The congress has been supported by generous contributions from:

NATIONAL COUNCIL FOR RESEARCH AND DEVELOPMENT

ISRAEL NATIONAL ACADEMY OF SCIENCES AND HUMANITIES

WEIZMANN INSTITUTE OF SCIENCE

TEL-AVIV UNIVERSITY

AGRICULTURAL RESEARCH ORGANIZATION

THE HEBREW UNIVERSITY OF JERUSALEM

TECHNION – ISRAEL INSTITUTE OF TECHNOLOGY

BEN-GURION UNIVERSITY

MINISTRY OF TOURISM

INTERNATIONAL ADVISORY COMMITTEE

M. AVRON, *Biochemistry Department, Weizmann Institute of Science, Rehovot* (Chairman)

H. BALTSCHEFFSKY, *Botaniska Institutionen, Lilla Frescati, 104 05 Stockholm 50*

N.K. BOARDMAN, *Commonwealth Scientific and Industrial Research Organization, P.O. Box 1600, Canberra City, A.C.T. 2601*

W.L. BUTLER, *University of California, Department of Biology, La Jolla, California 92037*

L.N.M. DUYSENS, *Biophysical Laboratory of the State University, Schelpenkade 14A, Leiden*

G. FORTI, *Instituto de Botanica della Universita, Via Foria 223, Napoli Napoli*

M. GIBBS, *Department of Biology, Brandeis University, Waltham, Massachusetts 02154*

D.O. HALL, *King's College, 68 Half-Moon Lane, London S.E. 24*

P. JOLIOT, *Institut de Biologie Physico-Chimique, 13, rue Pierre Curie, Paris Ve*

S. KATOH, *Department of Biophysics and Biochemistry, Faculty of Science, University of Tokyo, Hongo, Tokyo*

A. SAN PIETRO, *Botany Department, Indiana University, Jordan Hall 138, Bloomington, Indiana 47401*

A. TREBST, *Institut für Biochemie der Pflanzen, Ruhr Universität 463 Bochum*

LOCAL ORGANIZING COMMITTEE

M. AVRON, *Biochemistry Department Weizmann Institute of Science, Rehovot* (Chairman)

H. CARMELI, *Biochemistry Department, Tel-Aviv University, Tel-Aviv*

Z. GROMET-ELHANAN, *Biochemistry Department, Weizmann Institute of Science, Rehovot*

S. KLEIN, *Botany Department, The Hebrew University of Jerusalem*

N. NELSON, *Biology Department, Technion–Israel Institute of Technology, Haifa*

J. NEUMANN, *Botany Department, Tel-Aviv University, Tel-Aviv*

I. OHAD, *Department of Biological Chemistry, The Hebrew University of Jerusalem*

N. SHAVIT, *Biology Department, Ben-Gurion University, Beer-Sheva*

DAN WILLIAM REED
June 6, 1940 - September 8, 1974

(Dr. Reed was travelling from Tel Aviv to Portugal to attend a NATO Advanced Study Institute on membranes when a plane crash, after leaving Athens, led to his untimely death.)

Dan Reed grew up on a farm in southwestern Michigan, near the town of Constantine. The discipline, self-reliance, and industriousness that he learned in his early years remained a part of him throughout his scientific career. During his life, he retained a respect for the land, the attitudes and the life-style of the rural community.

Curiosity and a special ability in science led Dan to concentrate on chemistry and physics at Manchester College in Indiana. Upon completion of his undergraduate training in 1962, Dan recognized the potential for further integration of the principles of chemistry and physics into the biological sciences. The graduate research Dan undertook in Donald Hulquist's laboratory at the University of Michigan is a record of his initial effort to apply the principles of physical chemistry for a greater understanding of the structure and function of several complex proteins obtained from erythrocytes: a green heme-protein, a pink copper containing protein and cytochrome $\underline{b}_5$.

Dan began his work in photosynthesis as a postdoctoral fellow in the laboratory of Roderick Clayton at Cornell University in 1967. At that time, Clayton was interested in characterizing the pigment content of reaction centers in photosynthetic bacteria and Dan applied his talents in protein purification to the isolation of the reaction center complex from *Rhodopseudomonas spheroides.* His success resulted in the first published isolation and characterization of this important complex.

From Cornell, Dan moved to the Charles F. Kettering Research Laboratory in Yellow Springs, Ohio where he continued to study the chemical and physical nature of the photo-synthetic reaction center and initiated studies on the structure and composition of photo-synthetic membranes. This resulted in a macromolecular model of the photosynthetic membrane which was presented at the II International Congress on Photosynthesis Research in 1971. He continued analyzing the reaction center complex and extended it to the molecular level as evidenced by the paper he presented during this Congress. His work was primal in leading to the concept of four bacteriochlorophyll and 2 bacteriopheophytin molecules, strongly interacting in the complex.

At Kettering, Dan had a number of diverse research interests under investigation which were centered on the structure of membranes, including studies of the physical and chemical nature of photosynthetic membrane systems, an evaluation of the physical nature of the visual rod outer segments and structure-function relationship of nitrogenase in nitrogen-fixing microorganisms and legume root nodules.

In addition to this scientific interests, Dan maintained an active role in community affairs and in his church. Dan worked well with people. He was energetic, stimulating, motivated to get answers to his research problems and was optimistic about the future. Photobiologists have lost an esteemed colleague.

Donald Keister

INTRODUCTION

These volumes contain most of the contributions presented during the Third International Congress of Photosynthesis. To facilitate rapid publication, the manuscripts were photographed directly as presented, no page-proofs were sent to authors, and editorial corrections were minimal. It was the feeling of the organizing committee that the less aesthetic appearance of these volumes as a result of these procedures will be more than compensated by the added benefit due to early publication.

The Congress took place in Israel less than a year after the traumatic experience of the last war. Thus much of the necessary preparation work was done immediately following this disturbance and I wish to express my thanks to all participants and particularly to all those who helped in the day to day organization for their help and cooperation in trying circumstances.

Special thanks is due to the members of the organizing committee, to the members of our small photosynthesis group. Fanny Itzhak, Binah Silberstein, Uri Pick and Haim Hardt, to our visitors, David Cahen and Andres Binder, and our tireless secretaries, Ruth Bar-Adon, Zippora Kahana and Zippora Zimmerman.

M. Avron
Editor

CONTENTS

VOLUME I

PRIMARY REACTIONS

ELECTRON TRANSPORT

VOLUME II

BIOENERGETICS

CARBON METABOLISM

VOLUME III

DEVELOPMENT AND ORGANIZATION

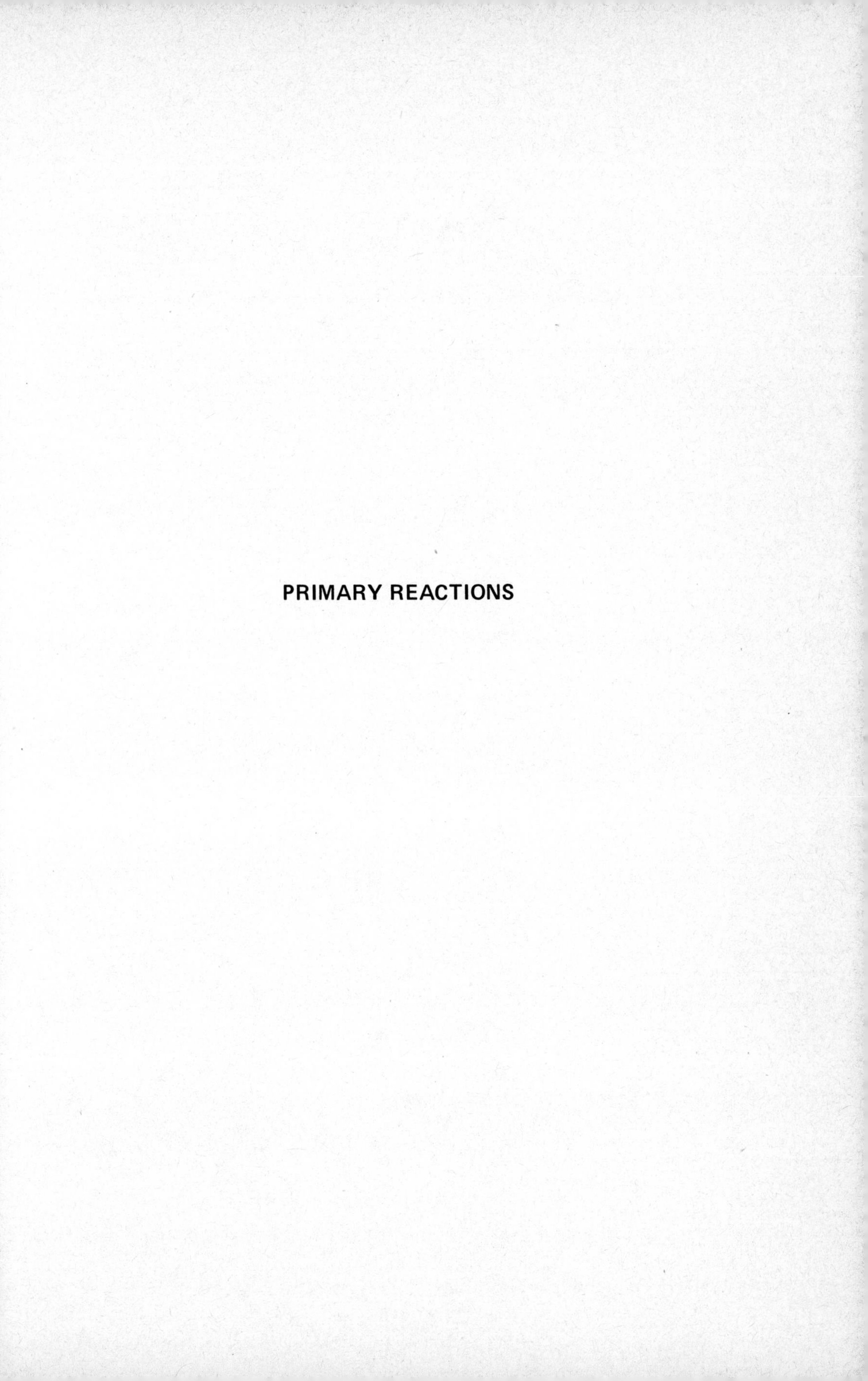

PRIMARY REACTIONS

M. AVRON, *Proceedings of the Third International Congress on Photosynthesis*
September 2-6, 1974, The Weizmann Institute of Science, Rehovot, Israel
Elsevier Scientific Publishing Company, Amsterdam, The Netherlands, 1974

RAPID REACTIONS OF PHOTOSYSTEM 2 AS STUDIED BY THE KINETICS OF THE FLUORESCENCE AND LUMINESCENCE OF CHLOROPHYLL a IN CHLORELLA PYRENOIDOSA

L.N.M. Duysens, G.A. den Haan and J.A. van Best

Biophysical Laboratory of the State University, P.O. Box 2120, Leiden, The Netherlands

1. Summary

The kinetics of the fluorescence yield and luminescence of chlorophyll a of system 2 were studied as function of the S-states and in the presence of hydroxylamine and DCMU in the time range from 0.4 to several hundreds of microseconds using a 30 ns ruby laser and xenon flash lamps as actinic and/or measuring light sources.

In the time range of 5 - 30 μs, in a flash sequence, under conditions in which the primary acceptor Q is in the reduced state, oscillations in luminescence occurred opposite to fluorescence oscillations. Also in the presence of 10^{-2} M hydroxylamine high luminescence was correlated with fluorescence quenching. These experiments indicated that the quencher was the oxidized primary donor P^+. The kinetics could be described in terms of 2 light reactions and 3 fluorescence quenchers: the oxidized primary acceptor Q, the quencher T (presumably a triplet), that is formed in a non-photosynthetic light reaction, which occurred with high efficiency after Q was reduced, and P^+, the oxidized primary donor. A transient initial quenching was observed during a time of the order of a microsecond which was not correlated with luminescence and occurred when the reaction centers were in states S_2 and S_3. This might be due to a quenching state or quencher, different from the quenchers mentioned above.

It was concluded amongst others that P^+ oxidized, at least in the first flash, the electron donor complex participating in oxygen evolution in less than 1 μs. In a fraction of the reaction centers in the S_2 and perhaps in the S_3 state, this donor complex does not reduce P^+, but another slower donor ($\tau \approx 15$ - 30 μs) does. In the presence of hydroxylamine only the reaction with this donor occurs.

2. Introduction

The fluorescence yield of chlorophyll a is affected by the state of intermediates of photosynthesis: specifically, the primary acceptor Q of photosystem 2 quenches the fluorescence of chlorophyll a in the oxidized state, Q, but not in the reduced state, Q^- (ref. 1).

For studying rapid phenomena in photosynthesis so far the fluorescence and luminescence method was unsurpassed in time resolution by other methods, such as ab-

Abbreviation: DCMU, 3-(3,4-dichlorophenyl)-1,1-dimethylurea

sorption difference spectroscopy; we were able to study the fluorescence and luminescence kinetics from 0.4 μs onward in a single experiment.

In the time range from about 100 μs to 0.1 s no other quencher than Q was required to interprete the fluorescence phenomena[2]. Q^- is oxidized by the plastoquinone pool with a half time of several hundreds of microseconds[3]. Thus, if the actinic intensity is such that the risetime of the fluorescence yield is 100 μs or less, the fluorescence yield as a function of absorbed energy is a (roughly exponentially) increasing function, which in the time range considered is independent of the intensity of the actinic light[2].

The fluorescence yield of Chlorella increased by roughly a factor 3 following a saturating flash. When exciting light flashes were used of such an intensity that an increase in fluorescence or reduction of Q occurred in less than 10 μs, the kinetics became more complex: the fluorescence yield plotted as function of absorbed energy became less steep or even bent downward after having increased by about a factor 2.

The kinetics could satisfactorily be described by the assumption of two light reactions and an additional quencher T. After the first light reaction, the reduction of Q, a second quantum caused the formation of a quencher T, presumably a carotenoid triplet[4,3]. At lower light intensities the quenching due to T cannot be observed since T rapidly reacts back in a dark reaction to the non-quenching form S.

In this paper we will give further evidence supporting the interpretation of the kinetics, and more precise data on the quantum yield and time constants of the reactions concerned. Also results will be given of studies on the oscillations of the fluorescence yield, discovered by Delosme[5], in successive flashes after a dark period, and on luminescence oscillations, which proved to be correlated with these fluorescence oscillations. An interpretation of these phenomena will be given in terms of reactions of reaction center 2.

2. Materials and methods

Algae

The algae used were Chlorella pyrenoidosa grown in continuous culture bubbled with air with about 5 % CO_2 (ref. 6). The algae were centrifuged and suspended in fresh growth medium and, while being illuminated, bubbled at room temperature for more than 15 minutes with air and 5 % CO_2. The algae were kept in the dark for at least 3 minutes prior to the measurements, and in general a new sample was taken after each series of flashes. If hydroxylamine was added, it was incubated in the dark during 10 minutes. The extinction of the samples was 0.1 at 680 nm (corrected for scattering at 720 nm) in a 1 mm cuvette, and 0.2 in the 5 mm cuvette used in the experiments with the laser.

Apparatus

Two apparatus were used, one applying xenon flash lamps ($\tau_{1/2} \approx 13$ or 1 μs), the other one was in addition provided with a Q-switched laser (λ = 694 nm; $\tau_{1/2} \approx 30$ ns) as exciting source.

Xenon flash lamp apparatus

A condensor in front of the flash lamps was imaged on the algae contained in a cuvette or deposited on a platinum electrode for oxygen measurements. Unless otherwise indicated, a set of filters isolated a band around 420 nm. By means of neutral density filters the intensity was adjusted. The intensity is given in percentage transmission of the neutral density filters. The fluorescence intensity at 681 nm, F(t), excited by the flash, and the intensity of the flash, I(t), were digitized at 256 equidistant points of the time interval and recorded in the memory system (disc, tape) of a PDP-9 computer[7]. A memoscope or a Calcomp plotter connected to the computer drew the fluorescence yield $\Phi(t) = F(t)/I(t)$ as a function of t or as a function of integrated energy: $E(t) = \int_0^t I(t')dt'$. The fluorescence was recorded in each of a series of flashes usually given with a time interval of 2.56 s.

By means of a second weak measuring flash, which was so weak as not to cause appreciable changes in the fluorescence yield, the fluorescence yield could be measured at certain times in the dark period after the exciting flash.

The algae could also be illuminated by continuous light, which was admitted by a shutter. A fluorescence band with maximum at 684 nm was isolated from the fluorescence spectrum by means of a filter set in front of a photomultiplier.

Laser apparatus

From the fluorescence spectrum excited by the 694 nm laser flash, a band was isolated with a maximum at 670 nm by means of a filter set in front of the measuring photomultiplier, which was connected to a memoscope. The intensity of the laser flash (which was rather variable) was simultaneously recorded. By means of a weak measuring xenon flash (risetime to the peak τ = 0.2 μs; energy 0.2 $\mu J/cm^2$; λ_{max} = 433 nm) the fluorescence yield (at 670 nm) was determined at the sharp peak of the flash at various times after and shortly before the laser flash (φ_0).

In addition to the laser a second xenon lamp ($\tau_{1/2}$ = 5 μs; broad band between 380 and 550 nm) could be used as actinic source. A number of saturating "preilluminating" xenon flashes could be given followed by the laser flash and the weak measuring flash. The time between the actinic flashes usually was 0.2 s.

Luminescence was measured with the same apparatus after the sensitivity of the photomultiplier had been increased. The luminescence was recorded by means of three memoscopes, set on different time bases, sensitivities and frequency responses.

When necessary, and this was always done when measuring luminescence, the voltage between kathode and first dynode was reversed during the laser flash, in order to avoid overloading of the photomultiplier. Sometimes further gates were used between photomultiplier and oscilloscope amplifiers to minimize transients from the photomultiplier.

4. Results and interpretation

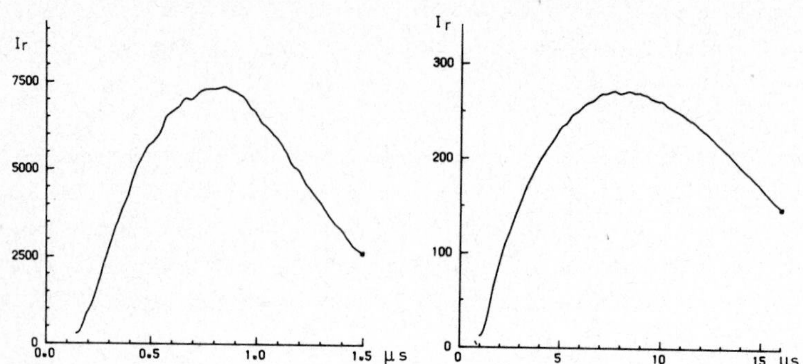

Fig. 1. Relative intensity of the xenon flashes as a function of time. Halfwidth 1.0 µs and 13 µs. The intensities are not comparable. A flash of 1 µs, intensity 100 %, was about saturating; a 3 %, 13 µs flash excited about two thirds of the reaction centers; its energy was 6 - 9 µJ/cm^2, λ = 420 nm.

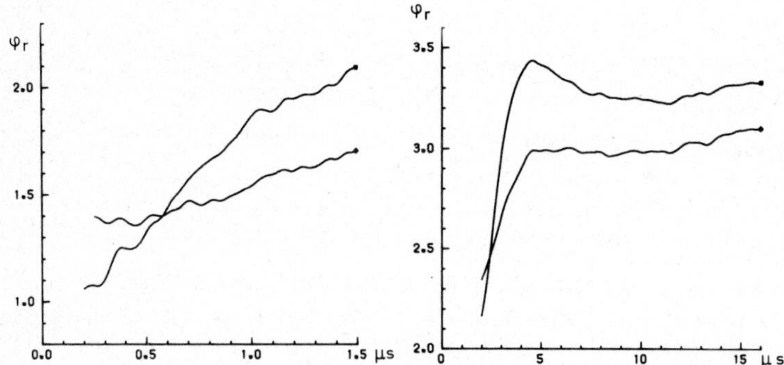

Fig. 2. a left, b right. Flash intensities 100 and 26.7 %, respectively. The steepest curves give the kinetics of the relative fluorescence yield of Chlorella in the first flash after a dark period of 3 minutes, the other curve that after the third flash. The φ_o values for fig. 2b, as determined from curves measured at a lower flash intensity, were 1.7 for the steep curve and about 2.0 for the other one.

Fig. 1 gives the intensity distribution in the 1 and 13 µs xenon flashes used as actinic and measuring light for part of the experiments. Fig. 2 shows the changes in fluorescence yield as a function of time in the first and third flash or a series given after a dark period. Note the initial rapid increase in fluorescence yield followed by a depression attributed to a second photoreaction (figs. 2b and 3a). In order to analyse these curves, in fig. 3a the fluorescence yield in various flashes is plotted as a function of energy absorbed (or rather of integrated flash intensity) in the 13 µs flash with the relative intensities of the flashes as parameters.

From fig. 3a it follows that the initial increase in fluorescence yield per unit

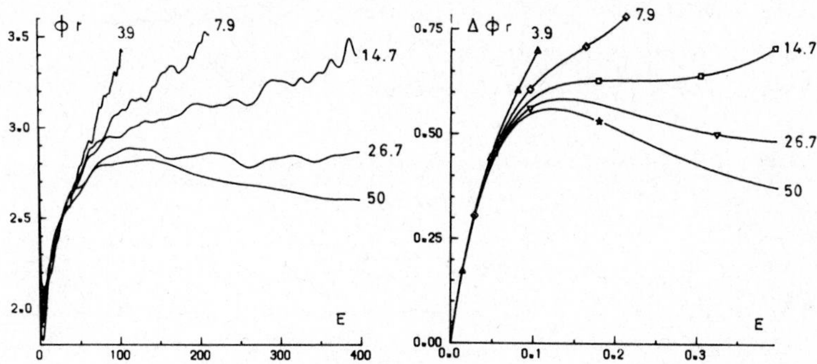

Fig. 3. a left, b right. Fig. 3a gives the kinetics of the relative fluorescence yield of Chlorella in 13 μs flashes of various intensities indicated as % of the maximum intensity. The algae were continuously illuminated with light of an intensity of about 0.5 mW/cm^2, with a wavelength centered at 420 nm, starting 8 minutes before the experiment. Fig. 3b gives the changes in fluorescence yield as computed from the equations in the text.

of time is roughly proportional to the momentary flash intensity. In a flash of a relative intensity of 50 given after a dark period of 3 minutes a 100 % increase in fluorescence yield occurred for the first flash after a dark period in about one μs, for an intensity of 3.9 in about 15 μs, leaving the first microsecond in which the flash intensity is low, out of consideration.

The kinetics of fig. 3a can be described by the following scheme:
(1) $QS + h\nu \rightarrow Q^-S$ (k_1I); (2) $Q^-S + h\nu \rightarrow Q^-T$ (k_2I); (3) $Q^-T \rightarrow Q^-S$ (k_3).

We assume for the sake of simplicity that system 2 consists of separate units, a unit being in this case a group of pigment molecules, which transfer excitation to one reaction center. Thus all reactions are assumed to be first order, the k's being the rate constants and I(t) the intensity of the flash.

Q is the primary acceptor, which is converted by the first light reaction (1) into the non- or weakly quenching form Q^-. By the second light reaction a weakly quenching substance S, contained in the same unit, is converted into a quenching form T, which like Q is assumed to quench the variable part of the fluorescence. In the dark reaction (3), T reverts to S. Q^-S or the increase in fluorescence $\Delta\varphi_r$ is calculated from the differential equations and is plotted (fig. 3b) as a function of the relative absorbed energy $E(t) = \int_0^t I(t')dt'$, in which $I(t')$ is the experimentally determined intensity of the flash. The life, decay or risetime is for exponential curves the time needed to reach a fraction (1 - 1/e) of the maximum amplitude of the change. This time was in some cases experimentally found by determining the intersection point of the linearly extrapolated initial slope of the curve with the final level. A satisfactory correspondence between computed and measured curves is found for $k_3 = 0.35\ (\mu s)^{-1}$, which means a lifetime of T, $\tau = 2.9\ \mu s$, $k_1 = -14$ (relative units, r.u.), k_2 = -6.5 (r.u.), k_2/k_1 = 0.46 (fig. 3b). Thus the second photoreaction S → T is only half as efficient as the first one, if we assume that one molecule T quenches the fluorescence of the unit almost completely. If T is a less efficient quencher than Q, the second light reaction may be as efficient as the first one.

In the presence of DCMU in the background of continuous light the fluorescence yield is maximal since DCMU strongly decreases the reoxidation rate of Q^-. A flash then only decreases the fluorescence yield. The kinetics of this decrease measured for the same batch could be satisfactorily simulated with the same values of the k's.

As mentioned before[4] the dip in the fluorescence was much more pronounced at low than at high partial pressures of oxygen, showing that k_3 was increased by oxygen.

In experiments with a 500 μs flash given after a few minutes of darkness it was found that for flash intensities corresponding to risetimes of 4 to 100 μs for an increase of 50 percent in φ, the curves representing φ as a function of energy all coincided within the precision of measurement. Delosme's experiments[2] provide the connection to the longer time ranges. It can be concluded from all these experiments that the fluorescence increase upon illumination in the time ranges of 1 μs to 10 ms is, in a first approximation, caused by the photoreduction of Q, modified by the quenching due to the formation of T for the shorter time ranges, and the dark reoxidation of Q^- for the longer ranges.

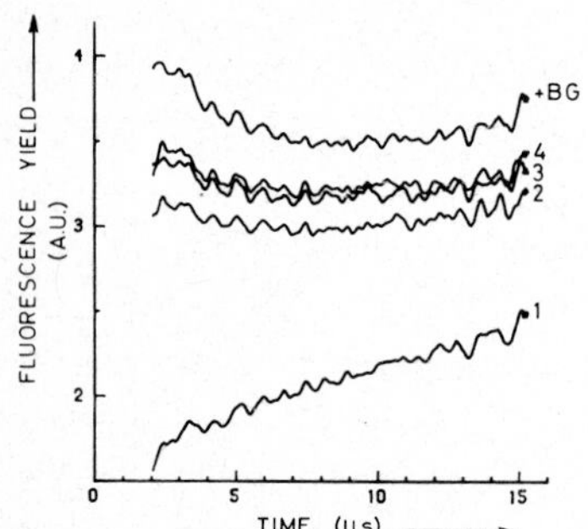

Fig. 4. Kinetics of fluorescence yield of Chlorella in the presence of 0.01 M hydroxylamine and 10 μM DCMU in a sequence of 13 μs, 7.9 % flashes following 10 min of incubation in the dark. The top curve is measured with a background of saturating continuous light. In the absence of hydroxylamine the curve of the fluorescence yield is much steeper and almost attains the maximum value in about 10 μs.

Quite different fluorescence kinetics were observed in the presence of 0.01 M hydroxylamine (fig. 4) and 10 μM DCMU. In the presence of hydroxylamine only, the kinetics in the first few flashes are similar to the kinetics in the first flash in the presence of hydroxylamine and DCMU. The fluorescence yield increases as if in all reaction centers after excitation, in addition to Q^-, a quencher is present which disappears in a dark reaction in about 30 - 40 μs (lifetime). After the flash, as was observed by weak measuring flashes, the fluorescence rise continues but does not quite reach the maximum fluorescence level reached in the absence of hydroxylamine. In the presence of DCMU, which prevents Q^- reoxidation, the level obtained in the presence of hydroxylamine is about 80 percent of that in its absence.

We confirmed Delosme's finding[5] that the maximum fluorescence yield obtained at the end of a flash carried out a damped oscillation as a function of the number of

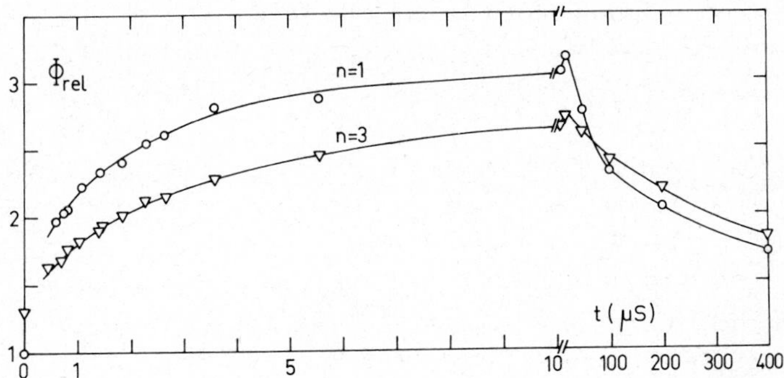

Fig. 5. Curve n = 1: fluorescence kinetics of Chlorella after a saturating (400 μJ/ cm^2) laser flash, given after a dark period of 3 min; n = 3: the same, but the laser flash is preceded by two saturating xenon flashes. The fluorescence yield is determined by means of a weak measuring xenon flash. The increase in yield from 0.6 to 10 μs for the curve marked n = 1 is probably caused by the disappearance of the quenching due to T. The lifetime of the quencher, as estimated from the n = 1 curve, τ = 2.3 μs, is somewhat smaller than that computed from the kinetics of fig. 3a.

the flashes with a period of four, the fluorescence being higher at the end of the first and fifth flash, and lower after the third flash, etc. Time courses of the fluorescence yield in the first and third 13 μs flash are shown in fig. 2a, and after a laser flash in fig. 5. An analysis of the fluorescence kinetics after the laser flash (fig. 5) was attempted as follows. We assume that the initial increase in fluorescence yield after the laser flash without preilluminating flashes (n = 1) is caused by the disappearance of the quencher T. The quenching factor d is then $d = (\varphi_1 - \varphi_o)/(\varphi_{max} - \varphi_o)$, φ_1 being the observed fluorescence yield in the first flash. If we further assume that the quenching factor is the same for the fluorescence φ_3, observed after the laser flash preceded by two preilluminating flashes, the corrected fluorescence difference is $\Phi_c = d(\varphi_3 - \varphi_1)$. The lifetime of the disappearance of Φ_c is roughly 20 μs (fig. 6).

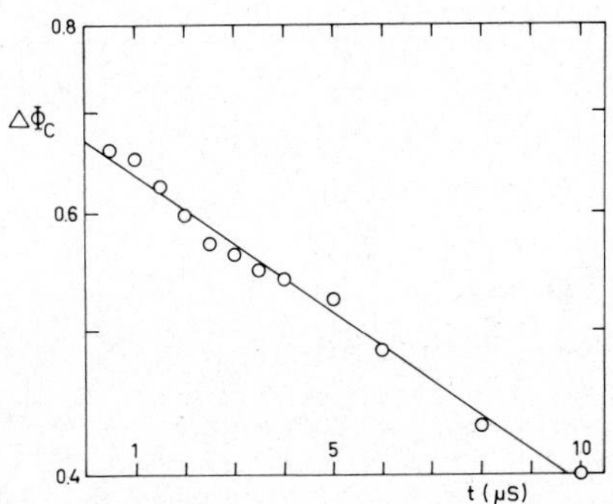

Fig. 6. The difference $\Delta\varphi_c$ between the fluorescence yield curves of fig. 5, corrected for quenching by T, plotted with a logarithmic ordinate as function of time (see text).

This experiment suggests that a quenching substance is formed in the first flash in part of the reaction centers and disappears in a time of about 20 μs, causing an additional increase in fluorescence. Since, mainly due to instability of the laser, so far this experiment could not be repeated with sufficient precision for constructing a Φ_c curve and in view of the uncertainty of the assumptions, we consider this conclusion as preliminary. For fluorescence yield kinetics measured in the first and third flash (fig. 2b), the curves approached each other between 5 and 10 μs, but much faster than expected if the disappearance of a quencher with a lifetime of 20 μs would be responsible. The higher rate of the light-induced formation of the quencher T for the top curve may explain the rapid phase.

Zankel[3] measured fluorescence curves analogous to those in fig. 5 but only from 80 μs onward. He estimated by a rather daring extrapolation that under certain conditions a rise in the fluorescence yield occurred in a time of about 35 μs. Our experiments give support to this conclusion, although the lifetime may be somewhat different.

Presumably an additional factor is responsible for the rise of the fluorescence curve in the third flash (fig. 2a). A quencher or quenching state may be formed in the light, which disappears with a lifetime of the order of one μs.

Luminescence kinetics were measured from 0.6 to 50 μs or more after a laser flash ($n = n_o$) preceded by $n_o - 1$ saturating xenon flashes. The curves were digitized, fed into a computer, and analyzed in terms of sums of exponentially decreasing functions with amplitude and decay time as parameters, by means of a parameter fitting program based on the method of Fletcher and Powell[8].

The luminescence contained "components" with lifetimes of about 0.8 μs, of about 20 μs, varying from 14 - 25 μs in different experiments, and 200 μs. We will concern ourselves with the 20 μs component. The variations may at least partly be caused by systematic errors. Fig. 7 shows the amplitude of the 25 and 200 μs component, and of the fluorescence yield at 14 μs as a function of the energy/cm^2 of the laser flash. Saturation of the three phenomena is reached for roughly the same energies/cm^2: two thirds of the changes in luminescence amplitudes and fluorescence yield is reached for laser energies of about 180 and 140 $\mu J/cm^2$, respectively, for this batch of algae. In other experiments it was found that the amplitudes of the oxygen pulses in 13 μs flashes as function of the flash energy saturated at roughly the same intensities as the fluorescence yields. This lends further support to the hypothesis that the phenomena observed originate from photoreaction 2.

Fig. 8 shows that the amplitude of a 23 μs component strongly oscillates as a function of the number of flashes preceding the luminescence measurement. These oscillations occur in opposite direction to those found in the fluorescence yield at the moment that Q is photoreduced. The quenching of the fluorescence yield at 16 μs in the nth flash (I = 3.9 %), $1 < n < 6$, with respect to the yield of the first flash, appeared to be proportional to the 23 μs component. In the presence of DCMU,

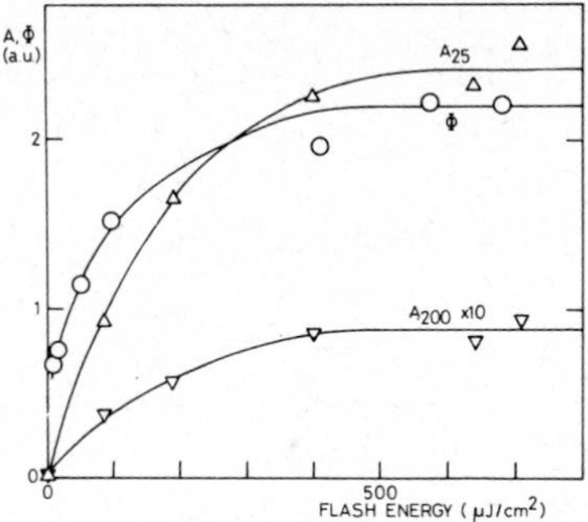

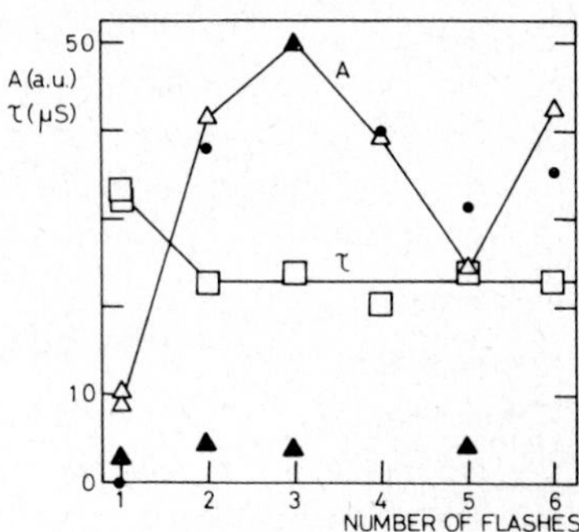

Fig. 7. Left: amplitudes in arbitrary units of 25 and 200 μs luminescence components, and of the fluorescence yield of Chlorella, plotted as a function of the energy of the laser flash. Energy saturation occurs for the three curves at about the same energies. For the luminescence measurements the laser flash was preceded by two saturating xenon flashes.

Fig. 8. Right: amplitudes (open triangles) and decay times (squares) as obtained by a computer analysis of the luminescence kinetics of Chlorella. Only the components with decay times around 20 μs are shown. Solid triangles: amplitudes in the presence of DCMU. The solid circles should be left out of consideration.

a 22 μs component remained small after any number of flashes (fig. 8). In an experiment with another batch of algae we found that in the presence of hydroxylamine the amplitude of a 20 μs component was about as high as that of the corresponding component after the third flash in the absence of hydroxylamine. After two flashes in the presence of hydroxylamine the 20 μs luminescence component almost doubled, and remained constant after a series of n flashes, spaced 0.2 s, $n \leqslant 6$. However, if the laser flash was given a few seconds after the series of xenon flashes, the 20 μs luminescence component did not increase. The fluorescence kinetics after the laser flash measured with the same batch, were the same after one or two flashes, whether these flashes were given with a spacing of 0.2 s or of a few seconds. The risetime of the fluorescence was 20 μs or a little less. This experiment indicates that in the presence of hydroxylamine in a flash an intermediate different from P^+ and Q^- is formed of about one second lifetime, which stimulates the amplitude of the 20 μs luminescence component, but does not affect the fluorescence yield or decay rate.

5. Discussion

As discussed, the initial light-induced increase in fluorescence yield is, with a single exception, probably caused by the photochemical reduction of the primary electron acceptor Q.

$DPQ + h\nu \rightarrow DP^+Q^- \rightarrow D^+PQ^-$

D is a secondary electron donor and is assumed to be part of the complex participating in water oxidation. From the observation that in the first flash after a dark period the fluorescence can rise in less than a microsecond, it follows that, if P^+ quenches the fluorescence yield of chlorophyll a, the oxidation of D by P^+ requires less than a microsecond in the first flash after a dark period.

In the presence of 0.01 M hydroxylamine a clearly time-dependent fluorescence increase, not caused by the T → S reaction, occurred in a 13 μs flash[9] (fig. 4) and after a laser flash. After the saturating laser flash the fluorescence yield increase occurred in about 20 μs or a somewhat shorter time. Under these conditions a strong luminescence component is present, which decays exponentially in 20 μs. The longer fluorescence risetime of 35 μs estimated by linear extrapolation of the initial slope in a 13 μs flash (fig. 4) might at least partly be caused by a depression of the fluorescence increase by the quencher T.

The oscillations observed in a series of flashes given after a dark period, both in fluorescence (in state Q^-)[4] and in the amplitude of a 20 μs luminescence component (fig. 8) were not observed in the presence of 0.01 M hydroxylamine (see also ref. 10). A lower fluorescence yield in state Q^- was correlated with a larger 20 μs luminescence component.

It is generally agreed[11] that luminescence is caused by the backreaction of the primary reaction $P^+Q^- \rightarrow PQ + h\nu_L$. Since a short time after excitation most reaction centers probably are in state Q^-, a stronger luminescence indicates, other things being equal, that more P^+ is present. The correlation of low fluorescence yield in state Q^- with high luminescence then indicates that P^+ is a quencher of fluorescence. The same suggestion has been made[21] on basis of fluorescence kinetics at $-196\ ^\circ$.

One may argue that the luminescence intensity may also be increased or decreased by the electric field due to charges on components of the reaction center, specifically by charges in the donor complex[12]. It is possible that positive charges in the states S^2 and S^3 stimulate luminescence. However, this does not explain the strong luminescence in the presence of hydroxylamine, since hydroxylamine reduces the secondary donor. It does not explain either the correlation of strong luminescence with quenched fluorescence. Therefore it is worthwhile to further discuss consequences of the assumption that P^+ quenches fluorescence and that the amplitude of the 20 μs luminescence components as observed by us are mainly determined by P^+ concentration.

Zankel[10] concluded from measurements of the luminescence kinetics of spinach chloroplasts between 68 and 800 μs that a luminescence component of 35 μs decay time was present which oscillated, when a series of flashes was given after a dark period, just like oxygen pulses. We found by means of a much faster apparatus that in Chlorella the luminescence component in the second flash was very high (see fig.

8), while the oxygen pulse in the second flash was negligible. Presumably extrapolation to shorter times did not lead to the correct result.

The oscillations in luminescence amplitude in subsequent light flashes were in fact very similar to oscillations in fractions of reaction centers, which were computed to be inactive in oxygen production. These fractions were computed by Delrieu[13] from the damping of the oscillations in oxygen pulses in a series of light flashes after a dark period.

The following model is an extension of the Joliot-Kok model[14,15] as modified by Delrieu[13]. Four positive charges or oxidizing equivalents have to be accumulated at the donor site of the reaction center in order to produce one oxygen molecule[15].

$$S^nPQ + h\nu \rightarrow S^nP^+Q^- \rightarrow S^{n+1}PQ^- \qquad (n = 0 \ldots 3)$$

$$S^4 + 2\,H_2O \rightarrow S^o + 4\,H^+ + O_2$$

S^n is a secondary electron donating complex containing n oxidizing equivalents. The fractions of S^o and S^1 were estimated[12] to be about 0.30 and 0.7, and an appreciable fraction, S_i^2, of S^2, was estimated to be inactive in oxygen production. We assume that S_i^n does not or only slowly donates electrons to P^+, and that another donor D_2 successfully competes with S_i^n. One might assume that $S_i^nP^+Q^-$ (n = 2,3) is the complex $M^{(n-1)+}Z^+P^+Q^-$ (see ref. 12) in which Z^+ is a secondary donor and M is a tertiary donor complex. Z can only donate one electron so that Z^+ is not able to transfer another electron to P^+. S^n is the complex $M^{n+}Z$ in which all n charges happen to be on M. Another possibility is that S^n and S_i^n are different conformational states. S^n and S_i^n are assumed to be in a kind of equilibrium[13].

$$\begin{array}{ccc} D_2PQ + h\nu \rightarrow & D_2P^+Q^- \rightarrow & D_2^+PQ^- \\ S_i^n & S_i^n & S_i^n \end{array}$$

The redox reaction between D_2 and P^+ occurs in about 20 µs. Our experiments (see also ref. 9) confirm the suggestion[16,17] that in the presence of hydroxylamine the reaction of P^+ with the S-complexes is inhibited, so that only the reaction with D_2 occurs.

The S-states present after illuminating spinach chloroplasts by 0 ... n preilluminating flashes can be frozen in by rapidly lowering the temperature down to -50 °C (ref. 18), which was concluded from the fluorescence kinetics at this temperature.

It was found that the initial rate of cytochrome $\underline{b}_{559}$ oxidation upon illumination with continuous light at -50 °C oscillated with the number of preilluminating flashes[19] in a similar way as the 20 µs luminescence component (fig. 8). This indicates that the donor D_2 is identical to cytochrome $\underline{b}_{559}$ or oxidizes this cytochrome. Cytochrome $\underline{b}_{559}$ might participate in cyclic electron transport, or in another oxidizing reaction not leading to evolution of oxygen.

An interesting phenomenon, also observed by Mauzerall[20], is the delay in increase in fluorescence in the submicrosecond region after preillumination with at least one flash (fig. 2a). This might be interpreted to indicate that reduction of P^+ by S^n ($n = 2,3$) requires a time of the order of a microsecond. This contradicts our finding that no increase in amplitude of a microsecond luminescence component occurred after a laser flash, preceded by two flashes. One explanation would be that P^+ is reduced within less than 1 μs, but that the donor complex in states S^n ($n = 2,3$) causes a transient quenching upon oxidation.

Acknowledgements

We are indebted to Mr. J.H.M. van de Water and Mrs. E.L. Klok-Huijser for making the computer programs.

The investigations were supported in part by the Netherlands Foundation for Biophysics, financed by the Netherlands Organization for the Advancement of Pure Research (ZWO).

References

1. Duysens, L.N.M. and Sweers, H.E., 1963, in: Studies on Microalgae and Photosynthetic Bacteria, special issue of Plant Cell Physiol. (University of Tokyo Press) p. 353.
2. Delosme, R., 1967, Biochim. Biophys. Acta 143, 108.
3. Zankel, K.L., 1973, Biochim. Biophys. Acta 325, 138.
4. Duysens, L.N.M., van der Schatte Olivier, T.E. and den Haan, G.A., 1972, Abstr. VI Int. Congr. Photobiology, Bochum, no. 277.
5. Delosme, R., 1971, C.R. Acad. Sci. Paris 272, 2828.
6. Hoogenhout, H. and Amesz, J., 1965, Arch. Mikrobiol. 50, 10.
7. den Haan, G.A. and Kooi, E.R., 1972, IEEE Transact. Instrumentation Measurement 21, 69.
8. Fletcher, R. and Powell, M.J.D., 1963, Computer J. 6, 163.
9. den Haan, G.A., Duysens, L.N.M. and Egberts, D.J.N., Biochim. Biophys. Acta, in the press.
10. Zankel, K.L., 1971, Biochim. Biophys. Acta 245, 373.
11. Lavorel, J., Review on Luminescence, in: Bioenergetics in Photosynthesis, ed. Govindjee (Academic Press) in the press.
12. van Gorkom, H.J. and Donze, M., 1973, Photochem. Photobiol. 17, 333.
13. Delrieu, M.-J., 1973, C.R. Acad. Sci. Paris 277, 2808.
14. Joliot, P., Barbieri, G. and Chabaud, R., 1969, Photochem. Photobiol. 10, 309.
15. Kok, B., Forbush, B. and McGloin, M., 1970, Photochem. Photobiol. 11, 457.
16. Cheniae, G.M. and Martin, I.F., 1970, Biochim. Biophys. Acta 197, 219.
17. Cheniae, G.M. and Martin, I.F., 1972, Plant Physiol. 50, 87.
18. Joliot, P. and Joliot, A., 1973, Biochim. Biophys. Acta 305, 302.
19. Amesz, J., Pulles, M.P.J., de Grooth, B.G. and Kerkhof, P.L.M., these proceedings.
20. Mauzerall, D., 1972, Proc. Natl. Acad. Sci. U.S. 69, 1358.
21. Butler, W.L., 1972, Proc. Natl. Acad. Sci. U.S. 69, 3420.

M. AVRON, *Proceedings of the Third International Congress on Photosynthesis*
September 2-6, 1974, The Weizmann Institute of Science, Rehovot, Israel
Elsevier Scientific Publishing Company, Amsterdam, The Netherlands, 1974

A TRIPARTITE MODEL FOR CHLOROPLAST FLUORESCENCE

W.L. Butler and M. Kitajima

Department of Biology, University of California, San Diego, La Jolla, California.

1. Introduction

A number of investigations have indicated that the photosynthetic apparatus of plants consists of, and can be physically separated into, two photochemical pigment systems, Photosystem I and Photosystem II, which operate at the reducing and oxidizing ends, respectively, of the photosynthetic electron transport chain. Out of this work and earlier work showing that chlorophyll molecules are organized into aggregates called photosynthetic units has grown the generalized concept of two photosystems each organized into units containing several hundred antenna chlorophyll molecules which absorb light energy from the environment and a special reaction center chlorophyll molecule which traps the excitation energy from the antenna chlorophyll and effects the conversion of the excitation energy to chemical free energy.

Attempts to study the primary photochemical reactions of either of the two reaction center chlorophylls in preparations free from large amounts of antenna chlorophyll (as has been achieved with reaction center preparations from photosynthetic bacteria) have failed: photochemically active particles enriched in the reaction center chlorophyll have been obtained but these preparations still contain large amounts of antenna chlorophyll relative to the amount of reaction center chlorophyll. Vernon et al.[1] suggested that both types of photosynthetic units of plants consist of an accessory complex which contains the accessory pigments and a major part of the chlorophyll a and a reaction center complex which contains the reaction center chlorophyll and the remainder (20 to 30%) of the chlorophyll a. The reaction center complexes can be separated from the accessory complexes by detergent treatment but the reaction center chlorophyll has not been separated from the chlorophyll a of the reaction center complex.

This concept of a bipartite organization of the photosynthetic apparatus into Photosystem I and Photosystem II has been modified recently by Thornber and coworkers. A chlorophyll protein complex containing equal amounts of chlorophyll a and chlorophyll b, which was ascribed earlier to Photosystem II[2], has been redesignated as a light-harvesting chlorophyll complex[3] which can feed excitation energy to either of the two photosystems. This chlorophyll a/b protein accounts for 50 to 60% of the total chlorophyll and all of the chlorophyll b. An active Photosystem I fraction[4] containing P_{700} and chlorophyll a, accounting for 10 to 20% of the total chlorophyll, has also been purified. The remaining chlorophyll a (approximately 30%) may be ascribed to Photosystem II complexes, analogous to the TS F-2a particles of Vernon *et al.*[1], which contain chlorophyll a and the reaction centers of Photosystem II. These results indicate a tripartite organization of the photosynthetic apparatus into Photosystem I and Photosystem II complexes, each containing appreciable amounts of antenna chlorophyll a in addition to their respective reaction center chlorophylls, and light-harvesting chlorophyll complexes which can transfer excitation energy to either of the two photosystem complexes. The purpose of the present paper is to consider the fluorescence properties of chloroplasts from the context of the proposed tripartite organization of the photosynthetic apparatus.

The discussion of the tripartite system will be based on a simple model for Photosystem II units (see Fig. 1) presented recently[5] to account for fluorescence measurements at 690 nm.

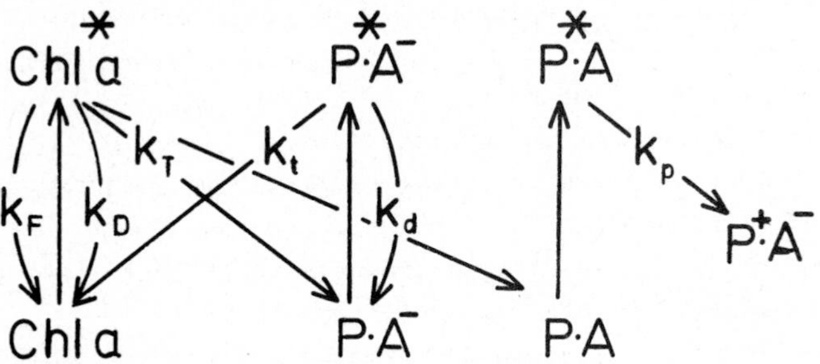

Fig. 1

That model for Photosystem II units can be summarized as follows:

1) There is only one origin of fluorescence at 690 nm, k_F, from the antenna chlorophyll of Photosystem II. (This assumption is in contrast to the usual assumption that a major part of initial fluorescence, F_o, comes from Photosystem I while the fluorescence of variable yield, i.e. the difference between the maximal and the initial fluorescence yields, $F_V = F_M - F_o$, comes from Photosystem II.)

2) Only two states of the Photosystem II reaction center complex need be considered: P·A represents the open centers and $P\cdot A^-$ or P(1-A) represents the closed centers. ("A" indicates the fraction of the primary electron acceptor molecules present in the oxidized state). The P^+ states (P^+A or P^+A^-) have too short of a lifetime to play a significant role under the conditions examined here.

3) Excitation energy trapped by a closed reaction center may be transferred back to the antenna chlorophyll, k_t, giving another opportunity for fluorescence or may be dissipated by a nonradiative decay process, k_d. A simplifying assumption (examined in some detail previously[5]) can be made that $k_p \gg k_t$ and k_d (i.e. $\phi_p = k_p/(k_p + k_t + k_d) = 1$ for open reaction centers).

4) The model for the Photosystem II unit can be applied rigorously for either extreme case of energy transfer between Photosystem II units:

a) Assuming no energy transfer between Photosystem II units (the separate package model) fluorescence can be taken as the sum of the fluorescences from the independent open and closed units

$$\phi_F = \frac{k_F A}{k_F + k_D + k_T} + \frac{k_F(1-A)}{k_F + k_D + k_T}\left[1 + \phi_T\phi_t + (\phi_T\phi_t)^2 + \ldots\right] .$$

Energy trapped within the closed units (the 1-A term) cycles into ($\phi_T = k_T/(k_F + k_D + k_T)$) and back out of the closed reaction centers ($\phi_t = k_t/(k_t + k_d)$) until it is dissipated by fluorescence, k_F, or a nonradiative decay process, k_D, in the antenna chlorophyll. The infinite series converges to $1/(1-\phi_T\phi_t)$ so that:

$$\phi_F = \phi_{F_o}\left[A + \frac{1-A}{1-\phi_T\phi_t}\right]$$

or written in terms of a constant, ϕ_{F_o} part, and a variable, ϕ_{F_v} part:

$$\phi_F = \phi_{F_o}\left[1 + \frac{\phi_T\phi_t(1-A)}{1-\phi_T\phi_t}\right] .$$

The yield of photochemistry which occurs only at open reaction centers is:

$$\phi_P = \phi_T \phi_p A = \phi_T A \quad (\text{since } \phi_p = 1) \; .$$

b) Assuming complete energy transfer between Photosystem II units (the matrix model) so that all reaction centers are available to all of the antenna chlorophyll of Photosystem II:

$$\phi_F = \frac{k_F}{k_F + k_D + k_T} \left[1 + \phi_T(1-A)\phi_t + \left[\phi_T(1-A)\phi_t\right]^2 + \ldots\ldots \right]$$

$$= \frac{k_F}{k_F + k_D + k_T} \left[\frac{1}{1 - \phi_T \phi_t (1-A)} \right]^{*}$$

which also can be written in terms of a constant and a variable part:

$$\phi_F = \phi_{F_o} \left[1 + \frac{\phi_T \phi_t (1-A)}{1 - \phi_T \phi_t (1-A)} \right] \; .$$

In the matrix model, energy transferred back from a closed reaction center may be trapped by another open center and used for photochemistry so that:

$$\phi_P = \phi_T \phi_p A \left[1 + (1-A)\phi_T \phi_t + \left[(1-A)\phi_T \phi_t\right]^2 + \ldots\ldots \right]$$

$$= \phi_T A \left[\frac{1}{1 - \phi_T \phi_t (1-A)} \right]$$

The two models predict different kinetic behavior[5] (in the separate package model ϕ_F and ϕ_P are linear with A, while in the matrix model they are nonlinear) but both models give the same expressions for ϕ_F and ϕ_P at the initial (A = 1) and final (A = 0) states. With either model:

$$\frac{F_V}{F_M} = \frac{\phi_{F_M} - \phi_{F_O}}{\phi_{F_M}} = \phi_T \phi_t \quad \text{and} \quad \phi_{P_O}(A = 1) = \phi_T \; .$$

---

* This equation was presented previously[5] in a different, but equivalent, form:

$$\phi_F = \frac{k_F}{k_F + k_D + k_T\left[A + (1-A)\phi_d\right]}$$

5) The simple model presented in Fig. 1 predicts two types of fluorescence quenching processes. Quenching in the antenna chlorophyll increases k_D and quenches both F_o and F_M, with F_M (or F_V) being somewhat more sensitive to the quenching than F_o. With this type of quenching the ratio F_V/F_M is decreased to the same extent as the initial yield of photochemistry, ϕ_{P_o}, since both are proportional to ϕ_T. Quenching at the reaction center chlorophyll, which increases k_d and thereby decreases ϕ_t of the closed reaction centers ($\phi_t = k_t/(k_t + k_d)$), quenches F_M (or F_V) but has no influence on F_o. With this type of quenching the ratio F_V/F_M may be markedly decreased with little or no influence of ϕ_{P_o} (to the extent that k_p remains large compared to k_d).

Both types of fluorescence quenching have been observed experimentally (with fluorescence being measured at 690 nm). Dibromothymoquinone (DBMIB) quenches both F_M and F_o as well as the yield of photochemistry (initial rate of photoreduction of C-550) at -196°C and the ratio F_V/F_M is decreased to the same extent as ϕ_{P_o}[6]. On the other hand, the addition of ferricyanide to chloroplasts prior to freezing causes a marked quenching of F_M (or F_V) at -196°C while F_o is unaffected[7] and ϕ_{P_o} is essentially unchanged[5]. DBMIB quenches by creating nonfunctional energy traps in the antenna chlorophyll which compete with the reaction centers for the excitation energy[6]: ferricyanide (at -196°C) induces (via the oxidation of cytochrome b_{559}[7]) the quenching of excitation energy at the reaction center chlorophyll specifically[5]. The quenching induced by UV radiation (at room temperature or at -196°C) is also at the reaction center chlorophyll. (F_V can be almost totally quenched with no effect on F_o and photochemistry supported by artificial electron donors for Photosystem II may persist to an appreciable extent[8].) Thus, the simple model presented in Fig. 1 accounts for observable phenomena which otherwise would appear anomalous, and no special assumptions need to be made as to different origins of F_o and F_V. The model for the Photosystem II units which was based on the "classical" bipartite organization of the photosynthetic apparatus can be expanded to consider the fluorescence from both Photosystem I and Photosystem II within the framework of a tripartite organization.

2. The Tripartite Model

A schematic diagram of the proposed tripartite organization of the photosynthetic apparatus is presented in Fig. 2.

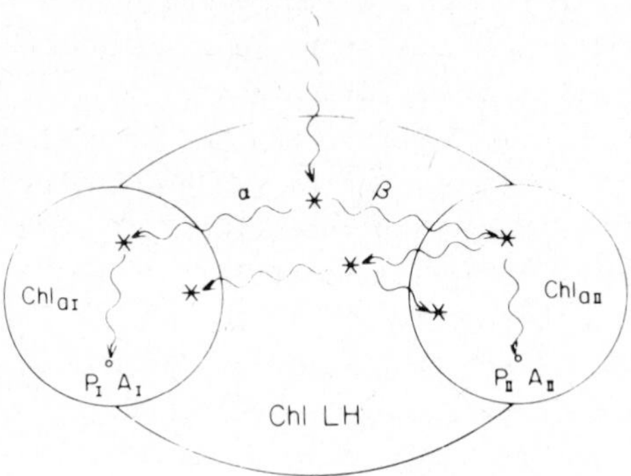

Fig. 2

Chl LH designates the light-harvesting chlorophyll complex which, in green plants, contains equal parts of chlorophyll a and chlorophyll b; Chl a_I designates the chlorophyll a of the Photosystem I complex and Chl a_{II} the chlorophyll a of the Photosystem II complex. P_I and P_{II} designate the reaction center chlorophylls of Photosystems I and II, respectively, and A_I and A_{II} the primary electron acceptors of the two photosystems.

The emission spectrum of chloroplasts at -196°C shows three bands of fluorescence at about 685, 695 and 730 nm: these will be ascribed to Chl LH, Chl a_{II} and Chl a_I, respectively. The association of the 730 nm emission band with Chl a_I is well supported by several lines of evidence but the identification of the source of the 695 nm emission band is more tentative. Experimentally, however, the 695 and 685 nm emission bands are coupled very closely. Murata[9] showed that the fluorescence induction curves for the 685 and 695 nm bands at -196°C were very similar. The half times for the light-induced fluorescence yield increase were the same at the two wavelengths but the ratio F_V/F_M was about 20% greater at 695 nm than it was at 685 nm. He suggested, on the basis of these results, that the 695 nm fluorescing form of chlorophyll was closer to the primary photochemical reaction of Photosystem II than was the 685 nm fluorescing form but that excitation energy was readily transferred back and forth between these two forms of chlorophyll.

The model for energy migration within the photochemical apparatus is shown in Fig. 3.

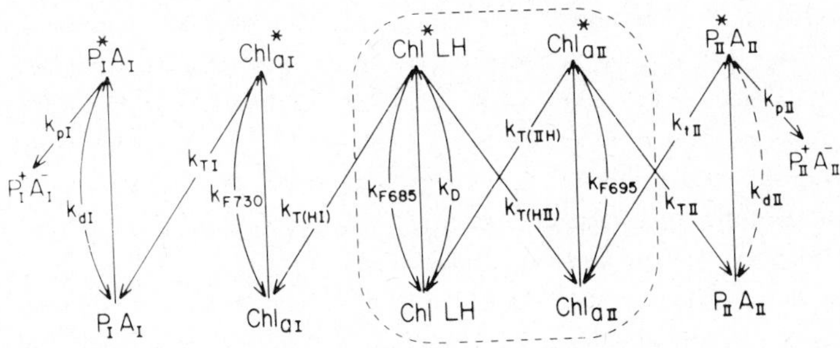

Fig. 3

The dashed line around Chl LH and Chl a_{II} indicates the tight coupling (rapid exchange of excitation energy) between these two complexes. No distinction will be made between the 685 and 695 nm fluorescence because of this tight coupling. Fluorescence at 690 nm will be assumed to be representative of both emissions and to emanate from both Chl LH and Chl a_{II}. The expression for the yield of 690 nm fluorescence is the same as that derived previously from the simpler model based on Photosystem II units with slight changes of nomenclature (assuming the separate package model).

$$\phi_{F690} = \frac{\beta k_{F690}}{k_{F690} + k_D + k_{T(II \to I)} + k_{TII}} \left[A_{II} + \frac{1 - A_{II}}{1 - \phi_{TII}\phi_{tII}} \right] \qquad (1)$$

In the previous model for Photosystem II units, it was recognized that a part of the k_D losses at the bulk chlorophyll of Photosystem II could be due to energy transfer to Photosystem I. In the present model, energy transfer from Photosystem II to Photosystem I is included as a specific term, $k_{T(II \to I)}$. However, a distinction should be made between a photon absorbed directly by Chl LH from the environment and an exciton transferred to Chl LH from Chl a_{II}. If the exciton has some degree of localization (i.e. does not spread uniformly throughout Chl LH) an exciton transferred out of Chl a_{II} to Chl LH should on the average be closer to the Photosystem II complex than to the Photosystem I complex and therefore have a greater probability of transfer back to Chl a_{II} than an exciton due to the direct absorption of a photon by Chl LH. For the initial absorption of light, β represents the probability of energy transfer to, or direct absorption by, Chl a_{II} and α, the probability of excitation of Chl a_I.

The introduction of α and β for the probabilities of a photon to excite Chl a_I and Chl a_{II}, respectively, shows that the formulation based on the tripartite model is not different in principle from that based on the bipartite organization. $k_{T(II \rightarrow I)}$ in equation 1 represents the process of energy transfer from Chl a_{II} to Chl a_I, probably via Chl LH, and as such incorporates both the $k_{T(IIH)}$ and the $k_{T(HI)}$ processes indicated in Fig. 3. $k_{T(II \rightarrow I)}$ is not defined rigorously in the model presented in Fig. 3 because the degree of coupling between Chl LH and Chl a_{II} is not expressable in a rigorous manner. k_{F690}, k_D and k_{TII} which compete with $k_{T(II \rightarrow I)}$ in Eqn. 1 may also be influenced to different degrees by this coupling and may represent a weighted average of these processes occurring within the Chl LH complexes and the Chl a_{II} complexes. In essence, the β part of the Chl LH complex and the Chl a_{II} complex function as a "classical" Photosystem II unit and the α part of the Chl LH complex and the Chl a_I complex function as a Photosystem I unit.

At room temperature the yield of fluorescence from Photosystem I is sufficiently low that it can be neglected for most purposes. The small amount of fluorescence measured at 730 nm at room temperature can be ascribed to the long wavelength tail of the Photosystem II emission on the basis that the ratio F_M/F_O is the same at 690 and at 730 nm. On cooling, however, a major emission band due to Photosystem I fluorescence appears at 730 nm. Measurements of fluorescence at 730 nm from chloroplasts frozen to -196°C also show a fluorescence of variable yield, although the ratio F_M/F_O at 730 nm is less than at 690 nm. Murata[9] proposed that the fluorescence of variable yield at 730 nm was due to energy transfer from Photosystem II because the kinetics of the fluorescence yield increase at -196°C were the same at 730 as at 690 nm. That proposal is also supported by the observation[5] that the fluorescence of variable yield at 730 nm occurs only in chloroplast particles which have Photosystem II activity: isolated Photosystem I particles show no variable yield fluorescence.

The observation that the fluorescence yield of Photosystem I particles at -196°C is invariable requires some discussion. It should be expected that the photooxidation of P_{700} at -196°C, which bleaches the 702 nm absorption band, would remove P_{700} as an energy trap, thereby increasing the yield of fluorescence from Chl a_I. The observation that the fluorescence yield of Photosystem I particles does not change on irradiation at -196°C indicates that the efficiency of trapping excitons in Chl a_I does not change with the photooxidation of P_{700}. It has been shown that the bleaching of the 702 nm band due to oxidation of P_{700} at -196°C is accompanied by an absorbance increase at 690 nm of approximately equal magnitude to the bleaching at 702 nm[10]. We suggest that the absorbance increase at 690 nm represents the appearance of nonfunctional trapping centers which are formed as P_{700} is oxidized (either P^+_{700} absorbs at 690 nm and traps excitation energy or the oxidation of P_{700} leads to the formation of another trapping center) and that the rate constant for trapping of excitation energy in Chl a_I, k_{TI}, is constant and independent of the redox state of P_I. The yield of fluorescence from Chl a_I should be:

$$\phi_{F730} = (\beta\phi_{T(II\rightarrow I)} + \alpha)\left[\frac{k_{F730}}{k_{F730} + k_{TI}}\right] \qquad (2)$$

where $\phi_{T(II\rightarrow I)}$ is yield of energy transfer from Chl a_{II} to Chl a_I.

$$\phi_{T(II\rightarrow I)} = \frac{k_{T(II\rightarrow I)}}{k_{F690} + k_D + k_{T(II\rightarrow I)} + k_{TII}}\left[A_{II} + \frac{1 - A_{II}}{1 - \phi_{TII}\phi_{tII}}\right] \qquad (3)$$

Thus, the yield of fluorescence at 730 nm from chloroplasts at -196°C has a variable component which depends on the Photosystem II reaction centers. The fluorescence contribution at 730 nm due to $\phi_{T(II\rightarrow I)}$ has the same ratio of constant and variable parts as the fluorescence at 690 nm.

The equations for the tripartite model are not as rigorous as the equations for the simpler Photosystem II model because $k_{T(II\rightarrow I)}$ is not rigorously defined. Nevertheless the tripartite model is useful for the conceptualization of energy transfer processes within the photosynthetic apparatus. Murata[11] showed that energy transfer from Photosystem II to Photosystem I ("spillover") was affected by the presence of divalent cations. The absence of divalent cations, which stimulates "spillover", is characterized by a marked quenching of F_V with very little effect on F_o. This characteristic of a specific quenching of F_V indicates a control on energy transfer from the reaction center chlorophyll of Photosystem II to Photosystem I. So far as Photosystem II is concerned such a transfer of energy out of Photosystem II could be included as a k_d process. Thus, two types of energy transfer from Photosystem II to Photosystem I were postulated[5]; energy transfer from the antenna chlorophyll analogous to a k_D process and energy transfer from the reaction center chlorophyll analogous to a k_d process.

The proposed tripartite organization of the photosynthetic apparatus presented in Fig. 3 suggests a reasonable mechanism for the control of "spillover". If the Photosystem I and Photosystem II complexes are brought closer together (a condition induced by the absence of divalent cations) $k_{T(II \rightarrow I)}$ should increase, relative to the other competing decay processes, so that, by Eqn. 3, $\phi_{T(II \rightarrow I)}$ should increase. It is also apparent from Eqn. 1 that an increase of $k_{T(II \rightarrow I)}$ could cause a slight quenching of F_o measured at 690 nm, but such quenching will be small because the denominator in Eqn. 1 should always be dominated by the k_{TII} term. In addition, excitons transferred out of the closed Photosystem II reaction centers should have a greater probability of being transferred to Photosystem I. This latter process decreases ϕ_{tII} thereby quenching F_V specifically.

With chloroplasts in the presence of divalent cations, which show a F_M/F_o ratio of 5.0, one may assume[5] that $\phi_{TII} = 0.8$ and $\phi_{tII} = 1.0$. In this case none of the excitons transferred out of closed Photosystem II reaction centers are lost to Chl a_I. (If the Photosystem I and Photosystem II complexes are far enough apart, an exciton transferred from a closed Photosystem II reaction center into Chl LH will be lost to fluorescence or to a nonradiative decay process before it reaches the Photosystem I complex or it will be returned to the Photosystem II complex.) If, in the absence of divalent cations, the Photosystems I and II are close enough together that 20% of the excitons transferred out of closed Photosystem II reaction centers reach Photosystem I, ϕ_{tII} will decrease to 0.8 and the ratio F_M/F_o (at 690 nm) will decrease to 2.8 . Studies of the effects of divalent cations on chloroplast fluorescence at 690 nm[11] (either at room temperature or at -196°C) generally show such a quenching of F_V with only a slight quenching (10 to 25%) of F_o.

The tripartite model predicts rather different results for the effects of divalent cations on the fluorescence yield at 730 nm at -196°C. The increase of $\Phi_{T(II \rightarrow I)}$ associated with an increase of "spillover" should cause an increase of fluorescence at 730 nm but the relative proportion of the variable component of the energy transferred to Photosystem I should be decreased because of the decrease of Φ_{tII} (see Eqn. 3). Preliminary results have shown that, in the absence of divalent cations (increased "spillover"), F_V at 730 nm is approximately the same magnitude as in the presence of divalent cations while F_O, comprized of both α and the invariable part of energy transferred from Photosystem II (see Eqn. 2), is appreciably larger. It should be possible to separate the α and the $\beta\Phi_{T(II \rightarrow I)}$ contributions to the 730 nm fluorescence at -196°C, since the latter will have the same ratio of variable to invariable parts as the fluorescence measured at 690 nm, and to determine experimentally α, β and $\Phi_{T(II \rightarrow I)}$ under various conditions. Such experiments may provide a basis for further experimental tests of the model.

The model may also be useful for the examination of chloroplast development. Thornber and Highkin[3] have shown that Chl LH is absent from the mutant of barley which lacks chlorophyll b but which, nevertheless, grows photosynthetically. Etiolated bean leaves greened in far-red light[12] or with a series of brief white-light flashes[13] (which are characterized structurally by unfused primary thylakoids) also lack chlorophyll b but have Photosystem I and Photosystem II activities. It is reasonable to suppose that such leaves also lack Chl LH. If energy transfer from Photosystem II to Photosystem I is mediated by Chl LH, leaves lacking Chl LH would not be expected to show a fluorescence of variable yield at 730 nm at -196°C since such fluorescence is due to energy transfer from Photosystem II to Photosystem I. Thus, a study of fluorescence induction curves at low temperature at 690 and at 730 nm under conditions where Chl LH is being formed in photosynthetically competent leaves (irradiating far-red-greened or flashed-greened bean leaves with white light induces the formation of chlorophyll b and the formation of grana) might show the appearance of "spillover" concomitant with the formation of chlorophyll b (i.e. Chl LH). The model presented in Eqns. 1, 2 and 3 appears promising in that it explains existing data on chloroplast fluorescence within a reasonably plausible and consistent framework and it appears to focus attention on new experimental approaches which hopefully will expand our understanding of the photosynthetic apparatus.

Acknowledgements

This work was supported by NSF grant GB-37938X to W.L.B. M.K. is on leave from the Fuji Photo Film Co., Ltd., Tokyo. Japan.

References

1. Vernon, L.P., Shaw, E.R., Ogawa, T., Raveed, D., 1971, Photochem. Photobiol. 14, 343.
2. Thornber, J.P., Gregory, R.P.F., Smith, C.A., Bailey, J.L., 1967, Biochem. 6, 391.
3. Thornber, J.P. and Highkin, H.R., 1974, Eur. J. Biochem. 41, 109.
4. Alberte, R.S., Thornber, J.P., Naylor, A.W., 1973, Proc. Natl. Acad. Sci. 70, 134.
5. Butler, W.L. and Kitajima, M., Biochim. Biophys. Acta (in press).
6. Kitajima, M. and Butler, W.L., Biochim. Biophys. Acta (in press).
7. Okayama, S. and Butler, W.L., 1972, Biochim. Biophys. Acta 267, 523.
8. Yamashita, T. and Butler, W.L., 1968, Plant Physiol. 43, 2037.
9. Murata, N., 1968, Biochim. Biophys. Acta 162, 106.
10. Lozier, R. and Butler, W.L. 1974, Biochim. Biophys. Acta 333, 465.
11. Murata, N., 1969, Biochim. Biophys. Acta 189, 171.
12. De Greef, J., Butler, W.L., Roth, T.F., 1971, Plant Physiol. 47, 453.
13. Strasser, R.J. and Sironval, C., 1972, FEBS L. 28, 56.

M. AVRON, *Proceedings on the Third International Congress on Photosynthesis*
September 2-6, 1974, The Weizmann Institute of Science, Rehovot, Israel
Elsevier Scientific Publishing Company, Amsterdam, The Netherlands, 1974

COMPARATIVE STUDY OF THE 520nm ABSORPTION CHANGE AND DELAYED LUMINESCENCE IN ALGAE

Pierre JOLIOT and Anne JOLIOT
Institut de Biologie Physico-Chimique
13, rue Pierre et Marie Curie
75231 Paris Cedex 05 France

1. Introduction

Delayed luminescence is generally interpreted as resulting from a back reaction occuring at the level of System II photocenters. Z^+ Chl Q^- → Z Chl Q (Arthur and Strehler[1], Lavorel[2]). This field has recently been reviewed by Lavorel[3]. The emission of delayed light implies that enough energy is available to reach the first excited singlet state of chlorophyll. The redox energy stored in the center after the photoact is not sufficient and an additional energy is required. Thermal activation of delayed luminescence has been studied and discussed by Arnold and Azzi[4]. Crofts suggested (quoted by Fleshmann[5]) that the activation energy could be lowered by a membrane potential.

If we accept the idea of Mitchell[6] that the donor and acceptor of both photosystems are localised on the inner and outer face of the thylakoid membrane respectively, we can predict an interaction between an electric field across the membrane and the dipole Z^+ Chl Q^-. This problem has been investigated by Miles and Jagendorf[7] who observed that a membrane potential induced by an ion gradient was able to increase the intensity of delayed light. This effect has been studied more quantitatively by Barber and Kraan[8], Kraan et al[9] and Hardt and Malkin[10]. Delayed luminescence can also be stimulated by various treatments, especially by an acid-base transition as shown by Mayne[11] (see also refs[8, 9, 10]). A direct effect of an electrical field on delayed luminescence has been reported by Arnold and Azzi[12] who showed that the intensity of delayed light produced by a chloroplast suspension could be modulated by an externally applied modulated electric field.

The theory proposed by Mitchell to explain the mechanism of the coupling between electron transport and phosphorylation supposes that the transfer of electrons induces both an electrical field and a proton gradient across a membrane. Extensive studies by the group of Witt (reviewed in ref[13]) and of Crofts[14, 15] established that this membrane potential induced a shift in the absorption of spectrum of the chloroplast pigments and especially carotenoids. In the case of chloroplasts, the maximum of the absorption change is observed at 515-520 nm. Thus, it appears that delayed luminescence and the 520 absorption change represent two independent probes able to detect a membrane potential. Several experimental results support this hypothesis. Firstly, Fleischmann and Clayton[16] reported a correlation between the kinetics of delayed luminescence and the carotenoïd shift in bacterial chromatophores. Secondly, Wraight and Crofts[17] observed a biphasic increase at the onset of continuous illumination. On the basis of the action of uncouplers and ionophores, this result was interpreted as an effect of the light induced electrical field and pH gradient on the luminescence emission.

2. Methods

Delayed luminescence is measured by a method described in ref[18] but without protecting the photomultiplier during the actinic flash. For times shorter than 10 msec, the actinic flash induces an artefact of higher intensity than the delayed luminescence. From 10 msec to 100 msec, this artefact is measured under conditions where the slow component of luminescence is inhibited (preilluminated algae in the presence of 10μM DCMU + 10mM Hydroxylamine. For these conditions, Bennoun[19] showed that the slow component of luminescence is completely inhibited. Using a fast phosphoroscope, Ducruet and Lavorel[20] established that under these conditions, only a very short component (< 500μsec) of the delayed luminescence is observed.

We checked that the actinic flash does not perturb the linearity of the response for dark times after the flash longer than 2 msec.

Absorption changes are measured with a spectrophotometer described in reference[21]. In this method, the absorption level is determined by short, weak monochromatic flashes.

All the experiments reported here were performed at 20°C. The algae (Chlorella or Chlamydomonas Mut 54)[25] are suspended in 50 mM phosphate buffer, pH 6.4 in presence of 10μM dichlorophenyl dimethylurea (DCMU).

3. Results

The aim of this work was a quantitative study of the correlation existent between delayed luminescence and the 520 nm absorption change. To minimize the contribution of the pH gradient, most of our experiments are performed with dark-adapted material illuminated by one or two flashes. The experiments reported here are performed on living algae in the presence of 10μM DCMU. The life time of the absorption change at 520 nm induced by flashes in dark-adapted algae is generally much longer than in isolated chloroplasts, the $t_{1/2}$ having values up to 5 sec depending upon the physiological state. It is thus possible to study the correlation between absorption change and delayed luminescence during a longer period.

Bennoun[22] demonstrated that in the presence of DCMU, the reoxidation of Q^- takes place only by a back reaction. Addition of Hydroxylamine, which rapidly removes the positive charges formed on the donor side, blocks both the back reaction and the delayed luminescence[22, 19]. On the other hand, in the absence of DCMU, Barbieri et al[23] observed that the intensity of luminescence depends on the number of positive charges accumulated on the donor side of System II. Delayed luminescence induced by salt effect or acid-base transitions also depends on the number of positive charges[10]. In the presence of DCMU, only one or two charges can be accumulated, depending upon the state of the center in the dark. It is noteworthy that during a series of saturating flashes given in the presence of DCMU, the state of the centers is the same immediately after each flash in the sequence. This property will be used in the experiments presented here.

To make the experimental section easier to understand, a summary of the essential features of the absorption change at 520 nm induced by a single flash on dark adapted algae is presented (Witt and Moraw[24], Joliot and Delosme[21]). One can observe 4 phases Fig.1.

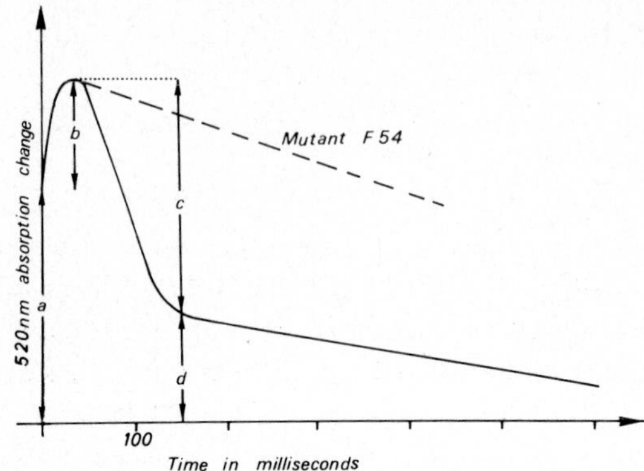

Fig. 1. Time course of the 520 absorption change induced by a single flash on dark adapted algae :
Chlorella or Chlamydomonas Wild type.
--- Chlamydomonas mutant F54.

1) a fast absorption increase (a) occurs in less than 1μsec. In our laboratory using Chlorella, we find 60 to 80% of this phase is linked to Photosystem I activity and 20 to 40% to System II activity.

2) a slower increasing phase (b) occurs in the time range 1 to 50 msec[24, 21]. Phase b depends only upon System I activity[21].

3) a fast absorption decrease (c) is observed between 50 and 200 msec. This phase appears to be related the phosphorylation activity [21].

4) a very slow decreasing phase (d) ($t_{1/2}$ 1.5 to 5 sec) probably corresponds to the decay of the field by leakage through the membrane[21]. The relative amplitude of phases c and d can be varied by a number of pretreatments. For instance, dark aerobic incubation (about one hour) increases the amplitude of the slow phase d. Often phase c is not observed after a single flash. On the other hand, phase c is always absent in Chlamydomonas mutant F54 isolated by Bennoun and Levine[25], which is known to be blocked at the level of the ATPase[26].

Addition of DCMU does not essentially modify the relaxation curve observed after a single flash. Nevertheless, long incubation in the presence of DCMU (1 hour) decreases the amplitude of phase c and increases phase d.

After a single flash in the presence of DCMU, the reoxidation of Q^-, which leads to restoration of System II activity, is a slow process ($t_{1/2}$ about 700 msec (Bennoun[22])). On the contrary, the primary donor of System I is rapidly reduced ($t_{1/2} \leqslant 100 \mu sec$)[21] by the secondary donor which is present in a reduced state in the dark. Thus, a second flash given a time t after the first induces an absorption increase proportional to the fraction of the primary donor of Photosystem I which has been reduced during the dark period t. In the case of repetitive flashes, a slower turnover is observed for Photosystem I (100 msec) which correspond to a cyclic electron transfer around this photosystem.

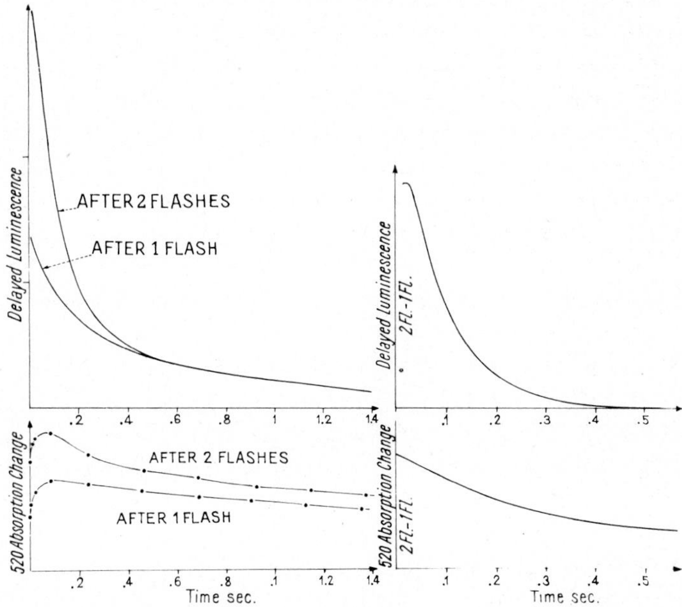

Fig. 2. Time course of the delayed luminescence and the 520 nm absorption change induced by one or two flashes 225 msec apart.
The vertical axis on the lower figure on the right hand side has been expanded by a factor of 2.

Figure 2 shows the time course of the absorption change at 520 nm and the delayed luminescence, after preillumination by one flash or by two flashes spaced 225 msec apart. One observe that, despite the fact that all the acceptor Q available has been reduced by the first flash (see Fig. 6), the delayed luminescence is much higher after the second flash than after the first. In the same way, the absorption level recorded after the second flash is higher than after the first. We represent (right of Fig 2) the difference between the effects produced by a group of two flashes and those produced by one flash, both for the delayed luminescence and for the absorption change. The lifetime of the luminescence stimulation is significantly shorter than the lifetime of the stimulation of the absorption change. Figure 2 is one of family of curves with different times between the two saturating flashes. Luminescence and absorption change can also be compared by considering the effect of the time interval between two saturating flashes, the time between the last flash and the measurement remaining constant (22 msec). Figure 3 shows the difference between the delayed luminescence measured after 2 flashes a time t apart and 2 flashes given synchronously. The same figure also shows similar differences for the absorption change. We subtract the effect of two synchronised flashes to allow for the fact that each flash is not perfectly saturating for the System I photoreaction. The correlation between the two types of measure remains excellent during a range of time of six orders of magnitude. The rising part of the curve gives the restoration time of Photosystem I activity (reduction of P_{700}). On a linear scale of time, one can observe a biphasic increase ($t_{1/2}$~ 70μsec and ~ 3msec) The decreasing part of the curve in Fig 3 is a measure of the decrease in absorption following the first flash, because in this time range the effect of the second flash is the same at each time and is added to the remaining effect of the first. The same type of correlation is obtained if luminescence and absorption are measured at dark times longer than 22msec after the last flash (up to 200msec). The time course of the decreasing part of the curve in Fig 3 depends strongly on the physiological state of the algae. The decay is slow ($t_{1/2}$ ~ 4sec) when the algae have been stored in the dark for one day and incubated with 10μM DCMU for more than one hour as in the experiment in Figure 3, but much faster ($t_{1/2} \simeq$ 200msec) for both luminescence and absorption, in the case of freshly harvested algae after a short incubation with DCMU (data not shown).

The experiments in Figure 4 were performed to check if the absorption increase which corresponds to phases a and b[18] produced the same effect on the delayed luminescence. It is important to answer this question because phase b remains difficult to interpret in terms of a membrane potential. Although the absorption spectrum of both phases is identical, it is difficult to rule out

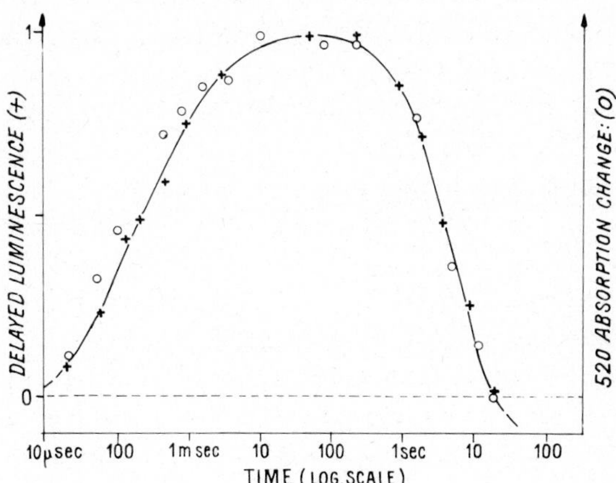

Fig. 3. Effect of the time interval between two saturating flashes on the delayed luminescence and the 520 nm absorption change.

Delayed luminescence (+) and absorption at 520 nm (o) were measured 22 msec after either the last flash of a group of two, or after two flashes given synchronously. The curve shows the differences between these two values normalised to the maximum value.

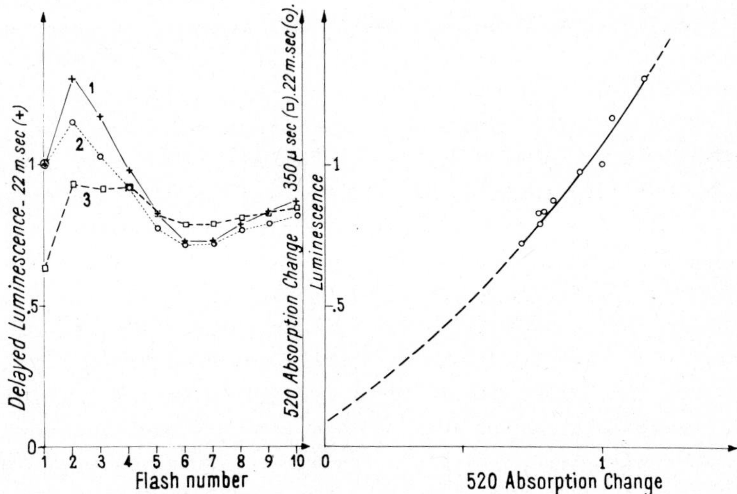

Fig. 4. Left. Delayed light and absorption change induced by a series of 10 flashes 450 msec apart. Delayed luminescence (+) was measured 22 msec, and the absorption at 520 nm at 350μsec (□) and 22 msec (o) after each flash of the sequence.

Right. Relationship between delayed luminescence and absorption change at 520 nm. The values were obtained from the sequence on the left.

the possibility that phases a and b correspond to different phenomena with similar spectra. Witt and Moraw[24] established that phase b is not observed with preilluminated material. Joliot and Delosme[21] observed that illumination by 5 to 6 flashes is sufficient to suppress phase b completely. In Figure 4, we plot the 520 nm absorption change measured 350μsec and 22 msec after each flash of a series of 10 flashes spaced 450 msec apart. The difference between the absorption change at 22 msec and 350μsec is a measure of the amplitude of the slow phase b which disappears after 5 flashes. On the same material, the delayed luminescence was measured 22 msec after each flash. It is clear that the delayed luminescence was correlated with the 22 msec component of the absorption change (phases a + b) and not with phase a alone. Thus, it is reasonable to assume that phases a and b correspond to the same property of the membrane as they are detected in the same way by two independent probes. In the right half of the figure, the intensity of delayed luminescence is plotted against the absorption level. In the relatively small range of absorption experimentaly analysed, the correlation curve is nearly linear, but a definite curvature, as shown in the figure, was observed in each of a number of experiments. It is difficult to obtain quantitative information on the lower part of the correlation curve. The only way is to decrease the energy of the flash but this causes a decrease in both the concentration of the precursor of the light emitting reaction (Z^+ Chl Q^-) and also the fluorescence yield which indirectly controls the delayed luminescence as shown by Lavorel[2]. Further information is provided by the observation that delayed luminescence can occur even when the absorption at 520 nm is equal to that in the dark. Thus the correlation curve does not go through zero but crosses the ordinate axis at a positive value. The dashed line gives an extrapolation of the experimental correlation curve. This extrapolation is certainly quite arbitrary ; however we can predicted a large dependence of luminescence upon the absorption change even when the latter is small.

The excellent correlation in Figs 3 and 4 between the absorption change and the delayed luminescence is difficult to reconcile with the data presented in Fig. 2 which shows that the stimulation of luminescence observed after a two flash preillumination disappears more rapidly than the stimulation in the absorption change. The lifetime of the stimulation of delayed luminescence and absorption has been studied under different experimental conditions (Fig. 5).

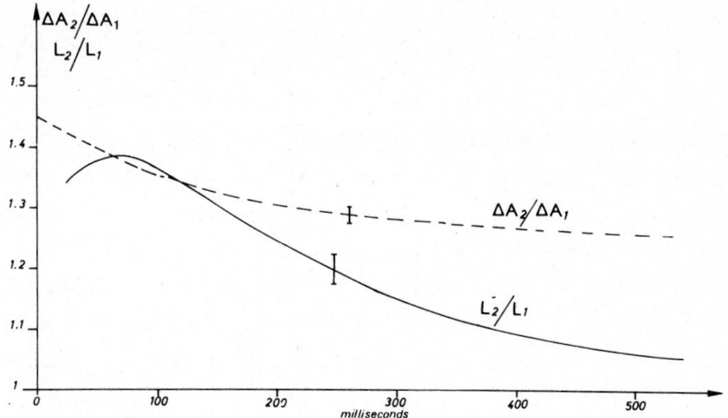

Fig. 5. Time course of the stimulation of the delayed luminescence and absorption change induced by a group of two flashes 9 sec apart compared to a single flash.

L_2 delayed luminescence measured after the second flash of a group of two 9 sec apart.

L_1 delayed luminescence measured after 1 flash.

ΔA_2 and ΔA_1 520 nm absorption change measured under the same conditions.

This experiment was performed with algae which presented an especialy long lived absorption change. The ratio of the delayed luminescence after two flashes 9 sec apart to the luminescence after one flash has been calculated as a function of the dark time after the last flash (L_2/L_1). The other curve ($\Delta A_2/\Delta A_1$) represents the same calculation for the 520 nm absorption change. It can be seen even more clearly than in Fig. 2 that the absorption presents a slower decay than the delayed luminescence.

Fig. 6 shows the time course of the decay of the prompt fluorescence after one or two saturating flashes 225 msec apart. A direct comparison of the slope of the two curves can be made even though the variable fluorescence yield is not proportional to the concentration of Q^- (see the insert from Bennoun[27]). The rate of the back reaction is seen to be faster after two flashes than after one flash. The stimulation in rate lasts for about 200 msec, a time comparable to the duration of stimulation of the delayed light (see lower curve). It is important to point out that the ratio of the initial slope of fluorescence decay after two flashes to that after one flash (1.55) is significantly lower than the ratio of delayed luminescence measured under the same conditions (2.2).

We can also observe that the prompt fluorescence yield measured immediately after each flash is practically equal, which means that the yield of fluorescence is not correlated with the absorption change. A similar experiment (Fig. 7) was carried out with mutant 54 subjected to a sequence of 4 flashes 450 msec apart which shows a slower decay of the absorption change and can thus integrate the effect of several flashes. The correlation between the rate of the back reaction and the intensity of delayed luminescence is good. As in Figure 6, we observe that the stimulation in the intensity of luminescence is higher than the stimulation of the rate.

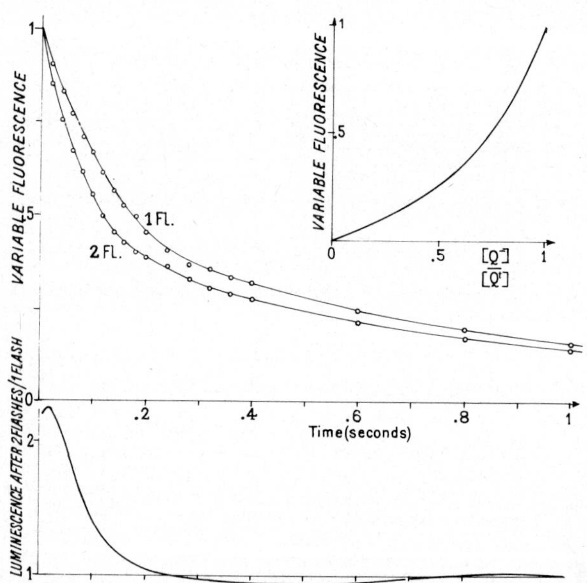

Fig. 6. Top. Time course of the prompt fluorescence measured after a group of two flashes 225 msec apart or after one flash.

Bottom. Ratio of the delayed luminescence measured after a group or two flashes 225 msec apart to the luminescence measured after one flash.

Insert. (from Bennoun[27]) relationship between the variable fluorescence yield and the concentration of Q^- (both normalized to their maximum value). This relationship was established by measuring the reoxidation of Q^- following continuous illumination.

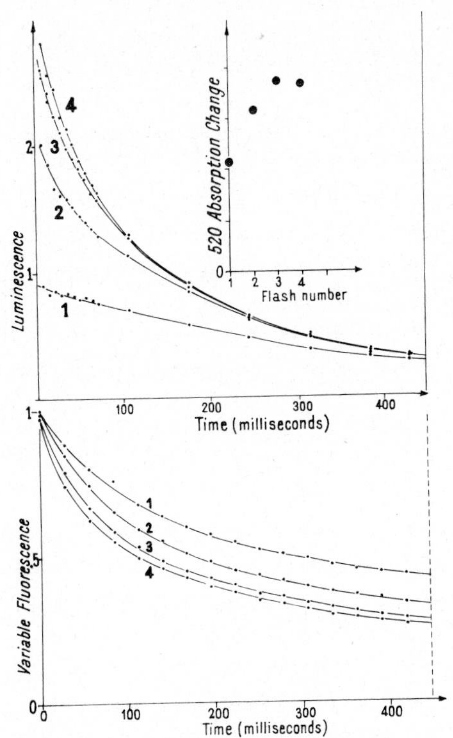

Fig. 7. Decay of the prompt fluorescence and delayed luminescence measured after each flash of a sequence of 4 flashes 450 msec apart in Chlamydomonas Mutant F54.

The number of preilluminating flashes is indicated on each curve. Insert. 520 nm absorption change measured 22 msec after each flash of the sequence.

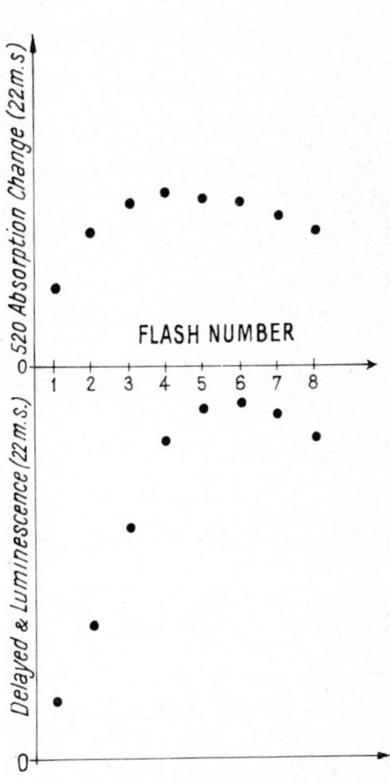

Fig. 8. Delayed luminescence and 520 nm absorption change measured 22 msec after each flash of a sequence of 8 flashes spaced 225 msec apart in Chlamydomonas M F54.

When mutant F54 is subjected to a series of flashes some discrepancy appears between the increase of the absorption change and the stimulation of the delayed light as a function of the number of the flash. One can observe in Figure 7 that the delayed luminescence is higher after 4 than after 3 flashes, while the absorption level is slightly lower. This discrepancy appears more clearly in Figure 8 where the flashes are spaced 225 msec apart. Thus, the intensity of delayed light is controled by other parameters in addition to that indicated by the absorption level at 520 nm.

4. Discussion

a) the results presented here demonstrate that, for a given concentration of the precursor of luminescence Z^+ Chl Q^-, the same property of the chloroplasts affects both the delayed luminescence and the 520 absorption change. This result could be predicted from the following assumptions which we discussed in the introduction :
1. Delayed light emission can be stimulated by a membrane potential
2. The 520 nm absorption change is a linear measurement of this membrane potential. Thus, further evidence has been provided for the idea that the 520 nm absorption change and the delayed luminescence represent two independent probes which measure the membrane potential.

b) The double flash experiment in the presence of DCMU shows that the electrical field generated by Photosystem I centers can be detected at the level of Photosystem II photocenters, less than 20 msec after its formation. This confirms the prediction of Mitchell that both photosystems are situated in parallel in the same membrane and that the electric field, initially localised as dipoles at the level of the photocenters, becomes delocalised in less than 10 msec, (the limit of resolution of our delayed fluorescence technique). The simplest interpretation is that suggested by Witt, namely that all variation of potential in the aqueous phase on either side of the membrane is rapidly dissipated by ion movements. Wraight and Crofts[17] previously reported experiments which they interpreted by assuming that a membrane potential generated by Photosystem I activity (in the presence of DCMU and a good electron donor to this photosystem) could stimulate delayed luminescence.

c) We show (Figs 6 and 7) that the increase in delayed light emission induced by a double flash compared to a single flash is directly correlated with an increase in the rate of the back reaction. On the basis of qualitative experiments, Kraan et al[9] have previously supposed that the stimulation of delayed light induced by various pretreatments (acid-base transition, salt effect) was also linked to an acceleration of the rate of the back reaction. On a more quantitative basis, we observe here that the stimulation of delayed light induced by repetitive flashes is higher than the stimulation of the rate of the back reaction (Figs 7 and 8). This discrepancy would be explained if there are several competitive pathways for the reoxidation of Q^- by the back reaction ; only a reaction passing via the excited state of chlorophyll can lead to luminescence emission. This problem has already been discussed by Lavorel[28] and Joliot et al[18] who showed that, in the absence of DCMU, the radiative pathway is predominant. Nevertheless, the intervention of non-radiative pathway is suggested by the experiments of A. Joliot[29]. Measuring the rate of the back reaction by the fluorescence decay in the presence of DCMU at temperatures between 0°C and -60°C, two independent back reactions were observed. One of these, a temperature dependent reaction , is associated with delayed luminescence emission ; another one, which becomes predominant below -40°C is temperature independent and does not seem associated with delayed light. At room temperature, the contribution of this temperature independent reaction is significant during the first 100 msec after a flash. Another hypothesis which is compatible with the first is that the extent to which the rate of the radiative back reaction depends on the membrane potential vary among the centers This hypothesis will be discussed in d.

d) More difficult to interpret is the fact that the dependence of the delayed luminescence and the rate of the back reaction on the membrane potential decreases with the time of observation (Figs 2 and 5). We do not think that this phenomenon can be interpreted only by the fact that the concentration of the light emitting precursor decreases faster after two flashes than after one flash. Actually, the concentration of Q^- measured at the same dark time after a group of two flashes or after a single flash never differ more than 8% (see Fig. 6). Such a small difference is not sufficient to explain the disappearence of the stimulation delayed luminescence for dark time longer than 300 msec.

This phenomenon can be interpreted if we postulate that, due to some structural heterogeneity, the dependence of the back reaction on the membrane potential is not identical for all the centers. In this case, the centers which are more sensitive to the field will disappear more rapidly. Such a hypothesis cannot be discussed in detail until more experimental data is available. We can just remark that some kinetic properties of Photosystem II centers can be explained in terms of a heterogeneity. For instance, the experiments of A. Joliot[29] show that the rate constant of the temperature-independent back reaction (studied at low temperature in the presence of DCMU) varies by several orders of magnitude between different centers.

e) We did not observe any direct dependence of the prompt fluorescence yield on the membrane potential. In the case of mutant F54, we were not able to observe an effect on the fluorescence yield measured immediately after each flash even after more than 10 flashes spaced 225 msec apart (data not shown. Under these conditions, we can expect that an appreciable proton gradient will have been formed, as well as a large membrane potential. These results are in contradiction with the data reported by Wraight and Crofts[30] who observed with continuous light in chloroplasts, a large quenching effect induced by the proton gradient. One cannot rule out the hypothesis that the acceleration of the back reaction both by the membrane potential and by the proton gradient can decrease the steady state level of fluorescence. However, a quantitative comparison of the amplitude of the proton gradient and of the photochemical turnover times under continuous and flashing light will be necessary before any final conclusions are possible.

References

1. Arthur, W.E. and Strehler, B. E., 1957, Arch. Biochem. Biophys. 70, 507.
2. Lavorel, J., 1968, Biochim. Biophys. Acta. 153, 727.
3. Lavorel, J., in Bioenergetics of Photosynthesis. ed : Govindjee, Academic Press, New York.
4. Arnold, W. and Azzi, J., 1968, Proc. Natl. Acad. Sci. US, 61, 29.
5. Fleischman, D.E., 1971, Photochem. Photobiol. 14, 277.
6. Mitchell, P., 1961, Nature, 204, 860.
7. Miles, C.D. and Jagendorf, A.T., 1969, Arch. Biochem. Biophys. 129, 711.
8. Barber, J. and Kraan, G.P.B., 1970, Biochim. Biophys. Acta 197, 49.
9. Kraan, G.P.B., Amesz,J., Velthuys, B.R. and Steemers, R.G., 1970, Biochim. Biophys. Acta. 223, 129.

10. Hardt, H. and Malkin, S., 1973, Photochem. Photobiol. 17, 433.

11. Mayne, B.C., 1967, Brookhaven. Symp. Biol. 19, 460.

12. Arnold, W. and Azzi, J., 1971, Photochem. Photobiol. 14, 233.

13. Witt, H.T., 1971, Quaterly Review, 4, 365.

14. Crofts, A.R. and Jackson, J.B., 1970, in Electron Transport and Energy Conversation. (Eds. S. Papa, J.M. Tager, E. Quagliariello and E.C. Slater, Adriatica Edictrice Bari, Italy). 383.

15. Jackson, J.B. and Crofts, A.R., 1960, FEBS Letters, 4, 185.

16. Fleischman, D.E. and Clayton, R.K., 1968, Photochem. Photobiol. 8, 287.

17. Wraight, C.A. and Crofts, A.R., 1971, Eur. J. Biochem. 19, 386.

18. Joliot, P., Joliot, A., Bouges, B. and Barbieri, G., 1971, Photochem. Photobiol. 14, 287.

19. Bennoun, P., 1971, C.R. Acad. Sci. Paris, 273, 2654.

20. Ducruet, J.M. and Lavorel, J., 1974, Biochem. Biophys. Res. Comm. In press.

21. Joliot, P. and Delosme, R., 1974, Biochim. Biophys. Acta. In press.

22. Bennoun, P., 1970, Biochim. Biophys. Acta. 216, 357.

23. Barbieri, G., Delosme, R. and Joliot, P., 1970, Photochem. Photobiol. 12, 197.

24. Witt, H.T. and Moraw, R., 1959, Z. Physik Chem. Nec Folge. 20, 254.

25. Bennoun, P. and Levine, R.P., 1967, Plant. Physiol. 42, 1284.

26. Sato, V.L., Levine, R.P. and Neumann, J., 1971, Biochim. Biophys. Acta. 253, 437.

27. Bennoun, P., 1971, Thèse de Doctorat d'Etat. Paris.

28. Lavorel, J., 1971, Photochem. Photobiol. 14, 261.

29. Joliot, A., 1974, Biochim. Biophys. Acta. In press.

30. Wraight, C.A. and Crofts, A.R., 1970, Eur. J. Biochem. 17, 319.

M. AVRON, *Proceedings of the Third International Congress on Photosynthesis*
September 2-6, 1974, The Weizmann Institute of Science, Rehovot, Israel
Elsevier Scientific Publishing Company, Amsterdam, The Netherlands, 1974

AN ANALYSIS OF THE PMS-EFFECTED QUENCHING OF CHLOROPLAST FLUORESCENCE

George Papageorgiou

Department of Biology, Nuclear Research Center "Democritus"
Athens, Greece

1. Summary

Oxidized and ascorbate-reduced PMS (probably a semiquinone, PMS-SQ) are shown to quench the fluorescence of isolated spinach chloroplasts by means of direct interactions with excited Chl a molecules. This effect manifests independently of the Q- and the X_E-mediated quenching effects, that are known to occur in the presence of PMS. Reduced PMS is a better direct quencher than oxidized PMS suggesting that the thylakoid membrane is more permeable to it. The permeation of these compounds is facilitated by the light-induced protonation of the membrane, and suppressed by added Mg^{++}.

At weak excitation, micromolar concentrations of ascorbate-reduced PMS stimulate the fluorescence of normal chloroplasts. This effect is attributed to the reduction of the primary electron acceptor (Q)of photosystem II by the reduced cofactor. At strong excitation, or in the absence of ascorbate, quenching is invariably observed.

2. Introduction

Addition of oxidized, or ascorbate-reduced PMS, to normal[1-3] and to DCMU-poisoned[4-9] photosynthetic preparations leads to quenching of Chl a fluorescence. When photosystem II is functional and the excitation intensity limiting, the fluorescence decrease is believed to reflect a PMS-mediated oxidation of Q.[2,3] With a nonfunctional photosystem II, the quenching reflects the buildup of the transmembrane electrochemical potential X_E[5], and in particular the protonation of the thylakoid membrane.[5,10] It is understood that X_E develops only in strong exciting light and as a result of PMS-catalysed electron transport around photosystem I.

As oxidants, the N-methylphenazonium cation (PMS) and its semiquinone (PMS-SQ; obtained by reduction with ascorbate) would be expected to quench also by means of direct interactions with excited chlorophylls. Electron deficient compounds are known to be quenchers of Chl _a_ fluorescence both in solution[11] and _in vivo_.[12-15] In order to interact directly with electronically excited chlorophylls, PMS must penetrate to the pigment layer of the membrane. On the other hand, in the X_E-mediated and in the Q-mediated quenching, PMS interacts with intermediates of the photosynthetic electron transport.

The present paper describes evidence in support of a direct quenching effect by oxidized and ascorbate-reduced PMS.

3. Methods and Materials

Broken chloroplasts were prepared in 300 mM sucrose, 50 mM potassium (N-morpholino)-ethanesulfonate, pH 6.4, according to a conventional procedure[14]. When present, DCMU and sodium ascorbate were introduced to final concentrations of 20 μM and 1 mM, respectively. The concentration of the PMS stocks was corrected on the basis of the molar absorptivity given by Zaugg[16] (E_{388}= 26,300 M^{-1} cm^{-1}). Pure Chl _a_ was prepared according to Strain and Sherma[17] from lyophilized cells of the blue-green alga _Phormidium luridum_, and it was transferred to methanol by evaporating the petroleum ether at 40° C under reduced pressure. Only freshly prepared methanolic solutions were used, as it was found that the absorption of Chl _a_ in this solvent changes on prolonged storage.

Fluorescence was measured essentially as described before.[14] A photographic shutter (Compur 1) permitted full intesity illumination of the sample within 1 ms, and the signal was displayed either by a Tektronix 549 storage oscilloscope or by a strip chart recorder. Excitation intensities were measured with a calibrated Bi-Ag thermopile. Absorption was measured with a Bausch and Lomb Spectronic 505 spectrophotometer, and with a Hitachi Model 356 dual wavelength spectrophotometer. The absorption of ascorbate-reduced PMS was obtained by adding microliter quantities of methanolic PMS to 6.9 mM sodium ascorbate in methanol (about 200-fold excess over PMS) directly in the spectrophotometer cuvette. Further spectral details are given in the legends to the figures.

4. Results

Figure 1 shows absorption spectra of methanolic solutions of oxidized and ascorbate-reduced PMS in the visible region. The spectrum

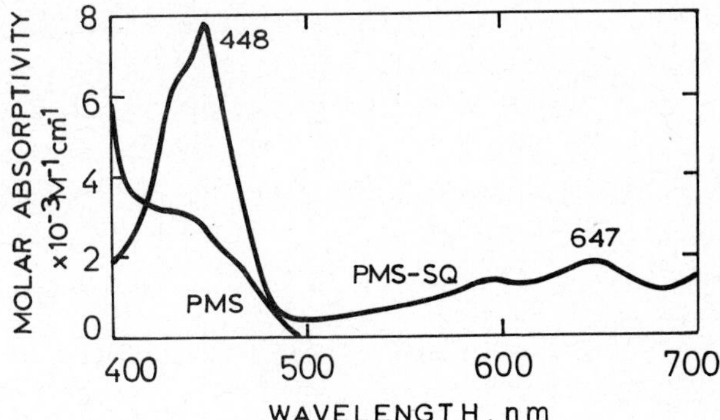

Fig. 1. Absorption spectra of oxidized PMS and ascorbate-reduced (PMS-SQ) in methanol.

of the reduced compound resembles that of the semiquinone[16]. It is characterized by an absorption band at 448 nM (E = 7,800 M^{-1} cm^{-1}) as well as by a weak absorption extending throughout the visible region. Fully reduced PMS (PMSH) was prepared by further reduction of the semiquinone with excess sodium dithionite. PMSH is insoluble in water, methanol, and petroleum ether, while in benzene solution it reoxidizes rapidly to the green semiquinone. On the basis of these observations we shall assume that, in the presence of excess ascorbate, PMS exists in the extrathylakoid space as a semiquinone.

Figure 2 shows that PMS is a quencher for excited Chl a in methanol. The quenching obeys a linear Stern-Volmer relation with a quen-

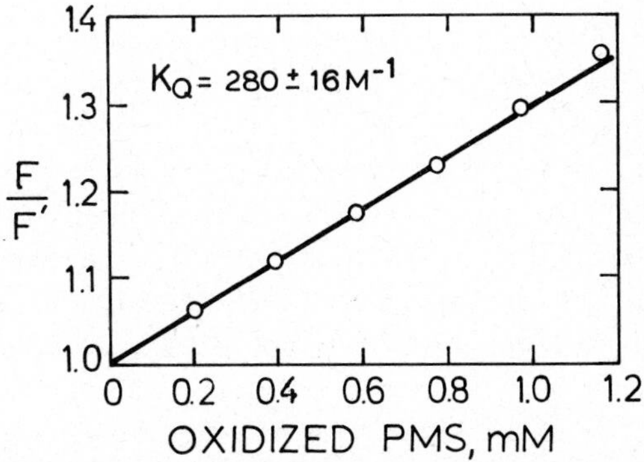

Fig. 2. Stern-Volmer plot of the quenching of the fluorescence of Chl a in methanol. Excitation, 620 nm; Δλ, 12 nm; 4.9 kergs. $cm.^{-2}$ s^{-1}. Detection, 675 nm; Δλ, 6.6 nm. F,F' unquenched and quenched fluorescence.

ching constant, K_Q = (F-F') /F'C, equal to 280 $\pm$ 16 M^{-1} (six determinations) On the basis of a fluorescence lifetime for Chl *a* in methanol of τ = 6.9 ns[18], this corresponds to a collision rate constant, $k = K_Q\tau^{-1}$, of 4.06 x $10^{10} M^{-1} s^{-1}$, i.e. in the range of the diffusion imposed limit for bimolecular collisions. Intermolecular transfer of excitation energy may contribute to the magnitude of the quenching constant, provided that the association of PMS and Chl *a* lasts longer than the exciton visit on the latter molecule[19]. Energy transfer, in other words, broadens the exciton trapping range of a quenching center. We consider, however, this possibility as unlikely in the present case because: (1) The average Chl *a* to Chl *a* distance of our samples was 965 Å, which is much greater than the distance for 50 % probability of excitation transfer (Ro = 69 Å; ref. 20). (2) A 10-fold concentrated solution of Chl *a* in methanol yielded the same quenching constant, although energy transfer is 100 times more probable in this instance.

Due to a strong "inner filter" effect (see fig. 1), we were unable to determine quantitatively the quenching constant for ascorbate-reduced PMS. Taking into account the "front face" fluorescence detection geometry of our fluorometer, we estimated that 20 µM PMS-SQ attenuates the true fluorescence signal by 5 %, and 50 µM by 12 %. Qualitatively, however, the measured fluorescence lowering is greater than the inner filter estimates, suggesting that the semiquinone quenches directly. At the concentration used in these experiments, ascorbic acid had no effect on the fluorescence of methanolic Chl *a*.

When PMS is introduced to strongly illuminated and DCMU-poisoned chloroplasts or algae, it causes a lowering of Chl *a* fluorescence[4-9] (hereafter denoted as ΔF). During a subsequent dark rest of the sample, the lost fluorescence is recovered but the recovery is never complete, irrespective of the duration of the dark rest. We shall call the maximal dark-recovered fraction of ΔF as ΔF_{REV}, and the unrecovered remainder as ΔF_{IRR}. We shall then assume, plausibly, that the light-induced ΔF_{REV} is the true measure of the X_E-mediated quenching, while ΔF_{IRR} represents the nonphotochemical direct quenching effect of PMS. In support of the notion that a fraction only of ΔF is due to X_E-mediated quenching, we found that the uncoupler CCCP, at 5 µM, inhibits the PMS-induced fluorescence lowering only 40 to 70 %. This uncoupler, however, exerts a complex influence, being itself a quencher of the fluorescence of Chl *a* *in vivo*.

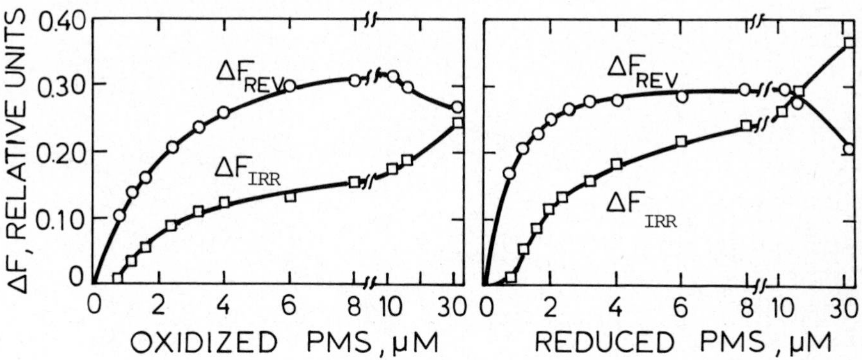

Fig. 3. Concentration curves of the dark-reversible (ΔF_{REV}) and the dark-irreversible (ΔF_{IRR}) portions of the fluorescence lowering induced by oxidized (left) and ascorbate-reduced PMS (right). Excitation through Corning filter C.S. 4-72 and 5 cm of 5 % $CuSO_4$; 29.1 kergs. cm^{-2}. s^{-1}. Detection, 680 nm.

According to fig. 3, the concentration curves of ΔF_{REV} follow a saturation pattern, while those of ΔF_{IRR}, although biphasic, they do not saturate. The biphasic character of the latter curves reflects the presence of quencher-sensitive (variable fluorescence) and quencher-resistant (constant fluorescence) fluorescence components[13]. After the saturation plateau, the concentration curves of ΔF_{REV} begin to decline, because direct quenching destroys increasingly larger fractions of variable fluorescence. Double-reciprocal plots (ΔF_{REV}^{-1} vs C^{-1}; not shown here) indicate maximal X_E-mediated quencing of 36 % both for oxidized and for ascorbate-reduced PMS. Below saturation, however, reduced PMS catalyses X_E-mediated quenching more efficiently than its oxidized counterpart.

Ascorbate-reduced PMS is more efficient as a direct quencher also. Quenching constants, $K_Q = (F-F')/F'C = \Delta F_{IRR}/(1-\Delta F_{IRR})C$, calculated from the initial slopes of the corresponding curves in fig. 3, are $6 \times 10^4\ M^{-1}$ for the oxidized and $16.4 \times 10^4\ M^{-1}$ for the reduced species of the cofactor. Two causes may be responsible for these extremely high values. (1) The extensive migration of the excitation among the Chl *a* molecules *in vivo*, which greatly magnifies the effectiveness of the PMS quenching centers; and (2) a preferential concentration of the cofactor in the pigment phase of the lamella.

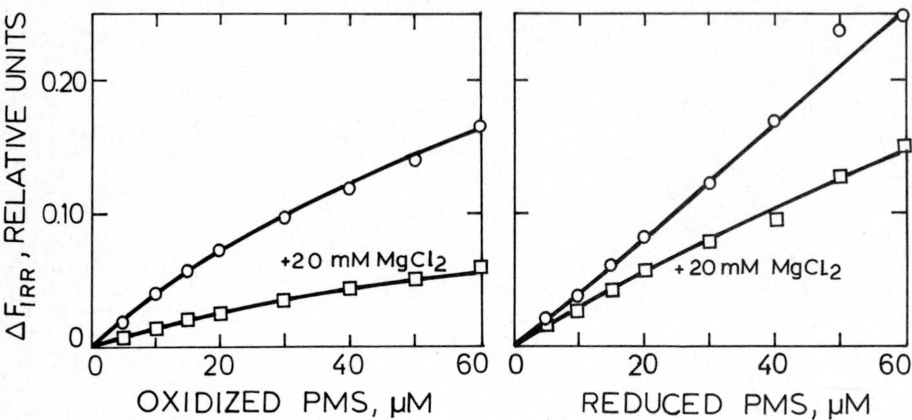

Fig. 4. Concentration curves of the dark-irreversible lowering (ΔF_{IRR}) of the fluorescence of weakly illuminated chloroplasts by oxidized (left) and ascorbate-reduced PMS (right). Excitation, 620 nm; $\Delta\lambda$, 12 nm; 0.35 kergs. cm^{-2}. s^{-1}. Detection 680 nm. The curves for ascorbate-reduced PMS are corrected for inner filter errors on the basis of the molar absorptivities of fig. 1, and by assuming non-scattering samples.

X_E-mediated quenching, in the presence of PMS, manifests only in strong exciting light[5]. The fluorescence lowering obtained when PMS is added to weakly illuminated and DCMU-poisoned chloroplasts is due, then, to direct quenching, which is not reversed by darkness. Fig. 4 presents quenching curves for oxidized and ascorbate-reduced PMS, obtained with chloroplasts excited with weak red light(λ=620 nm; I = 350 ergs. cm^{-1}.s^{-1}). With red excitation, the inner filter error is less in the case of ascorbate-reduced PMS, and completely absent in the case of oxidized PMS (cf. absorption spectra in fig. 1). A further, inner filter correction has been included in the quenching curves for reduced PMS. This was computed by assuming nonscattering samples, and on the basis of the absorptivities given in fig. 1 and the "front-face" detection geometry of our apparatus. At low co-factor concentrations, when the inner filter error is better corrected, both oxidized and ascorbate-reduced PMS yield the same quenching constant, namely K_Q= 3.5 x 10^4 M^{-1}. This constant is less than those calculated for direct quenching under strong illumination of DCMU-poisoned chloroplasts (vide supra). At higher concentrations, reduced PMS appears to be a better direct quencher than oxidized PMS, but this can also be due to uncorrected inner filter attenuation of the fluorescence.

Mg^{++}, which has been shown to have no effect on the X_E-mediated quenching in the presence of cyclic electron transport cofactors and strong exciting light[7,21], is shown in fig. 4 to inhibit quenching by both oxidation species of PMS. At 20 mM Mg^{++}, the inhibition is maximal, since exactly the same effect is obtained on increasing the Mg^{++} concentration to 100 mM. Mg^{++} is known to influence direct quenching of chloroplast fluorescence by nonreducible nitroaromatic compounds, an effect which has been attributed to cation-induced changes in the membrane ultrastructure[14].

Arnon et al[1] have demonstrated extensive quenching of normal chloroplast fluorescence in strong exciting light and in the presence of PMS. The quenching was attributed to an induced spillover of Chl *a* excitation from the highly fluorescent photosystem II to the weakly fluorescent photosystem I as a result of the cofactor catalysed cyclic electron transport. Murata and Sugahara[5], however, in discussing these results, suggested that the observed quenching is due to the oxidation of the photosystem II electron acceptor (Q) by the added cofactor. A Q-mediated quenching of normal chloroplast

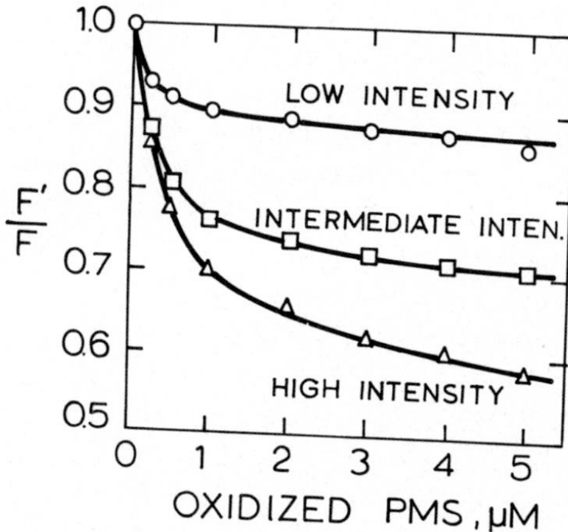

Fig. 5. Dependence of the ratio of unquenched to quenched fluorescence on the concentration of oxidized PMS. Excitation as in fig. 3; low intensity, 0.68 kergs. $cm^{-2}.s^{-1}$; intermediate intensity, 5.5 kergs $cm^{-2}.s^{-1}$; high intensity, 29.2 kergs $cm^{-2}.s^{-1}$ Detection, 680 nm.

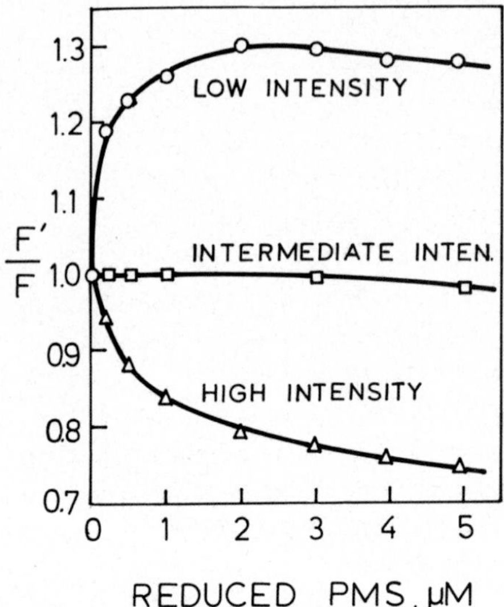

Fig. 6. Dependence of the ratio of unquenched to quenched fluorescence on the concentration of ascorbate-reduced PMS. Fluorescence excitation and detection as in fig. 5.

fluorescence was invoked also by Heath and Packer[2], and by Schwartz[3] to account for similar observations.

Figure 5 presents concentration curves of the ratio F'/F (quenched over unquenched fluorescence of normal chloroplasts) for oxidized PMS, at three different excitation intensities. Quenching is more prominent at higher excitation intensities and shows a quasi-saturation at about 1-2 μM PMS. This may suggest that PMS does not intervene as a reducible substrate but as a catalyst mediating electron transport to some other acceptor. Relevant to this is the inability of PMS to support oxygen evolution by chloroplasts up to a concentration of 50 μM.[22]

In contrast to oxidized PMS, ascorbate-reduced PMS is shown in fig. 6 to stimulate the emission of fluorescence by weakly illuminated chloroplasts, and to quench the fluorescence of strongly illuminated chloroplasts. At a suitably chosen intermediate excitation intensity, the stimulatory and the quenching effects balance out. Because of the existence of the stimulatory effect less quenching is observed in strong light with ascorbate-reduced PMS, than with oxidised PMS. The stimulation and the quenching of fluorescence saturate at about 1-2 μM of ascorbate-reduced PMS. Beyond this concentration, a slow decline of the Chl a fluorescence is observed.

5. Discussion

Our results indicate that in addition to the X_E- and the Q-mediated quenching effects, oxidized and ascorbate-reduced PMS suppress also the fluorescence of chloroplasts by means of direct quenching interactions with excited Chl a molecules. Arguments in support of direct quenching are as follows: (1) Oxidized and (most likely) ascorbate-reduced PMS are efficient quenchers of excited Chl a in methanol. (2) The dark-irreversible portion of the light-induced fluorescence lowering in the presence of PMS increases monotonously with the concentration of the cofactor. (3) PMS lowers the fluorescence of DCMU-poisoned chloroplasts at an excitation that is too weak for formation of X_E.

The bimolecular collision rate constant for Chl a and PMS in methanol (fig. 2) is in the range of the diffusion-imposed limit, suggesting that there is no preferred direction for an effective encounter between these molecules. This is in agreement with the delocalization of the positive change over the entire PMS cation, and implies that *in vivo* the quencher may interact with any one of its neighboring chlorophylls that happens to be visited by the migrating exciton. The linearity of the fig. 2 plot suggests further that the observed quenching is the result of a single process. Since we did not detect new absorption bands in mixtures of PMS and Chl a in methanol we are inclined to dismiss the possibility of "static quenching"[23]. It appears that PMS quenches "dynamically" interacting with excited molecules only. In conformity with this conclusion, Mohanty et al[7]. reported that PMS shortens the lifetime of Chl a *in vivo*.

In intense exciting light, ascorbate-reduced PMS (presumably a semiquinone free radical) is a better direct quencher than oxidized PMS (fig. 3). Since the quenching efficiency of the reduced species cannot exceed that of the oxidized species (which is the theoretical maximum, cf. fig. 2), stronger quenching by reduced PMS implies an easier permeation to the pigment phase. On the basis of the measured quenching constants for oxidized ($K_Q = 6 \times 10^4\ M^{-1}$) and for reduced

PMS ($K_Q = 16.4 \times 10^4 M^{-1}$), the reduced species appears to permeate at least 2.7 times faster than the oxidized species.

In weak exciting light, only dark-irreversible quenching is observed (fig. 4) which is less pronounced than dark-irreversible quenching in strong light (cf.fig. 3). Two possible causes are: (1) Less quencher-sensitive variable fluorescence in weak exciting light; and (2) easier permeation of PMS to the pigment layer of the lamella in strong exciting light. The latter possibility can be ascribed to the light-induced protonation of the thylakoid membrane which would facilitate penetration of positively charged molecules by suppressing the bound negative charge[23,24].

It has been shown that the PMS-induced fluorescence lowering in strong light is completely unresponsive to added Mg^{++}.[8] The true effect of this cation, however, should be obscured by its tendency to be expelled to the extrathylakoid space under strong illumination[25]. In weak exciting light (fig. 4), Mg^{++} inhibits direct quenching by both oxidized and ascorbate-reduced PMS, although the inhibition is more pronounced for the former species. Binding of Mg^{++}, therefore, influences the extent of direct quenching in the opposite direction relative to the effect of the light-induced protonation of the membrane, implying that different permeability barriers are affected in each case. Since the dissociation constants of free side chain carboxyl groups are smaller than those of their salts, we may surmise that protonation affects primarily an electrostatic permeability barrier. On the other hand, the effect of Mg^{++} should be mostly stereochemical, since as a divalent cation it can interact with two different nucleophiles.

PMS interferes with the electron transport system of normal chloroplasts both at the level of the primary electron acceptor Q of photosystem II, and at the level of the cyclic electron transport. The second process requires intense excitation, and on this basis it can be discriminated from the first. In the case of weakly illuminated chloroplasts, partial quenching by low concentrations of oxidized PMS should be attributed to a Q-mediated effect, while in strong light to combined Q-mediated and X_E-mediated effects. (fig. 5). The clearest indication of a cofactor interaction with Q is the stimulation of chloroplast fluorescence by ascorbate-reduced PMS (fig. 6). This phenomenon manifests only in weak excitation where the X_E-mediated quenching by PMS is not operating. The second, less steep declining portions of the quenching curves of fig. 5 and 6 represents probably a direct quenching of chloroplast fluorescence by PMS.

References

1. Arnon, D.I., Tsujimoto, H.V. and McSwain, B.D., 1965, Proc. Natl. Acad. Sci. U.S. 54, 927.
2. Heath, R.L. and Packer, L., 1968, Arch. Biochem. Biophys. 125, 1019.
3. Schwartz, M., 1967, Biochim. Biophys. Acta 131, 548.
4. Govindjee and Yang, L., 1967, cited by Govindjee, Papageorgiou, G. and Rabinowitch, E. in: Fluorescence, Theory, Instrumentation, and Practice, ed. G.G. Guilbault (Marcel Dekker, New York) p. 511.
5. Murata, N. and Sugahara, K. 1969, Biochim. Biophys. Acta 189, 182.
6. Papageorgiou, G., Isaakidou, J. and Argoudelis, C. 1972, FEBS Letters 25, 139.
7. Mohanty, P.K., Zilinskas-Braun, B., and Govindjee 1973, Biochim. Biophys. Acta 292, 459.
8. Mohanty, P.K., and Govindjee, 1973, Photosynthetica 7, 146.
9. Papageorgiou, G. 1974 in: Bioenergetics of Photosynthesis, ed. Govindjee (Academic Press, New York) chapter 6, in press.
10. Wraight, C.A. and Crofts, A.R. 1971. Eur. J. Biochem. 17, 319.
11. Livingston, R. and Ke, C.L. 1950. J. Amer. Chem. Soc. 72, 909.
12. Teale, F.W.J. 1960, Biochim. Biophys. Acta 42, 69.
13. Amesz, J. and Fork, D.C. 1967, Biochim. Biophys. Acta 143, 97.
14. Papageorgiou, G., and Argoudelis, C. 1973, Arch. Biochem. Biophys. 156, 134.
15. Etienne, A.L. and Lavergne, J. 1972, Biochim. Biophys. Acta 282, 268.
16. Zaugg, W.S. 1964, J. Biol. Chem. 239, 3964.
17. Strain, H. and Sherma, J. 1969, J. Chem. Educ. 46, 476.
18. Brody, S.S. and Rabinowitch, E. 1957, Science 125, 555.
19. Lavorel, J. 1967, J. Chem. Phys. 47, 2235.
20. Duysens, L.N.M. 1964, in: Progress in Biophysics and Molecular Biology, eds. J.A.V. Butler and H.E. Huxley, (Pergamon Press, Oxford) 14, 1.

21. Cohen, W.S. and Sherman, L.A. 1971, FEBS Letters 16, 319.

22. Reeves, S.G. and Hall, D.O. 1973, Biochim. Biophys. Acta 314, 66.

23. Dilley, R.A. 1968, Biochemistry 7, 338.

24. Murakami, S., and Packer, L. 1970, J. Cell Biol. 47, 332.

25. Dilley, R.A., and Vernon, L.P. 1965, Arch. Biochem. Biophys. III, 365.

M. AVRON, *Proceedings of the Third International Congress on Photosynthesis*
September 2-6, 1974, The Weizmann Institute of Science, Rehovot, Israel
Elsevier Scientific Publishing Company, Amsterdam, The Netherlands, 1974

SLOW CHLOROPHYLL FLUORESCENCE CHANGES IN ISOLATED INTACT CHLOROPLASTS: EVIDENCE FOR CATION CONTROL

J. Barber, A. Telfer, J. Mills and J. Nicolson.
Botany Department, Imperial College, London, S.W. 7.

1. Summary

Isolated spinach chloroplasts show dark reversible slow fluorescence quenching when illuminated which is not observed after removal of their outer membranes by osmotic shock. The rate of quenching is dependent on the ability of the chloroplasts to fix CO_2 and is speeded up by increasing the rate of coupled electron flow and/or by reducing the lag period between the initiation of the illumination and O_2 evolution. Reconstitution studies with osmotically shocked chloroplasts demonstrate that the quenching is due to an energy dependent net movement of cations from the granal to the stromal compartments. In a cation free medium Ionophore A23187 reverse the quenching observed with whole chloroplasts but nigericin and monensin had no effect suggesting that light induced divalent and not monovalent cation gradients are involved in the in vivo quenching process. Gramicidin D, even in the presence of added monovalent cations, does not relieve the quenched state with whole chloroplasts suggesting that this antibiotic allows cation leakage across the outer membranes as well as across the thylakoids. Correlation of the cation dependent fluorescence with millisecond delayed light emission indicates a direct relationship between the fluorescence changes and the chemical component of the high energy state (ΔpH). The results are discussed in terms of their possible significance to the State 1 - State 2 hypothesis[1].

2. Introduction

Intact cells show light induced slow fluorescence yield changes which are independent of the redox state of the S2 trap and seem to be associated with regulating the delivery of quanta to the two photosystems[1]. Indications of the mechanisms involved in this regulation come from studies of fluorescence yield changes induced by varying the cation levels of isolated broken chloroplast suspensions[2,3]. However until very recently the slow light induced fluorescence changes detected with intact organisms were not observed with isolated chloroplasts. Krause[4,5] and work from our laboratory[6,7] have demonstrated that if chloroplasts are isolated so that they retain their outer membranes (envelope) they show slow fluorescence changes comparable with those observed with intact leaves. In this paper we relate these light induced fluorescence changes with energy dependent cation fluxes between the granal and stromal compartments and discuss the possible significance of this process in regulating photosynthetic electron flow.

3. Materials and Methods

Intact chloroplasts were isolated from spinach leaves by the method of Stokes and Walker[8]. Preparations contained about 65% whole chloroplasts as determined by the ferricyanide method[9]. Illumination of the chloroplasts was via a 4 mm Schott BG 18 filter at an intensity of 70 kergs $cm^{-2}sec^{-1}$ and the resulting prompt fluorescence and msec delayed light were measured as previously described[10]. Ionophore A23187 was obtained from Eli Lilly, Indianapolis (Lot No. 361-066-275) as was the nigericin (Lot No. 189-380B-171-A) and monensin (Lot No. 370-589 AD-291). Gramicidin D was obtained from Sigma and phlorizin from R.H. Emanuel Ltd., London. Additional details are given in the figure legends.

4. Results

(i) General Properties of the Slow Fluorescence Changes

Fig. 1 summarises some of the properties of the slow quenching observed with isolated intact chloroplasts. The rate of quenching is dependent on the rate of electron flow and on the lag period between turning on the illumination and observing oxygen evolution, for example phosphoglycerate (PGA) versus HCO_3^- as electron acceptor. The quenching is reversed in the dark with a $t_{\frac{1}{2}}$ of about 1.5 min and is independent of the time taken to reach the low fluorescing state. By using oxaloacetate, which unlike PGA and HCO_3^- does not require ATP for its reduction, it has been shown that the quenching can be independent of the redox state of Q (ref. 6). Addition of DCMU relieves the quenched state at rates comparable with the dark reversal except for an initial fast phase which probably reflects a "Q effect". The energy transfer inhibitor phlorizin does not relieve the quenched state but uncouplers like nigericin (if K^+ is in the suspending medium) and carbonyl cyanide m-chlorophenyl hydrazone (CCCP) induce a relatively rapid return from the low to

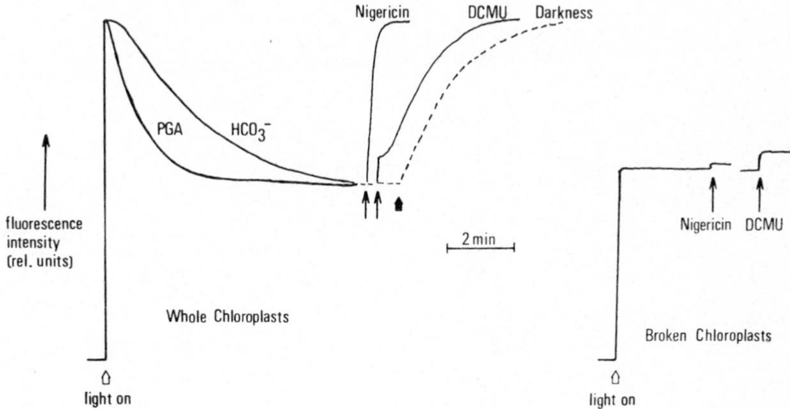

Fig. 1. An illustration of some of the properties of the slow fluorescence changes observed on illuminating a suspension of dark pretreated isolated whole spinach chloroplasts. The experiments were carried out in 3 mls of medium containing 0.33M sorbitol and 50mM HEPES adjusted to pH 7.6 with KOH with a chlorophyll concentration of about 30 μg/ml. The medium also contained either 5mM HCO_3^- or 1mM PGA and the injections were 0.1 μM nigericin or 3.3 μM DCMU. The broken chloroplasts were obtained by osmotic shock in 1½ mls H_2O followed by the addition of 1½ mls of the above medium at double strength.

the high fluorescence level ($t_{\frac{1}{2}}$ of about 2 sec). When whole chloroplasts are subjected to an osmotic shock the fluorescence level decreases and the slow induction effect is lost. These osmotically shocked chloroplasts did not reduce HCO_3^- or PGA, but did show endogenous electron flow (seen as a Mehler type reaction) and were able to establish a light induced pH gradient.

The above observations suggest that the outer membranes control the factors which give rise to the slow fluorescence quenching and that the mechanism involved is independent of phosphorylation products and in some way associated with the establishment of the high energy state (HES) generated by electron flow.

(ii) Cation Involvement

Experiments with osmotically shocked chloroplasts have helped in understanding the mechanism of the quenching phenomenon as have experiments involving the use of ionophores.

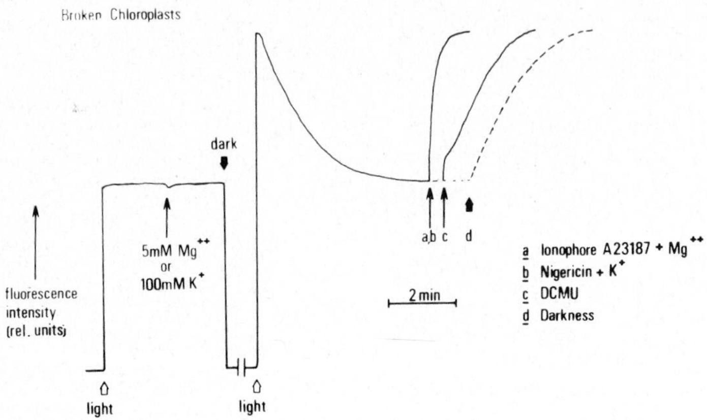

Fig. 2. An illustration of some of the properties of cation dependent fluorescence changes. The method of preparing the broken chloroplasts and the concentrations of nigericin and DCMU are the same as Fig. 1. Ionophore A23187 was added to give a final concentration of 1μg/ml (i.e. 2μM).

We have found, as did Krause[5], that the addition of about 5mM Mg^{++} or 100mM K^{+} to dark treated osmotically shocked chloroplasts not only induced a high fluorescing state as reported earlier[2,11] but also restored the ability of the chloroplasts to show the slow light induced quenching (see Fig. 2). Like whole chloroplasts, the quenched state can be relieved by darkness or DCMU treatment and also by the divalent cation ionophore, Ionophore A23187 when Mg^{++} is present[7] or by nigericin when K^{+} is present in the medium (see Fig. 2). Under similar conditions both ionophores uncouple electron flow. The results indicate that the slow fluorescence quenching observed with whole chloroplasts is related to net cation transfer from the granal to the stromal compartments powered by light driven uptake of protons into the grana. This would account for the lack of effect on fluorescence of cation additions to illuminated broken chloroplasts which, as mentioned above, were able to maintain a light induced pH gradient under these conditions. In fact as Fig. 2 shows it was necessary to give a dark pretreatment before observing the cation dependent light induced quenching.

When whole chloroplasts were suspended in a medium free of externally added alkali and alkaline earth cations Ionophore A23187 relieved the low fluorescing state but nigericin and monensin did not until K^+ or Na^+ respectively were added to the suspending medium (see Fig. 3). This result suggests that within the intact organelle the fluorescence changes are entirely associated with the energy dependent net movement of Mg^{++} (or possibly Ca^{++}) between the granal and stromal compartments. The effect of the above ionophores on electron transport with whole chloroplasts suspended in the same cation free medium also suggests that divalent but not monovalent cations are available for exchange with H^+ across the thylakoid membranes so as to bring about uncoupling.

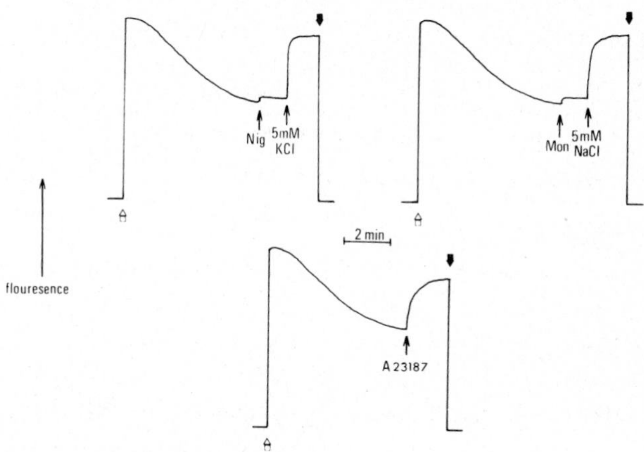

Fig. 3. Effect of the ionophores, nigericin (0.1 μM), monensin (0.1 μM) and A23187 (1.0 μM) on the light induced quenched state in whole chloroplasts suspended in 0.33M sorbitol and 10mM HEPES adjusted to pH 7.6 with tris base. The chlorophyll concentration was 10 μg/ml and other conditions as Fig. 1.

When the non-specific cationic ionophore gramicidin D was used the results were different to those obtained with the above mentioned uncouplers. As Fig. 4 shows when this antibiotic was added to illuminated whole chloroplasts suspended in a medium containing K^+ ions the low fluorescing state was not relieved. If DCMU was added after the gramicidin treatment it also had no significant effect. However, gramicidin did relieve the low fluorescence state created with illuminated broken chloroplasts treated with Mg^{++} or K^+ and so did DCMU in the usual way. Without sufficient

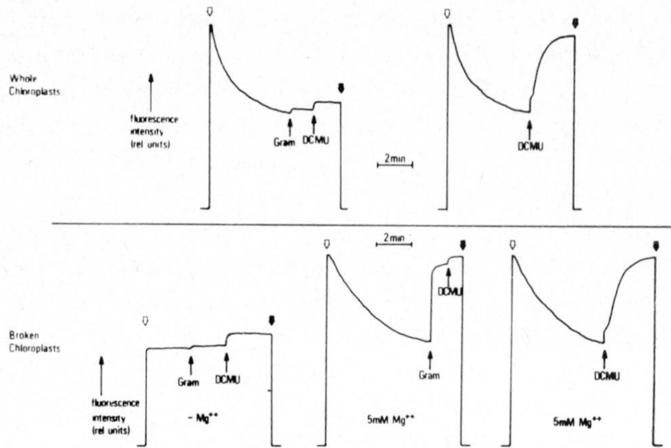

Fig. 4. Effect of 1μM gramicidin D on light induced fluorescence quenching observed with whole and broken chloroplasts. The chlorophyll concentration was 35 μg/ml and DCMU was added to give 3.3 μM. The suspensions contained 1mM PGA and other conditions as Fig. 1.

Mg^{++} or K^+ present to generate the high fluorescence state the effect of gramicidin and DCMU was similar to that observed with whole chloroplasts. These results suggest that gramicidin not only acts on the thylakoid membranes to induce uncoupling but also interacts with the outer membranes making them sufficiently leaky to lower the stromal cationic level such that the high fluorescing state can not be attained (comparable with removal of the outer membranes by osmotic shock). This effect is not normally seen with those compounds like nigericin, CCCP, NH_4Cl etc. which rely on the existance of a H^+ ion gradient for their action and therefore tend to exert their influence at the thylakoid membranes rather than at the outer membranes.

(iii) Relationship with HES and msec Delayed Light Emission

From above it is clear that cation dependent fluorescence quenching reflects the existence of a HES. In addition to this, the HES affects the intensity of one msec delayed light[12].

As Fig. 5a shows when nigericin is added to a suspension of illuminated whole chloroplasts which are at their fully quenched level, the fluorescence rises and at the same time msec delayed

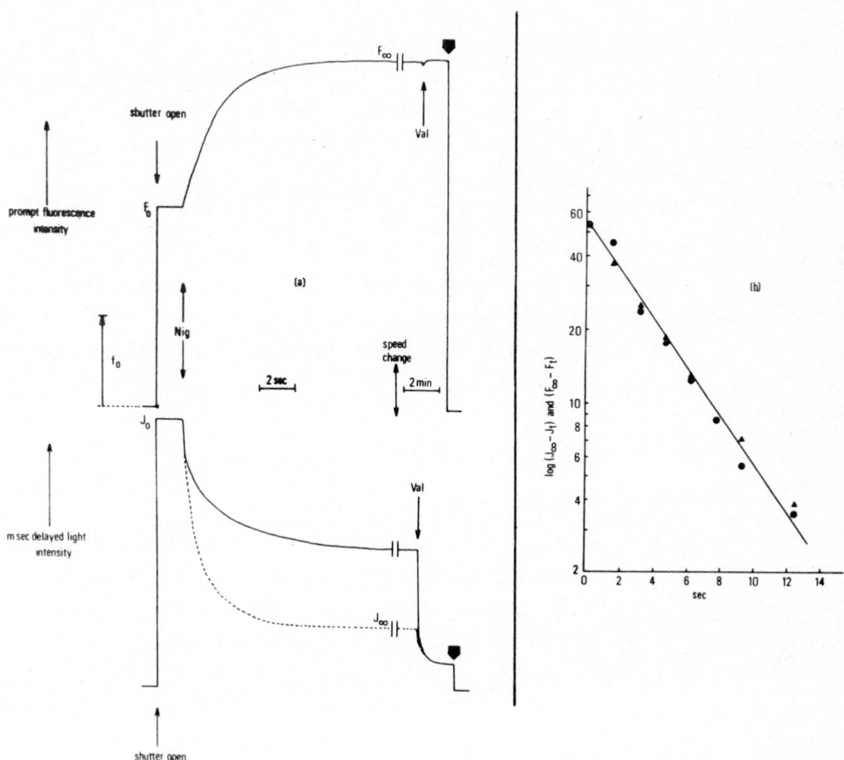

Fig. 5(a). Effect of 0.1 μM nigericin on prompt fluorescence and one millisecond delayed light from whole chloroplasts. The chloroplasts were preilluminated for 5 min in order to establish the low fluorescing state before rapidly injecting 0.1 μM nigericin into the cuvette. After about 7 sec the chart speed was reduced and 5×10^{-8}M valinomycin was injected into the cuvette. The chlorophyll concentration was 38 μg/ml and the f_o level was established on an identical dark treated suspension by recording the initial fluorescence level using a CRO. The observed delayed light (L) has been corrected for the time dependent fluorescence yield changes above f_o to give J (see text) which is shown as a dotted curve.

(b) Semilog plots of the time dependent nigericin induced fluorescence and msec delayed light changes. F_t is the fluorescence level above F_o at anytime t and F_α is the final fluorescence level. J_t is the corrected value of delayed light at anytime t below J_o and J_α is the final level. (see Fig.5a clarification of symbols).

light is inhibited. In the case of msec delayed light there is evidence that the HES effect is due to its ability to reduce the activation barrier for the emission process such that the intensity of emission (L) is exponentially related to Mitchell's HES (Δp)[13,14].

As Fig. 5b shows the nigericin sensitive fluorescence (F) and msec delayed light changes (J) were first order and logarithmically related. In this analysis delayed light (L) was corrected for live fluorescence yield changes: that is those changes which occurred above the f_o level[15].

Thus it seems that $F \alpha \log J$ (where $J = \phi L$ and ϕ is live fluorescence yield) and since $\log J \alpha \Delta p$ then $F \alpha \Delta p$. Furthermore it seems likely that only the chemical component of Δp (i.e. ΔpH) controls the fluorescence changes since as Fig. 5a shows the addition of valinomycin to the nigericin treated whole chloroplasts had no affect on the fluorescence but further decreased the msec delayed light to a lower level.

5. Discussion

The above findings show that to establish the high fluorescing state in intact chloroplasts there must be a sufficiently high concentration of Mg^{++}, or possibly Ca^{++}, within the granal interiors. The light induced fluorescence quenching can be explained as a net Mg^{++} efflux from the grana to the stroma induced by inward pumping of H^+ (i.e. establishment of the ΔpH component of the HES). On the other hand the dark reversal is dependent on sufficient Mg^{++} in the stroma so that this cation can leak back into the granal compartment as the pH gradient is dissipated. Clearly the outer chloroplast membranes must act as a barrier to Mg^{++} diffusion which is supported by the recent work of Pflüger[16]. In fact it has been estimated that the stromal Mg^{++} concentration may rise by 10mM in the light[17], a value which is consistent with the level of Mg^{++} required to establish the high fluorescing state with broken chloroplasts bearing in mind that activities and not concentrations should be compared.

Although Mg^{++} is necessary to establish the high fluorescing state it is not clear whether the quenching is due directly to Mg^{++} release from specific binding sites on the inner side of the thylakoid membranes or to protonation of these specific sites. Illuminated chloroplasts usually show energy dependent granal shrinkage indicating that the H^+ ions taken up became more firmly bound (less osmotically active) than were the effluxing counter-ions[18]. This is, of course, especially significant if the cation exchange *in vivo* is $2H^+/Mg^{++}$ which seems to be the case as indicated from our ionophore studies and also from the careful flux studies on broken chloroplasts by Hind et al[19]. The resulting fluorescence changes are probably due to conformational changes within the thylakoid membranes. These structural changes presumably bring about changes in the physical status of the pigments such as orientation and distance. There is evidence that conformational changes of this type do occur *in vivo*[20] but their relationship with chlorophyll fluorescence is as yet not clear[5,17].

The finding that an energy dependent net movement of Mg^{++} (or possibly Ca^{++}) between the granal and stromal space can control changes in fluorescence yield has important implications with regard to the State 1 - State 2 hypothesis for the regulation of quantal distribution to the two photosystems[1,21]. The above work with intact chloroplasts seems to bridge the gap between fluorescence observations with intact cells[1,22], and with broken chloroplasts[2,3]. The question now arises whether the high and low fluorescing levels seen with intact chloroplasts correspond to State 1 and State 2 respectively. Our preliminary investigations into this question have made use of the fact that both $chla_2$ and $chla_1$ fluorescence at liquid N_2 temperatures. In agreement with Murata[1] and others[23,24] we found that the addition of Mg^{++} to DCMU treated broken chloroplasts did seem to decrease spillover between $chla_2$ and $chla_1$ (i.e. caused a shift from State 2 to State 1). In contrast HES quenching of chlorophyll fluorescence from broken chloroplasts treated with DCMU induced by diaminodurene (DAD) mediated electron flow[25] caused an overall quenching of the $chla_2$ and $chla_1$ low temperature emission peaks. With intact chloroplasts the measurements were complicated by the fact that the quenching involves both Mg^{++} and HES effects. Nevertheless our measurements did suggest that the light induced fluorescence quenching may be partly due to an increase in spillover between the two photosystems.

At this stage our investigations are incomplete but if we assume that the proton driven Mg^{++} efflux from the grana brings about a transition from State 1 to State 2 we must think of a model which will also account for antagonism between the activity of the two photosystems. That is, how in the presence of a HES could the over activation of photosystem one increase the Mg^{++} level in the granal compartment so as to reduce transfer between $chla_2$ and $chla_1$ (seen as an increase in the fluorescence yield of $chla_2$) and so optimise electron flow through the two photosystems? One simple idea recently considered during discussions with Dr. J. Amesz is that when photosystem one is over activated compared to photosystem two there is an increase in the chloroplast ATP pool due to cyclic phosphorylation and that this stimulates the functioning of an ATP dependent Mg^{++} pump operating to oppose the primary light induced H^+/Mg^+ exchange. In this way changes in the granal Mg^{++} levels would regulate spillover and optimise net electron flow through the two photosystems.

Acknowledgements

The authors wish to thank the Science Research Council for financial support.

References

1. Bonaventura, C. and Myers, J., 1969, Biochim. Biophys. Acta 189, 366.
2. Murata, N. 1969, Biochim. Biophys. Acta 189, 171.
3. Briantais, J.M., Vernotte, C. and Moya, I., 1973, Biochim. Biophys. Acta 325, 530.
4. Krause, G.H., 1973, Biochim. Biophys. Acta 292, 715.
5. Krause, G.H., 1974, Biochim. Biophys. Acta 333, 301.
6. Barber, J. and Telfer, A., 1974 in: Ion transport in plant cells and organelles, eds. J. Dainty and U. Zimmerman (Springer Verlag, Berlin) in press.
7. Barber, J., Telfer, A. and Nicolson, J., 1974, Biochim. Biophys. Acta 357, 161.
8. Stokes, D.M. and Walker, D.A., 1971, Plant Physiol. 48, 163.
9. Heber, U. and Santarius, K.A. 1970, Z. Naturforsch. 25b, 718.
10. Barber, J., 1972, Biochim. Biophys. Acta 275, 105.
11. Homann, P.H., 1969, Plant Physiol. 44, 932.
12. Mayne, B.C., 1967, Photochem. Photobiol. 6, 189.
13. Wraight, C.A. and Crofts, A.R. 1971, European J. Biochem. 19, 386.
14. Hipkins, M.F. and Barber, J., 1974, FEBS Letters 42, 289.
15. Clayton, R.K., 1969, Biophys. J. 9, 60.
16. Pflüger, R., 1969, Z. Naturforsch. 28b, 11.

17. Hind, G. and McCarty, R.E., in: Photophysiology, ed. A.C. Giese (Academic Press, N.Y.) 8, 113.
18. Dilley, R.A., 1971, Curr. Top. Bioenerg. 4, 237.
19. Hind, G., Nakatani, H.Y. and Izawa, S., 1974, Proc. Nat. Acad. Sci. U.S. 71, 1484.
20. Murakami, S., and Packer, L., 1970, J. Cell Biol. 47, 332.
21. Duysens, L.N.M., 1972, Biophys. J. 12, 858.
22. Govindjee and Papageorgiou, G., 1971, in: Photophysiology, ed. A.C. Giese (Academic Press, N.Y.) 6, 1.
23. Mohanty, P., Braun, B.Z. and Govindjee, 1973, Biochim. Biophys. Acta 292, 459.
24. Gross, E.L. and Hess, S.C., 1973, Arch. Biochim. Biophys. 159, 832.
25. Wraight, C.A. and Crofts, A.R., 1970, Eur. J. Biochem. 17, 319.

M. AVRON, *Proceedings of the Third International Congress on Photosynthesis*
September 2-6, 1974, The Weizmann Institute of Science, Rehovot, Israel
Elsevier Scientific Publishing Company, Amsterdam, The Netherlands, 1974

DELAYED LIGHT STUDIES ON PHOTOSYNTHETIC ENERGY CONVERSION

IX. EVIDENCE FROM MSEC EMISSION FOR TWO DISTINCT SITES OF ACTION OF TREATMENTS THAT INHIBIT ON THE OXIDIZING SIDE OF PHOTOSYSTEM II

D. E. Cohn, W. S. Cohen, S. Lurie* and W. Bertsch

Department of Biological Sciences
Hunter College, New York, N.Y., U.S.A.
and
*Department of Botany
Hebrew University, Rehovot, Israel

1. Introduction

A number of treatments inhibit electron flow between the site of water oxidation and photoreaction II: tris-ageing[1], heat[2], ultraviolet irradiation[3], Triton X-100[4], hydroxylamine[5], uncouplers at alkaline pH[6,7], ageing at alkaline pH[7], and chloride depletion[8]. Treated chloroplasts can oxidize artificial electron donors, and this donor supported electron flow to Hill acceptors is sensitive to 3-(3,4-dichlorophenyl)-1,1-dimethylurea (DCMU)[1]. We have examined the effects of electron donors and of the inhibitor DCMU on the msec delayed light emission (DLE) from chloroplasts exposed to the above treatments. The treatments all result in rapid msec decay of DLE in the absence of an acceptor. Donors or DCMU slowed DLE decay kinetics after some treatments, while after other treatments the DLE kinetics remained rapid in the presence of these agents. This data suggests that different treatments affect at least two different sites between the reaction center and the site of water oxidation.

2. Materials and Methods

Chloroplasts were prepared from greenhouse-grown oat seedlings (Avena sativa var. Garry) or market spinach as previously described[9]. Chlorophyll was determined according to Arnon[10]. Tris-aged chloroplasts were prepared according to Yamashita et al[1]; heat-treated and ultraviolet-irradiated chloroplasts were prepared according to Lurie[11]. Alkaline-inactivated chloroplasts were prepared according to Cohn et al[7]. Triton X-100[3], trypsin[4], and hydroxylamine[5] treatments were performed as previously described. Chloride-depleted chloroplasts were prepared as described by Izawa _et al_[8]. Whole cells of _Euglena gracilis_ strain Z were depleted of manganese as described by Heath _et al_[12].

Photoreduction of 2,6-dichloroindophenol (DCPIP) was monitored with an Aminco-Chance dual wavelength spectrophotometer following the absorbance change Δ590 - 470 nm[7]. Oxygen evolution was monitored using the luminol method[13] in the laboratory of Dr. David Mauzerall. Msec DLE was examined between 0.8 and 3.2 msec after the centers of repeating flashes of white exciting light as previously described[9,14].

3. Results

Fig. 1 shows the effects of ageing oat chloroplasts for 15 min in 0.8 M Tris-HCl (pH 8.2) on msec DLE. As previously reported[14], the tris-aged chloroplasts are characterized by rapidly decaying DLE in the absence of an electron acceptor. The electron acceptor (DCPIP) had little or no effect on the decay kinetics in the Tris-treated chloroplasts, but lowered the overall intensity of the DLE slightly. Addition of DCPIP to untreated chloroplasts resulted in a more than 2-fold increase in emission at 1 msec and a rapid decay of the DLE as previously reported[15]. Fig. 1 also shows that DLE decay kinetics from tris-aged chloroplasts are slower in the presence of the artificial electron donor diphenylcarbohydrazide (which restored DCPIP reduction to 60% of the control). Diphenylcarbohydrazide (DPC) had little or no effect on the DLE or the rates of electron transport in untreated chloroplasts (data not shown). In tris-treated chloroplasts DCMU slowed the decay of the DLE in the absence of an acceptor in a manner similar to DPC, but also lowered the overall emission intensity. Addition of DCMU (2 x 10^{-6} M) to untreated chloroplasts lowered the overall intensity and prevented any effect of the acceptor DCPIP on the DLE (data not shown). Orthophenanthroline (2 x 10^{-4} M) had the same effect as DCMU on the DLE.

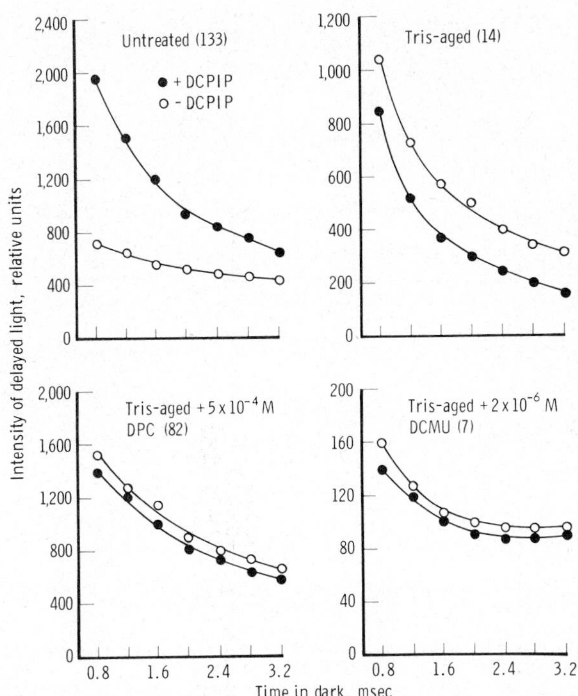

Fig. 1. Effect of DPC and DCMU on msec DLE from tris-aged oat chloroplasts. The reaction mixtures contained 20 mM NaCl, 25 mM tricine-NaOH (pH 7.8),5 mM $MgCl_2$, either .012 mM DCPIP (photoreduction) or .001 mM DCPIP (DLE), and chloroplasts equivalent to 10 ug/ml chlorophyll. Tris-aged chloroplasts were prepared by incubation of oat chloroplasts for 15 min in 0.8 M Tris-HCl, pH 8.2.

Fig. 2 shows the effects on msec DLE of irradiating chloroplasts with ultraviolet light for seven minutes. The DLE decay kinetics in the ultraviolet-irradiated chloroplasts were similar to that observed with tris-aged chloroplasts. The msec DLE decay kinetics and emission intensity were not affected in the absence of an acceptor by either the addition of DCMU or of DPC (which restored reduction of the acceptor DCPIP to 58% of the control). The lack of an effect of DCMU on the DLE intensity of ultraviolet-irradiated chloroplasts was unusual. Following all the other treatments examined (see Table I below) DCMU lowered the overall intensity of DLE whether it slowed the decay kinetics, as

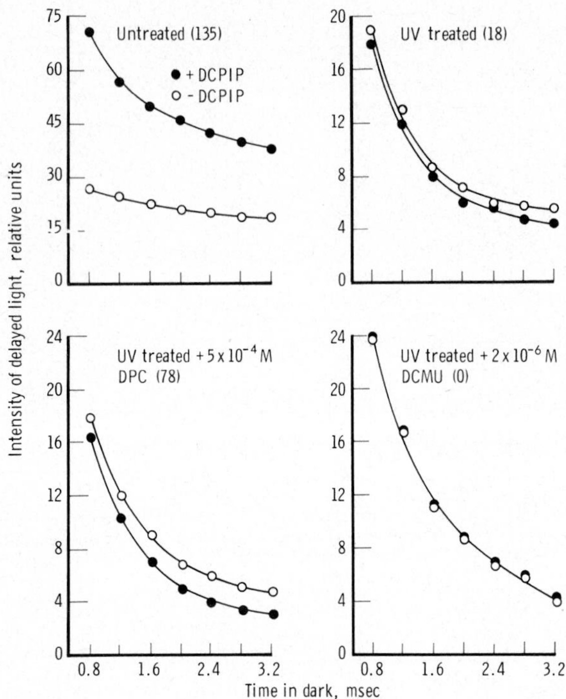

Fig. 2. Effect of DPC and DCMU on msec DLE from ultraviolet-irradiated spinach chloroplasts. The reaction mixtures were as in Fig. 1. Ultraviolet-irradiated chloroplasts were prepared by exposing spinach chloroplasts to ultraviolet light for seven min as described by Lurie[11].

in tris-aged chloroplasts, or whether the decay kinetics remained rapid. Ultraviolet irradiation itself lowers the overall DLE intensity (Fig. 2) in contrast to the other treatments which all result in an increase in DLE intensity at 1 msec when compared to DLE from untreated chloroplasts.

Fig. 3 shows that in chloroplasts inactivated by exposure to gramicidin at pH 8.9[7] DPC (which restored DCPIP reduction to 76% of the control) had no effect on the rapid DLE kinetics in the absence of an acceptor. On the other hand, DCMU slowed the rapid decay observed in the gramicidin-treated chloroplasts.

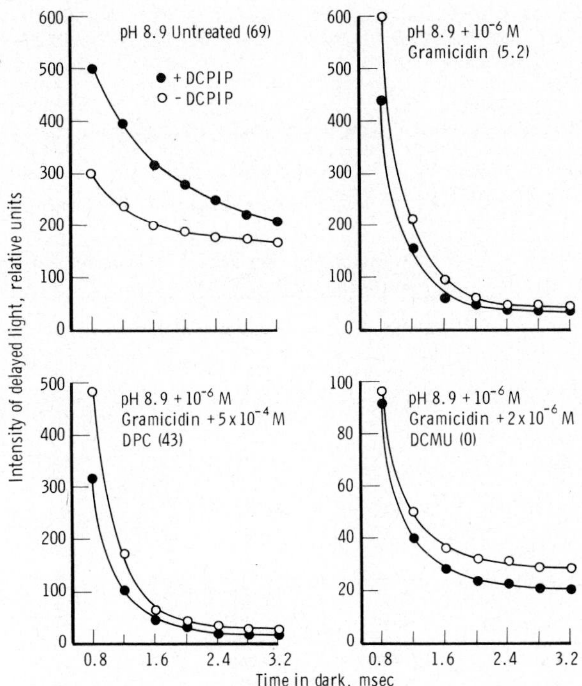

Fig. 3. Effect of DPC and DCMU on msec DLE from gramicidin-inhibited oat chloroplasts. The reaction mixtures were similar to those described in Fig. 1, except that the buffer was 50 mM Tricine-NaOH (pH 8.9).

Table I summarizes data from the three treatments described above, as well as similar experiments with choroplasts and algae inactivated by a variety of other treatments. In this table the ratio of DLE at 0.8 msec to that at 2.8 msec is used as a crude measure of the decay of the DLE in the time range examined in our experiments. With the exception of gramicidin-treated chloroplasts, the treatments appear to fall into two classes: those in which the rapidly decaying DLE of the treated systems is slowed by both donors and DCMU (Class "A"), and those in which the rapidly decaying DLE associated with the treatment remains rapid in the presence of either donors or DCMU (Class "B").

TABLE I

EFFECT OF ELECTRON DONORS AND DCMU ON DECAY KINETICS OF MSEC DELAYED LIGHT EMISSION FROM UNTREATED AND TREATED CHLOROPLASTS

Rates of DCPIP photoreduction (ueq/mg chl·hr) in untreated chloroplasts varied from 70 in coupled preparations to 400 in the uncoupled preparations used for chloride-depletion experiments. The whole cells of Euglena gracilis used as controls for the manganese-deficiency experiments evolved oxygen at 25 umoles/mg chl·hr. Where indicated Cl^- (NaCl) was employed at 10^{-2} M, DCMU at 2×10^{-6} M, DPC at 5×10^{-4} M, Mn ($MnSO_4 \cdot H_2O$) at 1.6×10^{-4} M, and $NH_2OH \cdot \frac{1}{2}H_2SO_4$ at 10^{-2} M.

	DLE intensity 0.8 msec / DLE intensity 2.8 msec			Rate of Electron Transport (% of control)	
	No Additions	+Donor	+DCMU	-Donor or cofactor	+Donor or cofactor
Untreated chloroplasts	1.58-1.70	1.58-1.79	1.35	100	90-110
Class "A" treatments					
Tris-ageing	3.00	2.06(DPC)	1.78	10	60
Cl^--depletion	3.90	2.13(Cl^-)	2.06	23	64
Mn^{++}-depletion*	4.37	2.19(Mn^{++})	-	2	92
Class "B" treatments					
UV-irradiation	3.80	3.80(DPC)	4.0	13	58
Heat	12.7	9.33(DPC)	10.5	27	63
pH 9.6 ageing	5.25	5.75(DPC)	4.0	39	93
Trypsin	5.77	5.50(DPC)	-	5	57
Triton X-100	5.00	4.00(DPC)	5.50	11	48
NH_2OH	15.0	10.0(NH_2OH)	9.50	19	36
Class "C" treatments					
Gramicidin/pH 8.9	15.3	20.0 (DPC)	3.42	8	62

*Untreated Euglena cells used as control had 0.8/2.8 ratio of 1.81.

4. Discussion

The wide variety of treatments that inhibit electron transport on the oxidizing side of PSII appear to have several characteristics in common: 1)electron transport through PSII is restoreable by addition of a suitable donor of electrons[1], 2)the variable fluorescence is severely attenuated in the treated systems[1], and 3)in the absence of an electron acceptor, the treatments induce a rapid decay in the msec DLE measured after repeating flashes of exciting light[14]. Although many of the treatments involve conditions that uncouple photophosphorylation, the DLE kinetics induced by the treatments were quite different from those seen in uncoupled chloroplasts[16].

In contrast to the common characteristics outlined above, there have been several reports that these treatments are not identical with respect to certain other parameters. Thus, after some treatments, the low variable fluorescence induced by the treatment is restored in the presence of an electron donor, while after other treatments it is unaffected[2,5,17]. Similarly, the addition of DCMU partially restores the variable fluorescence after some treatments, but not after others[2,5,17].

The present DLE data show a similar pattern. With one group of treatments (Class "A") the addition of an electron donor in the absence of an electron acceptor significantly slowed the decay of the DLE (see Table I). The decay kinetics under these conditions appear much more like those of untreated chloroplasts. In a second group of treatments (Class "B") the rapid decay of the DLE, in the absence of an acceptor, remained rapid on addition of a donor.

The effect of DCMU on the DLE kinetics correlated, with one exception, with the effects of the donors. After Class "A" treatments the rapid decay of the DLE was markedly slowed by DCMU. In Class "B" the decay kinetics in the presence of DCMU remained rapid. These differences presumably reflect a difference in the site of action of the treatments of the two classes. The single exception was found in chloroplasts exposed to gramicidin at pH 8.9; Class "C". In Class "C", the DLE decay kinetics in the absence of an acceptor were unaffected by electron donors, but slowed by DCMU.

One possible model for the reaction center of PSII has been elaborated by Van Gorkum _et al_[18]. All delayed light would arise from the recombination of P^+ with Q^-. This recombination normally would be prevented by the rapid reduction of P^+ by Z. On such a model the two broad classes of behavior shown by the treatments could result from blocks in the electron transport chain at the two sites between H_2O and the reaction center as shown below.

$$H_2O - - \underset{\text{"A"}}{\#} - Z \underset{\text{"B"}}{\#} P\text{-}Q - A - - PSI -$$

In Class "B" treatments the interaction between Z and P would be inhibited, leading to the recombination of the P^+ and Q^- generated in the light. This recombination should yield the rapidly decaying DLE observed in the msec time range. Added electron donors are assumed to reduce P^+ more slowly than the recombination with Q^-, and thus they should not affect the kinetics of the DLE decay to any significant extent. DCMU, which inhibits electron transport between Q and A[19], would not be expected to influence the DLE decay kinetics in such a treated system, since the reduction of A by Q^- is also quite slow compared to the recombination of Q^- with P^+.

In Class "A" treatments the rapid component of the DLE decay would arise from reaction centers that have accumulated Z^+ from previous flashes of exciting light, and therefore would be unable to stabilize the charge separated on a subsequent flash by reducing P^+. In this case, an electron donor should have a large effect on the DLE since the accumulated Z^+ would be reduced in the dark period preceding the next flash of exciting light. DCMU in such blocked systems would lead to an accumulation of closed traps in the state $\|Z^+PQ^-\|$ after a flash of exciting light. These traps could not further separate charge and would only yield delayed light by the regeneration of P^+ through a slow back reaction between Z^+ and P.

In the case of chloroplasts exposed to gramicidin at pH 8.9, the addition of an electron donor had little effect on the decay of the DLE, whereas DCMU slowed this decay considerably. Such "mixed" behavior is not predicted by the simple model presented above. This treatment was, however, the only one that did not fit into the simple classification. Furthermore, such "mixed" behavior was not shown when chloroplasts were aged at pH 9.6 for 5 minutes and then titrated to pH 7.5 for storage, a treatment we believe inactivates the chloroplasts in a way similar to exposure to gramicidin at pH 8.9[7].

We feel that the "mixed" behavior shown by the gramicidin-treated chloroplasts may result from the fact that this is the only treated system assayed at a high pH. We have observed (data not shown) that at pH 8.9, in the absence of gramicidin, the addition of DCMU causes the decay kinetics of DLE in the absence of an acceptor to become significantly <u>more rapid</u>. The DLE is similar to that seen in the presence of DCMU in the gramicidin-inhibited system (Fig. 3). Thus, it appears that at pH 8.9 DCMU may have an effect on the reaction center of PSII which obscures the effects of the gramicidin treatment. Possibly DCMU at pH 8.9 causes the withdrawal of holes from P^+ to some endogenous donor other than Z. Such an effect of DCMU could result in the unusual slowing of the DLE decay kinetics by DCMU that is found in the gramicidin-inhibited chloroplasts. In any event, the behavior of this single treatment, while not predicted by the simple model, may be a special case.

The effects on variable fluorescence and on DLE decay kinetics of treatments that act on the oxidizing side of PSII have also been ascribed to wasteful cyclic side reactions, induced by the treatments, in which oxidizing and reducing equivalents recombine in a non-luminescent fashion[14]. Using this approach, the different responses of the DLE from the treated systems to the presence of electron donors or DCMU could result from the induction of different cyclic pathways by the various treatments as shown below.

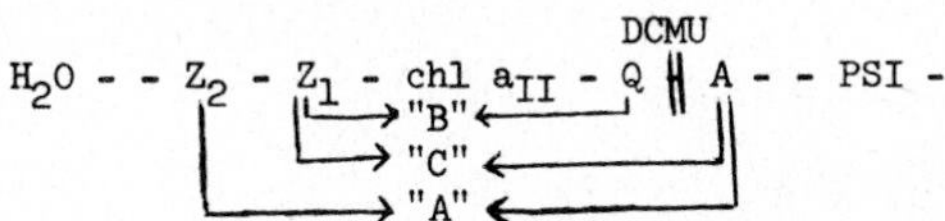

If the reduction of Z_1^+ by an added electron donor is assumed to be slower than the cyclic reactions that reduce Z_1^+, then the rapid DLE decay induced by the treatments of Classes "B" and "C" would not be significantly altered in the presence of an electron donor. Similarly, if it is assumed that Z_2^+ is more rapidly reduced by an added reductant, then the flow of charge through the cycle, and thus the DLE decay kinetics, would be modified by an electron donor. Only those cycles involving electron transport component A (Classes "A" and "C") would be affected by the presence of DCMU. This model also allows for the possibility of a cycle between Q^- and Z_2^+, in which case the addition of an electron donor, but not of DCMU, would be expected to slow the decay of the DLE. We have not observed this case.

We do not feel, as yet, that it is possible to distinguish between the linear model suggested by the scheme of Van Gorkum et al[18] and the cyclic model suggested by Bertsch et al[14]. These two models are not mutually exclusive. In any case, the observation that all the treatments fall into two classes of behavior (with one arguable exception) suggests that the treatments induce at least two different types of lesions in the photosynthetic apparatus.

5. References

1. Yamashita, T. and Butler, W. L., 1968, Plant Physiol. 43, 1978.
2. Yamashita, T. and Butler, W. L., 1968, Plant Physiol. 43, 2037.
3. Malkin, R., 1971, Biochim. Biophys. Acta 253, 421.
4. Selman, B. and Bannister, T., 1971, Biochim. Biophys. Acta 253, 428.
5. Katoh, S., Ikegami, I. and Takamiya, A., 1970, Arch. Biochem. Biophys. 141, 207.
6. Harth, E., Reimer, S. and Trebst, A., 1974, FEBS Letters 42, 165.
7. Cohn, D. E., Cohen, W. S. and Bertsch, W., 1974, Submitted to Biochim. Biophys. Acta.
8. Izawa, S., Heath, R. and Hind, G., 1969, Biochim. Biophys. Acta 180, 388.
9. Cohen, W. S. and Bertsch, W., 1974, Biochim. Biophys. Acta 347, 371.
10. Arnon, D. I., 1949, Plant Physiol. 24, 1.
11. Lurie, S., 1972, Doctoral Dissertation, Hunter College, C.U.N.Y., New York.
12. Heath, R. and Hind, G., 1969, Biochim. Biophys. Acta 189, 222.
13. Burr, A. and Mauzerall, D., 1968, Biochim. Biophys. Acta 153, 614.
14. Bertsch, W. and Lurie, S., 1971, Photochem. Photobiol. 14, 251.
15. Bertsch, W., West, J. and Hill, R., 1971, Photochem. Photobiol. 14, 241.
16. Wells, R., Bertsch, W. and Cohen, W. S., 1971, in: Proc. 2nd Int. Congr. on Photosyn. Res. Vol. I, eds. G. Fiorti, M. Avron and A. Melandri (Dr. W. Junk - the Hague) p. 207.
17. Katoh, S. and Kimimura, M., 1974, Biochim. Biophys. Acta 333, 71.
18. Van Gorkum, H. J. and Donze, M., 1973, Photochem. Photobiol. 17, 333.
19. Duysens, L. N. M. and Sweers, H. E., 1963, in: Studies on Microalgae and Photosynthetic Bacteria, ed. Jap. Soc. Plant Physiol. (University of Tokyo Press, Tokyo) p. 353.

M. AVRON, *Proceedings of the Third International Congress on Photosynthesis*
September 2-6, 1974, The Weizmann Institute of Science, Rehovot, Israel
Elsevier Scientific Publishing Company, Amsterdam, The Netherlands, 1974

FLUORESCENCE AND TRIGGERED LUMINESCENCE OF MUTANTS OF CHLAMYDOMONAS REINHARDI WITH LESIONS ASSOCIATED WITH PHOTOSYSTEM II.

by

H. Hardt*, S. Malkin* and B. Epel**

*Department of Biochemistry, The Weizmann Institute of Science, Rehovot, and

**Department of Botany, Tel Aviv University, Tel-Aviv, Israel

1. INTRODUCTION

Two families of mutants of the alga Chlamydomonas reinhardi with lesions associated with photosystem II (PS II) were recently isolated and investigated[1,2,3]. One family was called hfd and was shown to lack the photochemical activities of PS II but to retain the normal activities of PS I[3]. The other, called lfd, was shown to have functional PS II reaction center but to lack the oxygen evolving Hill reaction due to a probable block on the oxidizing side of PS II[1,2]. PS II mediated electron transport in this family could be restored by artificial electron donors.

In this study we investigated the fluorescence induction and the triggered luminescence properties of two representative mutants: hfd-49 and lfd-13 as compared to the properties of the wild type strain.

2. MATERIALS AND METHODS

The mutant strains were grown as previously described[3]. Chlorophyll content was determined by the method of Arnon[4]. The spectrophotometric activities were measured as previously described[3]. Chloroplast fragments were prepared from cells in the logarithmic phase of growth by a modification of the method of Levine and Gorman[5]. Methylviologen reduction and benzoquinone reduction were followed with a Clark type oxygen electrode.

Fluorescence induction was measured in an instrument which was described before[6]. The exciting light was passed through an 602 nm Schott interference filter, with half bandwidth of 16 nm. The light intensity was 1400 ergs cm^{-2} sec^{-1}. The fluorescence emission at 90° to the exciting beam entered an EMI 9558 (S-20) photomultipler after passing through a Schott RG 665 filter. The fluorescence induction was displayed on a Tektronix memory oscilloscope. A preillumination of far-red light (724 nm, 1 minute) was given standardly before each experiment of fluorescence induction[7]. We also checked that a dark period (2-3 minutes) instead of far-red gave identical results.

Triggered luminescence was measured by an instrument previously described[8]. A preillumination by saturating flashes, after a dark period of about 10 minutes was given as described before[9].

3. RESULTS AND DISCUSSION

The photochemical activities of chloroplast fragments prepared from cells of these algae are shown in Table 1. Both mutants lfd-13 and hfd-49 are unable to transfer electrons from water either to an acceptor of PS I, such as methylviologen (MV) or to an acceptor of PS II such as benzoquinone (BQ). The mutant blocked on the water side of PS II, lfd-13, shows photoreduction of MV using the ascorbate hydroquinone couple as an artificial electron donor. This is a PS II reaction as seen by its sensitivity to the addition of DCMU. This reaction is absent in the hfd-49 mutant which lacks PS II reaction center.

Photoreduction of methylviologen by reduced 2,6-dichlorophenolindophenol (DPIP) is shown by both mutants in presence of DCMU, indicating the normal activity of PS I in these chloroplast fragments.

TABLE 1

Photosynthetic electron transport reactions of lfd-13, hfd-49 and wild type C.reinhardi chloroplast fragments

(μmoles O_2/mg chlorophyll hr)*

Reaction	lfd-13	hfd-49	wild type
$H_2O \longrightarrow$ MV	O	O	-64
$H_2O \longrightarrow$ BQ	O	O	+90
DPIP/Asc/DCMU $\longrightarrow$ MV	- 216	- 480	- 318
HQ/Asc $\longrightarrow$ MV	- 137	O	-
HQ/Asc/DCMU $\longrightarrow$ MV	- 22	-	-

* - O_2 uptake
\+ O_2 evolved

Experimental conditions as described in ref. 1.

The spectrophotometric activities of both mutants are shown in Table 2. The hfd-49 lacks the high potential cytochrome b-559 and has only half of the low potential cytochrome b-559. This mutant also lacks the C-550 component which was recently suggested to act as the primary acceptor of PS II[10]. In the lfd-13 mutant there exist both C-550 and low potential cytochrome b-559, but it has only half of the high potential cytochrome b-559. Cytochrome -f and cytochrome b-563 are fully present in both mutants.

TABLE 2

Spectrophotometric activities of lfd-13, hfd-49, and wild-type C. reinhardi chloroplast fragments

Strain	C-550	Cyt b-559 High potential	Cyt b-559 Low potential	Cyt-f	Cyt b-563
Wild type	+	100 %	100 %	100 %	100 %
hfd-49	-	0	50 %	100 %	100 %
lfd-13	+	50%	100 %	100 %	100 %

C-550 - light induced change. All others, chemically induced changes. All changes were measured at liquid nitrogen temperature. Other experimental details as described in ref.2,3.

The photochemical and spectrophotometric activities summarized in Table 1 and Table 2 are in complete agreement with previously published studies on these mutant classes[1,2,3] (cf. Table 1 in ref. 1; Table 1 in ref. 3).

The chlorophyll fluorescence induction as exhibited by the wild type chloroplast fragments is shown in Fig.1,A and remains unchanged in the presence of 2 mM ascorbate with 0.5 mM hydroquinone. Addition of DCMU is accelerating the fluorescence rise as expected (Fig.1,B). The fluorescence induction (variable fluorescence) is completely missing in the mutant lacking PS II reaction center, hfd-49 (Fig. 1,C).

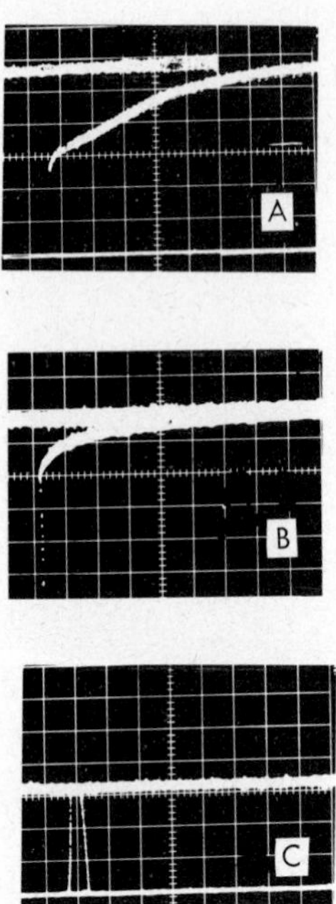

Fig. 1. Fluorescence induction of chloroplast fragments from the wild-type and the hfd-49 mutant.
The chlorophyll concentration with both strains was 7.5 μg/ml in a medium containing 0.1 M KCl, 0.2 M sucrose solution. (A) wild type; time scale 200 msec/div. (B) wild type with addition of 5 μM DCMU; 200 msec/div. (C) hfd-49 mutant; 10 msec /div. The left and right vertical lines in this picture correspond to light on and to light off, after a few seconds. Vertical sensitivity is the same for A-C. Exp. A was repeated in presence of 2 mM ascorbate and 0.5 mM hydroquinone with identical results.

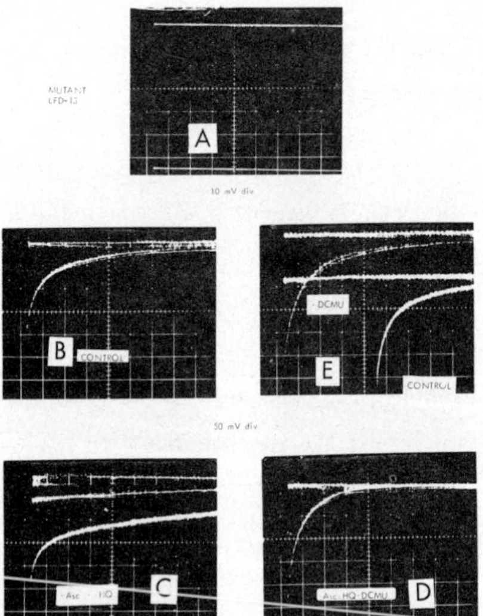

Fig. 2. Fluorescence induction of lfd-13 mutant chloroplast fragments. Chlorophyll concentration was 10.8 μg/ml. Time scale 200 msec/div. In picture A the vertical sensitivity scale is 5 times lower than in picture B, C, D, E. Other conditions as indicated. Experiments B, C, D were performed on the same sample.
Light intensity in Figs. 1 and 2 was the same, 1400 ergs cm^{-2} sec^{-1}.

The characteristcs of the fluorescence induction of the lfd-13 mutant, blocked on the water side of PS II are shown in Fig. 2. This mutant does show variable fluorescence starting from F_o (initial fluorescence level) at about 5/6 of F_m (final fluorescence level) as can be seen in Fig. 2,A. Fig. 2,B is showing the variable fluorescence on a 5 times larger scale. Fig. 2,C shows the effects of addition of 2 mM ascorbate with 0.5 mM hydroquinone to the same sample. F_o the initial level of fluorescence is decreased and at the same time the area above the fluoresdence curve is significantly increased (x 4 times). The full increase required a prior far-red or dark period of restoration as described in Materials and Methods. Addition of 5 µM DCMU to the same sample, as shown in Fig. 2,D accelerated the fluorescence rise which now appears the same as for the wild type plus DCMU (Fig. 1,B). The kinetics of variable fluorescence in another sample of this mutant is the same before and after addition of 5 µM DCMU as seen in Fig. 2,E.

These experiments show probably that this mutant is limited by the oxidizing side of PS II. A block of electron transfer from water limits the available pool for electron transfer to the reducing side. The fluorescence induction area is proportional in this case to the pool available on the oxidizing side. When an electron donor is added this limitation disappears, and the only limitation is now the pool on the reducing side, as is the usual case. The necessity for a far-red restoration period is also indicating that the rate limiting step is on this side. DCMU limits this pool by a considerable degree. If one assumes that because of the oxidizing side limitation and the possible fast reaction of Q^- with the pool A at the reducing side, the final reaction center state is Z^+Q in this mutant, a possible conclusion is that Z^+ affects the fluorescence in the same way as Q^- (i.e. Z^+Q is a non-quencher state). The initial level of fluorescence F_o which is lower in presence of ascorbate hydroquinone indicates that if normally some Z is oxidized at the beginning, it will become reduced, thus pointing again toward the roll of Z^+ as a non-quencher as Q^-.

The normal function of PS II reaction center is reflected by another activity; the possibility to trigger a pulse of luminescence by injecting an amount of acid to the preilluminated chloroplasts[11]. This activity is clearly exhibited by chloroplast fragments of the wild type strain.

No signal of luminescence or even of delayed light was detected in the mutant hfd-49, lacking PS II. The mutant lfd-13, which has an active PS II reaction center, shows delayed light and luminescence triggered by acid (Fig. 3) or by an organic solvent[12] (not shown).

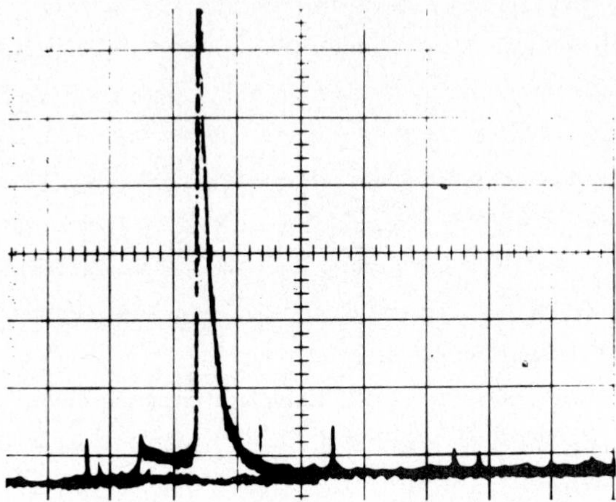

Fig. 3. HCl triggered luminescence from lfd-13 chloroplast fragments. 0.3 ml 0.1 N HCl was injected into 1 ml of chloroplast fragments (45 μg chlorophyll) in a 0.1 M KCl, 0.2 M sucrose solution. The preillumination was by 1 saturating flash. Time scale 0.5 sec/div.

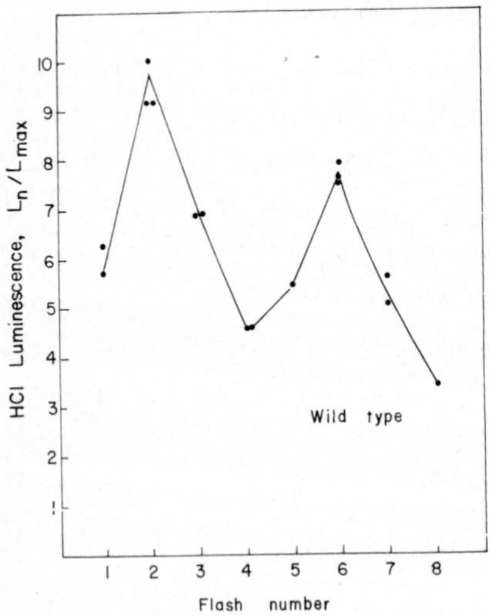

Fig. 4. Oscillations of the HCl triggered luminescence elicited by saturating flashes, plotted against the flash number, in wild-type chloroplast fragments. Before illumination the samples were dark adapted for about 10 minutes. Other experimental conditions as in Fig. 3.

The dependence of the luminescence peaks on the number of preilluminating flashes in the wild type chloroplast fragments is shown in Fig. 4. The oscillations with a period of four are similar to those found in lettuce chloroplasts[9]. These oscillations are indicative to the involvement of the S-states of the scheme of Kok et al[13] as reactants in the back reactions which lead to luminescence. In contrast to the wild type, no oscillatory pattern was seen with the chloroplast fragments of lfd-13 (Fig. 5) except for an initial increase from flash No. 1 to flash No. 2. Additional flashes resulted in no further changes. Fig. 6 shows that the presence of ascorbate and hydroquinone does not restore the pattern of oscillations in lfd-13, however, the initial rise is perhaps composed of 3 flashes instead of 2.

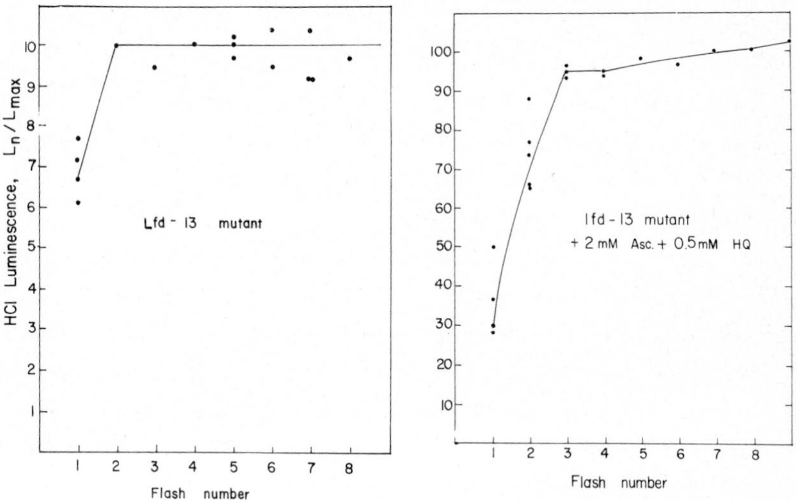

Fig. 5. Dependence of HCl-triggered luminescence on the number of preilluminating saturating flashes, in lfd-13 mutant chloroplast fragments. Experimental conditions as in Fig. 3.

Fig. 6. Dependence on HCl-triggered luminescence on the number of preilluminating saturating flashes, in lfd-13 mutant chloroplast fragments. Experimental conditions as in Fig. 3 except for addition of 2 mM ascorbate and 0.5 mM hydroquinone during the dark period.

It is therefore probable that the lesion in this mutant between water and PS II still enables the production of two States such as S_2^{+2} and S_3^{+3}, however, it does not enable to complete the cycle of the S-strates reactions which is accompanied by the production of oxygen. The presence of ascorbate and hydroquinone results probably with the reduction of the first S-state which is present in the dark, thus another flash is required to arrive to the steady state level of induced luminescence.

In summary the fluorescence properties of the mutant blocked on the oxidizing side of PS II are pointing toward the effect of Z^+ on the fluorescence yield. The pattern of dependence of luminescence on the number of preilluminating flashes in this mutant suggests that 2 or 3 S-states are still able to be produced on the Z side of PS II.

References

1. Epel, B. L. and Levine, R. P. 1971, Biochim. Biophys. Acta, 226, 154.
2. Epel, B. L., Butler, W. L. and Levine, R. P., 1972, Biochim. Biophys. Acta, 275, 395.
3. Epel, B. L. and Butler, W. L., 1972, Biophys. J., 12, 922.
4. Arnon, D. I., 1949. Plant Physiol. 24, 1.
5. Levine, R. P. and Gorman, D. S., 1966, Plant Physiol. 43, 1293.
6. Malkin, S. and Michaeli, G., 1971, in: 2nd. International Congress on Photosynthesis, Stresa, Vol. 1, eds. G. Forti, M. Avron and A. Melandri (Dr. W. Junk Publishers, N. V., the Hague), p. 149.
7. Malkin, S. and Kok, B. , 1966 , Biochim. Biophys. Acta 126 , 413 .
8. Malkin, S. and Hardt, H.,1971, in : 2nd. International Congress on Photosynthesis, Stresa, Vol. 1, eds. G.Forti, M. Avron and A. Melandri(Dr. W. Junk Publishers, N. V. , the Hague) p. 253
9 Hardt, H. and Malkin, S. 1973, Photochem. Photobiol. 17,433.
10. Erixon, K. and Butler, W. L., 1971, Biochim. Biophys. Acta 234, 381.
11. Miles, C. D. and Jagendorf, A. T., 1969, Arch. Biochem. Biophys. 129, 711.
12. Hardt, H. and Malkin S., 1972, Biochim. Biophys. Acta 267, 588.
13. Kok, B., Forbush, B. and McGloin, M., 1970, Photochem. Photobiol. 11, 457.

M. AVRON, *Proceedings on the Third International Congress on Photosynthesis*
September 2-6, 1974, The Weizmann Institute of Science, Rehovot, Israel
Elsevier Scientific Publishing Company, Amsterdam, The Netherlands, 1974

THE EFFECT ON STIMULATED LUMINESCENCE OF TREATMENTS INHIBITING PHOTOSYSTEM II

S. Lurie and S. Malkin

Department of Botany, Faculty of Agriculture, Hebrew University
Rehovot, and Department of Biochemistry, The Weizmann Institute
of Science, Rehovot, Israel

1. Summary

The effect on second delayed* light and stimulated luminescence of four treatments inhibiting photosystem II was examined: UV irradiation, heat inactivation, tris ageing and chloride depletion. All treatments inhibited the extent of both emissions without affecting their decay kinetics. DCMU which enhances these light emissions, did not enhance emission from treated chloroplasts.

The effect of UV irradiation was examined after increasing times of treatment. It was found that photochemistry (water to DCPIP) and delayed light were inhibited in a parallel fashion, whereas the stimulated luminescence was inhibited differently. Acid and salt induced luminescence are less sensitive to UV inhibition. They were not inhibited and in some experiments were increased slightly when delayed light was already 30% inhibited. Solvent and temperature jump stimulated luminescences are inhibited in a qualitatively similar fashion to the delayed light.

The results will be discussed in terms of mechanism of action of the treatments and their effect on the precursors of delayed light.

2. Introduction

Isolated chloroplasts exhibit light emission after preillumination. Delayed light emission, first seen by Strehler and Arnold[1], is measurable after the end of the exciting irradiation and persists for several minutes after illumination ceases. More recently light has been found to be emitted as the result of various triggering processes. These include pH transitions[2,3], salt addition[3,4], temperature jump[5,6] and organic solvent injection[7]. All of these triggering processes require prior irradiation.

Delayed light occurring as it does from microseconds to seconds after illumination, reflects several reaction types or different pools of precursors. The triggered luminescences are examined one or more seconds after illumination, and may induce emptying of different precursor pools, depending on the type of triggering. If these pools are components on the oxidizing and reducing side of photosystem II, then inhibiting electron flow at various points along the chain, should affect the triggered luminescence.

* Measured from 22 msec. after preillumination to about 1 sec.

There are a number of treatments which affect photosynthesis by inhibiting photosystem II on the oxidizing side. These include tris ageing[8], heating[9]. UV irradiation[9,10,11] and chloride deficiency[12]. All of these treatments inhibit oxygen production coupled to photochemical reduction, and appear to have different sites of action. Examining their effect on triggered luminescence may enable us to differentiate between precursors of the different types of triggered luminescence.

3. Materials and Methods

The experimental apparatus for monitoring triggered luminescence has been described previously[13,14]. The procedure for obtaining a measurement is the following: A one ml sample containing 50 μg chlorophyll in 1 mM Tris-HCl, pH 7.8 is dark adapted for at least one minute in a cuvette which can either be opened to illumination or to a photomultiplier. After dark adaptation the front shutter is opened and the sample is illuminated. When the front shutter closes the back shutter opens automatically 22 msec after the closure of the front shutter and delayed light can be monitored. At predetermined times there is the addition of the triggering reagents from syringes. Complete mixing is obtained 0.1 sec after the addition, and the triggered luminescence is then recorded on a memory oscilloscope.

Chloroplasts were prepared from lettuce as described by Avron[15]. The treatments, except for chloride depletion, were performed on control chloroplasts. For tris ageing, chloroplasts were aged in 0.8 M Tris-HCl, pH 8.0 for 10 minutes at 0°C, as described by Yamashita and Butler[8]. The chloroplasts were then collected by centrifugation. To obtain heat inactivated chloroplasts, 0.5 ml aliquots of chloroplasts at 1 mg/ml were heated for various periods of time at 50°C. The heated chloroplasts were immediately cooled to 0°C and stored on ice until use. For UV irradiation chloroplasts were suspended in the reaction mixture and placed in a cuvette 8 inches from a Osram 150 W mercury lamp. Between the lamp and the sample were UV filters transmitting between 250 to 350 nm. Irradiation was for varying lengths of time. Control chloroplasts were treated the same way but were not irradiated.

To obtain chloride depleted chloroplasts, leaves were ground in a medium containing 0.4 M sucrose, 0.025 M TES-NaOH, pH 7.4, and 0.005 M Na_2SO_4. The chloroplasts were depleted of Cl using the EDTA-washing procedure of Izawa et al[12] and were resuspended in 0.1 M sucrose, 0.005 M TES-NaOH, pH 7.4, and 0.005 M $MgSO_4$ at 0°C.

DCPIP reduction was assayed by absorbance changes at 600 nm. For UV inhibition we used weak 685 nm actinic light (light limiting conditions). For heat treatment 434 nm actinic light was used. It was found that UV inhibition is the same for light limiting or saturating conditions.

4. Results

Figure 1 shows oscilloscope pictures of the four types of triggered luminescence on untreated chloroplasts. Acid and salt triggered luminescence consistently gave a peak as high as, or higher than, the delayed light, while solvent and temperature jump triggered luminescences were lower than the delayed light. Figure 2 shows the effect of the four inhibitory treatments on the acid triggered luminescence. The delayed light emission is decreased by the treatments and the triggered luminescence is also inhibited by all four treatments. Salt, solvent and temperature jump triggered luminescences were also inhibited by the treatments in a parallel manner to that of acid triggered luminescence (Table I)

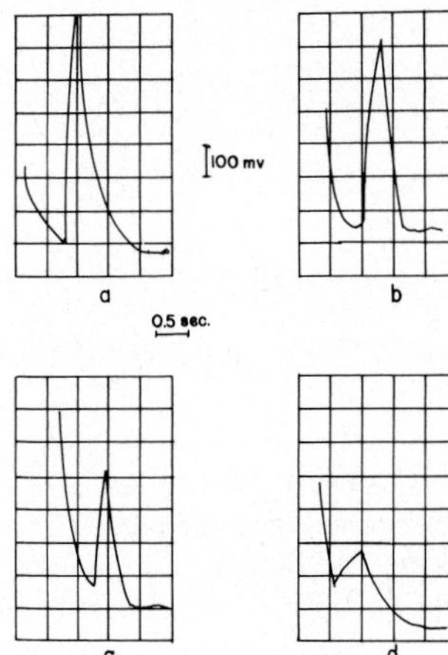

Figure 1. Four types of triggered luminescence from untreated chloroplasts. (a) HCl (0.25 ml 0.1 M)(b) sodium benzoate (0.25 ml 1.5 M) (c) dimethylsulfoxide; (d) temperature jump (1 ml 95°c water). Vertical scale is 20 mV/division and horizontal is 0.5 sec/division.

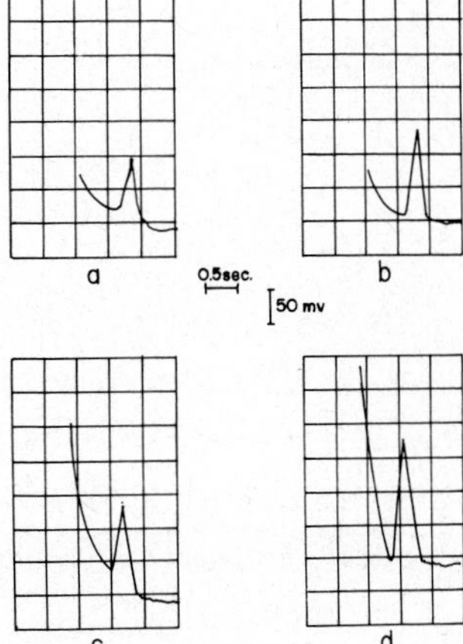

Figure 2. The effect of four inhibitory treatments on HCl triggered luminescence. a) tris treated; b) heating for 90 seconds at 50°C; c) chloride depleted; d) UV irradiated for three minutes. Vertical scale is 10 mV/div. and horizontal scale is 0.5 sec./div.

Treatment	Type of triggered luminescence							
	Acid		Salt		Solvent		Temperature	
	D.L. peak height	T.L. peak height	D.L. peak height	T.L. peak height	D.L. peak height	T.L. peak height	D.L. peak height	T.L. peak height
TRIS aged	150	0	120	60	140	40	130	40
Heated (1 min)	120	50	130	80	120	50	120	60
UV irradiated (1.5 min)	200	125	200	125	190	50	200	75
Chloride depleted	200	10	200	100	200	90	200	40
Untreated	450	700	500	500	450	300	450	250

Table I. The effect of treatments inhibition photosystem II on second delayed light (D.L.) and triggered luminescence (T.L.).

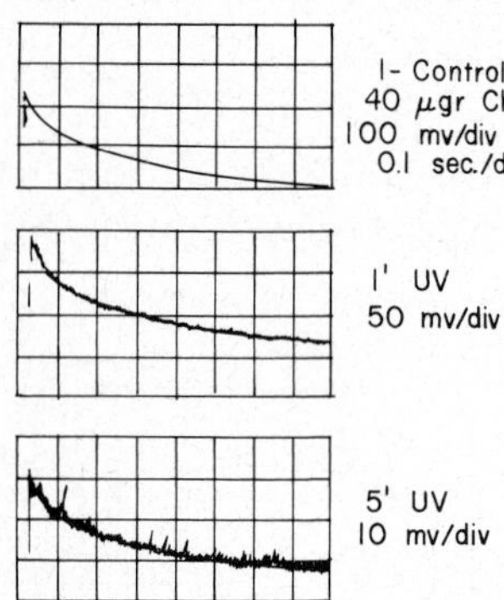

Figure 3. Comparison of the decay kinetics of untreated chloroplasts and UV irradiated chloroplasts, for delayed light. Similar results were obtained for the other luminescence types.

As shown in figure 3, the rate of decay of delayed light or stimulated luminescence was not affected by the treatments, only the amount of delayed light and extent of triggered luminescence was decreased. DCMU, which enhances the emission of second delayed light and triggered luminescence from untreated chloroplasts, does not enhance emission from treated chloroplasts.

Two of the treatments, heat and UV irradiation, can be examined sequentially after various times of inhibition. The effect of different times of UV irradiation was examined. It was found that photochemistry (water to DCPIP) and delayed light were inhibited in a parallel fashion, whereas the triggered luminescence of various types were inhibited differentially (figure 4). Acid and salt induced triggered luminescence are less sensitive to UV inhibition than the delayed light. They were not inhibited, and in some experiments were even increased slightly when delayed light was already 30 % inhibited. Solvent and temperature jump triggered luminescence are inhibited in a qualitatively similar fashion to the delayed light.

Chloroplasts heated for varying lengths of time showed an inhibition of delayed light which went together with an inhibition of photochemistry. The triggered luminescences were also inhibited in the same manner. There were no differences on the effect of heating on acid and salt triggered luminescence as compared to solvent and temperature jump. (figure 5).

5. Discussion

There appears to be an intrinsic difference between the effect on triggered luminescence of inhibitors on either the oxidizing or reducing side of photosystem II. An earlier paper[16] showed that DCMU, NQNO and DBHB which inhibit on the reducing side of photosystem II all enhanced triggered luminescence. The treatments which inhibit between water and photosystem II all decrease triggered luminescence. This may mean that it is the state of the pools on the oxidizing side of photosystem II (S states)[17] which determine the degree of triggered luminescence. It was found that the triggered luminescences showed an oscillatory pattern in flashing light experiments similar to that of delayed light. This is interpreted as reflecting differential charging of the S states and shows that the triggered luminescences are sensitive to alterations in these states.[18]

The mechanism of action of the inhibitory treatments is not known. There is some evidence that they induce different cyclic reactions around photosystem II, thus both competing with normal electron flow and with normal back reactions to delayed light.[12,19] In this case neither the S states nor Q would accumulate in a charged state, and so triggered luminescence would be inhibited. Other researchers think that the treatments initiate a block on the oxidizing side of photosystem II similar to the action of DCMU on the reducing side.[11] In this case the higher S states would not be formed. If these states are the luminescence precursors inhibition of luminescence would result.

We found that all treatments which inhibit photosystem II also cause inhibition of both

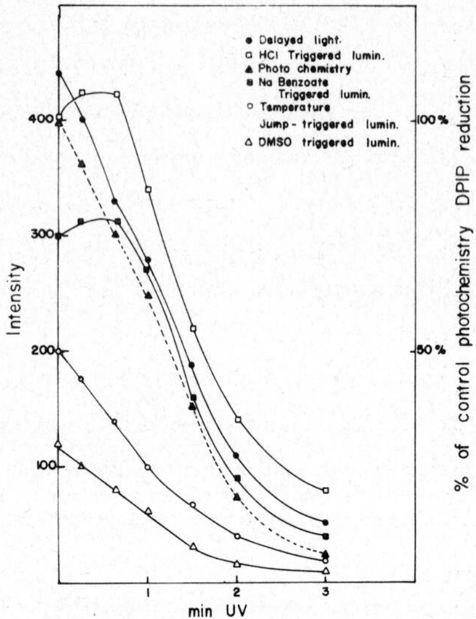

Figure 4. The effect of different times of UV irradiation on photochemistry, delayed light and triggered luminescence. Photochemistry is measured by DCPIP reduction by 685 nm light at limiting light conditions.

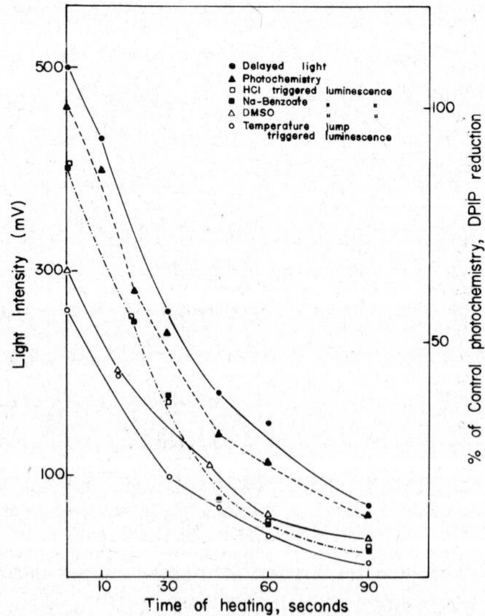

Figure 5. The effect of different times of heating ($50^{\circ}c$) on photochemistry, delayed light, and triggered luminescence. Photochemistry was measured by DCPIP reduction by 434 nm light.

delayed light and all types of triggered luminescences. In the case of heat treatment (figure 5) all the luminescent types were inhibited to the same degree, which was also the same degree of inhibition of the photochemical activity of photosystem II. This result shows that this type of inhibition affects photosynthetic units in an all or none fashion, so that all types of activities are proportional to the amount of surviving competent units.

The light emission in the second time range is coming only from the unaffected photosynthetic units, in contrast to msec delayed light which includes light emission from inhibited and uninhibited units together.[20] The inhibited photosynthetic units exhibit a very rapid decay of light emission in the msec range which does not extend out the second time range. Therefore, as we have seen (Fig. 3), the decay kinetics of second delayed light and triggered luminescence are not affected by the treatments, only the extent of light emission is lowered proportional to the number of unaffected photosystem II reaction centers.

With this perspective it is of interest to examine why the UV treatment acts in a much more selective way; the salt and acid luminescences being less affected than the other types of activity. Figure 4 shows the distinction between these two types. Salt and acid luminescence inhibition must be multiple photon processes (which is indicated by the lag in figure 4). The other activities are inhibited in apparently simple kinetics. It appeared that even in units in which electron transport is inhibited precursors for HCl and salt luminescences are still formed. Only as more UV photons are absorbed in the same unit will these luminescences be inhibited.

It has been suggested that the inhibition blocks electron transfer from water and enhances cyclic pathways from the oxidized and reduced moieties formed from the primary photoact. Activity can be restored by adding suitable electron donors instead of water. However, this restoration is limited and is less effective as the UV dose increases. Figure 6 shows the photochemical activity of the original system and the restored system (by addition of diphenylcarbazide) as a function of the time in UV. We can deduce that the initial effect of UV is to block electron transfer from H_20 but not to inhibit the primary photoact. As more UV photons are absorbed in the unit other types of destruction occur which may stop even the primary act.

It appears that at short times of UV irradiation the second delayed light, temperature jump and solvent luminescences are inhibited following the first step of UV inhibition, namely a block of electron transfer from H_20. A possible rationale for that is the fast recombination (directly, or by cyclic pathways) between Z^+ and Q^- which does not allow for storage of luminescence precursors for the above particular types of luminescence.

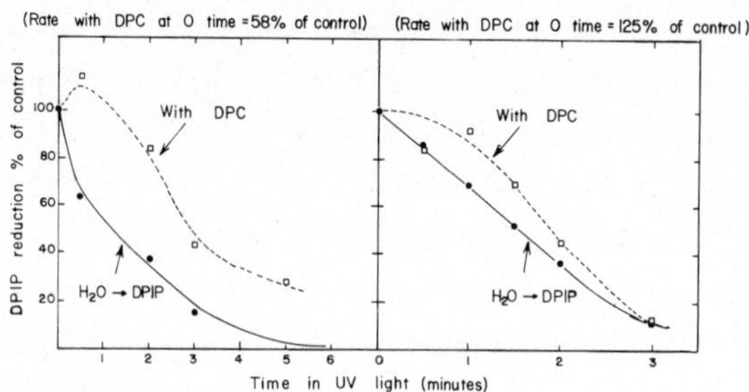

Figure 6. Restoration of photochemistry by diphenylcarbazide (DPC) in UV inhibited chloroplasts (2 Experiments). Photochemistry is measured by DCPIP reduction by 685 nm light at limiting light conditions. DPC concentration 0.5 mM. The rate of control and of control with DPC is normalized to 100 %. The rate of control with DPC compared to the control is indicated in the figure.

HCl and salt luminescences are inhibited following the second step of UV inhibition, namely the inhibition of the primary photoact itself, which is probably a multiphoton type. It appears that the rate of formation of precursors to these types of luminescence must be quite fast to compete with the recombination of Z^+ and Q^-.

One conclusion to be drawn is that the precursors for HCl and salt luminescence are different from the precursors of delayed light, temperature jump and solvent luminescence. Pursuing this idea further, luminescence precursors may be formed from the primary oxidized and reduced moieties and they need not be the primary products themselves in all cases, as is usually assumed.

These conclusions amplify our views expressed in other articles[13], which divided between different types of precursors according to kinetics. A distinction was made between delayed light, temperature jump, and HCl/salt types.

References

1. Strehler, B. L. and Arnold, W., 1951, J. Gen. Physiol., 34, 809.
2. Mayne, B. C. and Clayton, R. K., 1966, Proc. Nat. Acad. Sci., 55, 494.
3. Miles, C. D. and Jagendorf, A. T., 1969, Arch. Biochem. Biophys., 129, 711.
4. Barber, J. and Kraan, G. P. B., 1970, Biochim. Biophys. Acta, 197, 49.
5. Mar, T. and Govindjee, 1971, Biochim. Biophys. Acta, 226, 200.
6. Jurisinic, P. and Govindjee, 1972, Photochem. and Photobiol., 15, 331.
7. Hardt, H. and Malkin, S., 1972, Biochim. Biophys. Acta, 267, 588.
8. Yamashita, T. and Butler, W. L., 1968, Plant Physiol., 43, 1978.
9. Yamashita, T. and Butler, W. L., 1968, Plant Physiol., 43, 2037.
10. Malkin, S., 1971, Biochim. Biophys. Acta, 253, 421.
11. Mantai, K. M., Wong, J. and Bishop, N. I., 1970, Biochim. Biophys. Acta, 131,350.
12. Izawa, S., Heath, R. and Hind, G., 1969, Biochim. Biophys. Acta, 180, 388.
13. Malkin, S. and Hardt, H., 1971, II Intl. Congress on Photosynth. (Forti, ed.), Stresa, Italy, 253.
14. Hardt, H. and Malkin, S., 1971, Photochem. and Photobiol., 14, 483.
15. Avron, M., 1960, Biochim. Biophys. Acta, 40, 257.
16. Hardt, H. and Malkin, S., 1972, Biochem. Biophys. Res. Comm., 46, 668.
17. Barbieri, G., Delosme, R. and Joliot, P., 1970, Photochem. and Photobiol., 12,197.
18. Hardt, H. and Malkin, S., 1973, Photochem. and Photobiol., 17, 433.
19. Bertsch, W. and Lurie, S., 1971, Photochem. and Photobiol., 14.
20. Cohn, D.E., Cohen, W. S., Lurie, S. and Bertsch, W., 1975, III Intl. Congress on Photosynth. (Avron, ed.) Rehovot, Israel, in press.

M. AVRON, *Proceedings of the Third International Congress on Photosynthesis*
September 2-6, 1974, The Weizmann Institute of Science, Rehovot, Israel
Elsevier Scientific Publishing Company, Amsterdam, The Netherlands, 1974

FLASH NUMBER DEPENDENT LUMINESCENCE OF ISOLATED SPINACH CHLOROPLASTS AT DIFFERENT pH'S, LOW TEMPERATURE AND IN THE PRESENCE OF NH_4Cl

B.R. Velthuys

Biophysical Laboratory of the State University, P.O. Box 2120, Leiden, The Netherlands

1. Summary

The luminescence intensity at 40 ms after each of a series of short saturating flashes given to spinach chloroplasts was flash number dependent in special circumstances only. Lowering of temperature to -35 °C or increase of the pH to 10.0 before the last flash of a series made the flash number dependence similar to that of Zankel's[1] 200 µs component. This suggests that under these circumstances the decay of the precursor state of 200 µs luminescence was inhibited. In the presence of a high concentration of NH_4Cl a strongly enhanced luminescence was observed after the third and following flashes, suggesting that in the presence of NH_4Cl the oxidizing side of photosystem 2 can still accumulate four positive charges which, however, do not longer decay by water oxidation. DCMU, if added after preillumination, did not abolish the preillumination dependence of luminescence at -35° indicating that it did not accelerate the decay of accumulated positive charge.

2. Introduction

In the sub-ms region luminescence of green plants induced by short saturating flashes is strongly dependent on the so-called S-state, i.e. the number of positive charges accumulated in the pathway to water. According to Zankel[1] this dependence is most pronounced for a luminescence component decaying, at room temperature, with a halftime of about 200 µs. The intensity of this component is largest after the third flash, i.e. after the flash inducing the transition to the S_4 state, the state from which oxygen is produced. At longer times, i.e. at say, 10, 40 or 1000 ms, a different flash number dependence is observed under normal conditions. With Chlorella, luminescence is now strongest after the second flash[1,2]; with isolated spinach chloroplasts hardly any flash-number dependence whatsoever is observed now. In some special circumstances, however, a strong luminescence after the third flash is obtained with chloroplasts at much longer times than the sub-ms region. Here we want to describe some of these conditions, including (a) lowered temperature, (b) high or low pH, (c) presence of NH_4Cl.

Abbreviation: DCMU, 3-3,4(dichlorophenyl)-1,1-dimethylurea.

3. Materials and methods

Chloroplasts from spinach were prepared as described in ref. 3. They were suspended in a solution of pH 7.8, containing 0.2 M sucrose, 0.06 M KCl, 0.04 M NaCl and 0.025 M sodium morpholinopropane sulfonate (MOPS) buffer, and kept in a darkened vessel until used. For increase and decrease of pH (section b) use was made of sodium cyclohexylaminopropane sulfonate (CAPS) and citric acid respectively. The chlorophyll concentration was 0.10 mM. The measurements were performed using the apparatus of Kraan et al[4]. Actinic illumination was given in the form of short saturating flashes from a xenon flash tube, transmitted by a Balzers IR mirror and a Corning CS 4-76 glass filter. In a flash series the flashes were given at 1-s intervals, unless otherwise stated.

For the low-temperature experiments (section a) the measuring compartment was equipped with a double-walled perspex chamber. A brass cuvette holder could be placed inside. The cuvette, made of perspex, had a 1 mm path length and an area of 4 x 5 cm^2. Luminescence was detected from an area of approx. 3 x 2 cm^2 in the center. A thermocouple at the center of this area, and in direct contact with the sample, was used to measure the temperature. After preillumination the cuvette, plus holder, was put into liquid nitrogen, cooled to a temperature about 10 $^{\circ}C$ lower than the desired temperature, and transferred to the measuring compartment. When the temperature of the sample had increased to the desired temperature a flash was given and luminescence recorded.

4. Results

a. Subzero temperatures

As the temperature is lowered from room temperature to 0 $^{\circ}C$ the luminescence intensity at 40 ms after a flash decreases. Upon further lowering, below 0 $^{\circ}C$, however, the luminescence intensity increases again, as shown, for dark adapted chloroplasts, in fig. 1A. In fig. 1B the temperature dependence of luminescence is given for chloroplasts which had been preilluminated by one or two flashes before cooling. Note that the intensity scales of fig. 1A and fig. 1B differ by a factor of 10. It is evident that the increase of luminescence obtained with preilluminated chloroplasts at the lower temperatures is much larger than with dark-adapted chloroplasts. With preilluminated chloroplasts an optimum was obtained at about -35 $^{\circ}C$. A similar result was obtained by Tollin et al[5], 16 years ago, but in this early study the dependence on preillumination before cooling was not taken into consideration.

Fig. 1B (broken line) also shows the effect of atrazine, an inhibitor of Q^--reoxidation, added before cooling. Before the addition of atrazine the chloroplasts had been preilluminated with two flashes. As one might expect, luminescence at

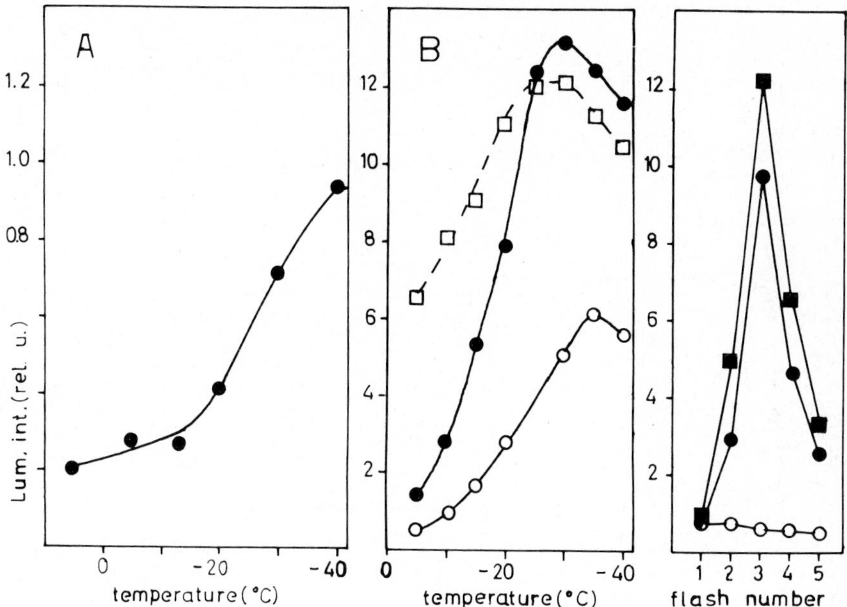

Fig. 1. Luminescence of spinach chloroplasts at 40 ms after a flash as function of temperature. (A) Dark adapted chloroplast. (B) Chloroplasts preilluminated before cooling by one (o) or two (• and ◻) flashes. The cooling was started 25 s after preillumination. ◻: 200 μM atrazine was added 5 s after preillumination.

Fig. 2. Effect of DCMU on luminescence of chloroplasts at -35 °C. Same conditions as for fig. 1B. Preillumination 0 to 4 flashes. The cooling was started 23 s after preillumination. •: 10 μM DCMU added 3 s after preillumination. o: 10 μ MDCMU was added before the preillumination.◻ : no DCMU added.

-5 °C was enhanced by atrazine. There was still an increase in luminescence, however, at lowering the temperature. This indicates that the increase obtained in the absence of inhibitor is only partly due to the inhibition of Q^- reoxidation which takes place in the temperature range from -10 °C to -30 °C (refs. 6,7); apparently the inhibition of some secondary electron transfer reaction on the oxidizing side of photosystem 2 is also partly responsible for the increase of luminescence in this temperature range.

The addition of an inhibitor of Q^- reoxidation (fig. 1B) had apparently only little affected the preillumination effect upon luminescence. This is confirmed by fig. 2. In this case 10 μM DCMU was added after preillumination, about 20 s before cooling to a final temperature of -35 °C. The controls without DCMU are also given, as are the results obtained when DCMU was added before illumination. One observes that, if DCMU was added after preillumination, the flash number dependence of luminescence was retained.

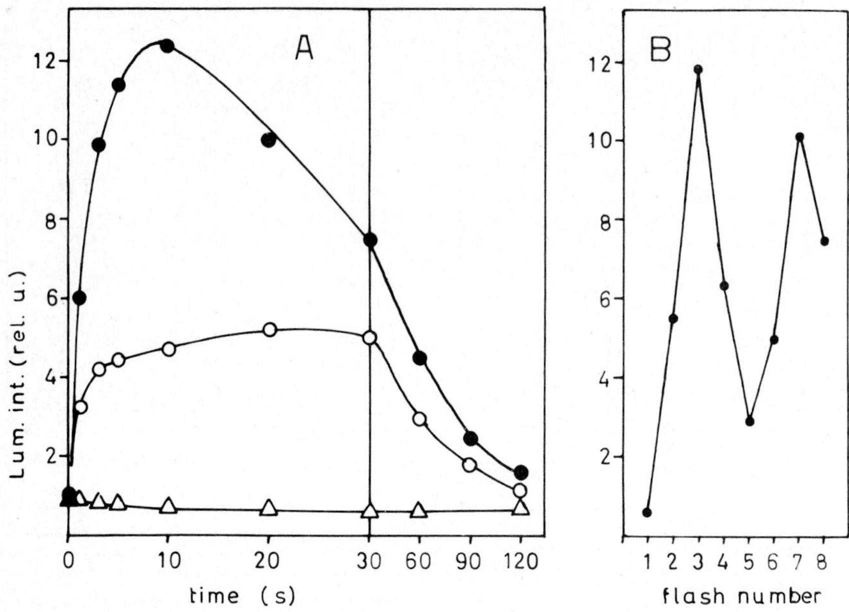

Fig. 3. (A) Luminescence of chloroplasts at 40 ms after a flash as function of incubation time at pH 10. Preillumination (5 s before the pH change from 7.8 to 10.0) zero (Δ), one (o) or two (•) flashes. (B) 40 ms-luminescence of chloroplasts at pH 10 as function of the flash number. Same conditions as for A (i.e. the last flash of a series was given at pH 10, the preceeding ones at pH 7.8). Incubation time at pH 10: 10 s.

b. pH changes

There are several reports about the interaction of pH with delayed light production. Wraight et al.[8] have studied the pH-dependence of luminescence of uncoupled chloroplasts illuminated, at the different pH's, with weak continuous light. Earlier the effect of an acid-base jump upon luminescence was studied by, amongst others, Mayne[9] and Kraan et al.[4] Our approach was different from both mentioned. Chloroplasts were preilluminated by a various number of flashes at pH 7.8. About 5 s later either base or acid was added. A variable time later an additional flash was given and luminescence recorded. The results obtained when base was added are shown in fig. 3. They strongly resemble those obtained in section (a): the luminescence intensity was strongly dependent on the number of preilluminating flashes, and was largest after two flashes preillumination. As fig. 3A shows the optimal effect needs some incubation time. Fig. 3B shows the flash number dependence after the optimal incubation time, 10 s. At longer incubation times the luminescence intensity obtained decreases again, presumably due to deactivation of higher S-states.

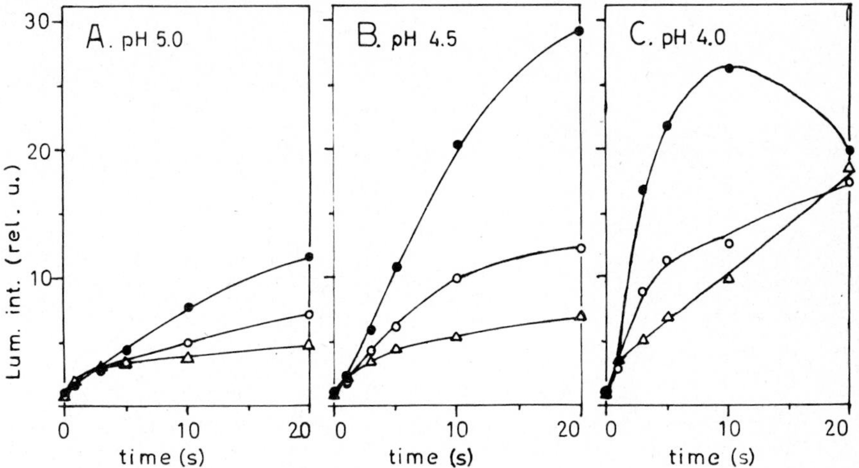

Fig. 4. Luminescence of chloroplasts at 40 ms after a flash as function of incubation time at pH 5.0 (A), 4.5 (B) or 4.0 (C). Preillumination (5 s before the acidification) zero (Δ), one (o) or two (•) flashes. 5 μM gramicidin D was present.

With dark-adapted chloroplast only a decrease was seen, at the shorter incubation times.

A preillumination dependence of flash-induced luminescence was also observed with chloroplasts subjected to acidification. Fig. 4A, B, C shows the results obtained after acidification to pH 5.0, pH 4.5 and pH 4.0 respectively. The chloroplasts were preilluminated by 0,1 or 2 flashes. Again, as with base, luminescence was largest with chloroplast preilluminated by 2 flashes. A notable difference, however, as compared with the effect of base, was that now luminescence with dark-adapted chloroplasts was enhanced also.

Both with acid addition and with base addition the incubation time needed for optimal enhancement of luminescence was not substantially decreased by substances like gramicidin D and nigericin (not shown). This indicates that the time dependence of the effect cannot be explained as being due to the time needed for equilibration of inside and outside pH.

c. Presence of NH_4Cl

A high concentration of NH_4Cl is known to inhibit oxygen evolution[10]. It was reported by Zankel[1] that luminescence is also affected: in the presence of 50 mM NH_4Cl the luminescence intensity at 0.5 ms, at room temperature, was strongly enhanced from the second flash on. We have observed that the affection of luminescence was still more pronounced at longer times. The luminescence intensity at 40 ms after flashes given in the presence of NH_4Cl was relatively little affected

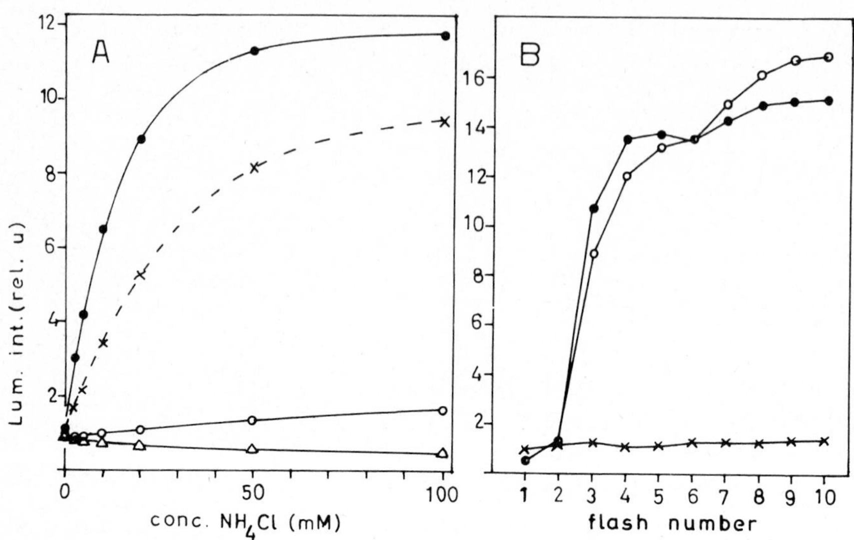

Fig. 5. (A) Luminescence at 40 ms after the first (Δ), second (o) and third (•) flash given in the presence of NH_4Cl. x: same conditions as for • except pH: pH 7.3 instead of 7.8. The flashes were given at 5 s intervals. (B) Luminescence at 40 ms after each of a series of 10 flashes given in the presence of 50 mM NH_4Cl. The flashes were given at 1-s (o) or 5-s (•) intervals. x: no NH_4Cl added (1-s flash intervals).

for the first two flashes; it was strongly enhanced from the third and following flashes on (fig. 5). Assuming that the predominant initial state is S_1 (see ref. 2),this result indicates that in the presence of NH_4Cl the oxidizing side of photosystem 2 can still accumulate four positive charges, of which the first three are stabilized but the fourth charge, which normally disappears in the reaction that gives oxygen, is not; apparently it decays at least partly by a recombination reaction with Q^-, giving an intense emission of luminescence. At pH 7.3 an about 3 times higher concentration of NH_4Cl was needed for stimulation of luminescence than at pH 7.8 (fig. 5A), indicating, in agreement with earlier reports about the action of NH_4Cl[10,11], that the unprotonated base, NH_3, is the active form. Remarkably, the optimal enhancement of luminescence after the third flash was obtained with the relatively long flash-interval of 5 s (fig. 5B), which may indicate that relatively slow changes are involved in the inhibiting action of NH_3.

5. Discussion

Under the usual conditions, i.e. at room temperature and near neutral pH, luminescence of isolated chloroplasts at longer times than the ms region is only little flash-number dependent. Under some special conditions, however, as we have shown above, the time region where a pronounced flash-number dependence can be observed is much extended. At subzero temperature and at pH 10 the flash number dependence at 40 ms strongly resembles that of the 200 μs component under normal conditions (compare figs. 2 and 3B with fig. 4A of ref. 1). This suggests that 40 ms luminescence under these conditions is produced from the same precursor state as that responsible for 200 μs luminescence, and that the extension of strong luminescence till much longer times than normal may be caused by an inhibition of the decay of this precursor state. A possible candidate for the precursor state of sub-ms luminescence appears to be the state $P680^+ - Q^-$, where P680 and Q are the primary donor and primary acceptor of photosystem 2 respectively. According to Gläser et al.[12] part of $P680^+$ ($P690^+$) decays with a halftime of 200 μs (the main part decaying more rapidly) which is also about the decay time of part of the Q^- (ref. 13) (the remaining part decaying more slowly). Our results then could indicate that at -35 °C and at pH 10 the decay of the slowly decaying part of the $P680^+$ may be much retarded, while the flash-number dependent amplitude of this component of $P680^+$ may be unaffected. From the same point of view the effects of acidification and NH_4Cl appear to be more complex. After acidification a large concentration of luminescence precursor is also produced by the first flash while normally little or no 200 μs luminescence is observed after this flash[1]. In the presence of NH_4Cl the decay of luminescence precursor only then appears to be substantially retarded when three positive charges have been accumulated in the pathway to water.

DCMU and atrazine, if added after preillumination, did not abolish the preillumination effect upon luminescence at -35 °C (section a). If, what seems to be a reasonable assumption, the preillumination effect is due to higher S-states, i.e. S_2 and S_3, then the results obtained indicate that atrazine and DCMU do not or only little accelerate the deactivation of these S-states. Yet, according to Rosenberg et al[14], the capacity for oxygen evolution by isolated chloroplasts is rapidly lost in the presence of DCMU, added after preillumination. Our results then indicate that this effect of DCMU is due to some inhibitory action of DCMU on reactions on the donor side of photosystem 2 rather then to destabilization of accumulated positive charge (see also refs. 15, 16).

Acknowledgement

This investigation was supported by the Netherlands Foundation for Chemical Research (SON), financed by the Netherlands Organization for the Advancement of Pure Research (ZWO).

References

1. Zankel, K.L., 1971, Biochim. Biophys. Acta 245, 373.
2. Joliot, P., Joliot, A., Bouges, B. and Barbieri, G., 1971, Photochem. Photobiol. 14, 287.
3. Amesz, J., Pulles, M.P.J. and Velthuys, B.R., 1973, Biochim. Biophys. Acta 325, 472.
4. Kraan, G.P.B., Amesz, J., Velthuys, B.R. and Steemers, R.G., 1971, Biochim. Biophys. Acta 223, 129.
5. Tollin, G., Fujimori, E. and Calvin, M., 1958, Proc. Natl. Acad Sci. U.S. 44, 1035.
6. Thorne, S.W. and Boardman, N.K., 1971, Biochim. Biophys. Acta 234, 113.
7. Malkin, S. and Michaeli, G., 1972, in: Proc. 2nd Int. Congr. Photosynthesis Research, Stresa, vol. 1, eds. G. Forti, M. Avron and A. Melandri (Dr. W. Junk N.V. Publishers, The Hague) p. 149.
8. Wraight, C.A., Kraan, G.P.B. and Gerrits, N.M., 1972, Biochim. Biophys. Acta 283, 259.
9. Mayne, B.C., 1968, Photochem. Photobiol. 8, 107.
10. Izawa, S., Heath, R.L. and Hind, G., 1969, Biochim. Biophys. Acta 180, 388.
11. Hind, G. and Whittingham, G.P., 1963, Biochim. Biophys. Acta 75, 194.
12. Gläser, M., Wolff, C., Buchwald, H.-E. and Witt, H.T., 1974, Photochem. Photobiol. 42, 81.
13. Zankel, K.L., 1973, Biochim. Biophys. Acta 325, 138.
14. Rosenberg, J., Sahu, S. and Bigat, T.K., 1972, Biophys. J. 12, 839.
15. Bouges-Bocquet, B., Bennoun, P. and Taboury, J., 1973, Biochim. Biophys. Acta 325, 247.
16. Etienne, A.L., 1974, Biochim. Biophys. Acta 333, 320.

M. AVRON, *Proceedings of the Third International Congress on Photosynthesis*
September 2-6, 1974, The Weizmann Institute of Science, Rehovot, Israel
Elsevier Scientific Publishing Company, Amsterdam, The Netherlands, 1974

ANALYSIS OF KINETICS AND TEMPERATURE SENSITIVITY OF DELAYED FLUORESCENCE FROM UNCOUPLED SPINACH CHLOROPLASTS

M.F. Hipkins and J. Barber

Botany Department, Imperial College, London SW7, England

1. Summary

The high-energy state of phosphorylation influences the intensity of delayed fluorescence in a way which is both time and temperature dependent. Previous analyses of the kinetics and temperature sensitivity of the emission process have failed to account for this. In this paper we have studied the above properties with spinach chloroplasts uncoupled with gramicidin. It has been found that the kinetics of the emission process between 3msec and 3.5sec, with a correction for fluorescence yield, are less complex than previously reported. Over this time range the decay seems to be equally well described by a $J^{-\frac{1}{2}}$ or a J^{-1} versus time relationship. In contrast to previous reports it appears from T-jump studies and the temperature sensitivity of the kinetics (assuming they follow the $J^{-\frac{1}{2}}$ versus time relationship) that the activation barrier for emission does not significantly change over the time range 3msec to 3.5sec, and is in the region of 0.5 to 0.7eV. Moreover, the monophasic nature of the decay suggests a single reaction mechanism in this time range.

2. Introduction

In order to understand the mechanism of delayed fluorescence it is important to have a knowledge of its decay kinetics and the temperature dependence of these kinetics. The mode of decay will give clues to the molecular processes involved, and the temperature dependence will give an idea of the magnitude of any activation energy barriers in the reaction.

It is becoming clear that delayed fluorescence may not appear as a simple function of time for at least two reasons. Firstly, the prompt fluorescence yield modifies in some way the delayed fluorescence intensity and secondly the high-energy state of phosphorylation is thought to act on the delayed fluorescence reaction.

The involvement of the fluorescence yield in the emission of delayed fluorescence was first suggested by Lavorel[1]. By analogy with the relation between incident light intensity I and the prompt fluorescence intensity F

$$F = \phi I \qquad 1$$

where ϕ is the prompt fluorescence yield, Lavorel suggested that delayed fluorescence L should be described by a similar relationship

$$L = \phi' J \qquad 2$$

where J is the rate of chlorophyll singlet formation via the delayed fluorescence reaction and ϕ' is the fluorescence yield of the chlorophyll molecules through which the delayed fluorescence exciton migrates before deexcitation. The exact relationship between ϕ and ϕ' is at present unclear.

The work of Mayne[2,3] and others [4,5] has shown that the high-energy state has an influence on the intensity of delayed fluorescence. The mechanism that leads to delayed fluorescence from oxygen-evolving photosynthetic systemsis thought to involve a back reaction between the reduced primary acceptor Q^- and oxidised primary electron donor Y^+ of photosystem two[1], and the high-energy state is envisaged as reducing the activation energy for the back reaction[6]. If the back reaction is assumed to be first order, then the rate of chlorophyll singlet formation may be expressed by [6,7]

$$J = [Y^+ChlQ^-]k' \, \nu \, \exp\{-(E_{ac} - \Delta p)/kT\} \qquad 3$$

where $[Y^+ChlQ^-]$ is the concentration of the charge transfer complex thought to act as the precursor for delayed fluorescence[8], k' is a constant containing entropy terms, ν is a frequency factor, E_{ac} is the activation energy for the back reaction and Δp is the high-energy state expressed as Mitchell's proton motive force[9] given by

$$\Delta p = \Delta\Psi + 2.303 \frac{RT}{F} \Delta pH \qquad 4$$

where $\Delta\Psi$ is the electrical gradient and ΔpH is the pH gradient across the thylakoid membranes, and the other symbols have their usual meaning.

Combining equations 2 and 3 gives

$$L = \phi' \, [Y^+ChlQ^-] \, k' \, \nu \, \exp\{-(E_{ac} - \Delta p)/kT\} \qquad 5$$

Clearly ϕ' and Δp are important parameters in this equation, which are time and temperature dependent, and will thus perturb both the measurement of E_{ac} and the kinetics of the decay process. These complications have not been fully realised in previous work so we have reinvestigated the temperature and kinetic properties of delayed fluorescence in the time range 3msec to 3.5sec. Previous studies in this time range have reported multiphasic decay kinetics (see for example Ruby[10]) and, in the millisecond region, a small or zero activation energy for delayed fluorescence [11,12]. The experiments described here were performed with fully uncoupled isolated chloroplasts (Δp reduced to zero) having no electron acceptor (Q fully reduced), and an attempt has been made to correct the measurements for fluorescence yield.

3. Materials and methods

Isolation of osmotically broken spinach chloroplasts was essentially the same as described previously[13] except that the chloroplasts were washed and suspended in a medium which contained 0.33M sucrose, 50mM KCl and 20mM N-Tris (hydroxy-methyl)-methyl-2-aminoethane-sulphonic acid (TES) brought to pH 7.0 with KOH. Chlorophyll concentrations were determined by the method of Arnon[14]. Prior to experimenting an appropriate quantity of stock chloroplasts was diluted with the above suspending medium to give a final chlorophyll concentration of 8 - 12 $\mu g\ ml^{-1}$.

(a) T-jumps. The intensity of 1msec delayed fluorescence and prompt fluorescence were measured as described earlier[13]. Temperature changes in the cuvette were monitored with a calibrated copper-constantan thermocouple connected to a Honeywell chart recorder. Addition of gramicidin was made by injecting 100μl of the appropriate stock through a light-tight diaphragm. Rapid addition of isothermal or hot suspending medium was accomplished with a syringe inserted through the same diaphragm. The addition of 1ml of suspending medium to 3ml of chloroplast suspension was found to give reproducible artifact-free mixing.

(b) Prompt and delayed fluorescence decays. The decay of prompt and delayed fluorescence was measured in a simple shutter apparatus. The sample, with uncoupler already added, was contained in a temperature controlled cuvette. It was dark adapted for 30sec and then illuminated with blue light (filter combination: 4mm BG18 and 5mm BG38) through an open camera shutter for a fixed time. At the end of this time the camera shutter closed and the photomultiplier was switched on by means of a high-voltage switch activated by a phototransistor. In order that the dead time be reduced to a minimum the photomultiplier was switched on by the addition of the final two dynodes to the dynode chain. In this way the dead time was reduced to 3msec. The signal from the photomultiplier was either fed into a storage oscilloscope for recording in the time range 3msec to 800msec, or into a chart recorder for the time range 0.3 to 3.5sec. For delayed fluorescence measurements the photomultiplier was protected with a 2mm RG695 filter. To monitor prompt fluorescence changes in the dark induced by the actinic beam a continuous weak blue light was used (filters: 7mm BG18 and 2mm BG38), and the measurements made under the same actinic light conditions as the delayed fluorescence. The intensity of the weak measuring beam was such that the prompt fluorescence signal (measured through 4mm BG665 and a 687nm interference filter) was not distorted by the delayed fluorescence due to the actinic beam. BG and RG filters by Schott.

4. Results

One possible way to estimate the activation energy for a process is to subject the process to a rapid change of temperature (T-jump) and to measure the change of rate of reaction. T-jumps have been carried out with the delayed fluorescence process both in the region of 0.5sec to 1sec[15], and in the region of 1sec to 10sec[16]. We turned our attention to the millisecond region because of reports of a very small activation energy for delayed fluorescence in this time domain[11,12] which appeared inconsistent with the marked effect on delayed fluorescence of the high-energy state and artificially produced electrical gradients across the thylakoid membrane[13]. Fig.1. shows the effect of an isothermal addition and a typical T-jump on prompt and millisecond delayed fluorescence from chloroplasts totally uncoupled with gramicidin D. The isothermal additions were used to correlate the relative decrease in prompt and delayed fluorescence so that the appropriate corrections could be made for the analysis of the T-jump experiments[7].

By substituting $\Delta p = 0$ in equation 5, and ignoring for the moment the temperature sensitivity of the parameters $[Y^{+}ChlQ^{-}]$, ν and k' then E_{ac} may be derived:

$$E_{ac} = 2.303\ k.\ \frac{T_1\ T_2}{T_1 - T_2}\ \log_{10} \frac{J_2}{J_1} \qquad 6$$

where J_1 and J_2 are the intensities of delayed fluorescence observed at the initial and final temperatures T_1 and T_2, corrected for

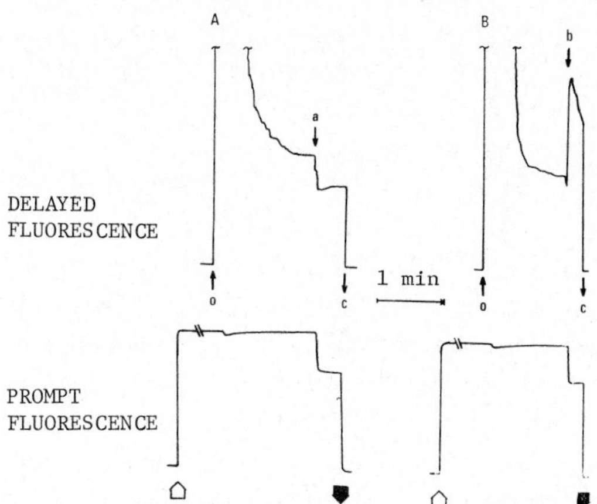

Fig.1. The effect of an isothermal addition (A) and a temperature jump (B) on the prompt and delayed fluorescence from uncoupled spinach chloroplasts with no electron acceptor. The suspension was illuminated for 1min before the shutter across the delayed fluorescence photomultiplier was opened. The chloroplasts were then uncoupled with 10^{-7}M gramicidin D (for the convenience of scale the initial level of delayed fluorescence and injection of gramicidin are not shown) and subjected to a rapid addition of 1ml of isothermal (a) or hot (b) suspending medium. The temperature change due to the hot addition was from 20° to 34.5°C while the isothermal addition was carried out at 20°C. Chlorophyll concentration 8μg ml^{-1}. Open arrows: light on, closed arrows: light off. The opening and closing of the delayed fluorescence photomultiplier shutter are indicated by o and c respectively.

fluorescence yield changes. A series of experiments with various initial and final temperatures was performed. The temperature ranges were kept fairly restricted to avoid problems of thermal denaturation and damage to the photosystem two reaction centres. The weighted mean of the results over all the temperature ranges was found to be 0.64 ± 0.01 eV .

The limitation of this approach is that it has assumed that all other processes are temperature independent. Our calculations have to some extent allowed for temperature induced changes in fluorescence yield and, since there was no net electron flow in our preparations, variations in fluorescence yield due to changes in the redox state of the system two traps were avoided. It has also been assumed that the formation and non-radiative decay of the precursor $Y^{+}ChlQ^{-}$ are temperature insensitive and that its concentration does not change during the T-jump. There is no way to check or correct for these effects because of the nature of the experiments and the methods used (in particular the creation and decay of the precursor more than once during the temperature jump), but Malkin and Hardt[15] have emphasised the necessity for such corrections which they were able to make at longer dark times.

Because of the possible changes in the concentration of $Y^{+}ChlQ^{-}$ with temperature another means of estimating the activation energy was employed. This involved measurement of the temperature sensitivity of the reaction constant. This method is independent of the sensitivity of the precursor concentration to temperature, but requires an analysis of the decay kinetics, and the formulation of a reaction mechanism before the activation energy can be estimated. Thus the decay of delayed and prompt fluorescence was studied in the time ranges 3msec to 800msec and 0.3sec to 3.5sec at temperatures between 5°C and 30°C. Fig. 2 shows typical successive measurements of the prompt and delayed fluorescence from the same sample of chloroplasts uncoupled with 5.10^{-7}M gramicidin, and without electron acceptor.

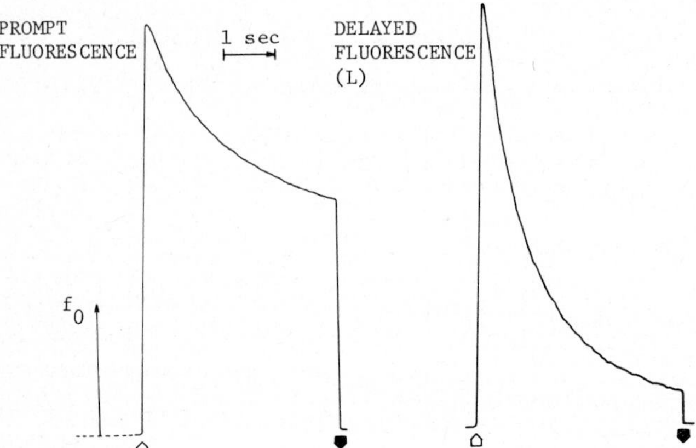

Fig. 2. The decay of prompt and delayed fluorescence in the time range 0.3 sec to 3.5 sec after a 25 sec period of continuous illumination. The open arrow denotes the start of the recording: the initial point of the decay was taken 0.3 sec later in order to allow for the response time of the recorder. The closed arrow denotes switching off the photomultiplier. The two decays were measured successively on the same sample, with a dark time of 90 sec between the two measurements. The f_o level was measured from the fluorescence induction on a CRO in a separate experiment. Temperature 18.3°C. Chlorophyll concentration 12μg ml^{-1}.

Following equation 2, the measured delayed fluorescence L was corrected with the prompt fluorescence yield in order to derive J. This was done both with the total prompt fluorescence yield and the variable prompt fluorescence (the part of the prompt fluorescence above the f_o level: see Fig. 2). The decay was then analysed in terms of three decay modes (i) lnJ versus time (ii) $J^{-\frac{1}{2}}$ versus time and (iii) J^{-1} versus time. Fig. 3 illustrates a typical set of plots from an experimental curve at one particular temperature. The linearity of each plot was measured by calculation of the correlation coefficient, and Table 1 summarises the results of this type of experiment carried out over two overlapping time ranges. As can be

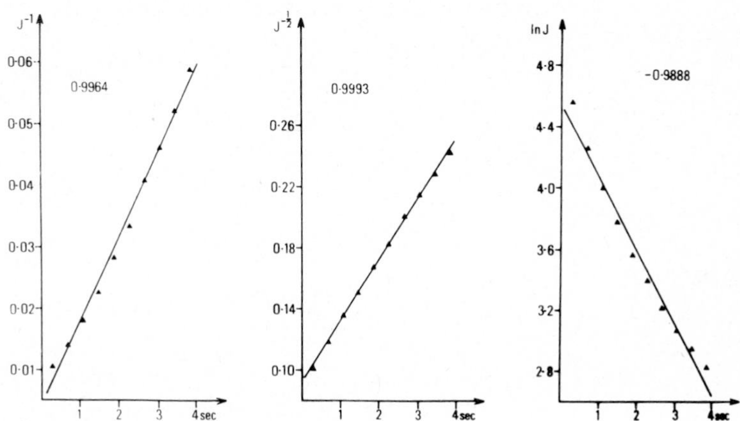

Fig. 3. Plots of three representations of the data of Fig. 2. The values of J were derived from equation 2 by using the total prompt fluorescence for ϕ'. The straight line fits were calculated by a least squares analysis, and the numbers on the plots denote the correlation coefficients of the particular set of points.

seen in Table 1, the more satisfactory straight line plots are obtained by correcting L with total prompt fluorescence, and that among the latter the lnJ representation is less satisfactory than either the J^{-1} or $J^{-\frac{1}{2}}$ fits. In general, both J^{-1} and $J^{-\frac{1}{2}}$ plots appear to give equally good straight lines, with neither of these consistently better than the other. If the delayed fluorescence is corrected with variable fluorescence then a different picture emerges. The J^{-1} plot gives a better fit for all the times and temperatures studied, but the fit is not as good as for the data corrected with total prompt fluorescence.

Table 1

Analysis of delayed fluorescence from uncoupled chloroplasts expressed as correlation coefficients of the stated kinetic laws

Time Range	Temperature °C	Total PF Correction			Variable PF Correction		
		ln J	J^{-1}	$J^{-\frac{1}{2}}$	ln J	J^{-1}	$J^{-\frac{1}{2}}$
0.3 sec -3.5 sec	29	-0.9893	0.9934	0.9997	-0.9877	0.9983	0.9982
	23	-0.9823	0.9969	0.9967	-0.9779	0.9965	0.9914
	18.3	-0.9888	0.9964	0.9993	-0.9890	0.9989	0.9974
	12.5	-0.9883	0.9884	0.9985	-0.9856	0.9995	0.9957
	6	-0.9819	0.9971	0.9915	-0.9606	0.9818	0.9722
3msec 800msec	29	-0.9932	0.9971	0.9975	-0.9908	0.9961	0.9946
	23	-0.9961	0.9975	0.9992	-0.9956	0.9977	0.9988
	18.3	-0.9906	0.9954	0.9950	-0.9861	0.9916	0.9907
	12.5	-0.9900	0.9958	0.9944	-0.9875	0.9942	0.9913
	6	-0.9872	0.9955	0.9921	-0.9844	0.9926	0.9894

5. Discussion

One trend emerges from the kinetic analysis of the decay of delayed fluorescence reported above and that is the decay does not seem to be described by a first order formulation. At first sight this would appear to argue against the postulated first order nature[8] of the back reaction thought to lead to delayed fluorescence[1]. However there are possible explanations for this.

It has been shown that both $J^{-\frac{1}{2}}$ and J^{-1} plots give relatively good fits to the decay over the 3msec to 3.5sec time range. Both decay modes have been reported before, for L (refs. 3, 15, 17) and for J (refs. 1, 16, 18) although in these cases no attempt was made to suppress the high-energy state effect.

What are the possible molecular mechanisms which would lead to these kinetic relations? Accepting for the moment that a first order process is involved then the rate of decay of the precursor through the delayed fluorescence back reaction is given by

$$J = -\frac{dC_{df}}{dt} \qquad 7$$

where C_{df} is that part of the precursor Y^+ChlQ^- that decays via the delayed fluorescence back reaction. Taking the special case of the delayed fluorescence back reaction being the sole route for the decay of the precursor (a 'deactivation' type decay of Lavorel[8]) equation 7 becomes:

$$J = - \frac{dC_{total}}{dt} \qquad 8$$

Under these circumstances the second order $J^{-\frac{1}{2}}$ plot gives support to a mechanism already suggested by others[19,16] in which delayed fluorescence results from the fusion of two triplet states to give a chlorophyll singlet state (s) in a second order reaction, so

$$J = \frac{ds}{dt} = k_2 \, C_T^2$$

where C_T is the concentration of triplet states and k_2 is a second order reaction constant. The triplet states are produced in the first order back reaction from the precursor Y^+ChlQ^-

$$\frac{dC_T}{dt} = k_1 \, C_{total}$$

If this reaction is rate-limiting then

$$\frac{ds}{dt} = k_1 \, C^2_{total}$$

and hence

$$J^{-\frac{1}{2}} = J_o^{-\frac{1}{2}} + k_1^{\frac{1}{2}}t \qquad 9$$

(see ref. 16), where J_o is the value of J at time t = 0. The involvement of triplet states in bringing about delayed fluorescence has yet to find rigorous experimental verification[20]. Nevertheless, this triplet model is attractive in view of the problem of understanding how delayed fluorescence excitons are able to diffuse into the bulk chlorophyll away from the open trap from which they originate, although recent considerations of Lavorel[21] and Wraight[22] do not rule out non-perfect trapping within a single photoactive unit.

The non-first order decay might also arise from other simultaneous processes competing with the first order back reaction for the delayed fluorescence precursor, in a 'leakage' type process as postulated by Lavorel[8]. These simultaneous processes could include the non-radiative decay of the precursor both on the oxidising or reducing side of the charge complex Y^+ChlQ^-. In this case the decay will reflect the complexity of the reactions deactivating the precursor. Consider a simple case where the competing reaction is second order; then

$$J = -\frac{dC}{dt} = k_1C \qquad 10$$

and

$$C^{-1} = C_o^{-1} + k_2t \qquad 11$$

By assuming that $k_2 >> k_1$, and combining equations 10 and 11,

$$J^{-1} = -(k_1C_o)^{-1} + (k_2/k_1)t \qquad 12$$

Such a model could explain the J^{-1} relationship, but in the above experiments it is likely that $k_2 \not> k_1$ since there was no net electron flow. Other simple explanations for the J^{-1} versus time plot are not clear. Lavorel[1,8] has suggested that the Elovich-Cope law[23] could be suitable for describing the decay of delayed fluorescence. This law, however, predicts that the slope of the J^{-1} plot should be independent of temperature. Such a property was not found here.

If the deactivation process was operative under our experimental conditions and there were no strongly competing reactions then we should consider the possibility that the relatively good fit of the data to the $J^{-\frac{1}{2}}$ law indicates that the delayed fluorescence back reaction is itself second order[15]. For example

$$J = k\,[Y^+]\,[Q^-] \qquad 13$$

and thus equation 9 would apply for the decay. The concept of a second order reaction receives some support from the work of Bennoun[24] in which he showed that the decay of (Q^-) in 3-(3,4 dichlorophenyl)-1, 1-dimethylurea (DCMU) blocked Chlorella (that is under apparently deactivation conditions) was linear in $(Q^-)^{-1}$. Overall the suggestion that the decay is due to a second order process means that at least one of the charged precursors of delayed fluorescence is diffusible and could give support to the general model of electron-hole interactions along the lines of that suggested by Bertsch et al[25].

In trying to clarify the mechanism of delayed fluorescence we have the serious problem of choosing between the two kinetic laws $J^{-\frac{1}{2}}$ and J^{-1}. As we have shown, a crucial factor in determining the shape of the delayed fluorescence decay is the emission factor ϕ' in equation 2. Its form and its relationship with the prompt fluorescence yield will depend on the model adopted for the photosynthetic unit. One may think of two extreme cases: firstly a separate model of the photosynthetic unit[26] in which an exciton cannot migrate from one unit to another. In this model the delayed fluorescence emission factor is equal to the prompt fluorescence yield of an open trap (assumed to be non-zero) and is thus independent both of time and of the macroscopic fluorescence yield which arises from the other photosynthetic units[26]. The other extreme is a totally connected model in which an exciton is free to migrate over a homogeneous array of photosynthetic units, with the exception of those chlorophyll molecules which give rise to the 'dead' fluorescence[18,26]. In this model it is not possible to differentiate between a delayed and a prompt fluorescence exciton, and ϕ' of equation 2 would be equal to $\Delta\phi$, the variable prompt fluorescence yield. The delayed fluorescence emission factor is thus dependent on the degree of connection between photosynthetic units. Although models have been proposed[26,8] the extent of the connection between units, and thus the exact measure of ϕ', is not clear. It seems likely from the work of Lavorel and Joliot[26] that the value of ϕ' is intermediate between the two extreme cases so that it is a function of time but less so than $\Delta\phi$. We thus feel that the total fluorescence yield changes probably give a better reflection of ϕ' than either $\Delta\phi$ or no correction at all, although in doing this we do not seek to give a meaning to the 'dead' fluorescence but simply to use it for conveniently reducing the time dependence of $\Delta\phi$. However, the computer simulation of the photosynthetic unit by Lavorel[21], in which he showed that an exciton might not visit a large number of units, could suggest that an even less strong function of time than total fluorescence yield should be used for ϕ'. In this case, the trend of the fit to become more like $J^{-\frac{1}{2}}$ than J^{-1} might continue, and we thus tend to favour at this stage the $J^{-\frac{1}{2}}$ law as the better representation of the kinetics, as has been done by others[16].

Therefore, using equation 9 to obtain k at various temperatures an Arrhenius plot may be obtained (Fig. 4) and E_{ac} estimated. The

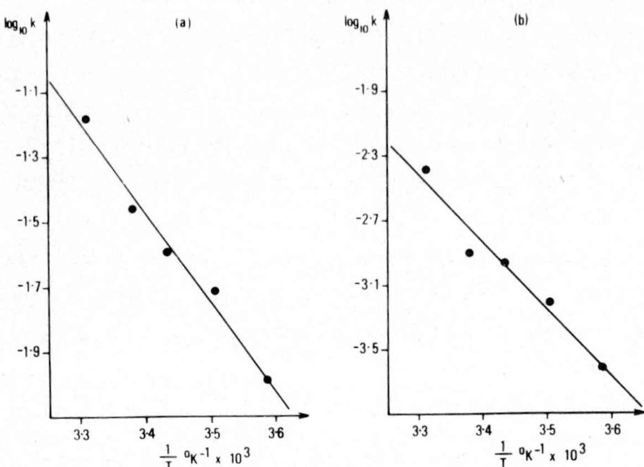

Fig. 4. Arrhenius plots of the reaction constants derived from the slopes of the $J^{-\frac{1}{2}}$ versus time plots, where L was corrected with total prompt fluorescence yield. (a) time range 3msec to 800msec (b) time range 0.3sec to 3.5sec.

errors on E_{ac} are taken from the standard errors of the slopes of the Arrhenius plots. Both the slopes and the errors were calculated from a least squares fit. Similar experiments, also using total prompt fluorescence to correct L have given values of E_{ac} which are shown in Table 2. The weighted mean of these values is 0.56 $\pm$0.03eV.

Table 2

Activation energies in eV of the delayed fluorescence back reaction from Arrhenius plots of the decay constant of the $J^{-\frac{1}{2}}$ representation

Expt.	Time Range	
	3 - 800msec	0.3 - 3.5 sec
A	0.47 $\pm$ 0.12	-
B	0.31 $\pm$ 0.14	0.55 $\pm$ 0.13
C	0.46 $\pm$ 0.09	0.76 $\pm$ 0.11
D	0.55 $\pm$ 0.04	0.82 $\pm$ 0.11

This value of E_{ac} is derived by making three assumptions: (i) the deactivation type of reaction exists, (ii) the total prompt fluorescence yield is the better measure of the delayed fluorescence emission factor and (iii) the $J^{-\frac{1}{2}}$ representation is the correct description for the decay of delayed fluorescence. It is in fairly good agreement with the value of 0.64eV found from T-jump studies on millisecond delayed fluorescence, and also with derivations of E_{ac} made by others[15,16]. The value for the millisecond activation energy contrasts with earlier reports[11,12], and is consistent with the high-energy state effect mentioned above. In conclusion, the relatively constant nature of the value of the activation energy, and the indication that there is no gross change with time in the decay mode suggest that there may be a single mechanism leading to delayed fluorescence in the time range studied, and that it could include a second order reaction. Accepting that this reaction is a back reaction in the photosystem two reaction centre and realising that more than one oxidation state of the oxygen-evolving system is probably involved[27,28] then a value for E_{ac} of about 0.6eV indicates that about 1.2eV can be extracted from the 1.8eV available from the singlet excited chlorophyll for driving photosynthesis, a value which is consistent with theoretical arguments (see ref. 6).

Acknowledgements

We thank the Science Research Council for financial support, and Mr. L. Hullis and Mr. C.T. Wright for technical assistance.

References

1. Lavorel, J., 1968, Biochim. Biophys. Acta 153, 727.
2. Mayne, B.C., 1967, Photochem. Photobiol. 6, 189.
3. Mayne, B.C., 1968, Photochem. Photobiol. 8, 107.
4. Wraight, C.A. and Crofts, A.R., 1971, Eur. J. Biochem. 19, 386.
5. Neumann, J., Barber, J. and Gregory, P., 1973, Plant Physiol. 51, 1069.
6. Crofts, A.R., Wraight, C.A. and Fleischman, D.E., 1971, FEBS Letters 15, 89.
7. Hipkins, M.F. and Barber, J., 1974, FEBS Letters 42, 289.
8. Lavorel, J., 1972, in: Bioenergetics of Photosynthesis, ed. Govindjee, in the press, Academic Press, New York.
9. Mitchell, P., 1966, Biol. Rev. 41, 445.
10. Ruby, R.H., 1968, Photochem. Photobiol. 8, 299.

11. Tollin, G., Fujimori, E. and Calvin, M., 1958, Proc. Natl. Acad. Sci. U.S. 44, 1035.

12. Sweetser, P.B., Todd, C.W. and Hersh, R.T., 1961, Biochim. Biophys. Acta 51, 509.

13. Barber, J., 1972, Biochim. Biophys. Acta 275, 105.

14. Arnon, D.I., 1949, Plant Physiol. 24, 1.

15. Malkin, S. and Hardt, H., 1973, Biochim. Biophys. Acta 305, 292.

16. Jursinic, P. and Govindjee, 1972, Photochem. Photobiol. 15, 331.

17. Malkin, S. and Hardt, H., 1972, in: 2nd Int. Congr. on Photosynthesis, Stresa, 1971, vol. 1, eds. G. Forti, M. Avron and A. Malandri (Dr. W. Junk Publishers N.V., The Hague) p. 253.

18. Clayton, R.K., 1969, Biophys. J. 9, 60.

19. Lavorel, J., 1969, in: Progress in Photosynthesis Research, vol. 2, ed. H. Metzner (T.H. Laupp, Jr., Tubingen) p. 883.

20. Stacy, W.T., Mar, T., Swenberg, C.E. and Govindjee, 1971, Photochem. Photobiol. 14, 197.

21. Lavorel, J., 1973, Physiol. Veg. 11, 681.

22. Wraight, C.A., 1972, Biochim. Biophys. Acta 283, 247.

23. Cope, F.W., 1964, Proc. Natl. Acad. Sci. U.S. 51, 809.

24. Bennoun, P., 1970, Biochim. Biophys. Acta 216, 357.

25. Bertsch, W., West, J. and Hill, R., 1971, Photochem. Photobiol. 14, 241.

26. Lavorel, J. and Joliot, P., 1972, Biophys. J. 12, 815.

27. Zankel, K.L., 1971, Biochim. Biophys. Acta 245, 273.

28. Barbieri, G., Delosme, R. and Joliot, P., 1970, Photochem. Photobiol. 12, 197.

M. AVRON, *Proceedings of the Third International Congress on Photosynthesis*
September 2-6, 1974, The Weizmann Institute of Science, Rehovot, Israel
Elsevier Scientific Publishing Company, Amsterdam, The Netherlands, 1974

STUDIES ON INDUCTION AND DECAY KINETICS OF DELAYED LIGHT EMISSION IN SPINACH CHLOROPLASTS.

Shigeru Itoh and Norio Murata.

Department of Biophysics and Biochemistry, Faculty of Science, University of Tokyo, Tokyo, Japan.

1. Summary

Induction-kinetics of the intensity of delayed light emission at 0.2 msec and 3.7 msec after repetitive flash light and decay kinetics from 0.2 to 150 msec were compared by a specially designed measurement system in spinach chloroplasts. It was shown that widely different induction patterns of millisecond delayed light ever reported can be understood as the changes in the relative yields of three exponentially decaying components with half decay times of 0.2 msec 0.8 msec and 35 msec depending on the redox state of the reaction centers of system II. The high energy state of phosphorylation in chloroplasts and experimental contitions also shown to affect their relative yields.

2. Introduction

Since the discovery of delayed light emission by Strehler and Arnold[1], many studies have been done on the induction kinetics of delayed light under various experimental conditions[1-8]. We have also investigated the induction kinetics of rapidly decaying components of delayed light (millisecond delayed light) using repetitive flash excitation[5,9], and have shown that the induction in the excitation of the delayed light is composed of two distinct phases ; The fast phase consists of an initial rapid rise in yield on onset of illumination followed by a fast decline to a lower steady state level, while the slow phase consists of a slow increase in yield from a lower level to a higher steady level. Suppression of the high energy state by addition of an uncoupler of phosphorylation elimanated the slow induction-phase and under this condition, rapid decline in the fast induction-phase described above is inversely related to the transitional change in the fluorescence yield during illumination. These findings led us to an inference that the millisecond delayed light originates from excitation of photosynthetic units of photoreaction II, in which the reaction centers are in an open state and that the subsequent development of high energy state in chloroplasts enhances the delayed light emission[9] thus giving rise to the slow induction-phase.

On the other hand, Clayton[3] suggested a parallel relationship between the yields of longer lived components of delayed light and fluorescence. Recently, Wraight[8] reported that the yield of millisecond delayed light also increased in parallel with that of the fluorescence during the induction period under low intensity of excitation light.

The discrepancy between their conclusion and ours has to be elucidated, which made the main purpose of the present study. In this study, the induction kinetics of delayed light emission at 0.2 and 3.7 msec after excitation with repetitive flash light under varied conditions were compared. Special experimental device was designed for the purpose. The decay kinetics of delayed light during a period of 0.2-150 msec after the excitation flash was also measured.

3. Materials and methods.

Chloroplasts were prepared from spinach leaves in a medium containing 0.4 M sucrose, 0.01 M NaCl and 0.05 M phosphate buffer pH 7.8 as described previously. In measurements of delayed light chloroplasts were diluted with the same medium to give a concentration, according to Arnon's method,[10] of 5-10 μg chlorophyll/ml.

Measurement of delayed light was performed with a Becquerrel-type phosphoroscope as described previously[5]. A 5.4 msec cycle of excitation and measurement was used: 0-0.9 msec for excitation, 0.9-5.4 msec for darkness. The time span between the full open and completely closed states of the sector combination was 0.1 msec. The delayed light during 0.2-3.7 msec after cessation of excitation flash was measured. To measure the decay kinetics of delayed light a combination of the phosphoroscope and a mechanical shutter was used.

Fig. 1. Induction of delayed light emission during repetitive flash excitation. (a) An oscilloscopic trace of the induction of delayed light. Upper and lower envelopes of the bright area correspond to the inductions of delayed light at 0.2 msec($L_{0.2}$) and 3.7 msec($L_{3.7}$) after the end of each excitation flash. The bright area corresponds to the decay of delayed light during each measurement cycle from 0.2 to 3.7 msec after the end of each 0.9 msec excitation flash. White excitation light used; 50,000 ergs/cm^{2}sec with the sector stopped. Temperature 22°. (b) Decay curves of delayed light at the I, D, D-S and S stages of the induction indicated in Fig. 1 (a).

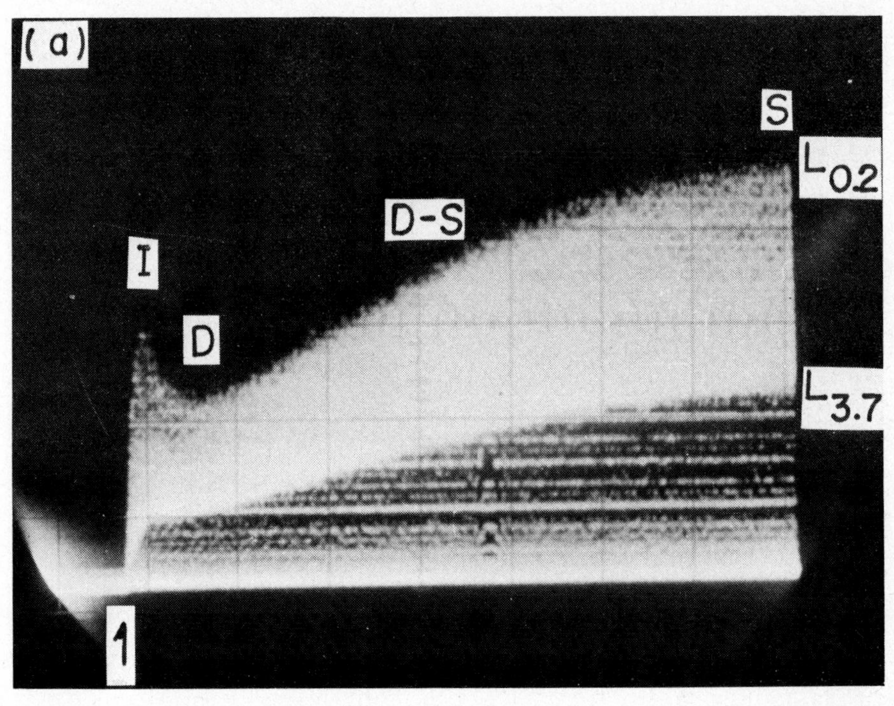
(a)
S
$L_{0.2}$
D-S
I
D
$L_{3.7}$
1
0 2 4 6 sec

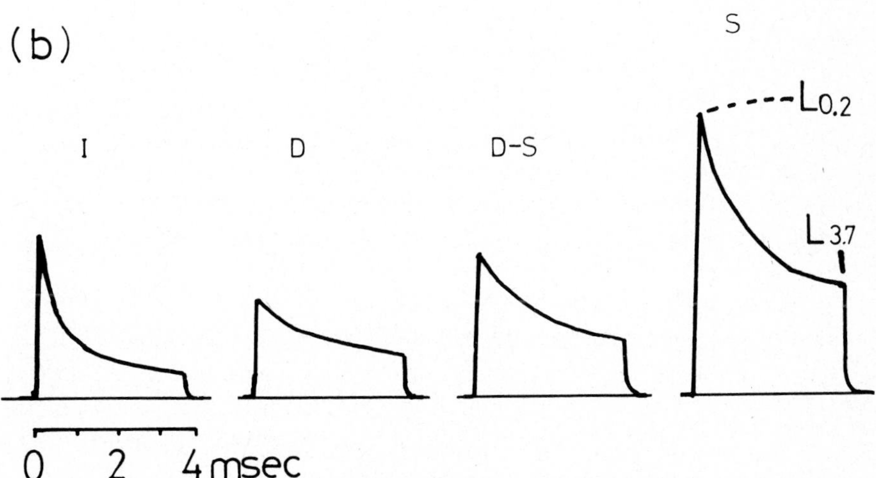
(b)
S
$L_{0.2}$
I
D
D-S
$L_{3.7}$
0 2 4 msec

White excitation light obtained from a 500 W xenon arc lamp was used as described previously[5]. The signal from a photomultiplier of the phosphoroscope was directly fed into a storage type oscilloscope and photographed or stored in a Transient Time Converter, from which records were drawn on a strip chart recorder.

4. Results

Induction and decay kinetics of delayed light emission.

Fig. 1(a) shows an oscilloscopic trace of the induction of delayed light emission during repeted excitation and measurement. The shutter was opened at time zero and closed after 15 sec. During this period an intermittent illumination was provided by repetitive flash light. The curves in Fig. 1(b) shows the time courses of each individual segment of delayed light emission responding to the previous excitation flash. It will be seen that the upper (lower) envelope of the bright area in Fig. 1(a) represents the change in intensity of the delayed light at 0.2 msec, $L_{0.2}$, (at 3.7 msec, $L_{3.7}$) after cessation of each excitation flash. As will be seen the decay pattern of delayed light widely differed with the duration of illumination with repeated flashes. In Fig. 1(a), $L_{0.2}$ showed a rapid increase on onset of illumination to the initial peak (I) which was followed by a rapid decline to a lower level, a dip (D); Fast induction-phase, then the $L_{0.2}$ level again increased gradually to reach a higher steady state level (S); Slow induction-phase. On the other hand, $L_{3.7}$ rapidly increased to a certain level; Fast induction-phase, and then, without showing any decline, it was followed by a slow increase to reach a steady state level (S); Slow induction-phase.

As shown in the following, the decay kinetics of delayed light was analysed. It was found that there are three exponentially decaying components of delayed light having half decay times of 0.2, 0.8 and 35 msec. For this computation, data were taken from traces of decay time courses of delayed light extending to a period of 0.2 to 150 msec after cessation of light. Considering the uncomparable slowness of decay rate of 35 msec component, it can be inferred that the three decay-components contributed to $L_{0.2}$, however, $L_{3.7}$ consisted mostly of the 35 msec decay-component. The value for $L_{0.2}$ minus $L_{3.7}$ will be practically equal to the sum of the 0.2 and 0.8 msec decay-components present at 0.2 msec after the flash excitation. Fig. 1(c) shows the change in $L_{0.2}$, $L_{3.7}$ and ($L_{0.2}$-$L_{3.7}$) computed

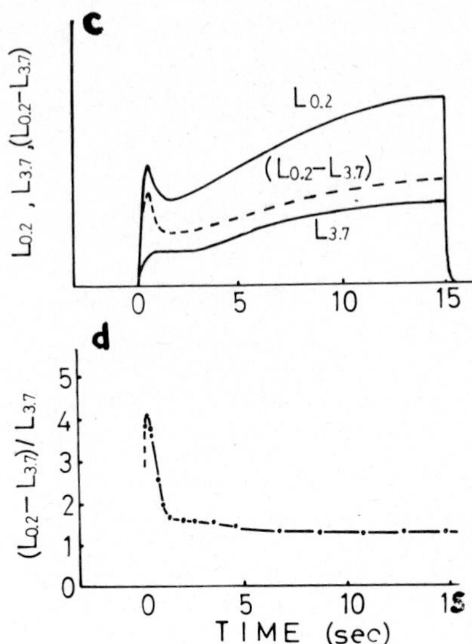

Fig. 1(c) Inductions of $L_{0.2}$, $L_{3.7}$ and $(L_{0.2}-L_{3.7})$. Curves are obtained from Fig. 1(a).
(d) Time course of the ratio, $(L_{0.2}-L_{3.7})/L_{3.7}$.

from the data in Fig. 1(a). The initial peak appeared sharper in $(L_{0.2}-L_{3.7})$ than in $L_{0.2}$ thus indicating that the decline from the I to D stage is exclusively due to the faster decay-components with 0.2 and 0.8 msec half decay-times.

In the decay time courses of delayed light emission at various induction stages presented in Fig. 1 (b), the decline from $L_{0.2}$ to $L_{3.7}$ was more marked at the I stage than in the D, D-S and S stages. The pattern of decay time courses, however, seemed to be unchanged at the latter three stages, although the levels in intensity increased with illumination time. Fig. 1 (d) shows the changes in ratio of the fast decay- to the slow decay-components during induction period; $(L_{0.2}-L_{3.7})/L_{3.7}$. The ratio was found to be maximum at the I stage declining to a steady state level and remaining almost unchanged during the subsequent slow induction-phase.

Decay curves of delayed light for longer dark periods (0.2-150 msec after cessation of excitation flash) at various induction stages are shown in Fig. 3 represented in logarithms of intensity of delayed light versus logarithms of dark time after cessation of light flash. It is shown that the mode of decay kinetics changes during the fast

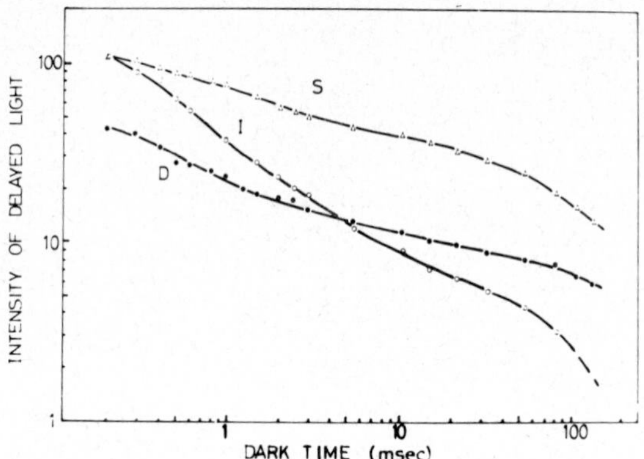

Fig. 2. Decay of delayed light from 0.2 to 150 msec after the end of the excitation flash at the I, D and S stages.

induction-phase, but remains almost unchanged during the slow induction-phase. All these decay curves were analysed into three exponentially decaying-components with half decay times of about 0.2, 0.8 and 35 msec. This result is consistent with some previous reports[11, 12]. Relative intensities of these 0.2, 0.8 and 35 msec decay-components were 54, 48 and 13 in the decay curve at the I stage, 11, 12 and 14 at the D stage, and 16, 48 and 48 at the S stage, respectively, when expressed as the intensities of each component at 0.2 msec after the cessation of excitation light. It will be clearly seen that the induction pattern of delayed light largely depends on the stage of induction and the choice of dark period after the flash excitation. The apparent discrepancies in induction pattern of delayed light emission reported in the literature will become, at least in part, comprehensible when we consider the circumstances shown in Figs. 1 and 2.

Effects of gramicidin S and methyl viologen on the induction and decay kinetics of delayed light emission.

An uncoupler of photophosphorylation, gramicidin S[13], is known to eliminate the slow induction-phase of the millisecond delayed light emission by suppressing the development of high energy state of phosphorylation[5,6]. On addition of gramicidin S, the slow induction-phase was eliminated in $L_{0.2}$, $L_{3.7}$ and $(L_{0.2}-L_{3.7})$ as shown in Fig. 3 (b), while the fast induction-phase was essentially unaffected.

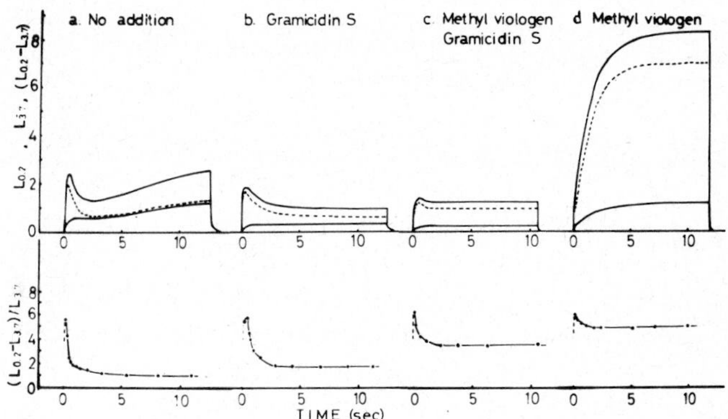

Fig. 3. Effects of gramicidin S and methyl viologen on the induction of $L_{0.2}$, $L_{3.7}$ and $L_{0.2}-L_{3.7}$, and on the ratio, $(L_{0.2}-L_{3.7})/L_{3.7}$. Upper traces; Relative intensities of $L_{0.2}$ (outer solid lines), $L_{3.7}$ (inner solid lines) and $L_{0.2}-L_{3.7}$ (dashed lines). Lower traces; ratio $(L_{0.2}-L_{3.7})/L_{3.7}$. Concentration of gramicidin S and methyl viologen, if added, were 6 and 10 μM, respectively.

This fact indicates that the slow induction-phase is due to the development of the high energy state.

An artificial electron acceptor, methyl viologen, stimulates the electron transport reaction, and is expected to cause an increase in a population of the reaction centers in the open state during the illumination. Fig. 3 (c) shows that a further addition of methyl viologen in addition to gramicidin S enhanced the steady state level of $(L_{0.2}-L_{3.7})$, but suppressed that of $L_{3.7}$. It is inferred that the higher population of the open state reaction centers increased the fast decaying components of delayed light but decreased the slow decay-component.

On addition of methyl viologen alone, both the rate and the extent of the slow induction-phase in $L_{0.2}$ were markedly stimulated (Fig. 3 (d)). $(L_{0.2}-L_{3.7})$ at the S stage became 5.4 times that of no addition. On the other hand, $L_{3.7}$ at the S stage was a little suppressed by the reagent. There occurred no peak in the fast induction-phase probably due to the enhancement of the later slow induction-phase.

The changes in ratios of $(L_{0.2}-L_{3.7})/L_{3.7}$ presented in the lower part of Fig. 3 for these cases showed a maximum at the I stage and declined to the steady state level during fast induction-phase and remained constant during the slow induction-phase thus indicating

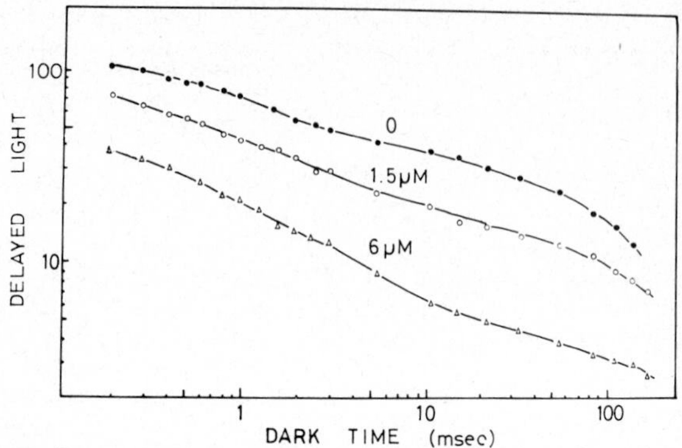

Fig. 4. Effects of gramicidin S on the decay kinetics of delayed light at the S stage (after 60 sec illumination). Excitation light; 110,000 ergs/cm^{2}sec. 22°.

that the decay kinetics remains unchanged during the slow induction-phase. It was also noted that gramicidin S had slight effects on this ratio, while methyl viologen enhanced its steady state level more than twice irrespective of the presence or absence of gramicidin S.

Fig. 4 shows the effects of gramicidin S on the decay kinetics of delayed light at the S stage. On addition of gramicidin S, the intensities of all the decay components of delayed light were suppressed to almost the same degree. The slow decay components seemed to be more sensitive than the faster ones to the higher concentration of the reagent. This is in part consistent with the findings by Wraight et al.[14] that delayed light from 40 to 1 sec was uncoupler sensitive.

In the presence of methyl viologen, the decay curves at the I and S stages showed similar decay kinetics each other (Fig. 5). A further addition of gramicidin S significantly suppressed the steady state level of delayed light, but in this case the slow decay-components seemed to be less sensitive to the uncoupler than the fast ones. The curves for the I and S stages in the presence of methyl viologen were analysed as described above in the same way, three decay-components having half decay times of about 0.2 msec, 0.8 msec and 20 msec were obtained. It is of interest that there was no difference in the half decay time of the faster decay-components of delayed light in the presence and absence of methyl viologen. The

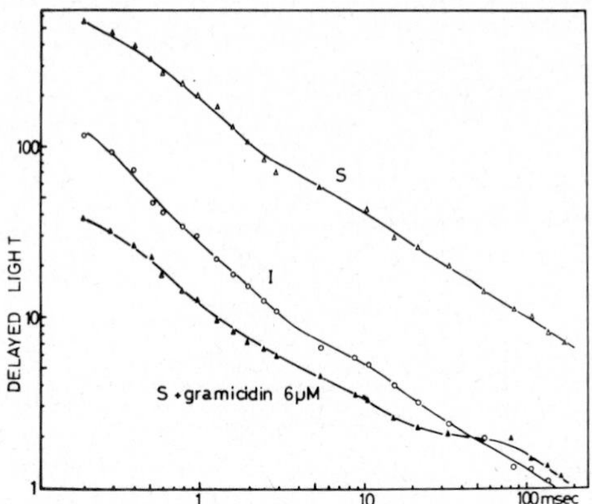

Fig. 5. Decay kinetics of delayed light at the I (after 0.1 sec illumination) and S (after 60 sec illumination) stages in the presence of 10 μM methylviologen, and at the S stage in the presence of 10 μM methyl viologen and 6 μM gramicidin S.

slow decay-component was found to be sensitive towards methyl viologen; the half time changed from 35 msec to 20 msec in the presence of methyl viologen.

Effects of excitation light intensity on the delayed light in the presence of gramicidin S.

Effects of excitation light intensity on the fast induction-phase and the decay kinetics of delayed light were investigated in the presence of gramicidin S. It was found that the ratio of the fast and the slow decay-components became higher at higher excitation light intensity. The characteristic features of the fast induction-phase, e.g., the initial peak, were observed only at higher light intensities(Fig. 6). The magnitude of $(L_{0.2}-L_{3.7})$ at the I

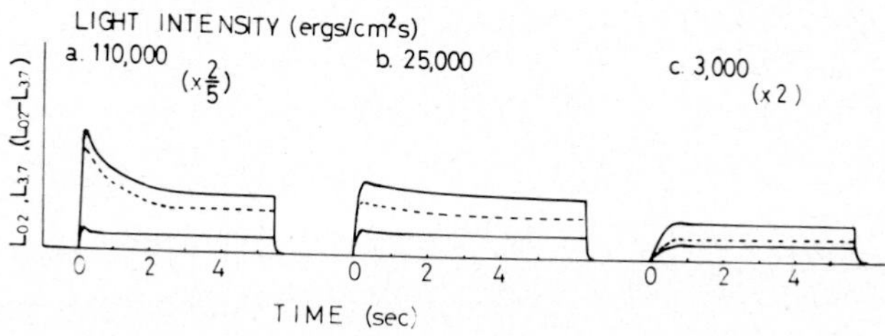

Fig. 6. Effects of excitation light intensity on the inductions of $L_{0.2}$, $L_{3.7}$ and $(L_{0.2}-L_{3.7})$ in the presence of 6 μM gramicidin S.

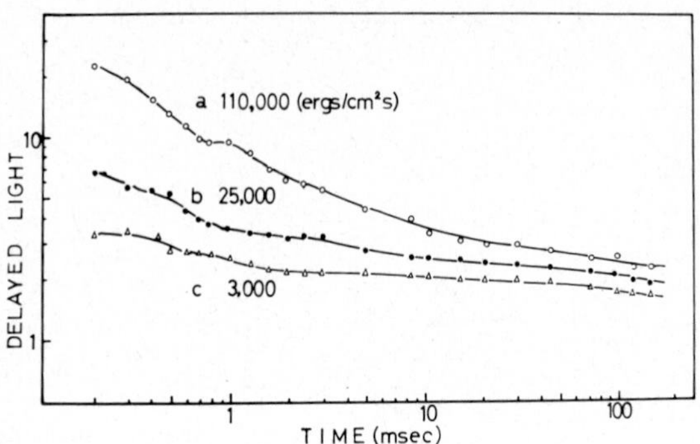

Fig. 7. Effects of the excitation light intensity on the decay kinetics of delayed light emission at the S stage in the presence of 6 µM gramicidin S.

stage was found to be proportional to the excitation light intensity, while that of $L_{3.7}$ at the S stage showed a saturation at a relatively low light intensity (at 21,000 ergs/cm^{2}sec).

Fig. 7 shows the decay kinetics of delayed light at the S stage in the presence of gramicidin S. It will be seen that the slow decay-components are less sensitive to the excitation light intensity. The intensity of delayed light emission at 0.2 msec was 16 times increased on 30 fold increase in excitation light intensity, while that at 50 msec was only 1.4 times increased under the same changes of excitation light. The analysis of the decay kinetics of delayed light emission gave values for half decay times of 0.2, 0.8 and 35 msec; and the respective relative intensities (at 0.2 msec after cessation of excitation), 11,6 and 3.4 at 110,000 ergs/cm^{2}sec and 1.3, 0.5 and 2.2 at 3,000 ergs/cm^{2}sec. These results indicate that the nature of the delayed light emission observed widely differ depending on the intensity of excitation light, even when the measurement was performed at a definite range of dark period. For instance the faster decay-components can only be measured at 0.2 msec by using higher intensities of excitation light.

5. Discussion

The results of the present study indicate that there are three major factors contributing to the induction kinetics of millisecond delayed light emission in chloroplasts. The first is the reaction centers of photosystem II in the open state, the excitation of which yields the fast decay-components having half decay times of 0.2 msec and 0.8 msec, respectively. The second is the reaction centers in the closed state, excitation of which yields a slow decay-component having half decay time of 20 to 35 msec. Widely different induction time courses as reported in literature, sometimes with and sometimes without an initial peak, may result from differences in relative contributions of these three components constituting the actual delayed light emission. The third is the development of the high energy state of photophosphorylation in chloroplasts which enhances the yields of all these components of delayed light emission. The electron acceptor such as methyl viologen affects the delayed light by stimulating the formation of the open state reaction centers; uncouplers such as gramicidin S affect the delayed light by eliminating the high energy state. Wraight at al.[15] and Barber et al.[16] had suggested that the development of high energy state induces membrane potential which, in turn, lowers the activation energy for the delayed light-producing reaction. The precise mechanism of such effect is still open to be further investigated. Experimental conditions such as excitation light intensity also affect the delayed light; higher excitation light intensities is in favor of the fast decay-components.

We found that the yields of the fast decay-components linearly respond to the excitation light intensity. We have also previously shown that low temperature favors the relative yields of the fast decay-components of delayed light. These results suggest that the fast decay-components of delayed light are more directly related to the photoreaction in photosystem II as compared to the slow decay-component.

Confusing statements have been made in the literature concerning the relationship between fluorescence and delayed light emission. Most experimental results thus far reported seems to support the view that delayed light is produced by a reverse process of photochemical reaction at the reaction center of system II. The results of the present study lead us to additional inferences that the fast decay-components of delayed light must arise directly from the reaction centers just after the trapping of photon quanta; the formation of the slow decay component is related to the closed state of reaction center, which may suggest a requirement of some dark reaction(s) in addition to the photon trapping reaction at the reaction center. It has to be noticed that it is the fast decay-components of delayed light that are inversely related to the fluorescence yield in chloroplasts while the slow decay component that behaves in a parallel manner with fluorescence yield.

Acknowledgements

The authors wish to express their thanks to Drs. A. Takamiya and S. Morita for suggestions made on the preparation of the manuscript.

References

1. Strehler, B.L. and Arnold, W., 1951, J. Gen. Physiol., 34, 809.
2. Goedheer, J.C., 1962, Biochim. Biophys. Acta, 64, 294.
3. Clayton, R.K., 1969, Biophys. J., 9, 60.
4. Joliot, P., Joliot, A., Bouges, B. and Barbieri, G., 1971, Photochem. Photobiol., 14, 287.
5. Itoh, S., Murata, N. and Takamiya, A., 1971, Biochim. Biophys. Acta, 245, 109.
6. Wraight, C.A. and Crofts, A.T., 1971, Eur. J. Biochem., 19, 286.
7. Fleishman, D.E. and Clayton, R.K., 1968, Photochem. Photobiol., 8, 287.
8. Wraight, C.A., 1972, Biochim. Biophys. Acta, 283, 247.
9. Itoh, S. and Murata, N., 1973, Photochem. Photobiol., 18, 209.
10. Arnon, D.I., 1949, Plant Physiol., 24, 1.
11. Itoh, S. and Murata, N., 1974, Biochim. Biophys. Acta, 333, 525.
12. Zankel, K.L., 1971, Biochim. Biophys. Acta, 245, 373.
13. Avron, M. and Shavit, N., 1965, Biochim. Biophys. Acta, 109, 317.
14. Wraight, C.A., Kraan, G.P.B. and Gerrits, N.M., 1972, in: 2nd international congress on photosynthesis research, edited by W. Junk (The Hague) vol. 2, p. 951.
15. Wraight, C.A., Kraan, G.P.B. and Gerrits, N.M., 1972, Biochim. Biophys. Acta, 283, 259.
16. Barber, J. and Kraan, G.P.B., 1970, Biochim. Biophys. Acta, 197, 49.

Present adress of S. Itoh:

Department of Biochemistry, Medical School, University of Bristol, Bristol BS 8 1TD, Great Britain.

M. AVRON, *Proceedings of the Third International Congress on Photosynthesis*
September 2-6, 1974, The Weizmann Institute of Science, Rehovot, Israel
Elsevier Scientific Publishing Company, Amsterdam, The Netherlands, 1974

STUDIES ON THE REACTION MECHANISM OF THE OXIDIZING EQUIVALENTS PRODUCED BY SYSTEM II

G. Renger

Max Volmer Institut für Physikalische Chemie und
Molekularbiologie der Technischen Universität Berlin

1. Introduction

In the photosynthetic electron transport pathway of higher photoautotrophic organisms electrons are transferred from water to $NADP^+$ by the action of two light reactions operating in series (for review s. ref. 1). Within this chain system II plays a central role because it generates by light excitation strong oxidizing equivalents which are able to use the ubiquitous water as the natural electron source for the reduction of $NADP^+$.

The oxidation of water to molecular oxygen requires the cooperation of four equivalents of sufficient oxidizing power:

$$2\ H_2O \quad + \quad 4\ \oplus \quad \longrightarrow \quad O_2 \quad + \quad 4\ H^+ \qquad (1)$$

Hence, the realization of this reaction in photosynthetic organisms needs a machinery which is able, firstly to use light quanta of the visible range of the spectrum for the generation of the indispensible holes and secondly to assure their cooperation.

In order to perform this process at highest degrees of efficiency nature has developed some two milliards years ago (2) a very complex machinery, commonly called system II.

2. The principal functional organization of system II

The principal functional organization of system II can be described in the following way (s. fig. 1): There exists a functional subdivision into 3 operational units:

A) The central operational unit is the generator system which transforms the energy of an exciton into the energy of an electron and a hole located at geometrical separated sites. According to this fundamental functional mechanism the central operational unit of system II will be designated as photoelectrical dipole generator II. The photoelectrical dipole generator II is connected with two other operational units regulating the influx and efflux of substrate and products, respectively.

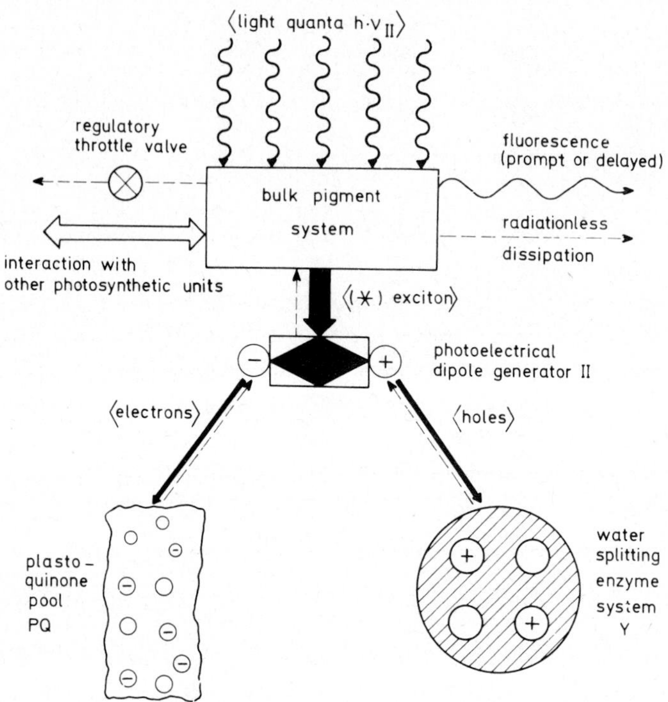

Fig. 1. Scheme of the principal functional organization of system II. Thick arrows indicate the main reaction pathway, thin broken lines indicate back fluxes of the electrical charges and excitons. For a detailed description see text.

B) The bulk pigment system consisting mainly of chlorophylla, chlorophyll b and carotenoids embedded in a protein (or lipoprotein) matrix (3,4) regulates the exciton influx into photoelectrical dipole generator II. This system provides an optimal adaptation on the environmental light quantum fluxes: If the quantum influx rate is slow in comparison to the turnover rate of photoelectrical dipole generator II the bulk pigment system acts as antenna which funnels practically all of the incoming light quanta into photoelectrical dipole generator II, probably via resonance energy transfer (5,6). On the other hand, at high quantum influx rates the dissipative exciton decay processes of the bulk pigment system (fluorescence, radiationless dissipation and regulatory throttle valve) prevent the accumulation of harmful excited metastable states (7,8).

C) Electron carrier systems regulate the efflux of the products (electron and hole) out of photoelectrical dipole generator II. The hole is captured by the reactor system which realizes the fundamental cooperation of four holes being indispensible for the oxidation of water to molecular oxygen. Hence, this system will be designated as the <u>watersplitting enzyme system Y</u> (9). The electron simultaneously produced with the hole is transferred into the plastoquinone pool.

For an understanding of the mode of action of the whole machinery of system II we have to analyze the molecular mechanisms of the operational units and their functional interrelationships. In order to attack the problem appropriate tools are required as well for the isolation and modification of the operational units as for the analytical detection of the functional operation of these units.

Because of the molecular dimensions of the machine,among the former types of tools effectors of molecular size, i. e. chemical agents with specific actions, seem to be most promising, whereas the latter types of tools include analytical techniques which allow the detection of molecular events (10). According to the functional organization of system II (s. fig. 1) in principle two classes of chemical effectors should be very useful:

a) chemicals which specifically interrupt the functional connections between the operational units without concomitant influence on their functional integrity.
b) chemicals which modify the mechanism of an operational unit in a well defined way.

In the present communication the investigation about the utility of chemical effectors will be restricted to those operational units which are indispensible for the photosynthetic water oxidation: the photoelectrical dipole generator II and the watersplitting enzyme system Y.

3. The isolation by DCMU of photoelectrical dipole generator II from the electron transfer systems

For the functional isolation of photoelectrical dipole generator II chemicals are required which interrupt specifically the functional connections to the other operational units.

Because the photoelectrical dipole generator II is structurally intimately connected with the bulk pigment system an interruption of the exciton flow is possible only by a geometrical separation of both operational units. By the use of detergents a partial disconnection between the bulk pigment system and photoelectrical dipole generator II accompanied by the destruction of the watersplitting enzyme system Y can be obtained (11,12). However, in contrast to photosynthetic bacteria (13,14) a complete separation of reaction centers from the bulk pigment system has never been achieved in chloroplasts. By contrast, the specific interruption of the functional connections between photoelectrical dipole generator II and the electrical charge efflux systems can be more easily realized.

Because photoelectrical dipole generator II can be operative only if both of its poles are free of electrical charges, one turnover is sufficient for its transformation into the functional inactive state. Hence, taking into account the pecularities of the watersplitting enzyme system Y (9,15,16) a functional isolation of photoelectrical dipole generator II from the electron transfer systems should be possible by chemicals which selectively block the electron efflux out of its negative pole. Under these circumstances the electron remains localized at the negative pole, whereas the hole can be transferred into the watersplitting enzyme system Y. But this hole is normally[x] unable to oxidize water because the systems Y have been found to act as functional independent operational units which prevent the statistical cooperation of holes of different photoelectrical dipole generators II (15,16). Therefore, in the absence of electron acceptors and donors which are able to react with the electron and the hole, respectively, both electrical charges (electron and hole) remain trapped in system II until they would react via an internal recombination. This back reaction regenerates the functional active state of photoelectrical dipole generator II.

It is now well established that DCMU type inhibitors (18-20) effec tively block the electron efflux out of photoelectrical dipole generator II. Hence, despite of its influence on the watersplitting enzyme system Y (21-23) DCMU will be used for the functional isolation of photoelectrical dipole generator II. However, this type of functional isolation forbids the indication of the functional active state by the commonly used methods based on the easily detectable secondary reactions leading to dye reduction (24-26) or of the holes giving rise to oxygen evolution.

On the basis of the functional organization of system II (fig. 1)

[x] If the watersplitting enzyme system Y is preloaded with 3 holes the next hole being transferred to Y leads to wateroxidation (17)

in principle two other techniques should be available:

a) Indirect methods which are based on the intimate functional connection between bulk pigment system and photoelectrial dipole generatorII. Because of this coupling the quantum efficiency of the exciton decay processes in the former operational unit should reflect the functional state of the latter operational unit. It has been shown by numerous investigators that the variable fluorescence arising practically exclusively from system II (27-29) can be often used as a suitable tool for the indication of the functional state of photoelectrical dipole generator II (30-32). However, as has been discussed recently by Joliot (33) this indirect technique fails under certain conditions.

b) Direct methods use the photoelectrical dipole generator II itself as an indicator of its functional state. The transition from the functional active state into the functional inactive state caused by a 1-exciton-induced single turnover leads to the generation of an electrical dipole. Hence, the number of electrical dipoles generated in system II by a short saturating flash is identical with the number of photoelectrical dipole generators II, which were functionally active before the flash. Therefore, the question arises: How can one determine the generation of these dipoles?

4. The electrochromic 515 nm absorption change as an indicator for the functional state of photoelectrical dipole generator II in DCMU blocked chloroplasts

Based on the rise time measurements of the field indicating absorption change at 515 nm Witt et al (34,35) have inferred that the photoelectrical dipole generators of systems I and II, respectively, are anisotropically arranged with respect to the plane of the thylakoid membrane. Therefore, the charge separation process at the photoelectrical dipole generators II and I necessarilly includes the simultaneous generation of an electrical field across the thylakoid membrane. This electrical field giving rise to the 515 nm absorption change should be a useful indicator of the functional state of photoelectrical dipole generator II if all photoelectrical dipole generators of system I are kept functionally inactive. This can be achieved in the following way (s. fig. 2). DCMU poissoned chloroplasts were illuminated in the presence of benzylviologen as system I elec-

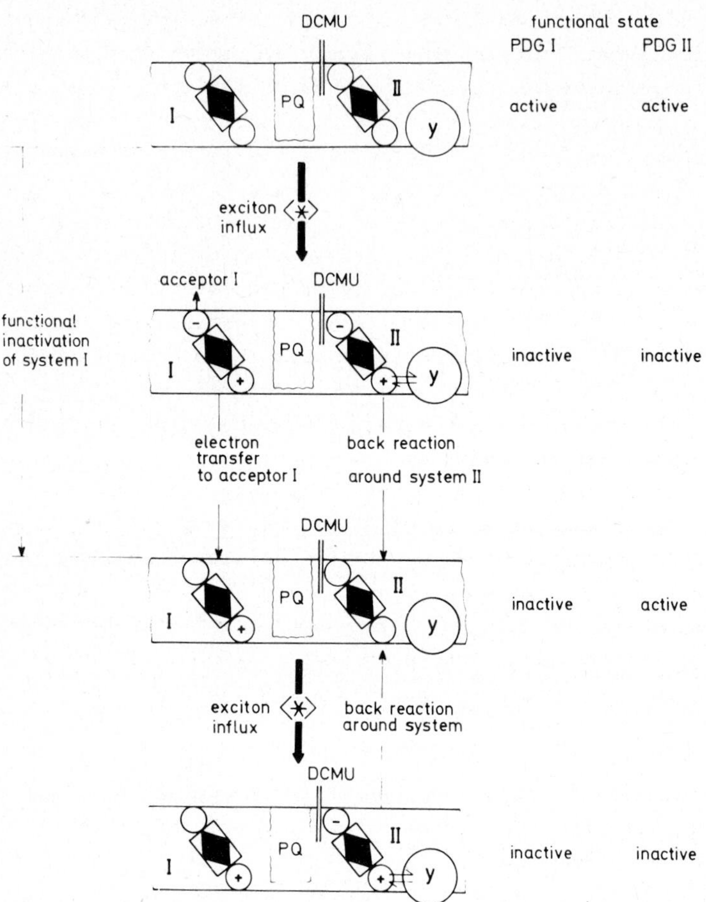

Fig. 2 Simplified scheme of the reactions of the photoelectrical dipole generators I and II in DCMU poisoned chloroplasts in the presence of benzylviologen as system I electron acceptor. For a detailed description see text.

tron acceptor. The electron produced by photoelectrical dipole generator I is transferred to benzylviologen whereas the hole of relative weak oxidizing power remains localized at the positive pole because the electron transfer from system II is prevented by DCMU and an internal cycle around system I is excluded by the action of benzylviologen (36). The following measurements have been carried out on DCMU poisoned chloroplasts whose systems I were functionally inactivated in the above described way.

The photoelectrical dipole generators II were transformed into their functional inactive state by a saturating preilluminating flash f_p followed by a dark time for allowing of a definite degree of recovery of the functional active state via the back reaction. At the time t_d after the preilluminating flash f_p an analyzing flash f_a of saturating intensity has been fired and the electrochromic 515 nm absorption change caused by this flash has been measured. After a dark time of 30 s which is sufficient for the completion of the back reaction the cycle was repeated and the contributions of 16 analyzing flashes were averaged (for excitation conditions s. insert of fig. 3). Because the electrochromic 515 nm absorption change has been found to act as a linear intrinsic voltmeter, the dependence on t_d of the amplitude of the 515 nm absorption change should reflect the kinetics of the back reaction around system II. The obtained results are depicted in the bottom curve of fig. 3.

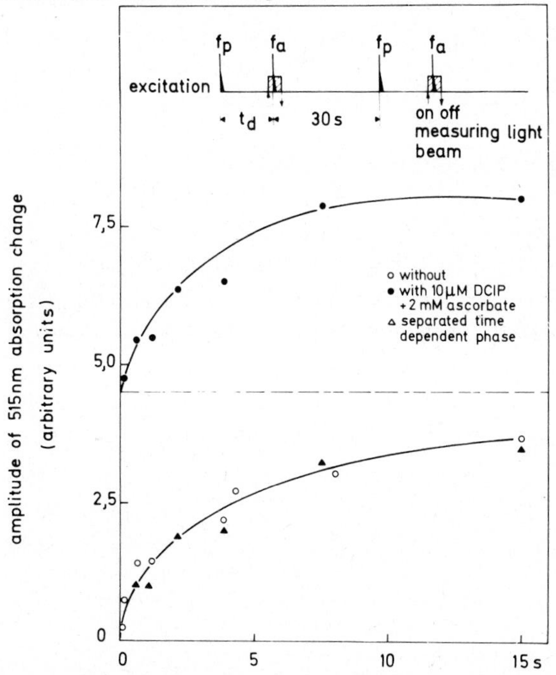

Fig. 3. Amplitude of the 515 nm absorption change as a function of the time t_d between the flashes f_p and f_a in spinach chloroplasts in the presence of 2 μM DCMU.
Chloroplast suspension: chloroplasts (10 μM chlorophyll), 0,1 mM benzylviologen, 10 mM KCl, 2 mM $MgCl_2$ and 20 mM N-tris(hydroxymethyl)-methylglycine-NaOH, pH = 7,5. For the details of the measuring conditions s. ref. 37.

It is seen that the recovery occurs with a half time of 2s. This value is in fair agreement with the experimental data for the half time earlier obtained by the use of variable fluorescence as indicator (32).

If the amplitudes of the 515 nm absorption change given in the bottom curve (open circles) would be caused practically exclusively by the regeneration of the functional active state of photoelectrical dipole generator II, then a different pattern should arise for the dependence on t_d of the 515 nm amplitudes under conditions where photoelectrical dipole generators I are functionally active, too. By the donor couple 2,6-dichlorphenolindophenol (DCIP) plus ascorbate the dark regeneration of the functional active state of photoelectrical dipole generator I is accomplished in a time at least two orders of magnitude smaller than the back reaction around system II. Hence, except for very short times t_d after the preillumination flash f_p ($t_d < 100$ ms) photoelectrical dipole generators I are fully active. Therefore, by the addition of DCIP + ascorbate, for $t_d > 100$ ms a two phase kinetical pattern should be observed for the dependence on t_d of the 515 nm amplitude: a time independent phase caused by the fully active photoelectrical dipole generators I and a time dependent phase reflecting the regeneration of the functional active state of photoelectrical dipole generators II via back reaction. Furthermore, the time course and the amplitude of the time dependent phase should be identical with the bottom curve obtained in the absence of DCIP + ascorbate. The experimental data given in the top curve of fig. 3 show that really a two phase kinetical pattern occurs and that by substraction of the time independent phase (indicated by dotted line) the separated time dependent phase (indicated by full triangles) is practical identical with the bottom curve.

Hence, our experimental data presented here (s. ref.37) prove that in DCMU poisoned chloroplasts the amplitude of the 515 nm absorption change is a direct indicator of the functional state of photoelectrical dipole generator II, if all systems I are functionally blocked.

5. The structural and functional organization of photoelectrical dipole generator II

It has been found (37) that the amplitude of the 515 nm absorption change caused by the functional active photoelectrical dipole generators II remains practically uneffected by DCMU. The decay of the electrical field generated by the functionally isolated photoelectrical dipole generator II can be accelerated by ionophores (valinomycin) in the same way as the decay of the electrical field settled up in normal chloroplasts by both photosystems (37). These results confirm the earlier assumptions about an anisotropic arrangement of photoelectrical dipole generator II itself in the thylakoid membrane (34,35). Furthermore, the independence of DCMU of the 515 nm amplitude caused by photoelectrical dipole generator II favors the suggestion that the negative pole and the positive pole are located near the outer and the inner phase, respectively, of the layer of the thylakoid membrane which acts as the barrier of low ion permeability.

Information about the functional organization of photoelectrical dipole generator II should be obtainable from the kinetics of the back reaction around system II. If there would exist only one type of photoelectrical dipole generator II acting as noncooperative operational unit, the recovery kinetics by internal back reaction have to be of first order. However, the experimental data of the bottom curve of fig. 3 can be described either by a two phase first order kinetics (37) or by a second order kinetics (37). A two phase first order kinetics could be explained by the existence of two types of noncooperative photoelectrical dipole generators II characterized by different internal recovery times in DCMU blocked chloroplasts. This assumption would be in agreement with earlier suggestions about the existence of two types of photoelectrical dipole generators II (38,39).

On the other hand, a second order kinetics should indicate any form of cooperation of electrons and holes of different photoelectrical dipole generators II. However, the cooperation of the electrons via the common plastoquinone pool (40) is interrupted in DCMU blocked chloroplasts, whereas the statistical cooperation of the holes of different photoelectrical dipole generators II has to be excluded according to Kok et al (15) and Joliot et al (16). A small range cooperation including only two photoelectrical dipole generators II has been proposed earlier on the basis of DCMU titration experiments (41). The present data do not allow an unequivocal decision about the details of the functional organization discussed above.

Because the rate of the back reaction around system II in DCMU blocked chloroplasts is slow in comparison to the rate of the back reaction at 77° K ascribed to the recombination of an electron and a hole which are assumed to be localized at the poles of the photoelectrical dipole generator II (42) it is inferred that in DCMU blocked chloroplasts the hole is not fixed at the positive pole, but can be transferred to secondary electron donors (system Y).

Until now the photoelectrical dipole generator II has been considered as an abstract operational unit. Now the chemical structure remains to be discussed. It has been found that the negative pole probably consists of a special plastoquinone molecule, designated as X 320 (43, 44). The situation is even more complicated for the positive pole. Recent experiments favor the assumption that a special chlorophyll a (designated Chl a_{II}, s. ref. 45-47) acts simultaneously as exciton acceptor and as electron donor, i. e. as positive pole. Hence, because of the anisotropic arrangement of photoelectrical dipole generator II it is inferred that X 320 and Chl a_{II} are located near the outer and the inner side of the impermeable part of the thylakoid membrane, respectively (s. also Wolff et al, this symposium). However, in chloroplasts X 320 is not freely accessible by agents from the outer aqueous phase, because until now no redox agent has been found which directly accepts electrons from X 320 (s. ref. 48).

6. The modification of the properties of the holes trapped in the watersplitting enzyme system Y, the ADRY effect

Though the holes produced by photoelectrical dipole generator II are of sufficient oxidizing power, a reaction with water resulting in the evolution of oxygen cannot occur until after trapping in the watersplitting enzyme system Y the accumulation of four trapped holes is reached (15,16). Because of its functional role the trapped holes can be suggested as to be the chemical intermediates of the photosynthetic wateroxidation. Generally, 3 different reaction types of the trapped holes can be distinguished:

A) <u>Fast trapping reaction</u> ($\tau_{1/2} < 1$ ms, s. ref. 49) in the watersplitting enzyme system Y produced by photoelectrical dipole generator II (symbolized by $\oplus_{II}$):

$$\oplus_{II} + S_i \xrightarrow{k_i} \bigcirc_{II} + S_{i+1} \qquad (2)$$

S_i represents the trapping state of system Y with i = number of trapped holes (i = 0,...3)

B) <u>Fast cooperative reaction</u> ($\tau_{1/2} \leq 1$ ms, s. ref. 49) with water leading to molecular oxygen:

$$S_4 + 2\,H_2O \xrightarrow{k_{O_2}} S_o + 4\,H^+ + O_2 \qquad (3)$$

C) <u>Slow dissipative discharge reaction</u> ($\tau_{1/2} > 1$ s, s. ref. 50,

51,52) of the trapped holes:

$$S_i \quad + \quad e \quad \xrightarrow{k_{D_i}} \quad S_{i-1} \qquad (4)$$

According to this scheme oxygen can be produced only if the turnover rate of the photoelectrical dipole generators II determined by the exciton influx rate is high enough in comparison to the rate of the deactivation.

Information about the nature of the trapped holes should be obtainable if one could modify the above described mechanism in a well defined way. In the last years a number of substances have been found to accelerate significantly the deactivation reactions in chloroplast (53-55). Therefore, this effect has been called ADRY[x]-effect (55). It has been shown (56) that the ADRY agents significantly enhance the rate of the dissipative discharge reactions of S_2 and S_3 (s. fig. 4), whereas S_1 characterized by a long lifetime (15,16) remains nearly uneffected. Furthermore, on the basis of earlier double flash group experiments (s.ref. 57, fig.5) the conclusion can be drawn, that ADRY agents do not significantly influence the rate of the fast trapping reaction leading from S_2 to S_3. So, it is reasonable to infer, that ADRY agents practically do not change the pattern of the fast trapping reactions. Recent measurements on μs luminescence confirm the assumption that there occurs practically no interference by ADRY agent on the transfer of holes from photoelectrical dipole generator II into the watersplitting enzyme system Y (58).

If the turnover rate of photoelectrical dipole generator II is fast in comparison to the ADRY agent catalyzed deactivation, the oxygen evolution remains practically uneffected by these substances. Hence, it can be assumed that ADRY agents do not interfer with the cooperative reaction of S_4 with water. However, at higher ADRY agent concentration an unspecific destruction of the functional integrity of the watersplitting enzyme system Y occurs.

Therefore, the ADRY agents (in a suitable concentration range) can be used as a very useful tool for labeling and for specific modification of the higher trapped hole accumulation states S_2 and S_3.

The dissipative discharge of trapped holes requires electrons. For stoichiometrical reasons the ADRY agents have been excluded to act as irreversible electron donors (59). This favors the ADRY agents to be acting as catalysts which accelerate the natural dissipative dischar-

[x] ADRY = Acceleration of the Deactivation Reactions of the watersplitting enzyme system Y

ge reactions. However, the strong inhibition of the delayed luminescence in the range of ms to s led us to assume, that the ADRY agent catalyzed deactivation of S_2 and S_3 differs in its mechanism in some way from that of the natural dissipative deactivation reactions (56).

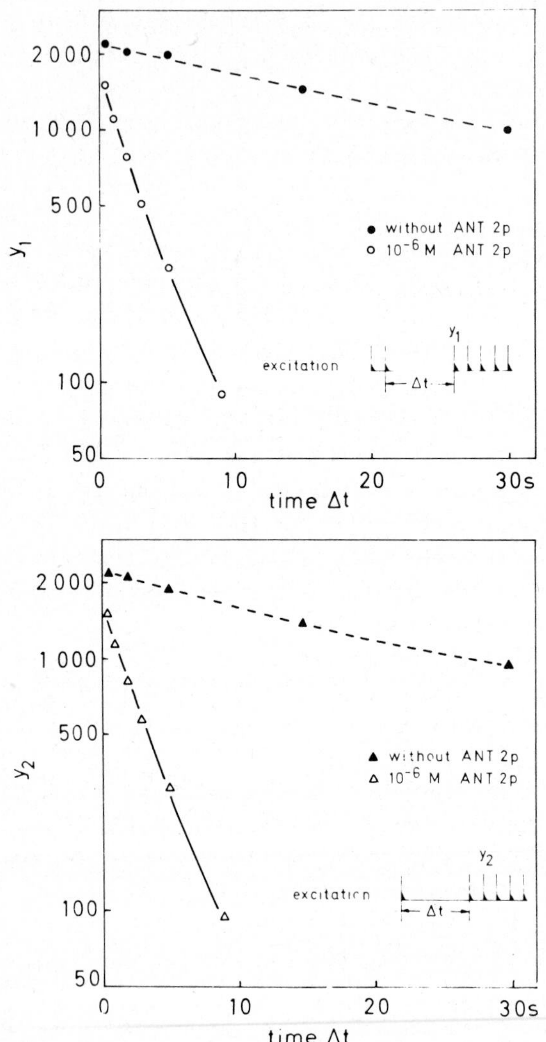

Fig. 4. Oxygen yield of the first flash (top) and the second Flash (bottom) in a sequence of short flashes after preillumination with one or two flashes, respectively. For the experimental details s. ref. 56.

The ADRY agents so far known (60) are more or less potent uncouplers. Hence, it could be possible, that the ADRY effect is correlated to the ability of these agents to accelerate the electrical field decay or the proton efflux rate. In fig. 5 (s. ref. 60) the effect of a few agents on the decrease of the relative average oxygen yield per flash reflecting the steady state concentration of S_3 (s. ref. 61) is compared with the relative Hill reaction rate indicating the uncoupling power.

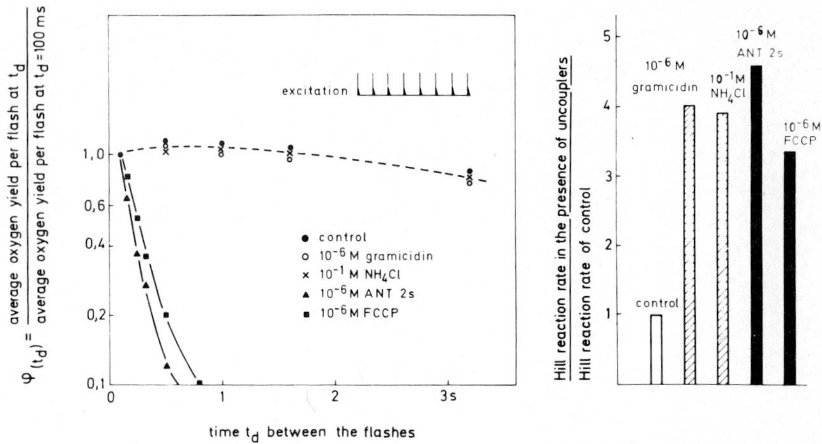

Fig. 5. Comparison between the ADRY effect (indicated by the decrease of the average oxygen yield per flash with increasing time t_d between the flashes) and the uncoupling power (indicated by the relative Hill reaction rate) for different substances in chloroplasts For details s. ref. 60.

It is seen that there exists no correlation between the uncoupling activity and the ADRY effect. Hence, it can be concluded that the transmembrane gradients of pH and of the electrical potential are not essential for the stabilization of S_2 and S_3. Therefore, the stability of S_2 and S_3 seems to be essentially determined by the structure of the watersplitting enzyme system Y itself.

In order to test this possibility the ADRY effect (expressed by the reciprocal half time of the decrease of average oxygen yield per flash, s. ref. 61) has been measured in dependence on temperature. The obtained results given in fig. 6 indicate that with respect to the activation energy of the ADRY effect two different ranges can be distinguished with a relative sharp transition temperature at 18° C. This result can be explained by a structural change in the water-

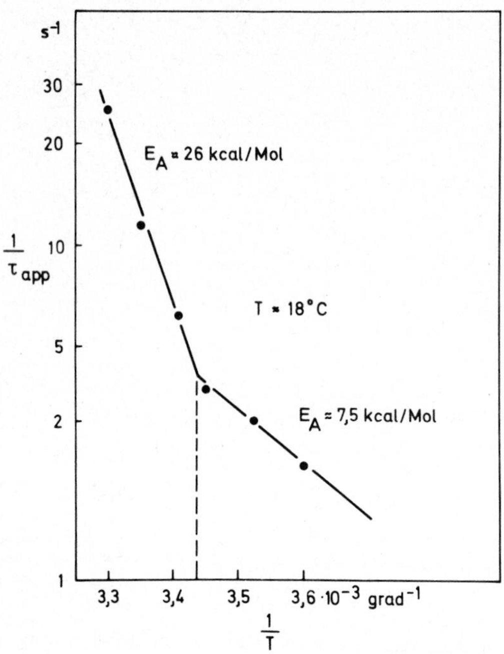

Fig. 6. Reciprocal half time of the decrease of the average oxygen yield per flash in dependence on the reciprocal temperature.

splitting enzyme system Y at the transition temperature. It is interesting to note that at nearly the same temperature a transition occurs also with respect to the permeability of the thylakoid membrane, which could be ascribed to a phase transition in the lipid matrix of the whole thylakoid membrane (62). Furthermore, the activation energies for the ion transport are nearly the same as those for the ADRY effect. Therefore, it could be possible, that the ADRY effect is kinetically limited by the transport of ADRY agents through the thylakoid membrane into the watersplitting enzyme system Y, which is assumed to be localized near the positive pole of photoelectrical dipole generator II, i. e. at the inner phase of the thylakoid membrane. This localization is in agreement with recent findings about the H^+ production of system Y (63,64) and of the Mn^{2+}-release in Tris-treated chloroplasts (65). However, because the electrical field decay in the presence of the ADRY agent ANT 2p occurs at least one order of magnitude faster than the deactivation reactions (G. Renger, unpublished) the ADRY effect is probably not kinetically limited by the transport rate of the ADRY agents. Hence, it is concluded that the results of fig. 6 reflect probably a structural change within the water-splitting enzyme system Y caused by a gross phase transition in the lipid phase of the thylakoid membrane.

With respect to its chemical composition the system Y has been proposed to be a manganese-protein-system (9,66-68), which covers up the positive pole of photoelectrical dipole generator II. In this way system Y not only realizes the oxidation of water, but in addition protects the sensitive organic material of the thylakoids against oxidative destruction by the strong oxidizing equivalents of system II.

A last point which remains to be shortly discussed is the role of cytochrome b 559 localized at the oxidizing side of system II (69-71). A direct participation of this enzyme in the process of wateroxidation has been shown to be very improbable (72,73). But, because some ADRY agents are known to transform the cytochrome b 559 from a high potential form into a low potential form (73-75) the possibility has to be considered that the latter species could be responsible for the ADRY effect. However, 3 lines of evidence point against the assumption that the ADRY effect is simply caused by a direct electron donor function of low potential cytochrome b 559 reduced by system II:

a) Because cytochrome b 559 is membrane bound a fixed place mechanism (61) has to be considered including only two types of system Y: Y connected with low potential cytochrome b 559 (fast deactivation) and Y connected with high potential cytochrome b 559 (slow deactivation). It has been shown, that such a simple mechanism is not compatible with the dependence on the ADRY agent concentration of the deactivation rate (61).
b) For stoichiometrical reasons an electron cycle around system II induced by ADRY agents should prevent the reduction of photooxidized P700. Recent experiments show that this does not occur for all ADRY agents (60).
c) A preincubation with ADRY agents should be sufficient for the full development of the ADRY effect. However, the very weak ADRY effect in CCCP washed chloroplasts (59) does not support this assumption.

Hence, the ADRY effect is not simply caused by an electron cycle around system II mediated by the low potential form of cytochrome b 559, but a participation of this cytochrome (via an indirect way) can not be totally excluded. The role of cytochrome b 559 for the in vivo reactions of the oxidizing side of system II still remains an open question.

References

1. Witt, H.T., 1971, Quart. Rev. 4, 365.
2. Cloud, P.E., 1965, Science 148, 27.
3. Cederstrand, C.N., Rabinowitch, E. and Govindjee, 1966, Biochim. Biophys. Acta 126, 1.
4. Borisov, A.Yu. and Ilina, M.D., 1969, Moleculyarnaya Biologia 3, 391.
5. Duysens, L.M.N., 1952, Utrecht
6. Bay, Z. and Pearlstein, R.M., 1963, Proc. Nat. Acad. Sci. US 50, 1071
7. Wolff, Ch. and Witt, H.T., 1969, Z. Naturforsch. 24 b, 1031
8. Wolff, Ch. (in preparation)
9. Renger, G., 1972, Physiol. Veg. 10, 329
10. Rüppel, H. and Witt, H.T., 1970, in: Methods in Enzymology, Vol. 16, ed. K. Kustin (Academic Press, New York and London) p. 316
11. Vernon, L.P., Shaw, E.R., Ogawa, T. and Raveed, D., 1971, Photochem. Photobiol. 14, 343.
12. Wessels, J.S.C., van Alphen-van Waveren, O. and Voorn, G., 1973, Biochim. Biophys. Acta 292, 741.
13. Mc Elroy, J.D., Mauzerall, D.C. and Feher, G., 1974, Biochim. Biophys. Acta 333, 261.
14. Jolchine, G. and Reiss-Husson, F., 1974, FEBS Letters 40, 5.
15. Kok, B., Forbush, B. and Mc Gloin, M.P., 1970, Photochem. Photobiol. 11, 457.
16. Joliot, P., Joliot, A., Bouges, B. and Barbieri, G., 1971, Photochem. Photobiol. 14, 287.
17. Duysens, L.M.N., 1972, in: Proc. 2nd Int. Congr. Photosynthesis Res. Stresa, Vol. 1, eds. G. Forti, M. Avron and A. Melandri, (Dr. W. Junk N. V. Publishers, The Hague), p. 19.
18. Duysens, L.M.N. and Sweers, H.E., 1963, in: Microalgae and Photosynthetic Bacteria, University of Tokyo Press, p. 353
19. Joliot, A., 1968, Physiol. Veg. 6, 235.
20. Van Rensen, J.J.S., 1971, Meded. Landbouwhogesch. Wageningen 71-9, 1.
21. Bouges-Bocquet, B., Bennoun, P. and Taboury, J., 1973, Biochim. biophys. Acta 325, 247.
22. Renger, G., 1973, Biochim. Biophys. Acta 314, 113
23. Etienne, A.L., 1974, Biochim. Biophys. Acta 333, 320.
24. Kelly, J. and Sauer, K., 1968, Biochemistry 7, 882.
25. Avron, M. and Ben Hayyim, B., 1969, in: Progress in Photosynthesis Research, Vol. 3, ed. H. Metzner (H. Laupp jr. Tübingen)
26. Sun, A.S.K. and Sauer, K., 1971, Biochim. Biophys. Acta 234, 399.
27. Clayton, R. K., 1969, Biophys. J. 9, 60.
28. Delosme, R., 1967, Biochim. Biophys. Acta 143, 108.
29. Lávorel, J. and Joliot, P., 1972, Biophys. J. 12, 815.

30. Malkin, S. and Kok, B., 1966, Biochim. Biophys. Acta 126, 413.

31. Forbush, B. and Kok, B., 1968, Biochim. Biophys. Acta 162, 243.

32. Bennoun, P., 1970, Biochim. Biophys. Acta 216, 357.

33. Joliot, P., Bennoun, P. and Joliot, A., 1973, Biochim. Biophys. Acta 305, 317.

34. Wolff, Ch., Buchwald, H.E., Rüppel, H., Witt, K. and Witt, H.T., 1969, Z. Naturforsch. 24 b, 1038

35. Schliephake, W., Junge, W. and Witt, H.T., 1968, Z. Naturforsch. 23 b, 1571.

36. Rumberg, B., 1964, Z. Naturforsch. 19 b, 707.

37. Renger, G. and Wolff, Ch., 1975, Biochim. Biophys. Acta (in press)

38. Vermeglio, A. and Mathis, P., 1973, Biochim. Biophys. Acta 314, 57.

39. Etienne, A.L., 1974, Biochim. Biophys. Acta 333, 497.

40. Siggel, U., Renger, G., Stiehl, H.H. and Rumberg, B., 1972, Biochim. Biophys. Acta 256, 328.

41. Siggel, U., Renger, G. and Rumberg, B., 1972, in:Proc. 2nd Int. Congr. Photosynthesis Re. Stresa, Vol. 1, eds. G. Forti, M. Avron and A. Melandri (Dr. W. Junk N. V. Publishers, The Hague), p. 753

42. Murata, N., Itoh, S. and Okada, M., 1973, Biochim. Biophys. Acta 325, 463.

43. Stiehl, H.H. and Witt, H.T., 1969, Z, Naturforsch. 24 b , 1588.

44. Witt, K., 1973, FEBS Letters 38, 116.

45. Döring, G., Stiehl, H.H. and Witt, H.T., 1967, Z. Naturforsch. 22 b, 639

46. Döring, G., Renger, G., Vater, J. and Witt, H.T., 1969, Z. Naturforsch. 24 b, 1139.

47. Gläser, M., Wolff, Ch., Buchwald, H.E. and Witt, H.T., 1974, FEBS Letters 42, 81.

48. Wolff, Ch., Büchel, K. H. and Renger, G., 1975, Biochim. Biophys. Acta (in press)

49. Bouges-Bocquet, B., 1973, Biochim. Biophys. Acta 292, 772

50. De Kouchkovsky, Y. and Joliot, P., 1967, Photochem. Photobiol. 6, 567.

51. Lemasson, C., 1970, C. R. Acad. Sci.(Paris) 270, 250.

52. Lemasson, C. and Barbieri, G., 1971, Biochim. Biophys. Acta 245, 386

53. Renger, G., 1969, Naturwiss. 56, 370.

54. Renger, G., 1971, Z. Naturforsch. 26 b, 149.

55. Renger, G., 1972, Biochim. Biophys. Acta 256, 428.

56. Renger, G., Bouges-Bocquet, B. and Delosme, R., 1973, Biochim. Biophys. Acta 292, 796.

57. Renger, G., 1972, in: Proc. 2 nd Int. Congr. Photosynthesis Res. Stresa, Vol. 1, eds. G. Forti, M. Avron and A. Melamdri (Dr. W. Junk Publishers, The Hague), p. 53.

58. Renger, G. and Lavorel, J. (in preparation)

59. Renger, G., 1972, Eur. J. Biochem. 27, 259.

60. Renger, G., Bouges-Bocquet, B. and Büchel, K. H., 1973, J. Bioenergetics 4, 491.

61. Renger, G., 1973, Biochim. Biophys. Acta 314, 390.

62. Gräber, P. and Witt, H.T., 1975, FEBS Letters (in preparation)

63. Junge, W. and Ausländer, W.,1974, Biochim. Biophys. Acta 333, 59.

64. Fowler, Ch. and Kok, B., 1974, Biochim. Biophys. Acta 357, 299.

65. Blankenship, R. E. and Sauer, K., 1974, Biochim. Biophys. Acta 357, 252.

66. Olson, J. M., 1970, Science 168, 438

67. Renger, G., 1970, Z. Naturforsch. 25 b, 966.

68. Cheniae, G. M. and Martin, I. F., 1970, Biochim. Biophys. Acta 197, 219.

69. Boardman, N. K. and Anderson, J. M., 1967, Biochim. Biophys. Acta 143, 187.

70. Knaff, D. B. and Arnon, D. I., 1969, Proc. Nat. Acad. Sci. US 63, 956.

71. Erixon, K. and Butler, W. L., 1971, Biochim. Biophys. Acta 234, 371.

72. Cox, R. P. and Bendall, D. S., 1972, Biochim. Biophys. Acta 283, 124.

73. Cramer, W. A. and Böhme, H., 1972, Biochim. Biophys. Acta 256, 358.

74. Cramer, W. A., Fan, H. N. and Böhme, H., 1971, J. Bioenergetics 2, 289.

75. Ben-Hayyim, 1974, Eur. J. Biochem. 41, 191.

M. AVRON, *Proceedings of the Third International Congress on Photosynthesis*
September 2-6, 1974, The Weizmann Institute of Science, Rehovot, Israel
Elsevier Scientific Publishing Company, Amsterdam, The Netherlands, 1974

ON THE SYSTEM II RECOMBINATION REACTION

J. LAVOREL

Laboratoire de Photosynthèse
C. N. R. S. Gif-sur-Yvette

According to the recombination hypothesis, luminescence is the radiative decay of a singlet exciton resulting from recombination of a pair of charges within the activated reaction center. Luminescence is thus the exact reversal of the early photochemical charge separation[1]. I shall discuss the importance of knowing the yield of the recombination reaction and survey several experimental results which may be explained by the recombination hypothesis. An important extension of the hypothesis will be contemplated.

The recombination yield is the fraction of activated centers $^{+}C^{-}$ which relax to the photoactive state C by recombination (C and $^{+}C^{-}$ are respectively equivalent to e.g. Z-YChlQ and Z-$^{+}$YChlQ^{-}). A quantitative knowledge of this yield would be of paramount importance in the interpretation of luminescence phenomena and more generally for our understanding of the System II photochemistry. Luminescence may be a priori qualified as a "leakage" or as a "deactivation" effect, with different kinetic properties, according to wether the recombination yield is low or high[2].

It would seem that the recombination yield could be simply deduced from the luminescence yield. The latter is defined as the number of photons emitted per activated center. According to Zankel[3], the luminescence yield is of the order of $10^{-4}h\nu$ center^{-1} ; no other experimental estimation is known. Its yield may be appreciably smaller than the recombination yield because several factors may modify the luminescence outcome of a recombination act. They are :

1) The exciton yield defined as the stoechiometric coefficient α in the recombination equation :

$$^{+}C^{-} \xrightarrow{k_L} C + \alpha\varepsilon_L \qquad (1)$$

(k_L = rate constant of recombination ; ε_L = concentration of luminescence excitons). Recent results[4] show that α can vary between large limits, in response to changes in the electroosmotic state of the thylakoid. Such large variations are understandable if $\alpha \ll 1$.

2) The emission yield (Φ) in the L relation L = (Φ)J, i.e. the yield of the radiative decay of ε_L. No decisive evidence has been produced yet to back the original proposal[5] of identifying (Φ) with the macroscopic System II fluorescence yield (which might itself be different from the total observed fluorescence yield !). I am inclined to believe that (Φ) is equal to Φ (0), the fluorescence

yield of an "open", weakly fluorescent System II unit[1].

It is therefore possible that both modulating factors α and (Φ) are rather appreciably smaller than 1.0, with the consequence that the recombination yield might be much larger than the luminescence yield and that it could have a sizeable effect on the System II photochemical yield. Several recent results from independent sources in effect point to the recombination reaction as a process competing with the normal utilization of the photochemical energy by the PS chain.

In Chlorella, the light responses observed in the presence of NH_2OH (high concentration) + DCMU are consistant with the following scheme :

$$C \underset{k_L}{\overset{k_I}{\rightleftarrows}} {}^{+}C^{-} \xrightarrow{k_+} C^{-} \qquad (2)$$

(k_+ = pseudo 1st order rate constant of NH_2OH oxidation). This scheme implies that, given enough light, the system will find itself in the "blocked" state C^-. However the process is rather inefficient : it has been shown by comparing luminescence, fluorescence and photochemical responses in flash experiments[6,7] that C^- accumulates exponentially as a function of flash number. This effect is explained by a strong competition of recombination (k_L) against NH_2OH oxidation (k_+). In this particular condition, it is estimated that $k_+/k_L \cong 10^{-2}$.

Low temperature experiments, as used by several authors, are well suited to demonstrate the occurrence of the recombination reaction. The activation energy of recombination is presumably smaller than that of the first stabilization step (such as or similar to k_+ in Eq. 2). It is significant that luminescence excited at 120°K by a single flash is much more intense than at ordinary temperature, but that the lifetime of the emission is about the same in the two conditions[8]. As a consequence of the unequal temperature coefficients of the two competing processes, the photochemical quantum yield is decreasing by lowering the temperature : from kinetics of fluorescence rise and C_{550} photoreduction, it is concluded that the relative quantum yield (taken as 1.0 at room temperature) is 0.4 at -100°C and 0.27 at -196°C[9]. At the latter temperature, short saturating flashes suppress only a small fraction of the variable fluorescence subsequently monitored by a continuous beam ; however complete suppression of the variable fluorescence is obtained by increasing the flash duration up to $\sim$ 1 sec[10]. This is readily explained by the above scheme (Eq. 2) where low temperature has the same effect as DCMU to stop the evolution of the System in state C^-. Incidentally, it is interesting to note that the same type of dependency of the kinetics on flash duration was also observed at room temperature in the above DCMU + NH_2OH system[4]. The fluorescence rise is dependent on the S states of the frozen material (i.e. S_0,S_1 for dark adapted material and S_2,S_3 with material given two flashes prior to freezing) : at -40°C, the variable fluorescence removed by a single flash is larger in S_0,S_1 than in S_2,S_3 and the fluorescence rise is faster in S_0,S_1 than in S_2,S_3[11]; similarly, a smaller number of flashes given at -40°C are required to raise fluorescence to its maximum level in S_0,S_1[12]. These results are in agreement with the recombination hypothesis if one assumes that the recombination reaction is more active in S_2,S_3 than in S_0,S_1 : luminescence in sequence experiments at room temperature may be explained by the same difference in recombination rates between the various S states.

Even at room temperature, the recombination reaction may influence the quantum yield of photoreaction II. A very interesting illustration is found in the work of Satoh _et al._[13]. They noticed that their kinetic data on Hill reaction with ferricyanide as acceptor could be fitted by the reciprocal plot :

$$\frac{1}{V} = \frac{1}{K_D} + \frac{1}{K_L} \cdot \frac{1}{I} \qquad (3)$$

(V = rate of Hill reaction ; I = light intensity ; K_D,K_L = constants). Furthermore, they observe that under normal conditions, the reaction being limited on the acceptor side of System II, chemical treatments acting on the acceptor side (e.g. DCMU) change K_D whereas those acting on the donnor side (e.g. FCCP) change K_L ; the situation is exactly opposite when the kinetic limitation is displaced to the donor side (decoupling and mild heat treatment). The basic scheme of Eq. 2 may be completed to express the first situation :

$$C \underset{k_L}{\overset{k_I}{\rightleftharpoons}} {}^{+}C^{-} \xrightarrow{k_+} C^{-} \qquad (4)$$
$$C^{-} \xrightarrow{k_-} C$$

where $k_+ > k_-$ because of kinetic limitation on the acceptor side. The steady-state solution of scheme 4 has the same form as Eq. 3 :

$$\frac{1}{V} = \left(\frac{1}{k_-} + \frac{1}{k_+}\right) + \left(1 + \frac{k_L}{k_+}\right)\frac{1}{k_I} \qquad (5)$$

Comparison with Eq. 3 yields $1/K_D = 1/k_- + 1/k_+ \cong 1/k_-$ and $1/K_L \sim (1 + k_L/k_+)$. This explains the above results of Satoh _et al._. The symetric situation is described by a scheme analogous to Eq. 4, where the roles of k_+ and k_- have been exchanged and C^- is replaced by ^+C. One must notice that the recombination hypothesis is essential in this explanation : the alternative hypothesis of a fast equilibrium between the active form C and an inactive form could not account for the specific effects on K_D or K_L according to which side of the reaction center is kinetically modified.

The recombination hypothesis is in my opinion important in the functioning of the linear four-step mechanism of the water splitting system as proposed by Kok _et al._[14]. Kinetic analysis of the O_2 emission in sequence of short saturating flashes shows that the quantum yield of the successive photochemical steps on the average does not exceed 0.8-0.9. The corresponding "misses" could be very well accounted for by recombination. More precisely, it is proposed that a close correspondence be established between the fast phases of luminescence and "misses" on the one hand, and the slow phases of luminescence and "deactivation" on the other hand. The separation between the two types of events would occur at a few milliseconds after the flash[4]. Several arguments may be produced in favor of this hypothesis : CCCP[6], methylamine[6], DNB[15] modify fast luminescence and misses in the same direction ; isolated chloroplasts show a lower probability of miss and a slower rate of deactivation than Chlorella and, correspondingly, the average light emission is smaller in chloroplasts than in Chlorella ; the analysis of many published O_2 sequences (as well as those obtained in my laboratory) discloses a systematic increase of the probability of misses at the begining of a flash sequence (Lavorel, unpublished) a similar increase in the submillisecond luminescence is also found during the first flashes of a sequence[4]. If this hypothesis happened to be verified, it would confirm the above assumed low value of the yield factors α and (Φ).

References

1. Lavorel, J., 1973, Biochim. Biophys. Acta 325, 213.
2. Lavorel, J., 1974, "Luminescence", in : Bioenergetics of Photosynthesis, ed. Govindjee (Academic Press, New York), in the press.
3. Zankel, K. L., 1971, Biochim. Biophys. Acta 245, 373.
4. Lavorel, J., 1974, submitted to Photochem. Photobiol.
5. Lavorel, J., 1968, Biochim. Biophys. Acta 153, 727.
6. Etienne, A. L., 1974, Biochim. Biophys. Acta 333, 497.
7. Ducruet, J. M. and Lavorel, J., 1974, Biochem. Biophys. Research Com. 58, 151.
8. Ruby, R. H., 1974, personnal communication.
9. Butler, W. L., Visser, J. W. M. and Simons, H. L., 1973, Biochim. Biophys. Acta 325, 539.
10. Murata, N., Itoh, S. and Okada, M., 1973, Biochim. Biophys. Acta 325, 463.
11. Joliot, P. and Joliot, A., 1972, in : Proceedings of the 2nd International Congress on Photosynthesis Research, eds. G. Forti, M. Avron and A. Melandri (Junk, The Hague) p. 26.
12. Amesz, J. Pulles, M. P. I. and Velthuys, B. R., 1973, Biochim. Biophys. Acta 325, 472.
13. Satoh, K., Katoh, S. and Takamiya, A., 1972, Plant and Cell Physiol. 13, 885.
14. Kok, B., Forbush, B. and McGloin, M. P., 1970, Photochem. Photobiol. 11, 457.
15. Etienne, A. L., Lemasson, C. and Lavorel, J., 1974, Biochim. Biophys. Acta 333, 288.

M. AVRON, *Proceedings of the Third International Congress on Photosynthesis*
September 2-6, 1974, The Weizmann Institute of Science, Rehovot, Israel
Elsevier Scientific Publishing Company, Amsterdam, The Netherlands, 1974

INVESTIGATIONS ON THE SLOW (200 μs) COMPONENT OF THE PHOTOACTIVE CHLOROPHYLL REACTION IN SYSTEM II OF PHOTOSYNTHESIS

Günter Döring

Max-Volmer-Institut für Physikalische Chemie und Molekularbiologie
Technische Universität Berlin

SUMMARY

1. The red maximum of the Chl-a_{II} (P 680) difference spectrum is shifted to shorter wavelengths after treating the chloroplasts with heat or digitonin.
2. The half-life time of the Chl-a_{II} absorbance changes is increased by heat-treatment. The addition of HQ + Asc to heat-treated chloroplasts leads nearly again to the half-life time in untreated chloroplasts.
3. The Chl-a_{II} activity in digitonin-treated chloroplasts and in the presence of histone resp. is low when the electron acceptor is BV and high when the electron acceptor is FeCy.
4. The half-life time of the Chl-a_{II} reaction depends on the temperature of the reaction mixture. The activation energy of the Chl-a_{II} reaction is approx. 7,5 kcal/Mole.
5. In heat-treated chloroplasts the linear electron flow in PS II seems to be replaced by a cyclic one, which is sensitive to DCMU.

INTRODUCTION

In 1967 the photoactive pigment of PS II has been identified by using the repetitive flash technique[1] as a chlorophyll of the type "a" because of its absorbance changes[1,2]. The absorbance changes of this "Chlorophyll-a_{II}" (Chl-a_{II}) are maximal in the regions of 435 nm and 685 nm. The half-life time of the Chl-a_{II} absorbance changes depends on the pretreatment of the chloroplasts; an average value is 200 μs.

From our difference spectrum and from an extinction coefficient for chl-a in the red of $\varepsilon = 7{,}5 \cdot 10^4\ l \cdot M^{-1} \cdot cm^{-1}$ we calculated that the ratio Chl-a_{II}:Chl-a_I (the active pigment in PS I) is 1:2. But, when both Chl-a_I and Chl-a_{II} are active centers and coupled in series between H_2O and $NADP^+$, a necessary condition is that the ratio within one electron transport chain is $Chl\text{-}a_{II}:Chl\text{-}a_{I} \approx 1:1$. Therefore, it seems that our $Chl\text{-}a_{II}$ absorbance changes are too small for unknown reasons.

Recently GLÄSER et al.[4] observed in PS II a new chl-a absorbance change with a half-life time of appr. 35 μs. The 35 μs-absorbance changes together with our 200 μs-absorbance changes are maximal at about 690 nm. GLÄSER et al. interprete the new 35 μs-component together with the 200 μs-component as the biphasic kinetics of one and the same $Chl\text{-}a_{II}$ in PS II. The sum of both absorbance changes corresponds to the extent of the absorbance changes of $Chl\text{-}a_{I}$. Assuming nearly the same ε for $Chl\text{-}a_{II}$ and $Chl\text{-}a_{I}$ it follows now from the extent of $Chl\text{-}a_{II}$ and $Chl\text{-}a_{I}$ that the ratio in one electron transport chain is $Chl\text{-}a_{II}:Chl\text{-}a_{I} \approx 1:1$.

Recently van GORKOM et al.[3] confirmed our results. In deoxycholate incubated chloroplasts they observed a light minus dark difference spectrum with a shape nearly identical with our $Chl\text{-}a_{II}$ difference spectrum in digitonin-treated chloroplasts[2].

In this paper some new results of the 200 μs-component of $Chl\text{-}a_{II}$ are reported.

MATERIALS AND METHODS

Stripped spinach chloroplasts have been isolated as described in l.c.5. Suspensions enriched with PS II have been prepared according to ANDERSON and BOARDMAN[6]. The spectroscopic measurements were performed by the repetitive flash technique described in l.c.1. Excitation: 385-500 nm (4 mm BG 23 + 2 mm KG 2 from Schott), 610-710 nm (4 mm RG 610 + 2 mm KG 2 from Schott), saturating flashes of 20 μs duration. The electrical band width ranged from 0,1 Hz to 30 kHz. The optical path length through the cuvette was 1,2 mm. The band width of the monitoring light (grating monochromator) was 5 nm, the intensity about 50 $ergs/cm^2s$. 10^3 $ergs/cm^2s$ $h\nu_I$-backgroundlight (716 nm) were irradiated. The temperature of the sample was 22°C.

For the measurements with histone all salts have been washed away from the chloroplasts before histone was added.

The chlorophyll-concentration was 10^{-4} M. The sample contained Tris-HCl-buffer 0,05 M pH 7,2, BV 10^{-4} M resp. FeCy $5\cdot10^{-4}$ M as electron acceptor, and NH_4Cl $2\cdot10^{-3}$ M as phosphorylation uncoupler.

Further details are given in the legends of the figures. Deviating conditions are noted.

RESULTS

1. Difference spectra of Chl-a_{II}

In fig.1 the difference spectra of Chl-a_{II} in untreated and in heat-treated chloroplasts are presented. These difference spectra

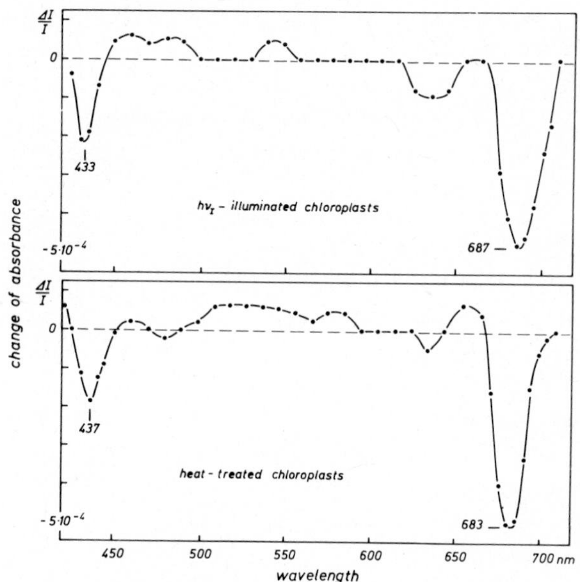

Fig. 1. Top: Absorbance changes with a life time of appr. 150 µs as a function of the wavelength in stripped spinach chloroplasts with BV as electron acceptor. Activity of the oxygen production: 151 moles O_2/mole Chl·h. Repetition rate 8 Hz. Intensity of the $h\nu_I$-backgroundlight: $5 \cdot 10^4$ ergs/cm^2s (720 nm). Bottom: Absorbance changes with a life time of appr. 250 µs as a function of the wavelength in heated spinach chloroplasts (5 min at 53 °C) with BV as electron acceptor. No activity of the oxygen production. Repetition rate 8 Hz.

are very similar to that measured in digitonin-treated chloroplasts[2]. During the measurements in untreated chloroplasts strong $h\nu_I$-background light ($5 \cdot 10^4$ ergs/cm^2s 720 nm) was irradiated in order to keep Cyt-f oxidized in the darktime between the exciting flashes. Therefore, no disturbing absorbance changes of Cyt-f were superimposed to the Chl-a_{II} absorbance changes in the blue region. The half-life time of the Chl-a_{II} absorbance changes is appr. 150 µs in untreated and appr. 250 μs in heat-treated (5 min at 53 °C) chloroplasts (in digitonin-treated chloroplasts it was appr. 200 μs). The difference spectra have maxima at 433 and 687 nm in untreated chloroplasts and at 437 and 683 nm in heat-treated chloroplasts (in digitonin-treated chloroplasts it was 435 and 682 nm). Obviously, pretreatment of the chloroplasts leads to a blue-shift of the red maximum of the Chl-a_{II} differences spectrum (this is in agreement with results of van GORKOM et al.[3]) and to an increase of the half-life time of the Chl-a_{II} absorbance changes.

The sensitivity of the Chl-a_{II} absorbance changes to DCMU does not depend on the pretreatment of the chloroplasts; it is always the same as for the relative oxygen yield per flash in untreated chloroplasts.

2. Chloroplasts in which the electron transport is damaged in PS II. Heat-treated chloroplasts

Whilst the amplitude of the Chl-a_{II} absorbance changes is not effected by a 5 min heat-treatment up to 50 °C[2], shows the half-life time a clear increase for temperatures higher than 45 °C (see fig. 2). The half-life time increases from 120 μs in untreated

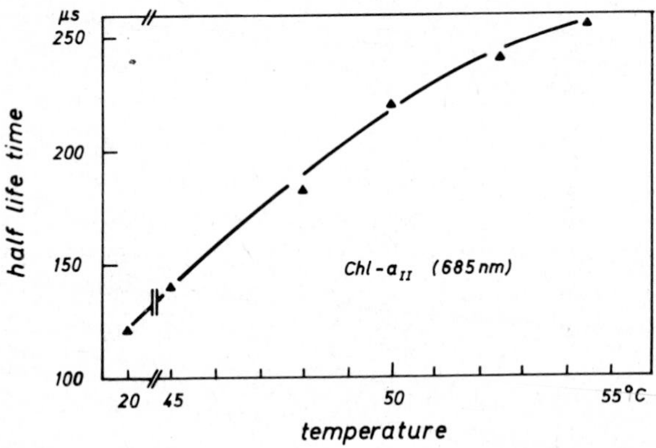

Fig. 2. Half-life time of Chl-a_{II} (685 nm) as a function of the heating temperature in stripped spinach chloroplasts with BV as electron acceptor. The chloroplasts had been exposed to the marked temperatures for 5 min before the measurement. during the measurement the temperature in the sample cuvette was 22 °C. Activity of the oxygen production (before heating): 108 moles O_2/mole Chl·h. Repetition rate 8 Hz.

chloroplasts up to 220 μs after treatment at 50 °C and to 250 μs at 54 °C.

The addition of benzo-hydroquinone plus ascorbate (HQ + Asc), which is an artificial electron donor system for PS II, to heat-treated chloroplasts (5 min at 52 °C) makes that the kinetics of the Chl-a_{II} absorbance changes get nearly as fast as before heat-treatment, but the amplitude is diminished down to 60% of the value in untreated chloroplasts. The amplitude of the Chl-a_I absorbance changes, which is diminished nearly completely by heat-treatment, is restored to 60% after the addition of HQ + Asc to heat-treated chloroplasts.

3. Chloroplasts in which the electron transport is damaged in PS I

3.1. Digitonin-treated chloroplasts

In the 10.000xg fraction of digitonin-treated chloroplasts (according to ANDERSON and BOARDMAN[6]) most of the electron transport chains are interrupted between the two photosystems, because most of the Cyt-f[7] and plastocyanine[8,9] is washed away. In the case that no artificial electron donor is used the behavior of these subchloroplasts depends drastically on the type of electron acceptor. When benzyl viologen (which is an electron acceptor for PS I[10]) is used, the activities of both PS I and PS II, indicated by the absorbance changes of Chl-a_I and Chl-a_{II}, are low (10-20% of the activity in untreated chloroplasts), indicating that in these subchloroplasts only 10-20% of the electron transport chains are intact. When potassium ferricyanide (which is an electron acceptor for PS I[10] and for PS II[11,12]) is used, the activity of PS II, indicated by the HILL-activity and the Chl-a_{II} absorbance changes, is 70-80%, the activity of PS I, indicated by the Chl-a absorbance changes, is 20-30% compared to the activities in untreated chloroplasts.

3.2. Histone-treated chloroplasts

From measurements of the $NADP^+$- and the FeCy-reduction BRAND et al.[13] concluded that histone inhibits the electron transport in PS I with greater efficiency than in PS II. We have confirmed these conclusions by measurements of the absorbance changes of Chl-a_I (705 nm) and Chl-a_{II} (685 nm). As shown in fig.3 for 50% inhibition of Chl-a_{II} about the 10-fold histone concentration is necessary as for Chl-a_I, when FeCy is used as electron acceptor. These measurements were carried out with a repetition rate of 3 Hz.

At higher repetition rates the inhibitory effect of histone is slightly less.

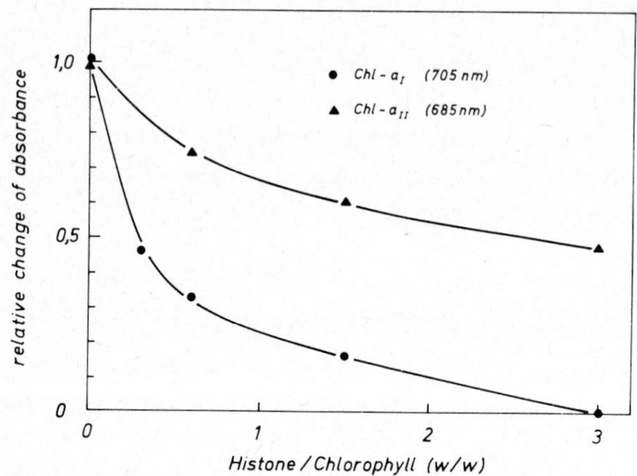

Fig. 3. Relative change of absorbance of Chl-a_I (705 nm) and Chl-a_{II} (685 nm) as a function of the ratio histone/chlorophyll (w/w) in salt free stripped spinach chloroplasts. Electron acceptor FeCy. Activity of the oxygen production (without histone): 97 moles O_2/ /mole Chl·h. Repetition rate 3 Hz.

When BV is used as electron acceptor, there is no difference in the inhibitory effect of histone for Chl-a_I and Chl-a_{II}: when the ratio histone/chlorophyll (w/w) is 3, the inhibition of both Chl-a_I and Chl-a_{II} is about 95%.

4. The activation energy of the Chl-a_{II} reaction

The half-life time of the Chl-a_{II} absorbance changes depends on the temperature in the sample cuvette. At room temperature (22°C) the half-life time is appr. 110 μs, but with decreasing temperature the Chl-a_{II} reaction gets slower, and at 2 $^{\circ}$C the half-life time is appr. 265 μs. From plotting the logarithm of the rate constant k as a function of the reciprocal absolute temperature (see fig.4) the activation energy of the Chl-a_{II} backreaction in the dark is estimated to be appr. 7,5 kcal/Mole.

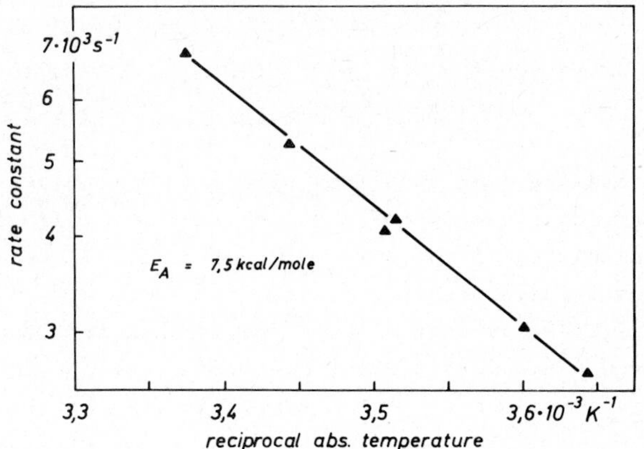

Fig. 4. Logarithm of the rate constant of the Chl-a_{II} reaction as a function of the reciprocal absolute temperature in stripped spinach chloroplasts. Electron acceptor BV. Activity of the oxygen production at 22 °C: 110 moles O_2/mole Chl·h. Repetition rate 8 Hz.

DISCUSSION

It is well known for a long time past that Chl-a_I is engaged in the electron transport chain with a redox reaction[14,15]. The decision whether the type of reaction is a redox reaction or a sensitizer reaction is much more difficult for Chl-a_{II}. Any reaction of Chl-a_{II} which leads to an electron transfer from water to PS I without an electron donation or acception by Chl-a_{II} itself, we call a sensitizer reaction. In the case of a sensitizer reaction the 200 μs of the Chl-a_{II} absorbance changes can not be a time constant of the electron transport, because the Chl-a_{II} molecules themselves do not accept or donate electrons.

For sensitizer reactions as defined above we can distinguish two cases:

1. The Chl-a_{II} reaction is active even without an electron transfer from Y to X, and
2. the Chl-a_{II} reaction is active only when an electron is transferred from Y to X (see Fig.5).

Reduced X has been identified as a special plastosemiquinone anion[21,22].

The first case is compatible with our findings that the Chl-a_{II} reaction is active in tris-washed[16], in heat-treated and in aged[2] chloroplasts. But our results in digitonin-treated and in histone-treated chloroplasts (high Chl-a_{II} activity with FeCy, low Chl-a_{II} activity with BV as electron acceptor) do exclude this possibility, because the Chl-a_{II} activity should be high, independent on the electron acceptor.

The second case is hard to distinguish from a redox reaction by experimental results. The experimental results reported so far can be interpreted as well if the Chl-a_{II} reaction is a redox reaction as if it is a sensitizer reaction which is coupled with an electron transfer from Y to X. An exception is the finding that the Chl-a_{II} absorbance changes become slower by heat-treatment, but faster again after the addition of the PS II donor system HQ + Asc. Because in the sensitizer model the half-life time of the Chl-a_{II} absorbance changes is no time of the electron transfers, it is hard to understand that the electron donor system HQ + Asc accelerates the Chl-a_{II} reaction in heat-treated chloroplasts. Therefore, we tend more to the assumption that Chl-a_{II} acts in a redox reaction.

But in the case of a sensitizer reaction which is coupled with an electron transfer from Y to X as well as in the case of a redox reaction our experimental results lead to the assumption that in tris-washed, in heat-treated and in aged chloroplasts the linear electron flow in PS II is replaced by a cyclic one including X, Chl-a_{II} (in the case of a redox reaction) and Y (see fig.5). This

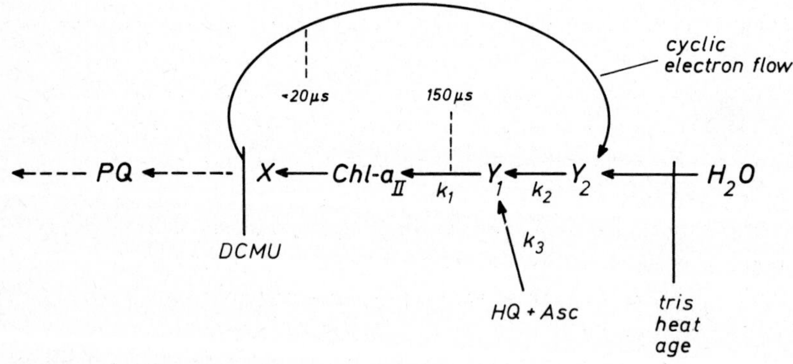

Fig. 5. Simplified scheme of the electron transport in tris-washed, in heat-treated and in aged chloroplasts. For details see text.

cyclic electron flow must be sensitive to DCMU. The existence of a DCMU-sensitive cyclic electron flow in PS II in tris-washed chloroplasts, which does not exist in untreated chloroplasts, has been postulated already by ROSENBERG et al.[17] (the reaction site of DCMU in this cyclic electron flow is assumed to be on the oxidizing side of Chl-a_{II}).

The electron transfer from X^- to Y_2^+ must be faster than 20 μs (this is the time resolution of our equipment), because we do not detect absorbance changes of X in tris-washed, in heat-trated and in aged chloroplasts.

In the case that the Chl-a_{II} reaction is a redox reaction, the half life time of the Chl-a_{II} absorbance changes depends on the rate constants k_1, k_2 and k_3. In untreated chloroplasts it is $k_2 \gg k_1$, and k_1 is leading to a half-life time of appr. 150 μs. During heat-treatment k_1 does not change, but k_2 becomes smaller and makes the half-life time of Chl-a_{II} increase. After the addition of HQ + Asc to heat-treated chloroplasts it is $k_3 \gg k_2$, k_1, and therefore, the rereduction of Chl-a_{II}^+ becomes faster again.

It should be noted that the reaction $S_1^* \rightarrow S_2$ in KOK's four-step-model of the oxygen evolution at room temperature has a first half-life time $\leq$ 200 μs[18], just as the half-life time of Chl-a_{II}. BOUGES has shown that the half-life time of S_1^* depends on the temperature in the reaction vessel[19,20]. Even though the temperature coefficient of the reaction $S_1^* \rightarrow S_2$ is slighly different from that of the Chl-a_{II} reaction, one can not exclude that the reduction of Chl-a_{II}^+ after flash excitation is limitting for the reaction $S_1^* \rightarrow S_2$.

REFERENCES

1. Döring,G., Stiehl,H.H. and Witt,H.T., 1967, Z. Naturforsch. 22b, 639

2. Döring,G., Renger,G., Vater,J. and Witt,H.T., 1969 Z. Naturforsch. 24b, 1139

3. van Gorkom,H.J., Tamminga,J.J. and Haveman,J., 1974 Biochim. Biophys. Acta 347, 417

4. Gläser,M., Wolff,Ch., Buchwald,H.-E. and Witt,H.T., 1974 FEBS Letters 42, 81

5. Winget,G.D., Izawa,S. and Good,N.E., 1965 Biochem. Biophys. Res. Commun. 21, 438

6. Anderson,J.M. and Boardman,N.K., 1966 Biochim. Biophys. Acta 112, 403

7. Boardman,N.K. and Anderson,J.M., 1967 Biochim. Biophys. Acta 143, 187

8. Wessels,J.S.C., 1966, Biochim. Biophys. Acta 126, 581

9. Sane,P.V. and Hauska,G.A., 1972, Z. Naturforsch. 27b, 932

10. Witt,H.T., Rumberg,B., Schmidt-Mende,P., Siggel,U., Skerra,B., Vater,J. and Weikard,J., 1965, Angew. Chemie 77, 821

11. Trebst,A., Harth,E. and Draber,W., 1970 Z. Naturforsch. 25b, 1157

12. Avron,M. and Ben-Hayyim,G., 1969, in: Progress in Photosynthesis Research (Metzner,H. ed.), Vol.3, 1185, Univ. of Tübingen Press

13. Brand,J., Baszynski,T., Crane,F.L. and Krogmann,D.W., 1971 Biochem. Biophys. Res. Commun. 45, 538

14. Kok,B., 1961, Biochim. Biophys. Acta 48, 527

15. Rumberg,B. and Witt,H.T., 1964, Z. Naturforsch. 19b, 693

16. Govindjee, Döring,G. and Govindjee,R., 1970 Biochim. Biophys. Acta 205, 303

17. Rosenberg,J.L., Sahu,S., and Bigat,T.K., 1972 Biophysical Journal 12, 839

18. Kok,B., Forbush, B. and McGloin,M., 1970 Photochem. Photobiol. 11, 457

19. Bouges,B., 1972, Biochim, Biophys. Acta 256, 381

20. Bouges-Bocquet,B., 1973, Biochim. Biophys. Acta 292, 772

21. Stiehl, H.H. and Witt,H.T., 1969 Z. Naturforsch. 24b, 1588

22. Witt,K., 1973, FEBS Letters 38, 116

M. AVRON, *Proceedings of the Third International Congress on Photosynthesis*
September 2-6, 1974, The Weizmann Institute of Science, Rehovot, Israel
Elsevier Scientific Publishing Company, Amsterdam, The Netherlands, 1974

IDENTIFICATION OF THE PRIMARY REACTANTS IN PHOTOSYSTEM 2

Hans J. van Gorkom

Biophysical Laboratory of the State University, P.O. Box 2120, Leiden,
The Netherlands

1. Summary

This is a discussion on the identification of the primary electron donor and acceptor of photosystem 2 on the basis of absorption difference spectra measured in subchloroplast fragments prepared with deoxycholate.

Some supplementary data are given, which confirm the conclusion that the primary acceptor is a plastoquinone and the primary donor is a chlorophyll a dimer.

2. Introduction

For an understanding of the primary reactions in photosynthesis it is essential to know the chemical identity of the primary reactants. We have been working on the identification of the primary donor-acceptor couple in photosystem 2. Most of the results have been published recently[1,2]. In this paper these results will be discussed in the light of some new data.

The primary electron donor of photosystem 2 is supposed to be the excited state of a chlorophyllous pigment with an absorption band at 680 nm, called P680. Until recently this notion was based more on the attractive analogy to P700 of photosystem 1 and P870 of bacterial systems than on experimental evidence. The difficulty in obtaining such evidence is commonly attributed to a very fast dark reduction of the oxidized $P680^+$ by a secondary donor.

3. Measurements with the repetitive flash technique

Fast measurements of absorption changes have been carried out with the repetitive flash technique by Döring and coworkers[3]. A reversible chlorophyll bleaching with a 200 μs dark decay was observed. The concentration of the pigment - if it was completely bleached - was much lower than the expected concentration of system 2 reaction centers. To explain this Döring suggested that P680 was acting as a sensitizer rather than as an electron donor. However, the change was inhibited by DCMU (3-(3,4-dichlorophenyl)-1,1-dimethylurea), which is known not to inhibit the photochemical reaction. As we have explained earlier[4], it is to be expected that most $P680^+$ was reduced faster than Döring could have measured, and that the 200 μs decay is seen only in a fraction of the reaction centers in which the secondary donor was in its oxidized form already. The DCMU sensitivity then merely confirms

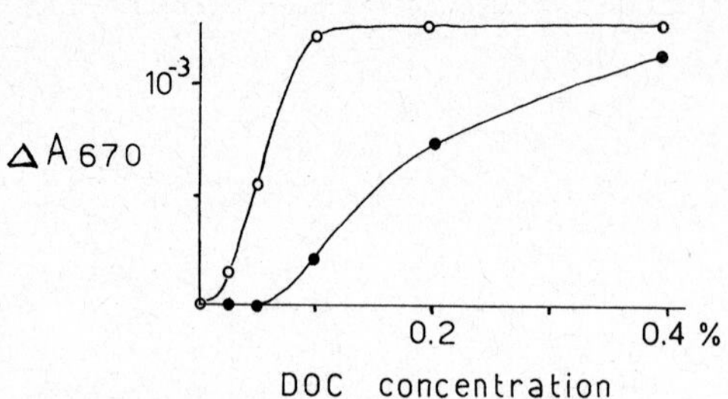

Fig. 1. DCMU sensitivity of the light-induced reversible chlorophyll bleaching in subchloroplast fragments prepared with deoxycholate (DOC). Optical path length 1 mm. Chlorophyll concentration 0.1 mM. Ferricyanide concentration 1 mM. Open circles: no DCMU. Solid circles: 10 µM DCMU added. Note: the DCMU sensitivity varies with the batch of spinach and is usually less pronounced.

the expectation that this situation occurs only in the higher S-states, which can not be formed if the reoxidation of the primary acceptor is blocked by DCMU.

In agreement with this interpretation Gläser and coworkers[5] have now reported a much larger 35 µs decay, preceding the 200 µs decay measured by Döring. However, this faster decaying change was also DCMU sensitive. If the 0.25 s dark time between flashes did allow an appreciable reoxidation of the primary acceptor we tend to conclude that even this 35 µs decay is observed only when the reduction of $P680^+$ is retarded by something which does not arise in the presence of DCMU, e.g. a higher S-state or membrane potential.

4. Measurements in subchloroplast fragments prepared with deoxycholate

We have shown that an absorbance change with the same difference spectrum as reported by Döring can be measured in chloroplasts treated with deoxycholate[1]. This change was not sensitive to DCMU and could be interpreted as the oxidation of a chlorophyll a dimer, present in a concentration of one to about 350 chlorophylls. This seems to confirm the interpretation of Döring's measurements as given above. We can now add that at the lowest deoxycholate concentrations where $P680^+$ can be seen it is sensitive to DCMU (see fig. 1), indicating that at least one secondary donor is still active. At higher deoxycholate concentrations the dark reduction of $P680^+$ presumably is caused by the back reaction with the reduced primary acceptor.

Fig. 2A shows the difference spectrum of the reduction of the primary acceptor, measured in subchloroplast fragments prepared with deoxycholate as reported in ref. 2. The ultra violet part is so similar to the difference spectrum caused by the reduction of plastoquinone to anionic plastosemiquinone (dashed line in fig. 2A, data from ref. 6) that it may be identified as such. Our spectrum is shifted to longer wavelengths, but this is due to the deoxycholate treatment, because in a concentrated chloroplast suspension immediately after addition of deoxycholate

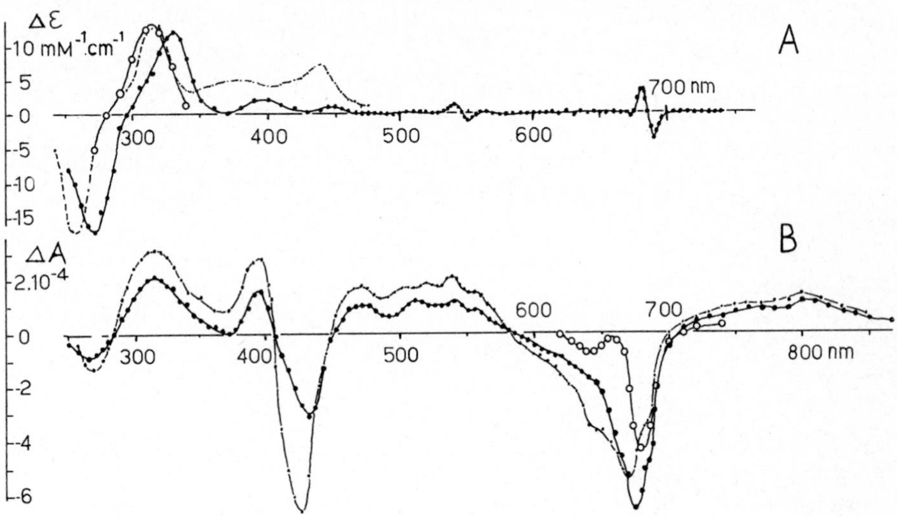

Fig. 2. Spectra of light-induced absorbance changes in subchloroplast fragments prepared with deoxycholate.
A) Reduction of the primary acceptor as reported in ref. 2. 250 μM chlorophyll after treatment with 0.4 % deoxycholate. 5 μM diphenylpicrylhydrazyl added. Open circles: similar measurements but with 2 mM chlorophyll immediately after addition of 2.5 % deoxycholate. (This spectrum was not calibrated to the Δε scale). Dashed line: plastosemiquinone anion minus plastoquinone difference spectrum given in ref. 6.
B) Oxidation of the primary donor as reported in ref. 1. Open circles: 2 mM chlorophyll, 2.5 % deoxycholate, 5 mM ferricyanide. Optical path length 0.1 mm. Solid circles: 0.1 mM chlorophyll, 0.4 % deoxycholate, 0.1 mM ferricyanide. Optical pathlength 1 mm. Dots: the same preparation after prolonged incubation with 5 % deoxycholate at room temperature.

no such shift was observed (open circles in fig. 2A). The band shifts around 545 nm (C550) and 685 nm may be atrributed to an indicator pigment in the neighbourhood of the primary acceptor. For the sake of simplicity (explaining both changes by one pigment) and because of the analogy to bacterial reaction centers we favor the hypothesis that the band shifts are both due to a bound or aggregated form of pheophytin _a_. This pigment has an absorption band at 536 nm which, if red-shifted by 10 nm, could account for C550.

Fig. 2B shows difference spectra attributed to the oxidation of the primary donor. If measured in a fresh and concentrated chloroplast suspension after addition of deoxycholate this appears to be a chlorophyll _a_ dimer (open circles, from ref. 1). Spectra obtained after more extensive deoxycholate treatment can be explained as a mixture of monomers and dimers (solid circles). Even more rigorous deoxycholate treatment resulted in a difference spectrum (dots) which seems to reflect a further change to a species with an absorption peak near 640 nm. The changes of the primary acceptor might well be superimposed on these spectra, as would be expected if the dark decay reflects the back reaction with the reduced acceptor.

Acknowledgement

This work was supported by the Netherlands Foundation for Chemical Research (SON), financed by the Netherlands Organization for the Advancement of Pure Research (ZWO).

References

1. van Gorkom, H.J., Tamminga, J.J., Haveman, J. and van der Linden, I.K., 1974, Biochim. Biophys. Acta 347, 417-438.
2. van Gorkom, H.J., 1974, Biochim. Biophys. Acta 347, 439-442.
3. Döring, G., Renger, G., Vater, J. and Witt, H.T., 1969, Z. Naturfosch. 24b, 1139-1143.
4. van Gorkom, H.J. and Donze, M., 1973, Photochem. Photobiol. 17, 333-342.
5. Gläser, M., Wolff, C., Buchwald, H.E. and Witt, H.T., 1974, FEBS Lett. 42, 81-85.
6. Bensasson, R. and Land, E.J., 1973, Biochim. Biophys. Acta 325, 175-181.

M. AVRON, *Proceedings on the Third International Congress on Photosynthesis*
September 2-6, 1974, The Weizmann Institute of Science, Rehovot, Israel
Elsevier Scientific Publishing Company, Amsterdam, The Netherlands, 1974

THE RELATIONSHIP BETWEEN THYLAKOID STACKING, STATE I AND STATE II PHENOMENA IN WHOLE CELLS AND THE CATION EFFECTS IN CHLOROPLASTS OF CHLAMYDOMONAS REINHARDI

Pierre BENNOUN [x] and Henri JUPIN [xx]

x Institut de Biologie Physico-Chimique, 13 rue Pierre et Marie Curie. 75231 - PARIS - Cedex 05

xx Laboratoire de Botanique de l'Ecole Normale Supérieure 24, rue Lhomond. 75231 - PARIS - Cedex 05

Introduction

When whole cells of algae are subjected to prolonged illumination with different types of light, they display slow adaptation to the intensity and the wavelength of the light. This was described by BONAVENTURA and MYERS [1] as a transition between two states (State I and State II). These are different in many respects : pattern of chlorophyll fluorescence [1, 2, 3], quantum yield of System II [1, 3, 4], amplitude of the EMERSON effect [1], pattern of deactivation in the dark of the oxygen precursor S_3 [5], apparent efficiency of energy transfer within the System II units [3]. This list is certainly not a complete one as much remains to be studied in this field.

In a previous publication, one of us [3] showed that similar differences were observed in isolated chloroplasts upon addition of magnesium ions. A correlation was shown to exist between State I of algae and the state of chloroplasts in the presence of magnesium and between State II of algae and the state of chloroplasts in the absence of magnesium. This correlation leads to the idea that the transition between State I and State II could be regulated by magnesium ions.

It has been shown that cations stabilize the association of thylakoid membranes into grana stacks [6, 7]. It was therefore of interest to investigate whether such variations of the membrane organisation occur <u>in vivo</u> and could therefore be an important feature of State I and State II phenomena. It is well known that stacking of two thylakoids leads to the appearance of an electron dense line in the area of contact. This presumably means that structural changes in the membranes occur at the point of contact of two thylakoids.

The changes in photosynthetic properties observed in the transition from State I to State II in algae and upon addition of cations in chloroplasts could be a result of these structural rearrangements.

Materials and Methods

a) Chamydomonas reinhardi wild type 137 C was grown mixotrophically. Chloroplasts displaying large effects of magnesium ions were prepared as described previously [3]. This preparation involves an osmotic shock by dilution into a buffer containing 0.2 M sorbitol. States I and II were developed as described previously [3]. Cells were resuspended in 10 mM phosphate buffer pH 6.8 and the suspension was stirred during illumination with light I or light II. Glutaraldehyde fixation was performed in the light and illumination with light I or II was continued for another five minutes before incubation in the dark at 4° for one hour. The chloroplast preparations were resuspended in 20 mM HEPES buffer pH 7.5 containing 10 mM Nacl, 10 mM KCl and 0.4 M sorbitol. Magnesium chloride (8 mM) was added in the dark and chloroplasts incubated in the dark before glutaraldehyde fixation.

b) Electron Microscopy. Fixation with 2.5 % glutaraldehyde at 4° for one hour was carried out as described above. Samples were embedded in Araldite following post-fixation with 0.5 % osmium tetroxide (1 hour at 4°) and with 0.5 % uranyl acetate (15 minutes). Micrographs were obtained with a Philips E 300 apparatus. The stacking profiles were measured according to TEICHLER-ZAHLEN [8] and GOODENOUGH et al [9]. A method similar to that described by WEIBEL [10] for the measurement of the surface of endoplasmic reticulum was used to estimate the percentage of thylakoid surface common to two thylakoids.

Plate 1 -

a) Wild type chloroplast of C. reinhardi whole cell in State I showing large stacks of thylakoids.

b) Wild type chloroplast of C. reinhardi whole cell in State II showing stacks of 2 or 3 thylakoids.

c) Chloroplast of C. reinhardi in the presence of 8 mM $MgCl_2$. The thylakoids are closely packed in a polygonal array.

d) Chloroplast of C. reinhardi in the absence of magnesium. The thylakoids have rounded outlines and little contact exists between them.

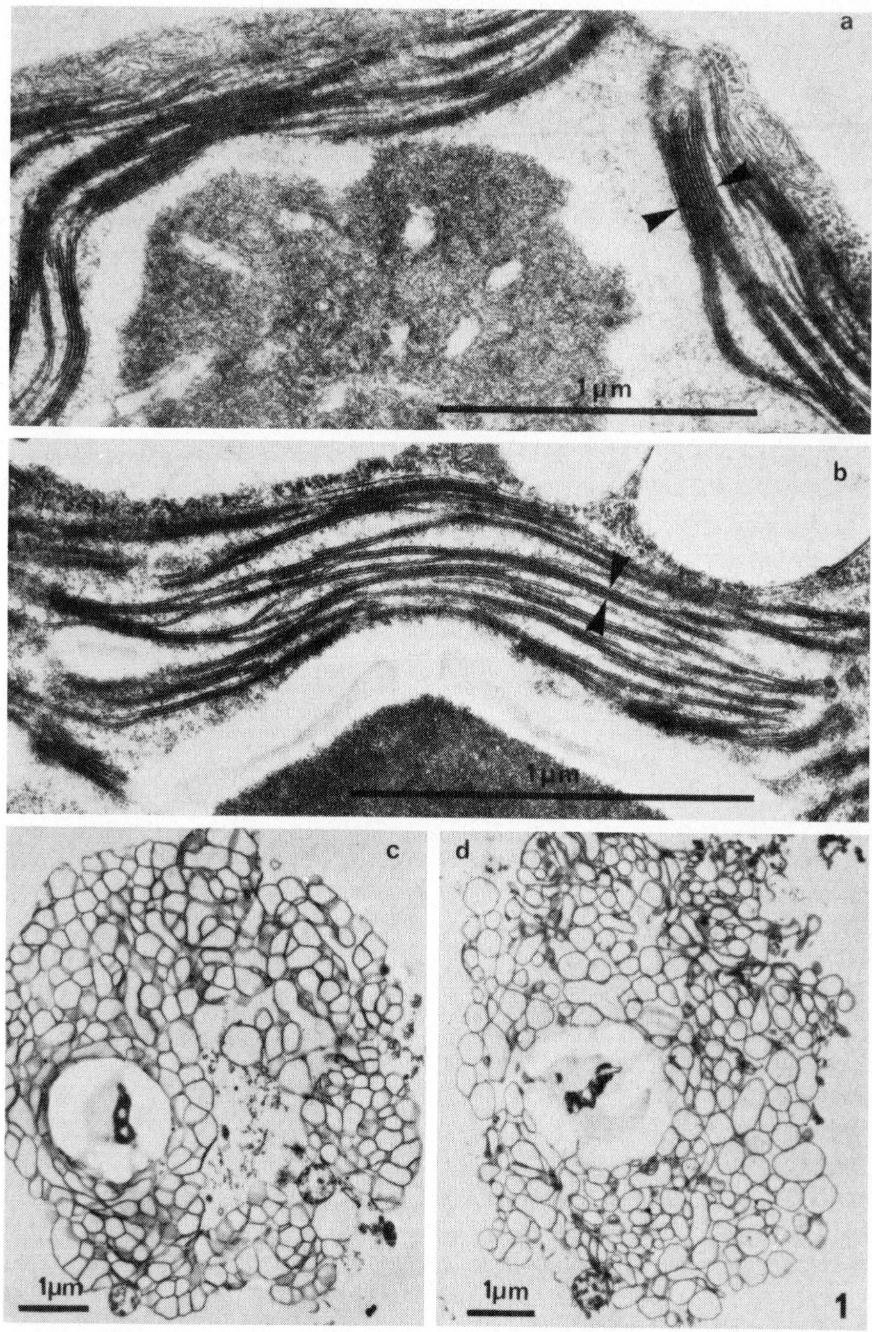
a
1 µm
b
1 µm
c
1µm
d
1µm
1

RESULTS

Plate 1 shows micrographs of whole cells of C. reinhardi fixed by glutaraldehyde in State I and in State II. It is apparent that in State I (a) the stacks are composed of more thylakoids than in State II (b). In order to examine this more quantitatively, stacking profiles of 17 cells taken at random were analyzed in each case. The data are shown in Figure 1. One observes that in State II the thylakoids are mainly grouped in pairs and that there is little spread in the distribution. In State I the distribution is much more broader and stacks of five thylakoids are the most common. It is interesting to check whether these differences represent a change in the proportion of stacked thylakoids. Measurements performed for 5 cells in each state show that 40 % of the total surface of the thylakoids is stacked in State II and 60 % in State I (see table 1).

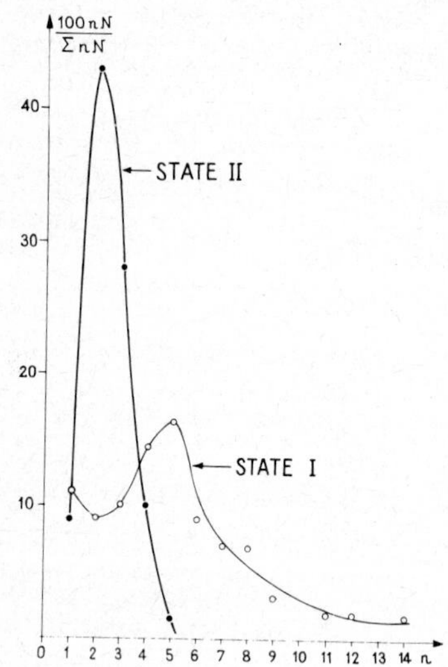

Fig. 1 - Stacking profiles of thylakoids in wild type cells of C.reinhardi in States I and II. A plot of $100\ nN/\Sigma nN$ versus n is given for each state where n = number of thylakoids in a stack and N = total number of stacks of size n. Measurements were performed on 17 cells for each state. In State I 542 thylakoids were counted and 376 in State II.

Plate I also shows micrograph of chloroplasts of C. reinhardi of the type that display large magnesium effects[3]. In the presence of 8 mM $MgCl_2$ (c) one observes that the thylakoids are in very close contact and arranged in a polygonal array. In the absence of magnesium ions (d) there is little contact between the thylakoids which have rounded outlines.

	STATE I			STATE II		
Cell number	Ratio of surface in contact with another thylakoid to total surface %	Total surface estimated by thylakoid length µm	Number of stacks analized per cell	Ratio of surface in contact with another thylakoid to total surface %	Total surface estimated by thylakoid length µm	Number of stacks analized per cell
1	59 ± 10	84	7	41 ± 10	86	6
2	65 ± 5	67	4	35 ± 9	61	4
3	60 ± 10	55	4	42 ± 7	98	6
4	47 ± 30	38	2	37 ± 14	74	4
5	64 ± 17	88	4	39 ± 30	30	2
Average percentage of membrane in contact	60% ± 5%			40 % ± 4 %		

Table 1 - Measurements of the percentage of membrane surface in contact with another membrane. C. reinhardi whole cells in States I and II.

The differences between the five samples of each state taken in pairs are not significantly different so that an average of the individual measurements is valid.

DISCUSSION

Chloroplast thylakoids are either free or in contact with one or two neighbours forming grana stacks. This paper shows that this pattern is not a permanent one ; illumination of algae with "light I" or "light II" leads to modifications in the degree of membrane stacking. On going from State I to State II a difference of 20 % in the proportion of stacked membranes is observed. It is noteworthy that some of the changes of photosynthetic properties observed between these two states are of comparable size.

For instance variations of 10 % in the quantum yield of System II, and of 25 % in the maximum fluorescence yield are observed [3, 4]. Another feature which differs in States I and II may be of some importance. In State II a majority of paired thylakoids are present, a situation in which each thylakoid has one face in contact with the outside. In State I stacks of 5 thylakoids are the most numerous class which implies that more than half of the thylakoids no longer have contact with the outside. Chloroplast preparations in the presence and absence of magnesium ions show similar but much more contrasted properties.

Although no precise measurements were made it is clear from the micrographs that magnesium induces a difference of much more than 20 % of the area of membranes in contact. As shown previously the magnitude of the changes in photosynthetic properties observed in this case is larger than that observed between State I and State II [3]. For instance, variations of 20 % in the quantum yield of System II and of 50 % in the maximum fluorescence yield are observed.

The results presented here provide additional evidence for the correlation between State I and State II phenomena and the effect of cations. Parallel variations of the photosynthetic properties and of the degree of stacking of the thylakoid membranes are observed in both cases. One may ask whether the changes in the stacking pattern may be considered as a primary or a secondary effect. In the first case, membranes would be held together by magnesium and structural rearrangements would occur as a consequence of stacking. These would be responsible for the observed modifications in the photosynthetic properties. In the second case conformational changes of the membranes would be induced directly by magnesium. The stacking of the membranes would be one of the results of these changes. In this case it should be possible to dissociate the stacking effect from the other parameters which are involved in these processes.

The relationship between State I and State II phenomena and the cation effects implies a mechanism for the regulation of the distribution of cations between the different phases of the chloroplast. The detailed mechanism of this is far from clear, but the observation that the pH rise observed upon illumination of chloroplasts is compensated both by Cl^- influx and by Mg^{2+} efflux [11] may be pertinent to this problem. However the control of magnesium in the different phases of the chloroplast may be rather complex as besides the effects described here this cation is involved in several other important chloroplast functions such as photophosphorylation, activity of enzymes in the carbon fixation pathway and even the activity of ribosomes.

REFERENCES

1. Bonaventura, C. and Myers, J., 1969, Biochim. Biophys. Acta, 189, 366-383.
2. Murata, N., 1969, Biochim. Biophys. Acta, 172, 242-251.
3. Bennoun, P., 1974, To be published in Biochim. Biophys. Acta.
4. Wang, R.T. and Myers, J., 1974, Biochim. Biophys. Acta, 347, 134-140.
5. Lemasson, C. and Barbieri, G., 1971, Biochim. Biophys. Acta, 333, 353-365.
6. Izawa, S. and Good, N.E., 1966, Plant Physiology, 41, 544.
7. Murakami, S. and Packer, L., 1971, Arch. Biochem. Biophys., 146, 337-347.
8. Teichler-Zallen, D., 1969, Plant Physiology, 44, 701-710.
9. Goodenough, U. and Levine R.P., 1969, Plant Physiology, 44, 990-1000.
10. Weibel, E.R., 1969, Int. Rev. Cytology, 26, 235-302.
11. Hind, G., Nakatani, H.Y. and Izawa, S., 1974, Proc. Nat. Acad. Sc. USA, 71, 1484-1488.

M. AVRON, *Proceedings of the Third International Congress on Photosynthesis*
September 2-6, 1974, The Weizmann Institute of Science, Rehovot, Israel
Elsevier Scientific Publishing Company, Amsterdam, The Netherlands, 1974

LIGHT-INDUCED VARIABLE PHOTOSYSTEM II CHLOROPHYLL FLUORESCENCE AS INDICATOR FOR PHOTOSYSTEM I ACTIVITY

(Fluorescence behaviour of Hydrogenase containing green algae under aerobic and anaerobic conditions)

R. Bauer and U.F. Franck

Institut für Physikalische Chemie der RWTH Aachen

Federal Republic of Germany

1. Summary

The well known chlorophyll fluorescence transient at the onset of illumination (KAUTSKY-effect) which is typical for all green plants and algae is drastically altered under anaerobic conditions.

Preferentially hydrogenase containing green algae (species of Scenedesmus, Chlorella and Ankistrodesmus) show after a distinct time of anaerobiosis (N_2/CO_2-gasing) in the dark at the onset of illumination a steep fluorescence drop starting from a maximum high initial fluorescence level.

Anaerobiosis causes in hydrogenase containing green algae strong reductive conditions. This results in the reduction of most of the redox systems in the photosynthetic electron transport chain.

There is evidence to show that the steep fluorescence drop reflects the rapid oxidation of anaerobically reduced primary Photosystem II electron acceptor pools via defect electrons (holes) which are produced in the light by the electron consuming light reaction I. This is supported by experimental results concerning the inhibition or accelaration of photosynthetic electron transport reactions.

2. Introduction

According to the concept of Kautsky and co-workers[1] the fluorescence changes 'in vivo' are brought about by energy acceptors being present in a very limited amount in the immediate neighbourhood of the light absorbing chlorophyll being able to fluoresce.

In terms of modern understanding of the primary processes of photosynthesis the energy acceptors in question are electron transferring redox systems in the oxidized state. According to Duysens and Sweers' hypothesis and denomination[2] the essential quencher of fluorescence is the primary electron acceptor of Photosystem II (PS II) : 'Q'. Q represents the first link of a chain of coupled redox cycles transferring electrons via an electrochemical gradient to Photosystem I (PS I) and defect electrons (holes) simultaneously in the opposite direction. Undoubtly the Kautsky-effect reflects the kinetic cooperation of both light reactions during their initial balancing caused by the onset of illumination. It is assumed that under normal conditions the variation of the concentration of oxidized Q is responsible for the temporal changes of fluorescence intensity. The fluorescence is quenched by Q only in its oxidized state. During the quenching process Q accepts an electron and turns into the reduced state loosing simultaneously the capability of quenching. If Q is reduced by endogenous or exogenous electron donors the reaction centers in PS II are 'closed' and fluorescence increases because the absorbed light energy can not be utilized by such centers containing electron saturated Q. Normally Q is in the oxidized state in the dark. Therefore the fluorescence yield has its minimum value at the onset of illumination.

During further illumination, however, fluorescence increases rapidly in the course of the first milliseconds because Q is reduced by the electron producing light reaction II.

Fluorescence decrease is only possible if light reaction I produces an excess of defect electrons by electron consuming reactions. These defect electrons are transmitted along the chain of coupled redox cycles to the site of Q and lead to its oxidation.

In this paper it will be shown that these fluorescence decreases can be used as a measure for the effectiveness and completeness of the electron transport between the two light reactions.

For instance green algae containing hydrogenase such as Scenedesmus obliquus, Ankistrodesmus braunii and Chlorella pyrenoidosa[3] can reduce Q nearly completely in the dark under anaerobic conditions. In this case the fluorescence yield is maximum at the onset of illumination. But it decreases rapidly on the normal stationary level under weak blue light or strong far-red light. Obviously this fast decrease of fluorescence represents preferentially the oxidation of Q by defect electrons which are produced during illumination by the electron consuming light reaction I.

This is supported by experimental results concerning the inhibition (DCMU, DBMIB) or accelaration (heating) of the electron transport and the effects of NH_2OH and DAD/ascorbate admixtures. These treatments influence the fast fluorescence decrease under anaerobic conditions in a very specific and well understandable way.

3. Materials and Methods

The fluorescence measuring apparatus was in principle the same as described previously[4] .
All experiments were carried out with unicellular green algae received from the culture collection at Göttingen(W-Germany).
The following strains were studied: with hydrogenase : Scenedesmus obliquus /D_3 276-6 ; Ankistrodesmus braunii 202-7c ; Chlorella pyrenoidosa 211-8b ; without hydrogenase : Chlorella vulgaris f.tertia 211-8m ; Chlorella saccharophila 211-1d and 211-9a .
The algae were grown homocontinously in aerated (3 % CO_2 in air) liquid cultures (inorganic medium as described by Kessler et al.[5]) at 20,000 Lux and 30°C .

Anaerobic conditions were obtained by gasing the samples for several hours with a mixture of 96 % purified nitrogen and 4 % carbondioxid. Traces of oxygen were eliminated of the gas mixture by applying an oxygen absorber (Oxisorb, type F, Messer Griesheim: Düsseldorf/ W-Germany). This instrument garantees an oxygen concentration below 0.1 vpm .
3-(3,4-dichlorophenyl)-1,1-dimethylurea (DCMU) and 2,5-dibromo-3-methyl-6-isopropyl-p-benzoquinone (DBMIB) were dissolved in pure ethanol and prepared at appropriate concentrations such that the final concentration of ethanol in the algae suspensions never exceeded 1 % . Unless noted otherwise all experiments were carried out at 28°C. The blue exciting light (beyound 460 nm) had an intensity of 3.10^4 $erg.cm^{-2}.s^{-1}$. DAD ≙ diaminodurene .

4. Results and Discussion

a) The Kautsky-effect (General remarks)

At the present time it is generally accepted that the Kautsky-effect reflects directly and without temporal delay the mechanism of the primary processes of photosynthesis during its adaptation period to sudden changes of illumination. This paper mainly deals with the kinetic interpretation of the special shape of the fluorescence curve and its alteration by exogenous influences. In particular it will be shown that although the fluorescence transients originate clearly in PS II, they give quantitative kinetic informations also about the activity of PS I and its cooperation with PS II .

b) The characteristics of the normal fluorescence induction curve

Referring to Fig.1 and applying Lavorel's terminology[6] the different sections of the normal fluorescence transient can be specified as follows :

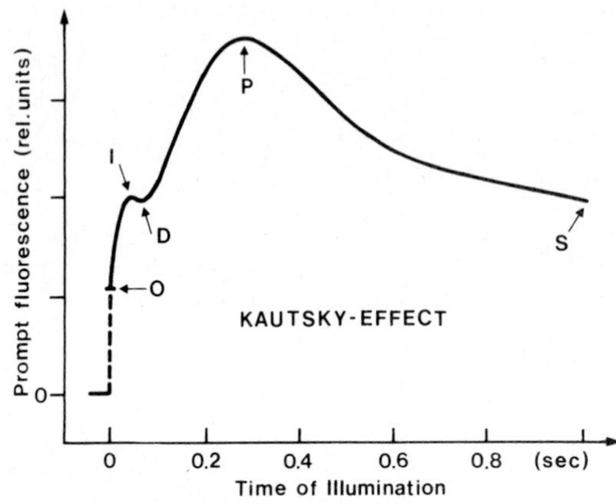

Fig.1. Typical chlorophyll fluorescence transient in aerobic Chlorella pyrenoidosa. Temperature 25°C; blue exciting light intensity 3.10^4 $erg.cm^{-2}.s^{-1}$. Notations : O = initial fluorescence measured after the shutter opening time of 0.5 msec, J = level of the initial peak, D = dip, P = peak, S = almost stationary fluorescence level.The different features of the induction curve are characteristic for all photosynthesizing green plants and algae.

o-O :The initial fluorescence

With the onset of illumination the initial value of fluorescence O is attained in less than 10^{-7} seconds[4]. It is the intrinsic starting point of the fluorescence curve. For healthy cells under normal conditions this value represents the state of lowest fluorescence yield during the entire period of constant illumination. This strongly quenched fluorescence obviously is emitted by the bulk of chlorophyll-a molecules transferring excitation energy (presumbly by inductive resonance[7])to the open, unoccupied reaction centers of the photosynthetic systems. This suggestion is supported by the findings of Lavorel that the initial fluorescence actually is built up by the fluorescence of PS I as well as PS II chlorophyll-a molecules[8].

Because the O-level is attained before the excitation energy being transferred by the bulk of chlorophyll molecules has reached the reaction centers, this level can be considered as a direct measure for the total amount of reaction centers available in the photosystem[9]. An occuring augmentation of the initial fluorescence by preillumination or by applying specific inhibitors etc. indicates that a fraction of the reaction centers has become closed or blocked by these influences.

O-J :The first fluorescence increase

The first increase in fluorescence reflects the beginning of the primary photochemical processes of photosynthesis. The first process in particular being indicated by the fluorescence behaviour consists in an 'activation' of the trapping PS II reaction centers[10].

According to this terminology the reaction centers are 'inactive' in the dark adapted state and incapable to transfer electrons through PS II. The light induced activation of this system obviously is related to alterations in the water-splitting enzyme system (distribution of S_0-S_4 states according to Fig.2), which causes an increase of fluorescence.

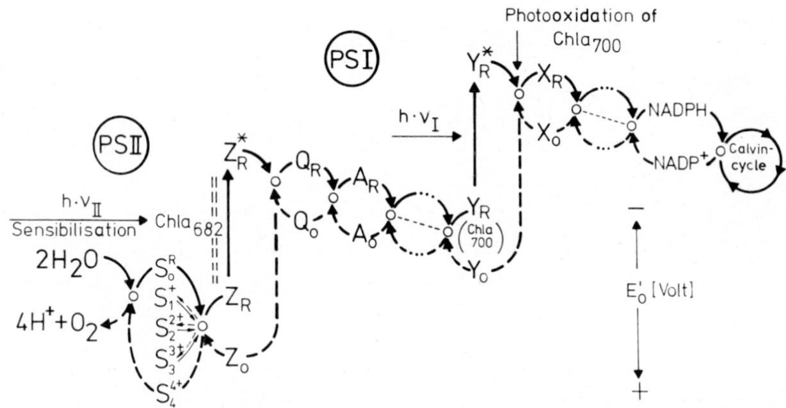

Fig.2. The photosynthetic electron transport chain represented as a mechanism of coupled redox cycles. S_0-S_4 symbolizes the different oxidation states of the oxygen-evolving enzyme system.

Immediately with the activation of PS II centers the photoreduction of Q (Q_o, Fig.2), the primary electron acceptor of PS II, sets in. Electron transfer from PS II to the Q-redox cycle is only possible if Q is available in the oxidized state (Q_o), i.e. only Q_o is capable to quench fluorescence. As a consequence photoreduction of Q by PS II activity leads to a diminution of the quenching power of the Q_o/Q_R cycle, resulting in the pronounced increase of fluorescence from O to J .

J-D : The first fluorescence decrease : the dip D

From experiments concerning inhibitors which interfere the electron transmission in the chain of redox cycles[11] as well as investigations of photosynthetic mutants of green plants lacking defined kinetic steps of the primary processes of photosynthesis[11,12], we have obtained strong evidence that the fluorescence decline indicates a first temporary, somewhat delayed oxidation of Q ($Q_R \rightarrow Q_O$, Fig.2) by defect electrons being transferred from PS I along the chain of redox cycles connecting both photosystems.

This PS I induced oxidation of Q, however, lasts only a short time probably as a result of kinetic limitations or exhaustions resp. which can occur in PS I, in the electron supply of PS II or in the connecting chain of redox cycles being limited subsequentially in their pools. At present it is still unknwown which of them is the actual site of this kinetic limitation giving rise to a consecutive increase in fluorescence.

D-P : The second fluorescence increase ; the peak P

The increase in fluorescence from D to P indicates that Q is reduced faster by PS II than it is simultaneously oxidized by the action of PS I. At the peak P, which represents a state of extensively reduced Q, a temporary balance of both reactions is attained. Here, the oxygen production is weak[10] according to the low efficiency of the photochemical electron transmission indicated by a corresponding high yield of fluorescence.

P-S : The second fluorescence decrease

In the same way the P-S decline can be interpreted kinetically as an increasing activity of PS I with respect to PS II, probably by the reactivation of electron carrying enzyme reactions.

The entire fluorescence transient from O to the final level S, which essentially represents the Kautsky-effect, in general can be understood as the manifestation of the delayed kinetic balancing of a magnitude of chemical and photochemical redox cycles. In the primary photosynthesizing apparatus these redox systems are present in different substantial quantities, different kinetical effectiveness and they are obviously well adjusted mutually with respect to their individual redox potentials whose concentration-dependences probably may be decisive for the substantial quantities of the different

redox systems actually found in green plants.
After a certain period of illumination depending on plant material, light intensity and temperature etc. the yield of fluorescence becomes approximately constant.

Its comparatively low value indicates that the photoreduction of Q_o by PS II is balanced by a considerable reoxidation rate by PS I, signifying an intensive steady state turn over of electrons by both light reactions. The real Kautsky-effect concerns the fluorescence behaviour during the first seconds of illumination. Slow fluorescence changes following this period have also been observed[13]. But obviously these changes are the result of light induced conformational alterations in the thylakoid membrane structure of the photosynthetic apparatus and do not depend on the oxidation-reduction level of Q[14].

c) Aspects concerning the kinetic interpretation of fluorescence changes.

Speaking in terms of reaction scheme of Fig.2 the following kinetical statements can be made :
1) The degree of fluorescence quenching indicates directly the quantity of the quencher Q_o present at the reaction centers of PS II as well as the reaction rate of the photochemical consumption of Q_o according to the redox reaction $Q_o + e \rightarrow Q_R$.
Hence a low yield of fluorescence corresponds to a high concentration of Q_o and a high rate of Q_o-consumption by the photochemical activity of PS I .
2) The slope of the fluorescence curve (e.g. Fig.1) gives direct informations about the temporal change of the quantity of Q_o at the reaction centers of PS II .Because Q_o is consumed on the one side by the electrons immediately supplied by the light reaction of PS II and produced on the other side by defect electrons supplied indirectly along the chain of redox systems $Y \rightarrow Q$ (Fig.2) by the light reaction of PS I, the slope of the fluorescence curve indicates at any point of time, which of the both light reactions is the more effective one. A temporal increase of fluorescence therefore corresponds to a predomination of PS II-and a decrease to a predomination of PS I-activity at the PS II centers.

d) Chlorophyll fluorescence induction of green algae under anaerobic conditions.

Under anaerobic conditions oxygen removal causes in hydrogenase containing green algae characteristic changes in the chlorophyll fluorescence induction. As already shown previously[11] the fluorescence induction curves changes with progressive anaerobiosis in three distinct phases (Fig.3).

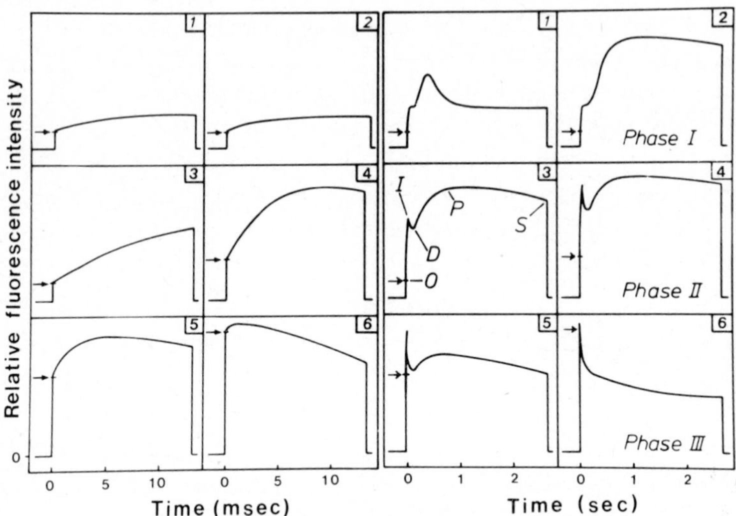

Fig.3. Change of the chlorophyll fluorescence transient in Scenedesmus obliquus during progressive anaerobiosis. Temperature 25 °C ; light intensity : 10^4 $erg.cm^{-2}.s^{-1}$. From ref.[11]. The three phases of anaerobiosis occur in the same manner also in Chlorella pyrenoidosa 211-8b and Ankistrodesmus braunii 202-7c. Adding air to the samples showning Phase III results in the restoration of the normal aerobic transient within 10 minutes. Note! There are two different time scales for the left and the right part of the figure.

In phase III the initial fluorescence level O increases up to a maximum level and the J-D transient appears more and more pronounced simultaneously with the gradually elimination of the D-P transient. This leads to an 'isolated fluorescence decline' (see Fig.3/6).

The isolated J-D decline as it occurs anaerobically in phase III was first reported by Schreiber et al.[11] in Scenedesmus obliquus. They found that this fluorescence decline depends on temperature, light intensity and the wavelength of the exciting light.

The J-D decline is most pronounced by temperatures around 30°C and low blue light (460 nm) (see Fig.4) or strong far-red (PS I)light. He could not be detected in the PS I-impaired mutant No.8 from Scenedesmus obliquus[11], but in mutant No.11 [11,15] which obviously is defective only in the oxygen-evolving reaction of PS II.

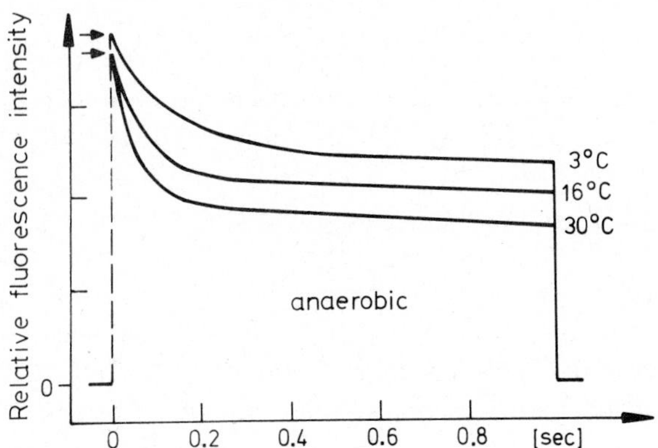

Fig.4. Scenedesmus obliquus; the influence of temperature on the anaerobic J-D decline.

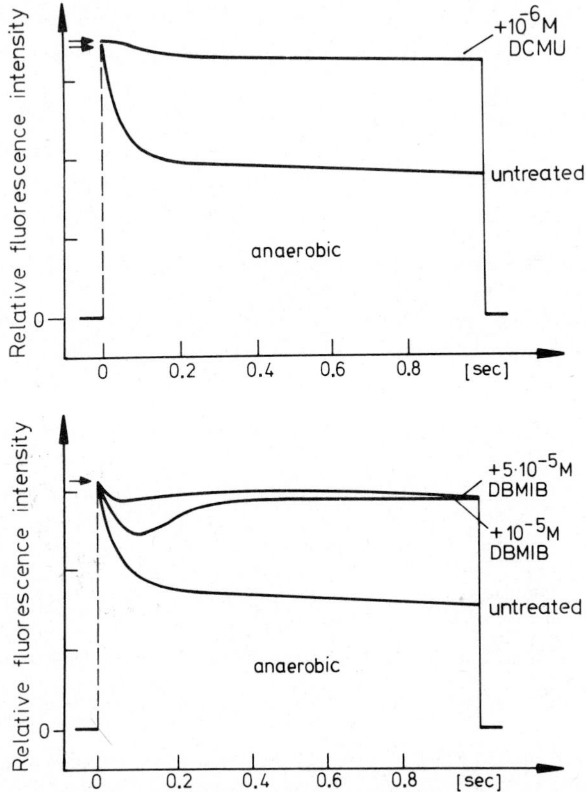

Fig.5. Scenedesmus obliquus; the effects of DCMU and DBMIB on the anaerobic J-D decline.

Fig.5 shows furthermore that the J-D decline is nearly totally diminished after the addition of appropriate amounts of DCMU or DBMIB - two inhibitors which block specifically the photosynthetic electron transport in the redox chain between the two light reactions[2,16].

On the other hand the J-D decline remains relative unaffected after specific inhibition of the water-splitting reaction by mild heat treatment (5 min 45°C) or the addition of hydroxylamine (up to 10^{-3} M). All these findings clearly demonstrate that the anaerobic J-D decline is mediated exclusively by the action of light reaction I. We assume that the formation of isolated J-D declines may be explained as follows: Anaerobiosis causes in plant cells strong reductive conditions. In particular this is known from hydrogenase containing green algae such as Scenedesmus obliquus, Chlorella pyrenoidosa and Ankistrodesmus braunii[3]. In these algae even a weak H_2-production (from endogenous organic compounds[17]) occurs in the dark or in weak light if they are incubated for a distinct time in N_2 or noble gases. The evolved molecular hydrogen - activated by the enzyme hydrogenase- serves as a strong reducing agent. In consequence this hydrogen is able to reduce all redox systems which are accessible in the chain of coupled redox systems between the two light reactions[18], in particular the primary and secondary electron acceptors of PS II, Q and plastoquinone, which are assumed to be located at the outer side of the thylakoid membrane[19].

Electron saturated PS II acceptor pools however result in the closure of PS II reaction centers. This causes that an electron transport immediately after the onset of illumination is not possible. The absorbed light energy is dissipated by the bulk of accessory chlorophyll-a molecules mainly as fluorescence which therefore has its maximum value at the onset of illumination. A renewed electron transport can not occur until defect electrons from PS I reactions are available. They are available as soon as the electron consuming NADP-reduction by light reaction I starts. Then the defect electrons are transported along the chain of coupled redox systems via the Q-cycle and cause its successive oxidation. Simultaneously with this oxidation of Q the fluorescence decreases rapidly until it reaches the normal stationary level.

The assumption that the anaerobic J-D decline reflects the oxidation of the quencher Q by PS I-produced defect electrons is once more manifested by experiments with DAD+ ascorbate. DAD + ascorbate at appropriate concentrations is a well known artificial redox system which can feed very effectively and specifically electrons into the electron transport chain between the two light reactions near PS I[20]. This system therefore should be able to 'capture' the PS I-produced defect electrons to an great deal before they can enter the redox cycles of the primary PS II electron acceptors.

Fig. 6 shows the anaerobic J-D decline after the addition of DAD + ascorbate. There is clear evidence that the J-D decline is strongly suppressed in the presence of DAD + ascorbate. This favours our assumption concerning the nature of the J-D decline.

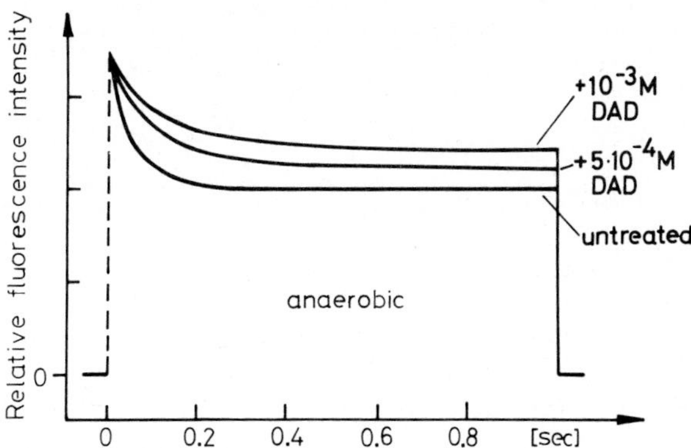

Fig.6. Scenedesmus obliquus; the effect of DAD on the anaerobic J-D decline. (DAD was reduced with a 3-fold excess of Na-ascorbate).

e) Concluding remarks

After the present studies we are inclined to believe that the anaerobic J-D decline in chlorophyll fluorescence ought to be used as a tool for monitoring the functioning and effectiveness of the photosynthetic electron transport between the two light reactions. For instance the action of various inhibitors as well as all physical and chemical influences which affect the electron transport between the two light reactions should affect the anaerobic J-D decline in a specific mode.

5. References

1. Kautsky, H. Appel, W. and Amann, H. , 1960, Biochem.Z. 332, 277.
2. Duysens, L. N.M. and Sweers, H. E,, 1963, in : Studies on Microalgae and Photosynthetic Bacteria, Jap.Soc. Plant.Physiol., Tokyo, p.177

3. Gaffron, H, 1940, Amer. J. Bot. 27, 273 .

4. Franck, U. F. Hoffmann, N. Arenz, H. and Schreiber, U., 1969 Ber.Bunsenges. Phys. Chem. 73, 871.

5. Kessler, E. Arthur, W. and Brugger, J. E., 1957, Arch.Biochem. Biophys. 71, 326 .

6. Lavorel, J. , 1959, Plant Physiol. 34, 204.

7. Förster, Th., 1948, Ann.Physik 2, 55 .

8. Lavorel, J. , 1962, Biochim. Biophys. Acta 60, 510.

9. Delosme, R., 1967, Biochim. Biophys. Acta 143, 108 .

10. Joliot, P., 1965, Biochim. Biophys. Acta 102, 116.

11. Schreiber, U. Bauer, R. and Franck, U. F. 1971, in : Proceedings of the IInd International Congress on Photosynthesis Research, Stresa, Vol.I, p.169

12. Aach, H. G. Franck, U. F. and Bauer, R., 1971, Die Naturwissenschaften 58, 525.

13. Kautsky, H. and Franck, U.F., 1943, Biochem.Z. 315, p. 139,156, 176, 209.

14. Mohanty, P. Papageorgiou, G. and Govindjee, 1971, Photochem. Photobiol. 14, 667 .

15. Schreiber, U. Bauer, R. and Franck, U.F., 1971, Z. Naturforsch. 26b, 1195 .

16. Trebst, A. and Harth, E. , 1970, Z. Naturforsch. 25b, 1157

17. Stuart, T. S. and Gaffron, H. , 1971, Planta(Berl.) 100, 228.

18. Kessler, E. , 1968, Planta(Berl.) 81, 264 .

19. Witt, H.T. Rumberg, B. and Junge, W. , 1968, in : Biochemie des Sauerstoffs, B.Hess and H. J. Staudinger eds. (Springer Berlin-Heidelberg-New York) p. 262 .

20. Trebst, A. and Pistorius, E. , 1965, Z. Naturforsch. 20b, 143.

M. AVRON, *Proceedings on the Third International Congress on Photosynthesis*
September 2-6, 1974, The Weizmann Institute of Science, Rehovot, Israel
Elsevier Scientific Publishing Company, Amsterdam, The Netherlands, 1974

COMPARISON OF EXCITONS TRANSFERS CHANGES INDUCED BY CATIONS AND BY ADAPTATION IN STATES I AND II

C. VERNOTTE*, J.M. BRIANTAIS* AND C.J. ARNTZEN**

*Laboratoire de Photosynthèse, C.N.R.S., 91190 Gif-sur-Yvette, France

**Department of Botany, University of Illinois, Urbana, Illinois, USA

Abstract

The excitons transfer from PS II to PS I is higher in State II than in State I adapted Chlorella cells, pea leaves and isolated intact spinach chloroplasts.These variations are identical to those observed in broken chloroplasts resuspended respectively in low and high concentration in cations.The two light adapted states are fixed by glutaraldehyde suggesting that they involve a structural change. Inhibition of State II adaptation by DCMU,NH_4Cl and valinomycin, its restoration by post addition of DCCD point out a correlation between State II and the energized membrane.

1. Introduction

There are now many data which indicate that an increase of cations concentration (up to 100 mM for monovalent, 10 mM for divalent) induces a decrease of the exciton transfer from Photosystem II to Photosystem I in a chloroplast suspension which is initially in presence of a concentration in monovalent cations of 10 mM[1-7].

Murata[8,9] observed that illumination of algae with System II light determines an increase of long wavelength maximum accompanied by a decrease of short wavelength maximum of the fluorescence emission spectrum. He suggested that in State II[10] adapted algae the intersystem transfer is larger than in dark adapted cells.

Here is reported a detailed comparison of salt effects previously published[6,7] to light adaptation, measuring in addition to emission spectra, amplitude and kinetics of the fluorescence induction and the sensitization of Photosystem I activity in States I and II. The results obtained agree with Murata's conclusion *i.e.*: the amplitude of the intersystem transfer in Chlorella cells, pea leaves and intact isolated spinach chloroplasts adapted to State II is close to the value reached in washed isolated chloroplast incubated in low cation medium. State I adaptation determines the same effect than an increase of cation concentration in the chloroplast suspension.

Transfer changes, induced varying the cation concentration,involve structural changes because it can be fixed by glutaraldehyde[11]. We tried succesfully the same fixation of State I and State II.

Krause[12] showed that the enhancement of Photosystem II fluorescence by Mg 5 mM addition is inhibited by illumination and that this inhibition is canceled by DCMU or FCCP. Then we checked the effect of NH_4Cl, valinomycin and subsequent addition of DCCD on State II adaptation. The data obtained point out that energized membrane and state II are closely correlated.

Hind *et al.*[13] measured a Mg^{++} and K^{+} exit corresponding to the photoinduced proton uptake in chloroplasts. Then we determined the pH dependancy of States I and II adaptations.

2. Materials and Methods

Chlorella pyrenoidosa cells are grown according to Delrieu and De Kouchkovsky[14]. The concentration of chlorophylls in the final suspension used is 50 µg/ml.

Chloroplasts are isolated as previously described[15] from pea, lettuce and spinach leaves. The grinding medium contains : Sorbitol 0.4 M and is buffered at pH 7.8 by Tricine 0.1 M. When required, chloroplasts are washed once in NaCl 0.01M. In some preparations Glutaraldehyde 0.5 % final concentration, is added to the grinding solution.

Preilluminations of algae, leaves and chloroplasts are performed using 647 nm $\pm$ 6.2 nm interference filter or a set of 2 filters (RG 8 cut off plus K 7 interference filter), a 710 nm $\pm$ 17 nm band is then obtained. The intensities used are respectively 2500 and 55000 ergs. cm^{-2}. s^{-1}. The suspensions, containing 50 µg chlorophylls /ml are 3 mm thick. In the case of leaves, one layer of them is floatting on a water bath. When it is indicated the suspension is preilluminated directly in the fluorimeter vessel, using the exciting monochromator with a half-bandwidth of 20 nm the thickness of the suspension then is 1 mm. The intensities received at 480, 650 and 710 nm are respectively : 50 000, 32 000, 25 000 ergs. cm^{-2}. s^{-1}. Emission spectra at liquid nitrogen temperature are determined using a Dewar flask in the bottom of which 1ml of the suspension is adsorbed on 2 layers of cheese-cloth as described by Cho and Govindjee[16]. The half bandwidth of the analytic monochromator is 1.6 nm.

Fluorescence inductions are recorded with an oscilloscope.

Photosystem I activity is determined by the rate of the oxygen uptake in limiting light of chloroplast suspensions containing 25 µg of chlorophylls/ml, DCMU 10^{-5}M, azide 10^{-4}M, Sodium Ascorbate 10^{-3}M, DPIP 10^{-4}M and methylviologen 2.5 10^{-4}M. Oxygen exchanges are detected using a Clark electrode. A 647 nm actinic light is obtained with an interference filter which half bandwidth is 12.4 nm, 710 light is obtained from K7 + RG8 filters, half bandwidth 35 nm.

3. Results

Emission spectra at 77°K of dark adapted and preilluminated materials

Figure 1 shows emission spectra of Chlorella cells preilluminated during 10 mn by 647 or 710 nm lights. The spectrum of dark adapted algae is intermediate between the two spectra presented. A decrease of 685 and 695 bands, an increase of 720 peak are induced by 647 nm preillumination. Opposite changes are determined by 710 nm preillumination.

Figure 2 points out the same qualitative phenomena in chloroplasts isolated from preilluminated pea leaves in presence of glutaraldehyde in the grinding medium.

Noticed that glutaraldehyde fixation yields a relative decrease of 730 nm band of all samples.

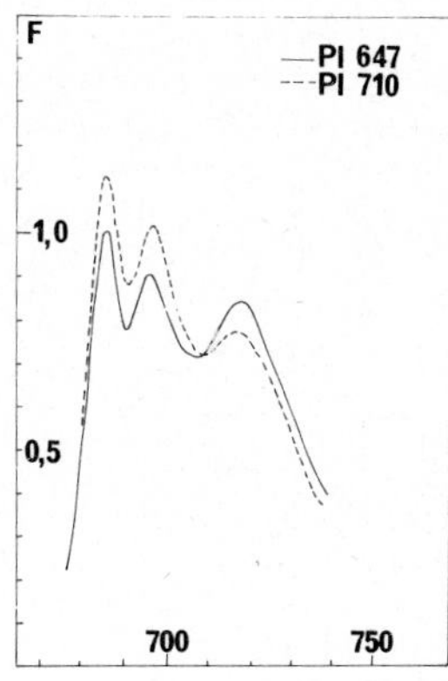

Fig. 1. Fluorescence emission spectra at 77°K of Chlorella cells preilluminated by 647 or 710 nm light during 10 mn. Excitation : 480 ± 10 nm. Analytic half bandwidth : 1.6 nm.

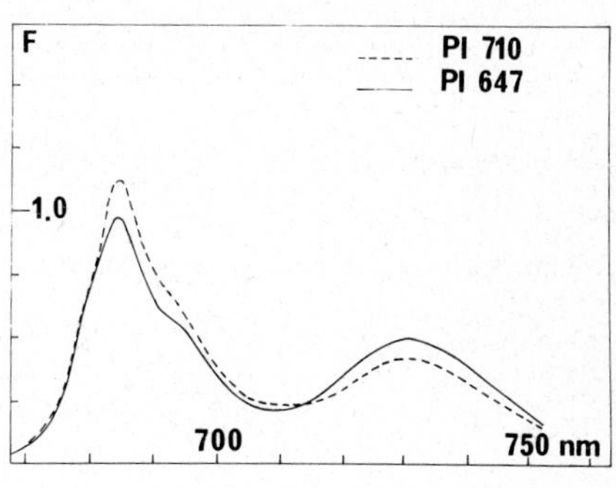

Fig. 2. Fluorescence emission spectra at 77°K of glutaraldehyde fixed chloroplasts isolated from preilluminated pea leaves and resuspended in a Tricine 100 mM, NaCl 10 mM, Sorbitol 400 mM, pH 7.8. Other conditions as for Figure 1.

As Krause[12] we observed that 650 nm light induces a slow decrease of 685nm fluorescence in intact isolated spinach chloroplasts. This decrease is accompanied by an increase of the 730/685 nm ratio of emission at 77°K (darkness : 0.72; 10 mn 647 light : 1.05).

Qualitatively those changes are the same as those induced by Mg^{++} on isolated chloroplasts[1].

Fluorescence induction at room temperature of dark adapted and preilluminated material.

In Figure 3a fluorescence inductions of dark adapted and preilluminated Chlorella cells are represented. DCMU 10^{-5}M is added just before induction measurements. It appears, as in cations effect, that it is mainly the variable part of the fluorescence which is affected by the preillumination. Its amplitude is decreased by 647 nm preillumination, increased by 710 nm light.

In Figure 3b, is plotted the amplitude of the variable yield of fluorescence versus oxidized Q; this last quantity is estimated by the complementary area according to Malkin[17]. This figure points out a less sigmoidal kinetics in State II than in State I adapted algae.

Figure 3. a) Induction of fluorescence at room temperature of Chlorella cells dark adapted or preilluminated (10 mn). DCMU 10^{-5}M is added just before the measurement. The algae are resuspended in Phosphate buffer 30 mM, KCl 100 mM, pH 6.5. Exciting beam 480 nm ± 10 nm. Analysis of fluorescence through a Corning filter 2-64 plus a Wratten filter 70. Oscilloscope recording.

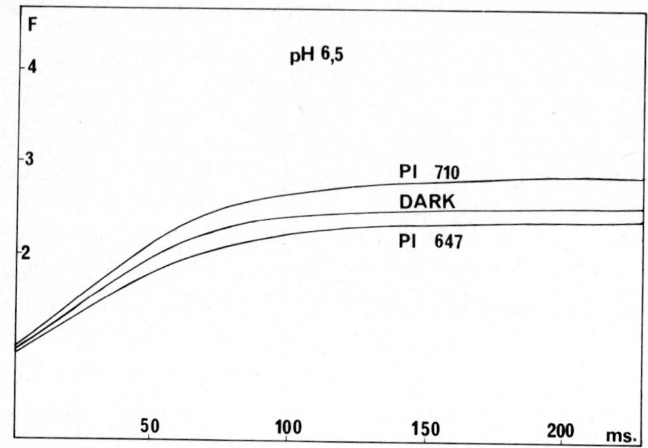

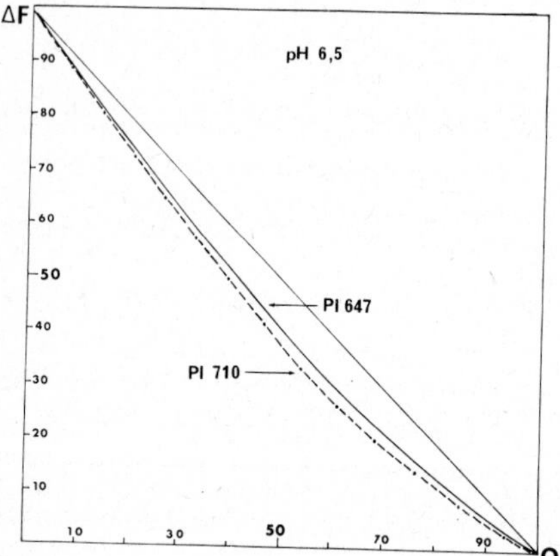

b) Variable yield of fluorescence as a function of Q oxidized plot deduced from Fig. 3a. Oxidized Q is measured by the complementary area of the induction according to Malkin[17].

In Table I, the relative amplitudes of variable fluorescence of glutaraldehyde fixed chloroplasts, isolated from preilluminated pea leaves are presented. The same type of variations than in Chlorella cells appears.

Furthermore, intact chloroplast isolated from dark adapted pea leaves show also a less sigmoidal kinetics when they have been subsequently preilluminated by 647 nm light.

TABLE I

Amplitude of variable fluorescence in glutaraldehyde fixed chloroplasts isolated from dark adapted and preilluminated pea leaves.

	Dark	PI 647	PI 710
Δ F/Fo	2.32	2.06	2.58

Effect of DCMU

Table II shows that, if DCMU is present during the preillumination of the algae, both 647 and 710 nm preilluminations produce an increase of variable fluorescence and a decrease of the 720/685 nm ratio of the liquid nitrogen emission spectra. Then in presence of DCMU no State II adaptation occurs whereas State I is still reached. Notice that DCMU added before cooling produces some changes.

TABLE II

Effect of DCMU 10^{-5}M on changes of 77°K emission spectra and variable fluorescence amplitude at room temperature induced by preillumination of Chlorella cells.

		Dark	PI 647	PI 710
$\frac{F\ 720}{F\ 685}$	0 DCMU	1.04	1.22	0.99
	+ DCMU 10^{-5}M	1.19	1.02	1.04
$F_{M_{685}}$ %	0 DCMU	100	82	110
	+ DCMU	100	106	110

The slow decrease of 685 nm fluorescence induced by 650 nm light in intact isolated chloroplasts is also inhibited by DCMU.

Besides, the increase of 685 nm emission, induced by $MgCl_2$ addition (5 mM) in washed chloroplasts, does not occur if the suspension is illuminated by a strong 650 nm light. This inhibition by light is suppressed by DCMU.

Electron chain and light adaptation in Chlorella.

Figure 4a represents the slow induction of dark adapted algae at room temperature, using a 480 nm exciting beam. The emission spectra at 77°K is determined at several places of the induction. The spectra are recorded when the steady state of fluorescence is reached, so when all Q is reduced[18]. Curve c shows that no change of the emission spectrum occurs before 15 seconds of light which intensity corresponds to 100 photons per second and per Photosystem II center.

In Figure 4d, changes of 77°K emission spectra are plotted versus the time of darkness which follows a 15 seconds 480 nm illumination.

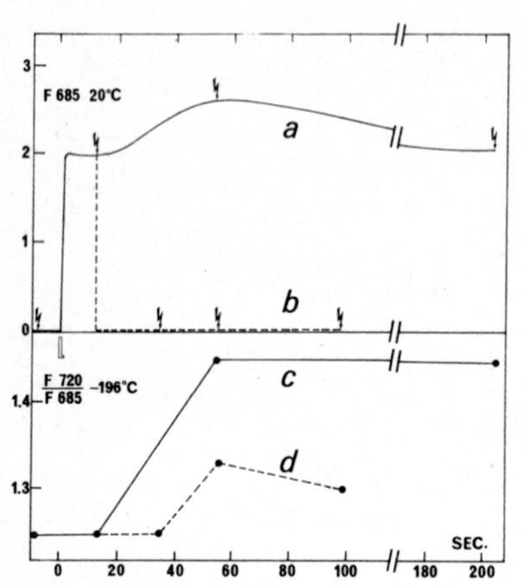

Figure 4a and b) Room temperature induction of fluorescence of dark adapted Chlorella cells using a 480 ± 10 nm exciting light which intensity corresponds to 100 photons per sec and per Photosystem II center. Emission at 685 nm is detected through a monochromator, the half bandwidth of which is 1.6 nm. 1 ml of the suspension, containing 50 μg of chlorophylls is adsorbed on two layers of 10 cm^2 cheese cloth discs placed on the bottom of a Dewar flask. c) Plot of 720/685 emissions at 77°K determined at the various places of the induction corresponding to the arrows of curve a. Liquid nitrogen is added at these times and the spectra is run when 77°K fluorescence induction is achieved. d) Value of 720/685 emissions at 77°K as a function of the time of darkness which follows a 15 sec 480 nm illumination, corresponding to arrows of curve b.

Sensitization of Photosystem I activity by 647 nm light.

As done previously for $MgCl_2$ effect[6,7] the relative importance of Photosystem I activity sensitization by 647 nm light has been determined in chloroplasts in State I or II. Table III shows the variations obtained in glutaraldehyde fixed chloroplasts isolated from preilluminated or dark adapted pea leaves. The preillumination by 647 nm induces a relative increase of the efficiency of 647 nm light sensitization of Photosystem I. In contrast 710 preillumination decreases it. Chloroplasts from dark adapted leaves can be in State I or II depending upon some uncontroled factor.

TABLE III

Variations of 647 nm sensitization of Photosystem I activity in glutaraldehyde fixed chloroplast isolated from dark adapted or preilluminated pea leaves.

	Dark	PI 647	PI 710
$\frac{647}{710}$	0.65	0.65	0.41
$\frac{647}{710}$	0.41	0.57	0.41

Effect of uncouplers

Krause[12] observed that CCCP suppress the inhibition by light of the stimulation of fluorescence obtained by $MgCl_2$ addition. Here the effect of NH_4Cl and Valinomycin have been checked on State II adaptation *i.e.* on the decrease of 685 nm fluorescence which occurs in intact chloroplasts isolated from spinach leaves when they are illuminated with 650 nm light. As depicted in Figure 5 the two uncouplers suppress the adaptation but the addition of ADP + P_i does not. Dicyclohexylcarbodiimide (DCCD) reverses partially the inhibition produced by uncouplers. We observed that DCMU is not efficient to block fluorescence decrease when DCCD is added to chloroplasts.

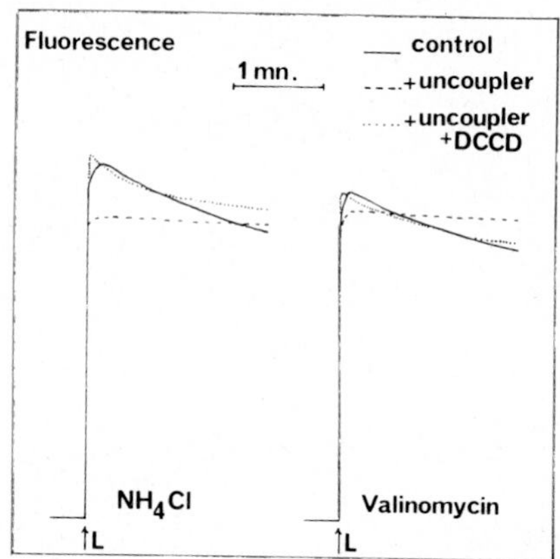

Figure 5. Effects of uncouplers and subsequent addition of an energy transfer inhibitor (Dicyclohexylcarbodiimide) on State II adaptation of intact chloroplasts. Room temperature fluorescence emission is detected at 685 ± 1.6 nm. The exciting beam is a 650 nm ± 10 light which intensity is 32 000 ergs cm^{-2} s^{-1}. Chloroplasts are isolated from spinach leaves and resuspended in the grinding medium (Tricine 0.1 M, Sorbitol 0.4 M, pH 7.8). The chlorophylls concentration is 50 μg/ml. NH_4Cl : 10^{-2}M, Valinomycin : 10^{-6}M, Dicyclohexylcarbodiimide : 10^{-3} M.

Effect of the external pH

In Figure 6 are represented the variations of 720 over 685 nm emissions at 77°K versus the external pH in dark adapted and preilluminated Chlorella cells. This ratio in dark adapted cells is maximum at pH 7. At pH 6 and 7, 647 and 710 nm preilluminations have antagonistic effects, whereas at pH 8 both lights determined a state II adaptation. If DCMU is present during the preillumination at pH 8 the ratio of dark adapted algae is maintained.

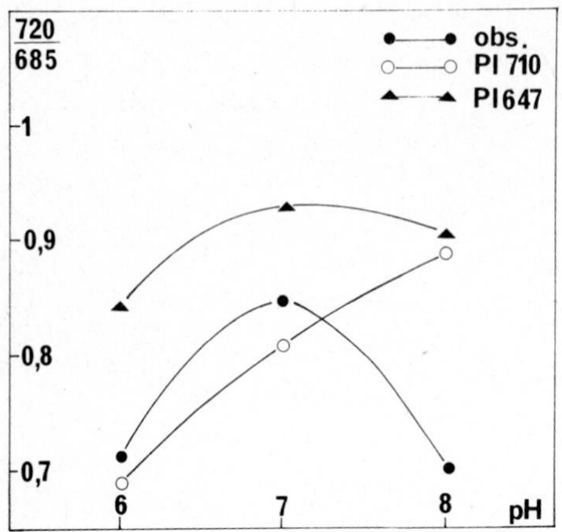

Figure 6. External pH dependency of 720/685 nm fluorescence emissions at 77°K in dark adapted or preilluminated Chlorella cells. Chlorophyll contain, excitation and analysis as in Figure 1.

In Figure 7 the effect of $MgCl_2$ addition (10 mM) on the fluorescence amplitude of washed (by NaCl 10 mM) isolated lettuce chloroplasts is plotted versus the pH of the resuspended medium. A smooth maximum of $MgCl_2$ effect is seen near pH 7.5 .

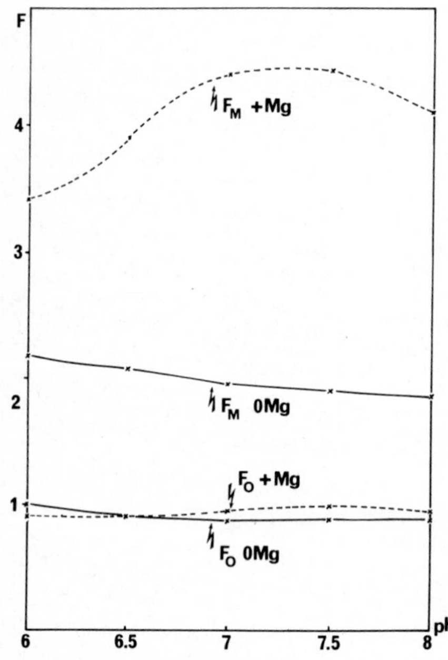

Figure 7. pH depency of $MgCl_2$(10 mM) effect on fluorescence levels (Fo : dead fluorescence, F_M: maximum fluorescence yield) of hypotonic (NaCl 10 mM) washed chloroplasts isolated from dark adapted lettuce leaves. DCMU 10^{-5}M is added just before the measurement. Incubation of chloroplasts in the darkness in presence of $MgCl_2$ is done during 10 mn. Chlorophylls concentration = 50 µg/ml. Excitation and analysis of fluorescence as in Figure 3.

Dependency of $MgCl_2$ effect in chloroplasts upon isotonicity of the washing medium

Table IV points out that isotonicaly washed chloroplasts present lower changes upon $MgCl_2$ addition than chloroplasts washed by NaCl 10 mM. Furthermore, the former preparation is characterized by a high variable fluorescence, a low sensitization of Photosystem I by 647 nm, and a decrease of 685 nm emission upon 650 nm strong illumination,as in NaCl 10 mM washed chloroplasts with $MgCl_2$ added .

TABLE IV

Characteristics of isolated pea chloroplasts depending upon the isotonicity of the washing medium. Limiting actinic lights for System I activity (R) are obtained from 647 nm interference filter and from a RG 8 cut off colored filter.

Chloroplasts washed by :	Tricine 0.1 M Sorbitol 0.4 M pH 7.8	NaCl 0.01 M
ΔF/Fo	1.42	1.07
$\frac{(\Delta F/Fo) + Mg}{(\Delta F/Fo)\ 0\ Mg}$	1.62	1.81
$\frac{R\ 647}{R\ 695}$	2.99	3.18

4. Discussion

The changes observed in light adapted algae and chloroplasts at pH 6 to 7, for emission spectra, induction of fluorescence, and 647 nm sensitization of Photosystem I activity, point out that State I is equivalent to isolated washed chloroplasts resuspended in a medium containing Mg^{++} and State II to chloroplasts in the absence of this cation.

In both magnesium or preillumination treatments, the adaptations involve structural changes since each state can be fixed by glutaraldehyde. We suggest that these structural changes determine variations of the distance between the two photosystems.

Summarizing : in State I the photosynthetic apparatus is characterized by a low Photosystem II → Photosystem I transfer as in chloroplasts incubated with $MgCl_2$. In State II the transfer amplitude is closed but smaller than in chloroplasts without $MgCl_2$. Then one can propose that State II adaptation corresponds to an exit of

Mg^{++} from the thylakoid to the stroma[19]. The main argument is given by Krause and our experiments : 647 nm illumination inhibits the fixation of Mg^{++} by washed chloroplasts. State I adaptation would corresponds to the functioning of the thylakoid as a Mg^{++} pump. According to this point of view, two questions arise 1 - How Mg^{++} can induce structural changes. 2- What process (es) monitor(s) Mg^{++} exchanges. To the first question no clear answer is given by our data, we can just say that both light adaptations and Mg addition effects are slow processes(Figure 4). We can also recall that Mg^{++} can be bound to several enzymes involved in Photosynthesis.

Concerning the second question, a larger discussion can be done on the basis of our results.

Considering the antagonistic effect of lights I and II in Chlorella, in the pH range from 6 to 7, in the one hand, taking care of the effect of pH on emission spectra of dark adapted algae in the other hand, it appears that light II effect is equivalent to an increase of the external pH and light I to a decrease (Figure 6). So let us propose the following simple hypothesis : the magnesium exit is linked to a proton absorption by the photosynthetic membrane as shown by Hind and coll.[13] and this proton absorption will correspond to the reduction of an electron carrier localized between the two photosystems. One candidate for a protonation monitored by antagonistic functioning of the two systems would be plastoquinones pool. Two main data agree with this scheme : State II but not State I adaptation is inhibited by DCMU and true ΔpH of Chlorella cells is suppressed by this inhibitor[20]

Antagonism between H^{+} and Mg^{++} is also suggested by pH dependency of the amplitude of Magnesium effect in washed chloroplasts.

This simple explanation cannot be used at pH 8 where the 2 lights produce a State II adaptation (no change compare to dark adapted would be at least expected in light I if the antagonistic scheme was valid) and where dark adapted Chlorella has the same emission spectra than State I adapted algae at pH 6. At pH 8 in Chlorella cells, it is necessary to look for another regulation or to revise the previous hypothesis suggested. For this purpose we would discuss separately the 2 light adaptations.

-State I adaptation

We observed that State I adaptation persists in presence of DCMU, Bedell[21] showed that PS I cyclic functioning occurs *in vivo* , so we can imagine that Magnesium fixation occurs through Photosystem I cyclic activity, instead to be linked to the ox-red state of plastoquinones pool. This cyclic Photosystem I would be inhibited at pH 8.

-State II adaptation

Besides the argument previously discussed : *i.e.* State II adaptation is prevented by DCMU, Mg^{++} fixation in washed chloroplasts is inhibited by light II and restored by DCMU or NH_4Cl; we checked in more detail State II adaptation using uncouplers . The DCMU effect previously discussed shows mainly that Photosystem II primary act is not sufficient to induce State II adaptation. So we try to investigate the role of later steps, investigating the effects of uncoupler on intact isolated chloroplasts. The inhibition of State II adaptation by Valinomycin as well as NH_4Cl and the restoration obtained by subsequent addition of the energy transfer inhibitor DCCD suggest that the energized membrane is unable to fixe Magnesium.

Experiments are in progress in order to elucidate the mechanisms of State I adaptation, structural modifications involved in transfer changes, with a peculiar attention to establish relationship between transfer changes and stacking process, and to know the specific site of Mg^{++} fixation in the desenergized membrane.

References

1. Murata, N., 1969, Biochim. Biophys.Acta 189, 171-181.
2. Murata, N., Tashiro, H. and Takamiya, H. 1970, Biochim. Biophys.Acta 197,250-256
3. Murata, N., 1971, Biochim. Biophys.Acta 245, 365-372.
4. Homann, P.H., 1969, Plant Physiol. 44, 932-936.
5. Gross, E.L. and Hess, S.C., 1973, Arch. Biochem. Biophys. 159, 832-836.
6. Briantais, J.M., Vernotte, C. and Moya, I.,1973, Biochim. Biophys.Acta 325, 530-538.
7. Vernotte, C., Briantais, J.M. et Bennoun, P.,1973, C.R. Acad. Sc. Paris 277, 1695-1698.
8. Murata, N., 1969, Biochim. Biophys. Acta 172, 242-251.
9. Murata, N., 1970, Biochim. Biophys. Acta 205, 379-389.
10. Bonaventura, C. and Myers, J., 1969, Biochim. Biophys. Acta 189, 366-383.
11. Mohanty, P., Braun, B.Z. and Govindjee, 1973, Biochim. Biophys.Acta 292,459-476.
12. Krause, G.H., 1974, Biochim. Biophys.Acta 333, 301-313.
13. Hind, G. Nakatani, H.Y. and Izawa, S., 1974, Proc. Natl Acad.Sci.71, 1484-1488.
14. Delrieu, M.J. and Kouchkovsky,Y., 1971, Biochim. Biophys.Acta 226, 409-421.
15. Arntzen, C.J., Armond, P.A., Zettinger, C.S. Vernotte, C. and Briantais, J.M., 1974 in press.
16. Cho, F. and Govindjee, 1970, Biochim. Biophys.Acta 205, 371-378.
17. Malkin, S., 1967, Biophys.J. 7, 629-649.
18. Butler, W.L., Visser, J.W.M. and Simons, H.J., 1973, Biochim. Biophys.Acta 293, 140-151.
19. Lin, D.C. and Nobel, P.S., 1971, Arch. Biochem. Biophys. 145, 622-632.
20. de Kouchkovsky, Y., Personnal Comm.
21. Bedell, G.W. and Govindjee,1973, Plant Cell Physiol. 14, 1081-1097.

M. AVRON, *Proceedings of the Third International Congress on Photosynthesis*,
September 2-6, 1974, Weizmann Institute of Science, Rehovot, Israel
Elsevier Scientific Publishing Company, Amsterdam, The Netherlands, 1974

COOPERATIVE PHENOMENA IN PHOTOSYNTHESIS

L. Tumerman
Chemical Physics Department
Weizmann Institute of Science
Rehovot, Israel.

1. Introduction

The question we were looking for an answer to was:

Are the electron transfer chains in chloroplasts independent from each other, or do they interact in a cooperative manner?

In the independent model each chain is supposed to have its own pool of intermediate electron carriers, so that electrons can be transferred from a given center of system II only to a definite center of system I, namely to the center which is connected to the donor center by this chain. On the contrary, in the cooperative model it is supposed that there exists a common pool of intermediate carriers for a number of chains. This pool obtains electrons from several centers of system II and dispense them to several centers of system I.

The cooperative model seems to be more plausible, because in the independent model a chain can function efficiently only if both centers it connects get excited simultaneously or in a very short interval of time. Such a coincidence is obviously a rare event, especially in the case of weak illumination. This restriction is many times less severe for the cooperative model.

The first experimental evidence in favor of the cooperative model was given by Witt, Rumberg and Junge[1], who have shown that at least ten electron transfer chains are combined by a strand of plastoquinones. This evidence was based on data bearing on the action of gramicidin D upon the time variation of absorption changes in chloroplasts.

Our way to solve the problem was different. We have studied the specific changes of chlorophyll fluorescence parameters, which occur during the "aging" of chloroplasts isolated from cells, and run in parallel with the loss of chloroplast capacity for the non-cyclic electron transport and the concomitant phosphorylation.

2. Experimental

The following five quantities were measured:

a) Duration (τ) and yield (ϕ) of fluorescence emitted by the chloroplasts. These quantities were measured for chloroplasts adapted to darkness ("dark" values τ_d and ϕ_d) as well as for chloroplasts in the steady state under illumination ("light" values τ_ℓ and ϕ_ℓ).

b) Duration of the "induction time" $t_{ind.}$, i.e. of the time required by the chloroplast to be transferred from the state of dark adaptation into the steady state under illumination.

All these quantities were measured as functions of the time of aging, i.e. of the time of storage after the chloroplasts had been isolated from cells.

Measurements were carried out on suspensions of chloroplasts isolated by the method of Wahtley and Arnon[2] from two week old pea seeds in the medium containing 0.35M NaCl and tris-buffer at pH 8.0. In some cases sodium ascorbate was added to the medium in concentration of 0.01M.

The instrument (a phase-shift fluorometer) and the "stop-flow" method used in these measurements were described previously[3]. Measurements were performed in the Institute of Molecular Biology in Moscow together with E. Sorokin.

3. Results and Conclusions

The measured values of τ_ℓ, τ_d and $t_{ind.}$ are plotted in Fig. 1 against the time of aging. Points for ϕ_ℓ and ϕ_d lie on curves similar to curves 1 and 2 of Fig. 1. In these experiments chloroplasts were stored at room temperature and no ascorbate was added to the suspension. They lost their activity in 8 to 10 hours. By lowering the temperature of storage and/or by adding sodium ascorbate the process of aging could be prolonged up to two weeks, but the pattern of fluorescence changes was the same. This proves that the observed changes in fluorescence are really due to the gradual loss of the activity of system II centers.

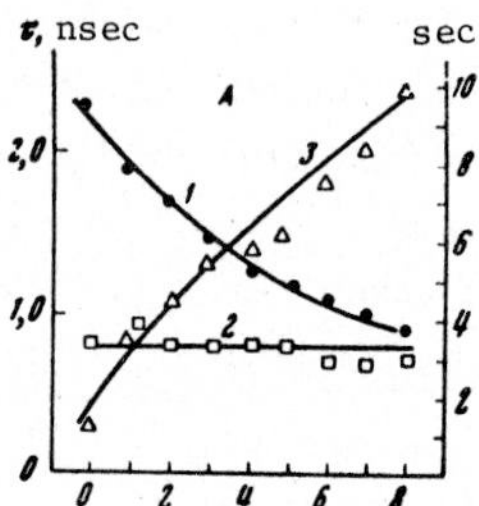

Fig. 1. Dependence of τ_ℓ (curve 1), τ_d(curve 2) and $t_{ind.}$ (curve 3), right ordinate) upon the time of aging (hours).

Three main facts were established by these measurements, and the following conclusions were drawn from them:

I. In freshly isolated chloroplasts the "light" values of fluorescence duration and yield (τ_ℓ and ϕ_ℓ) are 3 to 4 times higher then the "dark" ones (τ_d and ϕ_d), but during the aging the values τ_ℓ and ϕ_ℓ decrease gradually and approach τ_d and

ϕ_d respectively. The ratio τ_ℓ/ϕ_ℓ remains, however, constant during the whole process of chloroplast aging.

In our former paper[3] the proportionality was also shown to exist between the transient values of τ and ϕ during the induction time, and two important conclusions were drawn from this fact, namely that:

a) The "Photosynthetic Unit" contains several reaction centers which compete for the common pool of excitations. Excitation can be transferred in principle from anyone of the "photon-collecting" pigment molecules to each center although of course, transfer will probably take place to the nearest center.

b) Fluorescence is emitted by the bulk of "photon-collecting" molecules during the excitation migration, i.e. before the excitation is trapped in a center.

By applying the same resoning it is easy to show that the now established proportionality between the values of τ_ℓ and ϕ_ℓ gives an independent proof of the validity of both these conclusions[4].

Ergo: Duration and yield of chlorphyll fluorescence in vivo are determined only by the number of trapping centers in the Photosynthetic Unit.

II. The second experimental fact is that the "dark" values of τ and ϕ remain constant during the aging; this means that the total number of trapping centers in dark adapted chloroplasts remains the same during the aging. In other words, the centers which have lost their activity are converted into the "fluorescence quenching" ones, but continue to act as trapping centers.

Ergo: The number of active electron transfer chains decreases during the aging, but the rate of photon uptake into the centers that retain their activity and hence the rate of electron uptake in corresponding chains remain constant.

III. The third experimental fact is that the induction time $t_{ind.}$ increases during the aging. It can be seen from Fig. 1 that under our conditions $t_{ind.}$ is approximately 0.5sec. for freshly isolated chloroplasts and ≈10sec. for the aged ones.

In the independent model $t_{ind.}$ is the time required for filling the separate pool of intermediares in the chain. This time is determined only by the rate of electron uptake into the active chains and cannot depend upon the number of active chains. In the cooperative model $t_{ind.}$ is the time required to fill the common pool of intermediates for several chains. Therefore, at constant rate of electron uptake $t_{ind.}$ must increase when the number of active chains decreases.

Ergo: There exist a common pool of intermediate carriers for a number of chains, or in other words the cooperative model is the true one.

4. Discussion

Cooperative interactions were found to exist between the pigment molecules in the "Photosynthetic Unit" as well as between centers of photochemical systems

I and II. Now it was established that such cooperative interactions do exist also between the electron-transfer chains, which connect these centers.

Witt a.o.[1] have shown that the function unit of photosynthesis is one thylakoid. This unit consists of several hundreds of electron-transfer chains in the membrane of one thylakoid. It seems naturally to predict that the cooperativity will be found also on higher levels, such as interactions between thylakoids in a chloroplast and interactions between chloroplasts, mitochondria and other sub-cellular functional units.

References

1. Witt, H.T., Rumberg, B., and Junge, W. (1968) 19. Colloquium der Gesellschaft fuer Biologische Chemie (Springer-Verlag, Berlin).

2. Whatley, F.R., and Arnon, D.I., in "Methods in Enzymology", Vol. 6 (Academic Press, New York-London) (1963).

3. Tumerman, L., and Sorokin, E. (1967). Molekulyarnaya Biologiya (Russ.) 1, 628 (Engl. translation: Molecular Biology 1, 527).

4. Sorokin, E., and Tumerman, L. (1971). Molekulyarnaya Biologiya (Russ.) 5, 753 (Engl. translation: Molecular Biology 5, 603).

M. AVRON, *Proceedings of the Third International Congress on Photosynthesis*, September 2-6, 1974, Weizmann Institute of Science, Rehovot, Israel
Elsevier Scientific Publishing Company, Amsterdam, The Netherlands, 1974

ON THE COOPERATION BETWEEN PHOTOSYNTHETIC UNITS OF PHOTOSYSTEM II

S. Malkin

Department of Biochemistry

Weizmann Institute of Science, Rehovot, Israel

1. Summary

From studies on the flash yield vs. flash intensity of DCPIP reduction by isolated chloroplasts we conclude:

1. The photosynthetic unit is a physical concept.
2. In some instances there is no or little interaction between photosynthetic units.
3. In some instances there are large interactions which lead to exchange of excitation between different units.

2. Introduction

Many models have been suggested in the past for the type of cooperation of reaction centers with respect to the collection of excitation quanta. On the one extreme there is the independent units model, in which each reaction center has its own antenae pigments, and there is no or very little excitation exchange between the units[1-4]. On the other extreme there is the uniform matrix model, in which a reaction center can in principle "catch" excitations absorbed in any other part of the pigment system [1-4]. Truly, this last model is an idealization which can never be fully realized but approximated, since the life time of the exciton always limits the range of energy transfer. Tumerman, from fluorescence life time measurements, tends to believe the uniform pigment matrix model [2]. On the other hand fluorescence induction kinetics was analyzed by the independentunits model [1]. Other data are interpreted in terms of restricted energy transfer between photosynthetic units[5-8].

Our approach to this problem is quite classical: It is a quantitative and precise evaluation of the flash yield vs. intensity in a variation of Emerson Arnold photosynthetic unit determination [9-10]. In our case, we assayed DCPIP reduction in isolated chloroplasts which gives a direct information on Photosystem II.

A review of previous work and other experimental approaches to this problem has been documented in a previous report on the same subject [11]. The reader can consult this paper for more details.

3. Experimental

A series of μ-sec flashes were given to a sample of isolated chloroplasts with DCPIP which was placed in a dual-wavelength Aminco-Chance spectrophotometer. The flash was nearly monochromatic with wavelength at the window of absorption (560 nm; half band-width 20 nm). Chloroplasts and DCPIP concentrations were very low. This ensured a homogeneous distribution of intensity throughout the sample, which was further perfected by placing a mirror at the back of the cuvette. DCPIP reduction was monitored by absorbance changes at 600 nm relative to 730 nm.

It was checked that the distribution of intensity between different flashes of a series could give only negligible difference in the results compared to truly constant flashes [11].

The maximum flash intensity gave a photoreduction extent around 90–100% of saturation. To check for the saturation extent we replaced the monochromatic filter with a cut-off (>520 nm) filter which caused a true saturation of reduction, as checked by placing additional neutral density filters.

4. Theory

We shall discuss the results in terms of a simple theory for energy trapping by quenching centers in a uniform matrix pigment model. This theory was already outlined by Duysens,[12,13] and has the advantage over other theories by giving straightforward expressions for the flash yield and the quantum yield of the photoreactions.

We denote by $\bar{\rho}$ the average number of jumps between pigment molecules of an excitation during its life time.

The quantum yield of trapping, ϕ_{Trapping} is given then by [11]:

$$\phi_{\text{Trapping}} = \frac{\bar{\rho}T/N}{1+\bar{\rho}T/N}$$

ϕ_{Trapping} approaches 1 when $\bar{\rho} \to \infty$. T is the concentration (number) of trapping centers and N is the number of pigment molecules. The maximum value of ϕ_{Trapping} is achieved when all the centers are "open".

$$\max \phi_{\text{Trapping}} = \frac{\bar{\rho}T_o/N}{1+\bar{\rho}T_o/N} = \frac{1/\alpha}{1+1/\alpha}$$

To is the maximum concentration of reaction centers. $N/\bar{\rho}$ To can be condensed to a single parameter α. During the short flash the only reaction possible is the conversion of reaction centers from "open" form to "closed" form. If the initial concentration of traps is To and the final concentration is T_f, DCPIP reduction is equal to $T_o - T_f$. The differential equation for the conversion of traps is:

$$\frac{dT}{dn} = -\phi\, \varnothing_{Trapping}$$

where n is the number of quanta of the flash and ϕ is the quantum yield of the charge separation after trapping.

From this equation one finds for the flash yield Y the following results:

$$\frac{Y}{Y_{max}} - a \ln\left(1 - \frac{Y}{Y_{max}}\right) = \frac{\phi n}{T_o}$$

For two limiting cases this equation can be simplified:

(1) For $\alpha = o$ ($\bar{\rho} \rightarrow \infty$)

$$Y = \phi n \qquad \phi n \leqslant T_o$$

$$Y = To \qquad \phi n \geqslant T_o$$

This linear increase of Y and abrupt approach to saturation reflects the ideal uniform pigment matrix model. In this case $\varnothing_{Trapping}$ must be equal to 1.

(2) For $\alpha \longrightarrow \infty$ ($\bar{\rho} \rightarrow o$)

$$\frac{Y}{Y_{max}} \approx 1 - e^{-\frac{\bar{\rho}\phi}{N} n}$$

Y vs. n is then an exponential function.

Fig. l gives the expectations of Y vs. n for several values of α. For the sake of comparison a test point is devised. This is the point corresponding to the intersection of the tangent at n=o and the line $Y/Y_{max} = 1$. The value of Y/Y_{max} at this point is l for a linear behavior and $1-e^{-1} \approx 0.63$ for the exponential behaviour.

Depending on α various values of Y/Y_{max} at the test point are expected. This is shown on the right half of Fig. 1.

The independent units model predicts always an exponential dependence of Y/Y_{max} vs. n, no matter what is the value of $\varnothing_{Trapping}$. This can be seen in the most direct way by writing:

$$\frac{dT}{dn} = -\varnothing . \frac{T}{T_o}$$

assuming that at any moment the light is distributed evenly between open and closed reaction center units (cf. Reference 1). $\varnothing$ is a constant. It is the quantum yield to convert reaction centers from open to closed form.

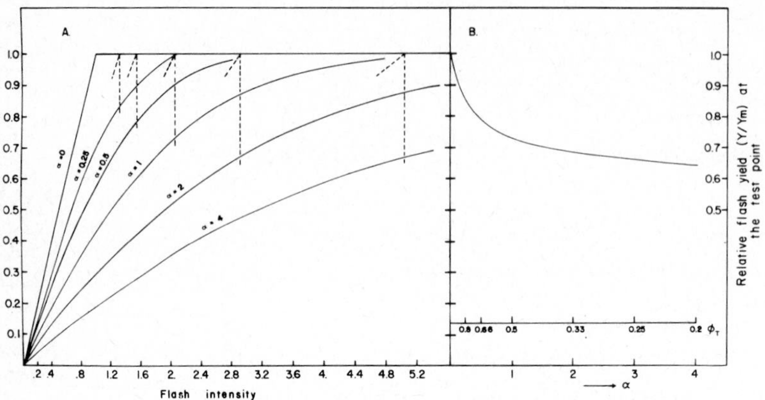

Fig. I. Theoretical predictions for the uniform pigment matrix model.
A. Relative flash yield vs. flash intensity for different values of the parameter α .
B. Relative flash yield at the test point vs. α . The abscissa shows also the corresponding values for max $\emptyset_{Trapping}$ (denoted $\emptyset_T$) vs. α .

5. Results

Previous results[11] showed many examples of exponential behaviour, while at the same time also giving high values of $\emptyset_{Trapping}$ (0.8). ($\emptyset_{Trapping}$ is estimated from the quantum yield of DCPIP reduction at steady-state limiting light conditions, λ = 685 nm. ($\emptyset_{Trapping}$ = 2 x quantum yield, to allow for light channelled into photosystem 1[14]). This showed that the concept of independent units is essentially correct. From the uniform pigment matrix we expect a value 0.82 at the test point while we actually got values around 0.6 - 0.63, in accordance with exponential behaviour.

We now checked systematically for medium effect and found some variations. Fig. 2 shows that chloroplasts in "low-salt" medium (0.01 M), behaved as if the photosynthetic units are independent. Addition of $MgCl_2$ (0.01 M) indicated some interaction, while in "high-salt" medium (0.1 M NaCl) the photosynthetic units seem to exchange excitations, as indicated by the approach to a linear behaviour. Whether this is a systematic effect should be checked by further experiments.

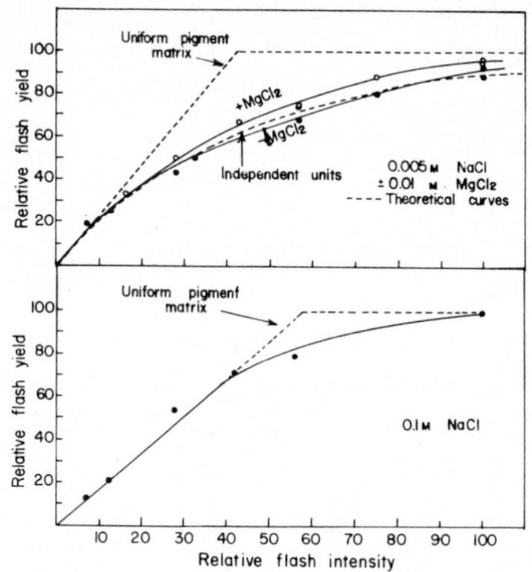

Fig. 2 Experimental relative flash yield vs. flash intensity.
Top: Reaction medium contained low concentration of NaCl with or without $MgCl_2$, as indicated.
Bottom: Reaction medium contained high concentration of NaCl, as indicated.
Otherwise the reaction medium contained: 7.5 μM DCPIP, 1mM TES buffer pH 7.5 and chloroplasts 10 μgm total-chlorophyll in 2 ml.

6. Discussion

The concept of the photosynthetic unit is probably not only statistical, relating the number of reaction centers to the number of pigments, but also has a physical significance. The fact that exponential behaviour is observed for the flash yield vs. flash intensity for some conditions proves the concept of independent units. We showed that by varying conditions the interaction between units can increase. This comes about presumably by decreasing the volume and therefore decreasing the distance between the units. Since the yield of resonance energy transfer is quite a sharp function of the distance, at some critical distance, the units can be transformed from a separated (independent) condition to a condition with large interactions by relatively a mild change of volume. A rough estimation shows that for R^6 dependence (Förster Resonance transfer mechanism) stretching the two dimensional membrane area by a factor of two is sufficient to cause the required transition. Whether this is what actually happens one has still to investigate.

A possible objection to the above considerations is the occurrence of photochemical reactions at units with closed centers. For example, the form, $D^+ A^-$ of the reaction center (D= doner A= Acceptor) can still be a quencher, as Butler and others[15, 16] have assumed. If, during the short flash time ($\angle$ 10 μsec) the reaction center remains in the quenching state* then even the ideal uniform pigment matrix model predicts a first-order reaction mechanism for the initial charge separation and hence an exponential dependence of Y vs. n.

This objection and other possible approaches to the problem will be investigated further.

References:

1. Malkin, S. and Kok, B., 1966, Biochim. Biophys. Acta 126, 413-432.
2. Tumerman, L. A. and Sorokin, E. M. , 1967, Mol. Biol. (English Transl.) I, 527-535.
3. Wang, R. T. and Meyers, J., 1973, Photochem. Photobiol. 17, 321-332.
4. Robinson, G. W., 1966, Brookhaven Symposia in Biology 19, 16-48.
5. Joliot, P. and Joliot, A., 1964, Compt. Rend. 258, 4622-4625.
6. Joliot, P. , Delosme, R. and Joliot, A., 1966, in: Currents of Photosynthesis pp. 359-366 (J. B. Thomas and J. C. Goodheer, eds.). Ad. Donker Rotterdam.
7. Joliot, P., Joliot, A., and Kok, B., 1958, Biochim. Biophys. Acta 153, 635-652.
8. Lavorel, J. and Joliot P., 1972, Biophys. J. 12, 815-831.
9. Emerson, R. and Arnold, W. J.,1932, Gen. Physiol. 16, 191-205.
10. Kohn, H. I., 1936, Nature 137, 706.
11. Malkin, S. Biophysical Chemistry in the press.
12. Duysens, L.N.M. (1952). "Transfer of excitation energy in photosynthesis" Thesis Utrecht.
13. Duysens, L.N.M. (1964) in: Progress of Biophysics Vol. 14 pp. 1-104 Pergamon Press.
14. Sauer, K. and Park, R. B., 1965, Biochemistry 4, 2791-2797.
15. Butler, W. L. , 1972, Proc. Natl. Acad. Sci. US, 69, 3420-3422.
16. Duysens, L. N. M., den Haan, G.A. and Van Best, J.A. , 1974 This proceedings.
17. H. J. van Gorkom. 1974 This proceedings.

* As van Gorkom has showed[17] P^+ has an absorption spectrum extending to the far red. Energy transfer from the bulk pigments to the P^+ form is possible.

M. AVRON, *Proceedings of the Third International Congress on Photosynthesis*,
September 2-6, 1974, Weizmann Institute of Science, Rehovot, Israel
Elsevier Scientific Publishing Company, Amsterdam, The Netherlands, 1974

INFRA-RED SPECTROMETRY OF ISOLATED CHLOROPHYLLS: ASSOCIATION CONSTANTS AND ISOTOPIC EFFECTS

J.P. LEICKNAM, M. HENRY, R. PLUS, R. GILET, J. KLEO
D.R.A. and D.B., C.E.N. SACLAY and GRENOBLE, FRANCE.

In the physico-chemical study of biological systems a parallel examination of simpler models, in which the structure and behaviour of the molecules are known exactly, must always be carried out for purposes of comparison. This applies to photosynthesis research where a thorough knowledge of the structure and behaviour of chlorophyll molecules (Ch) is necessary.

Although their chemical composition has long been well known [1], their structure continues to pose problems, in spite of the vast amount of research carried out on the subject [2]. This paper will be devoted to a quantitative study of self-association and interactions of Ch a with polar molecules, the aim being to determine how many electron donor-acceptor type intermolecular bonds can be formed, and how easily, per molecule of Ch a. From these results, hypotheses will be made concerning the position of Mg with respect to the tetrapyrrolic plane and its coordinence, then compared with those obtained in the case of Ch b when the central natural magnesium is replaced by ^{26}Mg. This work claims not to reach a final conclusion but rather to develop two methods of approach which we consider complementary. The first is used to determine the relative reactivity of the central Mg towards electron donors, the accepting nature of this atom certainly being linked not only to its electronic state but also to its position in relation to the tetrapyrrolic plane of the Ch molecule : the farther the centre of gravity of the positive charges inside the tetrapyrrolic nucleus is removed from the negative charges, the stronger should be the accepting nature of the Mg electron. As for the isotopic substitution method, it has often been used to determine the symmetry of molecules.

The techniques used, infrared spectrometry (i.r.) and Raman resonance spectrometry (R.r.), are particularly suitable for both the study of molecular interactions and the determination of molecular structure.

Ch infrared spectra have been obtained for some years, certain bands being attributed without question to well-defined vibrators

[3,4], but almost accurate quantitative study exists. R.r. spectra of Ch and of similar molecules containing the tetrapyrrolic ring have been obtained recently [5,6,7,8,9]. In i.r. spectrometry the spectral regions most thoroughly investigated include those where the vibrators C = 0 (valence vibration C = 0 : 1,600-1,800 cm^{-1}) and Mg - N (200-500 cm^{-1}) absorb, the latter region also showing, in the case of solids or apolar solvent solutions, one or more vibrations due to molecular interactions of Ch with itself or with certain ligands (∿ 300 cm^{-1}) [3,4]. For simplification purposes, these two spectral regions only will be examined here.

I - Quantitative study of the molecular interactions of anhydrous Ch a with itself and with other polar molecules (L).

1) - General remarks

Anhydrous Ch a self-associates to give polymers $(Ch\ a)_n$, a phenomenon caused by an electron donor-acceptor type interaction between the ketone carbonyl group of position 9 of one molecule and the central Mg atom of another [2,4,10]. The present study will thus yield information on the nature of both the ketone function as electron-donor and the Ch a Mg as acceptor.

In the solid state and in apolar solvents a band is observed at 1,698 cm^{-1}, assigned to the free C_9 = 0 groups of polymers and free molecules, together with one or several bands around 1,655 cm^{-1} due to complexed carbonyl vibrators (C_9 = O ...Mg) [2,3,4].

If polar molecules (L) are added to the medium in sufficient quantities (specified below) the band or bands around 1,655 cm^{-1} and the Mg ...0 band at about 300 cm^{-1} disappear : the intermolecular bonds of the (Ch a) polymers are broken [2,3,4].

To simplify again, we shall concentrate mainly on solutions in CCl_4 (concentrations between $0.2.10^{-3}$ and 35.10^{-3} mol. l^{-1}) and at room temperature : under these conditions vapour phase osmometry showed Ch a present as dimers in equilibrium with monomers [11].

The spectrometer is a model 225 Perkin Elmer with spectral slits of about 2 cm^{-1} ; a Perkin Elmer 180 (spectral slits ∿ 4 cm^{-1}) is used for low frequencies. The ICs cells are between 0.1 and 2.5 cm thick.

The Ch a samples are either extracts obtained by Mathis'method[12] or Sigma products. After purification and drying they are checked by i.r. spectrometry : they contain well below one molecule of water per hundred Ch molecules and no detectable amounts of pheophytine, allomer or other impurities.

^{26}Mg-substituted Ch is extracted in the usual way by growing chlorellae in $^{26}Mg\ SO_4$-containing media. $\left(\frac{^{26}Mg}{^{25}Mg + {}^{26}Mg} > 98\ \%\right)$.

Chromatographs on oil-impregnated Kieselguhr plates show that all these samples contain about 20 % epimer a'and 80 % a, in agreement with earlier work [13,14]. It should be emphasized that only epimer a is present "in vivo" : the characteristic a' band is absent on chromatograms of total chloroplast extracts and only appears after extraction of Ch its time of appearance depending on the temperature of the environment.

To our knowledge no technique has yet been found to extract a'-free samples of stereoisomer a in appreciable amounts. The simultaneous presence of a and a' in all samples of Ch extracted from green plants undoubtedly falsifies the results obtained by any physico-chemical technique, because although these epimers are described as differing only by the steric position occupied by the ester function fixed on the carbon 10 of ring V [13,14] their electron donor and acceptor behaviour is quite different, as shown below.

2) - Dimerization constant of Ch a in CCl_4 solution at room temperature

In general the polymer formation constant K_n corresponding to the equilibrium :

$$n\ \text{Ch}\ \underline{a} \rightleftharpoons (\text{Ch}\ \underline{a})\ n \qquad (1)$$

should be obtainable from optical density measurements at 1,698 cm^{-1} ($\nu\ C_9 = 0$ of the free ketone carbonyl groups) by application of the Beer Lambert and mass action laws [4]. If we call : $C° =$ nominal Ch (a + a') concentration $= C°\underline{a} + C°\underline{a}'$, C_1, C_c, C_p and C_m respectively the concentrations at equilibrium of free $C_9 = 0$, complexed $C_9 = 0$, degree n polymers and monomers, they are linked by the expressions :

$$C_c = C° - C_1, \quad C_p = \frac{C_c}{n-1} \quad \text{and } C_m = C_1 - C_p$$

$$K_n = \frac{Cp}{C_m{}^n} = (n-1)^{n-1} \cdot \frac{C° - C_1}{(nC_1 - C°)^n}$$

In CCl_4 where $n = 2$, $C_p = C_c$

$$\text{and } K_2 = \frac{C° - C_1}{(2C_1 - C°)^2}$$

Table 1 shows that K_2 varies regularly with the initial concentration C°, confirming our previous observations [4] ; on the other hand the hypothesis put forward at the time, of several simultaneous equilibria leading to the formation of polymers (Ch $\underline{a})_2$, (Ch $\underline{a})_3$ ... (Ch $\underline{a})_n$, was later invalidated by vapour phase osmometry [11] : only manomers and dimers exist in CCl_4 solution.

Because of the assumed equivalence of the species $\underline{a}$ and $\underline{a}'$ with regard to dimerisation [14,15] the existence of the epimers $\underline{a}$ and $\underline{a}'$ was not then taken into account. Strictly speaking equilibrium (1) should be written :

$$4\underline{a} + 4\underline{a}' \rightleftharpoons \underline{a}_2 + \underline{a}\text{-}\underline{a}' + \underline{a}'_2 + \underline{a}'\text{-}\underline{a} \qquad (2)$$

in which :

$\underline{a}$-$\underline{a}'$ refers to complexes with $\underline{a}$ electron donor and $\underline{a}'$ acceptor, $\underline{a}'$-$\underline{a}$ are complexes with $\underline{a}'$ electron donor and $\underline{a}$ acceptor.

According to the hypothesis used previously to estimate statistically the quantities of complexes $\underline{a}_2$, $\underline{a}'_2$ and ($\underline{a}$-$\underline{a}'$ + $\underline{a}'$-$\underline{a}$) [14], which have the same probabilities of formation, K_2 may be written :

$$K_2 = (K\underline{a}_2 \cdot K\underline{a}'_2 \cdot K\underline{a}\text{-}\underline{a}' \cdot K\underline{a}'\text{-}\underline{a})^{1/4}$$

$K\underline{a}_2$, $K\underline{a}'_2$, $K\underline{a}$-$\underline{a}'$, $K\underline{a}'$-$\underline{a}$ referring to the complex $\underline{a}_2$, $\underline{a}'_2$, $\underline{a}$-$\underline{a}'$, $\underline{a}'$-$\underline{a}$ formation constants respectively. The fact that K_2 is not constant shows that the two epimers behave differently.

Since only two experimental quantities are available ($C°$ and C_1) it is not possible to evaluate these four constants separately and to find out which is a weaker donor (and/or acceptor) than the other.

On the other hand it is theoretically possible to estimate separately the electron donor character of $\underline{a}$ and $\underline{a}'$, assuming these two epimers to have identical electron acceptor properties :

$$(Ka_2 \cdot Ka\text{-}a')^{1/2} = Ka - \Big|^{a}_{a'} \quad \text{and} \quad (Ka'_2 \cdot Ka'\text{-}a)^{1/2} = Ka' - \Big|^{a}_{a'}$$

- the first of these constants corresponds to equilibria where $\underline{a}$ alone is electron donor, both $\underline{a}$ and $\underline{a}'$ being acceptors,
- the second, to equilibria where $\underline{a}'$ alone is electron donor and both $\underline{a}$ and $\underline{a}'$ are acceptors.

It has been shown elsewhere [4] that whatever the concentrations $C°\underline{a}$ and $C°\underline{a}'$ before equilibrium we again have :

$$Ka - \Big|^{a}_{a'} = Ka' - \Big|^{a}_{a'} = \frac{C°-C1}{(2C_1-C°)^2}$$

This quantity is not constant as a function of C° (see Table 1) : the epimers have different electron acceptor properties.

It may seem therefore that one of the epimers has little or no donor or acceptor properties and that consequently the amounts of mixed dimers $\underline{a}$-$\underline{a}'$ and $\underline{a}'$-$\underline{a}$ are small or inexistent, in which case it is only necessary to consider one of the two equilibria :

TABLE I

CHLOROPHYLL AND PYROCHLOROPHYLL ☆ a IN CCl_4 AT ROOM TEMPERATURE

$C°.10^3$ mole. l^{-1}	$\frac{C°-C1}{(2C_1-C°)^2}$ l.mole^{-1}	Ka_2 l.mole^{-1}
35.1	180	7,000
13.0	410	8,000
8.0	750	13,000
6.0	760	9,400
5.7_5	875	13,100
3.0	1,500	18,700
2.5_7	1,600	16,900
2.0	1,700	11,800
1.6	2,100	14,800
1.5	2,200	16,000
0.8	2,250	21,000
0.5	4,300	19,000
0.2	6,000	18,000
5.0 ☆	14,000 ☆	$(1.4 \pm 0.7).10^4$

$$2\ a \rightleftharpoons a_2 \qquad (3)$$

$$2\ a' \rightleftharpoons a'_2 \qquad (4)$$

We propose to check whether one of the two constants is really independent of the initial concentration C°.

Knowing the concentrations of each epimer ($Ca° = 0.8\ C°$ and $C°a' = 0.2\ C°$) the quantities corresponding to the two formation constants Ka_2 and Ka'_2 of dimers $\underline{a}_2$ and $\underline{a}'_2$ are calculated from the C° and C_1 measurements.

It has already been shown [4] that :

$$Ka_2 = \frac{C° - C_1}{(2C_1 - 1.2C°)^2}$$

and

$$Ka'_2 = \frac{C° - C_1}{(2C_1 - 1.8C°)^2}$$

The results obtained show that only Ka_2 is practically constant as a function of concentration (Table 1). Moreover the difference ($2C_1$ - 1.8C°) represents the amount of monomer a' at equilibrium : this quantity is always negative.

It should be noted that the average of the Ka_2 values ($1.4.10^4$ $l.mole^{-1}$) is the same as that obtained with pyrochlorophyll a in the same solvent (Table 1). It will be recalled that with this molecule there is no possibility of isomerization at the carbon 10 level, the Ch ester being replaced on this position by a hydrogen atom.

The a isomer of Ch is therefore much more liable to form electron donor-acceptor complexes ; while it cannot be stated that a' is totally incapable of producing interactions of this type it has been estimated that, allowing for experimental error, the corresponding dimerization constant would be about 10 $l.mole^{-1}$ at the very most. The lack of reactivity of epimer a' towards electron donors and acceptors is confirmed by the quantitative study of Ch a - polar molecule interactions.

Thus the isomers a and a' are distinguishable not only by a different position of the ester function fixed on the carbon 10. This stereoisomery alone would not explain the lack of Mg reactivity of the species a' towards electron donors, suggestint that the electron densities on the Mg of the two species must be different. This is perhaps relevant to the fact that rings with five carbon atoms are known to exist in two stereoisomeric forms : the different steric positions on the carbon 10 of ring V distinguishing Ch a from Ch a' may be either due to, or responsible for, a deformation of this ring ; in either case the result should be an electronic change in the molecule as a whole, reflected on the Mg. It therefore seems likely that the accepting nature of the Ch would depend on this stereoisomery, and such effects have been observed [16,17].

3) - Influence of polar molecules (L) on the dimerization of Ch a. L-Ch a and L-Ch a' interactions.

The study of this type of interaction should give information on the reactivity of the central Mg as electron acceptor. In view of the above it will also be possible to compare a and a' with other electron donors.

The addition of varying amounts of L to Ch a dissolved in non-polar solvents has been shown to cause breakage of the intermolecular bonds of polymers a_n. The disappearance of these polymers is assumed to result from the formation of new L-a type complexes in which an L atom is an electron donor and the central Mg of Ch a an acceptor [2,3,4]. Proof of this hypothesis will be shown in the case of the ternary system tributylphosphate (TBP) Ch a dissolved in CS_2.

In CCl_4 the quantity of dimers not destroyed by addition of L can be determined for each solution from optical density measurements at 1,698 cm^{-1} (corresponding to free ketone C = 0 vibrators) or at 1,655 cm^{-1} (band assigned to C = 0 vibrators perturbed by an 0...Mg intermolecular bond).

For each L and each non-polar solvent, $\frac{C_p}{C°}.10^2$ is plotted against the molar fraction of L added : $x°_L = \frac{C°_L}{C°}$ with $C°_L$ = total concentration of L added to the solution. Figure 1 shows the curves obtained for a few typical systems : we no longer observe any detectable amount of polymer a_n ($C_p \sim 0$) for quite different $x°_L$ values, not only according to the nature of L but also according to that of the apolar solvent. Table 2 lists the approximate $x°_L$ values for which $C_p \rightarrow 0$, obtained by extrapolation, for different ternary systems donor + Ch dissolved in an apolar solvent. With water as polar molecule the study in CCl_4 is restricted to a small concentration range ($C°_{H_2O} = 8.10^{-3}$ mole.l.$^{-1}$ at most) because of the very low solubility of water in CCl_4. In this range the curve $\frac{C_p}{C°}\ 10^2 = f\ (x°_{H_2O})$ is similar to that obtained with alcohols.*

All these results can be interpretated by assuming that in CCl_4 or CS_2 solution ($n \sim 2$) the following equilibria, besides that leading to the formation of dimers a_2 (1), are in competition :

$$L + a_2 \rightleftharpoons L\text{-}a + a \qquad (5)$$

$$L + a \rightleftharpoons L - a \qquad (6)$$

$$L + a' \rightleftharpoons L - a' \qquad (7)$$

Generally speaking polar molecules (L) can be classified according to their ability to form L-a type complexes from polymers a_n in various apolar solvents, i.e. their more or less pronounced character as electron donors : in CCl_4 for instance :

acetone < ethanol, water < pyridine < TBP.

For ethanol therefore equilibrium (1) only shifts to dissociation if x° is large, whereas the same result is obtained with only 2.5 times as many pyridine as Ch a molecules, and 0.4 in the case of TBP. The equilibrium shifts observed with various donors L indicate that

---

* Note that here we do not obtain the $(Ch\text{-}H_2O)_n$ micelle described elsewhere [18], which require completely different preparation conditions

TABLE 2

$(x^\circ_L)_{C_p} \rightarrow 0$ FOR DIFFERENT TERNARY SYSTEMS ELECTRON DONOR + Ch a DISSOLVED IN APOLAR SOLVENT

Apolar Solvents	L	$(x^\circ_L)_{C_p \to 0}$	Apolar Solvents	L	$(x^\circ_L)_{C_p \to 0}$
CCl_4	Dioxane	10 ± 2	CCl_4	Pyridine	2,5±0,5
	Menadione (vitamine K_3	10 ± 2		Acetonitrile	50±10
				TBP *	0,4±0,04
	Vitamine K_1	100 ± 30		Ovolécithine	0,4±0,04
			CS_2	TBP *	0,6±0,1
	Tetrahydro-furan	10 ± 2	Cyclohexa-ne	TBP *	0,4±0,04
	Ethyl ether	50 ± 10			
	Butyl oxyde	50 ± 10	Hexane	TBP *	1,4±0,2
	Acetone	150 ± 30	Dodecane	TBP *	1,1±0,1
	Acetophenone	150 ± 30			
	Ethanol	10 ± 1	* Tributylphosphate		
	Methanol	10 ± 1			

as electron donors epimer a is stronger than pyridine but weaker than TBP, a' being deprived (or almost) of this property.

In the various apolar solvents the differences for a given L are less pronounced but still clear, as seen in the case of TBP-containing systems (fig. 1) : as shown below this must be related to the degree n of polymerisation of the chlorophyll before L is added.

Case of TBP - Ch a interaction

Since TBP is the most active electron donor investigated it was examined in CS_2 solution with the special aim of recording the absorption structure corresponding to P = 0 vibrators. As TBP is added to Ch a dissolved in CS_2 the following phenomena are observed simultaneously :

- disappearance of the band centred on 1.655 cm^{-1}, accompanied by an increase in the relative intensity of the band at 1.698 cm^{-1};
- disappearance of the ν Mg...0 band (308 cm^{-1}) characteristic of

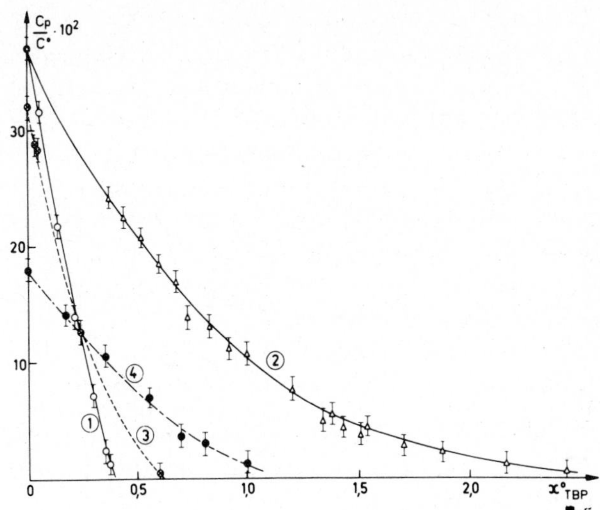

Fig. 1. $\frac{C_p}{C^\circ} \cdot 10^2 = f(x^\circ)$: 1 - Ch a + TBP in CCl_4. 2 - Ch a + Pyridine in CCl_4. 3 - Ch a + TBP in CS_2. 4 - Ch a + TBP in $C_{12}H_{26}$.

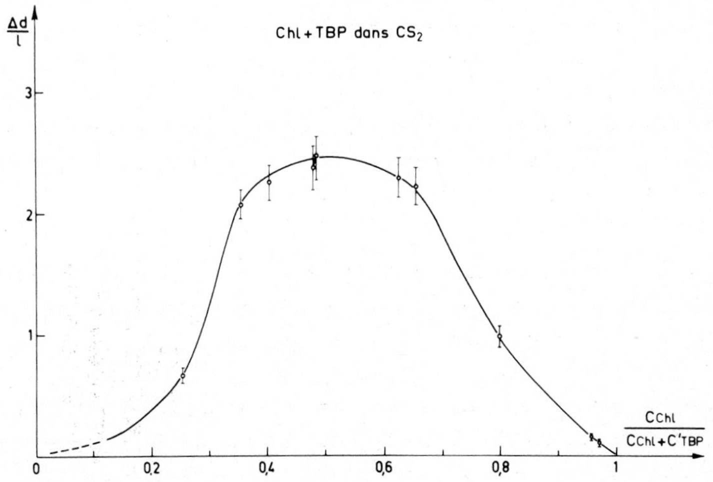

Fig. 2 - Job's curve (Ch a + TBP in CS_2)

chlorophyll dimers $\underline{a}_2$ and appearance of a band at 323 cm^{-1} ;

- reduction in the relative intensity of the structure in which is located the TBP $\nu_{P=0}$ band centred around 1,280 cm^{-1} and appearance of bands between 1,242 and 1,260 cm^{-1}, the latter being observed only at relatively high $x°_{TBP}$ values (about 0.7).

These observations fit in with the hypothesis stipulating the formation of L - Ch $\underline{a}$ type complexes in which a directed bond is created between the phosphate P=0 group and the Ch Mg.

It is thus possible using TBP to determine at most how many molecules of a strong ligand can be fixed to one Ch molecule, the standard method [19] being to plot $\frac{\delta d}{l} = f\left(\frac{C°}{C° + C°_{TBP}}\right)$ in which

$$\frac{\delta d}{l} = \frac{d° - dl}{l}$$

- With $\frac{d°}{l}$ the optical density per unit cell length at the maximum of the band centred on 1,280 cm^{-1} (absorption due to free P=0 vibrators in the absence of Ch) and $\frac{dl}{l}$ that measured at the same TBP concentration after addition of Ch, the absorption of the latter at this frequency being subtracted ;

- $C°_{TBP}$ the TBP concentration of the solution.

The curve obtained (fig. 2) is centred on

$$\frac{C°}{C° + C°_{TBP}} = 0.5$$

The complexes thus belong to the type 1 Ch $\underline{a}$ for 1 TBP molecule, and species $\underline{a}$ and $\underline{a}'$ both interact with this phosphate ; however the differences between the reactivity of dimers $\underline{a}_2$ and monomers $\underline{a}$ and $\underline{a}'$ should be emphasized.

The observations made are not sufficient to calculate the equilibrium constants of (5), (6) and (7), but they can nevertheless be classified with respect to the dimerization constant Ka_2. It is noticed first that in CCl_4 solution the $\frac{C_p}{C°} 10^2 = f(x°_{TBP})$ curve is very close to a straight line cutting the abscissa axis at about 0.4 ± 0.04, which means that each TBP molecule added reacts preferentially on a dimer $\underline{a}_2$, (present to the extent of 36 % before addition of TBP) rather than on a monomer $\underline{a}$ or $\underline{a}'$.

In CS_2 the results are much the same (fig. 1) although the average value of n is probably slightly above 2, as shown by the ordinate at the origin of the $\frac{C_p}{C°} = f(x°)$ curve.

Several P = 0 bands perturbed by the P = 0...Mg interaction have been shown to exist at around 1,242 and 1,260 cm^{-1}. The first appears as soon as TBP is added and corresponds to stronger P = 0...Mg interactions then the second. The very weak electron accepting nature of epimer a' (which reacts well only with strong donors such as TBP and hardly or not at all with "average" donors such as Ch a) suggests that TBP-a' type complexes form less easily and lead to a less perturbed P = 0...Mg vibration ($1,260\ cm^{-1}$) * than those of the TBP-a type (P = 0...Mg = 1.242). The equilibrium constants of (1) and (5-7) in CCl_4 therefore can probably be classified as follows :

$$K_5 > Ka_2 \quad \text{and}$$
$$K_5 > K_6 > K_7$$

In acyclic hydrocarbons the results differ from those obtained with CCl_4, due most likely to the existence of several polymerization equilibria :

$$n\ a \rightleftharpoons a_n \qquad (1)$$
$$a_n + a \rightleftharpoons a_{n+1} \qquad (8)$$
$$\cdots\cdots\cdots\cdots \qquad 11$$
$$\text{or} \quad m\ (a_2) \rightleftharpoons (a_2)_m \qquad (9)$$

In dodecane for example, under our concentration and temperature conditions, the degree of polymerization averages n = 4 as shown by vapour phase osmometru [11] and the ordinate at the origin of the $\frac{C_p}{C°} = f\ (x°)$ curve.

In these solvents the equilibria with TBP can be summed up by :

$$a_n + TBP \rightleftharpoons TBP - a + a_{n-1} \qquad (10)$$
$$a_{n-1} + TBP \rightleftharpoons TBP - a + a_{n-1} \qquad (10')$$
$$\cdots\cdots\cdots\cdots\cdots$$
$$\text{or} \quad (a_2)_m + TBP \rightleftharpoons TBP - a + a + (a_2)_{m-1} \qquad (11)$$

which probably explains the value $(x°_{TBP})\ C_p \rightarrow 0 \sim 1$ in dodecane.

In a dimer, and also apparently in a polymer, the Ch a molecule in which the central Mg is not involved in an intermolecular bond is therefore an especially active electron acceptor : it seems that the molecules at the end of the chain do indeed react first with donors L since according to the scheme of Katz et al [11], when suitable solvents are used (dielectric constant and long hydrocarbon chain), polymers would form according to equilibrium (9).

---

* This band appears alone on the spectra of TBP in the presence of allomer and Ch a, a molecule corresponding to Ch a with its ring V open. Like epimer a' it has little or no electron donor properties and polymerises hardly or not at all in apolar solvents and in the solid state.

To sum up this first part it has been shown that :

1) the electron donor and acceptor reactivities of epimers _a_ and _a'_ are quite different, which disproves the hypothesis [14,15] that a statistical proportion of dimers a_2, a'_2 and aa' exists in mixtures of _a_ and _a'_, solid or dissolved in apolar solvents ;

2) dimers and polymers $\underline{a}_2$ and $\underline{a}_n$ interact more easily with donors L than monomers _a_, and a fortiori _a'_. Hence the difference between _a_ and _a'_ lies not only in the position of the ester on the carbon 10 with respect to the ketone function [13,14], but also in electronic structural difference at the Mg electron acceptor site.

Four types of Ch _a_ molecules must therefore be taken into consideration, each with different properties towards electron donors : monomers _a_ and _a'_, molecules of polymers $\underline{a}_n$ in which the Mg is already involved in an intermolecular bond, and those at the end of the chains where the Mg is free of any intermolecular interaction.

The different reactivities of the Mg in these four cases are undoubtedly due to different hybridizations and thus to its position with respect to the tetrapyrrolic plane. We shall return to this point later ;

3) with an electron donor as powerful as TBP, no 1 : 2 (L - a - L) type complexes were observed even at high $x°_{TBP}$ values under our experimental conditions. This agrees with the a_2 or a_n models proposed earlier [11].

No L - a - OH_2 complexes were observed either, the very low water content of the systems investigated preventing their formation. In the case of CCl_4 it was also impossible to detect any L - a-apolar solvent complex ; the spectrum of CCl_4 impregnating solid Ch _a_ is identical with that of the solvent as a pure liquid film.

In the same way as for dimer or polymer self-association, even in the solid state, only one intermolecular bond is therefore formed on each Ch Mg bonded with a strong or relatively strong electron donor. Research in progress will show whether the same applies to weak donors. However the following two points suggest that these latter are unlikely to give rise to L - a - L complexes : if the model proposed by Katz et al. [18] for $(Ch - OH_2)_n$ micellae is correct there is always only one intermolecular bond per Ch, a single water molecule being fixed on its Mg ; in addition the isotopic shifts observed in the presence of strong donors (TBP for example) or weak donors (such as butyl oxide and acetone) lead to the same conclusions concerning the position of Mg with respect to the tetrapyrrolic plane, as shown below.

Note : with weak electron donors, unlike TBP, 2 : 1 types of complex perhaps can exist ; it is not impossible that equilibrium (10) includes an intermidiate stage and that when $x°_L$ is small we obtain the equilibrium :

$$a_2 + L \rightleftharpoons L - a_2 \qquad (12)$$

L would then be fixed without breaking the dimer (or polymer) and this would explain the relative stability of the complex H_2O - 2 Ch a the water of which is known to be very difficult to eliminate. This remark is consistent with all previous observations.

Finally the reactivities of the various Ch a molecules towards electron donors may be classified as follows :

- a < a' < a < a -

where - a are epimer a molecules of a_n or L - a complexes in which the Mg is involved in an intermolecular bond and a - those of the same complexes but with Mg free of such bonding, a' ans a referring as before to epimer a' and a monomers free of any directed interaction with polar molecules L. Lastly it should be stressed that because of the very high constant Ka_2 the systems examined all contain very small proportions of a monomers.

II - Isotopic effects.

For reasons of chemical stability only ^{26}Mg-substituted Ch b has been examined so far since it oxidises less easily than Ch a.

The effects due to substitution of natural Mg by ^{26}Mg in Ch b will be described according to whether the Ch is involved in a strong or a weak intermolecular bond : as mentioned above it is not yet possible to study these effects on isolated Ch monomers, apart from the case of Ch a or b allomers which are not examined here.

The results obtained with Ch a in part one will then be compared with those observed here on Ch b, in spite of its slightly different mode of self-association *.

1) - Ch b molecules involved in bonds with a "strong" electron donor (polymers b_n or L - b complexes).

In media where Ch b self-associates, the relatively strong i.r. band corresponding to Mg...O intermolecular bonds (297 cm^{-1} in the

---

* We know that in CCl_4 for example the latter molecules are mainly present in the form of trimer b_3, the aldehyde function being involved in these molecular interactions ; it seems unlikely that this will greatly modify the central Mg.

solid state) is lowered by 4 ± 2 cm^{-1}. The same applies with a strong electron donor such as TBP for the band at 323 cm^{-1}.

This confirms earlier work [2,3,4] assigning the Ch a or b band around 300 cm^{-1} to an Mg...0 vibration in which the Mg motion would be fairly large.

2) - Ch b molecules involved in bonds with a "weak" electron donor (L - b complex).

The above mentioned bands mask the Mg-N band or bands which in same cases should lie around 300 cm^{-1}. If a polar molecule with weak electron donor properties is added they disappear to reveal the underlying band structure (the ν Mg...L band being most certainly shifted towards the low frequency regions examined below). In this region the i.r. spectra of Ch b solutions in butyl oxide show a structure containing at least two low-intensity bands (314 and 302 cm^{-1}), one of which at least is assigned to Mg-N vibrators. The isotopic shift on the second is 5 ± 2 cm^{-1} ; the first has the form of a shoulder and a shift would be difficult to observe.

R.r. spectra[9] obtained with a Coderg PHO spectrometer (spectral slits about 5 cm^{-1}) and a Spectra Physics argon laser (power around 50 mw, set on the 457.9 nm line) in acetone solution also display several bands (a shoulder at about 320, a band centred on 303 and a slight shoulder around 295 cm^{-1}) ; the second emission, relatively strong, undergoes an isotopic shift of about 4 ± 2 cm^{-1}. No other isotopic substitution effect was observed. The magnitude of these isotopic shifts must be emphasized since they approximate the highest values found for Mg...N vibrations of the compound $^{26}Mg\ (NH_3)_6\ Cl_2$. [8]

This shows that Mg plays an important part in the motion corresponding to the 302, 303 cm^{-1} vibration and confirms the earlier assignment of this band [2,3,4].

It should also be remarked that the frequencies observed in R.r. seem to match those obtained by i.r., within the limits of experimental error.

III - Discussion.

These observations will need to be confirmed with ^{26}Mg-bound Ch a, in sufficient quantities to detect any other bands in the far i.r. corresponding to Mg motions. In addition the electron accepting reactivities of Ch molecules must be determined with a larger number of donors L.

The present results are used to consider all possible hybridizations and decide which best account for the Ch a complexing and Ch b isotopic effects reported here, the two phenomena being comparable since the reactivity of Mg towards electron donors is undoubtedly bound up with its position in relation to the tetrapyrrolic plane : its reactivity should be stronger as the centres of gravity of the + and - charges are farther apart.

The hybridization form most commonly accepted for Mg is Sp^3d^2 (octahedral structure Oh around Mg). This structure ought to be found in Ch, but the present study shows that in most cases this is not so ; since Mg accepts only one extra ligand the apparent coordinence is 5, as already observed [4] in Ch a and b dimers or polymers, or in L - a complexes.

This figure has sometimes been proposed for non-ionic Mg complexes. Examples are certain organic magnesium compounds, though these are described as relatively unstable [21,22], while pyridine-porphyrine complexes [23] and Ch complexes [2,10] have also been quoted. The corresponding hybridization, Sp^3d, would imply that the Mg is bound to three nitrogens and two ligands, which contradicts the results of part one.

This hybridization will therefore be rejected.

The same applies to hybridizations situating the Mg outside the plane, but bound to either three nitrogens (Sp^3 and p^3) or two nitrogens and two ligands (Sp^2d, perpendicular to the tetrapyrrolic plane (see part one).

The square plane Sp^2d structure (considered very stable) cannot apply except to monomers, which are not easy to study ; however the lack of reactivity of monomers a', and to a lesser extent a, suggests that the Mg of these molecules might approach this configuration. The same is true for Sp hybridization which cannot be rejected, work having shown that in certain basic porphyrines two central hydrogen atoms are localised on opposite nitrogens [24,25], which thus play a special part.

Finally it seems that for L - Ch complexes or $(Ch)_n$ molecules where the Mg is involved in an intermolecular bond the most suitable hybridization is Sp^2, the Mg being bound to two opposing nitrogens and one electron donor. This structure implies that the Mg is outside the tetrapyrrolic plane and explains both the impossibility of forming a second Mg - L bond with strong electron donors, as observed with TBP, and the relatively large isotopic shift of the ν Mg... O band.

This coordinance 3 for Mg may seem to contradict the value 4 in organic magnesium compounds, for example [21] :

$$\begin{array}{c} H_3C - H_2C \searrow \\ H_3C - H_2C \nearrow \end{array} O - \underset{R}{\overset{R}{Mg}} - O \begin{array}{c} \nearrow CH_2 - CH_3 \\ \searrow CH_2 - CH_3 \end{array}$$

However in the same way as for the 2 pyridine-porphyrine complex [23] it is found that one of the molecules, ether in the formes case and pyridine in the latter, is very easily removed whereas the bonds between Mg and the remaining ligand molecules are only broken with great difficulty. The polymar $(Ch)_n$ molecules in which the Mg is free of intermolecular bonding raise a special problem : they cannot have the Sp or Sp^2d structure but their strong reactivity towards electron donors suggests that in these hybridizations the Mg is in a special state, the electron orbitals being considerably deformed : the C=O complexing of these molecules undoubtedly leads to modifications in their geometry and in the electron distribution around the Mg which could then be removed from the tetrapyrrolic plane, in a hybridization state between Sp (or Sp^2d) and Sp^2 ; it would then interact easily with an electron donor to reach the Sp^2 configuration. However the D_4h symmetry hypothesis has been used to interpret the i.r. spectra of certain metallo-porphyrines [26, 27] but for porphyrine the free space between two opposing nitrogens (about 2.6 Å) is roughly equal to the diameter of Mg, which means that the slightest deformation of the tetrapyrrolic ring can expel Mg from the tetrapyrrolic plane. The many Ch substituants and especially the existence of ring V, together with the partial saturation of ring IV, must break this symmetry and in some degree separate the Mg from the plane, even in the absence of intermolecular bonding.

This theory is supported by the fact that the spectra of Ch a solid or dissolved (in CCl_4, CS_2, dodecane and cyclohexane) in the presence of TBP are identical with those of Ch a alone under the same conditions, apart from the modifications described above (due to breaking of the intermolecular bonds of polymers a_n and to the creation of TBP - a complexes). This point will be cleared up by the examination of Ch allomers, molecules possessing the chlorine ring and various substituants.

Finally the isotopic substitution of Mg has shown that an Mg - N vibration band is active in R.r., which means that the molecule possesses no symmetry centre. Three reasons for this are possible :
- the existence of the Mg...O intermolecular bond to the extent that it takes part in the resonance, a hypothesis which seems unlikely but will be checked ; - or else even at resonance the tetrapyrrolic

ring lacks a D_4h symmetry because of the many substituents ;- lastly the position of the Mg outside the plane. It is not impossible that all these effects are involved. The presence of this R.r-active Mg-N vibration can only confirm our previous conclusions.

Conclusion

Four types of Ch a molecules with quite different electron accepting reactivities have been detected and the following correlations are proposed :

- The weak reactivity of the species a' could correspond to a particularly stable Mg bonding state, sp^2 d (or Sp), placing Mg inside or very close to the tetrapyrrolic plane.
- In the case of monomers a, slightly more reactive, the Mg could lie a little farther outside the tetrapyrrolic plane, Sp(or Sp^2 d) with some orbital deformation.
- The Mg of L - Ch complexes and those involved in the bonds of polymers $(Ch)_n$ would have an Sp^2 type hybridization.
- Finally the Mg not involved in the intermolecular bonds of polymers $(Ch)_n$ are certainly right outside the tetrapyrrolic plane with a particularly deformed Sp (or Sp^2d) hybridization.

For $(Ch)_n$ and L - Ch complexes the isotopic effect observed does not disagree with these hypotheses.

These preliminary results show in any case that by a quantitative study of Ch complexing combined with the use of isotopically labelled molecules it should be possible to define more occurately the electronic state of the central Mg of these molecules. Meanwhile the hypotheses put forward on the basis of this work concerning the distorsion of the tetrapyrrolic ring, it repercussion on the electron structure [16] and the hybridizations of the central Mg seem to be corroborated by recent work on oxido-reduction reactions and ESR spectra of magnesium-containing tetrapyrrolic complexes and Ch itself [17].

Références :

1. Woodward, R.B. , 1960, Angew. Chem. 72, 651and 1961, Pure Appl. Chem. 2, 383.

2. Katz, J.J., 1968, Develop. Appl. Spectrosc. 6, 201.

3. for example : Katz, J.J., Closs, G.L., Pennington, F.C., Thomas, M.R. and Strain, H.H., 1963, J. Amer. Chem. Soc. 85, 3801.

4. Henry, M. and Leicknam, J.P., 1969, Colloque International du C.N.R.S. n° 191 "La nature et les propriétés des liaisons de coordi-

nation", Paris and Breton, J., Henry, M. and Leicknam, J.P., to be published.

5. Lutz, M., 1972, C.R. Acad. Sci. Paris, 275 B, 497.
6. Lutz, M. and Breton J., 1973, Biochem. Biophys. Res. Com. 53, 413.
7. Plus, R. and Lutz, M., 1974, Spectrosc. Letters 7, 73 and 133.
8. Plus, R. 1974 communications, IVth Intern. Conf. of Raman Spectrosc., Brunswick, Maine.
9. Lutz, M., 1974, communications, IVth Intern. Conf. of Raman Spectrosc., Brunswick, Maine, and private communication.
10. Katz, J.J. and Janson, T.R., 1973, Anals N.Y. Acad. Sci. 206, 579.
11. Ballschimter, K., Truesdell, K. and Katz, J.J., 1969, Biochim. Biophys. Acta 184, 604.
12. Mathis P., 1970, Thesis. Orsay and rapport CEA R - 4116 (1971).
13. Strain, H.H. and Manning, W.M., 1942, J. Biol. chem. 146, 275.
14. Katz, J.J., Norman, G.D., Svec, W.A. and Strain, H.H., 1968, J. Amer. chem. Soc. 90, 6841.
15. Houssier, C. and Sauer, K., 1970, J. Amer. Chem. Soc. 92, 779.
16. Petterson, R.C. and Alexander, L.E., 1968, J. Amer. Chem. Soc. 90, 3873.
17. Lexa, D. and Reix, M. 1974, J. Chim. Phys. 71, 511 and 517.
18. for example : Ballschmiter, K., Cotton, T.M., Strain, H.H. and Katz, J.J., 1969, Biochim. Biophys. Acta 180, 347.
19. Job, P., 1926, Ann. Chim. (Paris) 9, 113.
20. Toma, F., Villemin, M., Ellenberger, M. and Brehamet L., 1970, Magnetic Resonance and Related Phenomena, Proceedings of the XVIth Congress A.M.P.E.R.E., (Bucharest), p 317.
21. House, H.O. and Oliver, J.E., 1968, J. Org. Chem. 33, 929.
22. Swift, T.J. and Lo, H.H., 1967, J. Amer. Chem. Soc. 89, 3988.
23. Dorough, G.D., Miller, J.R. and Huennekens, 1951, J. Amer. Chem. Soc. 73, 4315.
24. Nelson, H.M., 1954, Ph. Thesis, Oregon State College.
25. Storm, C.B. and Teklu, Y., 1972, J. Amer. Chem. Soc. 94, 1745.
26. Ogoshi, H., Masai, N., Yoshida, Z. Takemoto, J. and Nakamoto, K., 1971, Bull. Chem. Soc. Jap. 44, 49.
27. Ogoshi, H., Saito, Y. and Nakamoto, K., 1972, J. Chem. Phys. 57, 4194.

M. AVRON, *Proceedings of the Third International Congress on Photosynthesis*,
September 2-6, 1974, Weizmann Institute of Science, Rehovot, Israel
Elsevier Scientific Publishing Company, Amsterdam, The Netherlands, 1974

FLUORESCENCE POLARIZATION MEASUREMENTS EXCITED BY HORIZONTALLY POLARIZED LIGHT

John Whitmarsh[+] and R. P. Levine

The Biological Laboratories, Harvard University
Cambridge, Massachusetts 02138, U.S.A.

SUMMARY

We have measured the fluorescence polarization for intact cells of Chlamydomonas reinhardi using horizontally polarized exciting light. These measurements were performed both in the presence and in the absence of DCMU at excitation wavelengths of 650, 660 and 670 nm. The polarization of the fluorescence in the absence of DCMU is greater than zero; which is indicative of a non-random orientation of the chlorophyll molecules. The addition of DCMU causes the polarization to increase slightly. In this paper we present an argument to show that the DCMU-induced increase is consistent with a Förster mechanism of excitation energy transfer. This interpretation does not, however, exclude the possibility of a delocalized process from being involved as well.

INTRODUCTION

Fluorescence polarization measurements made recently by Breton et al.[1] using spinach chloroplasts have revealed a non-zero polarization excited by horizontally polarized light. The authors account for this result in terms of an antenna model in which the chlorophyll transition dipoles are preferentially oriented parallel to the plane of the chloroplast membrane.

Using intact cells of the green alga C. reinhardi we have found that the polarization observed while exciting with horizontally polarized light is greater than zero and increases slightly upon the addition of DCMU (Table 1). The non-zero polarization and the fact that it increases as the excitation wavelength increases can be explained in terms of a preferred orientation of the chlorophyll molecules[1,2]. However, the DCMU-induced increase in the polarization is more difficult to account for. It is the purpose of this paper to present a qualitative argument, based upon specific assumptions concerning the local order of the chlorophyll molecules, that demonstrates that the increase in the fluorescence polarization is consistent with a Förster energy transfer mechanism[3,4]. We wish to emphasize that whereas the

---

Abbreviation: DCMU, 3-(3,4-dichlorophenyl)-1,1-dimethylurea.

[+]Present address: Centre d'Etudes Nucleaires de Saclay, Service de Biophysique, Departement de Biologie, BP No. 2. 91190 Gif sur Yvette, France.

argument does not rule the involvement of a delocalized energy transfer process it does provide support for the occurrence of a Förster transfer process in the antenna system.

MATERIALS AND METHODS

Intact cells of the wild-type strain (137c) of C. reinhardi were used in the experiments described here. Cells were cultured in 300 ml of tris-acetate-phosphate (TAP) medium under conditions previously described[5]. The cells were washed and suspended in fresh TAP. The resuspended cells were stirred and kept at room temperature in the light prior to the measurements. The exciting light intensity varied between 1 and $5 \cdot 10^{14}$ photons/cm^2-sec. The chlorophyll concentration, determined by a modification of the method of Mackinney[6,7], was 2 μg/ml.

Both the method of measurement and the instrument used to make the fluorescence polarization measurements will be described elsewhere[2].

RESULTS

The results of measurements of the polarization of the fluorescence observed at a right angle to the exciting beam are shown in table 1. Although the increase in the polarization induced by DCMU is slight, it is nevertheless statistically significant at 650 nm.

Table 1

The *in vivo* fluorescence polarization for *C. reinhardi* excited by horizontally polarized light

Excitation Wavelength	POLARIZATION % normal	POLARIZATION % with DCMU	FL with DCMU / FL normal
650 nm	0.42 ± .05	0.47 ± .05	2.1
660 nm	0.56 ± .05	0.58 ± .05	2.1
670 nm	0.85 ± .05	0.87 ± .05	2.2

The values shown are the average of four measurements. FL is the relative fluorescence yield.

DISCUSSION

The value of the fluorescence polarization observed using horizontally polarized exciting light depends primarily upon the orientation of the absorbing and emitting pigments and to a lesser extent upon the amount of energy transfer. For a randomly oriented group of pigments the fluorescence polarization should be zero[1]. The fact that the polarization observed for intact cells of C. reinhardi is greater than zero indicates that some relative and/or net order exists between the chlorophyll transition dipoles absorbing in the red[1,2].

In order to explain the action of DCMU we make use of the following assumptions:

1. The absorption and emission transition dipole of the chlorophyll molecules are either parallel or very nearly parallel in the region from 650 to 670 nm.

2. The red transition dipoles of the chlorophyll molecules tend to be aligned in the same plane on a molecular level, e.g., several molecules, grouped together, exhibit a preferred orientation of their red transition dipoles in the same plane (Figure 1). These planes of local order may be randomly oriented with respect to one another.

3. We assume that the antenna system for photosystem II can be made up of these planar groups of local order oriented perpendicular to one of three orthogonal axes (Figure 2).

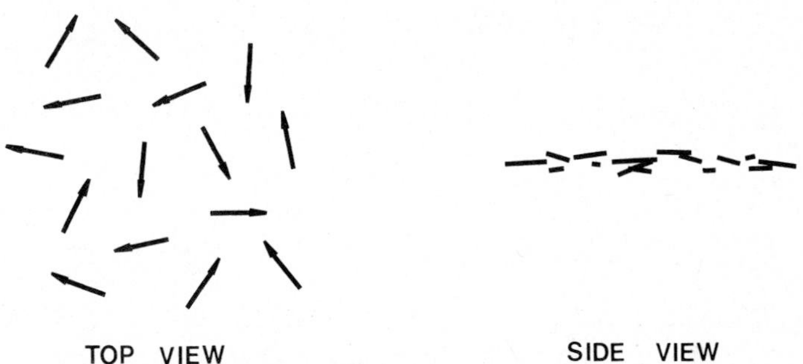

Figure 1. Local order. The arrows represent the red transition dipoles of the chlorophyll molecules.

The assumptions and the model are not intended to be consistent with all of the data relevant to the orientation of chlorophyll in vivo, but rather have been chosen in order to provide a simple explanation for the DCMU-induced increase in the polarization.

Consider first pigment groups oriented perpendicular to the x-axis (group A, figure 2). The incident light will be absorbed primarily by oscillators that are nearly parallel to the polarization of the exciting light. In the absence of energy transfer the fluorescence observed along the y-axis would be partially polarized parallel to the z-axis since when group A is viewed on edge most of the oscillators are nearly parallel to the yz-plane. If energy transfer occurs, the degree of polarization will increase because more oscillators oriented parallel to the z-axis will become excited and be able to fluoresce. We would expect the polarization to exhibit a minimum value in the absence of energy transfer and to

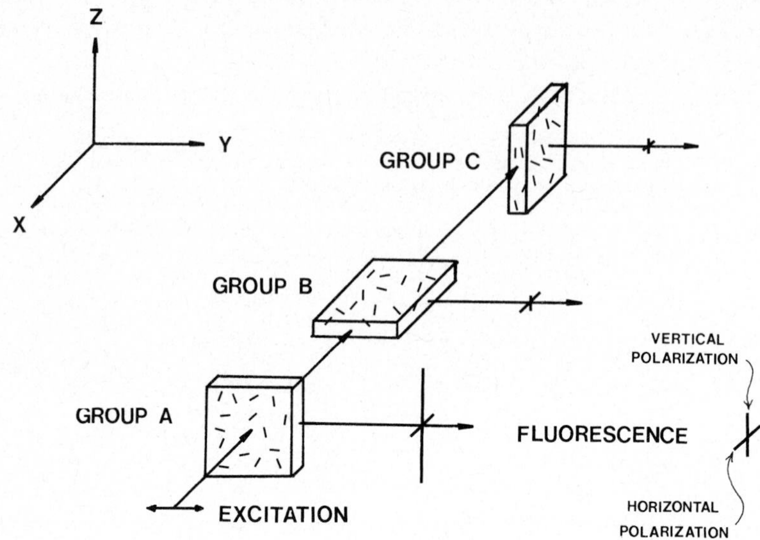

Figure 2. The lines within the squares represent the red transition dipoles of the chlorophyll molecules. The excitation (horizontally polarized) is incident along the x-axis and the fluorescence is observed along the y-axis.

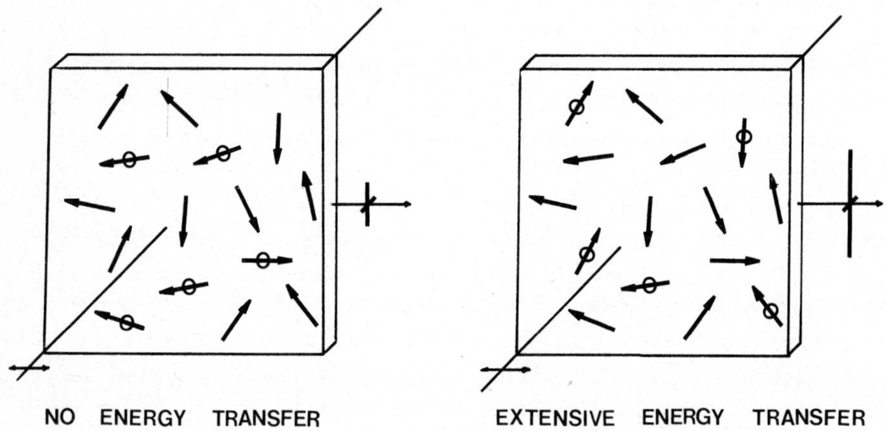

Figure 3. As excitation energy transfer occurs more chlorophyll molecules aligned parallel to the z-axis become excited and the polarization increases. The circles represent excitation energy.

increase to a maximum value if the excitation energy becomes randomized over the pigments before fluorescence occurs. Figure 3 shows the effect of energy transfer on the polarization for a localized transfer mechanism.

In group B, figure 2, many of the oscillators are aligned nearly parallel to the polarization of the exciting light so that they will absorb most of the incident light. There is, however, a major difference between group A and group B with respect to the relative amount of light absorbed by the two groups. Group B will absorb much less light than group A since its cross sectional area normal to the direction of the incident light is much smaller than that of group A. Therefore, although group B will emit fluorescence polarized parallel to the x-axis, its intensity will be much less than the fluorescence from group A.

Group C will absorb even less light than group B. Not only is the cross sectional area of group C small relative to group A, but, since the oscillators are aligned nearly perpendicular to the polarization of the exciting light there will be little absorption. The fluorescence from group C viewed along the y-axis would be unpolarized.

Therefore, since the primary contribution to the fluorescence will be from pigments in group A, we would expect the polarization of the fluorescence to be polarized parallel to the z-axis and and that the degree of polarization should increase as the excitation energy becomes more randomized over the molecular array.

If, in _C. reinhardi_, the excitation energy is not randomized over the chlorophyll molecules in photosystem II within the fluorescence lifetime, then a DCMU-induced increase in the fluorescence lifetime, which corresponds to a greater amount of enery transfer, may result in an increase in the polarization. If, however, the energy transfer process were due solely to a delocalized mechanism the migration of the energy would be so rapid that it would be randomized over the molecular array before fluorescence occurs and the addition of DCMU would have no effect upon the polarization. Therfore, an increase in the polarization upon the addition of DCMU is consistent with a Förster transfer process sufficiently slow so that the excitation energy is not randomized over the chlorophyll molecules before the fluorescence lifetime[2,8].

ACKNOWLEDGEMENTS

We wish to thank Professor N. Geacintov, Dr. J. Breton, Dr. F. van Nostrand, and Mr. J. Becker for helpful discussions. This research was supported by a grant from NSF (GB 18666 and GB 29203) and from the Maria Cabot Foundation for Botanical Research, Harvard University.

REFERENCES

1. Breton, J., Becker, J.F. and Geacintov, N.E. (1973) Biochim. Biophys. Res. Comm. 54, 1403.

2. Whitmarsh, J. and Levine, R.P., Biochim. Biophys. Acta (in press).

3. Förster, Th. (1968) in: Modern Quantum Chemistry, ed. Sinanglo (Academic Press, New York) p. 93.

4. Trosper, T., Park, R.B. and Sauer, K. (1968) Photochem. Photobiol. 7, 451.

5. Levine, R.P. and Gorman, D.S. (1966) Plant Physiol. 41, 1293.

6. Arnon, D.I. (1949) Plant Physiol. 24, 1.

7. Mackinney, G. (1941) J. Biol. Chem. 140, 315.

8. Mar, T. and Govindjee (1971) IInd International Congress on Photosynthesis, Stresa.

M. AVRON, *Proceedings of the Third International Congress on Photosynthesis*,
September 2-6, 1974, Weizmann Institute of Science, Rehovot, Israel
Elsevier Scientific Publishing Company, Amsterdam, The Netherlands, 1974

POLARIZED LIGHT SPECTROSCOPY ON ORIENTED SPINACH CHLOROPLASTS: FLUORESCENCE EMISSION AND EXCITATION SPECTRA

Jacques BRETON

Departement de BIOLOGIE.CEN-SACLAY,
BP. n°2. 91190. GIF-SUR-YVETTE - FRANCE.

The polarized light fluorescence properties of oriented chloroplasts are shown to be consistent with the data concerning the orientation of the absorption oscillators previously determined by linear dichroism (LD) measurements (Breton and Roux, Biochem. Biophys. Res. Comm., 1971, 45, 557). The emission is preferentially polarized parallel to the plane of the photosynthetic membrane, and exhibits a greater degree of polarization in the long wavelength edge of the emission spectrum. For those regions where the LD is positive (red band of $Chl_{\underline{a}}$ and carotenoid absorbing region), and with exciting light propagating along the plane of the oriented membranes, it is found that the fluorescence intensity is higher when the E vector of the exciting light is polarized parallel to the membrane plane than when it is perpendicular to this plane. An opposite effect is observed when exciting in the Soret region where the LD is negative.

It has been recently shown[1-5] through polarized light spectroscopy on oriented photosynthetic membranes that the degree of order of the pigments is substantially higher than has previously been supposed. In particular, linear dichroism (LD) spectra of oriented chloroplasts has revealed the orientation of chlorophyll and carotenoid molecules _in vivo_[1-4]. In this report we compare to previously published LD spectra[1] both the fluorescence emission and excitation spectra in polarized light on oriented spinach chloroplasts. These fluorescence data are shown to be consistent with the LD data.

The LD spectrum of oriented photosynthetic membranes from spinach chloroplasts is presented in fig. 1. Contribution of various artifacts (textural dichroism, selective polarized scattering, selective polarized reflection) are negligible as discussed elsewhere[4-5]. The sign and magnitude of the LD signals led us to reach the following conclusions concerning the orientation of the pigments[1-4] :

- The Y-polarized transitions (corresponding to absorption and emission in the red) of $Chl_{\underline{a}\ 680}$ and longer wavelength forms of $Chl_{\underline{a}}$ lie nearly parallel to the membrane plane. (It must be pointed out that this does not imply that the transition moments are parallel to each other).

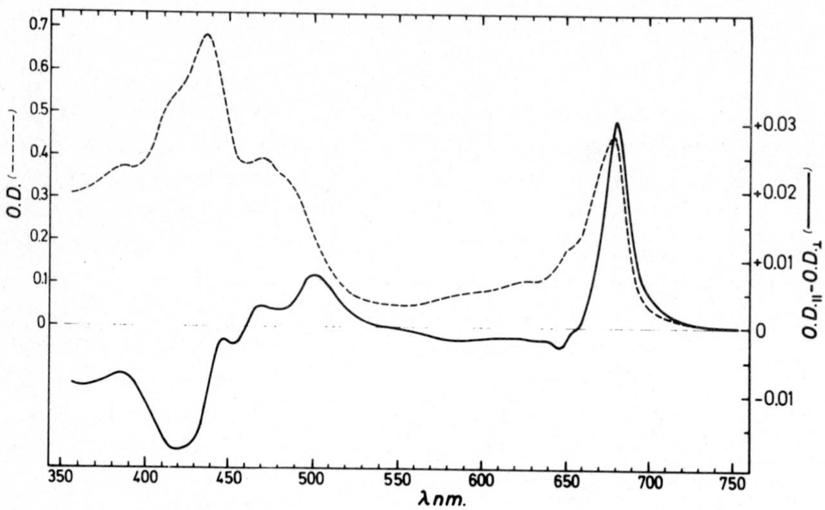

Fig. 1 : Oriented photosynthetic membranes of spinach chloroplasts (———) Linear dichroism spectrum ; (------) Absorption spectrum. (From Breton and Roux[1]).

- The Y-polarized transitions relevant to shorter wavelength forms of Chl $_a$ are either unoriented or oriented near the "magic angle" of 35° with respect to the plane of the membrane.
- The X-polarized transitions of Chl $_a$ are pointing out of the membrane plane at an angle estimated to approximately 45°.
- The main fraction of the carotenoid molecules are lying parallel to the membrane plane.

These conclusions obtained through LD measurements allow us to infer the expected behaviour of the polarized fluorescence from oriented spinach chloroplast membranes. Fig. 2 represents an idealized membrane in which the exciting light is incident along the X-axis and the fluorescence is viewed along the Y-axis.

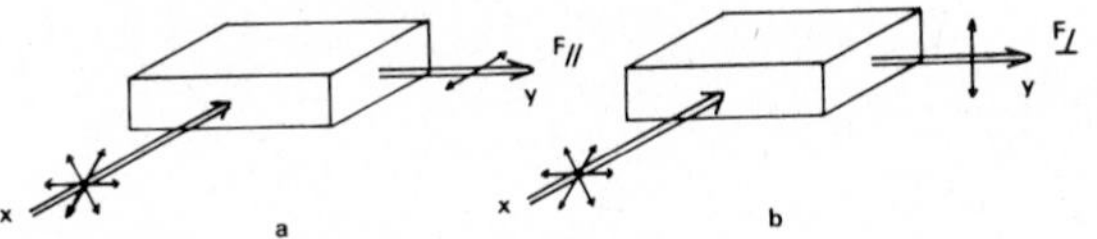

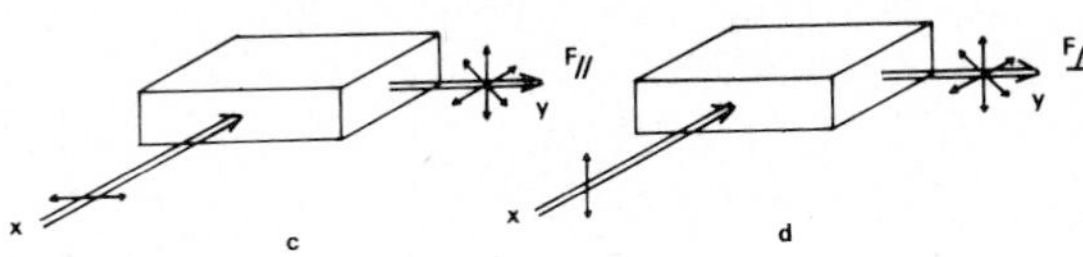

Fig. 2

Diagrammatic representation of the polarized light fluorescence experiments. The membrane plane is approximated by a rectangular slab ; a and b refer to polarized fluorescence emission. c and d to polarized excitation.

POLARIZED FLUORESCENCE EMISSION.

Owing to the preferred orientation of the emission oscillator close to the lamellar plane, the fluorescence emitted along the Y-axis will be preferentially polarized parallel to the membrane plane. Fig. 3 shows the result of the experiments depicted in Fig. 2a and 2b in which spinach chloroplasts, oriented by air-drying on a glass-plate in a 70 kG magnetic field, were excited at 436 nm and the fluorescence emissions spectra for the two polarizations were recorded. The expected behaviour is observed. This confirms the previous results of Geacintov et al.[3] In this experiment, the chlorophyll concentration was kept as low as possible in order to minimize the reabsorption of the fluorescence. Any reabsorption would be maximum for the component of the fluorescence polarized parallel to the membrane plane. Even at very low concentration it is difficult to avoid or even estimate the amount of reabsorption occuring within a single membrane when it is viewed parallel to its plane.

The ratio $\frac{F_{/\!/}}{F_{\perp}}$ is wavelength dependent (Table I) increasing towards longer wavelengths. This effect cannot be attributed to reabsorption as there is very little absorption above 710 nm.
Neither can it be explained by selective scattering since the wavelength dependence of the $\frac{F_{/\!/}}{F_{\perp}}$ ratio does not follow the scattering curves (Geacintov, personal communication). The wavelength dependence

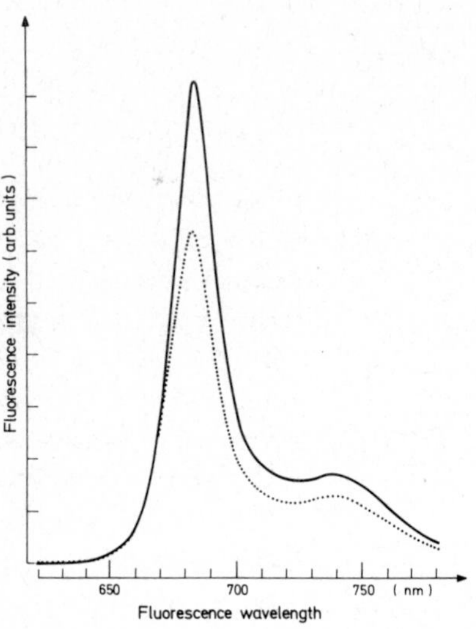

Fig. 3
Polarized fluorescence emission spectra of oriented spinach chloroplasts (——) $F_{\parallel}$; (.....) $F_{\perp}$. Excitation wavelength 436 nm. Half-band widths 7.5 nm.

of the $\frac{F_{\parallel}}{F_{\perp}}$ ratio is most probably related to the different spectroscopic forms of the Chl_a in vivo exhibiting different degree of order as demonstrated by LD spectroscopy[1].

Table I : Wavelength dependence of the polarized fluorescence emission of oriented spinach chloroplasts.

λ (nm)	660	670	680	690	700	725	750
$F_{\parallel}/F_{\perp}$	1.0	1.1	1.4	1.5	1.4	1.3	1.3

We wish to point out that the polarization parallel to the membrane plane of the fluorescence emitted along this plane may bring out an artifact in the measurement of energy transfers by the classical technique of the depolarization of fluorescence*. Making the hypothesis of a complete randomization of the initial polarization by energy

* However, this objection does not necessarily apply to relative polarization changes which may be indicative of the energy transfer processes.

transfers between chlorophyll molecules oriented parallel to the membrane plane (but mutually unoriented), we can expect that the polarization would be zero. It has been shown[3] that membranes excited with light propagating perpendicular to their plane absorb more light than the ones excited parallely. So a classical measurement of the fluorescence polarization using randomly oriented suspension of chloroplasts will photoselect for membranes oriented perpendicular to the light propagation vector and consequently give rise to a non-zero polarization of the fluorescence. Such an effect has been recently described for spinach chloroplasts and chlorella[6].

POLARIZED FLUORESCENCE EXCITATION

This experiment is depicted in Fig. 2c and 2d. Whenever the excitation takes place in a region of positive LD (Fig. 1) more mo-

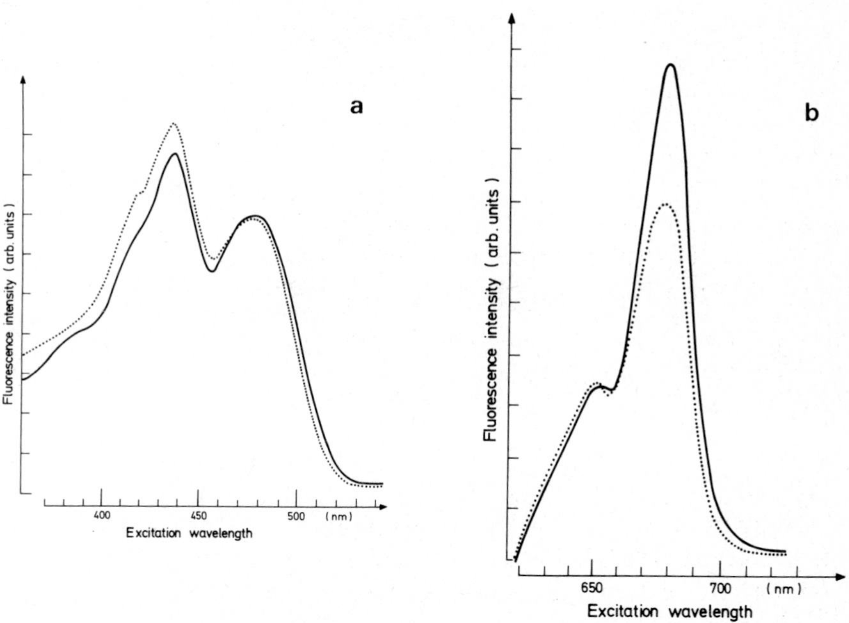

Fig. 4 : Polarized excitation spectra of oriented spinach chloroplasts. (———) F∥ ; (......) F⊥ . Fig. 4a : Emission wavelength 685 nm. Fig. 4b : Emission wavelength 730 nm. Half-band widths : 7.5 nm.

lecules are excited in the experimental arrangement shown in Fig. 2c than in that shown in Fig. 2d. Therefore when exciting in a region where the LD is positive, we can expect to find $F_{\parallel} > F_{\perp}$. An opposite effect is predicted in the regions of negative LD. Fig. 4a shows the polarized excitation spectra in the spectral region 350-550 nm for the emission at 685 nm. The changes of sign expected from the LD data are observed. Similar results have been obtained at a few representative wavelengths with spinach chloroplasts directly suspended in tris-sucrose buffer and oriented by a 13 kG magnetic field (unpublished). Fig. 4b shows the polarized excitation spectra in the red region for the emission at 730 nm. The polarization anisotropy observed in these spectra is smaller for the short wavelengths forms of $Chl_{\underline{a}}$ than for the longer wavelengths forms, a fact already suggested by LD measurements.

In general, the data obtained by fluorescence emission and excitation in polarized light using oriented chloroplasts are consistent with the conclusions drawn from the LD spectra. However, the technique of polarized excitation of the fluorescence is more selective than the LD technique wich necessarily takes into account all the oriented pigments that absorb light. It should then be possible to discriminate between the orientations of different sets of pigments giving rise to resolved emission bands. In this respect the use of low temperature might be fruitful.

REFERENCES

1 - BRETON, J. and ROUX, E., 1971, Biochem. Biophys. Res. Comm. 45, 557.

2 - MORITA, S., and MIYAZAKI, T., 1971, Biochim. Biophys. Acta, 245, 151.

3 - GEACINTOV, N.E., VAN NOSTRAND, F., BEKER, J.F., and TINKEL, J.B., 1972, Biochim. Biophys. Acta 267, 65.

4 - BRETON, J., MICHEL-VILLAZ, M., and PAILLOTIN, G., 1973 Biochim. Biophys. Acta 314, 42.

5 - BRETON, J., 1974, Biochem. Biophys. Res. Comm. 59, 1011.

6 - BRETON, J., BEKER, J.F., and GEACINTOV, N.E., 1973, Biochem. Biophys. Res. Comm. 54, 1403.

M. AVRON, *Proceedings of the Third International Congress on Photosynthesis*, September 2-6, 1974, Weizmann Institute of Science, Rehovot, Israel
Elsevier Scientific Publishing Company, Amsterdam, The Netherlands, 1974

A COMPARATIVE STUDY OF PORPHYRIDIUM CHL A BANDS AND CHL A BANDS IN VITRO

J.-C.Leclerc* and J.Hoarau

Laboratoire de Physiologie Cellulaire Végétale de l'Université de Paris-Sud,Associé au C.N.R.S.(LA40).91405 ORSAY,FRANCE.

1. Introduction

The use of a digitized spectrophotometer system allowed us to obtain fourth derivative of red algae Porphyridium absorption spectra. For cells obtained in well defined conditions the same peaks in the same positions were always present. Some of these peaks correspond to chlorophyll a forms positions defined in French's laboratory[1,2]. In order to know if the peak positions represented exact positions of elementary peaks in the bulk of chlorophyll a,it was necessary to test the Butler's fourth derivative method[3,4] with our apparatus. We used pure chlorophyll a and chlorophyll b the peaks positions of them being fairly known in several solvents. Fourth derivation of spectra of chlorophyll a and b mixtures gave always peaks corresponding (± 1nm) to chlorophyll a and b absorption maxima. Surprisingly pure chlorophyll a and b don't gave simple fourth derivative peaks at room temperature. At low temperature,far red absorption bands _in vitro_ may appear accordingly to Brody and Broyde[5] for instance. So a study of chlorophyll a _in vitro_ and a study of chlorophyll a in Porphyridium cells was undertaken.

2. Techniques

Spectrophotometric techniques are described in a paper submitted for publication in Photochemistry and Photobiology,they are only resumed here.

A single beam spectrophotometer is used. Signals from the sample or from the blank are given,at very well defined wavelength,to a digital voltmeter (Solartron). $\text{Log}_{10} \frac{I_0}{I}$ is calculated by a microcomputer (Hewlett Packard 9810A) and absorption spectra are drawn by a calculator plotter (Hewlett Packard 9362A) . The

---

* This work forms the principle article from a thesis of "Doctorat d'Etat ès Sciences Naturelles" submitted to the Faculté des Sciences d'Orsay in 1974 . C.N.R.S. registration N° AO 10 195.

fourth derivative of the curves are also computed. The positions and the importance of the peaks obtained with fourth derivative method lead us in the choice of positions, bandwidths and heights of Gaussian curves used for absorption curves reconstitutions. These reconstitutions are difficult because of many interactions between the fourth derivative of each natural component. The fourth derivative signal of a little component may be hidden by the negative part of the signal issued from an important neighbour component. A divergence of only $\pm 1.10^{-3}$ in optical density may be rapidly reached between the real and the reconstitued absorption spectra. The reconstitued fourth derivative obtained in this way may be very different from the natural one, when this occurs, it is necessary to modify some positions, heights and bandwidths. Finally, after several attempts, real and reconstitued absorption curves, with divergences less than $\pm 1.10^{-3}$ and fourth derivatives with identical main characteristics, are obtained (fig.1,2). The accuracy of individual components is also obtained with certain difference spectra in which only a few number of elementary components are involved. Some fourth derivations of difference spectra are also made.

Chlorophyll a samples were prepared from Spinach or Tomato leaves or Porphyridium cells by column or thin layer chromatography. Pheophytinization may be avoided: chlorophyll a is stable in high concentrated solutions or after drying under vacuum: dilute solutions in methanol are stable during several hours (but not during several days). For dilute solutions, a 1cm path-length rectangular glass cuvette is used. For concentrated solutions, 5 to 20 microns path-length cuvettes are made with two optical glass plates sealed with silicone fat on the edges.

3. Results

A. Porphyridium cells.

The wavelength positions of the gaussian curves used for the reconstitutions are defined after fourth derivative studies and also after several kinds of difference spectra obtained with cells grown under red or green light, iron-starved cells, French press cells extracts treated with 0.1% Triton X100. This led to the figure 1 curves system and to the comparison of figure 2. Because of the complexity of the fourth derivatives, some of the components don't appear on figure 2 and some other are displaced. The shoulder corresponding to the 682.4nm component in red light grown cells disappear with green light grown cells, where the 687nm component is very important.

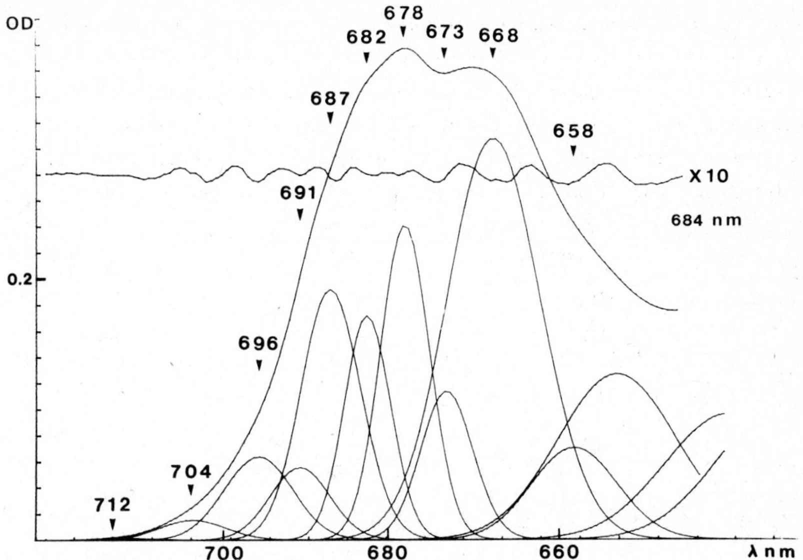

Fig.1. Reconstitued absorption curve of Porphyridium cells grown under intense 680nm monochromatic light. The sinuous line represents the ten times amplified divergence with the real spectrum.

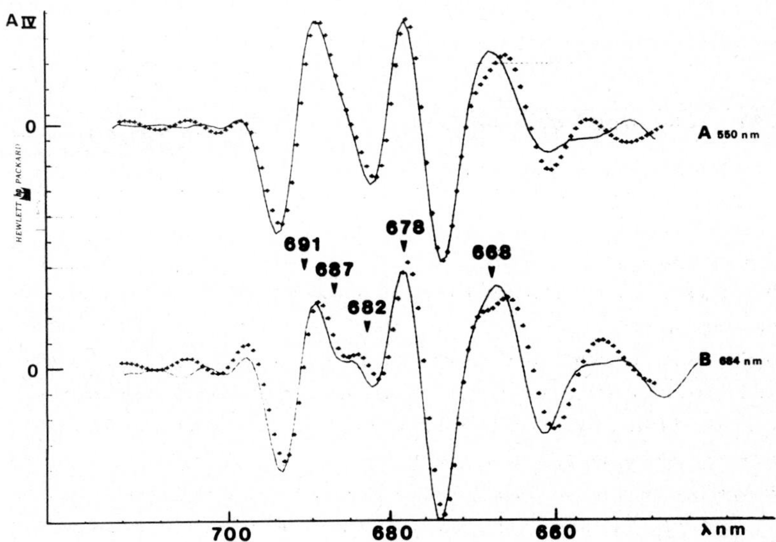

Fig.2. Fourth derivatives of the natural and reconstitued absorption curves. Values on ordinate are arbitrary.

The figure 3 shows a French press extract in phosphate buffer,the 678 and 668nm chlorophyll a forms are visible,the 683nm form is not clearly perceptible in French press extracts,but it is visible in whole cells spectra presented in a previous paper[6]. On figure 3,the dotted line represents the same French press extract after TritonX100 treatment: the 668nm form is increased,and with the decrease of 677 and 683nm forms,the 687 and 696nm forms appear clearly.

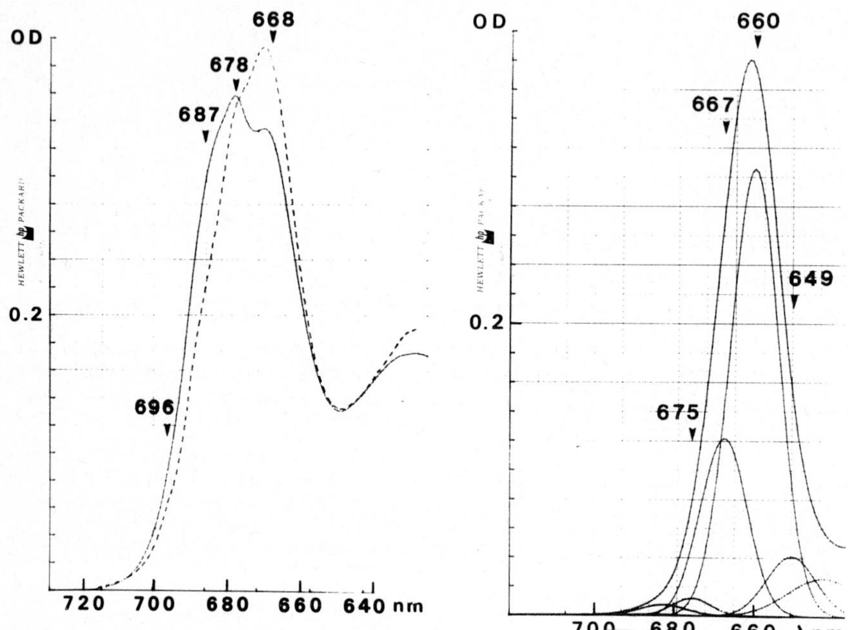

Fig.3. French press extract of Porphyridium; on dotted line: with 0,1% triton.

Fig.4. Chlorophyll a in acetone.

B. Chlorophyll a in vitro.

Two kinds of polar solvents are used: primary alcohols and ketones. The absorption peak in red region ($\Pi\ \Pi^*$ transition with electron of $a_{1\ u}$ orbital[7]) seems simple and give simple fourth derivatives with ketones (fig.4). In alcohols such as methanol and ethanol,the fourth derivatives show a dissymetrical peak and a low shoulder at about 675nm (fig.5). A curve reconstitution with a 675nm and a 663nm component is not good.

In frozen methanol or acetone solutions,the absorption maxima are slightly shifted towards the infra-red and important absorption bands appear at about 700nm (fig.6). The fourth derivations give four peaks at 687nm,675nm,668nm,660nm. An accurate curve reconstitution is made with these peaks positions,the results are shown in Table 1.

Table 1

Parameters used for chlorophyll a gaussian components,in dilute solutions and in Porphyridium cells

Methanol at -196°C			Methanol at 20°C			Acetone at 20°C			Porphyridium at -196°C		
712	12.5	230	—	—	—	—	—	—	712.3	8.4	22
703	12.5	240	—	—	—	—	—	—	703.9	9.9	158
696	12.5	200	698	15.5	6	697	17.5	10	695.7	10.4	642
689.5	12.5	540	689	17.5	10	—	—	—	690.7	9.2	535
—	—	—	685.5	15	150	—	—	—	686.8	9.2	1924
682.2	11.7	1360	682.4	14.2	140	682.2	14.8	90	682.4	7.1	1714
—	—	—	—	—	—	—	—	—	677.8	7.6	2405
674.7	11.7	2485	674.8	12.6	705	674.9	12.2	110	673.0	8.1	1144
667.8	12.4	2620	667.6	14.7	2535	667.5	14.5	1220	667.5	14.3	3095
659.1	12.7	1040	659.1	15.3	2165	659.2	15.2	3060	658.2	12.9	730
649	13.1	430	648.5	15	540	649	14.9	410	—	—	—
638	15.7	660	638	21	400	640	21.6	255	—	—	—

For each condition: wavelength,in nanometers,is on the left;bandwidth,in nanometers,is in the middle,relative height is on the right.

The same peaks positions are used with increased bandwidth and various height for curve reconstitutions of chlorophyll a in methanol at room temperature,the fourth derivatives of these reconstitutions are valid.

With concentrated solutions of methanol or acetone up to 25mM,there is only slight modification of the absorption. Bands in the 700-720nm region appear also in aqueous methanol (fig.7).

In non polar solutions,the peaks'positions are not changed at room temperature,or liquid nitrogen temperature. Solution in glycerol-trioleate for instance gives an important peak at 667nm. Concentrated solution in benzene shows a peak at 667nm and a shoulder at 660nm on the fourth derivative curve. In frozen benzene in addition to the 675nm,660nm and 668nm bands,a very important band appears at 640nm. Concentrated solutions (0.5 to 1mM) in heavy aliphatic hydrocarbon such as "white spirit" from Prolabo,D=0.74,rich in dodecane,give a very important peak in the 740nm region,shown in fig.8 after a formation time of about one hour. The curve analysis and the fourth derivation of this last system show between the 740 and the 665nm peaks several intermediate forms. If the concentrated solution in "white spirit" is diluted,the 740nm form slowly reverses in 665nm form.

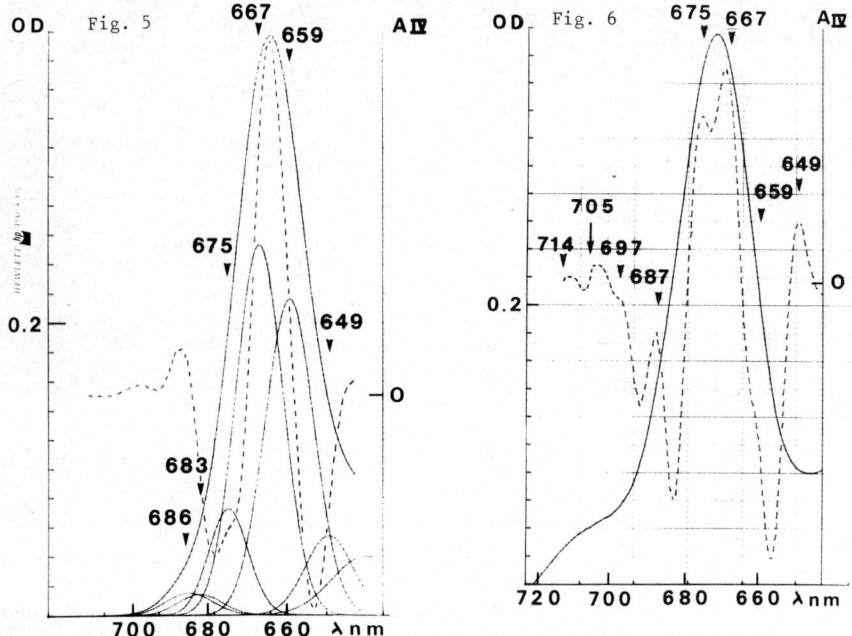

Fig.5 and Fig.6 : Chl a absorption spectra in methanol at 20°C (Fig 5) and -196°C (Fig 6). Dotted lines represent the fourth derivatives.

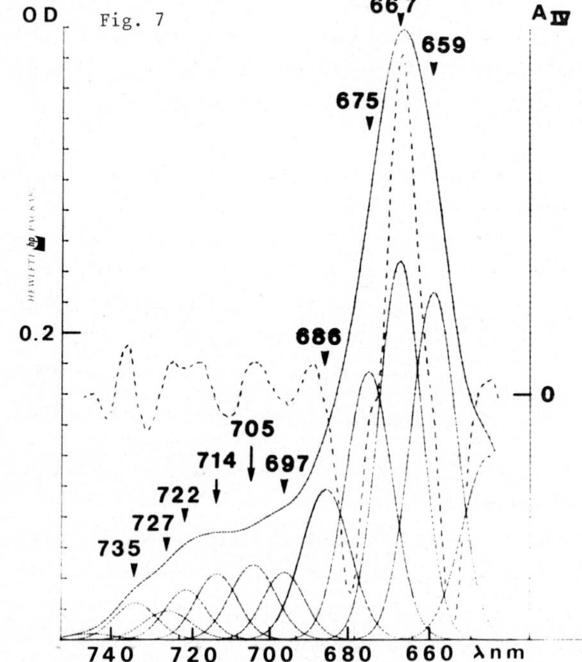

Fig. 7 : Chl a absorption spectra in methanol (67%) and water (33%).

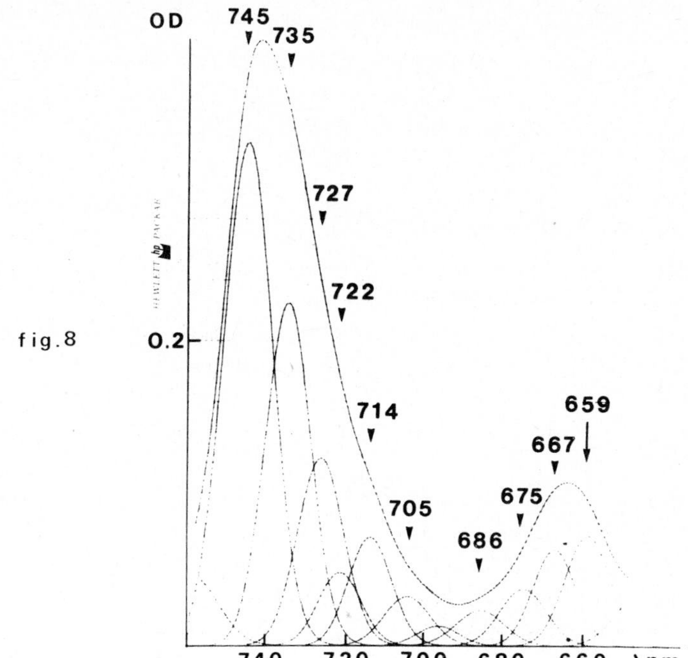

Fig. 8 : Concentrated solution of Chl a in "White Spirit"

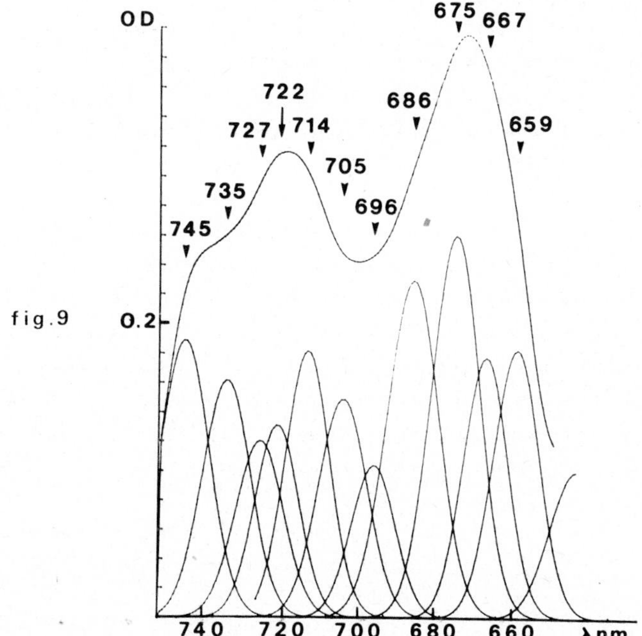

Fig. 9 : Film of dryed Chl a issued from a "White Spirit" and hexane mixture

Solid films of chlorophyll a are studied also, they are prepared with concentrated solutions, under vaccum or not. Solutions in acetone give films having important absorption at 675 nm, 687 nm and in 700-720 nm region; the difference spectra between a solution in acetone and the film give a fourth derivative showing a diminution of 659 nm form and an increasing of 675 nm and 668 nm forms. A similar system can be seen with films issued from solutions in butanol or benzene. Films obtained with solutions in aliphatic hydrocarbons always show near infrared peaks at 745 nm, 735 nm and 727 nm but never important with solutions in hexane for instance. Films obtained with "White Spirit" and hexane mixtures may have very important intermediate peaks : 705 nm, 714 nm and 722 nm as in figure 9.

4. Discussion

Numerous peaks are found in the 650-710 nm region in red algae Porphyridium cells but known photosynthesis don't require a great number of chlorophyll a forms. This may be generally explained if we consider the in vitro phenomena. Chlorophyll a in solution in simple solvent may have several important bands in the red region. If we admit that in Porphyridium thylakoids several physical environment exist, each of these environment may favour one electronic transition in single chlorophyll or one or several polymerized forms of chlorophyll.

Some of our peak positions don't correspond to the general system elaborated by French, Brown et al, perhaps because there is no chlorophyll b in Porphyridium. A study of chlorophyll b show very different wavelength positions not only for the single molecules but also for the polymers. In an early work Sironval et al[8] reported a 673 nm absorption form. The important 640 nm band found in certain conditions may be related to J.B. Thomas works[9]. The 687 nm band is clear in triton treated Porphyridium, amounts of 687 nm band vary under several culture conditions and this gives signals in difference spectra; it is the same with 696 nm band.

In vitro several of these positions are found with difference of one nanometer or less: 659 nm, 667 nm, 686 nm, 696 nm and 705 nm bands. But other such as 677 nm, 683 nm and 691 nm bands are specific of thylakoid structures. 683 nm band and perhaps 691 nm band are destroyed by triton treatment, 677 nm is only a few more stable, these two bands may result of chlorophyll a complexes with macromolecules such as proteins. The 658 nm and 667.5 nm absorption peaks of Porphyridium are probably due to free chlorophyll or chlorophyll bounded with small hydrophobic molecules, the absorption maxima of chlorophyll a in glycerol trioleate is found at 667.5 nm. The 687 nm and 696 nm absorption forms in Porphyridium correspond to the 675 nm, 686 nm and 696 nm group of absorption bands of chlorophyll a dryed from polar solvents, or in water colloids. The 675 nm band in vitro has no equivalent in vivo whereas the 691 nm absorption form has no equivalent in vitro. The 687 nm and 696 nm in vivo absorption bands seems to be the result of short chlorophyll polymers. The 705 nm and 714 nm absorption bands may be related to higher polymers of chlorophyll. The 745 nm band seen in Chlorella[10] but not in Porphyridium may be related to chlorophyll water polymers[11].

References

1. French, C.S., et al, 1970, Carnegie Inst. Year Book 69, 662.
2. French, C.S. and Brown, J.S., 1971, IInd International Congress on Photosynthesis Stresa, 291.
3. Butler, W.L. and Hopkins, D.W., 1970 a, Photochem. Photobiol. 12, 439.
4. Butler, W.L. and Hopkins, D.W., 1970 b, Photochem. Photobiol. 12, 451.
5. Brody, S.S. and Broyde, S.B., 1968, Biophysical J. 8, 1511.
6. Hoarau, J. and Leclerc, J.C., 1973, Photochem. Photobiol. 17, 403.
7. Lhoste, J.M., 1968, Bull. Soc. Franç. Physiol. Vég. 14, 379.
8. Sironval, C., et al, 1965, Biochim. Biophys. Acta 94, 344.
9. Thomas, J.B., 1974, Biochim. Biophys. Acta 333, 415.
10. Thomas, J., et al, 1970, Photochem. Photobiol. 11, 85.
11. Katz, J.J., 1973, Naturwissenschaften 60, 32.

M. AVRON, *Proceedings of the Third International Congress on Photosynthesis*, September 2-6, 1974, Weizmann Institute of Science, Rehovot, Israel
Elsevier Scientific Publishing Company, Amsterdam, The Netherlands, 1974

THE CONTRIBUTION OF PHYTOPLANKTON TO THE ATTENUATION OF LIGHT WITHIN NATURAL WATERS : A THEORETICAL ANALYSIS

J.T.O. Kirk
CSIRO Division of Plant Industry, Canberra, Australia.

Primary production by photosynthesis in aquatic ecosystems is limited by a number of physical and chemical parameters of the environment. Near the surface primary production may well be limited by the supply of phosphorus, nitrogen or other nutrients, but in almost all bodies of water, other than very clear or very shallow ones, at increasing depth the factor limiting primary production is the availability (intensity and spectral quality) of light in the photosynthetic waveband (350-700 nm). Attenuation of the light as it penetrates the water is a consequence of absorption of the light energy and its conversion, almost entirely, to heat. Water itself absorbs strongly at wavelengths above about 580 nm. Most inland, and many coastal, waters have in addition a strong absorption at wavelengths below 500 nm due to the presence of dissolved yellow plant breakdown products (gelbstoff). Turbidity due to suspended silt can greatly increase light attenuation but it does so mainly by scattering the light and increasing the pathlength of the photons.

The chloroplast pigments of the phytoplankton present in the water also absorb some of the light, but in waters of low or moderate productivity the algae make only a small contribution to the overall light attenuation[1]. As a consequence of nutrient enrichment resulting from the activities of man, a substantial and increasing proportion of the world's lakes, rivers, and to a lesser extent coastal waters, are in a eutrophic, highly productive state. In these waters phytoplankton can grow to densities such that they make a substantial contribution to light attenuation : the algal population can by self-shading limit its own photosynthesis and growth. While it is recognized that the structure of the phytoplankton canopy (cell size, shape, pigment content etc.) must affect light penetration through it, [2,3]the actual nature of the relationships between these factors and light attenuation has received little attention.

As part of a research project on the factors limiting primary productivity in aquatic ecosystems, I am engaged in a theoretical analysis of the contribution that suspended algal cells of varying types make to the attenuation of light within natural waters. By considering the properties of simple model systems, interesting conclusions concerning canopy structure and light penetration have been reached[4,5] and are summarized in this paper.

1. General treatment of cells of any shape and orientation

Duysens[6] presented a theoretical treatment of the absorption spectra of suspensions of pigmented particles which accounted for the fact that such spectra ($\log_{10} i_o/i$ versus λ) are distinctly flattened compared to the spectra of the corresponding pigment solutions. Some parts of Duysens' treatment have been made use of in the present work. Specifically, attention has been confined to the simplified case of non-scattering particles. Duysens showed that Beer's law may be applied to a suspension of such particles provided these are small compared to the total volume of the suspension : the applicability of Beer's law has been assumed in the present work.

It has been shown[4] that in a suspension containing n algal cells per ml illuminated with parallel monochromatic light at an angle (within the water) of β to the horizontal and intensity (on a horizontal surface) just below the water surface of I_o, the intensity, I_{sus}, at depth z metres is given by

$$I_{sus} = I_o\, e^{-(k + 100\, n\overline{Aa})z \text{ cosec } \beta} \tag{1}$$

where $\overline{Aa}$ is the mean value of the product, for each cell, of the effective cross-sectional area (A cm^2, the area of the cell projected onto a plane normal to the light beam) and that fraction (a, given as a proportion of 1.0) of the light incident upon the cell which is absorbed by it. Since the cells are randomly oriented and of any shape, A and a will vary from one cell to the other in the suspension. k is the absorption coefficient (m^{-1}) of the water in which the cells are suspended and is given by

$$k = w + g$$

where w is the absorption coefficient of pure water, and g is the absorption coefficient due to the gelbstoff present, at the wavelength in question.

It has also been shown that if the algal pigments were uniformly dispersed throughout the system as though in solution, the light intensity, I_{sol}, at depth z metres is given by

$$I_{sol} = I_o\, e^{-(k + 100\, Cn\bar{v}\gamma)\, z \text{ cosec } \beta} \tag{2}$$

where C is the intracellular concentration of pigment (g l^{-1}) in the (undispersed) cells, $\bar{v}$ is the average volume (cm^3) per cell, and γ is the (natural logarithm) specific absorption coefficient (l g^{-1} cm^{-1}) of the chloroplast pigment mixture at the wavelength in question. A system with uniformly dispersed pigments can be regarded as one extreme of phytoplankton canopy structure, equivalent to a very large number of very small cells. As a measure of the effects of changes in canopy structure on light attenuation it is useful to compare light penetration into a given suspension with that into an equivalent solution. Accordingly, a new parameter, the penetration coefficient, P_λ, has been introduced and is defined by

$$P_\lambda = \frac{I_{sus}}{I_{sol}} \qquad (3)$$

It follows that

$$P_\lambda = e^{(C\bar{v}\gamma - \bar{A}\bar{a})100nz \text{ cosec } \beta} \qquad (4)$$

It can be shown[4] that the penetration coefficient is always greater than zero and that if (at a constant set of A values) the values of a for all the cells are increased either by raising the intracellular pigment concentration, or by altering the wavelength to one more strongly absorbed, then P_λ is increased. Also, if a given amount of cellular material containing a specified pigment concentration is placed into a smaller number of larger cells (i.e. A and a are increased together) then P_λ is increased (I_{sol} is unaltered, I_{sus} increases); but if a given amount of pigment is distributed into a larger quantity of cell material (by increasing cell size or numbers) then P_λ decreases (I_{sol} is unaltered, I_{sus} decreases). It also follows, as may be seen from equation (4), that the value of P_λ rises if the number of cells per unit volume, or the depth, are increased, or if the light path is increased by diminishing the angle to the horizontal.

2. Calculation of spectral intensity distribution and vertical attenuation coefficients

For calculation of I_{sus} at any given wavelength all the parameters in equation (1) can readily be evaluated with the exception of $\bar{A}\bar{a}$. $\bar{A}\bar{a}$ for any cell is a complex function of its shape and orientation. $\bar{A}\bar{a}$ can fairly readily be evaluated for the unique case of spherical cells. For a sphere A and a are independent of orientation to the light beam : A is equal to πR^2 and a is given by

$$a = 1 - \frac{2[1 - (1 + 2\gamma CR)\, e^{-2\gamma CR}]}{(2\gamma CR)^2} \qquad (5)$$

where R is the radius (cm) of the sphere[6]. For any cells other than spherical ones the determination of $\bar{A}\bar{a}$ is possible but will generally involve multiple numerical integration. The value of k can vary enormously. It can not be less than w, the absorption coefficient of pure water, but can be much greater if there is a substantial level of gelbstoff in the water, as there commonly is in inland, and some coastal, waters. On the basis of data in the literature and also measurements in local waters a gelbstoff concentration sufficient to give a g value of 2.00 m^{-1} at 440 nm has been adopted as "typical" of inland waters of intermediate colour, and the more highly coloured coastal waters[4]. From absorption spectrum measurements on local water, a set of g values for different wavelengths, with g_{440} set to 2.00 m^{-1} has been obtained. The values of Smith and Tyler[7] for diffuse attenuation coefficients of the very pure water of Crater Lake, have been taken as good estimates of the values of w. At each wavelength of interest, the values of w and g, obtained in this way, have been added together to give k.

To relate the findings to photosynthesis it is desirable that light intensity be expressed in terms of quanta cm^{-2} s^{-1}, rather than in terms of energy cm^{-2}

s^{-1} (irradiance). To indicate light intensities expressed in quanta we replace I_o, I_{sus} and I_{sol}, in equations (1) and (2), with q_o, q_{sus} and q_{sol}, respectively. To provide a set of standard values of q_o for different wavelengths, measurements made in Australia[8] of the spectral energy distribution of direct sunlight for an air mass range of 1.25 - 1.50 have been converted to quanta cm^{-2} s^{-1} incident upon a horizontal surface just below the water surface, assuming a solar elevation of 45°, giving rise to an apparent solar elevation (β) of 58° within the water[4]. q_o has been calculated for each of thirty-five 10 nm bands, from 350 to 700 nm. From this set of values of q_o the spectral distribution of light intensity (q_{sus} or q_{sol} versus λ) at any given depth in the suspension or equivalent solution may be calculated. The total number of incident quanta cm^{-2} s^{-1} in the photosynthetic waveband (350 - 700 nm) just below the surface, Q_o, and at depth z m, Q_{sus} or Q_{sol}, are obtained by summing the values of q_o, q_{sus} or q_{sol}, respectively, for the thirty-five 10 nm wavebands. It should be noted that we have concerned ourselves only with direct sunlight : the contribution of diffuse skylight has been ignored.

It has been suggested that the most useful way of characterizing the attenuation properties of natural waters for photosynthetically active radiation is in terms of the vertical attenuation coefficient, K (m^{-1}), for the whole waveband between 350 and 700 nm[9,10]. The value of K changes somewhat with depth[9] : its average value down to depth z m is defined by the equation

$$ln \frac{Q_o}{Q} = Kz \qquad (6)$$

where Q and K can be either Q_{sus} and K_{sus} (for the suspension), or Q_w and K_w (for the water without algal cells). The value of K for a given water body will be a function both of solar elevation[10] and of the spectral quality of the sunlight. In the present work K values have been calculated from the Q_o and Q_{sus} (or Q_w) values, obtained as indicated above on the basis of a solar elevation of 45° and a corresponding spectral quality. A parameter of particular relevance for an understanding of self-shading in phytoplankton population is the average increment in vertical attenuation coefficient (for the 350 - 700 nm waveband), ΔK_{sus}, per unit increase in algal concentration (expressed as mg chlorophyll a m^{-3}). It is given by

$$\Delta K_{sus} = \frac{K_{sus} - K_w}{c} \qquad (7)$$

where K_{sus} is the vertical attenuation coefficient of an algal suspension containing c mg chlorophyll a m^{-3}.

3. Spectral distribution of light intensity in suspensions of spherical cells and colonies

In section 1 we arrived on theoretical grounds, at a set of qualitative predictions as to the consequences for light attenuation of various changes in the structure of the phytoplankton canopy. We shall now consider the results of some calculations, carried out as indicated in section 2, for model suspensions of spherical cells and colonies of green and blue-green algae, in order to determine the size of these effects[5]. All the algal cells were assumed to contain 2% (dry weight) chlorophyll. Sets of values of specific absorption coefficients for the green and blue-green algal pigment mixtures were calculated from data in the literature.

In the case of the green algae, a cell diameter of 8 μm was adopted to make the results of relevance to typical members of the *Chlorococcales* such as *Chlorella*. Three types of blue-green algae were considered : large colonies, 57.6 μm in diameter, as might be expected for *Microcystis* sp. (volume 100,000 μm^3 - ref[11]); colonies 8 μm in diameter for comparison with the green algal cells; small single cells, 0.8 μm in diameter, of the *Synechococcus* type. The spectral distribution of light intensity, and the vertical attenuation coefficients (350 - 700 nm), were calculated for a depth of 5 m within suspensions of these algae in water with the "standard" gelbstoff concentration (g_{440} = 2.0 m^{-1}). As may be seen from the spectral distribution of light intensity for the water alone (Fig. 1) the gelbstoff concentration is sufficient to absorb virtually all the light of wavelength below 500 nm. Absorption by water itself removes a substantial proportion of light at the red end of the spectrum. In fact nearly 80% of the light penetrating the water to 5 m is in the 545 - 655 nm waveband. This is a spectral region in which the green algal cells have comparatively small a values. The blue spectral region, where the cells have high a values has disappeared. Since the penetration coefficient is a function of a (section 1) we might anticipate that in this particular system

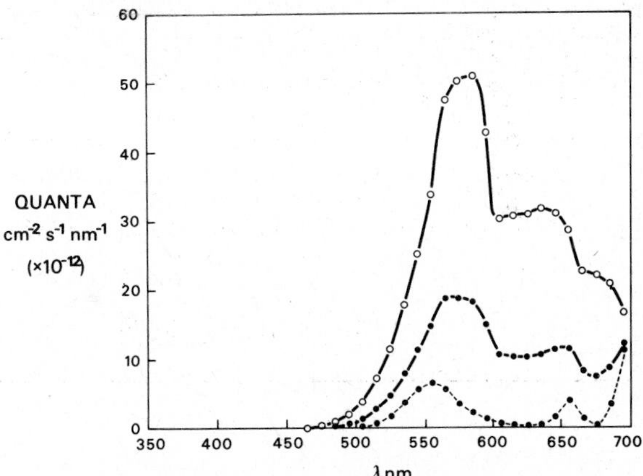

Fig. 1. Calculated spectral distribution of light intensity at depth 5 m in a hypothetical suspension of blue-green algal colonies (57.6 μm diameter; chlorophyll 2% of dry weight) at a chlorophyll a concentration of 30 mg m^{-3}. Water alone ——o——. Algal suspension ——●——. Equivalent solution of algal pigments ---●----.

light would penetrate the suspension to only a slightly greater extent than it penetrates an equivalent solution. This turns out to be the case : at 30 mg chlorophyll *a* m^{-3} the flux of total photosynthetic quanta at 5 m is only 1.7% higher in the suspension than it is in the solution (Table 1).

In the case of blue-green algae which have, owing to the presence of biliprotein pigments, substantial *a* values in the 545-655 nm region, the differences in light penetration between suspension and solution become significant. For colonies of the same size (8 μm diameter) as the green algal cells, at 30 mg chlorophyll *a* m^{-3}, the flux of total photosynthetic quanta at 5 m is 34% higher in the suspension than in the equivalent solution (Table 1). It was predicted in section 1 that the penetration coefficient would increase if a given amount of cellular material was disposed into a smaller number of larger particles. Fig. 1 and Table 1 confirm this for the blue-green algae. In the suspension of large (57.6 μm) colonies, at 30 mg chlorophyll *a* m^{-3}, the flux of photosynthetic quanta at 5 m is three times as high as that in the suspension of 8 μm colonies, and four times as high as in the equivalent pigment solution. By contrast the calculated spectral energy distribution at 5 m for the 0.8 μm cells (30 mg chlorophyll *a* m^{-3}) is virtually identical with that for the pigment solution, and the total flux of photosynthetic quanta is only 3.8% higher than in the solution.

Table 1. Attenuation of light in the 350 - 700 nm waveband in suspensions of spherical algal cells or colonies. 30 mg chlorophyll *a* m^{-3} in the suspension. 2% (dry weight) chlorophyll in the cells. Q_o is 99.41 x 10^{15} quanta cm^{-2} s^{-1}. The values of Q_{sus} and Q_{sol} are for depth 5 m. Q_w at 5 m is 5.59 x 10^{15} quanta cm^{-2} s^{-1}. K_w is 0.576 m^{-1}.

Algal type	Diameter μm	Q_{sus}	Q_{sol} quanta cm^{-2} s^{-1} x 10^{15}	Q_{sus}/Q_{sol}	K_{sus} m^{-1}	ΔK_{sus}	z_{eu} m
Green	8.0	2.98	2.93	1.017	0.701	0.0042	6.57
Blue-green	57.6	2.15	0.53	4.057	0.766	0.0063	6.01
Blue-green	8.0	0.71	0.53	1.340	0.987	0.0137	4.67
Blue-green	0.8	0.55	0.53	1.038	1.040	0.0155	4.43

The penetration coefficients increase, as predicted, with cell numbers. For the 57.6 μm blue-green colonies the ratio of flux of photosynthetic quanta (5 m) in the suspension to that in the solution is 1.804 at 10 mg chlorophyll *a* m^{-3}, compared to 4.057 at 30 mg chlorophyll *a* m^{-3}. When the depth is reduced to 2 m (at 30 mg chlorophyll a m^{-3}) this ratio falls to 1.984 : equation (4) predicts that the penetration coefficient should increase with depth.

4. Vertical attenuation coefficients

As a consequence of their strong biliprotein absorption in the 545-655 nm region the blue-green algal suspensions have higher vertical attenuation coefficients than similar suspensions of green algae. Comparison of the K_{sus} values for the three different blue-green colony/cell sizes in Table 1 shows how the attenuation coefficient increases as particle size diminishes. The increment in attenuation coefficient per unit algal concentration is more sensitive both to the pigment composition, and to the physical structure, of the phytoplankton canopy. ΔK_{sus} is much higher for the blue-green, than for the green, algae, and increases markedly as particle size decreases.

For algae of a given size alterations in pigment content or composition which decrease the vertical attenuation coefficient of the suspension, increase the depth of medium (euphotic depth, z_{eu} : corresponds approximately to Q_{sus} = 1% Q_o) which can be exploited for algal photosynthesis and growth. Similarly, for a given pigment content and composition, an increase, in cell or colony size (at constant biomass concentration) will also increase the euphotic depth (Table 1).

If it is wished to specify a particular euphotic depth, z'_{eu}, e.g. one that corresponds to the full depth of a particular algal culture pond, sewage oxidation pond etc., then the standing crop of algae, S_w, (expressed as g dry weight m^{-2}) that can be supported in that euphotic depth is given by

$$S_w = \frac{4.605 - K_w z'_{eu}}{10p\ \Delta K_{sus}} \qquad (8)$$

where p is the chlorophyll a content of the cells as a percentage of the dry weight[5]. It follows that to maximize the standing crop for a given euphotic depth ΔK_{sus} should be as small as possible, e.g. the cells or colonies should be as large as possible (at a given pigment concentration per unit algal weight). For the blue-green algae, with a euphotic depth of 5 m, standing crops of 13.7, 6.3 and 5.6 g m^{-2} could be supported with cell/colony diameters of 57.6, 8.0 and 0.8 μm respectively. It also follows from equation (8) that to obtain the maximum possible standing crop the depth of the water body (if this is to coincide with z_{eu}) should be the minimum practicable.

SUMMARY AND CONCLUSIONS

A mathematical treatment of light attenuation within suspensions of phytoplankton in natural waters has been developed which can be used to calculate the spectral distribution of light intensity at a given depth, the vertical attenuation coefficient for photosynthetically active radiation, and the increment in the attenuation coefficient for a given increase in total algal concentration.

The treatment can also be used to predict the effect of changes in phytoplankton canopy structure (cell size, shape, pigment concentration etc.) on light attenuation within the suspension, on the depth of the euphotic zone, and on the sustainable standing crop of algae. The limitations of the theory in its present form are that it has been derived for non-scattering, internally homogeneous cells, and that it is concerned only with the attenuation of direct, parallel sunlight and ignores the contribution of diffuse sky light.

The theory has been applied to hypothetical model suspensions of spherical green and blue-green algal cells and colonies. Plausible assumptions as to cell/colony size, and pigment content and composition have been made. Values of the specific absorption coefficients for the algal pigment mixtures have been obtained from the literature and used to calculate the absorption characteristics of the individual cells and colonies. The water in which the cells are suspended is assumed to contain levels of dissolved yellow substances (gelbstoff) characteristic of inland waters of intermediate colour : the actual values for the absorption coefficients of this water at different wavelengths have been based on measurements carried out on various local waters. On the basis of the light absorption characteristics of the cells/colonies, and the water, determined in this way, and assuming a solar elevation (above the surface) of 45°, calculations of the spectral distribution of light (quanta) intensity, and of the total photosynthetically active quanta (350 - 700 nm), for various depths, have been carried out.

It is found that for green algae in water in which effectively all the blue light is absorbed (depth 5 m) by dissolved yellow substances (gelbstoff) the light attenuation properties of the suspension are rather similar to those of an equivalent pigment solution : this is because the individual cells have relatively low absorption values in the spectral region (545 - 655 nm) in which most of the transmitted light occurs. In the case of blue-green algae, which have a strong absorption in this region, there can be marked differences between the attenuation properties of suspensions and of the equivalent pigment solutions. The suspensions transmit substantially more photosynthetically active radiation than the solutions : the effect increases with cell/colony size, with algal concentration and with depth, and can amount to a difference of several-fold. Vertical attenuation coefficients (for the 350 - 700 nm waveband), the increments in the attenuation coefficients per unit increase in algal concentration, euphotic depths, and the maximum substainable standing crops of algae, have been calculated, and their dependence on cell/colony size, and pigment composition, is demonstrated.

REFERENCES

1. Talling, J.F., 1960, Wett. u. Leben 12, 235.
2. Talling, J.F., 1970, In, Prediction and measurement of photosynthetic productivity (Pudoc, Wageningen) 431.
3. Talling, J.F., 1971, Mitt. Internat. Verein. Limnol. 19, 214.
4. Kirk, J.T.O., 1974a, submitted for publication
5. Kirk, J.T.O., 1974b, submitted for publication
6. Duysens, L.N.M., 1956, Biochim. Biophys. Acta 19, 1.
7. Smith, R.C. and Tyler, J.E., 1967, J. Opt. Soc. Amer. 57, 589.
8. Collins, B.G., CSIRO, Division of Atmospheric Physics, 1971, personal communication.
9. Smith, R.C., 1967, Limnol. Oceanogr. 13, 423.
10. Jerlov, N.G. and Nygard, K., 1969, Rep. Inst. Fys. Oceanogr., Copenhagen. 10, 19pp.
11. Findenegg, I., 1969, In, Primary production in aquatic environments (Blackwell, London) 18.

M. AVRON, *Proceedings of the Third International Congress on Photosynthesis*
September 2-6, 1974, The Weizmann Institute of Science, Rehovot, Israel
Elsevier Scientific Publishing Company, Amsterdam, The Netherlands, 1974

MONOMOLECULAR FILMS OF OXIDIZED AND REDUCED CYTOCHROME f

P. Chin and S. S. Brody
New York University, Washington Square, New York, N.Y.

1. Introduction

In the lamellar membrane of the chloroplast cytochrome f is closely associated with System I chlorophyll. The latter are oriented in a closely packed monolayer between layers of lipid and protein[1]. Pigment monolayers on an aqueous surface closely stimulate the *in vivo* state and serves as a useful model to study the physical and chemical properties of endogenous materials in an oriented, well defined, system. While the properties of cyt f in solution are well known, the properties in a membrane situation are not[2].

Surface properties of monomolecular films of ox and red cyt f are presented in this paper. Properties studied include area/molecule (A) and surface potential of the film (ΔV) as a function of pH. Changes in A and ΔV reflect conformational changes of the protein on the surface. Before interaction between cyt and chlorophyll can be evaluated, the surface properties of cyt monolayers must be determined.

2. Materials and Method

Monolayer studies are carried out using a Wilhelmy plate surface balance similar to that described previously[3]. The entire apparatus is housed in an environmental chamber. The sensitivity of the balance is $\pm$ 0.2 dyne/cm. A constant temperature of 15°C is maintained throughout the experiments using a thermostatically controlled, water cooling system imbedded in the trough.

The concentration of the aqueous solution of cytochrome (cyt) is determined spectrophotometrically. For cyt f a molar extinction coefficient of 2.6×10^4 is used for the reduced form and 1.1×10^4 for the oxidized form.

The area/molecule, A_{π} and surface potential ΔV_{π} are measured at a given surface pressure, π dyne/cm. The accuracy of A_{π} is $\pm$ 2% unless otherwise noted. Surface potential, ΔV, measurements are made as described previously[3]. The accuracy of ΔV is $\pm$ 10 mV.

π - ΔV and π - A curves are measured repeatedly until reproducible values are obtained. Usually after three isotherms consistant values are obtained.

Since the π - A curves for cyt do not have a linear portion an extrapolation to determine A_{π} at π = 0 dyne/cm is very uncertain, therefore A_{π} is measured at π = 6 dyne/cm. ΔV is equal to the difference in potential of the surface with the film (V) and the clean surface (V_{H_2O}) ie. ($V - V_{H_2O}$). More reproducible values of ΔV are obtained by calculating V_{H_2O} from V as a function of concentration of molecules on the surface, see results[4].

The cyt f is a gift from Prof. N. Bishop of the University of Oregon, Corvallis. Chl a is prepared as described by Aghion et al (1969). Addition of excess sodium ascorbate (Nutritional Biochemical Corporation, Cleveland, Ohio) is used to completely reduce the cytochromes. Addition of excess potassium ferricyanide is used to completely oxidize the cytochromes (Fischer Chemical Co.). Phosphate buffer, ionic strength 0.6 is used in all the experiments. These experiments are performed in air and under fluorescent light.

3. Results and Discussion

Surface isotherms for ox and red cyt f at pH 7.8 are shown in Fig. 1; A_6 for ox cyt f is 2640 $Å^2$ and for red cyt f is 2900 $Å^2$.

A graph of ΔV as a function of 1/A (molecules/$Å^2$) at pH 7.8 is shown in Fig. 2. Extrapolating to 1/A equal to zero gives the potential of water under the film, i.e. V_{H_2O}; for red cyt f, V_{H_2O} is -137 mV, for ox cyt f it equals -231 mV. These values are used to calculate ΔV.

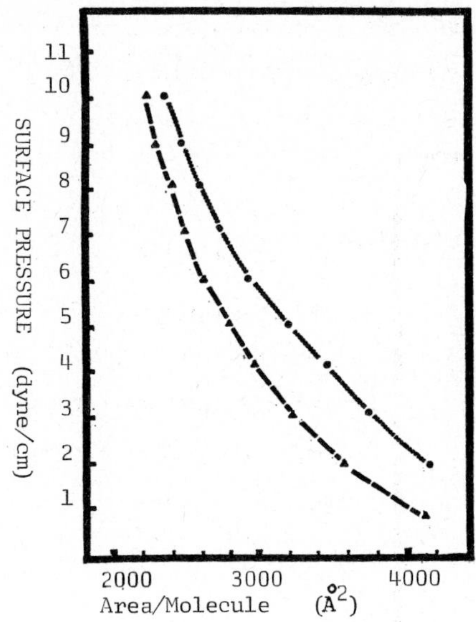

Figure 1: Area/molecule of oxidized (— —) and reduced (....) cytochrome f as a function of surface pressure. The aqueous subphase contained phosphate buffer pH 7.8, ionic strength 0.6 at 15°C.

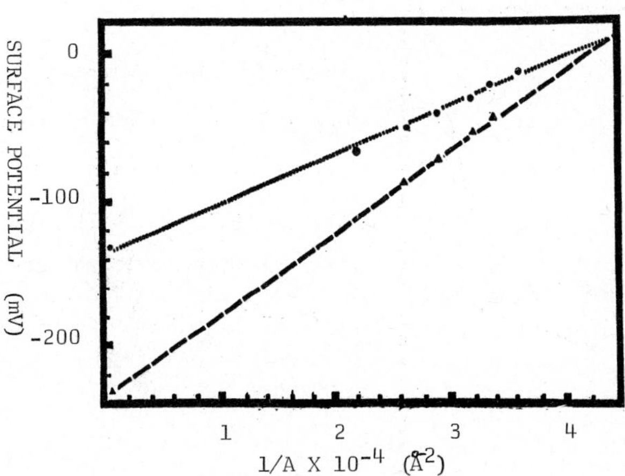

Figure 2: Surface potential as a function of molecules/area of oxidized (— —) and reduced (....) cytochrome f. Subphase same as in Fig. 1.

The ΔV_6 and A_6 of ox and red. cyt f as a function of pH are shown in Fig. 3 & 4. Ox cyt f has a maximum for both ΔV_6 and A_6 at about pH 7.5. The maximum value for ΔV_6 is 200 mV, the maximum for A_6 is 2600 $\mathring{A}^2$.

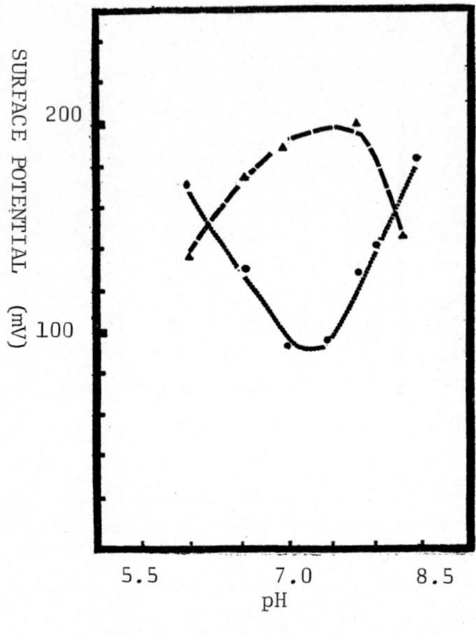

Figure 3: The surface potential of oxidized (— —) and reduced (....) cytochrome f as a function of pH. Subphase contained phosphate buffer ionic strength 0.6 at 15°C. The surface pressure is 6 dyne/cm.

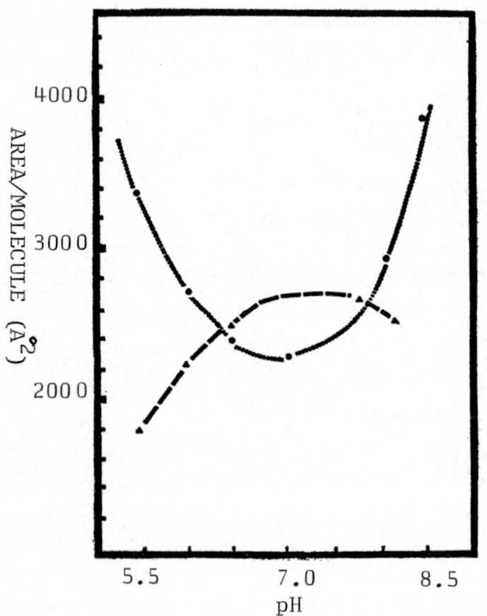

Figure 4: Area/molecule of oxidized (— —) and reduced (....) cytochrome f at surface pressure of 6 dyne/cm as a function pH. Subphase same as in Fig. 3.

Red cyt *f* on the other hand, has a minimum for both ΔV_6 and A_6 at about pH 7.3. The minimum value for ΔV_6 is 95 mV; the minimum A_6 is 2200 $Å^2$.

The curves for ox and red cyt *f* as a function of pH have equal areas at pH of 6.8 and 7.8 (Fig. 3); they have equal ΔV's at pH of 6.2 and 8.2 (Fig. 4). Once the film stabilizes on the surface, constant reproducible isotherms can be measured for several hours. There is no evidence of time dependent denaturation.

As can be seen in Figs. 3 and 4, there are two pH's where both the area/molecule and surface potential are approximately the same for the oxidized and reduced forms of cyt. At these pH's a redox reaction (of cyt) might be carried out with minimal energy since no significant change in protein conformation appears to be involved.

At the isoelectric point a maximum or minimum area for a protein is expected. The present study was not carried out at the isoelectric point of cyt *f* which occur at a pH of 4.7. The isoelectric point, as usually determined, is an average neutral charge for the entire surface of the protein molecule. The present monolayer study of A and ΔV as a function of pH is an effective isoelectric point for that portion of the protein surface in contact with, or submerged in, the aqueous phase. The fact that the pH maximum and minimum are about the same for the red. and ox cyt *f* indicates that the same portions of the cyt molecule are in contact with the water in the two redox states. That is, the same group of amino acids are being titrated. Reduction of cyt *f* results in a decrease of ΔV at a pH of 7.5. This is the direction of change to be expected for ΔV for the addition of an electron.

When outside the physiological pH range of 6 to 8 there is a reversal in the relative size and potential of the ox and red. forms of cyt (Fig.3). *In vivo*, these changes in A and ΔV could well be related to the position or retention of a particular redox state of cyt in the membrane. Changes in shape of the protein molecule as a result of redox state has previously been proposed[6,7].

References

1. Kreutz, W. 1970, Adv. in Bot. Res. (R. D. Preston, ed.), 3, 53.
2. Davenport, H. L. and Hill, R. 1952, Proc. Roy. Soc., B139, 327.
3. Brody, S. S. 1971, Zeit f. Natur forschg, 26b, 922.
4. Brody, S. S. 1973, Zeit f. Natur forschg, 28c, 157.
5. Aghion, J., Broyde, S. B. and Brody, S. S. 1969, Biochemistry, 8,3120.
6. Butt, W. D. and Keilin, D. 1962, Proc. Roy. Soc., B156, 429.
7. Hagihara, B., Morikawa, I., Sekuzu, I. and Okunuki, K. 1958, J. Biochem. (Tokyo), 45, 551.

M. AVRON, *Proceedings of the Third International Congress on Photosynthesis*, September 2-6, 1974, Weizmann Institute of Science, Rehovot, Israel
Elsevier Scientific Publishing Company, Amsterdam, The Netherlands, 1974

ENERGY TRANSFERS FROM PROTOCHLOROPHYLLIDE TO CHLOROPHYLLIDE AT ROOM TEMPERATURE IN LYOPHILIZED BEAN LEAVES.

E. Dujardin

Laboratoire de Photobiologie
Université de Liege, Sart-Tilman
4000-Liege, Belgium.

1. Introduction

In another paper[1] we show that energy transfers occur at 77^0K in lyophilized primary bean leaves from the $P_{645-639}$ complex to the $P_{657-647}$ complex when the leaves are etiolated and kept in the dark, and from the $P_{645-639}$ and $P_{657-647}$ complexes to the $P_{688-676}$, or to the $P_{685-675}$ complexes when the leaves have been irradiated by a 1 ms flash of white light.

We now give evidences for the occurence of an energy transfer from the $P_{645-639}$complex to the $P_{688-676}$ complex at room temperature. It is again stressed that part of the $P_{645-639}$ and of the $P_{657-647}$ complexes is not photoreducible in lyophilized leaf. In lyophilized etiolated leaves, analysis of extracts show that only 50 % of the protochlorophyllide molecules are reduced by light.

2. Material and methods

Beans of the variety "Commodore" were grown in complete darkness and their 15-day-old primary leaves were lyophilized as described in ref. 2.

The apparatus used for the registration of the fluorescence emission spectra in liquid nitrogen is described in ref. 3. The same device has been used for registrating emission spectra at room temperature.

The method for the measurement of the excitation of the fluorescence emission at liquid nitrogen temperature has been described in ref. 4. The same method was used for measurements of the excitation spectra at room temperature (the leaves were simply not frozen).

The exposure of the leaves to light occurred at room temperature. As a rule, a 1 ms polychromatic "Multiblitz Report Porba" electronic flash was used. The characteristics of this flash and the way it was used for obtaining different irradiation intensities are described in ref. 1.

3. Results and discussion

The comparison of the fluorescence emission spectrum with the excitation spectrum of the main emission band[1] shows that at 77°K, lyophilized etiolated bean leaves contain the protochlorophyllide-lipoprotein complexes $P_{645-639}$ and $P_{657-647}$, and that $P_{645-639}$ transfers energy to $P_{657-647}$ in these leaves.

Energy transfers occur at 77°K also after irradiation by light[1]. For instance, when a lyophilized leaf has received at room temperature a low-intensity polychromatic 1 ms flash, just before being frozen in liquid nitrogen, the emission at 688 nm of the chlorophyllide which is produced is strongly enhanced as a result of an energy transfer from protochlorophyllide. The efficency of this transfer is marquedly higher in the lyophilized leaf than in the fresh leaf[1].

If half of the initial protochlorophyllide has been reduced at room temperature by a medium-intensity flash, the fluorescence emission at 77°K comes essentially from chlorophyllide, but the excitation spectrum of this emission shows a substantial contribution of the energy absorbed by the protochlophyllide-lipoprotein complexes between 625 and 647 nm (fig. 1, a and b).

When the intensity of the 1 ms flash given at room temperature is increased up to saturation of the photoreduction, the protochlorophyllide contribution to the chlorophyllide fluorescence at 77°K decreases, but it never desappears completely.

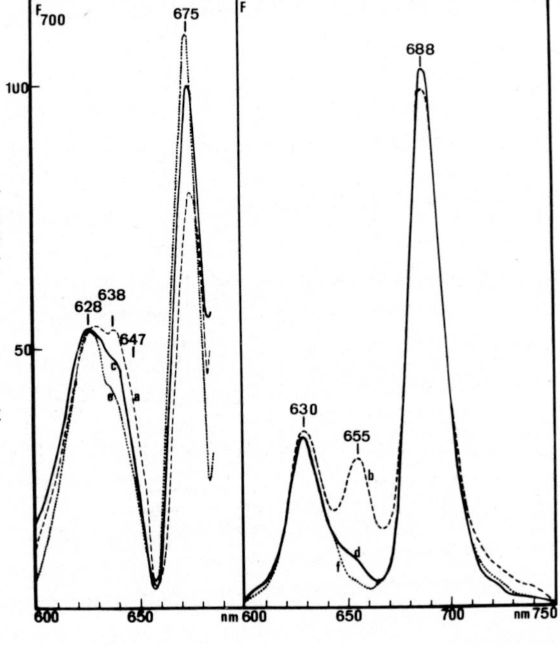

Fig. 1. 77°K excitation spectra of the fluorescence at 700 nm of lyophilized primary bean leaves after irradiation by a 1 ms flash of polychromatic light at room temperature, and the 77°K fluorescence emissions of the same leaves. Curves a and b : the flash was positioned 1 m from the leaf; curves c and d : a single saturating flash at 60 cm from the leaf was given. Curves e and f : two flashes were given, separated by a 7 sec dark interval (from Dujardin[1]).

A shoulder remains around 640 nm in the excitation spectrum of the chlorophyllide emission at 77°K even after a very intense flash has been repeated twice at room temperature.(fig. 1, e).

In the lyophilized bean leaf the spectral characteristics of the $P_{688-676}$ complex formed by a saturating flash remain stable for a very long time at room temperature; shifts to longer or shorter wavelengths are not seen[2]. This feature permits registration of both fluorescence emission and fluorescence excitation spectra at room temperature (at saturation of photoreduction).

After an intense flash the fluorescence emission spectrum of the lyophilized leaf at room temperature is similar to that at 77°K (fig. 2). Both comprise two bands : one band is emitted around 630 nm by the non-photoreducible $P_{633-628}$ complex; the second is emitted by the chlorophyllide-complex at 688 nm; as expected, the half-band width is increased at room temperature. A weak band is often seen around 655 nm at 77°K (fig. 1, d, f). There is no distinct emission at 645 nm at room temperature, although some $P_{645-639}$ complex molecules remain present as shown by the 640 nm band in the excitation spectrum of chlorophyllide emission; the excitation spectrum of chlorophyllide emission is the same at room temperature as at 77°K (fig. 1, c, e).

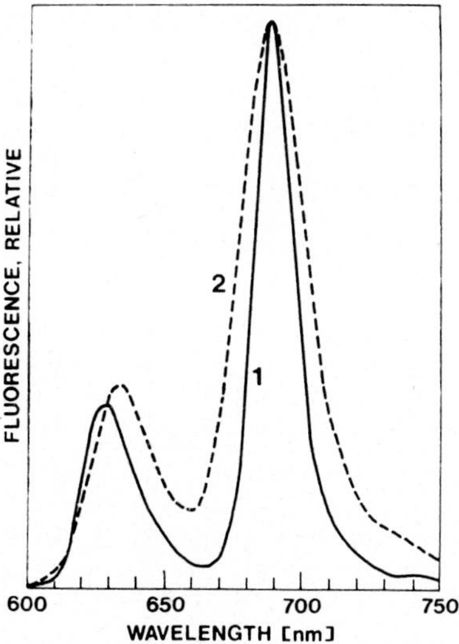

Fig. 2. Fluorescence emission of a lyophilized leaf irradiated with an intense flash as recorded : 1) at liquid nitrogen temperature (full line);
2) at room temperature (dashed line).
(from Dujardin[2])..

It is concluded that : a) some protochlorophyllide molecules belonging to the $P_{645-639}$ and $P_{657-647}$ complexes are not photoreduced at room temperature when a lyophilized etiolated leaf is irradiated by one or even two intense polychromatic flashes; b) the $P_{645-639}$ molecules which remain unreduced transfer energy to chlorophyllide at room temperature as well as at $77^{o}K$.

An energy transfer at $77^{o}K$ from protochlorophyllide to chlorophyllide in the lyophilized leaf is observed at the beginning of the irradiation when chlorophyllide belongs to the complex emitting at 685 nm ($P_{685-675}$), and after a more intense irradiation when it belongs to the complex emitting at 688 nm ($P_{688-676}$). We do not know however if energy transfer occurs also at room temperature from protochlorophyllide to chlorophyllide at an early stage of the irradiation.

References

1. Dujardin, E., 1974, Photosynthetica, submitted for publication.
2. Dujardin, E., 1973, Photosynthetica 7, 121.
3. Sironval, C., Brouers, M., Michel, J. M. and Kuiper, Y., 1968, Photosynthetica 2, 268.
4. Brouers, M., Kuiper, Y. and Sironval, C., 1972, Photosynthetica 6, 169.

M. AVRON, *Proceedings of the Third International Congress on Photosynthesis*, September 2-6, 1974, Weizmann Institute of Science, Rehovot, Israel
Elsevier Scientific Publishing Company, Amsterdam, The Netherlands, 1974

CHLOROPHYLL CATION RADICAL IN PHOSPHOLIPID VESICLES

M. Tomkiewicz and G. Corker
IBM Thomas J. Watson Research Center
Yorktown Heights, New York 10598

1. Introduction

Chlorophyll a dissolved in sonicated phospholipid vesicles might serve as a good model system for understanding some of the physical processes that occur during early stages of photosynthesis, a process which efficiently converts electromagnetic energy into useful chemical energy. With the proper electron acceptor, one might achieve in the phospholipid system a similar efficient photon convertion. With this in mind, we initiated a study of the model system using the detectability of the photochemically produced chlorophyll cation radical as evidence of photochemical charge separation. We present preliminary results of that study.

2. Results and Discussion

Structure

The phospholipid that we used was egg lecithin and the primary electron acceptors were both organic and inorganic. However, we have worked primarily with $K_3F_e(CN)_6$, $SmCl_3$ and $EuCl_3$. The first one was used for the purpose of studying the behavior of the chlorophyll radical in the vesicles, while the other two were used with the hope that in their reduced form they will be able to reduce water to form hydrogen gas.

Transmission Electron Microscopy of this system, with $SmCl_3$ as the negative stain, shows a mixed population made of small, mostly single wall vesicles with a radius around 250Å and big, r=0.5μ, multilamellar structures. At this stage, we made no attempt to separate those structures.[1] The electron acceptor was added after sonication. As a result, if the concentration of the acceptor was kept below a critical concentration ($\sim 5 \times 10^{-2}$M) the acceptor remained outside the vesicles. Absorption spectrum of the chlorophyll in the vesicles shows maximum at 670nm without any red shifted components. The Circular Dichroic spectrum is typical of the monomer. The ESR line-width (∿8g) also does not show chlorophyll aggregation. It is very significant to note that although on the average we have about 500 molecules of chlorophyll per vesicle, which is equivalent to 5×10^{-2}M in concentration, we did not observe aggregation of the chlorophylls.

Charge Separation

The light induced charge separation in the system was detected by following the behavior of the chlorophyll radical with EPR. The EPR spectrum of the chlorophyll radical is shown in Fig. 1.

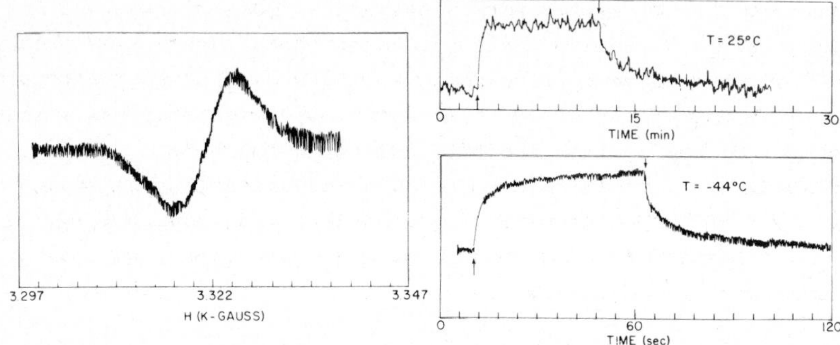

Fig. 1. ESR spectrum (left), time evolution at 25° (upper right) and time evolution at -44°C (lower right) of the light induced chlorophyll radical in phospholipid vesicles. The ESR spectrum and the time evolution at -44°C were obtained with 10^{-2}M $K_3Fe(CN)_6$ as the electron acceptor while the time evolution at 25°C was recorded using 10^{-2}M $SmCl_3$ as the acceptor.

The linewidth and the g value were invariant under x band and q band measurements. This indicates that we are dealing with one species. The steady state concentration of the radical and its time behavior at temperatures below freezing were identical with $SmCl_3$ and $K_3Fe(CN)_6$ as acceptors. The steady state concentration was much lower using $EuCl_3$ as acceptor. At room temperatures, the amplitude of the chlorophyll signal with $SmCl_3$ as acceptor was higher than the amplitude with $K_3Fe(CN)_6$.

Starting at liquid He temperature and going up to -100°C the EPR signal of the chlorophyll radical was formed upon illumination but did not decay for at least the first few hours. A measurable decay started to appear around -100°C. A representative trace of the time behavior in the temperature range -100°C to freezing is given in Fig. 1.

Both the rise and the decay have a very characteristic shape, both have a fast component and a slow component. The slow decay component is much more dramatically illustrated in the experimental trace in Fig. 2. This time behavior can be explained by invoking space and time dependence of the concentration of the reduced acceptor. That is, the diffusion of the reduced acceptor (or the electron) is probably one of the rate limiting steps at low temperatures.

A mathematical model which is based on those premises was developed. Its essential points are summarized in the Appendix. A detailed account of this theory will be given elsewhere. It will be sufficient to note here that the approximations that we made in developing the model were to neglect interactions between vesicles, to assume single-wall, average-size vesicles and to assume that the acceptor is located on the outside of the vesicle. The first approximation seems to be valid with the slow diffusion at subfreezing temperatures. However, it is invalid at temperatures above freezing where diffusion is more rapid. The least mean square fit of the experimental results at T = $-52^{o}C$ to this theoretical model is shown in Fig. 2.

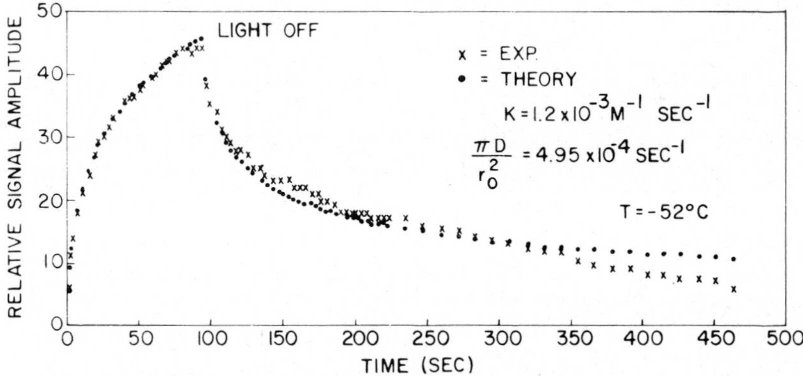

Fig. 2. Least mean square fit of the diffusion model (see Appendix) to the experimental results obtained at $-52^{o}C$. $10^{-2}M$ $K_3F_e(CN)_6$ was used as the electron acceptor.

For this fit, we adjusted two parameters, one was the recombination rate constant and the other was the ratio of the diffusion coefficient to the square of the radius of the vesicles. We have no independent estimate for the recombination rate constant. However, we can a priori estimate the diffusion coefficient at this temperature and from that derive the radius of the "average size" vesicle. The estimated size is well within the range observed by Electron Microscopy.

The misfit at the end of the curve can be theoretical or experimental in nature. Further work is needed to determine its origin. As mentioned before, the model completely fails when the rate of diffusion is much larger than the recombination rate. This seems to be the case at room temperature (see Fig. 1). We cannot explain in detail the time behavior at room temperature. It is worth mentioning that from the rise portion of the curve, it does not seem that diffusion is a rate determinating step. The steady state amount of the radical at room temperature is about 5% of the steady state amount produced at -52°C.. Comparing the radical signal amplitude produced at room temperature to the signal height of the reaction-center radical signal in Chlorella, when both samples had the same chlorophyll concentration and subjected to identical illumination, suggests that the yield with the vesicles is 20% of the algae.

Reducing Power

When ubiquinone (or benzoquinone) was incorporated into the system, no EPR signals of the quinone radical or the light induced chlorophyll were observed. In the presence of other electron acceptors, the ubiquinone had no influence. This is an interesting observation in conjunction with the work of others[2] with the corresponding model systems in organic solvents.

When spinach ferrodoxin was used as a possible acceptor in the system, optical studies were inconclusive while no evidence for the reduced ferrodoxin was found by EPR at liquid He temperatures, when the samples were illuminated at room temperature and frozen. However, when the iron sulfur protein which was identified by Shanmugan and Arnon[3] as Ferrodoxin II from R. rubrum was used, the protein was reduced by illumination of the vesicles. The EPR signal observed in illuminated vesicles containing this protein was identical to the signal observed when pure protein is reduced by dithionite. Some details about this protein, together with the

EPR of the reduced and oxidized form are given by Corker and Sharpe.[4]

The system with $SmCl_3$ and $EuCl_3$ as acceptors was checked for hydrogen evolution. Both metals are known to reduce water in their reduced form. The reaction is thermal in the case of Sm. However, there is some doubt whether the Europium reaction is thermal or photochemical. With our system, hydrogen was not detected when we used Europium as the acceptor, while hydrogen gas was observed using $SmCl_3$ as the acceptor. However, the evolution of hydrogen was a very irreproducible phenomena. We are now in the process of establishing the experimental factors which govern its irreproducibility.

3. Summary

Excited molecules of chlorophyll, solubilized in sonicated lecithin vesicles, are capable of undergoing reversible donation of an electron by interaction with a suitable partner molecule. $K_3F_e(CN)_6$, $SmCl_3$, $EuCl_3$, ferrodoxines and quinones, were used as electron acceptors.

The kinetics of the charge separation was measured at various temperatures and appeared to be characteristic to this particular heterogenous environment. Diffusion of the reduced acceptor, away from the vesicle surface, is stipulated as one of the rate limiting steps at subfreezing temperatures. A model which describes the diffusion process was developed and it can account well for the observed kinetics.

When $SmCl_3$ was used as an acceptor, it was found that the reduced samarium can reduce water and the system can produce hydrogen gas.

4. Appendix

We are assuming the chlorophyll radical to be formed according to the following mechanism:

$$\mathrm{Chl} \xrightarrow{h\nu} \mathrm{Chl}^*$$
$$\mathrm{Chl}^* \xrightarrow{k\text{-}1} \mathrm{Chl}^+ + \mathrm{A}^-$$
$$\mathrm{Chl}^* \xrightarrow{k\text{-}2} \mathrm{Chl}$$
$$\mathrm{Chl}^+ + \mathrm{A}^- \xrightarrow{k} \mathrm{Chl} + \mathrm{A}$$

where A is the electron acceptor and A^- is the reduced electron acceptor. We assume that these reactions occur on the surface of the vesicles. Thus, a concentration gradient of A^- will be established between the surface and the bulk. We further assume that the reduced acceptor will obey the diffusion equation expressed in spherical coordinates.

$$\frac{\partial c}{\partial t} = D\left(\frac{\partial^2 c}{\partial r^2} + \frac{2}{r}\frac{\partial c}{\partial r}\right)$$

with the boundary conditions:

$$\begin{array}{lll} c = c(r,t) & r \geq r_o & t \geq 0 \\ c = x_1(t) & r = r_o & t \geq 0 \\ c = f(r) & r \geq r_o & t = 0 \end{array}$$

when r_o is the external radius of the vesicle, c is the concentration of the reduced acceptor and $x_1(t)$ is the concentration of the reduced acceptor at the surface of the vesicles. Assuming a single vesicle in an infinite media, we can solve the equation[5] to obtain

$$C = \frac{1}{2r\sqrt{\pi Dt}} \int_{r_o}^{\infty} r'f(r')\{e^{-(r-r')^2/4Dt} - e^{-(r+r'-2r_o)^2/4Dt}\}dr'$$

$$+ \frac{2r_o}{r\sqrt{\pi}} \int_{\frac{r-r_o}{2\sqrt{Dt}}}^{\infty} x_1\left[t - \frac{r-r_o}{4D\mu^2}\right] e^{-\mu^2}\, d\mu$$

Assuming only chlorophyll and acceptor interactions and using the following relationships

$$y_1 = \int_{r_o}^{\infty} 4\pi r^2 c\, dr,$$

where y_1 is the number of chlorophyll radicals per vesicle, integration of these equations yields the following results:

A. For the rise time:

$$f(r) = 0$$

$$\frac{dy}{dt} = \frac{k_{-1}[A]I_a}{k_{-1}[A]+k_2} - ky\, x_1(t)$$

$$y = \sqrt{a}\int_0^t \frac{1 + \sqrt{a(t-\xi)}}{\sqrt{t-\xi}}\, x_1(\xi)d\xi$$

where I_a is the intensity of the absorbed light and [A] is the concentration of the acceptor, assumed to be space and time independent because of its large access, and

$$a = \pi D/r_o^2$$

and

$$y = y_1/4r_o^3$$

B. For the decay:

We assumed that at the time the light was turned off, the spatial distribution of the reduced acceptor will be Gaussian such that:

$$f(r) = x_1^{off}\left[e^{-\pi(r-r_o)^2/4\alpha^2 r_o^2}\right]$$

where x_1^{off} is the reduced acceptor concentration on the surface of the vesicle when the light is turned off; and α should obey the following cubic equation:

$$y^{off} = x_1^{off}(\pi\alpha + 4\alpha^2 + 2\alpha^3)$$

where y^{off} is the concentration of the chlorophyll radical when the light is turned off. The concentration of the chlorophyll radical will be given by the following equations:

$$\frac{dy}{dt} = -ky\, x_1(t)$$

$$y = g(t) + \sqrt{a}\int_o^t \frac{1+\sqrt{a(t-\xi)}}{\sqrt{t-\xi}}\; x_1(\xi)d\xi$$

$$g(t) = x_1^{off}\left[\pi\alpha + 2\alpha^2 + 2\alpha^3 + \frac{2\alpha^2}{\sqrt{1+\frac{at}{\alpha^2}}} - 2\alpha\arctang\left(\frac{\sqrt{at}}{\alpha}\right)\right]$$

where t=0 is the time when the light was turned off. The rise and decay kinetics involve a couple of equations; one is a differential equation and one an integral equation. Both equations have to be satisfied simultaneously. The midpoint product integration method[6] was used for a numerical solution of those equations. The Fletcher and Powel[7] minimization routine was used for the curve fitting.

References

1. Sheetz, M. P. and Chan, S. I., 1972, Biochem. 11, 4573.

2. Harbour, J. R. and Tollin, G. 1974, Photochem. Photobiol. 19, 69.

3. Shanmugan, K. T. and Arnon, D. I. 1972, Biochem. Biophys. Acta. 256, 487.

4. Corker, G. A. and Sharpe, S., this symposium.

5. Carslaw, H. S. and Jaeger, J. C. 1959, in: Conduction of Heat in Solids (Oxford) p. 247.

6. Weiss, R. and Anderson, R. S. 1972, Numer. Math. 18, 442.

7. Fletcher, R. and Powell, M. J. D. 1963, The Computer Journal 6, 163.

M. AVRON, *Proceedings of the Third International Congress on Photosynthesis*, September 2-6, 1974, Weizmann Institute of Science, Rehovot, Israel
Elsevier Scientific Publishing Company, Amsterdam, The Netherlands, 1974

STRUCTURAL FEATURES OF THE PRIMARY CHARGE SEPARATION ACROSS THE FUNCTIONAL MEMBRANE OF PHOTOSYNTHESIS IN GREEN PLANTS

Wolfgang Junge

Max-Volmer-Institut für Physikalische Chemie und Molekularbiologie
Technische Universität Berlin

1. Introduction

Triggered by Mitchell's hypothesis[1,2] evidence was accumulated that stimulation of the photosynthetic electron transport chain generates an electrochemical potential of the proton across the thylakoid membrane. It was demonstrated that light causes the uptake of protons from the outer aqueous phase[3], the release of protons into the inner one[4-6] and the generation of an electric potential difference across the membrane[7]. The structural properties of the electric generator in photosystem I are subject of this communication. First the vectorial and the protolytic properties of the electron transport chain and the layer structure of the thylakoid membrane are reviewed. Then evidence for the location of chlorophyll-a_I (chl-a_I) at the inner side with a small inclination to the membrane is presented. This structure is discussed with respect to the mechanisms of energy transfer and electron transport across the thylakoid membrane.

Vectorial and protolytic properties of the electron transport chain

Experiments which are reviewed below confirmed the following scheme for the generation of the electrochemical potential difference (Fig.1): Light Reaction II translocates one electron from the inner to the outer side of the thylakoid membrane. One proton per electron is released into the inner phase. Plastoquinone (PQ) on reduction at the outer side binds one proton per electron. The oxidation of PQ by plastocyanin (PC) at the inner side is followed by the release of one proton per electron into the inner phase. Light Reaction I drives the electron from the inner to the outer side of the membrane where the reduction of the terminal electron acceptor $NADP^+$ causes the binding of 1/2 proton per electron. Fig.1 incorporates two features which are only presented to this congress, the exclusive role of PC (W.Haehnel) and the position of chl-a_I (this comm.). The evidence for the vectorial electron-hydrogen transport system illustrated in Fig.1 is as follows: Flash spectrophotometric studies revealed that the electric potential across the thylakoid membrane can be measured via electrochromic absorption changes of chloroplast bulk pigments[7,8]. It was demonstrated that both light reactions contribute about equal amounts to the electric potential[9]. The extremely rapid rise of the electric potential[10] which is paralleled by the extremely rapid photooxidation of chl-a_I[11] led to the conclusion that the electric potential generation is closely coupled to the primary photochemical act (for details see Ref.12). Four protolytic reaction sites were identified which according to stoichiometric[13] and kinetic[14] evidence were attributed to the oxidation of water at the inner side, the reduction of PQ at the outer and its subsequent oxidation at the inner side and the reduction of the terminal electron acceptor at the outer side[13] (see Ausländer & Junge).

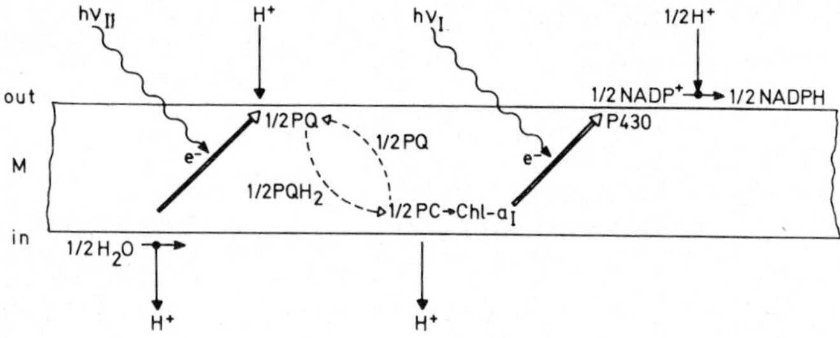

Fig.1. Scheme for the vectorial electron-hydrogen transport across the membrane

B i o c h e m i c a l studies with antibodies against electron acceptors of chl-a_I revealed the location of this acceptor system at the outer side of the membrane[15-17] and of plastocyanin at the inner side[18]. Studies with artificial redox compounds which were chemically modified for different lipid solubilities could be interpreted only under the assumption of a vectorial electron transport mediated by Light Reaction I across the membrane[19,20]. Studies on the influence of a transmembrane potential on the d e l a y e d l i g h t emission[21] led to the conclusion of vectorial electron transfer at Light Reaction II[22].

Two comments as to the vectorial electron transport scheme may be worthwhile to avoid misconception. The first concerns the question whether the electric field across the membrane is localized or delocalized. Delocalization over a functional unit containing at least 10^5 chlorophyll molecules at the millisecond time scale was demonstrated by a titration experiment with the ionophorous antibiotic gramicidin at extremely low concentrations[7]. Thus the electric field although originating from a localized charge separation at both light reactions is smeared out over the membrane because the ions redistribute in the aqueous phases. An estimate on the velocity of this redistribution can be based on recent experiments on electric induction phenomena with chloroplast suspensions excited with flashlight at nonsaturating energy[23,24]. From the observed time for the redistibution of ions around one thylakoid or more likely one grana stack which was about 10 μsec[24] we conclude that the redistribution between neighbouring reaction centers which are about 20 nm apart[25] occurs about two orders of magnitude more rapid. We expect that the electric field delocalizes over the thylakoid membrane in less than a microsecond at room temperature.

The second comment is related to the evolution of oxygen at the inner side of the membrane which may cause a critical oxygen concentration in the inner phase. It has to be questionned whether the oxygen permeability of the thylakoid membrane is large enough to avoid this. A rough estimate can be based on model studies on gas diffusion across lipoid monolayers at an air-water interface[26]. Even whith effective diffusion constants as low as $D=10^{-8} cm^2/sec$ [27] the diffusion time for a gas molecule across a layer of 10nm according to Einstein-Schmoluchowski[28] will below 0.1msec, that is two orders of magnitude more rapid than the relaxation of the electron transport chain.

The layer structure of the thylakoid membrane

Electron microscopy and X-ray small angle scattering (for reviews, see Ref.s 29,30) produced conflicting results on the layer structure of the thylakoid membrane as well as on its plane structure. It is not even settled whether the dielectric core of the membrane is a single or a double layer of lipids. Recent kinetic studies on the protolytic reactions with the outer aqueous phase revealed the existence of a shielding structure covering the reducing sites of both light reactions at the outer side of the membrane[14] (see Ausländer & Junge). The shielding structure is kinetically distinct from the main dielectric core of the membrane. While a pH-difference across the former relaxes within 60msec[14] it relaxes within several seconds across the whole membrane. It is obvious that the electric potential, which relaxes slower than with 60msec in freshly prepared chloroplasts (Fig.4) is generated across the core layer of the membrane.

Properties of chlorophyll-a_I

Photochemical activity of Light Reaction I is accompanied by negative going absoption changes at 700nm[32], 438nm[33] and a minor peak at 682nm[34]. These absorption changes were attributed to the photooxidation [35] of a special chlorophyll, chlorophyll-a_I[33]. Three independent types of observations gave evidence that chl-a_I is at least a dimer of chlorophyll-a molecules. The appearance of a double band at 682 and 700nm[34] can be interpreted by exciton interaction in an oblique chl-a dimer[36]. The circular dichroism spectrum of the absorption changes around 700nm fullfills the expectation for a dimer under exciton interaction[37]. The ESR line width of oxidized chl-a_I corresponds to the expectation for an unpaired electron delocalized over two porphyrin rings[38]. Despite of model studies with chlorophyll-a aggregates in vitro [38,39] little is known on the internal structure of chl-a_I. We use this term to denote the species which produces the above absorption changes on excitation.

2. Materials and Methods

Broken chloroplasts isolated from market spinach were used. The experiments in Fig.4 were carried out with freshly prepared chloroplasts selected for a low intrinsic membrane conductivity as previously[31]. For the remainder experiments we used broken chloroplasts which were stored under liquid nitrogen until use as in Ref.41. The chloroplasts were suspended to an average chlorophyll-concentration of 10μM in the following reaction medium:
tricine pH8, 3mM; KCl, 10mM; $MgCl_2$, 1mM; benzylviologen, 30μM.
In the photoselection experiments NH_4Cl at 3 mM was present.
The suspension was filled into a 2cm absorption cell which was mounted into a rapid kinetic spectrophotometer[43]. Photosynthesis was stimulated by a Xenon-flash (halftime 15μsec). The exciting light was tuned by absorption filters (Schott&Gen.), RG1/4 ($\lambda >$ 600nm) and BG28/4 (380-490nm) in Fig.4 and RG8/3($\lambda >$ 690nm) in the experiments in Fig.s 5 and 6. The wavelength of the interrogating light was dialed by a grating monochromator with a bandwidth of 5nm. The intensity of the interrogating light was below 100erg/cm^{2}sec in Fig.4 and 400 erg/cm^{2}sec in Fig.s 5 and 6. Changes in transmission on excitation were recorded. The signals were excited repetively by periodical flashes and averaged in a CAT 1000 on-line computer for improvement of the signal-to-noise ratio[43]. The electrical bandwidth of the detection system was limited by the dwell time per adress of the CAT which was chosen 62μsec (Fig.4) and 125μsec (Fig.s 5 and 6), resp..

Photoselection experiments were carried out in a set up the principle of which is illustrated in Fig.3[40]. An exciting and an interrogating beam impinge perpendicular to each other on an isotropic chloroplast suspension. The E-vector polarization of the exciting light was in the vertical, the E-vector polarization of the interrogating beam was varied, either in parallel or perpendicular to the former.

The exciting beam was polarized by a polaroid sheet,the interrogating one by a rotatable Glan-Thompson which were placed between the respective light source and the absorption cell. Experimental imperfections as well as check-up procedures for possible artifacts were discussed elsewhere[41]. The energy of the exciting flash was attenuated to yield about 20% saturation.

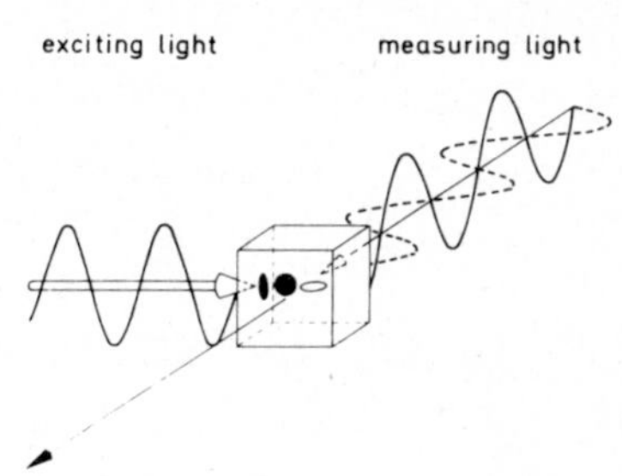

Fig.2. The optical set up in photoselection experiments. Full circles indicate excited thylakoids.

When exciting an isotropic chloroplast suspension with a linearly polarized flash of nonsaturating energy at wavelength greater than 680nm one excites by preference those orientations of thylakoid membranes which are illustrated by full circles in Fig.2[40,41]. This is evident from three physical properties of the system: 1. The probability for the absorption of light is proportional to the square of the skalar product between the E-vector of the exciting light and the transition moment of the molecular system. 2. Linear dichroism studies on thylakoid membranes oriented by spreading on surfaces[44,45] or by high magnetic field strength[46,47] revealed that the transition moments of those antennae pigments which absorb at wavelength greater than 680nm are oriented almost parallel to the thylakoid membrane[45]. 3. Resonant energy transfer which governs the action of the antennae pigments proceeds always "towards longer wavelength" no matter whether by the Förster[48] or by the weak exciton mechanism[49]. If the antennae transition moments were oriented at random in the thylakoid membrane no photoselection by a linearly polarized flash would occur. The information on the original polarization of the exciting light were lost during the resonant energy transfer within the antennae system. However, if the exciting light is tuned to the wavelength of the inplane oriented transition moments of the antennae ($\lambda > 680$nm) resonant energy transfer distributes the original polarization in the plane of the membrane but not isotropically (40,41). This makes the excitation conditions illustrated in Fig.2 select for those membrane orientations with their plane parallel to the E-vector of the exciting light. Absorption changes will result from the same photoselected ensemble. Their extent, in general, will differ depending on the polarization of the interrogating beam to the exciting one. We observed a linear dichroism by photoselection for the absorption changes of chl-a_I at 705 and at 430nm[40,41]. Information as to the orientation of chl-a_I in the membrane was inferred therefrom.

3. Results and Discussion

The location of chlorophyll-a_I in the membrane

ESR studies revealed that the unpaired electron in chl-a_I^+ is delocalized over two porphyrin rings[38]. We ask as to the location of these porphyrin rings with respect to the dielectric core layer of the thylakoid membrane. The gross location can be deduced from a kinetic argument[52]: The electric potential rises in less than 20nsec on flash excitation[10]. The photooxidation of chl-a_I occurs with the same high velocity[11] while its reduction is several orders of magnitude slower. Thus the oxidation of chl-a_I to chl-a_I^+ is accompanied by the electric potential generation. From the polarity of the electric potential which is positive inside[7] it follows that chl-a_I is located at the inner side of the thylakoid membrane.
Let us furthermore specify the term "at the inner side". If chl-a_I^+ were located at the inner side of the membrane but still within the hydrophobic core of the membrane as illustrated by positive charges in Fig.3, an inhomogenous electric field were formed in the membrane. This field pattern (Fig.3,left) should transform into the homogenous pattern illustrated in the right of Fig.3, when chl-a_I is reduced via plastocyanin which according to Ref.18 is likely to be in contact with the inner aqueous phase. This produces a decrease of the average field strength in the space between chl-a_I and the outer side of the membrane and an increase between chl-a_I and the inner side. This was subjected to an experimental test.
That the electric field after the charge transfer between chl-a_I and the acceptor extends into the space between chl-a_I and the inner side is due to electrostatic induction between the positive charges and the inner aqueous phase (distance between reaction centers:about 20 nm, membrane thickness: less than 7 nm[30]). The electric field would not reach the inner side of the membrane if the positive charge were delocalized in a stratum at the same depth as the positive point charges in the left of Fig.3.
We asked whether there is any component in the time course of the electric field kinetically corresponding to the reduction of chl-a_I as to be expected if chl-a_I were located within the core of the membrane. The experimental result is shown in Fig.4.

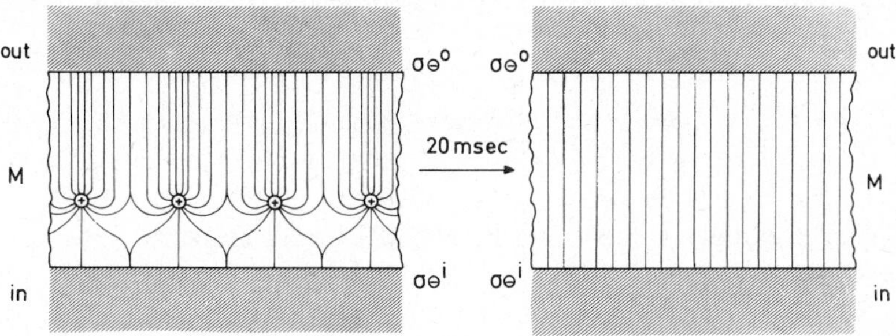

Fig.3. Scheme for the electric field if chl-a_I were located within the membrane, disproved by the experiments in Fig.4. left: shortly after the exciting flash right: after reduction of chl-a_I

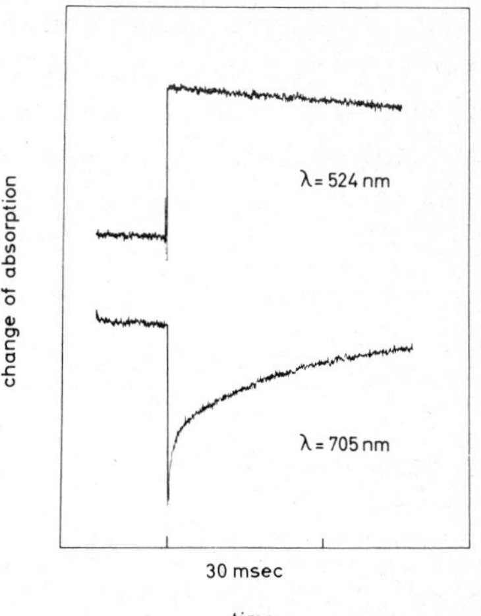

Fig.4. Time course of the electrochroic absorption change at 524nm (above), time course of the absorption change of chl-a_I at 705nm (below).

The upper trace in Fig.4 shows the time course of the electrochromic absorption change at 524nm on excitation with a short flash at t=0. The absorbency changes around 520nm are due to the response of carotenoids to the light induced electric field in their microenvironment in the membrane[7,8,55]. The lower trace shows the absorption change of chl-a_I at 705nm under the same conditions. It is evident that the field indicating absorption change at 524nm has not the slightest component matching kinetically the components of the chl-a_I reduction (half times of relaxation: 20μsec, 200μsec, 20msec [54]). Thus it has to be concluded that the porphyrin rings of chl-a_I are located at the inner interface of the thylakoid membrane if not in contact with the inner aqueous phase. The reaction path between plastocyanin and chl-a_I has no vector component which is perpendicular to the membrane and lies within the badly conducting dielectric core of the membrane.

The orientation of chlorophyll-a_I in the thylakoid membrane

We studied the orientation of chl-a_I in the membrane by the photoselection technique described under Materials and Methods. The linear dichroism of the absorption changes of chl-a_I on excitation into the in-plane oriented antennae is illustrated in Fig.4[56]. Detailed experimental conditions were given elsewhere[40,41]. The larger signal was obtained for parallel the smaller one for perpendicular orientation of the E-vectors of the exciting and the interrogating light (Fig.2). A dichroic ratio ($\Delta A_{\parallel}/\Delta A_{\perp}$) greater than 1 qualitatively indicates that the transition moment under interrogation is "more or less parallel" to the transition moment interacting with the exciting light. A quantitative description was presented elsewhere[41]. The spectrum of the dichroic ratio by photoselection is shown in the upper part of Fig.5, and the spectrum of the extent of the absorption changes in the lower.

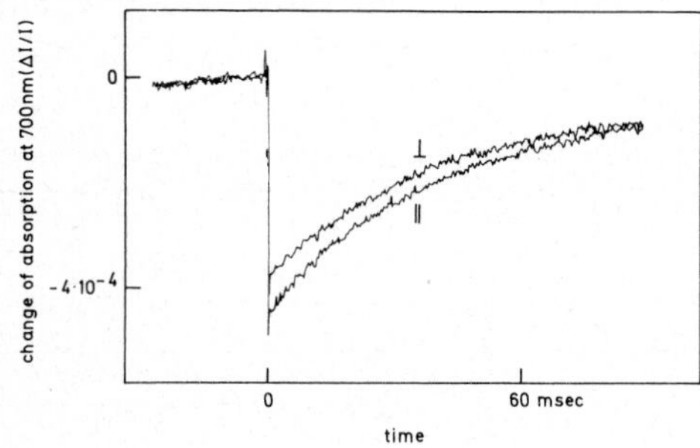

Fig.4. Linear dichroism by photoselection of the absorption changes of chl-a_I on excitation at $\lambda > 690$nm in the presence of valinomycin (1μM)[56].

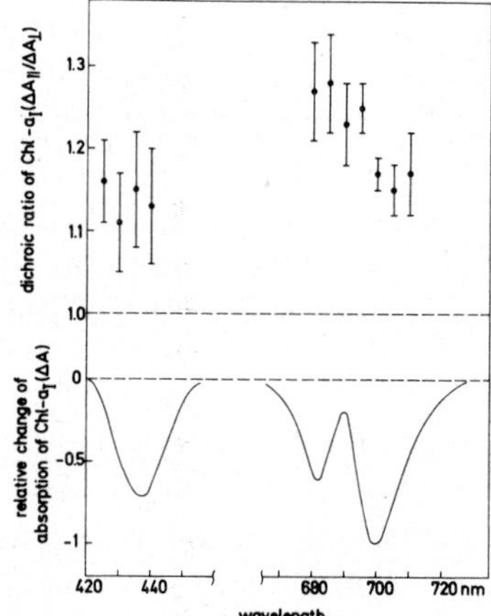

Fig.5. Spectrum of the dichroic ratio by photoselection of chloroplasts in the presence of valinomycin (above), spectrum of the extent of the absorption changes under the same conditions (below)[56].

The spectrum of the absorption changes in the lower part of Fig.5 closely resembles the spectrum for the photooxidation of chl-a_I as obtained from photosystem I particles[34], since the electrochroic absorption changes in the same wavelength range were apparently eliminated by accelleration of the field decay by valinomycin[42]. As obvious from the upper part of Fig.5 the dichroic ratio is greater than 1.11 within all three bands of chl-a_I. For the minor band at 682nm it rises up to 1.28. It is important to note that the variance of the dichroic ratio within each band is rather small. The relative large error in the dichroic ratio, indicated by the bars in Fig.5, is due to the fact that the dichroic difference ($\Delta A_{\parallel}-\Delta A_{\perp}$) in experiments as depicted in Fig.4 is in the order of only 10^{-5} (see Ref.41). As an illustration for the order of magnitude to be expected for the dichroic ratio in photoselection experiments it should be mentionned that the ratio is 3, if the excited and the interrogated transition moments are non-degenerate and in parallel to each other, and 1.333 if one of them is circularly degenerate[57].

A quantitative interpretation of the above dichroic ratios to yield the orientation of chl-a_I in the thylakoid membrane can be based on the above mentionned evidence for the in-plane orientation of the antennae transition moments above 680nm[44-47]. However a rigorous interpretation requires information on two further items: 1. the angular distribution of the original polarization of the exciting light via resonant energy transfer among the antennae, 2. the orientation of the respective transition moments of chl-a_I within the coordinate system of the dimer.

We asked for the symmetry properties of the in-plane oriented part of the antennae system. At the extremes two possibilities are to be distinguished: a circular distribution of the original polarization in the plane (circular degeneracy) and the conservation of polarisation during resonant energy transfer within a linear antennae sytem. To find out which of these possibilities is closer to reallity we carried out photoselection experiments with oriented chloroplasts. Orientation occurred by sedimentation and drying of chloroplast fragments on glass slides as in Ref.44, however, to keep the dried films photochemically active the sediment was dried at constant humidity (94.7%) controlled by $ZnSO_4 x 7H_2O$ at 5^oC[58]. The spreading technique produces an almost parallel orientation of thylakoid lamellae to the glass slide[44,45]. The optical conditions of the photoselection experiment with slides are illustrated in Fig.7. On variation of the polarization of the exciting beam (see Fig.7) one expects different dichroic ratios depending on which of the two extreme cases from above is real. While the dichroic ratio should be 1 if the antennae are circular degenerate, it should be up to 3 if they are linear[57]. The experiment revealed a ratio of 1 ($\Delta A_{\parallel}/\Delta A_{\perp}$ =1.1±5%)[59]. This gives evidence for a circular distribution of the original polarization of the exciting light in the plane of the circular degenerate antennae at wavelength greater than 680nm.

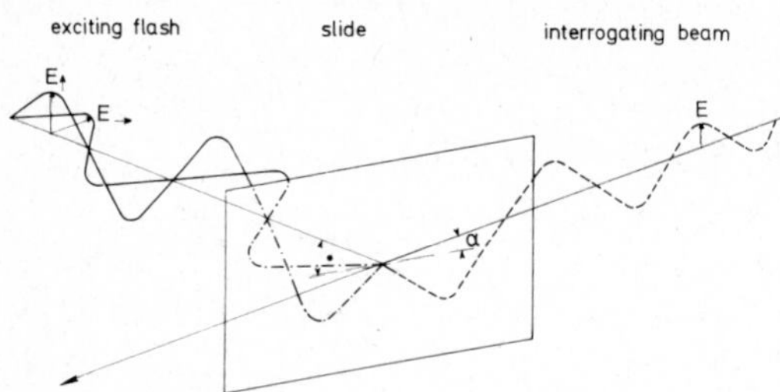

Fig.7. The geometry in photoselection experiments with chloroplast fragments oriented with their lamellae coplanar on a glass slide. The slide was tilted at 10° out of the plane formed by the E-vector and the propagation-vector of the interrogating beam.

We interpreted our earlier photoselection experiments under the assumption of a circularly degenerate antennae system[40,41]. According to Fig.4 in Ref.41 a dichroic ratio of 1 in photoselection with isotropic suspensions implies an inclination of 35.26° of the respective transition moment to the plane of the circular degenerate antennae system. A ratio of 1.28 as for the 682nm-band of chl-a_I corresponds to an inclination of only 13.5°. This inclination represents an upper limit already, since experimental imperfections tend to shift the dichroic ratio towards 1, equivalent to a larger inclination[41].

From the dichroic ratios in the upper part of Fig.6 we conclude that all transition moments which characterize the transformation of chl-a_I into chl-a_I^+ are inclined at less than 30°, some at even less than 13° to the plane of the membrane.

We ask as to the relevance of these findings for the orientation of the porphyrin rings of chl-a_I. The interpretation is complicated by the fact that little is known about the internal structure of the chl-a_I-dimer and, moreover, by our observation of absorption changes which relates our results to differences between transition moments of chl-a_I in the two states. However, a qualitative argument can be based on the observation that all transition moments in the difference spectrum of chl-a_I are inclined at less than 30° to the membrane. This argument is specified below: The transition moments of a chl-a-monomer are oriented in the plane of the porphyrin ring[60-62]. The Q-band transition moment is directed along one diagonal of the ring, while the Soret is mixed from two subtransitions one in parallel the other one perpendicular to the transition moment of the Q-band[62]. The resulting transition moments of the dimer are linear combinations of the monomer transition moments[36]. As yet it is unknown whether the absorption changes of chl-a_I reflect the bleaching of the dimer band, only, or its bleaching together with the appearance of the monomer band. Anyhow, the resulting transition moments for the difference of absorption of chl-a_I will be linear vectorial combinations of the

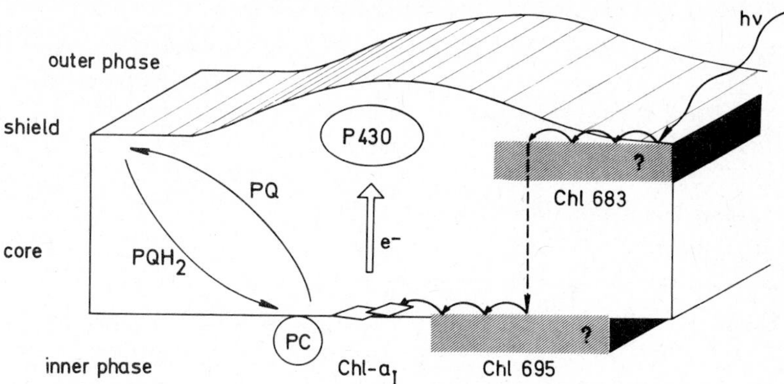

Fig.8. Model for the structural conditions of the energy transfer and the vectorial electron transport in Photosystem I

respective transition moments of the two monomers. The mixing coefficients for the linear combination will vary strongly depending on the wavelength. The experimental observation that the dichroic ratio was greater than 1.11, corresponding to an inclination of less than 30°, at *any* wavelength is not compatible with an almost perpendicular orientation of any of the transition moments of the chl-a monomers (a quantitative evaluation will be presented elsewhere[59]).

In conclusion, we find that the porphyrin rings of chl-a_I ly at the inner interface of the thylakoid membrane oriented rather parallel to the membrane plane. This configuration is illustrated in Fig.8.

4. Conclusions

The energy transfer in Photosystem I

The question whether the antennae chlorophylls are located at the outer or the inner side of the thylakoid membrane cannot be answered yet by experimental evidence. They were placed at any side in speculative membrane models[30,63-65]. We ask which position is compatible with the location of chlorophyll-a_I at the inner side of the membrane with regard to energy migration in Photosystem I. Resonant energy transfer from the antennae into the trap (chl-a_I) is characterized by a relaxation time of less than 30psec, which under the assumption of about 100 transfer steps corresponds to an average transfer time of 0.3 psec[50]. This transfer time depends by the third power on the distance between chromophores under exciton interaction[36,49].
The order of magnitude can be inferred from model studies on

bimolecular lipid membranes. They revealed a shift in the center wavelength of the Q-band of 10 nm if the average distance between chlorophyll molecules was 2 nm[66]. The wavelength shift is related to the transfer time by the following relationship (see Ref.s 36,49):

$\Delta\tau = h/\Delta\varepsilon = \lambda^2/\Delta\lambda \times c_o$ wherein $\Delta\varepsilon$ is the energy of exciton interaction between neighbouring chromophores, λ is the wavelength (662nm), h and c_o are as usual. The above data from model membranes thus imply a transfer time in the order of 0.1 psec at a chl-chl distance of 2nm. From the dependence of the transfer time on the third power of the distance it follows that the transfer time is still in the order of 1 psec if chlorophyll-a molecules are 4 or 5 nm apart, which is the thickness to be expected for the dielectric core of the thylakoid membrane[30]. Thus any position of the antennae chlorophylls which is illustrated by the hatched patterns in Fig. 8 is compatible with the location of chlorophyll-a_I at the inner side of the membrane.
If some of the antennae chlorophylls (say chl-683) were located at the outer side of the core layer the efficiency of resonant energy transfer into the trap would be increased if others (say chl-695) were placed at the inner side in"head-to-tail" configuration with the red transition moment of chl-a_I as tentatively shown in Fig.8.

Vectorial electron transport in Photosystem I

As reviewed above,excited chlorophyll-a_I transfers an electron across the thylakoid membrane within less than 20 nsec. We provided kinetic evidence that the electron transport between chl-a_I and its primary acceptor (P430[67]) crosses the full thickness of the dielectric core of the membrane. This thickness is below 7 nm according to electron microscopy[29] and X-ray scattering[30]. It is questionable whether special devices are necessary to bridge this distance for electronic conduction between chl-a_I and P430. Carotenoids by overlap of their π-electron orbitals[68-70] and special proteins acting as injection semiconductors[71] were discussed as possible candidates.
Such a role of the carotenoids becomes less probable by recent reports on a rather planar orientation of these linear molecules in the thylakoid membrane[45,51,53]. Concerning the necessity of a semiconducting protein it is noteworthy that the electric mobility of electrons photoinjected into pure hydrocarbon phases is very high ($>1\ cm^2/V\ sec$)[56]. This at a voltage of 100mV across the membrane makes an electron cross a distance of 10nm within 10psec. A voltage sufficient to direct the motion of photoinjected electrons across the membrane is provided by the difference in the redox potentials between excited chl-a_I and its

primary electron acceptor. It is an open question, however, whether the relatively low energy of a light quantum processed by chl-a_I, which is 1.77eV (λ=700nm), is sufficient to photoinject an electron into an undisturbed lipid layer. This deserves further study.

Aknowledgements

I am very much indepted to Armin Eckhof and Friedemann Graef for their collaboration, to Ilse Columbus for technical assistance and to Waltraud Falkenstein for the graphs. Financial support from the Deutsche Forschungsgemeinschaft is gratefully aknowledged.

References

1. Mitchell,P., 1961, Nature 191,144
2. Mitchell,P., 1966, Biol.Rev. 41,445
3. Neumann,J. and Jagendorf,A.T., 1964, Arch.Biochem.Biophys.107,109
4. Hager,A., 1969, Planta 89,224
5. Rumberg,B. and Siggel,U., 1969, Naturwissenschaften 56,130
6. Rottenberg,H.,Grunwald,T. and Avron,M., 1971, FEBS Lett.13,41
7. Junge,W. and Witt,H.T., 1968, Z.Naturforsch. 23b,244
8. Emrich,H.M.,Junge,W. and Witt,H.T., 1969,Z.Naturforsch.24b,1144
9. Schliephake,W.,Junge,W. and Witt,H.T.,1968,Z.Naturforsch.23b,1561
10. Wolff,Chr.,Buchwald,H.E.,Rüppel,H., 1969,Z.Naturforsch. 24b,1038 Witt,K. and Witt,H.T.
11. Witt,K. and Wolff,Chr., 1970, Z.Naturforsch.25b,387
12. Witt,H.T., 1971, Quart.Rev.Biophys. 4,4 ,365
13. Junge,W. and Ausländer,W., 1974, Biochim.Biophys.Acta 333,59
14. Ausländer,W. and Junge,W., 1974, Biochim.Biophys.Acta -in press-
15. Berzborn,R.J.,Menke,W.,Trebst,A. 1966, Z.Naturforsch.21b,1057 and Pistorius,E.
16. Berzborn,R.J., 1968, Z.Naturforsch.23b,1096
17. Regitz,G.,Berzborn,R.J. and Trebst,A., 1970, Planta 91,8
18. Hauska,G.A.,McCarty,R.E.,Berzborn,R.J. 1971, J.Biol.Chem.246,3524 and Racker,E.,
19. Trebst,A. and Reimer,S., 1973, Biochim.Biophys.Acta 305,129
20. Trebst,A., 1974, Ann.Rev.Plant Physiol. 25,423
21. Barber,J. and Kraan,G.P.B., 1970, Biochim.Biophys.Acta 197,49
22. Crofts,A.R.,Wraight,C.A. and Fleischmann,D.E.,1971,FEBS Lett.15,89
23. Kok,B., 1972, presented to the VI.Intern.Congr.Photobiol.,Bochum
24. Witt,H.T. and Zickler,A., 1973, FEBS Lett. 37,307
25. Kreutz,W., 1970, Z.Naturforsch.25b,88
26. Blank,M. and Britten,J.S.,1970, in:Physical Princ.Biol.Membranes, eds.F.Snell,J.Wolken,G.J.Iverson and J.Lam, (Gordon&Breach, New York) p.143
27. Blank,M. and Britten,J.S., 1965, J.Colloid.Sci. 20,353
28. Moelwyn-Hughes,E.A., 1965, Physical Chemistry, 2nd ed. (North Holland, amsterdam)
29. Mühlethaler,K., 1972, Chem.Phys.Lipids 8,259
30. Kreutz,W., 1970, Adv.Bot.Res. 3,53
31. Junge,W.,Rumberg,B. and Schröder,H.,1970,Europ.J.Biochem.14,575
32. Kok,B.,1956, Biochim.Biophys.Acta 22,399
33. Rumberg,B. and Witt,H.T., 1964, Z.Naturforsch.19b,693
34. Döring,G.,Bailey,J.L.,Kreutz,W., 1968, Naturwissenschaften 55,219 Weikard,J. and Witt,H.T.,

35. Kok,B., 1961, Biochim.Biophys.Acta 48,527
36. Kasha,M., 1963, Radiat.Res. 20,55
37. Phillipson,K.D.,Sato,V.L. and Sauer,K., 1972, Biochem.11,4591
38. Norris,J.R.,Uphaus,R.A., 1971, Proc.Natl.Acad.Sci.US 68,625 Crespi,H.L. and Katz,J.J.
39. Katz,J.J. and Norris,J.R., 1973, in:Current Topics in Bionergetics eds. D.R.Sanadi and L.Packer (Academic Press, New York) p.41
40. Junge,W. and Eckhof.A., 1973, FEBS Lett. 36,207
41. Junge,W. and Eckhof,A., 1974, Biochim.Biophys.Acta 357,103
42. Junge,W. and Schmid,R., 1971, J.Membrane Biol. 4, 179
43. Rüppel,H. and Witt,H.T., 1969, in:Meth.in Enzymology, vol.XVI ed.K.Kustin (Academic Press, New York) p.316
44. Breton,J. and Roux,E., 1971, Biochem.Biophys.Res.Commun.45,557
45. Breton,J.,Michel-Villaz,M. and Paillotin,G., 1973, Biochim.Biophys. Acta 314,42
46. Geacintov,N.,VanNostrand,F., 1971, Biochim.Biophys.Acta 226,486 Pope, M. and Tinkel,J.B.
47. Geacintov,N.,VanNostrand,F., 1972, Biochim.Biophys.Acta 267,65 Becker,J.F. and Tinkel,J.B.
48. Förster,Th., 1947, Z.Naturforsch. 2b,147
49. Hochstrasser,H.M. and Kasha,M., 1964, Photochem.Photobiol.3,317
50. Borisov,A.Yu. and Ilina,M.D., 1973, Biochim.Biophys.Acta 305,368
51. Geacintov,N.E.,VanNostrand,F. 1974, Biochim.Biophys.Acta 347,443 and Becker,J.F.,
52. Junge,W.,Ausländer,W. and Eckhof,A., 1974, in:Membrane Transport in Plants, eds.U.Zimmermann and J.Dainty (Springer, Berlin)
53. Breton,J. and Mathis,P., 1974, Biochem.Biophys.Res.Commun.58,1071
54. Haehnel,W., 1973, Biochim.Biophys.Acta 305,618
55. Schmidt,S.,Reich,R. and Witt,H.T., 1971, Naturwissenschaften 8,414
56. Minday,R.M.,Schmidt,L.D. and Davis,H.T., 1971, J.Chem.Phys.54,3112
57. Albrecht,A.C., 1961, J.Molec.Spectrosc. 6,84
58. Handbook of Chemistry and Physics, 53rd Ed.,ed.R.C.Weast,(CRC Press, Cleveland)
59. Junge,W. and Graef,F. 1974, in preparation
60. Gouterman,M. and Streyer,L., 1962, J.Chem.Phys.37,2260
61. Gouterman,M.,Wagniere,G. and Snyder,L.C.,1963,J.Mol.Spectrosc.11,108
62. Goedheer,J.C., 1957, Dissertation, University of Utrecht
63. Wolken,J.J. and Schwertz,F.A., 1953, J.Gen.Physiol. 37,111
64. Calvin,M., 1959, Brookhaven Symp.Biol. 11,160
65. Frey-Wyssling,A., 1960, Nova Acta Leopoldina, 22,147
66. Steinemann,A.,Alamuti,N.,Brodmann,W., 1971,J.Membrane Biol.4,284 Marschall,O. and Läuger,P.
67. Ke,B., 1973, Biochim.Biophys.Acta 301,1
68. Dartnall,H.J.A., 1948, Nature 162,122
69. Jahn,T.L., 1962, J.Theoret.Biol.2,129
70. Ilani,A. and Berns,D.S., 1973, Biophysik 9,207
71. Tributsch,H., 1972, Photochem.Photobiol. 16,261

M. AVRON, *Proceedings of the Third International Congress on Photosynthesis*, September 2-6, 1974, Weizmann Institute of Science, Rehovot, Israel
Elsevier Scientific Publishing Company, Amsterdam, The Netherlands, 1974

THE ELECTRICAL GENERATOR AND THE LAYER STRUCTURE OF THE FUNCTIONAL MEMBRANE IN PHOTOSYNTHESIS

Winfried Ausländer and Wolfgang Junge

Max-Volmer-Institut für Physikalische Chemie und Molekularbiologie
Technische Universität Berlin

1. Summary

In this communication it is demonstrated that the rate of proton uptake from the outer phase of the functional membrane of photosynthesis is slowed down by a diffusion barrier for protons which shields the redox reaction sites at the outer side of the membrane against the outer aqueous phase. This barrier can be lowered by sand grinding of chloroplasts, by digitonin treatment and by uncoupling agents. At the extreme this barrier can be practically eliminated to yield rates of proton uptake matching the rates of the corresponding redox reactions.This gives conclusive evidence that the electrochemical potential difference which light induces across the functional membrane of photosynthesis is generated by a vectorial electron-hydrogen transport system, as postulated by Mitchell[1].

2. Introduction

In a preceding paper[2] we attributed the four protolytic reactions at the outer and the inner side of the functional membrane of photosynthesis to the protolytic properties of the redox components water, plastoquinone and the terminal electron acceptor.The experimental evidence presented was conclusive except for one argument left,the rate of the protolytic reactions as detected by the dye cresol red after a short flash of light was considerably slower than the rate of the corresponding redox reactions.A similar discrepancy between the rates of redox and protolytic reactions in chromatophores has led to the suggestion that both are only indirectly linked to each other via a Membrane Bohr effect[3].If so, our concept of the electrochemical generator in photosynthesis (for detail, see our preceding papers[2,4]) has to be reevaluated.

We have demonstrated earlier, that sand grinding of chloroplasts leads to an acceleration of the rate of proton binding from the outer phase[4]. This led us to propose that a diffusion barrier for protons shields the redox reaction sites at the outer side of the membrane against the outer aqueous phase.In this communication we demonstrate that this barrier is lowered also by detergent treatment and by proton permeability increasing agents. At the extreme this barrier can practically eliminated as to yield rates of proton binding matching the rates of the respective redox reactions.

3. Materials and methods

Three types of chloroplasts were used in the experiments.Control chloroplasts were of the broken type. Ground chloroplasts and detergent treated chloroplasts were prepared from the control type by grinding with sea sand or by digitonin treatment.The preparation methods for the above types of chloroplasts are specified in our recent paper[5].After preparation, chloroplasts were stored under liquid nitrogen until use.Thawed samples were suspended in a 20-mm spectrophotometer cell containing the following standard reaction medium: KCl,20mM; $MgCl_2$,2mM; cresol red,30μM.The average chlorophyll concentration in the cuvette was 10μM.Electron acceptors, uncoupling agents and buffer were added as indicated in the Figures.
Photosynthesis was excited with a saturating "single-turnover flash". The absorption changes were measured by a rapid spectrophotometer. For details and specific references,see our preceding paper[2]. The frequency of the repetitive flashes was chosen to allow for relaxation of the light induced pH difference across the functional membrane. Since the treatment of chloroplasts by grinding or digitonin accelerated the relaxation rate the repetition rate was increased to 0.5 sec^{-1} for treated chloroplasts in comparison with 0.1 sec^{-1} for the control.

As in our preceding paper[2] pH changes in the outer aqueous phase of the inner chloroplast membrane were measured with the pH-indicating dye cresol red at a pH around 8.The rise of absorption at 574 nm indicates alkalisation.The pH indicating "absorption changes of cresol red at 574 nm" which are represented in the Figures were measured as follows: First the absorption changes at 574 nm were measured for a chloroplast suspension without buffer.Repetitive signals were summed up in the averaging computer.Then the same number of signals from chloroplasts in the presence of buffer (tricine,1mM (pH 8)) was subtracted.

In the Figures the ordinate label "different scale factors" indicated that we used different number of repetitions to obtain the documented signals.This was for convenience only,to represent signals comparable in extent and signal to noise ratio.There are several reasons for variations in the extent of the pH indicating absorption changes in the experiments:

1. The buffer capacity of the treated chloroplasts was greater in comparison with the control chloroplasts.
2. We studied the protolytic reactions in treated chloroplasts which are enriched in photosystem I with different electron acceptors. It is evident that proton binding from the outer phase at LR I will be to a different extent depending on whether BENZ (noncyclic) or DCIP (cyclic electron transport) is used as acceptor.In the first case the reaction is limited by the small number of active LR II centers, while in the latter it is not.
3. In the presence of FECY proton uptake from the outer phase is due to LR II, only.
4. Uncouplers contribute to the buffer capacity of the chloroplast suspension.

4. Results

(A) Factors affecting the kinetics of proton binding from the outer phase

It is obvious from Fig.1 that treatment of chloroplasts with sea sand or digitonin accelerates the rate of proton uptake (the half-times are written to the traces).Moreover, there was a further acceleration if DCIP instead of BENZ was used as terminal acceptor. In the control chloroplasts DCIP accepts electrons behind LR I without changing the activity of both light reactions under linear electron transport conditions[5].The kinetics obtained with both BENZ and DCIP yields a good approximation to a first order process, as checked by kinetic analysis.That indicates, that the proton uptake at two different reaction sites was accelerated, only one of which is linked to the reduction of the terminal acceptor, which was exchanged in the experiments depicted in Fig.1.This suggests that the additional acceleration of proton binding by DCIP might be due to its uncoupling properties[6] than to its redox properties.
The dependence of the kinetics of proton binding from the outer phase for ground chloroplasts on electron acceptors belonging to different classes is documented in Fig.2 According to the classification of Saha et al.[7] Class II acceptors can serve both as electron acceptors and as uncouplers in contrast to Class I and Class III acceptors.

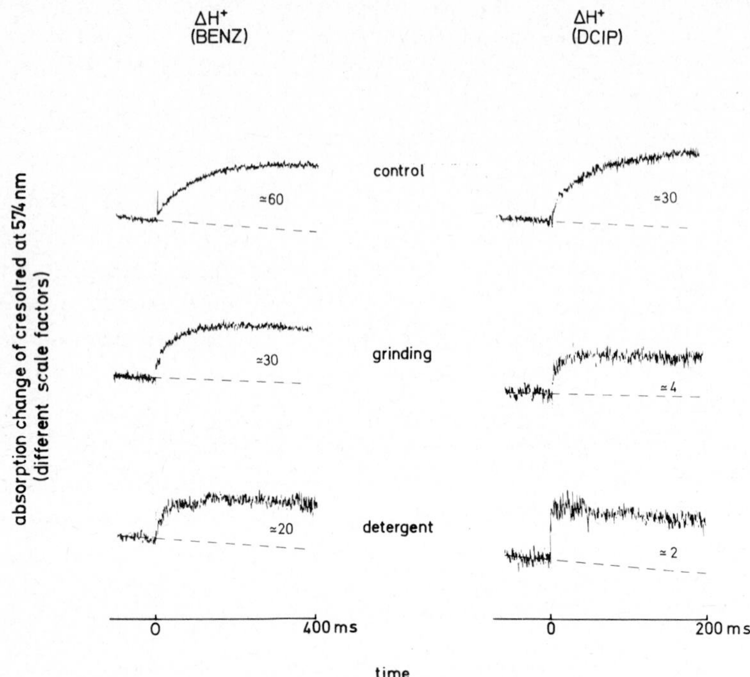

Fig.1: Absorption change of cresol red at 574 nm induced by a single-turnover flash of light at t=0. (standard reaction medium)
left: benzylviologen (60μM) as terminal acceptor
right: 2,6-dichlorophenol-indophenol (10μM) as terminal acceptor.

It is obvious from Fig.2 that the kinetics obtained with Class II acceptors are the most rapid ones. Experiments with uncoupling agents (ammonium-chloride,FCCP) together with electron acceptors of Class I (BENZ,FECY) yielded the same high rates of proton uptake, as measured with Class II acceptors[5].

The experimental evidence presented so far suggests that a hydrophobic diffusion barrier for protons shields the primary proton binding sites of the electron transport at the outer side of the functional membrane from the outer aqueous phase. The height of this barrier can be lowered by mechanical and chemical treatment of chloroplasts and by uncoupling agents.

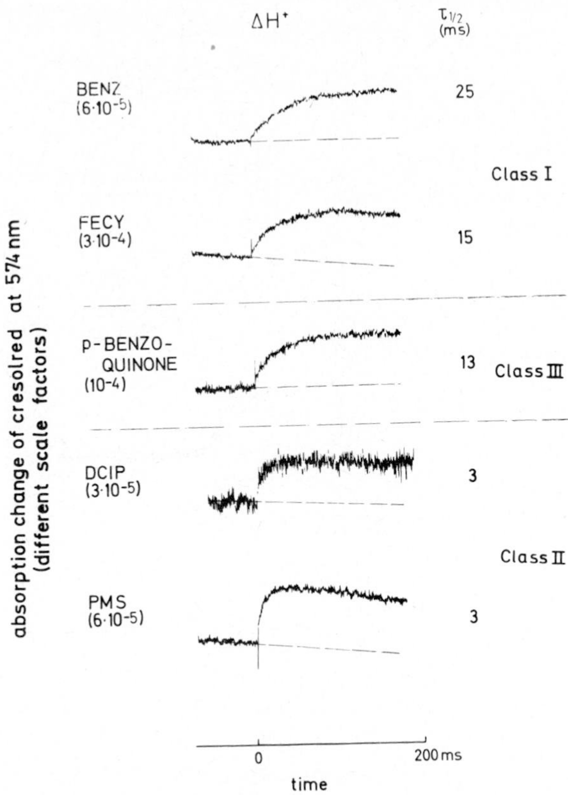

Fig.2: Dependence of the proton uptake kinetics on different types of acceptors (standard reaction medium, ground chloroplasts). The molarity of the electron acceptors is given in parenthese.

(B) Conditions for parallel kinetics of proton binding and redox reactions

We have asked whether this barrier can be lowered as to eliminate the delay between proton binding and redox kinetics of both light reactions. As depicted in Fig.3, in digitonin treated chloroplasts the rate of proton binding at LR I follows the rate of DCIP-reduction at LR I without detectable delay. Due to the low activity of LR II in our digitonin treated chloroplasts (below 20%), both kinetics were attributable to LR I. Variation of the DCIP concentration over more than one order of magnitude leaves the rates of proton binding and of the DCIP-reduction in reasonable agreement[5]. Next we studied the correlation between the rate of proton binding at LR II with the rate of the reduction of PQ (half-time 0.6 msec[8]) to which it was attributed in our preceding paper[2]. With control chloroplasts we have measured the proton uptake kinetics with FECY (0.1 mM) as terminal

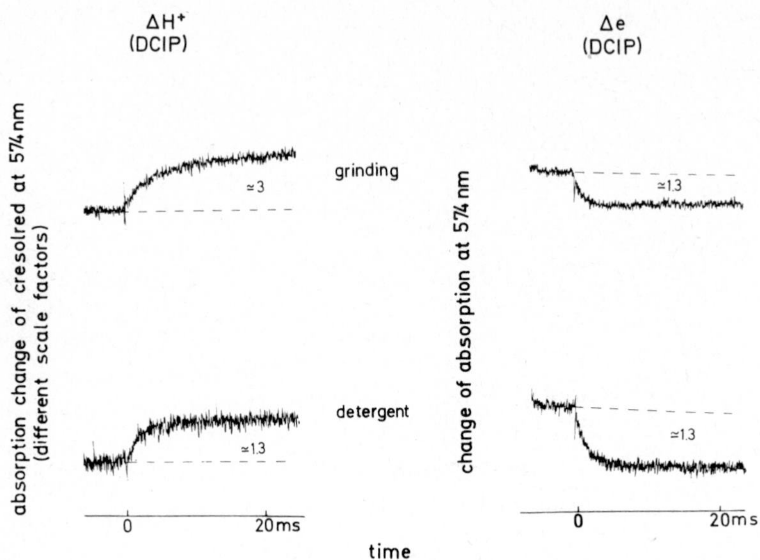

Fig.3: Correlation of the kinetics of proton binding and DCIP-reduction (DCIP 10μM).
left traces:"abs. changes of cresol red at 574 nm" (see Materials and methods), standard reaction medium.
right traces: abs. changes of DCIP measured in the absence of cresol red and in the presence of buffer (tricine 1mM,pH 8).

acceptor. Due to its low pK-value of 4.3, FECY doesn't bind a proton when it is reduced at LR I, so that the remaining proton uptake is attributable to LR II.We accelerated the rate of proton uptake at this site by adding the uncoupler FCCP.The result,which is documented in Fig.4 reveals a rapid proton binding with a half-time of about 2 msec. The time course of the pH indicating absorption changes of cresol red at the low time resolution (left traces) was discussed in detail in our preceding paper[2]. The maximum rise time of the alkalisation of 2 msec in the presence of FCCP is still in disagreement with the 0.6 msec rise time of the PQ-reduction. This may be due to a residual proton-permeability barrier, resistant to the proton carrier FCCP, delaying proton binding against the PQ-reduction. It might be argued that the time lag in the electron transfer between PQ and chlorophyll-a_I, which has been reported by Haehnel[9], reflects the time elapsing between the electron transfer to PQ and its proto-

nation, which might be a prequisite for its reaction with chlorophyll-a_I.
Although,the rates of proton binding and PQ-reduction could not be brought to a strict coincidence the above experiments have demonstrated, that the discrepancy between these rates in untreated chloroplasts is due to a diffusion barrier for protons which can be overcome e.g. by proton carriers as FCCP.

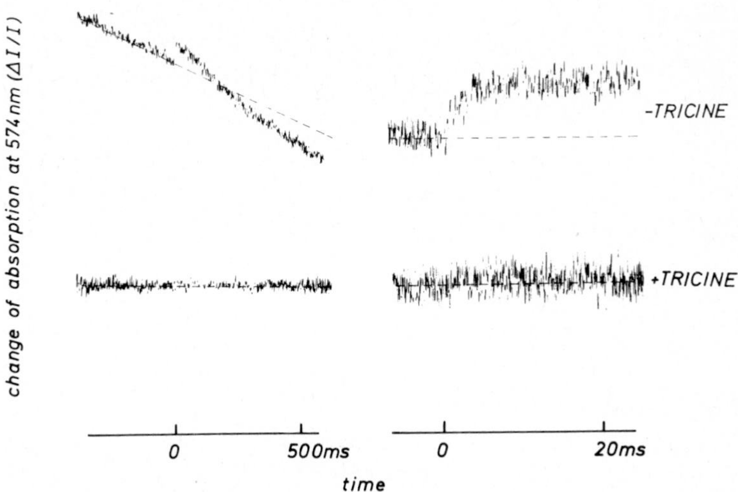

Fig.4: Absorption changes at 574 nm at control chloroplasts with FECY (0.1 mM) and the uncoupler FCCP (1μM);standard reaction medium
above: no buffer below: with buffer (tricine 1mM,pH 8)

5. Discussion

Our experimental results imply that the rapid electron transfer at both light reactions from a donor at the inner side of the membrane to an acceptor at the outer side, takes place across a membrane core which is covered by a shielding structure for protons.It is an open question whether this shielding structure is formed by isolated knobs covering each light reaction center or by a continous layer covering the whole membrane. This shielding structure is kinetically well distinct from the membrane core. While a pH-difference across the shielding structure relaxes within 60 msec it relaxes within several seconds across the total membrane.
From our experiments with the uncouplers and from the fact that the

rate limiting step of the electron transport amounts to 20 msec in comparison to the 60 msec relaxation time for protons across the shielding structure, we have to postulate a proton reservoir between the shielding structure and the membrane core. This reservoir rapidly supplies protons to the reduced electron acceptors and is slowly refilled from the outer aqueous phase. Independent evidence for such a proton storage capicity inside the membrane has been provided by Nishizaki[10].

Based on our earlier results[2,4] together with the experimental evidence presented in this paper, we can see no objection against a Mitchellian mechanism for the electrochemical generator in the functional membrane of photosynthesis of green plants.

Abbreviations: BENZ,benzylviologen; FECY,ferricyanide; DCIP,2,6-dichlorophenolindophenol; PQ,plastoquinone; FCCP,carbonycyanide-p-trifluormethoxyphenylhydrazone; LR I,light reaction I; LR II,light reaction II.

6. References

1. Mitchell,P.,1966, Biol. Rev. 41,445.
2. Junge,W. and Ausländer,W.,1974, Biochim.Biophys. Acta 333,59.
3. Chance,B.,Mc Cray,J.A. and Bunkenburg,J.,1970, Nature 225,705.
4. Schliephake,W.,Junge,W. and Witt,H.T.,1968,Z. Naturforsch.23b,1561
5. Ausländer,W. and Junge,W.,1974, Biochim. Biophys. Acta-in press-.
6. Elhanan,Z.G. and Avron,M.,1964, Biochemistry 3,365.
7. Saha,S.,Ouitrakul,R.,Izawa,S. and Good,N.E.,1971, J. Biol. Chem. Vol.246,3204.
8. Vater,J.,Renger,G.,Stiehl,H.H. and Witt,H.T.,1968, Naturwissenschaften 55,220.
9. Haehnel,W.,1973, Biochim. Biophys. Acta 305,618.
10. Nishizaki,Y.,1972, Biochim. Biophys. Acta. 275, 177.

M. AVRON, *Proceedings of the Third International Congress on Photosynthesis*,
September 2-6, 1974, Weizmann Institute of Science, Rehovot, Israel
Elsevier Scientific Publishing Company, Amsterdam, The Netherlands, 1974

STUDIES ON THE PHOTOCHEMICAL ACTIVE CHLOROPHYLL-a_{II} IN SYSTEM II OF PHOTOSYNTHESIS

Ch. Wolff, M. Gläser and H.T. Witt

Max-Volmer-Institut für Physikalische Chemie und Molekularbiologie,
Technische Universität Berlin

1. Introduction

In photosynthesis the electron transport from H_2O to $NADP^+$ is driven by two photochemical reactions which are coupled in series[1-4]. These light reactions occur at the two reaction centers C_I and C_{II} (Fig. 1) where the light energy is transformed into chemical energy. Both the two light reactions lead to a charge separation across the thylakoid membrane[5]. But it is open so far in which way this charge separation is realized in center II.

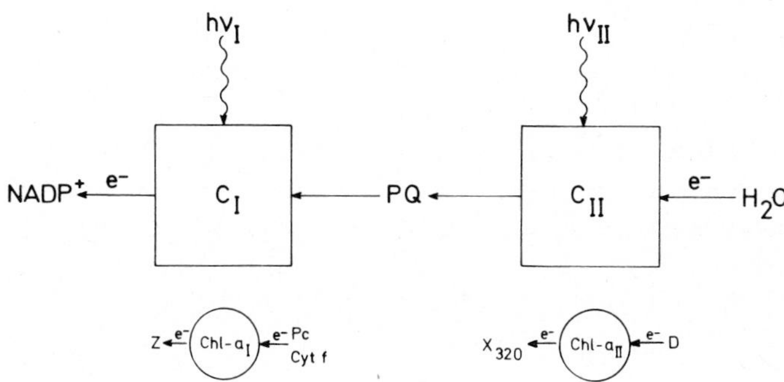

Fig. 1. Simplified Scheme of the electron transport chain in photosynthesis. For details see ref. 42.
C_I, C_{II}: photochemical active reaction centers, PQ: plastoquinone-pool, Chl-a_I, Chl-a_{II}: photochemical active chlorophylls in center I and II, Z: primary acceptor in center I, Pc: plastocyanine, Cyt f: cytochrome f, X_{320}: special plastoquinone molecule as primary acceptor in center II[14-17], D: primary donor in center II.

The primary step in center I is known to be the photooxidation of Chl-a_I[6-8] in less than 20 ns[9]. The released electron reduces lastly $NADP^+$. Chl-a_I^+ is rereduced in the dark by three electron donors Pc, Cyt f, PQ being located between Chl-a_I and Chl-a_{II}[10,11].

Chl-a_{II} was identified as a photoactive chlorophyll-a molecule in center II with an absorption band in the range 680 - 690 nm[12,13]. However, it is not yet known whether Chl-a_{II} acts as a redox-component (similar to Chl-a_I) or only as a sensitizer which mediates the electron transport from a donor D to the known primary acceptor X_{320}. X_{320} has been identified in its reduced form as a semiquinone-anion [14-17]. In any case the primary charge separation in center II takes place within less than 20 ns, i.e. as fast as the electrical field is set up across the thylakoid membrane by light reaction II[18,5]. The half time of the observed Chl-a_{II} reaction at room temperature was found to be ~200 µs. A similar reaction time has been observed in an intermediate step engaged with the oxidation of water[19]. However, it is unsatisfying that according to the extent of the 200 µs-reaction of Chl-a_{II} a concentration of Chl-a_{II} molecules was estimated which is 4 times smaller than that of Chl-a_I [12]. This calculation was made under the assumption that both components have the same extinction coefficient at maximal absorption.

The experiments reported as follows show, however, that there is a faster reacting component (τ = 35 µs) of a special chlorophyll-a which has to be correlated with the electron transport chain of photosystem II [20]. The extent of both the new 35 µs-component plus the already known 200 µs-component now corresponds to that of Chl-a_I. It may be that the new 35 µs-component represents together with the 200 µs-component the biphasic kinetics of one and the same Chl-a_{II} in reaction center II.

2. Results and Discussion

With the aid of a repetitive flash photometer[21] with high frequency modulated detecting light[22] and the use of ultra short flashes[23] (duration 0,4 µs) we were able to extend the time resolution to 1 µs for the observation of chlorophyll absorption changes in the red spectral region (see Materials and Methods).

(1) With this method we measured the time course of absorption changes at 690 nm in spinach chloroplasts in the presence of far-red background light[20]. The rise time of the changes is less than 1 µs. The decay kinetics are three-phasic: besides the well-known decay kinetics of Chl-a_I (~ 20 ms) and the 200 µs-Chl-a_{II}-component observed by Döring et al.[13] there is a new component with a half lifetime of τ = 35 µs.

(2) The spectrum of the new fast phase (35 µs) in the red region

can be derived from Fig. 2. Between 600 and 720 nm the sum of the 35 µs- and 200 µs-component is compared with the 200 µs-component alone which was obtained by Döring[24] under the same measuring conditions, i.e. whole spinach chloroplasts in the presence of far-red background light. Under these conditions the 200 µs-component has its absorption maximum at 687 nm instead of 682 nm observed in system II-enriched particles[13]. Also the main maximum of the sum spectrum (35 µs- plus 200 µs-component) is located around 690 nm. Thus the 35 µs-component alone has its red absorption maximum around 690 nm, too. Because no other known chloroplast pigments have absorption bands in the red region between 660 and 710 nm than chlorophyll-a the result shows that the 35 µs-component is caused by a reaction of a chlorophyll molecule of type "a". The corresponding blue band of absorption changes with a decay of 35 µs was found around 435 nm.

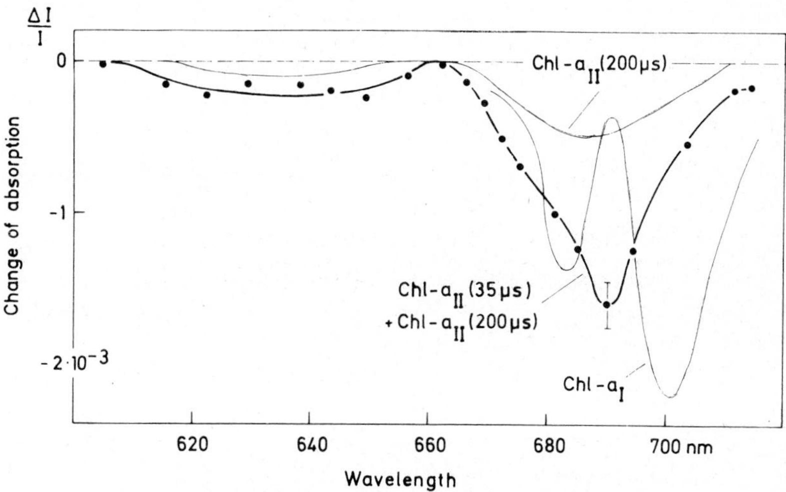

Fig. 2. Transient difference spectrum of the 35 µs-component plus 200 µs-component of Chl-a_{II} induced by saturating actinic flash light. For comparison it is also depicted the spectrum of the 200 µs-component alone[24] and of Chl-a_I[25].

(3) With increasing flash intensity the extent of the 35 µs-component reaches a saturation level. This fact indicates that the new chlorophyll-a component is a special one and differs from the normal bulk chlorophylls.

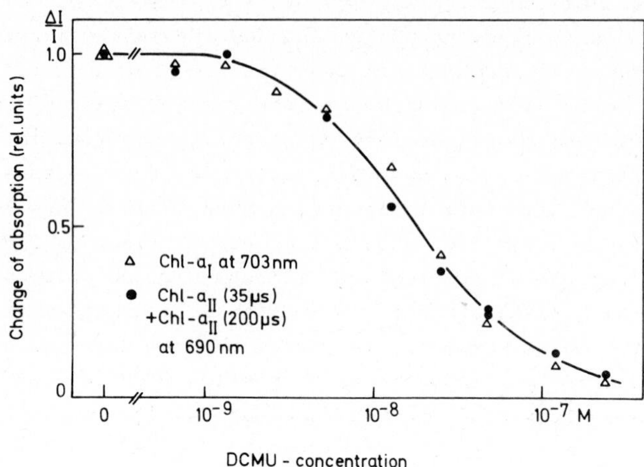

Fig. 3. Relative changes of absorption of Chl-a_{II} (35 μs-component plus 200 μs-component) at 690 nm and of Chl-a_I at 703 nm in spinach chloroplasts as a function of the DCMU-concentration.

(4) Fig. 3 shows the extent of the 35 μs- plus 200 μs-component in dependence of the DCMU-concentration. The DCMU-sensitivity is the same as that of Chl-a_I (P700) as well as of the 200 μs-component alone[13]. Therefore the shown dependence is valid also for the 35 μs-component which means that the 35 μs-component should be associated with the electron transfer reactions.

(5) All measurements were carried out in the presence of far-red background light (720 nm) which mainly excites Chl-a_I. Because we obseived that the 35 μs-component is insensitive to this light the 35 μs-component does not belong to photosystem I.

(6) Its belonging to photosystem II was shown as follows: the 35 μs-absorption change was measured in the presence of DBMIB plus ferricyanide $K_3[Fe(CN)_6]$.as described in ref. 20. From DBMIB it is known to be an agent which blocks the electron transport from plastoquinone to system I [26]. This was checked by observation of the Chl-a_I-absorption change being drastically decreased. In combination with ferricyanide the substance DBMIB acts at higher concentrations as electron acceptor for system II [27]. The fact that photosystem II was full active was checked by measuring the oxygen production which was not changed by the addition of DBMIB. Because the 35 μs-component (together with the 200 μs-component) at 690 nm was not diminished

at all by addition of DBMIB it can be concluded that the new 35 μs-component as well as the already known 200 μs-component is active in system II. Therefore both components can be designated as Chl-a_{II}. For distinguishing both components we symbolize them as

Chl-a_{II}(35 μs) and Chl-a_{II}(200 μs).

(7) The 35 μs-component and the 200 μs-component could represent the reactions of one and the same Chl-a_{II} molecule with two decay kinetics after excitation (similar to the tree decay kinetics of Chl-a_I). This is supported by the location of the maxima of the absorption changes of both components at approx. similar wavelengths around 690 nm. Both components can therefore be designated as P 690.

(8) Assuming a similar extinction coefficient for Chl-a_{II}(35 μs) plus Chl-a_{II}(200 μs) as for Chl-a_I it can be estimated from the saturation value of the absorption changes that about one molecule Chl-a_{II} is active per electron transport chain - a necessary condition if both Chl-a_I and Chl-a_{II} are coupled in series.

(9) Chl-a_{II} with its absorption band at 690 nm is assumed to be engaged as a redox-component in the primary photochemical act in reaction center II. This is supported by trapping experiments performed by Ke et al.[28] at low temperatures. In system II-particles Ke et al.[28] observed a chlorophyll species absorbing at 689 nm which was irreversibly photobleached at 77 $^{\circ}$K. This chlorophyll being trapped in its photobleached form was correlated to a light-induced EPR-signal (at 34 $^{\circ}$K) which probably indicates the oxidized form of reaction center II-chlorophyll[28,29]. Thus the primary photochemical step in reaction center II seems to be the

photo-oxidation of Chl-a_{II} (P 690).

(10) A possible photoreduction of Chl-a_{II} as the primary photochemical act can be excluded because of the following reasons:

(a) If $$X_{320}\ \text{Chl-a}_{II}\ D \xrightarrow{h\nu_{II}} X_{320}\ \text{Chl-a}_{II}^{-}\ D^{+}$$
is the primary step, then the electron transfer

$$X_{320}\ \text{Chl-a}_{II}^{-}\ D^{+} \longrightarrow X_{320}^{-}\ \text{Chl-a}_{II}\ D^{+}$$

must occur faster than 10 μs because the full amount of X_{320} was found to be reduced (to a semiquinone) faster than 10 μs[15]. Thus the decay kinetics of the full Chl-a_{II} should also be faster than 10 μs. However, this requirement is in contradiction to our measurements on Chl-a_{II}.

(b) Furthermore, under the assumption that only the primary photochemical step is trapped at liquid nitrogen temperature a possible photoreduction of Chl-a_{II} would lead only to trapping of Chl-a_{II}^{-} D^{+}

but not of the reduced acceptor X_{320}^{-}. However, the irreversible trapping of X_{320}^{-} at liquid nitrogen temperature was shown by K. Witt[16]. Therefore the photoreduction of Chl-a_{II} is improbable except if at liquid nitrogen temperatures an electron transfer to X_{320} is still possible in the dark.

Summarizing the results we may tentatively suggest the following scheme for the electron transport in system II (Fig. 4). Chl-a_{II} with maximal absorption at 690 nm is the reaction center chlorophyll molecule which is oxidized in the light. The primary electron transfer $X_{320} \xleftarrow{e^-}$ Chl-a_{II} occurs in less than 20 ns according to the measurements of the primary charge separation[18,5,23]. In the dark Chl-a_{II}^{+} is rereduced in the two observed phases: 35 µs and 200 µs. The different decay times are explained by the release of electrons from two electron donors D and Y acting in series. D and Y could belong to the water splitting enzyme system with its charge accumulating states S_o ... S_3[30-32,19]. The extent of the 35 µs-decay depends on the concentration D existing preoxidized (D^+) in the dark before flash excitation. When D is in its uncharged form Chl-a_{II}^{+} will be rereduced by D within 35 µs. However, when D is preoxidized to D^+ the decay kinetics of Chl-a_{II} is determined by the electron delivery from Y to D within 200 µs. Assuming a charge equilibrium between Y and D the amount of D^+ will be determined by the extent of the accumulated positive charges in the watersplitting device according to the S-states of Kok's model[19]. Under our repetitive measuring conditions all S-states are present to the same concentrations. Thus the preoxidized form D^+ of the primary donor possibly can be enriched to some degree in order to explain the extent of the 200 µs-component of Chl-a_{II}.

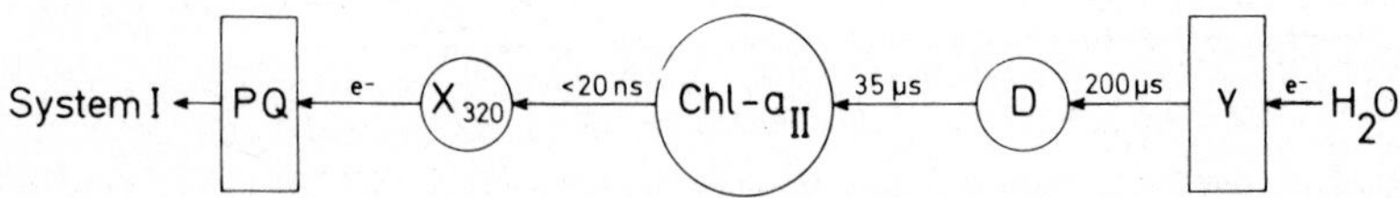

Fig. 4. Scheme of the electron transport chain in photosystem II. Chl-a_{II}: photochemical active chlorophyll in center II, X_{320}: special plastoquinone molecule as primary acceptor[14-17], PQ: plastoquinone pool, D and Y: unknown primary and secondary donors, resp., belonging to the water-splitting device.

It is of interest to note that in photosynthetic delayed light emission[33,34] as well as in fluorescence changes[35,36] two decay kinetics have been observed around 35 μs and 200 μs similar to the kinetics of $Chl\text{-}a_{II}$. Fluorescence and luminescence are known to be strongly related to the reaction center state of photosystem II. However, further results on $Chl\text{-}a_{II}$ would be necessary in order to compare the kinetics of our absorption changes with those of fluorescence and luminescence.

Reversible absorption changes around 680 nm with decay kinetics of about 30 μs in spinach chloroplasts were observed by Floyd et al.[37], however, at 77 °K. These changes were attributed to our reaction center chlorophyll $Chl\text{-}a_{II}$. But according to our conclusions -see point (9) - such reversible absorption changes of $Chl\text{-}a_{II}$ at 77 °K are just not expected. Besides the results of Floyd et al.[37] seem to be ambigious because it was not shown whether these changes belong to the main peak of $Chl\text{-}a_{II}$ or to the minor peak of $Chl\text{-}a_{I}$ (P 700) at 682 nm (see Fig. 2) which decays (without far-red background light) also with a component of 20 - 30 μs[11].

Concerning the structural organization of both the photochemical active reaction centers the following facts can be discussed:
The primary light-induced steps of the reaction centers lead to a charge separation at each center across the thylakoid membrane. Thereby the negative charges are transferred from the inner side to the outer side of the membrane[5]. Further the primary reaction of $Chl\text{-}a_{I}$ is known to be a photooxidation[6-8]. The same fact is probably valid for $Chl\text{-}a_{II}$ as was shown above. Thus the positive charges of $Chl\text{-}a_{I}^{+}$ and $Chl\text{-}a_{II}^{+}$ have to be localized close to the inner side of the membrane. Because it is known that a chlorophyll-cation is formed by the release of a delocalized π-electron of the porphyrin ring, the porphyrin rings of $Chl\text{-}a_{I}$ and $Chl\text{-}a_{II}$ have to be assumed to be placed at the <u>inner</u> plane of the thylakoid membrane.

3. Materials and Methods

All measurements were carried out on whole spinach chloroplasts at room temperature. For chloroplast preparation and detailed experimental conditions see ref. 20.

Concerning the elimination of fluorescence changes usually disturbing any detection of absorption changes in the red spectral region the following remarks should be made:

(a) By the use of the flash photometer with high frequency modulated detecting light[21,22] the absorption changes were discriminated exactly from fluorescence and luminescence emission originating in the flash light energy.

(b) The modulated detecting light itself produces also some fluorescence and luminescence light which will be changed when the photochemical active reaction centers II are excited by the flash light. Especially there are fluorescence and luminescence changes with kinetics around 35 μs and 200 μs[33-36]. Naturally, such changes would not be eliminated by the modulation technique[22] because these changes are modulated with the frequency of the detecting light. However, in a similar way as has been described by Döring and Witt[38] we have checked that the intensity of fluorescence plus luminescence (both originating in the detecting light) was less than 10^{-6} of the detecting light intensity itself (all intensities measured at the photomultiplier input). Further it is known that any change of fluorescence yield (including the much lower luminescence yield) caused by excitation of the reaction centers ranges maximal by a factor of five[39-41, 35,36]. Thus in our measurements it were possible that there are intensity changes $\Delta I/I < 5\cdot10^{-6}$ due to fluorescence changes (plus luminescence changes) which, however, are negligible compared with the observed intensity changes $\Delta I/I > 10^{-4}$. Therefore these observed intensity changes doubtlessly indicate absorption changes of Chl-a_{II}.

4. Summary

The electron transport in system II is driven by light reaction center II being a complex X_{320} - Chl-a_{II} - D with Chl-a_{II} the photochemical active chlorophyll-a molecule, X_{320} the primary electron acceptor[14-17] and D an unknown primary donor.

A new fast reaction (35 μs) of a special chlorophyll-a molecule was found. Its belonging to Chl-a_{II} (with the already known 200 μs-reaction) is probable because of the following reasons:

1) The spectral maximum of the transient absorption changes indicating the 35 μs-reaction was found to be located at a similar wavelength around 690 nm as for the 200 μs-reaction of Chl-a_{II}.

2) The sensitivity of the 35 μs-reaction to DCMU being the same as of Chl-a_{II}(200 μs) and Chl-a_{I} indicates its relation to the electron transfer chain.

3) The insensitivity of the 35 μs-reaction to far-red background light as well as to addition of DBMIB plus ferricyanide indicates

that it is localized in system II.

4) From the amplitude of the 690 nm-absorption change (35 µs-component together with the 200 µs-component of system II) a ratio of about one chlorophyll molecule per electron transport chain can be estimated.

Chl-a_{II} with its absorption band at 690 nm is probably photooxidized in the primary photochemical act of reaction center II. According to the primary charge separation (set up of an electrical field across the thylakoid membrane)[18,5] the photooxidation of Chl-a_{II} occurs faster than 20 ns.

The decay kinetics of Chl-a_{II} (with τ = 35 µs and 200 µs) are explained as its rereduction by two different electron donor states of the water-splitting system.

According to the primary transfer of a negative charge (electron) from the inner side to the outer side of the thylakoid membrane at both the reaction centers I and II it has to be assumed that the porphyrin rings of the photochemical active chlorophyll molecules Chl-a_I and Chl-a_{II} are placed at the inner plane of the thylakoid membrane.

Abbreviations: DCMU = 3-(3,4-dichlorophenyl)-1,1-dimethyl-urea
DBMIB = 2,5-dibromo-3-methyl-6-isopropyl-1,4-benzoquinone

Acknowledgement

The financial support of the Deutsche Forschungsgemeinschaft is gratefully acknowledged.

References

1. Hill,R. and Bendall,F. (1960). Nature 186, 136-137.
2. Duysens, L.N.M., Amesz, J. and Kamp, B.M. (1961). Nature 190, 510.
3. Kok, B. and Hoch, G. (1961). In: Light and Life (McElroy, W.D. and Glass, B., eds.), p.397-416, The John Hopkins Press, Baltimore.
4. Witt, H.T., Müller,A. and Rumberg, B. (1961). Nature 191, 194.
5. Schliephake, W., Junge, W. and Witt, H.T. (1968). Z. Naturforsch. 23b, 1571.
6. Kok, B. (1961). Biochim. Biophys. Acta 48, 527.
7. Rumberg, B. and Witt, H.T. (1964). Z. Naturforsch. 19b, 693.
8. Witt, H.T., Müller, A. and Rumberg, B. (1961). Nature 192, 967.

9. Witt, K. and Wolff, Ch. (1970). Z. Naturforsch. 25b, 387.
10. Haehnel, W., Döring, G. and Witt, H.T. (1971). Z. Naturforsch. 26b, 1171.
11. Haehnel, W. and Witt, H.T. (1972). In: Proc. 2nd intern. Congr. Photosynthesis Res., Stresa 1971 (Forti, G., Avron, M. and melandri, A., eds.), Vol. 1, p.469-476, Dr. W. Junk, The Hague.
12. Döring, G., Stiehl, H.H. and Witt, H.T. (1967). Z. Naturforsch. 22b, 639,
13. Döring, G., Renger, G., Vater, J. and Witt, H.T. (1969). Z. Natur forsch. 24b, 1139.
14. Stiehl, H.H. and Witt, H.T. (1968). Z. Naturforsch. 23b, 220.
15. Stiehl, H.H. and Witt, H.T. (1969). Z. Naturforsch. 24b, 1588.
16. Witt, K. (1973). FEBS Letters 38, 116.
17. Bensasson, R. and Land, E.J. (1973). Biochim. Biophys. Acta 325, 175.
18. Wolff, Ch., Buchwald, H.E., Rüppel, H., Witt, K. and Witt, H.T. (1969). Z. Naturforsch. 24b, 1038.
19. Kok, B., Forbush, B. and McGloin, M. (1970). Photochem. Photobiol 11, 457.
20. Gläser, M., Wolff, Ch., Buchwald, H.E. and Witt, H.T. (1974). FEBS Letters 42,81.
21. Rüppel, H. and Witt, H.T. (1970). In: Methods in Enzymology, Vol. 16 'Fast Reactions' (Colowick, S.P. and Kaplan, N.O., eds.) Academic Press New York, 1969, p.316.
22. Buchwald, H.E. and Rüppel, H. (1968). Nature 220, 57.
23. Wolff, Ch. and Witt, H.T. (1969). Z. Naturforsch. 24b, 1031.
24. Döring, G. (1974). 3rd Intern. Congress on Photosynthesis, Rehovot/Israel.
25. Döring, G., Bailey, J.L., Kreutz, W., Weikard, J. and Witt, H.T. (1968). Naturwiss. 55, 219.
26. Trebst, A., Harth, E. and Draber, W. (1970). Z. Naturforsch. 25b, 1157.
27. Gould, J.M. and Izawa, S. (1973). Eur. J. Biochem. 37, 185.
28. Ke, B., Sahu, S., Shaw, E. and Beinert, H. (1974). Biochim. Biophys. Acta 347, 36.
29. Malkin, R. and Bearden, A.J. (1973). Proc. Nat. Acad. Sci. USA 70, 294.
30. Renger, G. (1970). Z. Naturforsch. 25b, 966.
31. Renger, G. (1972), Physiol. Veg. 10, 329.
32. van Gorkom, H.J. and Donze, M. (1973). Photochem. Photobiol. 17, 333.

33. Zankel, K.L. (1971). Biochim. Biophys. Acta 245, 373.
34. Lavorel, J. (1973). Biochim. Biophys. Acta 325, 213.
35. Mauzerall, D. (1972). Proc. Nat. Acad. Sci. USA 69, 1358.
36. Zankel, K.L. (1973). Biochim. Biophys. Acta 325, 138.
37. Floyd, R.A., Chance, B. and DeVault, D. (1971). Biochim. Biophys. Acta 226, 103.
38. Döring, G. and Witt, H.T. (1972). In: Proc. 2nd Intern. Congr. Photosynthesis Res., Stresa 1971 (Forti, G., Avron, M. and Melandri, A., eds.), Vol.1, p. 39-45, Dr. W. Junk, The Hague.
39. Duysens, L.M.N. and Sweers, H.E. (1963). In: Microalgae and Photosynthetic Bacteria, pp. 353-372, University of Tokyo Press, Tokyo.
40. Malkin, S. and KoK, B. (1966). Biochim. Biophys. Acta 126, 413.
41. Delosme, R. (1967). Biochim. Biophys. Acta 143, 108.
42. Witt, H.T. (1971). Quart. Rev. Biophys. 4, 365.

M. AVRON, *Proceedings of the Third International Congress on Photosynthesis*, September 2-6, 1974, Weizmann Institute of Science, Rehovot, Israel
Elsevier Scientific Publishing Company, Amsterdam, The Netherlands, 1974

PHOTOREACTIONS OF CHLOROPLASTS AT -20 TO -50 °C

J. Amesz, M.P.J. Pulles, B.G. de Grooth and P.L.M. Kerkhof

Biophysical Laboratory of the State University, P.O. Box 2120, Leiden, The Netherlands

1. Summary

It is shown, that the absorbance change at 518 nm (P518) in spinach chloroplasts at -50 °C consists of two components, associated with photosystems 1 and 2, respectively. The first one appears to reflect the redox level of P700, the other one that of the primary electron acceptor of system 2 associated with C-550. Measurements in the ultraviolet indicate that this acceptor is plastoquinone, which is reduced to the semiquinone anion. In chloroplasts preilluminated with two flashes before cooling, the system-2 changes were reduced by about 40 %, and a slow photooxidation of cytochrome $\underline{b}_{559}$ occurred. At -20 °C, the photooxidation was slower than at -50 °C and preillumination was not necessary. Photooxidation of cytochrome $\underline{f}$ by system 1 was observed at -20, but not at -50 °C.

The results are discussed in terms of two different photoreactions of system 2 occurring in preilluminated chloroplasts. For one of these reactions the electron acceptor is plastoquinone; for the other one it might be an oxidized donor of system 2.

2. Introduction

Experiments with spinach chloroplasts cooled to temperatures between -40 and -90 °C have shown[1-4] that the photochemical reactions of system 2 of photosynthesis are affected by the so-called "S-states"[5-6] of the reaction centers. Dark-adapted chloroplasts, which are in States S_0 or S_1 showed upon illumination at low temperature a rapid increase in the yield of chlorophyll $\underline{a}$ fluorescence and phototransformation of C-550 (reflecting the reduction of the primary electron acceptor[7]). Chloroplasts that had been converted to States S_2 and S_3 by preillumination with two short, saturating light flashes before cooling, showed a slower fluorescence increase, and less transformation of C-550. The extent of absorbance increase at 518 nm (due to "P518") was also less in these chloroplasts[3]. Cytochrome $\underline{b}_{559}$ oxidation, however, occurred much more extensively in preilluminated than in dark-adapted chloroplasts[4].

The present paper reports the results of experiments on the effect of excitation of photosystem 1 and 2 respectively on these phenomena, on light-induced absorbance

Abbreviation: DCMU, 3-(3,4-dichlorophenyl)-1,1-dimethylurea

changes in the ultraviolet region and on cytochrome reactions at temperatures above -40 °C. The results support the hypothesis that two different photoreactions, with different efficiencies occur in preilluminated chloroplasts. The more efficient one consists of the photoreduction of plastoquinone to the semiquinone anion, accompanied by a band shift of C-550 and the oxidation of an unknown electron donor. The other one results in the oxidation of cytochrome $\underline{b}_{559}$; the electron acceptor for this reaction is not known.

3. Material and methods

Spinach chloroplasts were prepared as described elsewhere[3] and stored on ice in the dark. Just before each measurement they were mixed with glycol and transferred to a thermostated quartz cuvette of 1.4 mm optical path length. Unless otherwise indicated, the final chlorophyll concentration was 0.5 mM. The apparatus used[8] enabled simultaneous measurement of absorbance and fluorescence changes or of absorbance changes at two different wavelengths. Chlorophyll $\underline{a}$ fluorescence was measured at 692 nm; for further details see ref. 3.

4. Results and interpretation

4.1 Absorbance changes in the region 500 - 565 nm at -50 °C

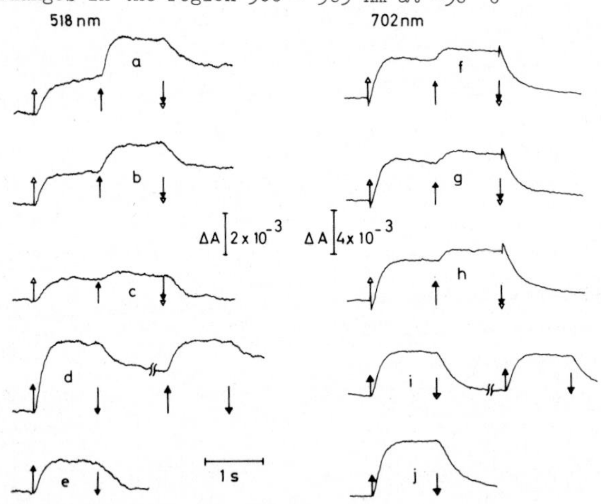

Fig. 1. Absorbance changes of chloroplasts at -50 °C at 518 nm (a - e) and 702 nm (f - j). Upward and downward pointing arrows mark the beginning and the end of illumination, respectively. Open arrows: 727 nm (17 nEinstein $cm^{-2}s^{-1}$), solid arrows: 630 nm (3.5 nEinstein $cm^{-2}s^{-1}$). At 518 nm an upward moving trace indicates an increase in absorbance, at 702 nm a decrease in absorbance. Conditions: a, d, f and i, dark-adapted chloroplasts; b and g, preillumination two saturating 8 µs flashes at 3 °C; c, e, h and j, 20 µM DCMU, 1 mM hydroxylamine, preillumination 12 flashes. For d and i, the second recording was made after 8 s darkness. Recordings at 518 and 702 nm were made simultaneously. Glycol concentration 53 % v/v.

Fig. 1 shows measurements of absorbance changes at 518 nm (due to P518) and 702 nm (due to P700, the primary electron donor of system 1) at -50 °C brought about by far-red light, almost exclusively absorbed by photosystem 1, and by red light, absorbed by both photosystems. The results show that, like at room temperature[9], systems 1 and 2 both cause a photoconversion of P518. The photosystem 1-induced absorbance change was independent of preillumination (recording b), whereas the photosystem 2-induced change at 518 nm was about 40 % smaller in chloroplasts preilluminated with two flashes before cooling than in dark-adapted chloroplasts. The absorbance change caused by system 2 was absent in chloroplasts preilluminated with NH_2OH and DCMU, as could be expected[10].

From the recordings a and c and a and f it can be calculated that in our conditions in dark-adapted chloroplasts about 40 % of the absorbance change at 518 nm was due to system 1 and 60 % to system 2.

The system-1 absorbance changes of P518 reversed rather rapidly in the dark. They showed similar kinetics as P700. In the presence of added oxidants (methylviologen or ferricyanide) both the P700 and P518 changes were almost irreversible (fig. 2). Upon longer illumination than shown in fig. 1, both the P700 and the P518 changes decreased slowly in the light, probably because of a reaction of P700 with a reductant different from the reduced primary acceptor. These results suggest that the absorbance change of P518 reflects the oxidation-reduction level of P700, but is not or little affected by the redox level of the primary electron acceptor of system 1.

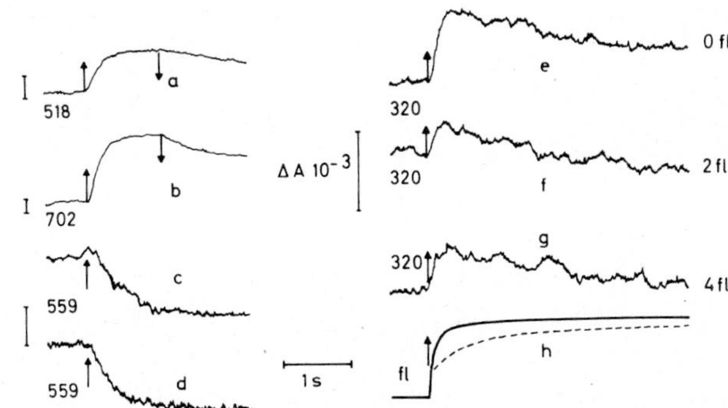

Fig. 2. Recordings a - g: absorbance changes at the wavelengths indicated. The vertical bars denote an absorbance increase of 10^{-3}. Conditions: a and b, as fig. 1, e, but with 0.1 mM methylviologen; c, as fig. 1, b (two flashes preillumination); d, as c, but five times lower actinic intensity and paper speed than the other recordings; e - g, no additions, preillumination as indicated, chlorophyll concentration 0.25 mM. Recording h: fluorescence of chlorophyll, measured at 692 nm; solid line: no preillumination, broken line: preillumination two flashes. Actinic illumination: 630 nm. Recordings a - d, h, -50 °C; e - g, -40 °C. See fig. 1 for further explanation.

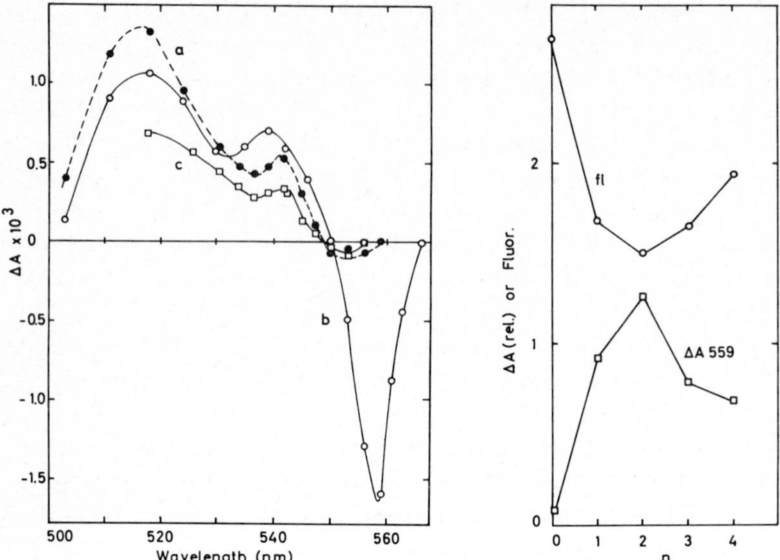

Fig. 3. Spectra of absorbance changes caused by system 2. Solid circles: absorbance changes brought about by 630-nm light upon a background of far-red light (conditions as for fig. 1, curve a (no preillumination)). Open circles: preillumination two flashes. Squares: preillumination two flashes with 3 mM ferricyanide. The spectra were obtained with three different batches of chloroplasts.

Fig. 4. Light-induced increase at -50 °C in chlorophyll a fluorescence (measured after 0.2 s illumination) and decrease in absorbance at 559 nm (measured after 1.5 s illumination) plotted as function of the number of preilluminating flashes, n. Actinic illumination: 630 nm. The intensity of initial fluorescence was taken equal to 1 unit.

The system-2 absorbance change of P518 reversed very slowly in the dark, with a half-time of several min. Difference spectra are shown in fig. 3. Spectrum a was obtained with dark-adapted chloroplasts, and shows a maximum near 518 nm. The band shift due to C-550 is clearly visible in the region 540 - 550 nm on the slope of the P518 nm band. The difference spectrum of the system-1 change (not shown) likewise showed a maximum near 518 nm, but C-550 was absent in this spectrum. After two flashes preillumination (b), the band of P518 was smaller, and the spectrum now showed an additional negative band at 558 nm, due to oxidation of cytochrome $\underline{b}_{559}$. This photooxidation was clearly slower than the photoconversion of C-550 or P518 (fig. 2), and proceeded at a similar rate as the increase in yield of chlorophyll a fluorescence. Fig. 4 shows that the 559-changes were dependent upon the number of preilluminating flashes with period four, with a maximum after two flashes. The fluorescence behaved approximately in a complementary way. The amount of cytochrome $\underline{b}_{559}$ that was photooxidized in chloroplasts preilluminated with two flashes was one per about 650 chlorophyll molecules.

Because of interference by the overlapping absorbance changes of P518 and cytochrome $\underline{b}_{559}$ it was impossible to decide from spectrum b if a conversion of C-550 occurred. In the presence of ferricyanide, however, the cytochrome band was absent (see also ref. 3) and the spectrum thus obtained (c) again showed the bands of C-550, reduced in about the same proportion as the band of P518, compared to dark-adapted chloroplasts. As reported earlier[2,3] a slow fluorescence increase is also observed in the presence of ferricyanide.

4.2 Absorbance changes in the ultraviolet region

Light-induced absorbance changes, recently reported by van Gorkom[11], in chloroplasts treated with the detergent deoxycholate, indicate that plastoquinone may act as primary electron acceptor of system 2. The difference spectrum suggested that the reduction product is the semiquinone anion.

Absorbance changes at 320 nm in intact chloroplasts at -40 °C (fig. 2) showed the same preillumination dependence, and similar kinetics as C-550 and the system-2 component of P518. The amplitude was about 40 % smaller after two flashes than without preillumination. After four flashes the changes were larger again. The absorbance difference spectrum (fig. 5) showed a maximum at 310 - 315 nm and a minimum near 270 nm. It is similar to the difference spectrum obtained *in vitro* upon reduction of plastoquinone to the semiquinone anion[14], but in the region below 280 nm it appears to be distorted, perhaps due to interference by another compound. In some preparations, the absorbance changes below 275 nm were larger than shown here. Compared to our spectrum and that of Bensasson and Land[14] the spectrum of van Gorkom[11] seems to be shifted towards longer wavelengths, perhaps due to the detergent used. The spectra reported by Stiehl and Witt[15] and Witt[16] show a maximum near 325 nm; and the first one shows only small absorbance changes below 290 nm. The reason for these differences is not clear yet. The size of the absorbance changes indicated reduction of one plastoquinone per about 300 chlorophyll molecules in dark-adapted chloroplasts. No reduction of plastoquinone occurred in chloroplasts preilluminated in the presence of hydroxylamine and DCMU, confirming that the reduction is driven by photosystem 2.

In conclusion, it may be stated that the experiments reported here, indicate that a plastoquinone molecule associated with the reaction center acts as primary electron acceptor of photoreaction 2. The plastoquinone is distinct from the large pool which acts as secondary electron acceptor[17]; reduction of the primary acceptor is accompanied by band shifts of P518 and C-550.

4.3 Photoreactions at temperatures above -40 °C

Photooxidation of cytochrome $\underline{b}_{559}$ was also observed at temperatures above -40 °C. The oxidation by light, however, became slower at increasing temperatures, and preillumination before cooling was no longer a prerequisite for the reaction. In most

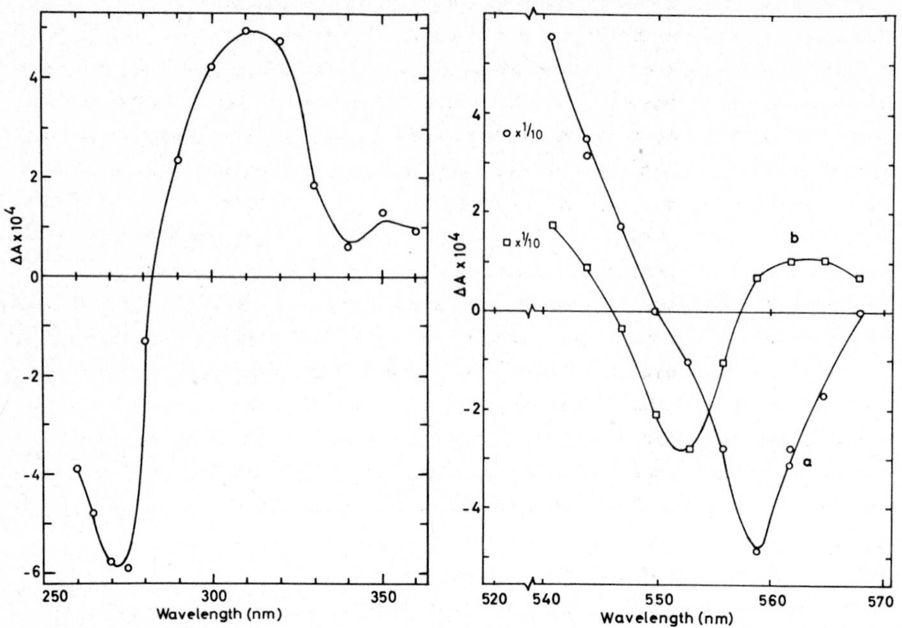

Fig. 5. Absorbance difference spectrum (light minus dark) in the ultraviolet region at -40 °C. Conditions as for fig. 2, e (no preillumination) except chlorophyll concentration, 0.12 mM. A correction for particle flattening was applied (see refs. 12,13) which varied between 1.37 (at 260 nm) and 1.18 (at 320 nm). Only the first, rapid absorbance change was plotted.

Fig. 6. Absorbance difference spectra at -20 °C. Glycol concentration 36 % v/v. Circles: induced by 630-nm light upon a far-red background (preceding dark time 3 min); squares: 727-nm illumination (dark time 0.5 min). No preillumination.

samples, however, the rate of reaction at -20 °C was larger after preillumination with two flashes, than with dark-adapted chloroplasts. At -20 °C, the photooxidation was sensitive to DCMU, especially with dark-adapted chloroplasts, indicating that the secondary pool of plastoquinone acted as electron acceptor. At -50 °C the photooxidation of cytochrome $\underline{b}_{559}$ was completely irreversible, even after 14 min darkness, but at -20 °C photooxidation of part of the cytochrome was again observed after 3 min darkness. A difference spectrum obtained in this way by illumination with red light upon a background of far-red light (fig. 6, spectrum a) showed, in addition to the cytochrome $\underline{b}_{559}$ band a relatively large increase at 520 nm. The increase was about 2.5 times larger than at -50 °C, and may have been partly due to secondary reactions affecting P518 at this temperature.

Spectrum b of fig. 6 was obtained upon excitation of system 1. It shows a smaller increase at 520 nm, and in addition a negative band at 552 nm, presumably due to photooxidation of cytochrome $\underline{f}$. As far as we know, photooxidation of cytochrome *f* at temperatures below 0 °C (see ref. 18) has not been observed so far. Some years ago, Knaff and Arnon[19] reported a photooxidation of cytochrome *f* in chloroplasts cooled to -189 °C, but this observation has not been confirmed by others (*e.g.* refs. 20,21). Remarkably, the amplitude of the absorbance decrease at 552 nm upon illumination was larger after 0.5 min than after 3 min of darkness, whereas the absorbance decrease of P700 showed the opposite behavior. The reason of this phenomenon is not clear yet.

5. Discussion

The results reported here show that in spinach chloroplasts at -50 °C the 518-nm change consists of two components, associated with systems 1 and 2, respectively. The first one appears to reflect the redox level of P700; the second one is correlated with C-550. The absorbance changes of C-550 reflect the redox level of a primary electron acceptor of system 2; evidence is given in section 4.2 that this acceptor is a plastoquinone molecule which is reduced to the semiquinone anion upon illumination.

Measurements of the kinetics of plastoquinone, C-550, P518 and cytochrome $\underline{b}_{559}$ support the notion[3] that in chloroplasts converted to States S_2 and S_3 by preillumination with one or two flashes two different photoreactions of system 2 occur at -50 °C. The first one is relatively efficient and consists of the reduction of plastoquinone, accompanied by band shifts of C-550 and P518 and coupled to the oxidation of an unknown electron donor. The other photoreaction is less efficient and results in the photooxidation of cytochrome $\underline{b}_{559}$. The fluorescence yield of chlorophyll *a* appears to be largely controlled by this reaction. The electron acceptor for the reaction is not known; the kinetics suggest that it is not plastoquinone. The reaction seems to be even less efficient at -20 °C than at -50 °C (section 4.3).

One may speculate that in States S_2 and S_3 in some reaction centers at -50 °C the electron acceptor is an oxidized donor produced by a previous photoact, and that, instead of the "normal" reaction (1)

$$D_1^+D_2^+D_3D_4PQ\ (a) \xrightarrow{h\nu} D_1^+D_2^+D_3D_4P^+Q^-\ (b) \longrightarrow D_1^+D_2^+D_3D_4^+PQ^-\ (c),$$

in some reaction centers the reaction (2)

$$D_1^+D_2^+D_3D_4PQ\ (a) \xrightarrow{h\nu} D_1^+D_2D_3D_4P^+Q\ \ (b) \longrightarrow D_1^+D_2D_3D_4^+PQ\ \ (e)$$

occurs. P and Q are P680 and plastoquinone, respectively, and D_1 - D_4 are electron donors on the pathway to water; (a) symbolizes a reaction center in State S_2. In reaction (2), D_2^+, only present in States S_2 and S_3, acts as electron acceptor instead of plastoquinone. In some reaction centers P^+ might oxidize cytochrome $\underline{b}_{559}$ rather than D_4. Such a reaction might occur upon excitation of reaction centers (a) and (e) and perhaps also of reaction centers (c), with D_4^+ as electron acceptor, causing a relatively slow accumulation of oxidized cytochrome $\underline{b}_{559}$. However, a detailed model to describe the various photochemical reactions that might be possible after the first photoact has occurred, would be too speculative on basis of the present evidence.

The investigations reported here were supported in part by the Netherlands Foundations for Biophysics and for Chemical Research, financed by the Netherlands Organization for the Advancement of Pure Research (ZWO).

References

1. Joliot, P. and Joliot, A., 1972, in: Proc. 2nd Int. Congr. Photosynthesis Research, vol. 1, eds. G. Forti, M. Avron and A. Melandri (Dr. W. Junk N.V. Publishers, The Hague) p. 149.
2. Joliot, P. and Joliot, A., 1973, Biochim. Biophys. Acta 305, 302.
3. Amesz, J., Pulles, M.P.J. and Velthuys, B.R., 1973, Biochim. Biophys. Acta 325, 472.
4. Vermeglio, A. and Mathis, P., 1973, Biochim. Biophys. Acta 314, 57.
5. Kok, B., Forbush, B. and McGloin, M.P., 1970, Photochem. Photobiol. 11, 457.
6. Joliot, P., Joliot, A., Bouges, B. and Barbieri, G., 1971, Photochem. Photobiol. 14, 287.
7. Butler, W.L., 1973, Accounts Chem. Res. 6, 177.
8. Amesz, J., Pulles, M.P.J., Visser, J.W.M. and Sibbing, F.A., 1972, Biochim. Biophys. Acta 275, 442.
9. Fork, D.C., Amesz, J. and Anderson, J.M., 1967, Brookhaven Symp. Biology 19, 81.
10. Bennoun, P., 1970, Biochim. Biophys. Acta 216, 357.
11. van Gorkom, H.J., 1974, Biochim. Biophys. Acta 347, 439.
12. Duysens, L.N.M., 1956, Biochim. Biophys. Acta 19, 1.
13. Amesz, J., 1964, Thesis, University of Leiden.
14. Bensasson, R. and Land, E.J., 1973, Biochim. Biophys. Acta 325, 175.
15. Stiehl, H.H. and Witt, H.T., 1968, Z. Naturforsch. 23b, 220.
16. Witt, K., 1973, FEBS Letters 38, 116.
17. Amesz, J., 1973, Biochim. Biophys. Acta 301, 35.
18. Floyd, R.A., 1972, Plant Physiol. 49, 455.
19. Knaff, D.B. and Arnon, D.I., 1969, Proc. Natl. Acad. Sci. U.S. 63, 956.
20. Boardman, N.K., Anderson, J.M. and Hiller, R.G., 1971, Biochim. Biophys. Acta 234, 126.
21. Erixon, K. and Butler, W.L., 1971, Photochem. Photobiol. 14, 427.

M. AVRON, *Proceedings of the Third International Congress on Photosynthesis*, September 2-6, 1974, Weizmann Institute of Science, Rehovot, Israel
Elsevier Scientific Publishing Company, Amsterdam, The Netherlands, 1974

FLUORESCENCE RISE FROM 36μs ON FOLLOWING A FLASH AT LOW TEMPERATURE (+ 2° . - 60°)

Anne JOLIOT

Institut de Biologie Physico-Chimique
13, rue Pierre et Marie Curie
75231 - Paris Cedex 05 France

Summary

The variation of the fluorescence yield 36μs after a short saturating flash has been studied in the +2° -60° temperature range on chloroplasts suspension. It has thus been possible to observe an increase in the fluorescence yield previously suggested by the works of Zankel[4]. The amplitude of this increase is temperature dependent and DCMU insensitive.

This increase is ascribed to the transfer of the positive charge stored on the chlorophyll during the primary photoact towards a secondary donor. The half-time value (~ 70μs at +2° - 220μs at - 30° ~ 400μs at - 60°) extrapolated to liquid nitrogen temperature fits nicely with the one expected by Murata et al.[18] for this reaction (35ms).

1. Introduction

The fluorescence yield increase observed upon illumination of chloroplasts or algal suspension is related to the reduction of a quencher Q , the Photosystem II primary electron acceptor[1]. More recently, several authors[2, 3, 4, 5], have observed a dependence of the fluorescence yield on the donor side of the photocenter, i.e. on the S states as defined in the model of Kok et al.[6] .

At room temperature, the quencher Q is reoxidized by a secondary quencher A, which is presumably a Plastoquinone. Low temperatures are now well known[7, 8] to slow down the Q to A reaction, resembling strongly the action of 3-4 dichlorophenyl dimethylurea (DCMU) on the fluorescence induction kinetics. Furthermore, it has been established [2, 5] that a short saturating flash given to a frozen sample is never able to totally reduce the quencher Q present. Only a part of the quenching ("Q_F") is suppressed with high quantum efficiency and its variations in amplitude are related to the charge separation leading to O_2 formation. The remaining fraction "Q_S" can only be reduced with low quantum efficiency, either by several flashes or continuous illumination, and has been related to the state of oxidation of Cytochrome b 559[5, 9].

In a previous paper[10], the variations of the fluorescence yield were studied in the -30° to -60° temperature range from 20ms to several minutes following the flash given at low temperature. The technique employed has been now improved by using weak detecting flashes instead of a continuous beam and allows us to study the variation of the fluorescence yield in a shorter time range following the flash.

2. Material and Methods

Chloroplasts are prepared according to Avron[11] and stored at -70° in the presence of 5% dimethylsulfoxide. Prior to use, the chloroplasts are suspended in 0.05 M Tris buffer (pH 7.5) with 0.4 M saccharose and 0.01M NaCl at a concentration of about 220 μg Chl/ml.

The experiments are performed with a previously described technique [5, 10]. The sample can be illuminated :

- by short saturating actinic flashes (Xenon, Verre and Quartz Co, model VQ.X CAD22, 1.3J, 6μs half band width) ;
- by a continuous beam (about 3000 ergs cm^{-2} sec^{-1}) used to analyse the fluorescence yield from 20ms after the actinic flash ;
- by weak detecting Xenon flash (General Radio Stroboslave). An electronic delay permits a variation of the dark time between the actinic flash and the first detecting flash. The frequency of the detecting flash may also be variable. The actinic flash causes an artifact which perturbs the signal up to 1ms. This artifact is measured and substracted from the detecting signal for each time shorter than 1ms. The intensity of the weak detecting beam has been adjusted to obtain a reasonable signal to artifact ratio for times as shorter as 36μs. This intensity corresponds to 0.5 to 1% of the centers hit per flash.

3. Results

The fluorescence yield is analysed at -50° at increasing dark times following a short saturating flash. The shortest dark time was ~ 40μs to avoid interference by reactions occuring in shorter times [12, 13, 14, 15].

From 36μs after the flash given at -50° two phases can be observed as depicted in Fig. 1 : first, an increase of the yield in the

36μs - 5ms range (Fig.1, part a), then a decrease beyond 5 ms. (Fig.1 part b).

The decrease in the fluorescence yield arises from the dark reoxidation of Q^- and has been previously studied[10] for times longer than 20ms after the actinic flash in the -30° to -60° range. The conclusions obtained here are in agreement with the ones previously demonstrated and are summarized in the Discussion section.

The results presented in this communication concern <u>an increase in the fluorescence yield</u> observed in shorter times following the actinic flash. The existence of this increase has been previously suggested by Zankel[4] at room temperature. However, Zankel was limited to times longer than 80μs after the actinic flash. The better resolution of our method combined with the fact that we work at lower temperature permits us to describe this phenomenom from 36μs on in more detail. This increase, still existant at low temperature, presents the following characteristics :

1. In dark adapted chloroplasts, its amplitude represents about 40% of the part of the quenching destroyed by a flash at -50° (Fig. 2, curve 1, black circles) and is DCMU insensitive (Fig.2, curve 1, open circles). But, as depicted in Fig.3 both amplitude and rise time of this increase are temperature dependent. The estimated half times rises are ~ 70μs at +2° , ~ 220μs at -30° and ~ 400μs at -60°.

When chloroplasts, in the presence of 20μM DCMU + 10mM Hydroxylamine, are illuminated by several flashes prior to cooling, an additionnal flash given at low temperature cannot induce any more charge separation[15]. Under these conditions, one can only observe at -50° a very small rise in the fluorescence yield, which represents not more than 10-12% of the total variable fluorescence and which is entirely terminated in 100μs. (Fig.2, curve 3).

2. When the chloroplasts received two flashes prior to cooling,the increase still occurs after a flash given at low temperature (Fig.1, part a, curve 3, and Fig. 2, curve 2). However, its amplitude is smaller (30% of the total quenching destroyed by a flash at -50° under the same conditions) and the initial yield measured at 36μs is lower than the one observed in dark adapted chloroplasts. Furthermore, the rise is much faster and at -50° goes to completion in ~ 1ms.

In Fig.4 is shown the fluorescence yield measured 4.95 ms (open circles) and 66μs (black circles) after a flash given at -50° to chloroplasts which received 0 to 5 flashes prior to cooling. The black squares sequence gives the difference between the two preceeding

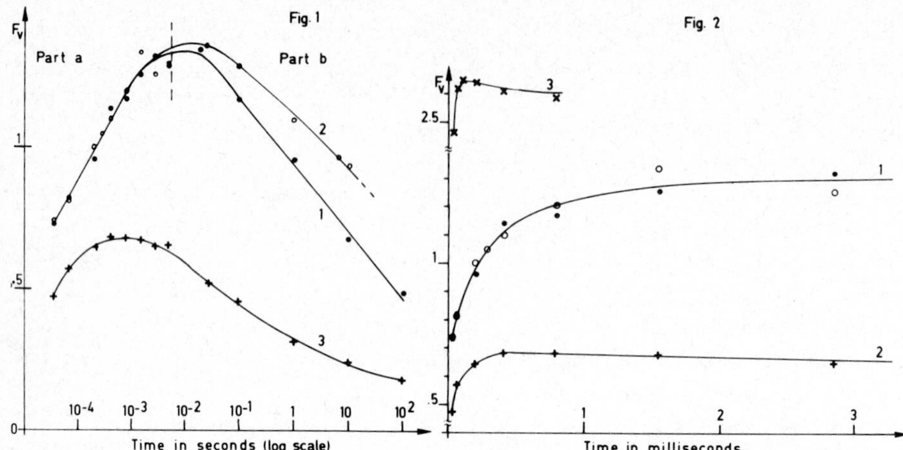

Fig.1. Variable fluorescence in function of dark time following a flash. T=-50°. Dark adapted chloroplasts: 1. ● No DCMU - 2. o 20μM DCMU. 3. Chloroplasts preilluminated by 2 flashes before cooling.

Fig.2. Fluorescence rise observed from 36μs on following a flash. T=-50°. 1. Dark adapted chloroplasts ● No DCMU - o 20μM DCMU.
2. Chloroplasts preilluminated by 2 flashes prior to cooling.
3. Chloroplasts with 20μM DCMU + 10mM Hydroxylamine preilluminated by several flashes prior to cooling.

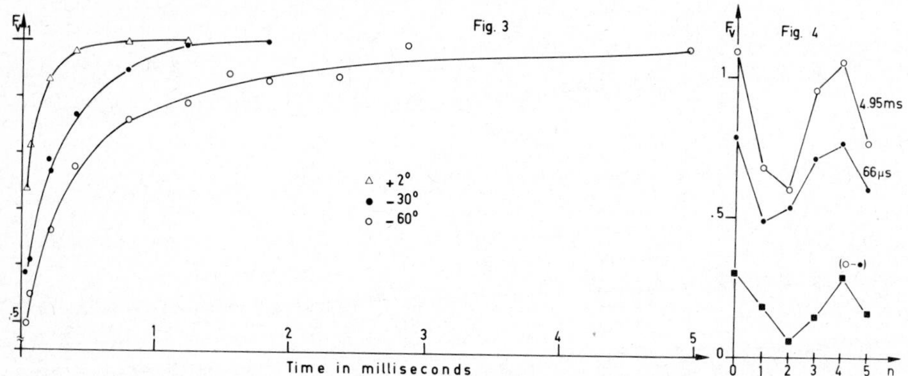

Fig.3. Temperature dependence of the fluorescence rise from 36μs on following a flash. Chloroplasts + 20μM DCMU. The curves have been normalized to the maximal level of fluorescence reached after a flash given at the indicated temperature.

Fig.4. Variable fluorescence at 4.95 ms and 66μs after a flash given at -50° in function of the number n of preilluminating flashes given to chloroplasts at +2° before cooling.
■ Amplitude of the fluorescence rise between 66μs and 4.95 ms, i.e. difference between the two preceeding sequences.

ones : the variations of the fluorescence yield oscillate in phase at the minimum and at the maximum of the increase. The dependence on the S states already established by Joliot et al.[2, 5] for longer dark times after the flash or by Delosme[3] during the flash are confirmed during the phenomenom described here : if States S'_n designate the state of the donor right after the additionnal flash given at low temperature[2] , S'_0 and S'_1 (after 0 or 4 preilluminating flash at +2°) are linked to a low quenching level whereas S'_2 and S'_3 (after a two flash - preillumination at +2°) are characterized by a high quenching level.

4. Discussion

1) Kinetics of the variation of the fluorescence yield

The time-course of the fluorescence yield measured during and after a short saturating flash has been studied by several authors and presents the following variations.

During the flash (2-15μs) two light-driven reactions occur : The first, increasing the fluorescence yield, is the charge separation and the second, decreasing the yield, is generally attributed to the formation of a carotenoïd triplet state in a carotenoïd-chlorophyll complex.

A subsequent fast increase is then observed in the 16-40μs time range corresponding to the relaxation time of the triplet state. The time-course of this increase is not modified by the initial state of quencher Q (oxidized or reduced) and from measurements at room temperature[12, 16] and at liquid nitrogen temperature[14] appears to be temperature independent.

Following this fast increase, a slower one occurs and has been studied here. It disappears in conditions where Q stays totally in the reduced state and thus cannot be ascribed to the same phenomenom as the first fast increase. (The small residual rise observed at -50° up to 100μs in preilluminated chloroplasts in the presence of DCMU + Hydroxylamine may be due to the "tail" of the 16-40μs rise).

In dark adapted chloroplasts, this second increase is temperature dependent which is another distinction with the preceeding one. From measurements at +4°, Zankel[4] proposed the existence of such a rise and the estimated half-times he supposed at +4° (65μs) and 22° (35μs) are coherent with the half-times values measured in this work. The absence of a DCMU effect also in agreement with Zankel's conclusions, may suggest that this increase corresponds to the transfer of the positive charge on the donor side of Photosystem II center.

From spectrophotometric studies, Gläser et al.[17] pointed out the existence of a fast $Chla_{II}$ absorption change which τ = 35µs at 20°C. It is very likely that this $Chla_{II}$ (35µs) absorption change and the increase in the fluorescence yield occuring in the same time range both denote the same reaction e.g. the transfer of the positive charge from the chlorophyll to a secondary donor.

The half-time values measured here at different temperatures gives an average Q_{10} of about 1.3. If one extrapolates the half-time value to liquid nitrogen temperature it would thus give a half-time of 20-40 ms. This value fits nicely with the one proposed by Murata et al.[18] for the same reaction. These authors suggested that the reduction of Chl^+ may occur by two competitive reactions, one is the stabilisation of the positive charge on a secondary donor, the other is a back reaction of the photochemical reaction. The temperature dependence we measured supports the hypotheses of Murata et al.[18] and Butler et al.[19] who argued that lowering the temperature favors the back reaction rather than the stabilisation of the positive charge on a secondary donor.

The slow decay (Fig. 1, part b) which follows the rise in the fluorescence yield corresponds to the reoxidation of the quencher Q^- reduced by the flash. By the time this decrease occurs, the positive charge has already migrated from the chlorophyll to a secondary donor. We previously studied this reoxidation[10], whose characteristics are the following : in the -30° to -60° range, it is a temperature dependent DCMU sensitive reaction. We showed that while the electron is transferred towards the pool A, the corresponding + charge is stored on the donor side of the center, Z, thus leading to an advance in the S states.

In the presence of DCMU, where the Q to A step, is blocked, two back reactions between Z^+ and Q^- participate in the reoxidation of Q : a temperature dependent one described by Bennoun[15] and a non-exponential temperature independent one which is the only one seen at -50°.

When the chloroplasts are in the S_2 + S_3 states before cooling, we established that the dark reoxidation following a flash at low temperature (Fig. 1, part b, curve 3) is principally due to a charge recombinaison and essentially no further advance in the S states can be observed. Below -30° this back reaction is temperature independent and is somewhat faster than that observed on dark adapted chloroplasts in the presence of DCMU. Nevertheless, it remains slower than the charge recombination between Chl^+ and Q^-. One must admit that the positive charge already migrated form Chl^+ towards a secondary donor.

It is surprising to notice that the fluorescence increase for chloroplasts in the $S_2 + S_3$ states is faster than the one observed on dark adapted chloroplasts if the same origin for the increase is supposed in both cases.

2) Quenching levels

Our results show a decrease of the fluorescence yield at 40μs when the temperature is lowered. (This difference still exists if the values are extrapolated to time zero). It seems reasonable to admit, as previously stated by Butler et al.[20] that Chl^+ is a quencher. Several hypotheses then may be proposed to take into account the variation of the quenching level with the temperature : if Chl^+ is a low efficiency quencher, one may suppose that its quenching efficiency is temperature dependent and increases when the temperature decreases. On the other hand, if Chl^+ is a high efficiency quencher, one can imagine the existence of an equilibrium between Chl^+ and an intermediary donor Y. This equilibrium should occur in a time shorter than 40μs and should be displaced in favor of Chl^+ when the temperature is decreased. This hypothesis could take into account the results obtained by Mauzerall[13] who observed a very low fluorescence yield 10 ns after the flash. The reaction which controls the fluorescence rise would be the reduction of Y^+ by Z, where the positive charges are stored. Due to the low equilibrium constant between Y and Chl, this step is associated with the reduction of Chl^+.

The quenching level observed for chloroplasts in the $S_2 + S_3$ states cooled to -50° is even higher. This result is in agreement with what Delosme[3] observed at room temperature in the μs range, and can be interpreted, as already proposed[2,5], by the intervention of a second quencher on the acceptor side of the center.

References

1. Duysens, L.N.M. and Sweers, H.E., 1963, in Microalgae and Photosynthetic Bacteria (Univ. of Tokyo Press, Tokyo), 353.

2. Joliot, P. and Joliot, A., 1971, in Proceedings of IInd International Congress on Photosynthesis Research (ed. W. Junk), 26.

3. Delosme, R., 1971, in Proceedings of IInd International Congress on Photosynthesis Research (ed. W. Junk), 187.

4. Zankel, K.L., 1973, Biochim. Biophys. Acta., 325, 138.

5. Joliot, P. and Joliot, A., 1973, Biochim. Biophys. Acta. 305, 302.

6. Kok, B., Forbush, B. and MacGloin, M., 1970, Photochem. Photobiol., 11, 457.

7. Joliot, P., 1965, Biochim. Biophys. Acta. 102, 135.

8. Malkin, S. and Michaeli, G., 1971, in Proceedings of IInd International Congress on Photosynthesis Research (ed. W. Junk), 146.

9. Vermeglio, A. and Mathis, P., 1973, Biochim. Biophys. Acta. 292, 763.

10. Joliot, A., 1974, Biochim. Biophys. Acta. (in press).

11. Avron, M., 1960, Biochim. Biophys. Acta. 40, 257.

12. Duysens, L.N.M., Van der Schatte Olivier, T.E. and Den Haan, G.A. 1972, Abstr. VI. International Congress on Photobiology. Bochum. 277.

13. Mauzerall, D., 1972, Proc. Natl. Acad. Sci. US., 69, 1358.

14. Den Haan, G.A., Warden, J.T. and Duysens, L.N.M., 1973, Biochim. Biophys. Acta. 325, 120.

15. Bennoun, P., 1970, Biochim. Biophys. Acta. 216, 357.

16. Delosme, R., personnal communication.

17. Gläser, M., Wolf, Ch., Buchwald, H.E. and Witt, H.T., 1974, FEBS Letters, 42, 1, 81.

18. Murata, S., Itoh, S. and Okada, M., 1973, Biochim. Biophys. Acta. 325, 463.

19. Butler, W.L., Visser, J. and Simons, J.L., 1973, Biochim. Biophys. Acta. 325, 539.

20. Butler, W.L., Visser, J. and Simons, J.L., 1973, Biochim. Biophys. Acta. 292, 140.

M. AVRON, *Proceedings of the Third International Congress on Photosynthesis*,
September 2-6, 1974, Weizmann Institute of Science, Rehovot, Israel
Elsevier Scientific Publishing Company, Amsterdam, The Netherlands, 1974

LIGHT-INDUCED ABSORPTION CHANGES IN SPINACH CHLOROPLASTS / A COMPARATIVE STUDY AT -50° AND -170°C

A. VERMEGLIO and P. MATHIS

Département de Biologie , CEN Saclay
BP n° 2 - 91190 Gif sur Yvette , France

Summary

1. Two Photosystem-2 (PS-2) photochemical reactions are studied :
   - Photooxidation of cytochrome b_{559}. Its magnitude (but not its kinetics) is dependent upon both the temperature and the S state.
   - Photoreduction of C-550. It is a more efficient reaction and its rate is about 4x smaller at -170° than at -50°, because of a back-reaction ($t_{1/2}$=4.2 ms). At -50°, neither the kinetics nor the magnitude of reduction are influenced by the S state. The reduction of C-550 is accompanied by a 518-nm effect which, however, exhibits a different sensitivity to ferricyanide (-170°) and to the S state (-50°).

   These two photochemical reactions are tightly coupled together.
2. A 518-nm absorption change due to PS-1 is large at -50° and much smaller at -170°. At -50° it reverses with a half-time of about 100 ms.
3. It is possible to photooxidize about two molecules of Cyt b_{559}HP (high-potential) by the way of alternate illuminations and temperature changes.

Introduction

Recently, low temperatures have been extensively used in the study of the reaction center of PS-2 in chloroplasts. Liquid N_2 temperature is specially convenient as it provides a condition in which most secondary dark reactions are blocked and several components have better resolved absorption bands [1]. However the photochemical routes may be different from what occurs in normal photosynthesis. In this respect, a study at -50° may be more valuable, since the photochemistry retains most of its room-temperature behaviour and most dark reactions (including the electron transfer from the primary to the secondary acceptor) are slowered [2].

In this article, in continuation of our previous work [3-7], we report on absorption changes measured at -50° and -170°, in an effort to elucidate the photochemistry of PS-2 . Our major conclusions are the existence at both temperatures (as we already proposed for -170°) of two separate PS-2 photochemical reactions [5] and of a 518-nm absorption change (field effect ?)[6] related to the reduction of a primary acceptor, C-550. Moreover we show that it is possible to photooxidize two Cyt b_{559} molecules at the reaction center of PS-2 . We also find several differences between the behaviour of PS-2 at -170° and in the S_2 or S_3 states at -50°, two cases which share several common properties : photooxidation of

Cyt b_{559}[3,4], slow fluorescence induction[2,8-10] and small proportion of the reaction centers that are left in a photoconverted state after excitation by a saturating flash[3,4,11-15].

Because of the overlap of the absorption changes, we must also discuss some of our data on PS-1 , in which we only study the 518-nm absorption change.

Material and Methods

Spinach chloroplasts, prepared by a standard technique, were diluted with 2 volumes of glycerol and stored in liquid N_2[3].In some specified cases we used freshly prepared chloroplasts,in a buffer containing 0.4 M sucrose , 0.02 M Tris buffer (pH 7.8) and 0.01 M NaCl .

Differential spectra were measured with a double-beam spectrophotometer[3]. The cuvette (optical path 1 mm) was cooled either by dipping into liquid N_2 (fig.7) or by a flow of cold N_2 gas set at the required temperature (fig.2 and intermediate steps in fig.7).The kinetics of the absorption changes were measured as previously described [5]. The copper frame of the cuvette (optical path 1 mm) was dipping in liquid N_2 (temperature : -170°) or in a mixture ethanol-dry ice (-50°). We used 630-nm exciting light , with an irradiance of 5 $mW.cm^{-2}$ at the surface of the cuvette (excepted when otherwise mentioned).

Results

1. Photosystem-1 reactions

The kinetics and magnitude of absorption changes (ΔA) due to PS-1 were measured with chloroplasts in which PS-2 activity was blocked by the addition of DCMU and of hydroxylamine followed by flash illumination before cooling[16]. Under these circumstances, neither photoreduction of C-550 nor photooxidation of Cyt b_{559} was observed , but another absorption change remained , whose difference spectrum is reported in fig.1. At -170°, ΔA are small,with a broad maximum around 530 nm.The signals, which are partly reversible, might be due to P_{700} which is known to be photooxidized at that temperature[17] in a partly reversible reaction[12]. At -50°,ΔA are much larger and the difference spectrum is similar to the room-temperature 515-nm absorption change[18].The signals are fully reversible (fig.1,insert) and can be induced identically by successive illuminations. These ΔA can also be observed in untreated chloroplasts if we use far-red illumination (data not shown ; see also ref.19) and they disappear with chloroplasts pretreated with 10 mM ferricyanide (FeCN).

From the difference in the PS-1 ΔA observed at -50° or at -170°,we can conclude that ,in the "field effect" hypothesis,[18] this effect is not a mere consequence of the primary photochemical reaction , as we also noticed for PS-2 (see below and [6]).From the possible reactivation of the signal at -50°,we

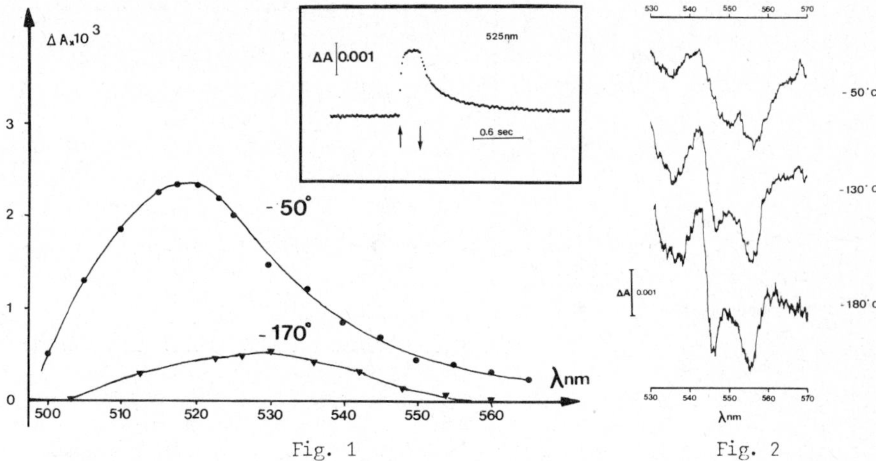

Fig. 1

Fig. 2

Fig.1 - Light-induced ΔA in chloroplasts at -50° (chlorophyll concentration :0.5 $mg.ml^{-1}$; glycerol added) and -170° (chlorophyll : 0.25 $mg.ml^{-1}$; no glycerol). Addition of 20 µM DCMU and 0.1 mM NH_2OH . Preillumination by one flash at room temperature before cooling. At -50° the signals are reversible , as indicated by the insert .

Fig.2 - Differential absorption between a reference cuvette and a cuvette that has been illuminated at -50° . Successive spectra were recorded at -50° and after lowering the temperature . Chloroplasts suspension (chlorophyll : 0.3 $mg.ml^{-1}$) dark-adapted before cooling.

conclude that the reversibility is not due to an electron migration but to a back reaction.

The remaining part of this report is devoted to PS-2 . In experiments performed at -50° the PS-1 signal has been taken in account by subtracting from the ΔA due to a first illumination the signal due to a second illumination given about 20 s after the first one and presumed to be due only to PS-1 . That subtraction was not performed in experiments done either at -50° on suspensions to which FeCN had been added (P_{700}oxidized) or at -170°,since the PS-1 signal is very small.

2. Photosystem-2 reactions : reduction of C-550 and the 518-nm effect

A major difficulty in this study is that the difference spectrum due to the reduction of C-550 , which is sharp and large at -170°,is much broader and smaller at -50° , as seen in fig.2 . The true difference may be even larger than it appears in fig.2 because of the possibility of a partial reoxidation of C-550 during the measurement of the successive difference spectra . In the PS-2 difference spectra obtained from kinetic measurements C-550 appears as a shoulder

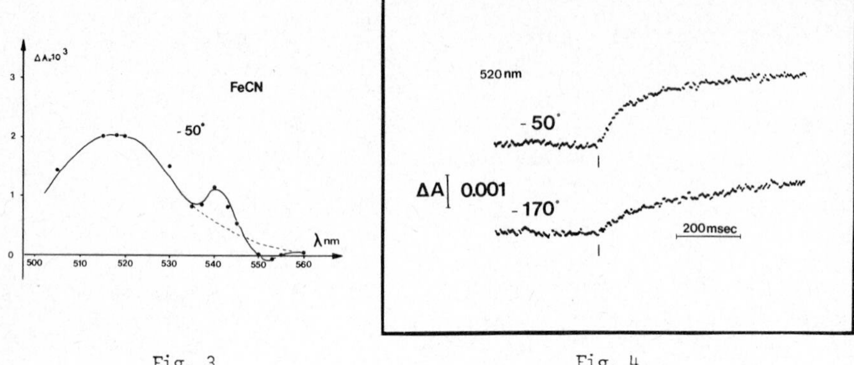

Fig. 3 Fig. 4

Fig.3 - Spectrum of light-induced ΔA measured at -50° with dark-adapted chloroplasts (chlorophyll : 0.5 $mg.ml^{-1}$). Addition of 20 mM of potassium ferricyanide.

Fig.4 - Kinetics of light-induced ΔA at 520 nm , at -50° and -170°, with dark-adapted chloroplasts (chlorophyll : 0.5 $mg.ml^{-1}$).In each case the signal due to a second illumination (20 s after the first) has been subtracted.

on a dominant 518-nm effect (fig.3) whereas both effects are of similar magnitude at -170° [6].

We already reported that the kinetics of the 518-nm effect and of the reduction of C-550 were the same at -170° [6].They are slow,owing probably to a back reaction ($t_{1/2}$=4.2 ms)[7].At -50° the rate of the reaction is about 3-4x greater, as shown in fig.4 for the 518-nm effect , and both phenomena continue to exhibit the same rate (fig.5). This rate is not influenced by the S state (fig.5),as determined by sequential flash excitation at room temperature before cooling.At 541 nm the absorption increase due to the reduction of C-550 is not affected by the S state (fig.5), in agreement with our previous report that , after a saturating illumination at -55° , all the C-550 can be trapped in a reduced state by rapid cooling to -196° [4]. On the other hand , the maximum ΔA due to the 518-nm effect is about 40% smaller in the states S_2 and S_3 than in the states S_0 and S_1 (fig.5).These results are in general agreement with those reported by Amesz et al [20], excepted that these authors find a smaller reduction of C-550 at -40° with chloroplasts preilluminated by two flashes than with dark-adapted chloroplasts.

We studied the effect of light intensity on both the rate of reduction of C-550 and the rate of appearance of the 518-nm effect . In the range studied ($t_{1/2}$ varied from 40 ms to 2 s) both rates are nearly identical and proportional to the irradiance . An identical result has been obtained at -170° for C-550 [5]. Addition of FeCN (up to 20 mM) has no influence on the reduction of C-550 in the

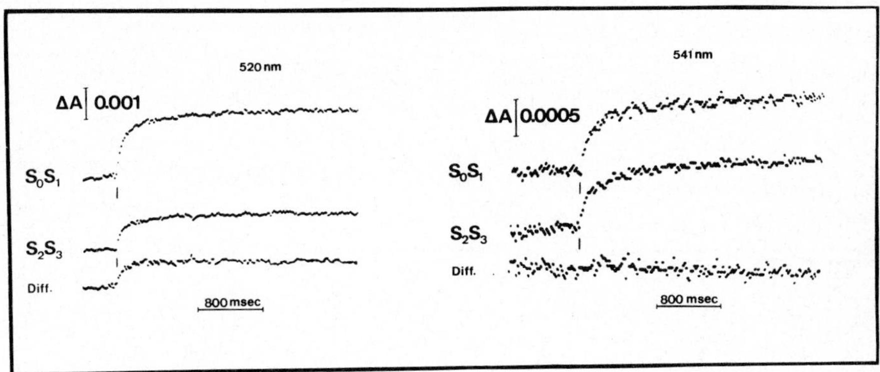

Fig. 5 - Kinetics of light-induced ΔA at -50° , at 520 and 541 nm , with chloroplast suspensions (chlorophyll : 0.5 mg.ml^{-1}) that were dark-adapted (S_0S_1) or received two flashes (S_2S_3) before cooling . The bottom traces represent the differences between the upper and the middle traces . As in fig.4 , the signals due to a second illumination have been subtracted .

time range studied . It also has no effect on the PS-2 518-nm effect occuring at -50° , in contrast with the experiments performed at -170° in which 10 mM FeCN completely abolishes the 518-nm effect [6].

3. Photosystem-2 reactions : oxidation of Cyt b_{559}

The amount of Cyt b_{559} that can be trapped in the oxidized state by rapid cooling to -196° of chloroplasts illuminated at -55° is dependent upon the S state of PS-2 [3,4]. This result is confirmed by kinetic measurements of ΔA at -50° . We find that the maximum absorption decrease at 558 nm is about 2x smaller with dark-adapted chloroplasts than after a two-flash preillumination , although the kinetics of absorption change are not affected (data not shown) . The rate of oxidation of Cyt b_{559} at -50° is about 6x smaller than the rate of reduction of C-550 . It is proportional to the irradiance (fig.6,left) and is identical to its rate of oxidation at -170° (fig.6,right) . These results differ from those obtained by Amesz et al [20] who did not detect a photooxidation of Cyt b_{559} in either S state at -40° , and by Knaff and Malkin [21] who found a variation with light intensity in the extent of Cyt b_{559} photooxidation at -50° .

We found the same amount of Cyt b_{559} photooxidized at -170° in either S state and at -50° in the S_2 and S_3 states [3,4]. This amount is usually evaluated to

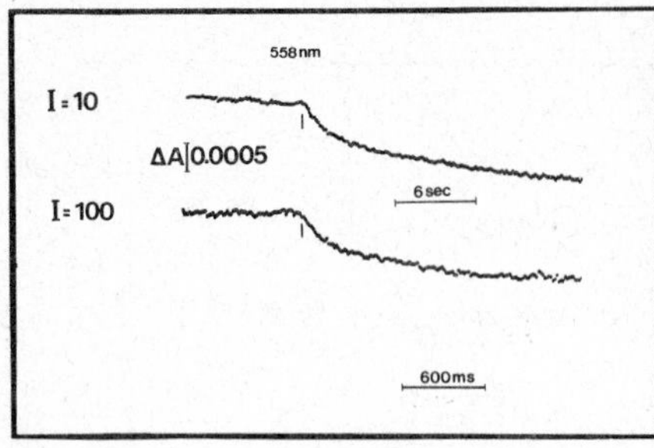

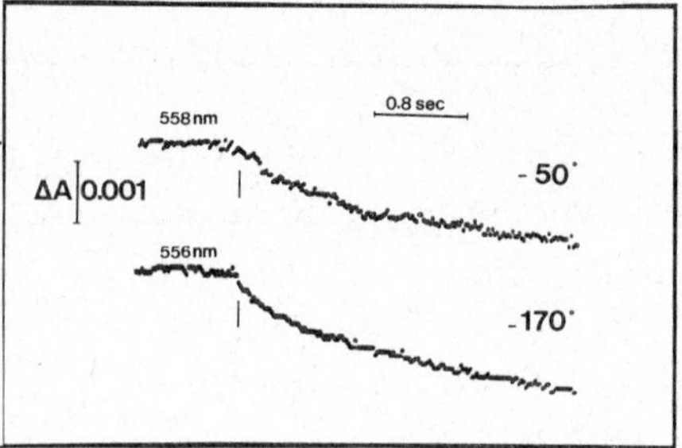

Fig. 6 - Kinetics of light-induced oxidation of Cyt b_{559} with chloroplast suspensions (chlorophyll : 0.5 $mg.ml^{-1}$) that received two flashes before cooling. Left traces : -50° , two light intensities (irradiance : 0.5 $mW.cm^{-2}$ for I = 10 and 5 $mW.cm^{-2}$ for I = 100) . Right traces : -50° and -170°, same light intensity.

be one molecule of Cyt b_{559} per C-550 , i.e. per PS-2 reaction center [22,23] and it represents about one half of the total amount of Cyt b_{559} present in the chloroplasts [24,25]. With fresh chloroplasts in sucrose buffer , in which we found that practically all the Cyt b_{559} (80-90%) was in the high-potential form , we wanted to check whether all the pool of Cyt b_{559} were located at the PS-2 reaction center and could be photooxidized . The experiments described in fig.7 show that this is indeed possible by a few cycles of alternated illuminations at -50° and rewarmings at -30° (this allows the partial reoxidation of C-550 , as we checked by differential absorption spectra) . The amount of Cyt b_{559} that is found oxidized after the previous treatment is nearly double the amount that is oxidized by illumination at -196° , i.e. nearly equal to the total amount of high-potential Cyt b_{559} (fig. 7d).

We believe that this last experiment demonstrates that practically all the pool of Cyt b_{559}HP is located at the reaction center of PS-2 and is photooxidizable at low temperature . The oxidation of all the pool of Cyt b_{559}HP is in agreement with experiments of Erixon and Butler [22] who found that C-550 was the compound present in limiting amount at the reaction center of PS-2 . The pool of Cyt b_{559}HP is however probably not homogeneous in view of the results obtained by Epel et al [23] ; moreover we found that , for chloroplasts suspended in glycerol , about one half of the Cyt b_{559}HP undergoes a rapid conversion to a lower-potential state (A.Vermeglio and P.Mathis , unpublished) .

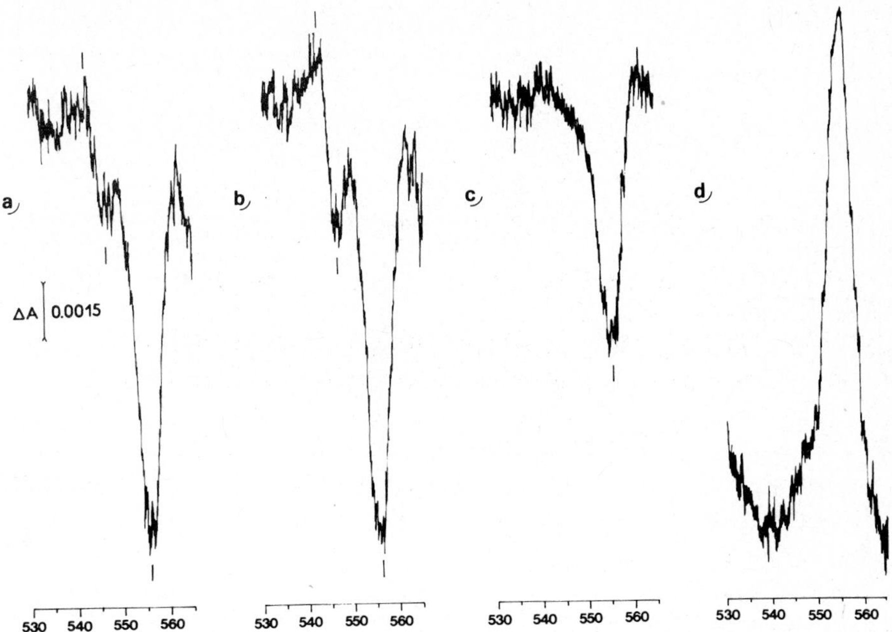

Fig. 7 - Differential absorption spectre recorded at -196° with chloroplasts (chlorophyll : 0.33 $mg.ml^{-1}$, no glycerol) that received the following treatments :

a,b,c : the sample cuvette received 2 flashes at room temperature before to be cooled at -50° , at which temperature it received a saturating illumination (10 s of white light) . It was then rewarmed for 2 min at -30° , and submitted to two additional cycles of cooling to -50° , illumination , warming to -30° . It was then cooled to -196° and spectrum a/ was recorded . The sample cuvette received an additional saturating illumination at -196° before recording spectrum b/ . Then the reference cuvette (previously protected during the illuminations of the sample cuvette) was illuminated at -196° and spectrum c/ was recorded .

d : 20 mM of potassium ferricyanide were added to the reference cuvette (for oxidation of Cyt f and of Cyt b_{559}HP). The sample cuvette has been illuminated for 15 s at 10° with far-red light (for a complete oxidation of Cyt f) and cooled to -196° . This represents a measurement of the total amount of Cyt b_{559}HP .

Discussion

Many properties of the reaction center of PS-2 at -196° are accounted for by the simple schema [10,11,13] :

$$A - Chl - D \underset{\text{back reaction}}{\overset{\text{light}}{\rightleftharpoons}} A^- - Chl^+ - D \xrightarrow{\text{dark}} A^- - Chl - D^+$$

(A is C-550 and D is Cyt b_{559}) . This schema accounts for the parallel amount of C-550 that is reduced and of Cyt b_{559} that is oxidized by continuous light [22,3].It is also convenient for explaining the small part of the fluorescence induction that can be accomplished by one saturating flash [11-15] and the small proportion of C-550 that is measured in a reduced state after a single flash [3,11,12]. We recently demonstrated the occurence of a back reaction [7] and found a half-time of 4.2 ms, a value very close to that obtained by Murata et al [13] in a precise study of the fluorescence induction . From the effect of one saturating flash (about 20% of the C-550 is left in a reduced state), it can be inferred that the half-time of the forward dark reaction is 20-30 ms .

However , in a previous study we reported that , in a large domain of exciting light intensity I , the rate of reduction of C-550 is about 2x greater than the rate of oxidation of Cyt b_{559} , both rates being nearly proportional to I .[5] This observation is not in agreement with the previous schema , which predicts a quasi-identical rate for both reactions at low light intensity (i.e. half-times greater than 0.5 s) . This discrepancy led us to propose [5] that two photoreactions were occuring at -196°, either in series or in parallel . In order to explain our present results on the rate of several reactions observed at -50° and -170° , we propose that these two photoreactions also occur at higher temperatures (-50°) for PS-2 in the S_2 and S_3 states .

The oxidation of Cyt b_{559} , whose magnitude varies with temperature [10] and with the S state at -55° [3,4], always proceeds at the same rate . This rate is proportional to the exciting light intensity , indicating a photochemical reaction of low quantum yield . As for C-550 , Butler et al [10] found that its rate of reduction is little affected by temperature between -196° and -100° . We report here that it is 3-4x smaller at -170° than at -50° . We conclude that it varies rapidly between -100° and -50° , a temperature range in which there is no change in the properties concerning Cyt b_{559} . Therefore we assume that the photoreaction leading to the reduction of C-550 has a rate which is affected by a temperature-dependent competition between a dark back reaction and a forward dark reaction , independently of the oxidation of Cyt b_{559} . We assume that , in the temperature range studied, the two photoreactions have different apparent quantum yields , the photoreduction of C-550 being always the more efficient reaction . Apart from accounting for the results regarding C-550 and Cyt b_{559} , the two-photoreaction hypothesis also provides a straightforward rationale to the parallelism between fluorescence induction and oxidation of Cyt b_{559} in conditions in which that oxidation is important [2,10,11,3] . The time course of both properties is determined by the slower photochemical reaction.

The two suggested photoreactions are not in competition since it is possible to fully photoreduce C-550 at -50° in either S state (fig. 5 and ref.3,4) and at -196° [22]. However there are reasons for supposing that they are tightly coupled. First , the oxidation of Cyt b_{559} and the reduction of C-550 occur at the same extent in a large variety of conditions of pretreatment [22] and of illumination[3,4]. Second , we found a nearly identical half-time for the back reoxidation of C-550 and for the oxidation of Cyt b_{559} after one flash at -170° [7]. The coupling can be realized by the way of a unique molecule of sensitizer (chlorophyll) for the two photoreactions , that would operate in series , e.g. :

$$\text{C-550} - \text{Chl} - \text{Cyt } b_{559} \underset{\text{back reaction}}{\overset{\text{light}}{\rightleftharpoons}} \text{C-550}^- - \text{Chl}^+ - \text{Cyt } b_{559}$$

$$\downarrow \text{light}$$

$$\text{C-550}^- - \text{Chl} - \text{Cyt } b_{559}^+$$

This type of model will account for most of the results , but it presents several limitations. For instance , it predicts a lag (that we could not observe [5]) for the photooxidation of Cyt b_{559} in continuous light at -170° . Moreover it is not able to simultaneously account for both the low yield of photoconversion after one flash at -196° [3,11,12] (leading to the assumption of a forward dark reaction with a half-time in the 10 ms range) and the possibility to photooxidize Cyt b_{559} by submillisecond flashes [3,4,7,27]. Considering mostly their results on fluorescence induction around -50° , Joliot and Joliot proposed a model which offers similarities with our "in series" model , as one photoactive chlorophyll can realize successively two different photoreactions [28].

In a second class of models , the two photoreactions occur "in parallel" and the coupling is realized by the way of dark redox reactions . We did not find a reasonably simple model that would account for all the present data .As an example of possible models , we present the following , in which it is assumed that cytochrome b_{559} will be oxidized if both D and Chl_2 are simultaneously in the oxidized state :

C-550 — Chl_1 — D ⟵ photoreaction 1 , rate aI
Cyt b_{559}

A — Chl_2 ⟵ photoreaction 2 , rate bI (a = 6 b)

After one saturating flash , the reaction center is in the following situation :

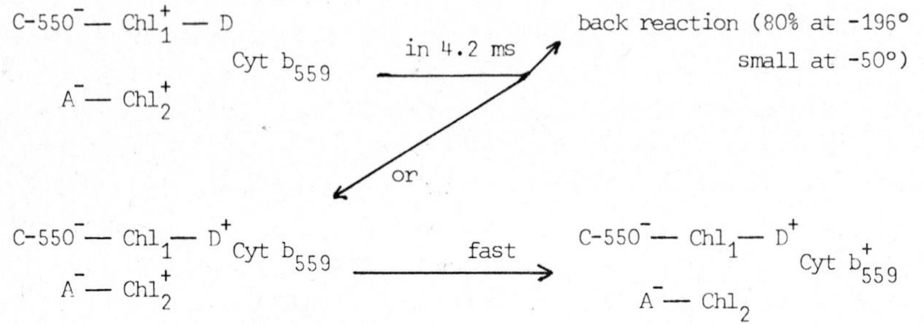

In any model we have to include two types of back reactions : a fast one ($t_{1/2}$=4.2 ms) , between $C\text{-}550^-$ and its oxidized primary donor ; and slower ones , for example between $C\text{-}550^-$ and D^+ . A slow back reaction has been reported by Knaff and Malkin [21], in the presence of ferricyanide , at intermediate temperatures (around -50°) . This reaction might explain why we were not able , after a saturating illumination of chloroplasts with continuous light at -55° , in the presence of FeCN , to trap all the C-550 in the reduced state , in contrast with what happens without FeCN [4]. Another slow back reaction has been reported by A. Joliot [26] who found it bo be much faster for chloroplasts in the S_2 and S_3 states than in the S_0 and S_1 states . The occurence of this last dark reaction might solve a discrepancy between some of our results ; at -50° , by the kinetic technique, we found the same rate and the same extent of reduction of C-550 in either S state . However , with chloroplasts that received one flash at -55° and were then rapidly cooled to -196° , we measured that 30% of the C-550 was reduced for chloroplasts in the S_2-S_3 states and 75% for chloroplasts in the S_0-S_1states [4]. On the basis of many similarities between the photochemical behaviour of PS-2 at -196° (in either S state) and at -50° (in the S_2-S_3 states) , it seems reasonable to assume that the low yield in flashes at -55° (S_2-S_3 states) [2,4] is due to a fast back reaction , as at -196° . However this proposal does not hold since C-550 is photoreduced at the same rate in either S state at -50° .

In a model with two photoreactions in parallel , a question remains as to the identity of the primary acceptors . Two candidates have been reported as primary acceptors : X-320[29] and C-550 , but there are several for considering that X-320 and C-550 reflect the semi-reduction of a unique plastoquinone molecule [30,31]. It remains a possibility that X-320 and C-550 represent two different acceptors functioning at low temperature since X-320 is fully and irreversibly reduced by one flash at -160° (according to Witt [32]) whereas C-550 is largely reoxidized after one flash [7]. At the present time , we are not able to make a definitive choice between different models for the reaction center of PS-2 at low temperature . However we feel that , at a given degree of complication , the model with two coupled photoreactions in parallel accounts for a greater number of expemental data .

Acknowledgement

We would like to thank A.Peronnard for constant and excellent technical assistance .

References

1. Butler,W.L.(1973) Acc.Chem.Res. 6,177-184
2. Joliot,P. and Joliot,A.(1973) Biochim.Biophys.Acta 305,302-316
3. Vermeglio,A. and Mathis,P.(1973) Biochim.Biophys.Acta 292,763-771
4. Vermeglio,A. and Mathis,P.(1973) Biochim.Biophys.Acta 314,57-65
5. Mathis,P.,Michel-Villaz,M. and Vermeglio,A.(1974) Biochem.Biophys.Res.Comm. 56, 682-688
6. Vermeglio,A. and Mathis,P.(1974) Biochim.Biophys.acta , in press
7. Mathis,P. and Vermeglio,A.(1974) Biochim.Biophys.Acta , in press
8. Thorne,S.W. and Boardman,N.K.(1971) Biochim.Biophys.Acta 234,113-125
9. Malkin,S. and Micheali,G.(1972) in Proc.2nd Intern.Congress on Photosynthesis Res. (Forti,G.,Avron,M. and Melandri,A. eds),pp.149-168,Dr Junk N.V. Publ., The Hague
10. Butler,W.L.,Visser,J.W.M. and Simons,H.L.(1973) Biochim.Biophys.Acta 292,140-151
11. Butler,W.L.,Visser,J.W.M. and Simons,H.L.(1973) Biochim.Biophys.Acta 325,539-545
12. Lozier,R.H. and Butler,W.L.(1974) Biochim.Biophys.Acta 333,465-480
13. Murata,N.,Itoh,S. and Okada,M.(1973) Biochim.Biophys.Acta 325,463-471
14. Butler,W.L.(1972) Proc.Natl.Acad.Sci.U.S. 69,3420-3422
15. Den Haan,G.A.,Warden,J.T. and Duysens,L.N.M.(1973) Biochim.Biophys.Acta 325, 120-125
16. Bennoun,P.(1970) Biochim.Biophys.Acta 216,357-363
17. Witt,H.T.,Müller,A. and Rumberg,B.(1961) Nature 192,967-969
18. Emrich,H.M.,Junge,W. and Witt,H.T.(1969) Z.Naturforsch. 24b,1144-1146
19. Vermeglio,A., Thesis, in preparation
20. Amesz,J.,Pulles,M.P.J. and Velthuys,B.R.(1973) Biochim.Biophys.Acta 325,472-482

21. Knaff,D.B. and Malkin,R.(1974) Biochim.Biophys.Acta 347,395-403
22. Erixon,K. and Butler,W.L.(1971) Biochim.Biophys.Acta 234,381-389
23. Epel,B.L.,Butler,W.L., and Levine,R.P.(1972) Biochim.Biophys.Acta 275,395-400
24. Boardman,N.K.,Anderson,J.M. and Hiller,R.G.(1971) Biochim.Biophys.Acta 234, 126-136
25. Anderson,J.M.,Than-Nyunt and Boardman,N.K.(1973) Arch.Biochem.Biophys. 155, 436-444
26. Joliot,A. Biochim.Biophys.Acta , in press
27. Floyd,R.A.,Chance,B. and DeVault,D.(1971) Biochim.Biophys.Acta 226,103-112
28. Joliot,P. and Joliot,A.(1972) in Proc.2nd Intern.Congress on Photosynthesis Res.(Forti,G.,Avron,M. and Melandri,A eds),pp.26-38, Dr Junk N.V. Publ.,The Hague
29. Stiehl,H.H. and Witt,H.T.(1969) Z.Naturforsch.24b,1588-1598
30. Van Gorkom,H.J.,Tamminga,J.J. and Haveman,J.(1974) Biochim.Biophys.Acta 347, 417-438
31. Van Gorkom,H.J.(1974) Biochim.Biophys.Acta 347,439-442
32. Witt,K.(1973) FEBS Lett. 38,116-118

M. AVRON, *Proceedings of the Third International Congress on Photosynthesis*,
September 2-6, 1974, Weizmann Institute of Science, Rehovot, Israel
Elsevier Scientific Publishing Company, Amsterdam, The Netherlands, 1974

EFFECTS OF CARBONYL CYANIDE m-CHLOROPHENYL HYDRAZONE (CCCP) AND OF 3-(3,4 DICHLOROPHENYL)-1,1-DIMETHYLUREA (DCMU) ON PHOTOSYSTEM II IN THE ALGA " *CHLORELLA PYRENOIDOSA* "

A.L. ETIENNE

Laboratoire de Photosynthèse, C.N.R.S., Gif-sur-Yvette (France)

Summary

The effects of CCCP and DCMU on photosystem II in algae are studied. They are well explained by a model in which the long lifetime fluorescent state is attributed to a photoinactive form of the center P^+Q. This photoinactive form results from an equilibrium between the primary and secondary donors with an equilibrium constant increasing with the number of positive charges stored. Its concentration therefore oscillates in synchrony with that of $S_2 + S_3$ and this explains the oscillations of the initial fluorescence yield during a flash sequence. The photoinactive form of the centers contributes to the misses and this implies an uneven distribution of the misses, more important for the S_2, S_3 states than for S_0, S_1. CCCP accelerates the reduction of P^+. DCMU in addition to the blocking between Q and A, enables a partial reduction of Q.

1. Introduction

After the discovery of the oxygen oscillations[1] and the linear model proposed for their explanation[2], the origin of their damping remains to be understood. Following Kok's proposal[2], two parameters can be defined : α percentage of "misses", β percentage of "double hits". The double hits are clearly due to the fact that some centers perform two charge-separations during the flash. The misses express a net deficit per flash in the production of positive charges useful for the oxygen production.

Two situations can be considered :

1/ Either all centers are in a photoactive state regardless of the stored charges and, after each flash, some positive charges disappear through a back reaction[3] or through a reduction of P^+ (P : photoactive chlorophyll) by a donor not involved in the oxygen production[4].

2/ Or, some centers are in photoinactive form and are unable to produce an electron-hole pair during a flash[5,6].

Such an inactive form may correspond either to P^+Q (Q : primary acceptor) or to PQ^-. The mechanisms proposed to explain the misses in 1/ can still apply in situation 2/ .

Up to now there has been no direct measurement of the number of photochemical charge separations occuring per flash in System II. The fluorescence oscillations[7] and 518 nm absorption change oscillations[8] might be taken as evidence for the second mechanism to occur.

In that respect it is important to determine the nature of the long lifetime fluorescent state responsible for the increase in the initial fluorescence yield correlated to the S_2 and S_3 states. The effects of CCCP and DCMU on photosystem II enable us to bring some additional information on that subject.

2. Material and Methods

Algae (*Chlorella pyrenoidosa*, Chick, Emerson Strain) were cultivated as previously described[6]. They are used in their culture medium (pH : 5.4) or in phosphate buffer.

Experiments of figures 1 and 4 were made with a stopped flow apparatus already described[6].

Experiments of figure 3 were made with a laser phosphoroscope built by Lavorel[9] and the oxygen flash sequences were measured with the polarographic method described by Joliot[1].

All experiments are performed at room temperature.

3. Results

A - DCMU effect on fluorescence and luminescence after preilluminating flashes.

The long lifetime fluorescent state of unknown nature correlated to the S_2 and S_3 states is still present shortly after the mixing with DCMU[6].

This is shown by the increased initial fluorescence yield, after two flashes and addition of DCMU, compared to the initial level for dark-adapted algae shortly after the addition of DCMU.

If the dark time after the DCMU addition is increased, the φ level increases for dark adapted algae, while the φ level of preillumination algae rapidly decreases and within 3 seconds reaches a value close to that of dark adapted algae (with no DCMU present) (fig. 1, a). This decrease is associated with an increased luminescence emission (fig. 1, b).

It has been recently demonstrated that there was an electron-carrier B which equilibrates with A in a two electron-transfer reaction and that the DCMU action can be explained by assuming that it lowers the mid-point potential of B[10]. The present results can therefore be interpreted in the following way : if at the time of DCMU addition there are some centers in a QB^- form, the addition of DCMU produces a partial reduction of Q, hence an increase of fluorescence if Q^- does not disappear as soon as it is formed through a back reaction. This is apparently the case for dark adapted algae and the increase of the φ level should be a way to know the amount of QB^- stable in the dark. If the QB^- stable in the dark belonged to centers in a S_1 state, the addition of DCMU should give rise to a luminescence emission : $Z^+ Q^- \xrightarrow{h} ZQ$. No luminescence was detected in that case and one has therefore to assume that QB^- stable in the dark belong to centers in a S_0 state.

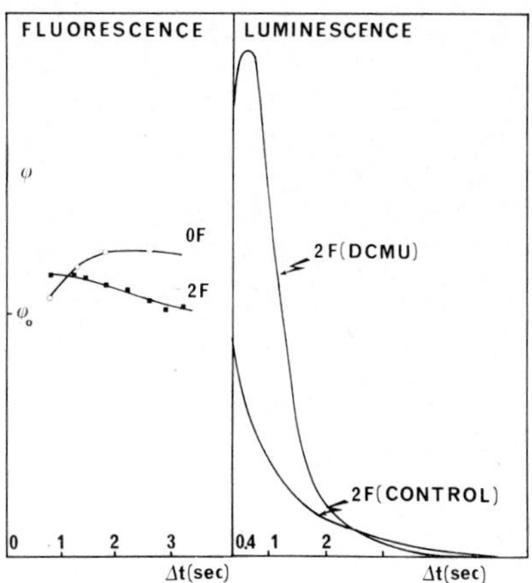

Fig. 1. Fluorescence yield and luminescence of dark adapted and preilluminated algae as a function of time after the addition of DCMU 5.10^{-5} M.

Dark adapted algae were preilluminated by 2 short saturating flashes 300 msec. apart., DCMU was added 400 msec. after the last flash. The fluorescence yield plotted in relative units is the initial fluorescence induced by a detecting beam level at variable time (Δt) after the addition of DCMU, the luminescence yield was observed with no detecting beam. pH : 5.7; chlorophyll concentration : 10 $\mu g.\ ml^{-1}$; temperature 20°C.

In the case of preilluminated algae, the increased luminescence emission triggered by the addition of DCMU and its fast decay indicate that the Q^- produced, rapidly recombines with positive charges stored on the donor side. The fast decrease of φ shows that the long lifetime fluorescent state cannot correspond to some PQ^-. It has another origin which could be a photoinactive form P^+Q.

B- CCCP effects

CCCP at high concentrations is known to act on the donor side of photosystem II[11,12,13]. This is well documented in chloroplasts and we found it to be true also in algae[14].

In a previous work[14] we have shown that in algae CCCP inhibits the fluorescence decay resistant to DCMU and that it also inhibits the luminescence emission. Therefore at high concentration, CCCP is a good inhibitor of the back reaction and this results from an interaction with the donor side.

The effect of CCCP 10^{-5}M on oxygen evolution during a flash series is shown in fig. 2.

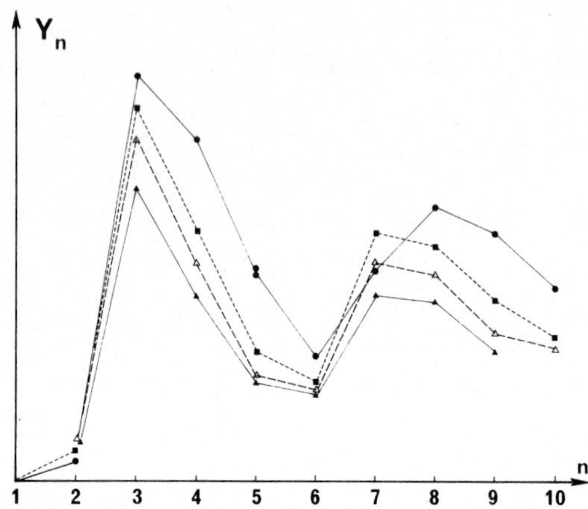

Fig. 2. Amount of oxygen produced per flash : Yn during a series of short saturating flashes ● No addition, 15 min darkness before the flashes, 100 msec between flashes.

■ Algae in the presence of CCCP 10^{-5}M for 15 min in the dark, 100 msec between flashes.

△ Same as ■ expect 300 msec between flashes.

▲ Same as ■ expect 600 msec " "

Phosphate buffer pH 6.5 + KCl 0.1M, temperature 20°C.

When the spacing between flashes is increased, the inhibition increases as could be expected from the accelerating effect of CCCP on the deactivation reactions (C. Lemasson, unpublished results).

More intriguing is its effect on α. This parameter can be computed independently of any curve fitting procedure[15].

There is a definite decrease of α in the presence of CCCP. For the experiment of fig. 2 (with 100 msec between flashes), the mean value of α is 0.2 for the control and 0.09 in the presence of CCCP. Therefore CCCP depresses one (or more) of the mechanisms responsible for the misses : it might decrease the amount of photoinactive centers, the probability for a back reaction to occur before the charge stabilization or suppress a donor not involved in the oxygen evolution (cyt. b-559 high-potential for example[16]).

Zankel has already shown that the correlation between the fluorescence oscillations and those of $S_2 + S_3$ concentration is not rigid[17]. While not preventing the formation of S_2 and S_3, CCCP suppresses also the oscillations of the fluorescence yield reached during each flash of a series (fig. 3).The oscillations of the initial fluorescence no longer exist and instead the φ level increases after the first three flashes then stays constant.

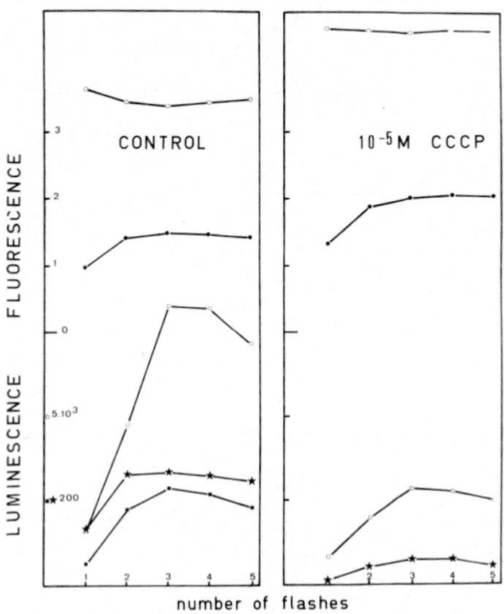

Fig. 3. Fluorescence and luminescence oscillations during a serie of flashes.

This experiment was done in collaboration with Dr. Lavorel.

Flash duration 180 μs. Time between flashes 400 msec.

Fluorescence plotted in relative units :

- ● initial fluorescence yield.
- ○ fluorescence yield reach at the end of the flash.

Each point is averaged over 20 experiments.

Luminescence plotted in arbitrary units.

- □ Integral of luminescence over the first 6.25 μsec after the flash.
- ■ " " from 6.25 to 12.5 μsec after each flash.
- ★ " " from 12.5 to 400 μsec " " " .

Each point is averaged over 100 experiments, the PMT is inhibited during the flash to prevent artefacts coming from the saturation of the amplifiers. Algae in culture medium pH 5.4, chlorophyll concentration : 5 $\mu g.ml^{-1}$.

As expected[14], the luminescence emission after each flash was lessened by CCCP but some oscillations remain and the fast component (characterized by the integral over the 6 first μsec is still maximum after the third flash).

In paragraph A, we have shown that the action of DCMU on the initial fluorescence yield was well explained if the long lifetime fluorescence state (responsible for the last part of the fluorescence decay) was attributed to a P^+Q form.

Since CCCP is known to interact with the donor side, its inhibitory effect on the back reaction can result from an increase in the rate of P^+ reduction by processes in competition with the back reaction : increase in the rate of $Z^{n+}P^+ \longrightarrow Z^{n+1}P$ or catalysis of another irreversible reduction patway (as demonstrated in chloroplasts[11,13]).

The fluorescence decay after some flashes has been decomposed[18] in three distinct phases : a first phase due to the fast reoxidation of a main part of Q^- by secondary electron acceptors, a medium phase which seems to correspond to the reoxidation of some Q^- via a back reaction and a slower phase already spoken of as " the long lifetime fluorescent state ".

In order to study the effect of CCCP on the different phases, we added CCCP before the flashes and after at a time sufficient for the two first phases of the decay to be achieved. The results of these experiments are shown in fig. 4.

When added after the complete reoxidation of Q^- (that is after the completion of the two first phases of the fluorescence decay), CCCP accelerates the last part of the decay (fig. 4a) as DCMU does (fig. 1) but in the present case, the fast fluorescence decay is associated to an inhibition of the luminescence emission. This shows that, if the long lifetime fluorescent state is due to P^+Q, CCCP accelerates the reduction of P^+ .

If added before the flashes, CCCP inhibits a part of the fluorescence decay over a long period (fig. 4b). The experimental set-up for the fluorescence measurement does not allow the detection of the fast component of the fluorescence decay. However since De Kouchkovsky (unpublished results) has shown that in algae the oxygen burst was increased in volume and rate by CCCP, we assume that CCCP does not inhibit the fast fluorescence decay. Therefore the observed inhibition is necessarily that of the medium phase corresponding to a back reaction and was to be expected[14].

In the presence of CCCP, the increase of φ during a flash sequence is especially large after the first flash, then φ remains practically constant. Since the inhibition of the fluorescence decay lasts longer than the total duration of the flash series, if the centers reopened by a back reaction were randomly distributed, each flash should yield, in the presence of CCCP, an additional increase of φ . The constancy of φ after the two first flashes contradicts that prediction and this may indicate that the centers reopened by a back reaction (and therefore not connected to A[19]) are different from those involved in the oxygen production and that the medium phase of the fluorescence decay does not necessarily contribute to the misses.

It is interesting to note that, in contradiction with the above statement, the mean α (0.21) computed by Joliot[18] from the amplitude of the medium phases : Δ after the first 4 flashes is in very good agreement with the mean α computed by Lavorel from the oxygen sequence (fig. 9, ref. 18) α = 0.19. However the oscillations of φ[7], ressemble those of Δ[18] and with an ad-hoc estimation of the fluorescenceyield of P^+Q, a correct α value might be computed (see for example ref. 20, fig. 4, although in that paper the nature of the photoinactive form is assumed to be P^+Q^- a completly fluorescent state).

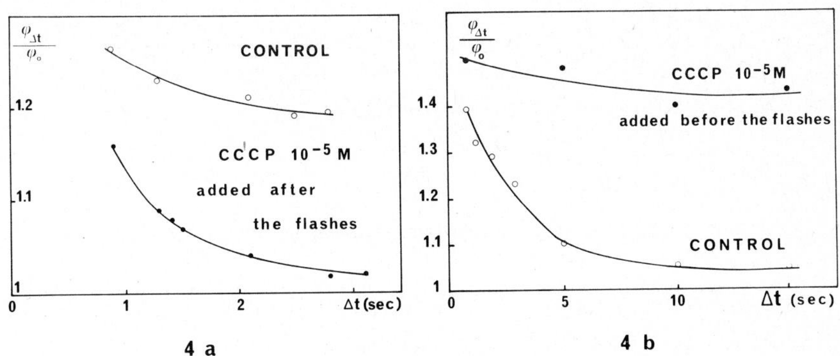

Fig. 4. Fluorescence decay in the presence of CCCP 10^{-5}M.

Fig. 4a : Dark adapted algae are preilluminated by two short saturating flashes 300 msec apart, CCCP 10^{-5}M final concentration is added 400 msec after the last flash and the fluorescence yield plotted in arbitrary units corresponds to the initial fluorescence excited by a blue detectory beam at varying times after the CCCP addition.

o Control : phosphate buffer alone added after the flashes.

● CCCP 10^{-5}M : CCCP is added 400 msec after the last flash.

Chlorophyll concentration : 10 μg.ml^{-1}, pH 5.7, Temperature : 20°C.

Fig. 4b : Dark adapted algae are mixed with CCCP 10^{-5}M for several minutes, then illuminated by two flashes, the initial fluorescence yield is detected at variable times after the last flash.

o Control no CCCP added

● CCCP 10^{-5}M added before the flashes.

4. Discussion and Conclusion

Different authors[18,19,21] have given experimental evidence against an attribution of the long lifetime fluorescent state to some remaining Q^-.
The behaviour of this state is better explained if one assumes that it corresponds to a photoinactive form P^+Q less quencher than the photochemical trap PQ. The oscillations in the concentration of that state during a flash sequence show that this state is favored when 2 or 3 positive charges are stored.
This means that there is a shift in the equilibrium :

$$Z^{n+}\ P \underset{k_n}{\overset{k_{-n}}{\rightleftharpoons}} Z^{(n-1)+}\ P^+ \qquad \text{eq (1)}$$

resulting from an increase of the mid-point potential of Z in its different oxidative states[5,6].

We propose therefore that :

1. The number of photoactive centers varies with the S states. It is maximum in $S_0 S_1$, minimum in S_2, S_3.

2. The photoinactive state, coming from centers in a P^+Q form, is responsible for the oscillations of φ and contributes for a large part to the misses.

3. The misses are therefore unevenly distributed, larger on S_2S_3 than on S_0S_1.

Various authors[14,22] now claim that P^+Q^- may be a fluorescence quencher. Therefore the fluorescence yield reached during the flashes will depend at least on two factors : the amount of closed traps : P^+Q when the flash is fired and the rate of reduction of P^+ which depends on k_n.

The fast luminescence decay reflecting stabilization processes[23] will also depend on k_n (see also ref. 24) and will contribute for a part to the misses[3]. Since it oscillates in synchrony with the O_2 sequence [24], it will contribute to the uneven distribution of α over the S states.

All the effects of a high concentration of CCCP (excluding the acceleration of the deactivation reactions) can be explained by assuming that in the presence of CCCP the reduction of P^+ is accelerated. In this way, CCCP prevents the stabilization of centers in a P^+Q state. This accelerating effect is also seen when Z is destroyed and replaced by others donors[14,25].

References

1. Joliot, P. Barbieri, G and Chabaud, R. 1969, Photochem. Photobiol. 10, 309.
2. Kok, B., Forbush, B. and McGloin, M. 1970, Photochem. Photobiol. 11, 457.

3. Lavorel, J., 1974, III[d] Congress on Photosynthesis, Rehovot.
4. Babcok, G.T., Sauer, K., 1973, Biochim. Biophys.Acta 325, 483.
5. Van Gorkom, H.J. and Donze, M. 1973, Photochem. Photobiol. 17, 333.
6. Etienne, A.L., 1974, Biochim. Biophys.Acta 333, 320.
7. Delosme, R. 1972, the II[d] Int. Congr. Photosynthesis, Stresa (Forti, G., ed.) p. 187.
8. Amesz, J. Pulles, M.P.J., Velthuys, B.R., 1973, Biochim. Biophys.Acta 325, 472.
9. Lavorel, J., 1973, Biochim. Biophys.Acta 325, 213.
10. Velthuys, B.R., Amesz, J., 1974, Biochim. Biophys.Acta 333, 85.
11. Homann, P., 1971, Biochim. Biophys. Acta 245, 129.
12. Renger, 1973, J. of Bioenergetics 4, 491.
13. Renger, G., Bouges-Bocquet, B. Delosme, R. 1973, Biochim. Biophys.Acta 292,796.
14. Etienne, A.L., 1974, Biochim. Biophys.Acta 333, 497.
15. Lavorel, J. 1974, to be published in J. of Theor. Biol.
16. Cramer, W.A., Fan, H.N., Böhme, H., 1971, Bioenergetics 2, 289.
17. Zankel, K.L., 1973, Biochim. Biophys.Acta 325, 138.
18. Joliot, P., Joliot, A., Bouges-Bocquet, B., Barbieri, G., 1971, Photochem. Photobiol. 14, 287.
19. Lavergne, J., to be published in Photochem. Photobiol.
20. Delrieu, M.J., 1973, C.R.Acad. Sci. Paris, 277, 243.
21. Forbush, B., Kok, B., 1968, Biochim. Biophys. Acta 162, 243.
22. Okayama, S., Butler, W.L., 1971, Biochim. Biophys.Acta 234, 381.
23. Lavorel, J., 1974, Bioenergetics of Photosynthesis (Govindjee, ed.).
24. Zankel, K.L., 1971, Biochim. Biophys. Acta 245, 373.
25. Haveman, J., 1973, Thesis, Leiden, The Netherlands.

M. AVRON, *Proceedings of the Third International Congress on Photosynthesis*, September 2-6, 1974, Weizmann Institute of Science, Rehovot, Israel
Elsevier Scientific Publishing Company, Amsterdam, The Netherlands, 1974

MONOVALENT AND DIVALENT CATION-INDUCED CHANGES IN CHLOROPHYLL a FLUORESCENCE AND CHLOROPLAST STRUCTURE

E. L. Gross,[†] T. Wydrzynski,[*] D. VanderMeulen[*] and Govindjee[*]

[†]Department of Biochemistry, The Ohio State University, Columbus, and

[*]Departments of Botany, Physiology and Biophysics, University of Illinois, Urbana, Illinois 61801 (U.S.A.)

Abstract

Monovalent cations were found to decrease both the constant and variable chlorophyll a fluorescence. The change in the constant fluorescence may imply a direct effect of the cations on the bulk chlorophyll a molecules. The monovalent cations were also found to cause changes in chloroplast structure as measured by light scattering, fluorescent probes and electron microscopic techniques. However, a correlation between larger structural changes and changes in chlorophyll a fluorescence was observed only in spinach but not in other plant species (oats, peas and lettuce). It is suggested that the cations may regulate energy distribution between the two photosystems, as reflected by the chlorophyll a fluorescence, through small conformational changes in certain key chlorophyll-protein complexes, perhaps in conjunction with larger chloroplast structural changes.

Introduction

Murata[1-3], Murakami and Packer[4], Mohanty, *et al.*[5], and others[6-8] have suggested that divalent cations such as Mg^{2+} and Ca^{2+} control the transfer of excitation energy between the two photosystems in isolated chloroplasts. Murata[3] and Murakami and Packer[4] found that the divalent cation-induced changes in energy transfer could be correlated with divalent cation-induced changes in chloroplast structure.

Gross and Hess[9] and Gross and Prasher[10] found that, in spinach, low concentrations (2-10 mM) of salts of monovalent cations such as Na^+ caused a decrease in the steady state chlorophyll a fluorescence (measured at 680 nm) at room temperature and an increase in the ratio of fluorescence emitted at 735 nm to that emitted at 685 nm at 77° K. The effects could be prevented or reversed by low concentrations of salts of divalent cations such as Mg^{2+} or Ca^{2+}. The divalent cation effects on fluorescence were shown to be the result of divalent cation binding to the chloroplast membranes[11]. Assuming that spillover proceeds from Photosystem II (PS II) to Photosystem I (PS I)[12], these results indicated that monovalent cations at low concentrations promote spillover and divalent cations or monovalent cations at much higher concentration can reverse this process. If, on the other hand, spillover proceeds from PS I to PS II[13] via the short wavelength form of chlorophyll a, then the opposite argument can be used. Alternately, the salt effects can be described as influencing the energy distribution in the photosynthetic membrane in favor of either PS I or PS II (cf. ref. 14). In this paper, we summarize observations that include species differences in cation effects on chlorophyll a fluorescence, effects of cations on fluorescence transients and the possible role of structural changes in the regulation of spillover (see VanderMeulen and Govindjee[15] and Wydrzynski *et al.*[16]).

Results

Comparison of the effects of cations on chlorophyll a fluorescence in oats and spinach

Gross and Hess[9] and Gross and Prasher[10] showed that, in spinach, divalent cations had no effect on chlorophyll a fluorescence when added alone in the absence of monovalent cations, indicating that spillover was maximally inhibited under these conditions so that addition of the divalent cations could have no further effect. Also, addition of the monovalent cations after the divalent cations had no effect.

VanderMeulen and Govindjee[15] examined the effects of salts on chlorophyll a fluorescence changes in other species including lettuce, peas and oats and found that although mono- and divalent cations excite opposite effects, these effects are somewhat different than observed for spinach. Namely, divalent cations added

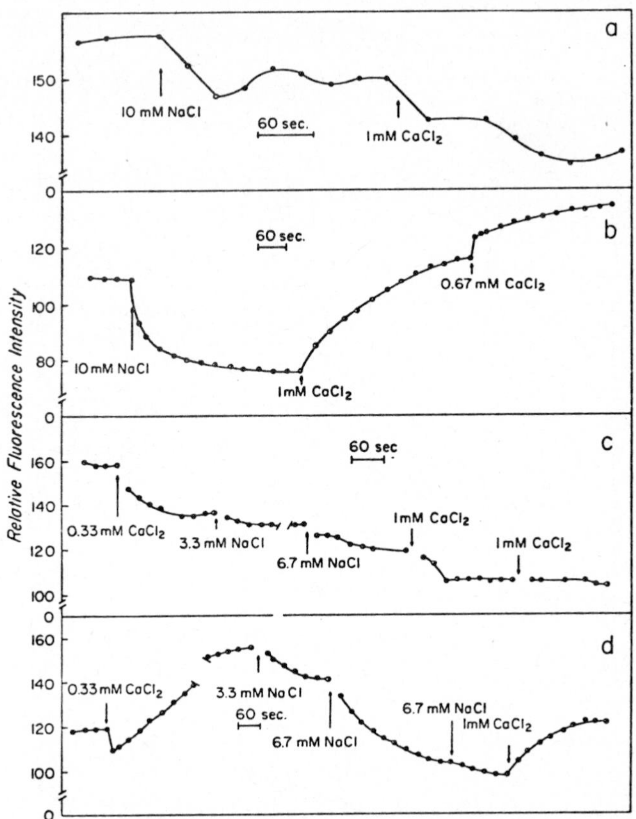

Fig. 1. Time course of cation-induced changes in fluorescence of N-phenyl-1-napthylamine (NPN) (a,c) and chlorophyll a (b,d) in washed oat chloroplasts. Chloroplasts were incubated in 100 mM sucrose, sufficient Tris base (0.17-0.2 mM) to titrate to pH 8, and 7.5 μM DCMU. The probe concentration was 5 μM (after VanderMeulen and Govindjee[15]).

alone cause a transient decrease in chlorophyll a fluorescence which is followed by an increase (Fig. 1d) and monovalent cations can reverse the divalent cation effects. These differences can be explained if spillover is not completely inhibited under low salt conditions in oats and peas as it is in spinach. (For a more complete discussion of these results see VanderMeulen and Govindjee[15].)

The effect of cations on chlorophyll a fluorescence transients

Chlorophyll a fluorescence can be divided in two parts, one of which is constant and independent of the state of the traps (O level[17]) and another, the so-called variable fluorescence (P minus O level), which depends on the state of the traps and is high when the traps are closed (Q reduced) and low when they are open (Q oxidized). Previous workers[3,5,6] found that when Mg^{2+} was added to chloroplasts incubated in 50 mM Tricine buffer or 1 mM Tris-Cl, there was a large increase in the variable fluorescence but very little effect on the constant fluorescence.

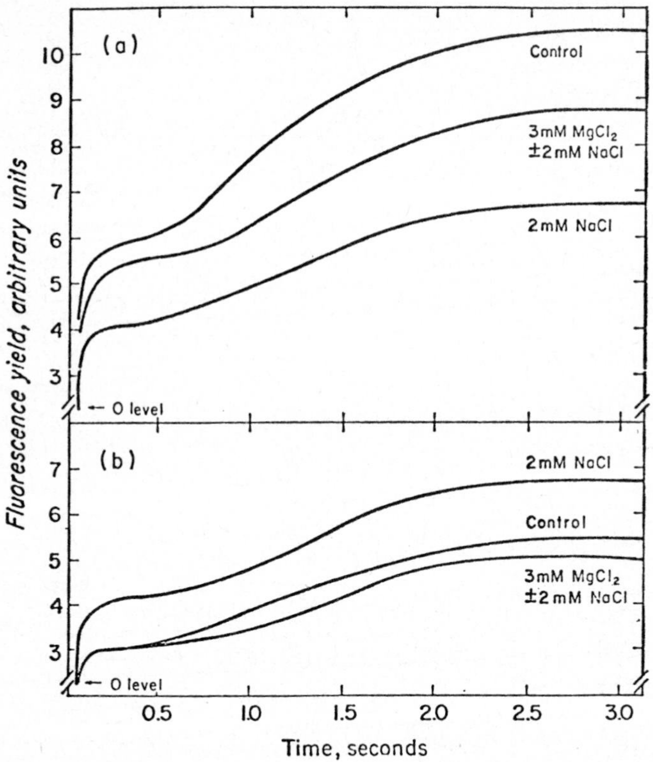

Fig. 2(a). Time course of chlorophyll a fluorescence in sucrose-washed spinach chloroplasts in the presence and absence of sodium and magnesium. Fluorescence was measured at 685 nm (half-band width, 6.6 nm), excitation, broad band blue light (CS 4-96 and CS 3-73); intensity, 1.2×10^4 ergs cm^{-2} sec^{-1}. The 4 ml reaction mixture consisted of 5 μg chlorophyll ml^{-1} suspension, 0.2 mM Tris buffer, pH 7.8-8.2, and 100 mM sucrose. Appropriate amounts of 100 mM stock salt solutions were added to give final concentrations of 2 mM NaCl and 3 mM $MgCl_2$; samples were kept dark adapted until the time of measurement. 2(b). Fluorescence transients from Fig. 1 normalized at the O level (after Wydrzynski et al.[16]).

Gross and Hess[9] studied the effects of monovalent cations only on the final steady-state level of fluorescence observed in the presence of DCMU. It was of interest, therefore, to examine the effects of monovalent cations on the fluorescence transients to determine whether they also only effect the variable fluorescence. It was found (Fig. 2) (see Wydrzynski <u>et al.</u>[16]) that both the constant and variable fluorescence was affected. This suggests that the bulk chlorophyll molecules[17] as well as the traps may be involved in the monovalent cation regulation of spillover. Moreover, it shows beyond any doubt that light and light-induced structural changes[18] are not required for the monovalent cation effects. $MgCl_2$ added alone caused a decrease in the P level and a very slight decrease in the O level. When $MgCl_2$ was added in the presence of 2 mM NaCl, the fluorescence transient was exactly the same as when $MgCl_2$ was added alone. When compared with the NaCl trace, this meant an increase in both O and P levels. The change in the O level apparently disagrees with the results of other workers[3,5] but the difference may be due to the actual salt conditions employed. Also, it appears that Mg^{2+} ions do not really "reverse" the NaCl effect but rather "override" it by superimposing their own effect.

An NaCl concentration curve for both the O and P levels[16] (Fig. 3) shows that in both cases the effect is saturated at 2 mM with half-maximal effects at 0.5 mM. In each case, addition of 3 mM $MgCl_2$ "reversed" the NaCl effect to the extent of producing its own effect on the fluorescence level.

The effect of cations on chlorophyll <u>a</u> fluorescence spectra

Emission spectra of chloroplasts at 77° K show three emission peaks at 685, 695 and 735 nm[19]. The 685 and 695 nm peaks are thought to be emitted from the light-harvesting antenna and molecules close to the trap of Photosystem II respectively whereas the 735 nm peak is thought to be emitted by Photosystem I[19-21]. The results[16] (Table I) are presented as the ratio of the fluorescence emitted at either 735 or 695 to that emitted at 685 nm with the values for the control chloroplasts (no salts added) normalized to 1.0. Addition of NaCl caused an increase in the F735/F685 ratio both with and without preillumination

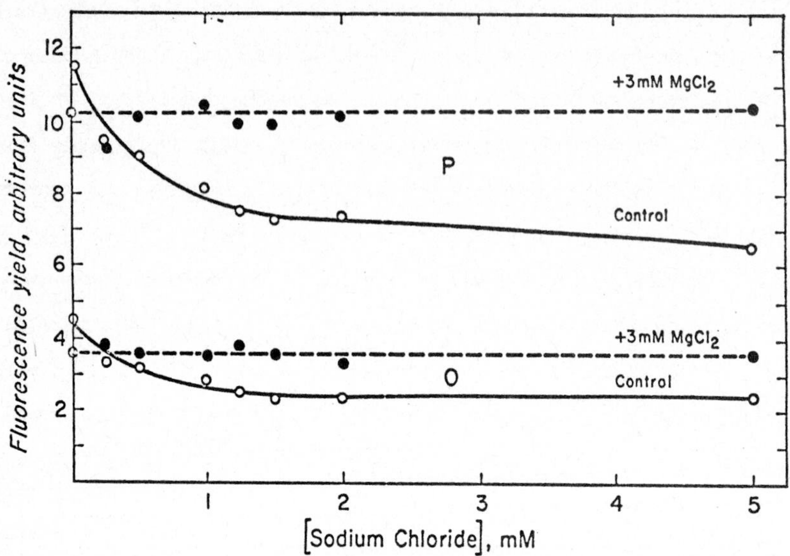

Fig. 3. Concentration curves showing the changes in the O and P level fluorescence as a function of NaCl concentration. O level was measured after 30 msec of illumination. Experimental conditions as in Fig. 2 (after Wydrzynski et al.[16]).

confirming the results of the fluorescence transients that the light harvesting chlorophylls may be involved in causing changes in energy distribution between PS I and PS II. No effect was observed on the F695/F685 ratio indicating that the Photosystem II traps may not be affected.

The Mg^{2+} ions effects are more complex. First, there is an increase in the F695/F685 ratio both in the dark and after preillumination suggesting that Mg^{2+} may affect the transfer of excitation energy from the light harvesting chlorophyll to molecules that are indicators of the Photosystem II traps or increase the number of traps[22,23].

Secondly, the effect on the F735/F685 ratio depends on whether the chloroplasts were preilluminated before freezing or not. In the dark adapted samples the F735/F685 ratio decreased whereas it increased in the preilluminated samples. These results can be explained on the basis that Mg^{2+} ions may have, at least, two effects. One effect is to decrease energy transfer to PS I. This effect could predominate in the dark adapted samples causing the observed decrease in

Table I. Emission Peak Ratios Normalized to a Control Value of 1.00 for Sucrose-Washed Spinach Chloroplasts at 77° K in the Presence and Absence of Sodium and Magnesium (after Wydrzynski *et al.*[16]).

	F735/F685		F695/F685	
	Dark*	Light**	Dark	Light
Control	1.00	1.00	1.00	1.00
2 mM NaCl	1.22	1.25	1.04	1.02
3 mM $MgCl_2$	0.81	1.17	1.15	1.11
2 mM NaCl + 3 mM $MgCl_2$	0.90	1.14	1.11	1.05

* Samples were kept dark-adapted at all times. Chloroplasts were preincubated with the salts for 3-5 minutes at room temperature before freezing. Exciting light, 435 nm, 6.6 nm half-band width, plus CS 4-96 filter; intensity, 30 ergs cm^{-2} sec^{-1}. Measuring wavelengths, variable, 6.6 nm half-band width; CS 2-61 filter before the analyzing monochromator. Chlorophyll concentration, 10 μg/ml^{-1} suspension. Each value represents an average of 6-11 different samples.

**Same condition as above except chloroplasts were preilluminated for 1 minute in strong white light (200 watt incandescent bulb) prior to freezing in the light.

the F735/F685 ratio. The second effect, perhaps, reflected by the change in the F695/F685 ratios, may involve an increase in the energy transfer from bulk Chl *a* to energy traps of PS II. This effect could predominate in the preilluminated samples causing the observed increase in the F735/F685 ratio.

The relation between structural changes and changes in chlorophyll *a* fluorescence

Murata[3] and Murakami and Packer[4] showed that structural changes involving stacking of the thylakoid membranes could be correlated with the chlorophyll *a* fluorescence changes. This appears reasonable since energy transfer between pigment molecules is determined by the distance between them and their mutual orientation[24]. A change in chloroplast structure which could alter the spatial distribution of pigment molecules could change the degree and direction of energy transfer. Alternately, a very minor change in the orientation of a few pigment molecules, such as has been proposed by Seely[25,26], could produce changes in spillover. Such small changes would not necessarily be reflected in large amplitude membrane alterations. To determine whether structural changes are the cause of the monovalent cation-induced decreases in chlorophyll a fluorescence, three methods for monitoring structural changes were used. These were light-scattering, fluorescent probes and electron microscopy.

The kinetics of 90° light-scattering changes and the corresponding changes in chlorophyll a fluorescence are shown for spinach and oat chloroplasts in Figure 4[15]. In both cases, the addition of NaCl caused a decrease in 90° light-scattering as well as in chlorophyll a fluorescence. The kinetics of the decrease in 90° light-scattering appear to be faster than for chlorophyll a fluorescence. Also, the correlation between the extent of 90° scattering and Chl a fluorescence is

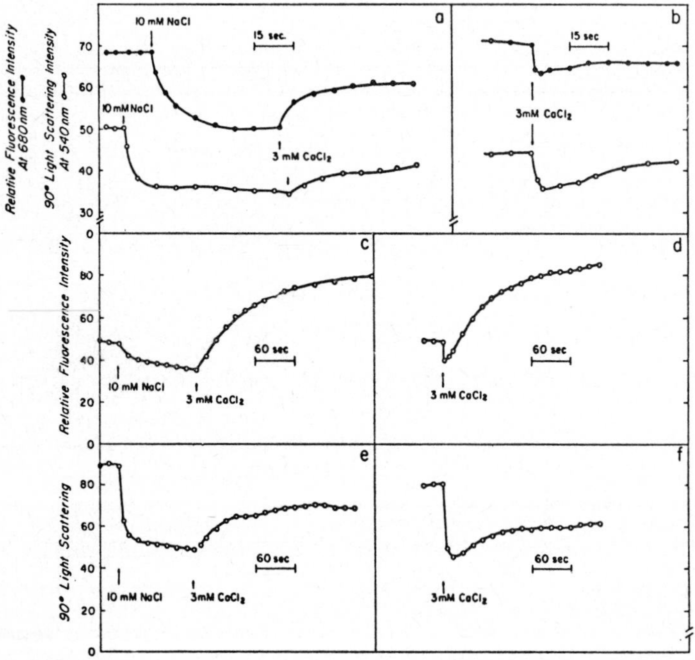

Fig. 4. Time course of cation-induced changes in chlorophyll a fluorescence and 90° light-scattering in washed spinach chloroplasts (a,b) and washed oat chloroplasts (c,d,e,f). Other conditions were as for Fig. 1 (after VanderMeulen and Govindjee[15]).

apparent for spinach (Fig. 4a & 4b) but in oats there are obvious differences (Fig. 4c-4f). Lettuce and pea chloroplasts behaved as oats. In spinach the kinetics of turbidity changes (ΔA_{540}) and chlorophyll a fluorescence changes (Fig. 5) are the same to within the limit of error of making either measurement.

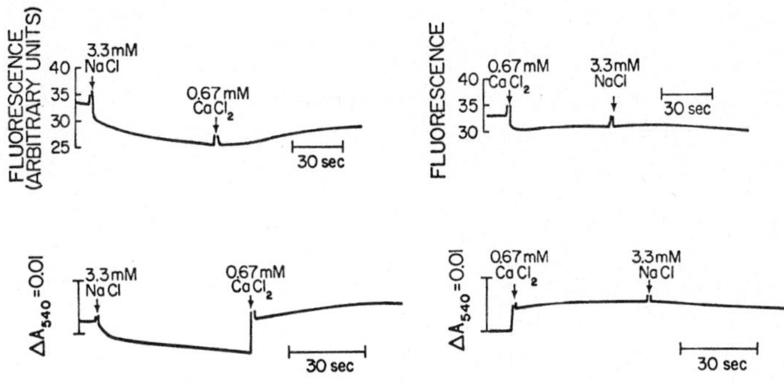

Fig. 5. Time course of cation-induced changes in chlorophyll a fluorescence and turbidity (ΔA_{540}) in spinach chloroplasts. Conditions were as for Fig. 1 (after E. L. Gross and Prasher[10]).

Moreover, in spinach, the NaCl concentration curve is the same for the decreases in fluorescence and turbidity (Fig. 6). These results with spinach may indicate that there is a better correlation between NaCl-induced chlorophyll a fluorescence changes and turbidity measurements whan with 90° light-scattering measurements. However, no data are available for oats at the time of the writing of this report.

Turbidity and 90° light-scattering measure different aspects of chloroplast structure. Turbidity is thought to measure large changes in size and shape[4] whereas 90° light-scattering measures smaller changes in the membranes themselves. It could be that the monovalent cations begin by promoting small, localized changes in the membranes observable by 90° light-scattering. However, these may not be enough to cause the fluorescence changes. As enough of the small amplitude changes take place they might cause the necessary changes in the chlorophyll bed that lead to fluorescence changes as well as to the overall changes in the shape of the chloroplasts as monitored by turbidity changes. Either the large or small amplitude changes may, in turn, regulate spillover. Alternately, the difference in kinetics of 90° scattering may result from the ambiguity of making light-scattering measurements on heterogeneous samples of large pigmented particles[27,28]. It is also likely that neither turbidity nor 90° scattering change measure the microevents that lead to changes in chlorophyll fluorescence. New approaches may be needed to solve this problem.

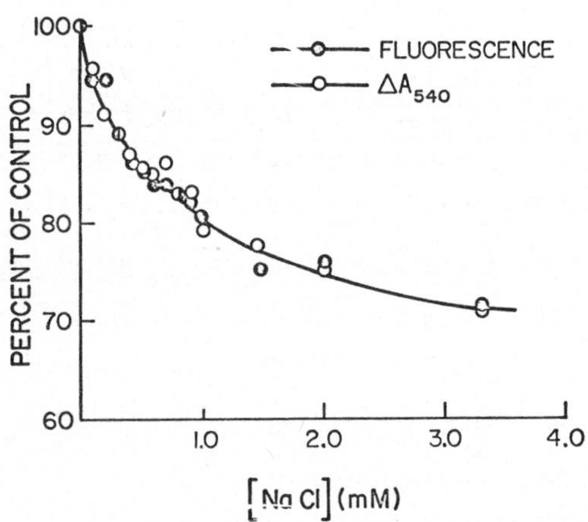

Fig. 6. The NaCl concentration dependence of changes in turbidity and chlorophyll a fluorescence for spinach chloroplasts. Various concentrations of NaCl were added to chloroplast suspensions and the effects on turbidity (ΔA_{540}) and chlorophyll a fluorescence were determined. The final fluorescence or turbidity levels obtained after salt addition were calculated as percent of the initial control level prior to salt addition (after E. L. Gross and Prasher[10]).

For both oat and spinach chloroplasts addition of divalent after monovalent cation increases light-scattering and turbidity as well as fluorescence. Again there are differences in kinetics between the different types of measurements. The fluorescence traces show a slow monotonic increase. The 90° light-scattering trace also increases but it is slower than the fluorescence trace. The turbidity trace, however, shows a rapid phase which is too fast to measure which is not reflected in the fluorescence trace. It could be that only the slow phase can be "correlated" with the fluorescence changes.

The similarities between the light-scattering (and turbidity) data and the chlorophyll a fluorescence changes leads us to postulate that structural change does cause the observed changes in spillover. The differences may be due to the complexities involved in light-scattering studies. On the other hand, direct and quantitative correlations cannot be made.

The fluorescent probes auramine-O, 1-anilinonapthalene-8-sulphonate (ANS^-) and N-phenyl-1-napthylamine (NPN) were also used to monitor changes in dielectric constant in the vicinity of the probe[4,29]. ANS^- is thought to reflect changes on the membrane surface such as those caused by ion binding[4,30-33], whereas NPN is thought to reflect events in the interior of the hydrophobic part of the membrane[34]. In contrast to their effects on chlorophyll a fluorescence changes (see Figs. 1 and 7), both mono- and divalent cations caused an increase in ANS^- fluorescence. The oppositely charged probe auramine-O responds in an opposite, but analogous manner. Moreover, both caused a decrease in NPN fluorescence (Fig. 1a,c). These results do not agree with those obtained for either chlorophyll a fluorescence or light-scattering but do agree with the cation-induced changes in pigment absorption which are due to absorption flattening[10,35,36] and may reflect cation binding to the membrane surface. Evidence supporting this interpretation include the observations that the cation binding sites on the chloroplast membrane[11] bind both mono- and divalent cations. Also the cation concentrations required for saturation of the probe fluorescence changes (namely 1 mM for divalent cations and 10 mM for monovalent cations[15]) are the same as those required for saturating the binding sites[11].

Electron microscopy was done[10] to clarify the nature of the cation-induced structural changes. It was found (Fig. 8) that spinach chloroplasts incubated in the absence of salts appeared slightly swollen due to the hypotonic medium but

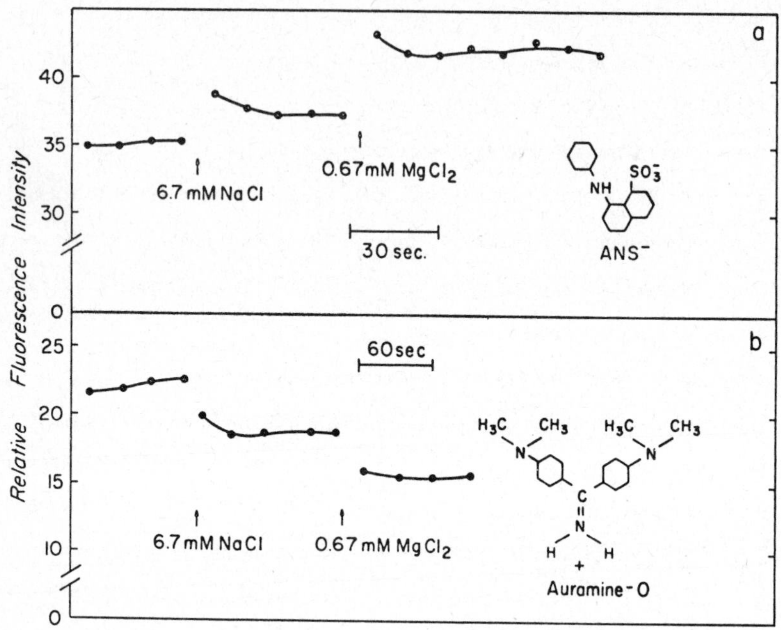

Fig. 7. Time course of the cation-induced changes in 1-anilinonapthalene-8-sulphonate (ANS^-) and auramine-O fluorescence in washed spinach chloroplasts. The probe concentration was 5 µM (after VanderMeulen and Govindjee[15]).

had visible grana stacks. This was surprising since it has been postulated[37] that unstacking of thylakoids occurs in a low ionic strength medium due to charge repulsion.

Addition of 3 mM NaCl caused the grana to unstack; it also promoted spillover. Addition of $CaCl_2$ (either together with (Fig. 9) or after the NaCl (not shown)) caused a restacking of the thylakoids. Unstacking of the thylakoids may bring the two photosystems together, thereby promoting spillover. The differences observed between species may be due to differences in the amount of thylakoid

---

Fig. 8. The effect of NaCl on the ultrastructure of chloroplasts. Chloroplasts were incubated in 100 mM sucrose + 0.2 mM Tris Base (A) and in the same medium + 3.0 mM NaCl (B) prior to fixation with glutaraldehyde. The magnification was 25,000 x.

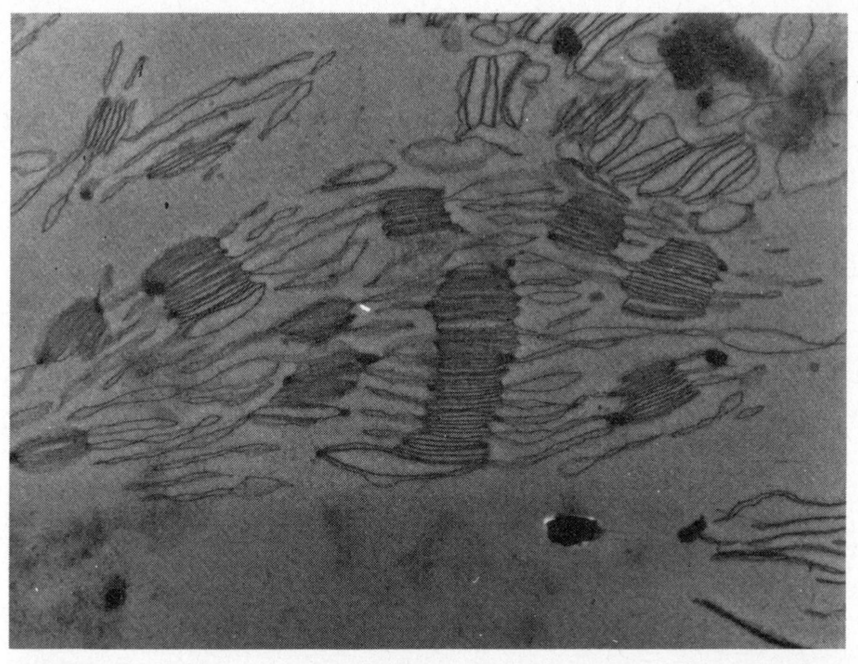

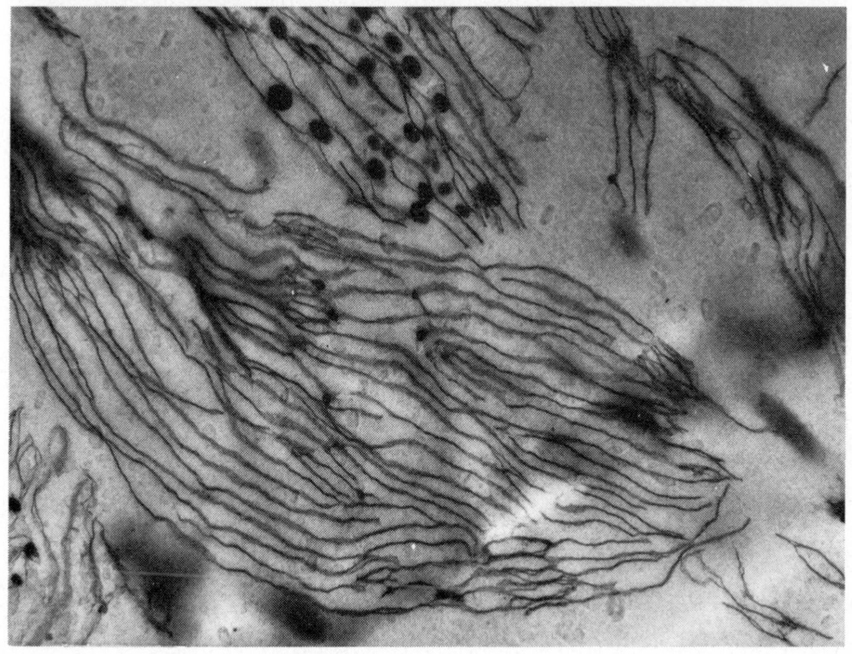

stacking observed under zero salt conditions. Alternately, the mechanism may be more subtle involving slight changes in orientation such as those postulated by Seely[25,26] which may be sufficient to regulate energy distribution. Further work is required to settle this problem.

Concluding Remarks

Since the discovery of Mg^{2+} effect on Chl a fluorescence yield by Homann in 1969 (ref. 6), large numbers of experiments have been done and several ideas have been proposed to explain them. It is now obvious that divalent ions have various effects on chlorophyll a fluorescence, some directly on PS II, and on PS I and some on energy distribution among photosystem I and photosystem II. The work presented here suggests that monovalent ions at low concentration decrease the "O" level, and this effect must be due to effect through bulk chlorophylls, not trap chlorophylls. As it has been difficult to obtain evidence for the motion of PS I and PS II closer or farther from each other to cause changes in energy spillover (in spite of recent evidence for motion of particles upon Mg^{2+} treatment[38]), it may be useful to probe an alternative picture in which light harvesting chlorophyll a molecules (common to both PS I and PS II) feed energy to both PS I and PS II. Monovalent and divalent cations may affect energy distribution by causing conformation changes on chlorophyll-protein complexes that lead to changes in orientation of few key chlorophyll molecules that control energy transfer to PS I and PS II[15].

Fig. 9. The effect of $CaCl_2$ on the ultrastructure of chloroplasts. Chloroplasts were incubated in the presence of 1 mM $CaCl_2$ (A) and 1 mM $CaCl_2$ + 3 mM NaCl (B) prior to fixation. The magnification was 25,000 x.

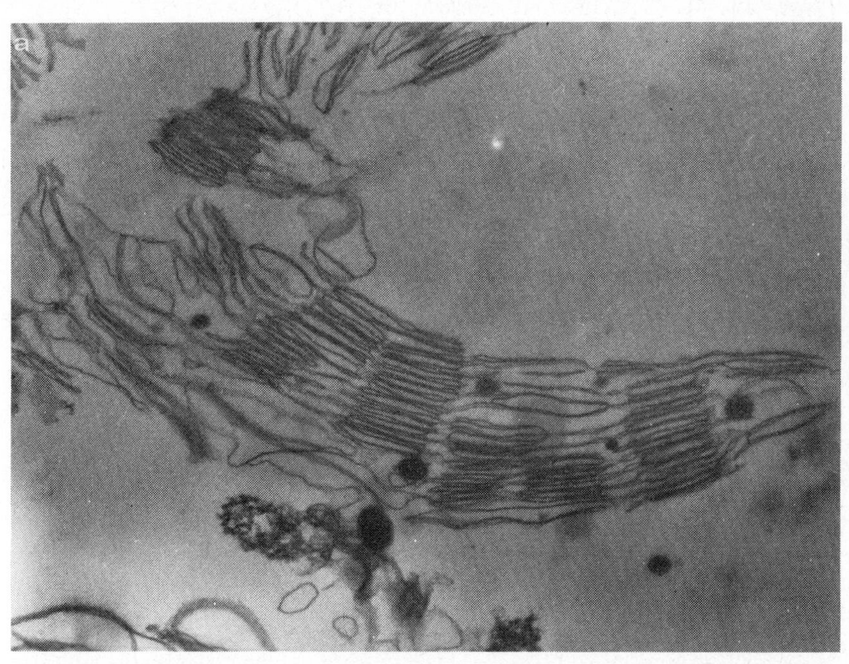
a

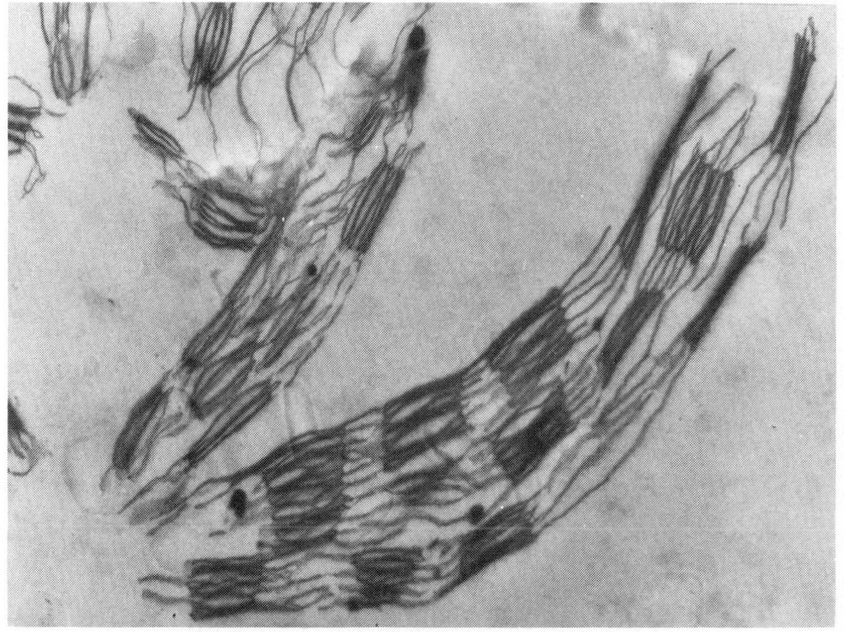

References

1. Murata, N., 1969, Biochim. Biophys. Acta 189, 171-181.

2. Murata, N., Tashiro, H. and Takamiya, A., 1970, Biochim. Biophys. Acta 197, 250-256.

3. Murata, N., 1971, Biochim. Biophys. Acta 245, 365-372.

4. Murakami, S. and Packer, L., 1971, Arch. Biochem. Biophys. 146, 337-347.

5. Mohanty, P., Braun, B. Z. and Govindjee, 1973, Biochim. Biophys. Acta 292, 459-476.

6. Homann, P., 1969, Plant Physiol. 44, 932-936.

7. Briantais, J. M., Vernotte, C. Moya, I., 1973, Biochim. Biophys. Acta 325, 530-538.

8. Marsho, T. V. and Kok, B., 1974, Biochim. Biophys. Acta 333, 353-365.

9. Gross, E. L. and Hess, S. C., 1973, Arch. Biochem. Biophys. 159, 832-836.

10. Gross, E. L. and Prasher, S. H., 1974, Arch. Biochem. Biophys. (in press).

11. Gross, E. L. and Hess, S. C., 1974, Biochim. Biophys. Acta 339, 334-346.

12. Murata, N., 1971, Biochim. Biophys. Acta 226, 422-432.

13. Sun, A. S. K. and Sauer, K., 1973, Biochim. Biophys. Acta 245, 409-427.

14. Bonaventura, C. and Myers, J., 1969, Biochim. Biophys. Acta 189, 366-383.

15. VanderMeulen, D. L. and Govindjee, 1974, Biochim. Biophys. Acta (in press).

16. Wydrzynski, T., Gross, E. L. and Govindjee, 1974, Biochim. Biophys. Acta (in press).

17. Govindjee and Papageorgiou, G., 1971, in: Photophysiology, vol. 6, Academic Press, New York, p. 1-46.

18. Murakami, S., Torres-Pereira, J. and Packer, L., 1975, in: Bioenergetics of Photosynthesis, ed. Govindjee, Academic Press, New York (in press).

19. Govindjee and Yang, L., 1966, J. Gen. Physiol. 49, 763-780.

20. Murata, N., Nishimura, M. and Takamiya, A., 1966, Biochim. Biophys. Acta 126, 234-243.

21. Murata, N., 1968, Biochim. Biophys. Acta 162, 106-121.

22. Rurainski, H. J. and Hoch, G., 1971, Proc. 2nd International Congress on Photosynthesis Research, Stresa, Italy, p. 133-141.

23. Jennings, R. C. and Forti, G., 1973, Biochim. Biophys. Acta 347, 299-310.

24. Hoch, G. and Knox, R. S., 1968, in: Photophysiology, vol. 3, Academic Press, New York, p. 225-250.

25. Seely, G. R., 1973, J. Theoret. Biol. 40, 173-187.

26. Seely, G. R., 1973, J. Theoret. Biol. 40, 188-198.

27. Bryant, F. D., Seiber, B. A. and Latimer, P., 1969, Arch. Biochem. Biophys. 135, 97-108.

28. Bryant, F. D., Latimer, P. and Seiber, B. A., 1969, Arch. Biochem. Biophys. 135, 109-117.

29. Radda, G. K. and Vanderkooi, J., 1972, Biochim. Biophys. Acta 241, 509-549.

30. Lesslauer, W., Cain. J. and Blasie, J. K., 1971, Biochim. Biophys. Acta 241, 547-566.

31. Zingsheim, H. and Haydon, D. A., 1973, Biochim. Biophys. Acta 298, 755-768.

32. Rubaclava, R., de Munoz, D. M. and Gitler, C., 1969, Biochim. 8, 2742-2747.

33. Vanderkooi, J. and Martinosi, A., 1969, Arch. Biochem. Biophys. 133, 153-163.

34. Trauble, H. and Overath, P., Biochim. Biophys. Acta 307, 491-512.

35. Duysens, L. N. M., 1956, Biochim. Biophys. Acta 19, 1-12.

36. Otoh, M., Izawa, S. and Shibata, K., 1963, Biochim. Biophys. Acta 69, 130-140.

37. Kirk, J. T. O., 1971, Ann. Rev. Biochem. 40, 161-196.

38. Ojakian, G. K. and Satir, P., 1974, Proc. Nat. Acad. Sci. 71, 2052-2056.

M. AVRON, *Proceedings of the Third International Congress on Photosynthesis*, September 2-6, 1974, Weizmann Institute of Science, Rehovot, Israel
Elsevier Scientific Publishing Company, Amsterdam, The Netherlands, 1974

A CRITICAL ROLE OF BICARBONATE IN THE RELAXATION OF REACTION CENTER II COMPLEX DURING OXYGEN EVOLUTION IN ISOLATED BROKEN CHLOROPLASTS

Govindjee*, A. J. Stemler* and G. T. Babcock†
*Departments of Botany and Physiology and Biophysics
University of Illinois, Urbana, Illinois 61801 USA
and †Department of Chemistry, Lawrence Berkeley Laboratory
University of California, Berkeley, California 94720 USA

Abstract

In bicarbcnate (HCO_3^-) depleted chloroplasts exposed to brief saturating light flashes, period 4 oscillations in O_2 yield per flash are damped within 3 cycles. Readdition of HCO_3^- to these preparations restores the oscillatory pattern to higher flash numbers, indicating that HCO_3^- reduces the probability of "misses" in the photosystem II reaction center. To explain our data, we must also propose that a certain percentage of reaction centers are completely inactive in the absence of HCO_3^- and that this, even more than an increase in the "miss" rate, lowers the steady state yield about 50 percent. Furthermore, the rate of the dark relaxation reaction $S_n' \text{----}{>} S_{n+1}$ (where S refers to the oxidation state of the oxygen evolving mechanism and n = 0, 1 or 2), following a photoact in the PS II reaction center, is retarded by about ten-fold in HCO_3^- depleted chloroplasts compared to the rate for this reaction in depleted chloroplasts to which HCO_3^- has been resupplied. However, HCO_3^- has no effect on the dark deactivation of the higher oxidation states (S_2 and S_3) of the positive charge accumulating system. We propose that the relaxation of reaction center II complex ($Z^{n+}Chla_{II}^{+}Q^{-}$, after a photoact, to $Z^{(n+1)+}Chla_2 \cdot Q$, where Z is the charge accumulating secondary electron donor, $Chla_{II}$ is primary electron donor and Q is primary electron acceptor) in dark requires HCO_3^-. This recovery reaction is affected either through the recovery of Q^- to Q or to the transfer of positive charges from $Chla_2$ to Z or both.

Abbreviations: Chl, chlorophyll; DCMU, 3-(3,4-dichlorophenyl)-1,1-dimethyl urea; DCPIP, dichlorophenol indophenol; DPC, diphenyl carbazide; PS II, photosystem II; Z Chl a_2 Q, reaction center of PS II, Z and Q being electron donor and acceptor, and Chl a_2 being the reaction center chlorophyll (the primary electron donor) of PS II.

Introduction

Recent investigation of the role of HCO_3^- in the Hill reaction indicates that this ion plays a critical role in the oxygen evolving mechanism[1-4]. Evidence is available which suggests[1] that HCO_3^- may act on the oxygen evolving side of photosystem (PS) II. Electron flow from the artificial electron donor diphenyl carbazide (DPC) to dichlorophenol indophenol (DCPIP) via PS II is insensitive[1] to HCO_3^-. Effects of HCO_3^- on chlorophyll (Chl) a fluorescence transients and on delayed light emission in the 0.5 - 5 sec time period were also interpreted to suggest a site of action of HCO_3^- on the oxygen evolving side of PS II . This latter work led Stemler and Govindjee[2] to speculate that HCO_3^- somehow stabilized higher oxidation states of the PS II reaction centers (referring to the kinetic model of oxygen evolution of Kok *et al.*[5]). We therefore studied[4] the effects of HCO_3^- on oxygen evolution in response to both continuous light and to brief light flashes. Our results[4] show that HCO_3^- reduces the frequency of reaction center "misses" and thus maintains oscillations in oxygen yield per flash for a much greater number of flashes. In addition, HCO_3^- was shown[4] to accelerate the relaxation reactions S_n--->S_{n+1}, where n = 0, 1 or 2) following a photoact. The present paper summarizes some of the results reported in reference 4.

Materials and Methods

Chloroplast Preparation. Maize (*Zea mays*) chloroplasts were obtained in a manner already described[1]. While even under optimum conditions maize chloroplasts usually do not perform the Hill reaction at very high rates compared to chloroplasts from other sources, we used maize to minimize precipitation of the chloroplasts during the HCO_3^- depletion procedure[3]. Bicarbonate depletion of pea (*Pisum sativa*) chloroplasts, under milder conditions, produced 4- 10-fold HCO_3^- stimulation of oxygen evolution with total yield equal to untreated controls (T. Wydrzynski and Govindjee, unpublished data). The HCO_3^- depletion procedure therefore does not necessarily result in gross chloroplast damage thereby accounting in some way for the HCO_3^- effect. To deplete them of HCO_3^- the chloroplasts were suspended in a solution containing 0.25 M NaCl, 0.04 M Na acetate, 0.05 M Na phosphate buffer at pH 5.0. The suspension was stirred slowly for 30 minutes at room temperature while the gas above the suspension was continuously flushed with nitrogen. After depleting the chloroplasts of HCO_3^- they were centrifuged in capped test tubes, previously flushed with N_2 and resuspended in reaction mixture. All vessels and reaction mixtures were carefully sealed or otherwise handled to avoid contamination with atmospheric CO_2 prior to deliberate addition of $NaHCO_3$.

O_2 Evolution in Flashing Light. The apparatus used for measuring oxygen evolution in response to brief light flashes was described by Weiss and Sauer[6] and

modified according to Babcock[7]. The xenon lamp pulses were 10 μsec in duration and were filtered through Corning 1-69 and 3-74 filters before being focused on the electrode surface. All flashes used in these experiments were of saturating intensity. The solution flowing above the membrane holding the chloroplasts to the surface of the platinum electrode contained 0.25 M NaCl, 0.04 M Na acetate, 0.05 M Na phosphate buffer at pH 6.8 and was either HCO_3^- free or supplied with 0.01 M $NaHCO_3$ in bicarbonate readdition experiments. The electrolyte was gassed continuously with 80 percent N_2, and 20 percent O_2.

Results

We have shown[4] that the stimulation of O_2 evolution caused by resupplying 0.01 M $NaHCO_3^-$ to the chloroplasts depends on light intensity. At saturating intensity stimulation is nearly 5-fold. This stimulation is clearly present at the lowest intensity used although it is reduced to about 2-fold.

1. Effects of HCO_3^- on Oscillations in Oxygen Yield. The kinetics of oxygen evolution in response to brief light flashes have already been described in detail for normal systems by Joliot *et al.*[8] and Kok *et al.*[5] (also see Mar and Govindjee[9]). From the evidence presented in Figure 1 (bottom), HCO_3^- depleted chloroplasts, when normalized to the same total oxygen yield (under steady state), show damped oscillations in oxygen evolution as a function of flash number compared to those chloroplasts resupplied with 0.01 M HCO_3^-. (In HCO_3^- depleted chloroplasts resupplied with HCO_3^-, oscillations were very similar to untreated controls; data not shown.) The damping of oscillations in the HCO_3^- depleted chloroplasts is suggested to be due to a greater number of "misses". Reducing the number of misses is not the only function of HCO_3^-, however, as will be seen from other results to be discussed below.

It is also evident from the recorder traces presented in Figure 1 (top) that total oxygen yield induced by light flashes spaced one second apart is nearly 2-fold greater in the presence of HCO_3^-. Under high intensity continuous light these same chloroplasts showed a 4- 5-fold greater rate of oxygen evolution. Thus brief flashes of high intensity light spaced one second apart produce the same reduced HCO_3^- stimulation as seen with low intensity continuous light. However, a 50 percent decrease in steady oxygen yield in the absence of HCO_3^- cannot be attributed to misses alone. If the miss rate were indeed 50 percent, the yield on the 3rd flash would be much less than that of the steady state yield. Since it is not, a miss rate of less than 20 percent is implied. We must therefore propose that a certain percentage of reaction centers are completely inactive in the absence of HCO_3^- and that this, even more than an increase in the miss rate, lowers the steady state yield about 50 percent.

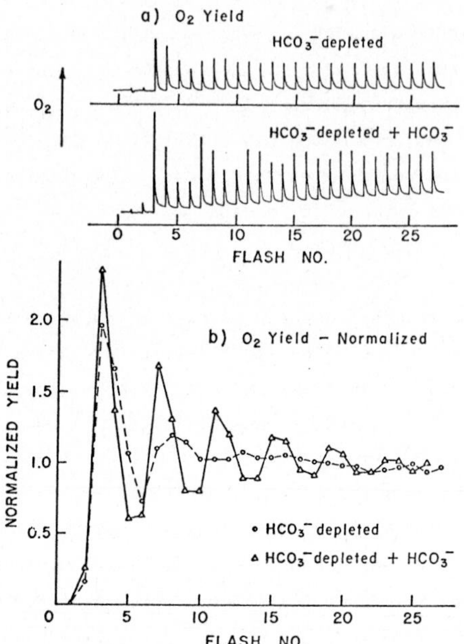

Fig. 1. Oxygen evolution in flashing light in the presence and absence of 0.01 M Na HCO_3^- following 5 minutes dark time. Top: recorder traces; saturating 10 μsec flashes spaced 1 sec apart were used to stimulate oxygen evolution. Bottom: oxygen yield as a function of flash number, from experimental traces in (a), normalized to the same total steady state yield of oxygen. The chloroplast suspension injected onto the platinum electrode contained 0.25 M NaCl, 0.04 M Na acetate, 0.05 M Na phosphate buffer pH 6.8, 20 μg ferredoxin ml^{-1} , 0.5 mM $NADP^+$ and 0.3 mg Chl ml^{-1} suspension. The chloroplasts used were previously depleted of HCO_3^-. To resupply HCO_3^- to the chloroplasts, HCO_3^- (to 10 mM) was added to the electrolyte flowing over the membrane holding the chloroplasts to the platinum. Other conditions are described in materials and methods (after Stemler et al.4).

2. Effect of HCO_3^- on Relaxation Reactions ($S_n' \dashrightarrow S_{n+1}$). Studies on the rates of the dark relaxation reactions occurring between photoacts have been made by Kok *et al.*[5] and particularly by Bouges-Boucquet[10]. The rates are measured by varying the time between the flashes and measuring the effect on the final yield of O_2. The half-times of the reactions $S_0' \xrightarrow{dark} S_1$, $S_1' \xrightarrow{dark} S_2$ and $S_2' \xrightarrow{dark} S_3$ are all in the order of 200-600 μsec in normal chloroplasts[10]. In HCO_3^- depleted chloroplasts, however, the half-times of these reactions are dramatically extended, while resupplying HCO_3^- restores the normal rates.

Figure 2 shows the effect of HCO_3^- on the reaction $S_2' \dashrightarrow S_3$ (this process proceeds biphasically and may involve two or more reactions[10]). The half-time for this

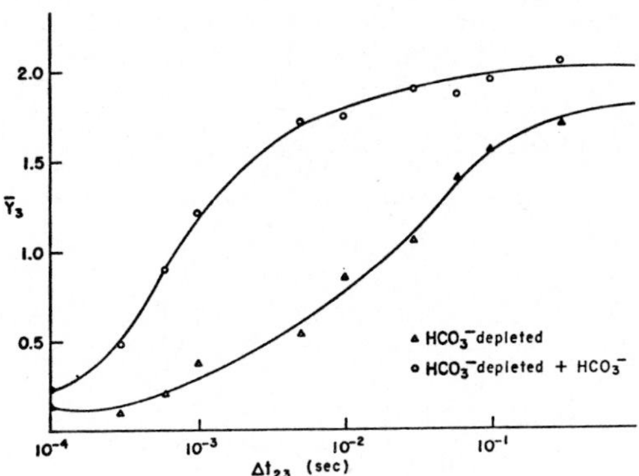

Fig. 2. Time course of the relaxation reaction $S_2' \dashrightarrow S_3$; oxygen yield on the third flash, $\bar{Y}_3$, in the presence and absence of 0.01 M Na HCO_3 as a function of the time between the second and third flash (Δt_{23}). $\bar{Y}_3$ is the yield normalized with respect to the steady state yield. Other flashes 1 sec apart. Other conditions as in Fig. 1 (after Stemler et al.[4]).

reaction is approximately 11 msec in HCO_3^- depleted chloroplasts and about 700 μsec in HCO_3^- depleted chloroplasts resupplied with HCO_3^-. Thus, HCO_3^- speeds the rate of this reaction, $S_2' \dashrightarrow S_3$ by more than 10-fold. The rate of the reaction $S_1' \dashrightarrow S_2$ was affected in a manner similar to that for $S_2' \dashrightarrow S_3$ (not shown here). Again the half-time was approximately 10 msec in HCO_3^- depleted chloroplasts and about 600 μsec in chloroplasts resupplied with HCO_3^-. Likewise the rate of the reaction $S_0' \dashrightarrow S_1$, calculated by the method of Bouges-Bocquet[10] is comparably reduced in HCO_3^- depleted chloroplasts (data not shown).

3. Effect of HCO_3^- on the Deactivation of the States S_2 and S_3. If the time between the first and second light flash, or between the second and third, is extended beyond about 1 sec, deactivation of the states S_2 or S_3, respectively, can be observed[11,12].

Figure 3 shows that the decay of the S_3 state in chloroplasts depleted of HCO_3^-, and those resupplied with HCO_3^-, follow the same time course, although there is a difference in the amount of O_2 evolved. Likewise the decay of the S_2 state (data not shown) is the same in the presence and absence of HCO_3^-. It is clear, therefore, that HCO_3^- has no effect on the stability of the higher oxidation states (i.e., S_2 and S_3) of the PS II positive charge accumulating system, but

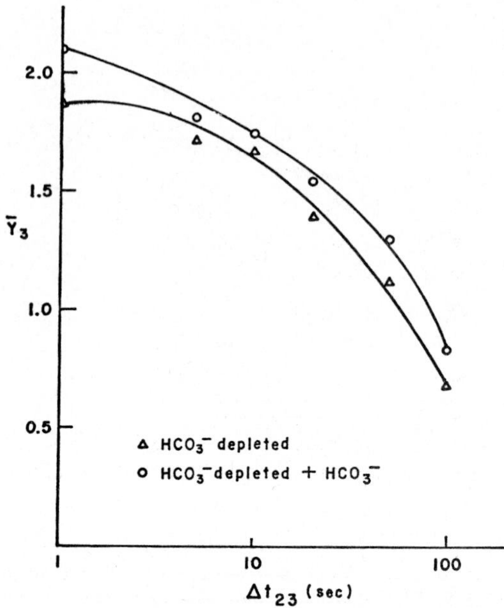

Fig. 3. Decay of the S_3 state; oxygen yield on the third flash $\bar{Y}_3$ in the presence and absence of 0.01 M Na HCO_3 as a function of the time between the second and third flash (Δt_{23}). Other conditions as in Fig. 2 (after Stemler et al.[4]).

it affects only on the rate of formation of these states following a photoact.

Discussion

The observation that HCO_3^- speeds about 10-fold the relaxation reactions between photoacts (Fig. 2) explains why less HCO_3^- stimulation is seen at low light intensity[4] and when saturating flashes are given spaced one second apart. Under these conditions the reaction centers have enough time to undergo relaxation (even at the lower rate imposed by HCO_3^- depletion) before another photon arrives. Thus HCO_3^- has less observed effect. The small stimulation in O_2 yield per flash (Fig. 1, top) that is still observed under these conditions must be due primarily to a greater number of active reaction centers and secondarily to the reduced number of "misses" that occur in the presence of HCO_3^-. Further quantitative analysis is needed to explain how the HCO_3^- effects reported here account for the 10-fold effect observed under saturating light intensities.

The ability of HCO_3^- to speed relaxation reactions can be interpreted to indicate either that this ion accelerates the reoxidation of the primary electron acceptor for PS II (Q) by the pool of intersystem intermediates or that HCO_3^- is acting on the electron donor side of PS II. The first interpretation, as mentioned in the introduction, appears to be inconsistent with our previous work[1-3]. For example, if HCO_3^- depletion imposed a block between Q and A, one would predict that the chlorophyll _a_ fluorescence transient would rise to maximum (F^∞) very fast, as it does in the presence of 3-(3,4-dichlorophenyl)-1,1-dimethylurea (DCMU). Although the fluorescence rise is, at first, faster, it slows down with time in the HCO_3^- depleted chloroplasts[2]. As HCO_3^- depletion may not cause a perfect block as does DCMU, new fluorescence experiments are being performed by T. Wydrzynski and Govindjee with different concentrations of DCMU to provide further data. Further suggestion that HCO_3^- may not be acting on the reducing side of PS II is provided by long-term delayed light emission, which is thought to reflect back reactions following light induced charge separation. If HCO_3^- accelerated the reoxidation of Q by A, less Q^- should be available to back react. Hence we would expect less delayed light emission in the presence of HCO_3^-. Instead, more delayed light emission is observed[2] in the presence of HCO_3^-. These results, the absence of an HCO_3^- effect when PS II electron flow is from DPC to DCPIP, and others already discussed[1-3] imply that HCO_3^- is acting on the electron donor side of PS II. However, since all our arguments which support this view are admittedly based on various assumptions, we plan to continue to test this hypothesis. New experiments are in progress in our laboratory (Urbana) to monitor the effect of HCO_3^- on membrane phenomena, on fluorescence decay after a flash of light, on the electron flow from H_2O to electron acceptors that may accept electrons directly from Q, on ΔpH, and on the $^{18}O/^{16}O$ ratio when chloroplasts are injected with $HC^{18}O_3^-$. Experiments are also planned to test the effect of HCO_3^- on the recovery of $Chla_2^+$ to $Chla_2$. Meanwhile it appears reasonable to consider how HCO_3^- may be influencing the oxygen evolving mechanism.

The kinetic model for oxygen evolution advanced by Kok _et al._[5] (see ref. 13) can now be extended in several possible ways to include the action of HCO_3^-. Ignoring for our purposes the reducing (Q) side of PS II, we may represent a photoact II as: $Chla_2 \xrightarrow{h\nu} Chla_2^+$, where $Chla_2$ is the primary electron donor to Q[14]. $Chla_2^+$ undergoes reaction with the first secondary electron donor Z: $Chla_2^+ + Z \longrightarrow Chla_2 + Z^+$. Z^+ in turn undergoes the HCO_3^- mediated reaction: $Z^+ + S_n \xrightarrow{HCO_3^-} Z + S_{n+1}$, where S is the charge accumulating enzyme or system in the nth state (n = 0, 1 or 2) and $Z^+ + S_n$ corresponds to the S_n' state mentioned earlier in the results section. We can imagine that O_2 evolution occurs simply as: $Z^+ + S_3 \longrightarrow S_4$; $S_4 + O_2$ precursor(s) $\longrightarrow S_0 + O_2$. A second possibility is that Z^+ and S_3 cooperate as: $Z^+ + S_3 + O_2$ precursor(s) $\longrightarrow Z + S_0 + O_2$. In the above model, HCO_3^- may control the transfer of the first 3 electrons from the positive charge accumulating mechanism to oxidized Z.

An alternative explanation of the HCO_3^- effect is also possible. In this second model Z is eliminated as an intermediate entirely, or rather it is equated with the positive charge accumulating system. Thus the reaction sequence can be written as: $Chla_2 \dashrightarrow Chla_2^+$; $Chla_2^+ + S \xrightarrow{HCO_3^-} Chla_2 + S_{n+1}$, where n is again equal to 0, 1 or 2. In this model HCO_3^- controls the rate of transfer of the first 3 electrons from the positive charge accumulating system (called S in the Kok *et al.* model) directly to the oxidized reaction center $Chla_2^+$. An entirely different view is to suggest that HCO_3^- accelerates electron flow from Q^- to the intersystem intermediate A. Experiments are in progress in our laboratory to decide between the models presented here.

Besides accelerating the relaxation reactions, HCO_3^- also reduces the number of misses which are apt to occur in reaction centers (Fig. 1). While these are clearly different effects, they are not necessarily independent. We propose that if the relaxation reactions following a photoact are accelerated by HCO_3^-, less time might be available for a back reaction of $Chla_2^+$ and Q^-. Such a back reaction, occurring in the msec time period or earlier after a flash could constitute a miss. It follows that this reaction would have less time to occur in the presence of HCO_3^-. If this is the case, we might expect greater amounts of delayed light emission (reflecting more misses) from HCO_3^- depleted systems in the msec and μsec time range following a flash.

The possible role of HCO_3^- on the actual O_2 evolving step must be seriously analyzed. Does it work as an allosteric effector on the O_2 evolving enzyme? New experiments are needed to understand how the absence of HCO_3^- inactivates reaction center II, increases "misses", and slows down the relaxation reactions of reaction center II -- all at the same time. Lastly, we wish to point out that the action of HCO_3^- described by us here and elsewhere[1-4,15] may have regulatory significance. When CO_2 (or HCO_3^-) is lacking in the microenvironment of chloroplasts so that they cannot manufacture carbohydrates, it would be unnecessary and perhaps harmful for the chloroplasts to produce the reducing power ($NADPH_2$) and ATP. Thus, nature may have evolved a system to "shut off" (or decrease drastically) the early reactions of photosynthesis -- the operation of reaction center II and consequently O_2 evolution. On the other hand, HCO_3^- may have some more direct role in the chemistry of O_2 evolution.

Acknowledgments

This research was conducted and completed with the aid of a grant GB 36751 from the National Science Foundation. We thank Professor Kenneth Sauer for allowing us the use of the facilities at the Lawrence Berkeley Laboratory and for helpful comments on the manuscript. Dr. Bruce Diner (at Dr. P. Joliot's laboratory in

Paris, France) has kindly collaborated with us and obtained independent data for us confirming the results reported here. We are higly grateful to him for his efforts. We also thank Dr. Bessel Kok for making the preliminary experiments for us, shown in Fig. 1, under his experimental conditions.

References

1. Stemler, A. and Govindjee, 1973, Plant Physiol. 52, 119-123.
2. Stemler, A. and Govindjee, 1973, Photochem. Photobiol. 19, 227-232.
3. Stemler, A. and Govindjee, 1974, Plant and Cell Physiol. 15,
4. Stemler, A., Babcock, G. and Govindjee, 1974, Proc. Nat. Acad. Sci. (in press).
5. Kok, B., Forbush, B. and McGloin, M, 1970, Photochem. Photobiol. 11, 457-475.
6. Weiss, C. and Sauer, K., 1970, Photochem. Photobiol. 11, 495-501.
7. Babcock, G., 1973, Kinetics and Intermediates in Photosynthetic Oxygen Evolution, Ph.D. Thesis, University of California, Berkeley. LBL-2172.
8. Joliot, P., Barbieri, G. and Chabaud, R., 1969, Photochem. Photobiol. 10, 309-329.
9. Mar, T. and Govindjee, 1972, J. Theor. Biol. 36, 427-446.
10. Bouges-Bocquet, B., 1973, Biochim. Biophys. Acta 292, 772-785.
11. Joliot, T., Joliot, A., Bouges, B. and Barbieri, G., 1970, Photochem. Photobiol. 14, 287-307.
12. Forbush, B., Kok, B. and McGloin, M., 1971, Photochem. Bhotobiol. 14, 307-321.
13. Joliot, P. and Kok, B., 1974, in: Bioenergetics of Photosynthesis, ed. Govindjee, Academic Press, New York (in press).
14. Butler, W. L., 1973, Accounts of Chemical Research 6, 177-184.
15. Stemler, A. and Govindjee, 1974, in: Procedings of the International Symposium on Biomembranes, Madurai, India, ed, L. Packer, Academic Press, New York (in press).

M. AVRON, *Proceedings of the Third International Congress on Photosynthesis*,
September 2-6, 1974, Weizmann Institute of Science, Rehovot, Israel
Elsevier Scientific Publishing Company, Amsterdam, The Netherlands, 1974

SOME COMMENTS ON THE PRESENT STATUS OF THE PRIMARY ELECTRON ACCEPTOR OF PHOTOSYSTEM I

Bacon Ke
Charles F. Kettering Res. Lab., Yellow Springs, Ohio 45387, U.S.A.

SUMMARY. Light-induced changes of EPR signals in photosystem-I subchloroplast particles at temperatures between 225° and 13°K showed that the rates of onset of photooxidation of P700 and photoreduction of iron-sulfur protein(s) (ISP) are rapid and identical within the limits of resolution of our instruments, and that $P700^+$ and ISP^- always recover in parallel, as expected for a charge-recombination reaction. Light-induced absorption changes due to P700 photooxidation at low temperatures monitored at 700 nm showed a similar kinetic pattern.

Redox titration by monitoring the attenuation of light-induced absorption changes due to P700 photooxidation at 86°K as a function of the redox potential imposed on the subchloroplast system revealed a midpoint potential of -530 mV for the photosystem-I primary acceptor. Experimental evidence available to date suggests that the bound ISP with an E_m of -530 mV is the most likely candidate for the photosystem-I primary acceptor.

Since the report in 1971 of observations of the spectral species "P430" by kinetic spectroscopy (1) and chloroplast-bound "ferredoxins" by EPR spectroscopy (2), and the suggestions that each may function as the primary electron acceptor in photosystem I, the continuing task in this field of research has been a quest for the chemical identity of P430, a possible correlation between P430 and the bound iron-sulfur proteins (ISP's), and/or additional kinetic evidence for the primary-acceptor role for the bound ISP's.

The primary-acceptor role of P430 appears to have been more definitively established from evidence of a kinetic correspondence between the reoxidation of the reduced primary acceptor and the reduction of a secondary acceptor (3-5). Further support was provided by the high quantum yield and the rapid rise time of the photosystem-I charge-separation reaction (6). However, it should be noted that photoreduction of the bound "ferredoxin" at cryogenic temperatures is a necessary but not sufficient condition for the assignment of bound "ferredoxin" the role of a primary acceptor of photosystem I. Indeed, questions have arisen recently on the alternative possibility that the bound proteins may be secondary acceptors (7,8). A definitive answer to this question has not yet been obtained because of the difficulty that detection of the reduced form of bound ISP's by EPR spectroscopy is possible only at very low temperature (<25°K). A kinetic correlation between the bound ISP's and P430 would require very rapid cycling of sample temperatures between that desired for the reaction and that necessary for observation (9).

This note summarizes recent studies designed to answer some of these questions, together with a discussion of the significance of some recent debates over the role of the bound ISP's.

SOME KINETIC STUDIES. Of the five reaction pathways taken by the charged primary donor and acceptor following their formation in the light, a kinetic correlation by optical and EPR spectroscopies for three of them, namely, charge recombination, the noncyclic and cyclic pathways, are subject to the technical difficulty stated above. However, the remaining two reactions, namely, the accumulation of $P700^+$ or $P430^-$, in which the accumulated species remain stable for relatively long time, may be examined by EPR spectroscopy in an attempt to clarify the chemical identity of P430 and a possible correlation between P430 and bound ISP's.

Primary photochemical charge separation followed by a "one-way" electron discharge by $P430^-$ to an autooxidizable secondary acceptor results in a difference spectrum of exclusively $P700^+$ (10). Illumination of photosystem-I subchloroplasts in the presence of an efficient reductant and under anaerobic conditions results in the reduction of both P700 and P430 (5). Subsequent EPR measurements of these reactions confirmed that exclusively $P700^+$ free radicals (11) or reduced ISP's (12) were formed in the two respective reactions. Considering the peculiar character of these reactions, namely, the rapid electron discharge by $P430^-$ and the accumulation of $P700^+$, or the rapid reduction of $P700^+$ and the accumulation of $P430^-$, and the fact that ferredoxins show an EPR signal only in their reduced forms, we think the evidence is significant.

The kinetic courses involving P430 in the first three reaction pathways at room temperature are well established, and they provide the most definitive basis for assigning a primary-acceptor role to this spectral species. A kinetic correlation between P430 and the ISP's would provide an equally definitive basis for assigning a primary-acceptor role to the latter. In principle, to perform the kinetic spectroscopic experiments at room temperature followed by examination of the decay course by EPR spectroscopy would require adapting and modifying the rapid-mixing/rapid-quenching technique which has been successfully used for studying enzyme kinetics (13). We are currently developing instrumentations for adapting this technique to our investigations.

It should be noted that certain background information, especially that regarding the decay of photosystem-I reactions at low temperatures, is also necessary for performing these rapid-quenching experiments. Unfortunately, background information on the low-temperature reactions of P700 is often contradictory and that on P430 is totally lacking at present. Some workers found P700 photooxidation practically irreversible between 110 and 77°K (e.g. 14), while others reported a partial reversibility (e.g. 15). We recently began to reexamine this problem, in an attempt to clarify this question and to obtain the background information necessary for the rapid quenching experiments. The reverse reaction (presumably recombination of the charged primary reactants) after light-induced charge separation in photosystem I appears to be quite complex. While the charge separation appears to be temperature independent, the kinetic course of the reverse reaction is very temperature sensitive. Details of this work will be reported elsewhere (16). Typical light-induced absorption changes accompanying photoreactions and their dark decay at two low temperatures (monitored at 700 nm for P700 photooxidation) are shown in Figure 1, left. As reported earlier, photosystem-I recombination occurs at room temperature with a $t_{1/2}$ of approx. 45 ms (3); the $t_{1/2}$ value increases as temperature is lowered. Beginning at 200°K, a small fraction of the absorption change becomes irreversible, and the reversible portion is apparently biphasic, and consists of equal portions of a rapid decay phase and a slow decay phase (Fig. 1, left, upper trace). As the temperature is lowered further, the irreversible fraction becomes greater, but the biphasic pattern of the reversible portion remains the same (Fig. 1, left, lower trace).

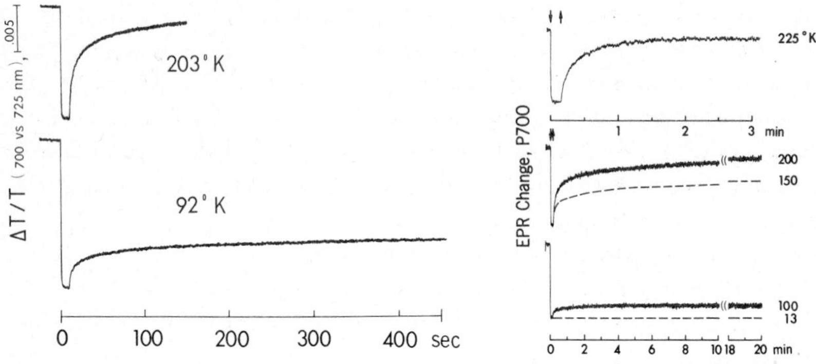

Figure 1. (left) Light-induced absorption changes due to P700 photooxidation at low temperatures. Triton-fractionated PS-I subchloroplast particles were suspended in a pH 8, 0.1 M Tricine buffer medium containing 2.3 M sucrose, 1 mM ascorbate and $5x10^{-5}$ M TMPD, at a Chl concn of 0.05 mM. Optical pathlength was 1 nm. Saturating actinic light between 400 and 460 nm was isolated with interference filters; illumination period was 10 sec.
(right) Light-induced EPR signal changes due to P700 photooxidation (monitored at the high-field derivative peak of the free-radical signal at g=2.0026) at various temperatures. Sample composition was the same as those used in the absorption changes at left, except at a Chl concn of 0.55 mM, and without sucrose. Microwave power was 1 μW at 13°K, 10 μW at 75°K, 30 μW between 100 and 175°K, and 100 μW between 200 and 225°K. Modulation frequency was 100 kHz, and amplitude, 2 G. Incident intensity (white light) at the sample tube was approx. 10^6 ergs/cm^2.sec, and was above saturation. Downward arrow, light on; upward arrow, light off.

From these results, it would seem desirable to examine the decay of the charged species at low temperatures also by EPR spectroscopy. The kinetics of the EPR free-radical signals representing $P700^+$ have almost identical characteristics to the absorption-change signals, as shown in Fig. 1, right. The formation of $P700^+$ free radicals was within the combined time resolution of the instruments (~0.25 sec). Upon cessation of illumination, the decay at 225°K appeared to follow first-order kinetics, with a half time of 7.7 sec. However, the signal decay was not complete; at 225°K, about 10% of the change was irreversible at the end of 20 min. At 200°K, the decay appears to be biphasic: the rapid and slow phases, which are approximately equal in magnitude, had decay half times of ~ 3 and 75 sec, respectively. At 175°K, the irreversible fraction was ~ 30%. At lower temperatures, the same biphasic-decay pattern remained, but the magnitude of the reversible portion became less and less. At 13°K, no visible decay could be observed at the end of 20 min. Although not shown in the Figures, the reversible portion of the absorption change and the EPR signal can undergo reversible bleaching again upon a second illumination.

If an ISP is indeed the reaction partner of P700 in the primary photochemical act of photosystem I, the onset time for the formation of the two charged species and their decay kinetics should be identical. This was found to be the case for 13°K, namely, the onset time for P700 photooxidation and that for ISP photoreduction (monitored at g = 1.86) are both rapid and not resolvable with the instrumentation used, and both changes are irreversible.

Because the EPR signals of reduced ISPs are only detectable at low temperatures, continuous monitoring of the course of decay at other temperatures cannot be made. Considering the fact that the EPR signals of $P700^+$ and ISP^- are practically completely stable at

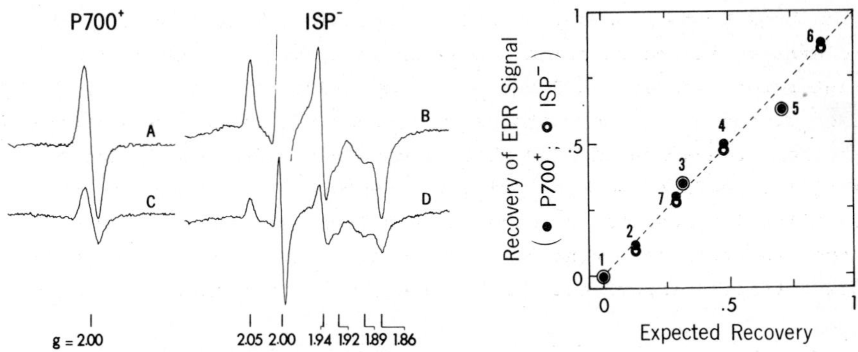

Figure 2. (left) EPR spectra of $P700^+$ (A) and ISP^- (B) after the subchloroplast sample was illuminated at 13°K for 20 sec. (C) and (d) are the corresponding spectra (recorded at 13°K) after the sample was maintained at 175°K for 6 min.
(right) Plot of the actual fraction of decay of the EPR signal of $P700^+$ (solid dots) and ISP^- (empty circles) vs. the expected extent of decay (derived from Fig. 1, right). Point sets 1, 2, and 3 are for samples maintained for 10 min at 13, 75 and 125°K, respectively; point sets 4 and 5, 6 min at 150 and 175°K, respectively; point set 6, 90 sec at 225°K; point set 7, 25 sec at 150°K.

13°K, one could study the decay of both $P700^+$ and ISP^- by exposing the EPR samples to other temperatures at which partial decay was allowed, and then rapidly refreezing the sample to 13°K. Fig. 2, left shows the EPR spectra of $P700^+$ and ISP^- produced in a subchloroplast sample after illumination at 13°K for 20 s (top row) and after dark decay at 175°K for 6 m (bottom row). Both signals recovered 63%, as calculated from these spectra, which is in good agreement with the expected value derived from the continuous decay course of $P700^+$ as shown in Fig. 1, right.* Additional determinations were made for exposures to a number of other temperatures ranging from 75 to 225°K for times ranging from 25 s to 10 m. In Fig. 2, right, the percentages of recovery derived from the $P700^+$ and ISP^- signals are plotted versus the expected values derived from the continuously monitored kinetic course. Detailed results and experimental procedures of this work will be reported elsewhere (9). These data show that not only are the measured decays in good agreement with the expected values, but the extent of decay of $P700^+$ and ISP^- also agrees very well with each other.

REDOX POTENTIAL OF THE PHOTOSYSTEM-I PRIMARY ELECTRON ACCEPTOR. Redox titrations (12) of the photosystem-I subchloroplasts monitored by EPR spectroscopy showed that possibly three bound ISP's are present in the reaction center of photosystem I, whose midpoint potentials are -530 and $\leqq$ -580 mV, respectively. Earlier redox titration of the photosystem-I primary acceptor by monitoring the attenuation of P700 photooxidation yielded a midpoint potential of -475 mV for the primary acceptor (3). As subsequent EPR-titration results indicated a much lower potential, it was suspected that the potential value derived from the room-temperature titration experiments might have been, for technical reasons,

*After the completion of this work, a report by A.J. Bearden and R. Malkin (Federation Proceedings, 33, abstract 378(1974) appeared, which described a similar experiment as reported here, in which they found a parallel decay of the EPR signals of P700 and "bound ferredoxin" approx. 40% at 150°K.

underestimated (12). It was soon confirmed that under the conditions of the room-temperature titration experiments, when a multitude of redox mediators is present in the reaction system, the reduced forms of these mediators under the required anaerobic conditions also served as electron donors to $P700^+$. These interceptions by the mediators consequently caused a premature attenuation of the magnitude of the light-induced absorption changes, and thus led to an underestimated redox-potential value.

It became obvious that the interference by the redox mediators with the $P700^+$ back reaction can be avoided only by observing the light-induced absorption changes at lower temperatures. As the background information shown in Fig. 1, left became available, we recently re-investigated the redox potential of the photosystem-I primary acceptor by measuring the attenuation of the light-induced absorption changes at 86°K, based on the same premise set out for the room temperature experiments (6). The magnitude of the light-induced absorption changes plotted as a function of the redox potential imposed on the reaction system is shown in Fig. 3. The attenuation of the absorption changes began a little before -500

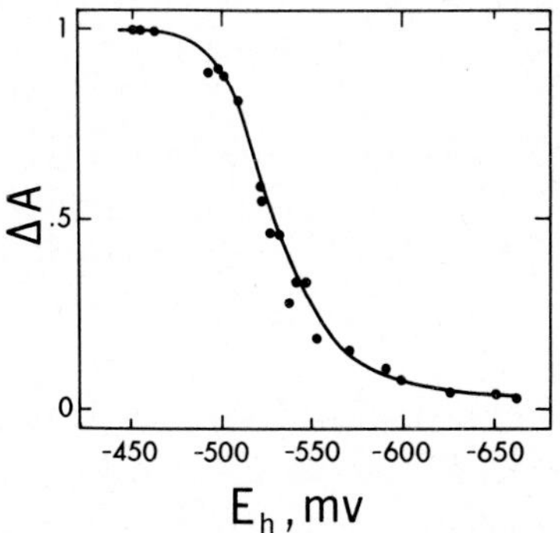

Figure 3. Light-induced absorption changes due to P700 photooxidation at 86°K vs. imposed redox potential. The redox mediators used and the experimental procedures were the same as described in refs. (5) and (12). The suspending medium was 0.1 M glycine-NaOH buffer, pH 11, containing 60% glycerol, the appropriate redox mediators, and subchloroplast particles at a Chl concn of 0.05 mM. Optical-pathlength and actinic-light conditions were the same as for Figure 1, left.

mV, then the light-induced absorption change decreased more promptly as the redox potential was lowered, and the residual (~5%) of the original change persisted beyond -600 mV. The shape of the titration curve also fits better with a one-electron change. Recently, Lozier and Butler (7) reported a similar redox titration using hydrogen/hydrogenase as the complementary redox couple and examined the attenuation of the light-induced absorption change due to P700 photooxidation by measuring the difference spectra of samples poised at different potentials before and after illumination at -196°C, and they also obtained a midpoint potential of -530 mV for the photosystem-I primary acceptor.

CONCLUDING REMARKS. Note that the midpoint potential reported here and that reported by Lozier and Butler (7) agree well with that of the first-stage EPR titration of the bound ISP's of photosystem I (12). These results are also consistent with the fact that illumination of chloroplasts at low temperature only leads to reduction of ISP's corresponding to the chemical reduction to the high-potential stage (12). These facts taken together seem to suggest that the less negative of the bound ISP's is the more likely candidate for the primary electron acceptor of photosystem I.

As shown in Fig. 1, light-induced absorption and EPR changes are partially reversible even at liquid-nitrogen temperature. The difference spectra reported in ref. (7) would then only consist of the contribution by the irreversible portion of the absorption change. Consequently, such spectra are subject to an alternative interpretation that the irreversible change of $P700^+$ resulted from an irreversible trapping of electrons by the secondary electron acceptor, presumably the bound ISP's (7,8). Although our titration results obtained at low temperature (Fig. 3) are essentially the same as that reported by Lozier and Butler (7), we actually measured the onset of P700 photooxidation and its attenuation, which should reflect more directly the fraction of the primary acceptor present in the oxidized state, and thus available to participate in the primary charge-separation reaction.

From a purely energetic viewpoint, it might be proposed that the more negative (or more reducing) ISP revealed by redox titration (12) could be the true primary acceptor, while the more positive (or less reducing) ISP could function as a "secondary" acceptor, presumably lying intermediate between the primary acceptor and soluble ferredoxin. However, available evidence appears not to support this contention. Kinetic spectroscopic studies with soluble ferredoxin as the secondary acceptor showed that soluble ferredoxin apparently receives electrons directly from P430 (4). In the previous titration experiments (12), we showed that at pH 9, the redox potential of the subchloroplast system reached approx. -560 mV, and the first-stage reduction corresponding to that of the higher-potential protein appears complete. If the proposal that the more negative ISP is the primary acceptor is correct, then subsequent illumination of the subchloroplast poised at -560 mV should cause a photoreduction of the more negative ISP in a photochemical charge-separation reaction. However, EPR spectra taken after illumination of such samples showed that no reduction of the more negative ISP occurred (12). More recently, we have further confirmed that no EPR signal corresponding to the more negative ISP (i.e., at $\underline{g}$=1.92 and 1.89) appeared in the subchloroplast sample poised near -560 mV and under continuous illumination (unpublished experiments).

Another argument against the suggestion that the more negative ISP functions as the primary acceptor is provided by the fact that EPR spectra recorded under continuous illumination of chloroplasts at all low temperatures yielded only the more positive ISP, and that the reduced ISP recovered in parallel with $P700^+$. Such a recombination is unlikely to occur at such low temperatures if the more positive ISP were a secondary acceptor.

ACKNOWLEDGEMENT. The author wishes to acknowledge with thanks the collaboration of Dr. H. Beinert and his colleagues in the EPR experiments discussed here. This work was supported in part by a National Science Foundation Grant GB-29161. Contribution No. 522 from the Charles F. Kettering Research Laboratory.

REFERENCES

1. Hiyama, T. and Ke, B. (1971) Proc. Nat. Acad. Sci., USA 68, 1010-1013.
2. Malkin, R. and Bearden, A.J. (1971) Proc. Nat. Acad. Sci., USA 68, 16-19.
3. Hiyama, T. and Ke, B. (1971) Arch. Biochem. Biophys., 147, 99-108.
4. Hiyama, T. and Ke, B. (1972) Proc. 2nd Intern. Congr. Photosynthesis Research, Stresa, 1971, Vol. I, pp. 491-497. Dr. W. Junk N.V. Publishers, The Hague.
5. Ke, B. (1973) Biochim. Biophys. Acta, 301, 1-33.
6. Ke, B. (1972) Arch. Biochem. Biophys., 152, 70-77.
7. Lozier, R.H. and Butler, W.L. (1974) Biochim. Biophys. Acta, 333, 460-464.
8. Chu, W., McIntosh, A.R., and Bolton, J.R. (1974) Abstract, 2nd Ann. Meeting of Am. Soc. Photobiology, Vancouver.
9. Ke, B., Sugahara, K., Shaw, E.R., Hansen, R.E. Hamilton, W.D., and Beinert, H. (1974) Biochim. Biophys. Acta, in press.
10. Ke, B. (1972) Biochim. Biophys. Acta, 267, 595-599.
11. Ke, B. and Beinert, H. (1973) Biochim. Biophys. Acta, 305, 689-693.
12. Ke, B., Hansen, R.E., and Beinert, H. (1973) Proc. Nat. Acad. Sci., USA 70, 2941-2945.
13. Bray, R.C. (pp. 195-203) and Palmer, G. and Beinert, H. (pp. 205-214), in Rapid Mixing and Sampling Techniques in Biochemistry (B. Chance et al., eds.) Acad. Press (1964).
14. Witt, K. (1973) FEBS Letters, 39, 112-115.
15. Mayne, B.C. and Rubinstein, D. (1966) Nature, 210, 734-735.
16. Sugahara, K. and Ke, B., in preparation.

M. AVRON, *Proceedings of the Third International Congress on Photosynthesis*,
September 2-6, 1974, Weizmann Institute of Science, Rehovot, Israel
Elsevier Scientific Publishing Company, Amsterdam, The Netherlands, 1974

PROPERTIES OF THE PHOTOCHEMICAL REACTION CENTRE OF PHOTOSYSTEM I IN SPINACH CHLOROPLASTS

M.C.W. Evans
Department of Botany & Microbiology
University College, London, U.K.

R. Cammack & S.G. Reeves
Department of Plant Sciences
King's College, London, U.K.

1. Introduction

The primary event in the energy conversion process of photosystem I (PSI) is generally thought to be the photooxidation of a chlorophyll molecule, P700, followed by the transfer of an electron to a primary acceptor and ultimately to NADP. This photooxidation of P700 is irreversible at cryogenic temperatures under the conditions normally used to investigate this reaction. It has been shown, using electron paramagnetic resonance (EPR) spectroscopy, that PSI particles contain two iron-sulphur centres which can be photoreduced at cryogenic temperatures[12]. The reduction of these centres is linked to the oxidation of P700[3] and it has been proposed that they are the primary electron acceptors of PSI[1,2]. Recently it has been reported that in detergent-prepared PSI particles, in which some disruption of the reaction centre may occur, the photooxidation of P700 is partially reversible[4], suggesting that a component other than the iron-sulphur centres is the primary electron acceptor. We have investigated the properties of the iron-sulphur centres in PSI particles prepared by mechanical disruption, and the effect of the redox state of these centres on the reversibility of P700 photooxidation at low temperatures.

2. Materials and Methods

PSI particles were prepared using the French press as described previously[2]. The particles were suspended in 0.1M Tris-Cl pH 8.0 and the pH adjusted with 2M NaOH as required. Oxidation-reduction potential titrations and EPR spectrometry were carried out as described previously[5].

3. Results and Discussion

PSI particles frozen in the dark in the presence of dithionite show little or no EPR signals in the $g = 2.00$ region (Fig.1). After illumination at low temperature the same sample will show a signal at $g = 2.00$ due to oxidized P700 and two signals due to reduced iron-sulphur centres (reduced P700 and oxidized iron-sulphur centres give no EPR signal). One iron-sulphur centre, A, has $g_z = 2.05$ $g_y = 1.94$ and $g_x = 1.86$ and the other, centre B, has $g_z = 2.05$ $g_y = 1.92$ and $g_x = 1.89$. In freshly prepared particles these centres show little or no reduction by dithionite at pH 8.0, but preparations which have been repeatedly frozen and thawed or exposed to alkaline pH show a variable reduction by dithionite. Samples illuminated at low temperature show extensive reduction of centre A and only slight reduction of centre B. Complete reduction of both centres can be obtained by illumination at room temperature in the presence of dithionite and freezing in the light (Fig.4) or by reduction in dark by dithionite in the presence of viologen dyes at alkaline pH.

Using a modification of the oxidation-reduction titration procedure developed by Dutton[6] we have determined midpoint potentials of centres A and B. Fig. 2 shows potential titration curves for each of the components of the spectrum of these centres. The midpoint potentials are very low and the titrations shown

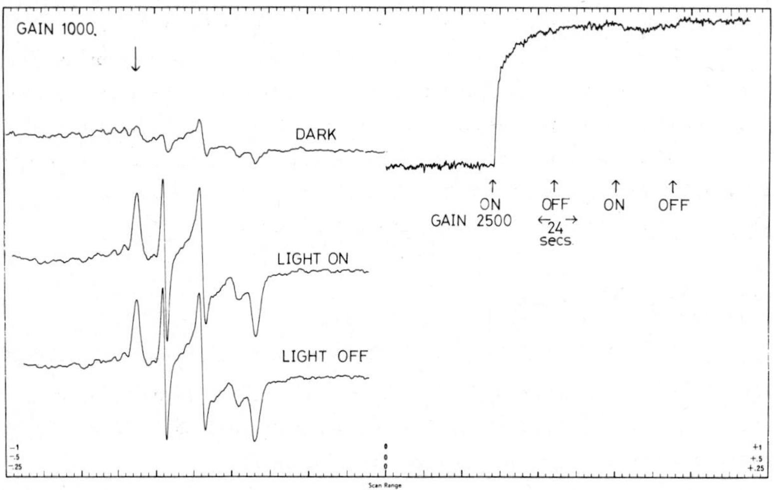

Fig. 1. The effect of illumination on the EPR spectrum (at 20°K) of PSI particles frozen in the dark. The time course of changes in the $g = 2.04$ signal and the effect of repeated light and dark intervals.

were done at pH 10.0. At this pH the potential of the hydrogen electrode is -592 mV and the over potential of the platinum electrode allows the measurement of potentials to about -640 mV. The curves fitted to the experimental data indicate that centre A has Em_{10} = -553 ± 20 mV and centre B has Em_{10} = -595 ± 20 mV. The titration curves fit the theoretical curve for a single electron donor. Ke et al[7] have carried out similar experiments. They obtained midpoint potentials for centre A (Em_{10} = -530 mV) and centre B (Em_{10} lower than -585 mV). Our results are in reasonably good agreement with this. However their interpretation of the titration curves was that the centres were two-electron acceptors, and that the g = 1.86 signal which appears with Em_{10} = -560 mV and then disappears again with Em_{10} = -600 mV was a third centre, requiring two electrons for the first reduction and two more electrons for the second stage reduction. These proposals imply that the centres are of unknown chemical composition, as all known iron-sulphur centres accept single electrons. It would also require eight

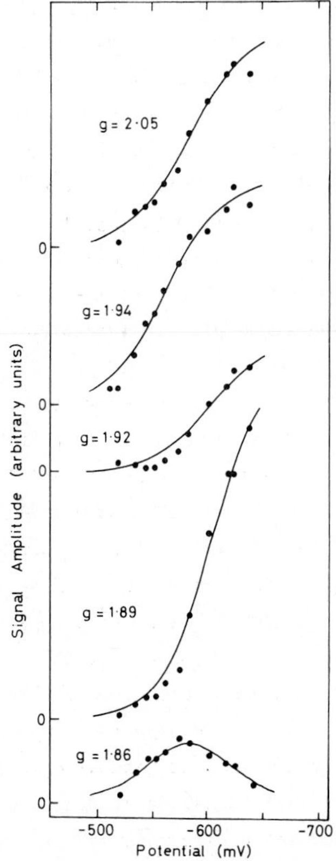

Fig. 2. Oxidation reduction potential titrations of the component of the iron-sulphur protein EPR signals in PSI particles. The curves shown are calculated assuming n = 1 for each of the centres.

electrons for complete reduction of the centres. We prefer the simpler interpretation that the centres are conventional iron-sulphur centres, and our results show that they are single electron acceptors. We propose an alternative hypothesis, that the disappearance of the g = 1.86 signal is not due to further reduction but to a change in lineshape resulting from interaction between the centres during reduction of centre B. A more detailed discussion of this problem is published elsewhere[8].

We have investigated the effect of illumination at low temperature on the EPR spectra of samples prepared under different conditions. Fig. 1 shows that in a sample prepared in the dark in the presence of a reducing agent, illumination at 20°K results in the formation of iron-sulphur protein signals chiefly corresponding to reduced centre A with a small contribution from reduced centre B. If the formation of one of the components (g = 2.05 in Fig. 1) is floowed it can be seen that on illumination the signal increases to a maximum, and if the light is turned off there is no decrease in signal size. Further illumination does not change the signal size. The reduction of the iron-sulphur centres is irreversible. On the other hand if the formation of the P700 signal under the same conditions is followed (Fig.3), it can be seen that a small part of the signal is reversible. If a sample is prepared so that the iron-sulphur centres are reduced before freezing (Fig. 4) it is found that all of the P700 change observed on illumination at 20°K is reversible. In the experiment shown in Fig. 4 the iron-sulphur centres were reduced by illumination at room temperature; under these conditions a large part of the P700 is already oxidized, and the reversible part of the P700 signal is about one-third of the "frozen-in" signal. Under these conditions, there is no change in the iron-sulphur signals on illumination.

Our interpretation of these results is that in the reaction centre complex at low temperature, an electron is transferred from the excited state of P700 to a primary acceptor X, by a reversible process. If centre A is in the oxidized state, it will tend to accept this electron from X in preference to centre B, but if A is already reduced, then B will accept the electron. These transfers are irreversible. It is possible that centre B is less accessible to X than centre A; but since B has a lower redox potential than A, it is also possible that there is free transfer of the electrons between A and B so that an equilibrium is set up. On our hypothesis, centres A and B are closely associated, since the reduction of centre B affects the EPR signal of centre A[8].

These results, in agreement with those of Bolton[4], indicate that iron-sulphur centres A and B are secondary rather than primary acceptors, which function even at these low temperatures as an electron sink preventing the reversal of the photooxidation of P700. Our experiments give no indication of the chemical

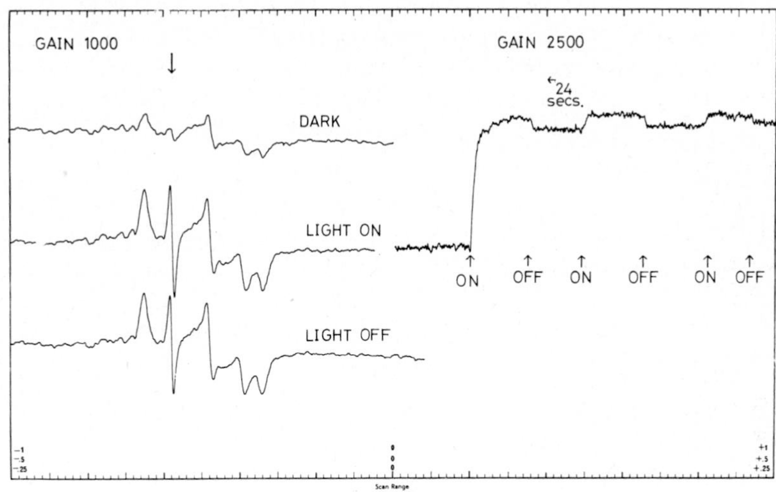

Fig. 3. The effect of illumination on the EPR spectrum of PSI particles frozen in the dark. Time course of changes in the g = 2.00 signal and the effect of repeated light and dark intervals.

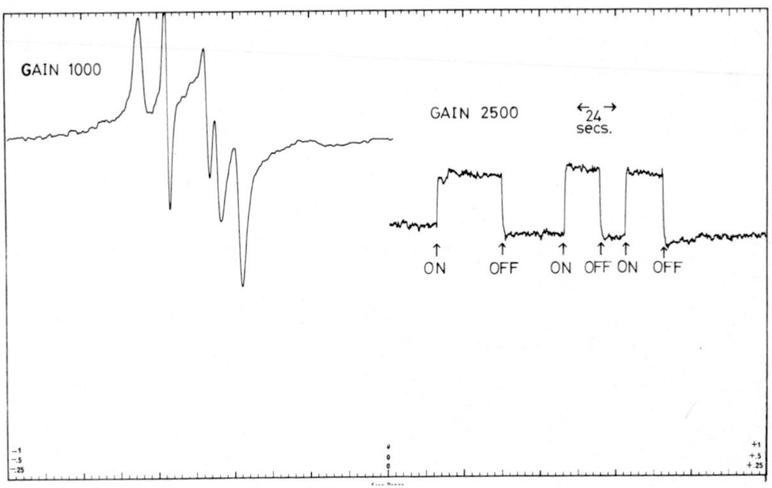

Fig. 4. The EPR spectrum (at 20°K) of PSI particles frozen in the light in the presence of $Na_2S_2O_4$. The time course shows changes in the g = 2.00 signal on illumination and in subsequent dark and light intervals.

nature of the primary acceptor X.

It now appears that there are at least five difference electron carriers between photosystem I chlorophyll P700 and the terminal acceptor, NADP. The reason for such a large number is not clear at present. It may be noted that the iron-sulphur centres have the useful ability to store electrons at potential below the hydrogen electrode, without evolving hydrogen.

References

1. Malkin, R. and Bearden, A.J., 1971, Proc. Natl. Acad. Sci., 68, 16-19.
2. Evans, M.C.W., Telfer, A. and Lord, A.V., 1972, Biochim. Biophys. Acta, 267, 530-37.
3. Bearden, A.J. and Malkin, R., 1972, Biochim. Biophys. Acta, 283, 456-468.
4. Bolton, J.R., 1974, Biochim. Biophys. Acta., In press, and proceedings of this meeting.
5. Evans, M.C.W., Lord, A.V. and Reeves, S.G., 1974, Biochem. J. 138, 177-183.
6. Dutton, P.L., 1971, Biochim. Biophys. Acta, 226, 63-80.
7. Ke, B., Hansen, R.E. and Beinert, H., 1973, Proc. Natl. Acad. Sci., 70, 2941-2945.
8. Evans, M.C.W., Reeves, S.G. and Cammack, R., 1974, Biochem. J. In press.

M. AVRON, *Proceedings of the Third International Congress on Photosynthesis*,
September 2-6, 1974, Weizmann Institute of Science, Rehovot, Israel
Elsevier Scientific Publishing Company, Amsterdam, The Netherlands, 1974

FLASH PHOTOLYSIS ELECTRON SPIN RESONANCE STUDIES OF THE LIGHT REVERSIBLE COUNTERPART TO SIGNAL 1 AT LOW TEMPERATURES IN SPINACH SUBCHLOROPLAST PARTICLES - THE TRUE PRIMARY ACCEPTOR OF PHOTOSYSTEM 1?

A. R. McIntosh, M. Chu and J. R. Bolton*

Photochemistry Unit, Department of Chemistry,
University of Western Ontario,
London, Ontario, Canada.

Abstract

The light induced electron spin resonance signals of Photosystem I spinach subchloroplast particles have been studied at ~6K. Using the technique of flash photolysis-electron spin resonance with actinic illumination at 647 nm, a kinetic analysis of the previously observed bound ferredoxin ESR signals was carried out. The kinetics of initial formation of these signals from dark incubated samples were studied as well as their possible light reversible behaviour in flashing light. Signal I ($P700^+$) exhibits a partial light reversible behaviour at 6K so it was expected that if the bound ferredoxin is the primary acceptor of Photosystem I, it should also exhibit a partial reversible behaviour. However, none of the bound ferredoxin ESR signals showed any such light reversible behaviour. Furthermore an analysis of the relative quantum yields of production of $P700^+$ vs. that of the ferredoxin signals shows that the yield of reduced ferredoxin is at least five times less than the yield of $P700^+$. This finding proves that the ferredoxin components are not the primary acceptor of Photosystem I.

A search to wider fields revealed two components which did exhibit the expected characteristics of the primary electron acceptor. A model is presented to account for the reversible and irreversible photochemical changes in Photosystem I. The possible identity of the primary acceptor responsible for these two new components is discussed in terms of the available information. The primary acceptor may be an iron-sulfur protein, but not of the type characteristic of the bound or water-soluble ferredoxins found so far in chloroplasts.

1. Introduction

Recently there has been a considerable debate as to the identity of the primary acceptor of Photosystem I (PSI) in green plant and algal photosynthesis. It is well known that a water-soluble ferredoxin is involved as an electron-transfer component on the reducing side of PSI[4,5]; however, since membrane fragments enriched in PSI which contain no soluble ferredoxin can still undergo the primary photochemistry of PSI, there must be one or more electron transfer components which precede the soluble ferredoxin.

* To whom correspondence should be sent.

In 1971, Malkin and Bearden[6] showed that spinach chloroplasts, when illuminated with red light at ~77K, exhibit a new electron spin resonance (ESR) signal at temperatures below ~20K. Since this new component is formed in parallel with the formation of Signal I ($P700^+$), they suggested that this new component is a ferredoxin which functions as the primary acceptor of PSI. This proposal was given further support by quantitative intensity comparisons and the fact that the production of this component is possible with far-red light (λ>700 nm)[7,8]. Yang and Blumberg[9] have also observed ferredoxin-like ESR signals at low temperatures and have made quantitative comparisons. Evans et al[13] detected two ferredoxin-like ESR signals, both of which could be photoproduced, and suggested that the two components might be two active centers of the same protein.

Recently, Ke[10,11] has suggested that an optical absorbing component P430 arises from the primary acceptor of PSI. On the basis of similarity of the light-minus-dark optical spectra he assigns this optical component to a bound ferredoxin. He has determined that the redox potential of P430 is -470 mV[10].

More recently Ke et al[12] have measured the intensities of the ferredoxin ESR components as a function of redox potential. They found three distinct components and claimed that each was produced by a two-electron reduction. However, this finding has been disputed by Evans et al[13] who claim that only two components are produced each by a one-electron reduction. One component has a redox potential (E_m' at pH10) of -563 mV with g factors of 2.05, 1.94 and 1.86, and the other component has a redox potential (E_m' at pH10) of -604 mV with g factors of 2.05, 1.92 and 1.89. We tend to believe that the results of Evans et al are more reasonable in view of the fact that only one electron is available in each reaction center in the photochemical reaction.

The primary acceptor of PSI should exhibit the same formation and decay kinetics as Signal I ($P700^+$) at low temperatures if we assume that the only decay process is direct return of the electron from the primary acceptor to $P700^+$. It has been known for some time that P700 exhibits a partial reversible photooxidation at temperatures ~77K[14]. We have determined that this reversible component of P700 has a decay half life of ~0.8s which is temperature independent from 5K to ~150K[3]. The primary acceptor should also exhibit this decay behaviour and thus we initiated a study of the kinetics of each of the ferredoxin ESR components of PSI using the flash photolysis electron spin resonance technique. The results of this study are presented in this paper.

2. Materials and Methods

Spinach subchloroplast particles were prepared according to the method of Vernon et al[1,15] with the detergent Triton X-100. The PSI particles were suspended in a 0.05 M phosphate buffer (pH7.7) with 0.5 M sucrose, 0.01 M NaCl and 50% glycerol*. The particles were stored in the dark at -20°C prior to use.

The actinic illumination was a laser beam operated at 647.1 nm interrupted with a programable electronic shutter as described earlier[3]. The ESR signals were accumulated on a computer of averaged transients[3]. Low temperatures were maintained with an Air Products Model LTD-3-110 "Helitran" liquid helium dewar and transfer system.

3. Results

The frozen-in ESR signal at 6K for the Triton subchloroplast particles is shown in fig. 1. Little or no ESR signal is seen for samples frozen in the dark prior to illumination. It appears that both components of Evans et al[13] are formed,as the g factors we obtain are in good agreement with theirs.

The formation kinetics of the ESR signals from dark incubated Triton subchloroplast samples were investigated. The $P700^+$ (Signal I) and bound ferredoxin line positions were individually monitored during a single four second flash for dark incubated samples, which had little or no dark ESR signals. Comparisons were made between different ESR lines derived from the same sample which was thawed in the dark to room temperature after each flash.

As can be seen from fig. 2, the rise kinetics of the P700 and the g = 1.94 ferredoxin line from one Triton Photosystem I sample are quite different at the beginning of their respective flashes. The initial rise kinetics of the P700 and ferredoxin ESR lines can be analyzed in a quantitative manner. From the results of Bearden and Malkin[7], we assume that the light induced ferredoxin components (g = 2.05, 1.94, 1.86) have a nearly one to one molar ratio relative to the light induced $P700^+$ signals. In order to compare the initial rise kinetics of $P700^+$ to those of the equimolar ferredoxin components, it is also necessary to take into consideration the peak heights of the respective peaks after saturating irradiation. Assuming no power saturation effects, the initial slopes of the kinetic rise curves for each ESR peak divided by the respective peak heights after saturating irradiation should give the relative quantum yields of formation for each ESR signal position. Such an analysis of the g = 2.05, 1.94, 1.92, 1.89 and 1.86 positions relative to the P700 position, revealed that the quantum yield of formation of P700 was greater by a factor of about five than that of the bound ferredoxin at 6K.

---

* The glycerol is added so that the preparation will form a glass at low temperatures. The glycerol does appear to have an effect on the behaviour of these particles at low temperatures as a higher proportion of the reversible component is found when glycerol is used. However, there is no change in the kinetics.

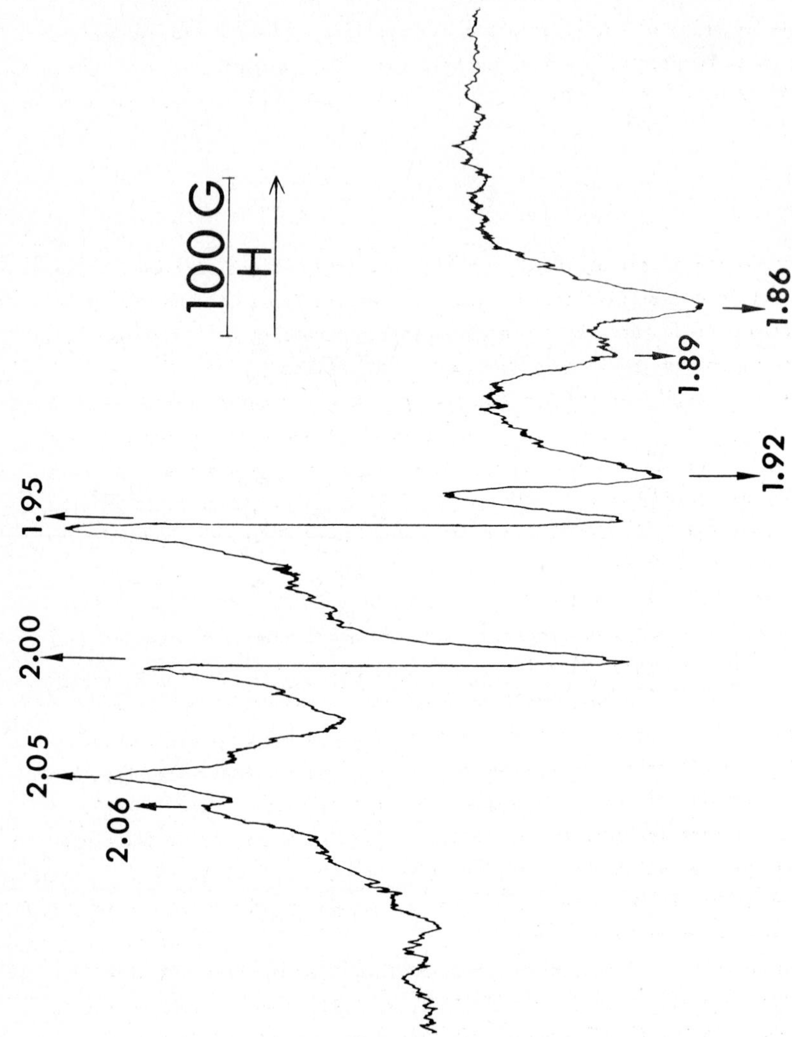

Fig. 1 ESR spectrum frozen in at 6K after irradiation at 647 nm of Triton PSI sub-chloroplast particles in 0.05 M phosphate buffer (pH = 7.7) with 0.5 M sucrose, 0.01 M NaCl and 50% glycerol.

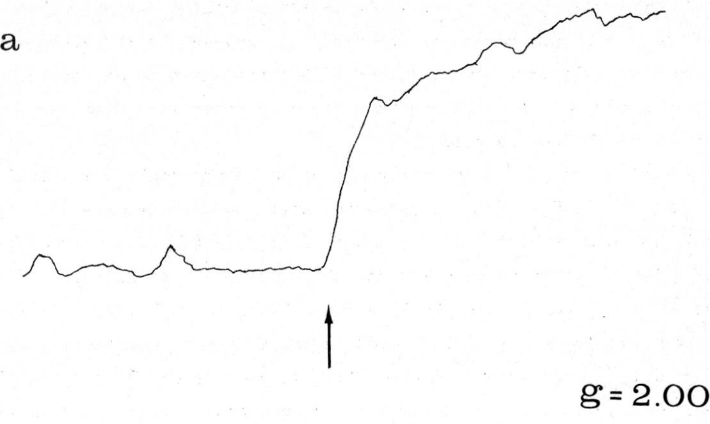

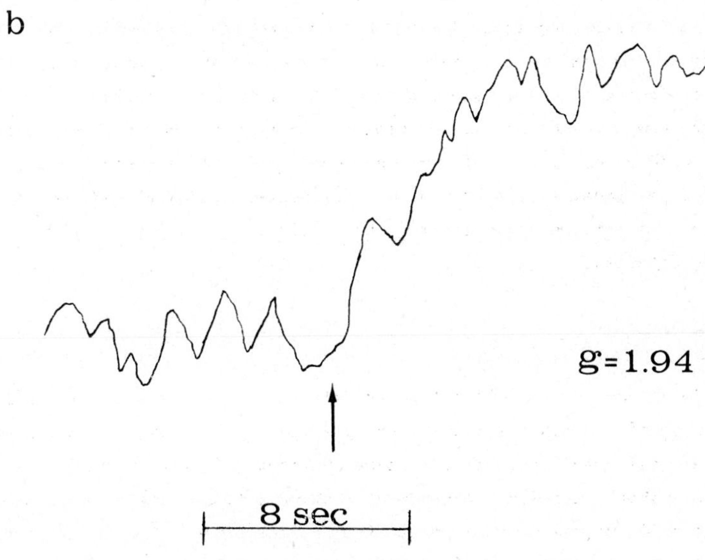

Fig. 2 Rise kinetics at 6K during one flash at 647 nm of Triton PSI subchloroplast particles in 0.05 M phosphate buffer (pH = 7.7) with 0.5 M sucrose, 0.01 M NaCl, and 50% glycerol.

(a) The high field peak position of Signal I at 0.1 mW power and 8G modulation amplitude.

(b) The low field peak position of the ferredoxin g = 1.94 component at 50 mW power and 20G modulation amplitude.

An intensive search was made for light reversible behaviour of each of the components of the two bound ferredoxin ESR signals. However, this search was totally unsuccessful. Our signal-to-noise ratio is such that we can, on the basis of these results, completely rule out the two ferredoxin components of Evans et al[13] as candidates for the primary acceptor of PSI.

We then began a search outside the range of most ferredoxin ESR signals. To higher field at g ~1.75 we found a resonance which did exhibit the same reversible characteristics as does Signal I. It can only be detected with very extreme ESR conditions (200 mW of microwave power and 25 G modulation amplitude). This component has a large linewidth (60-80 G) and an absorption-like shape in the negative mode. To lower field at g ~2.07 we found another component which also has a large linewidth (60-80 G) and an absorption-like shape but in the positive mode. It can be seen from fig. 3 that these two components exhibit the same kinetic behaviour as Signal I as is expected for the primary acceptor. Because of the very broad linewidths, the signal-to-noise ratio for these new components is rather poor and necessitates the accumulation of many hundreds of kinetic runs. We are certain that these new signals are not due to an artifact of heating by the light flash as no kinetic signal is seen on any of the ferredoxin peaks.

Since the new g = 2.07 and 1.75 components appear in reverse phase to each other, we would expect a derivative-like signal somewhere between these two components if this signal arises from an S = 1/2 state. We have searched for such a component but as yet, we have not found one. Either it is not there or it is too broad to be detected.

4. Discussion

It is clear that these new light reversible ESR signals are strong candidates for components of the ESR spectrum of the primary acceptor of PSI. As these g factors are considerably outside the range of known g factors for ferredoxins, it is not likely that the primary acceptor (which we will call X) is a normal ferredoxin although it very well could be an iron-sulfur protein[20]. For instance, a 4-methoxybenzoate o-demethylase from pseudomonas putida has g values[21] of g = 2.01, 1.91 and 1.78.

On the basis of our results we propose the following model to explain the fact that both reversible and irreversible photochemical processes occur at low temperatures:

We assume that some of the reaction centers are damaged, either in sample preparation or in freezing, such that the electron from P700 can reach only the primary acceptor X.

$$\text{P700} - \text{X} \not\rightarrow \text{Y}_1 - \text{Y}_2 \xrightarrow{h\nu} \text{P700}^+ - \text{X}^- \not\rightarrow \text{Y}_1 - \text{Y}_2$$

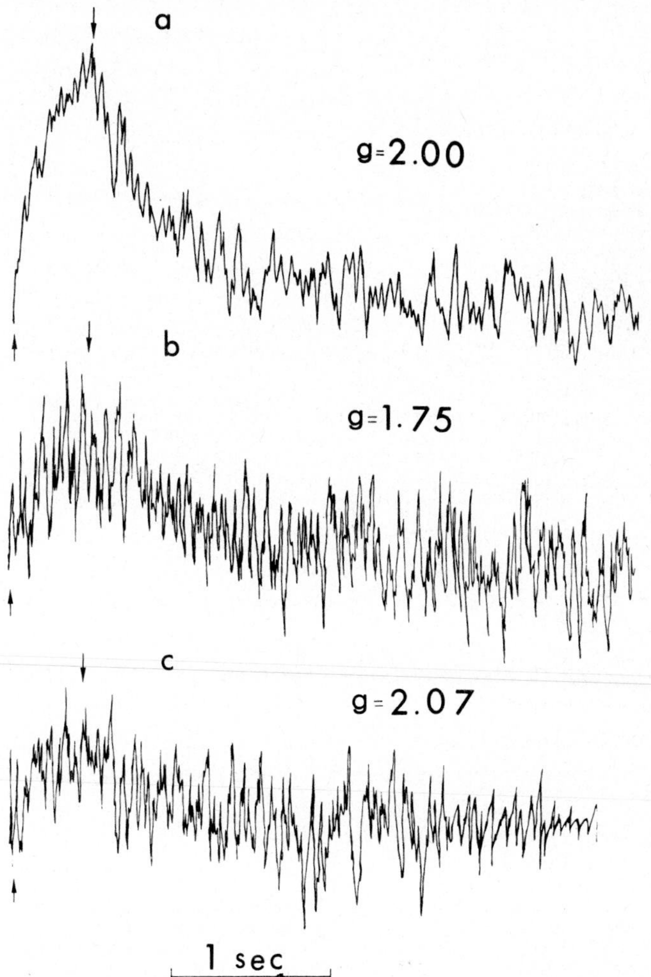

Fig. 3 Time course of the formation and decay of ESR signals at 6K during flash photolysis at 647 nm of Triton PSI subchloroplast particles in 0.05 M phosphate buffer (pH = 7.7) with 0.5 M sucrose, 0.01 M NaCl, and 50% glycerol.

(a) The high field peak position of Signal I at 1 mW power and 5G modulation amplitude.

(b) The g = 2.07 field position at 200 mW power and 25G modulation amplitude. The display is reversed in phase from the actual signal.

(c) The g = 2.07 field position at 200 mW power and 25G modulation amplitude.

The electron X^- may now return to $P700^+$ via a quantum mechanical tunneling process[3]. Support for this is found in the fact that the reversible portion of Signal I varies from 10-50% depending on the sample or species from which the sample was prepared[3].

For other reaction centers, the system is not damaged as much; and the electron can reach the secondary component Y_1.

$$P700 - X - Y_1 \nrightarrow Y_2 \xrightarrow{h\nu} \begin{array}{c} P700^+ - X^- - Y_1 \nrightarrow Y_2 \\ \downarrow \\ P700^+ - X - Y_1^- \nrightarrow Y_2 \end{array}$$

This process is irreversible once it occurs and accounts for the observation of one of the ferredoxin components* of Evans et al[13].

For other reaction centers, the system is still less damaged such that the electron can reach component Y_2.

$$P700 - X - Y_1 - Y_2 \xrightarrow{h\nu} \begin{array}{c} P700^+ - X^- - Y_1 - Y_2 \\ \downarrow \\ P700^+ - X - Y_1^- - Y_2 \\ \downarrow \\ P700^+ - X - Y_1 - Y_2^- \end{array}$$

This would account for the irreversible formation of the other ferredoxin component of Evans et al[13].

An alternative explanation for the irreversible formation of components Y_1 and Y_2 is a parallel scheme in which either Y_1 or Y_2 can be reduced by X^-:

$$P700 - X\begin{matrix} \nearrow Y_1 \\ \searrow Y_2 \end{matrix} \xrightarrow{h\nu} P700^+ - X^-\begin{matrix} \nearrow Y_1 \\ \searrow Y_2 \end{matrix}$$
$$\swarrow \qquad\qquad \searrow$$
$$P700^+ - X\begin{matrix} \nearrow Y_1^- \\ \searrow Y_2 \end{matrix} \quad \text{or} \quad P700^+ - X\begin{matrix} \nearrow Y_1 \\ \searrow Y_2^- \end{matrix}$$

Lozier and Butler[16] have determined the redox potential of -0.53V for the disappearance of the photo-induced $P700^+$ ESR signal. They claimed this to be the redox potential of the primary acceptor but, of course, if the electron can return from the primary acceptor at ~77K then they were not determining the redox potential of X but of Y_1 or Y_2. Evans et al[13] have measured the redox potentials of Y_1 and Y_2 and find -0.60V for Y_1 and -0.56 for Y_2. Clearly the redox potential of X must be more negative than that of either Y_1 or Y_2.

---

* Component Y_1 is probably the one with the more negative redox potential and therefore is to be compared with component B of Evans et al[13].

An enticing possibility for the identity of X would be a reaction-center protein similar to the one found in photosynthetic bacteria[17]. Dutton and Leigh[18,19] have described a 'photoredoxin' component in this protein which exhibits all the characteristics of a primary acceptor for bacterial photosynthesis. This 'photoredoxin' species has light reversible components at g = 1.82 and g = 1.68. The 1.68 component is similar to our 1.75 component but as yet, we have found nothing around g = 1.82 in our PSI particles.

5. Conclusions

It would appear from our results that the ferredoxin components, which have been proposed as the primary acceptor of PSI, are in fact secondary acceptors and a new ESR signal with g factors of 1.75 and 2.07 should be assigned to the primary acceptor X. Although X may be an iron-sulfur protein, it does not have g factors in the range of normal ferredoxins. Further experiments are in progress, such as a potentiometric titration to characterize the nature of the unknown primary acceptor we have observed, and will be reported in later communications.

References

1. Warden, J. T. and Bolton, J. R., 1974, Photochem. Photobiol. 20, 251-262.
2. Warden, J. T. and Bolton, J. R., 1974, Photochem. Photobiol. 20, 263-270.
3. Warden, J. T., et al, 1974, Biochem. Biophys. Research Commun. 59, 872-878.
4. Arnon, D. I., 1965, Science 149, 1460-1470.
5. Vernon, L. P. and Ke, B., 1966, in The Chlorophylls, eds. Vernon, L. P. and Seely, G. R. (Academic Press, New York) pp 569-609.
6. Malkin, R. and Bearden, A. J., 1971, Proc. Natl. Acad. Sci. U.S. 68, 16-19.
7. Bearden, A. J. and Malkin, R., 1972, Biochem. Biophys. Res. Commun. 46, 1299-1305; 1972, Biochem. Biophys. Acta 283, 456-468.
8. Evans, M.C.W. et al, 1972, Biochem. Biophys. Acta 267, 530-537.
9. Yang, C. S. and Blumberg, W. E., 1972, Biochem. Biophys. Res. Commun. 46, 422-428.
10. Ke, B., 1972, Arch. Biochem. Biophys. 152, 70-77.
11. Ke, B., 1973, Biochem. Biophys. Acta 301, 1-33.
12. Ke, B. et al, 1973, Proc. Natl. Acad. Sci. U.S. 70, 2941-2945.
13. Evans, M.C.W. et al, Biochem. Biophys. Acta. (In press)
14. Mayne, B. C. and Rubinstein, D., 1966, Nature 210, 734-735.

15. Vernon, L. P. et al, 1966, J. Biol. Chem. 241, 4101-4109.

16. Lozier, R. H. and Butler, W. L., 1974, Biochem. Biophys. Acta 333, 460-464

17. Feher, G., 1971, Photochem. Photobiol. 14, 373-387.

18. Dutton, P. L. et al, 1973, Febs Letters 36, 169-173.

19. Dutton, P. L. et al, 1973, Biochem. Biophys. Acta 292, 654-664.

20. Blumberg, W. E. and Peisach, J., 1974, Archives of Biochem. and Biophys. 162, 502-512.

21. Bernhardt, F. H. et al, 1971, Hoppe-Seyler's Zeit. Physiol. Chem. 352, 1091-1099.

M. AVRON, *Proceedings of the Third International Congress on Photosynthesis*,
September 2-6, 1974, Weizmann Institute of Science, Rehovot, Israel
Elsevier Scientific Publishing Company, Amsterdam, The Netherlands, 1974

REVERSIBILITY OF LIGHT-INDUCED REACTIONS OF PHOTOSYSTEM 1 AND 2 AT TEMPERATURES BELOW 200 °K

J.W.M. Visser and C.P. Rijgersberg

Biophysical Laboratory of the State University, P.O. Box 2120, Leiden,
The Netherlands

1. Summary

The reactions in the photosynthetic systems 1 and 2 at low temperatures were studied with spinach chloroplasts by use of electron paramagnetic resonance (EPR) measurements.

The photooxidation of the primary electron donor of photosystem 1, P700, was found to be partly reversible at temperatures between 10 and 200 °K. The photoreduction of a ferredoxin molecule with EPR signals at $g = 2.05$, 1.94 and 1.86 was also partly reversible, and the kinetics were identical to those of P700. These results support the hypothesis of Malkin et al. (1971, Proc. Natl. Acad. Sci. U.S. 68, 16-19) that this ferredoxin is the primary electron acceptor of photosystem 1.

In the presence of ferricyanide the photooxidation of P680, the primary electron donor of photosystem 2, was observed to be partly reversible after 0.5 s of saturating illumination at 110 °K. During prolonged irradiation P^+680 was reduced by a chlorophyll molecule in a secondary reaction stabilizing the charge separation under these conditions.

2. Introduction

The available data concerning the reversibility of the photoconversion of P700 and of ferredoxin at low temperatures are contradictory. Some authors[1-4] have observed a partial reversibility of the photoconversion of P700, others[5-9] have reported a completely or nearly completely irreversible photooxidation at temperatures near 80 °K. If P700 is photooxidized reversibly, one would expect a reversible photoreduction of the primary electron acceptor. However, bound ferredoxin, which has been proposed to be the primary electron acceptor of photosystem 1, has been reported to be photoreduced practically irreversibly at 77 °K (ref. 10).

Also with respect to a backreaction in the reaction center of photosystem 2 conflicting models have been proposed. Butler et al.[11] and Murata et al.[12] suggested models involving a backreaction between the primary reactants with a halftime of 4.5 ms at 80 °K. In these models a secondary donor (*e.g.* cytochrome $\underline{b}_{559}$)may also reduce P^+680 to stabilize the charge separation. Therefore, one would expect that it is possible to observe the backreaction if no secondary donor is available, *e.g.* if the samples are frozen in the presence of an oxidant. However, Malkin and coworkers[13,14] and Ke et al.[15] have observed a light-induced signal near g = 2.00 in the presence of ferricyanide, which was irreversible at temperatures below 170 °K. They suggested that this signal was due to irreversibly photooxidized P680.

In the present paper we report, firstly, optical and EPR measurements of the kinetics of the photoconversion of P700 and ferredoxin at temperatures between 10 and 200 °K. Secondly, we report measurements of the kinetics of the light-induced EPR signal near g = 2.00 in the presence of ferricyanide at 110 °K.

3. Materials and methods

Chloroplasts were prepared from market spinach leaves and suspended in a medium as described elsewhere[16]. For optical measurements samples were suspended in glycerol (final concentration 55 % v/v) shortly before freezing.

Absorbance changes were measured with a splitbeam spectrophotometer as described in ref. 17, equipped with a sample holder that could be cooled to any desired temperature above 80 °K. Appropriate combinations of a Schott RG cut-off and a Balzers B40 interference filter (bandwidth 12 nm) were placed in front of the photomultiplier in order to diminish artifacts due to changes in the fluorescence yield of pigment system 2. For the same purpose absorbance measurements were performed in the presence of dibromothymoquinone (DBMIB), which quenches fluorescence[18].

Fluorescence yield changes were measured using an apparatus described previously[11].

EPR measurements were recorded using a Varian E-9 spectrometer, operating near 9.1 GHz, that was calibrated as described in ref. 16. Samples were illuminated through the slotted front side of the cavity by an Aldis slide projector (500 W); the intensity at the sample position was about 100 $mW.cm^{-2}$, except where indicated otherwise.

4. Results and discussion

4.1 Photosystem 1 reactions

At the first illumination period at 110 °K with spinach chloroplasts which were frozen in the dark, 60 - 70 % of the absorbance change at 703 nm was found to be reversible. At subsequent illumination periods the extent of the change was equal to that of the reversible part at the first illumination period. Fig. 1 shows the spectrum of these reversible changes between 660 and 720 nm. The spectrum closely resembles that of the photooxidation of P700 (cf. refs. 1,4). At several wavelengths we also measured the changes at the first illumination period with spinach chloroplasts which were illuminated in the presence of 10^{-5} M DCMU and 10^{-2} M hydroxylamine shortly before freezing (fig. 1). By this treatment photoreactions of photosystem 2 are blocked, due to accumulation of the reduced electron acceptor[19].

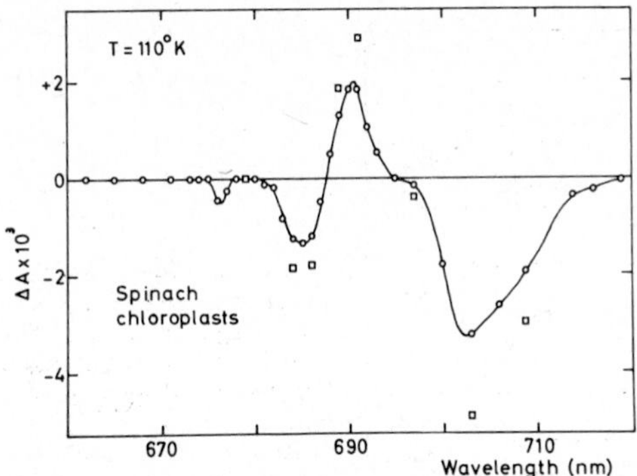

Fig. 1. Absorbance difference spectra at 110 °K with spinach chloroplasts (0.22 mg chlorophyll/ml in 1-mm cuvette), which were frozen in the dark in the presence of DBMIB (50 μM). □ , changes upon first illumination period with samples illuminated for 5 s shortly before freezing in the presence of DCMU (10^{-5} M) and NH_2OH (10^{-2} M). 0, changes at subsequent illumination periods (1 min dark, 5 s light). Actinic illumination: 420 - 500 nm, 10 nEinstein.$cm^{-2}s^{-1}$.

The spectrum obtained is similar to that of the reversible changes. The kinetics at 703 nm were the same with and without hydroxylamine and DCMU. This indicates that both the reversible and irreversible changes are due to photooxidation of P700.

The most obvious way to explain the partial irreversibility of P700 oxidation would be by means of a dark reaction that stabilized the charge separation (cf. ref. 4). For the reaction center of photosystem 2 it has been suggested[11,12] that the charge separation is irreversible at low temperatures because of secondary electron donation to P^+680. Similarly, irreversibility of the photooxidation of P700 at 110 °K may be explained by secondary electron acceptors of photosystem 1. Interestingly, it has been reported[10] that a bound ferredoxin molecule with EPR signals at g = 2.05, 1.94 and 1.86 is photoreduced irreversibly at 77 °K and at lower temperatures. Therefore, this ferredoxin possibly functions as secondary electron acceptor at low temperatures in the reaction centers with irreversibly photooxidized P700. Similar models have been discussed by Yang et al.[3] and by Lozier et al.[4]

We studied this hypothesis by measuring the kinetics of the EPR signals of P^+700 and reduced ferredoxin with the magnetic field fixed at the position of the lower field lines of the signals at g = 2.00 and 1.94, respectively. Fig. 2 shows these kinetics at various temperatures for P700 and at 11 °K for ferredoxin. The kinetics of the reactions of ferredoxin could only be observed at temperatures below 40 °K for reasons of EPR techniques (cf. ref. 20). It is clear from fig. 2 that the fraction of reversibly photooxidized P700 is temperature dependent. More detailed experiments indicated that between 200 and 170 °K this fraction was 90 %, between 170 and 40 °K the fraction decreased gradually with temperature from 90 until 20 %, and between 40 and 10 °K it remained 20 %. The bottom traces of fig. 2 show that

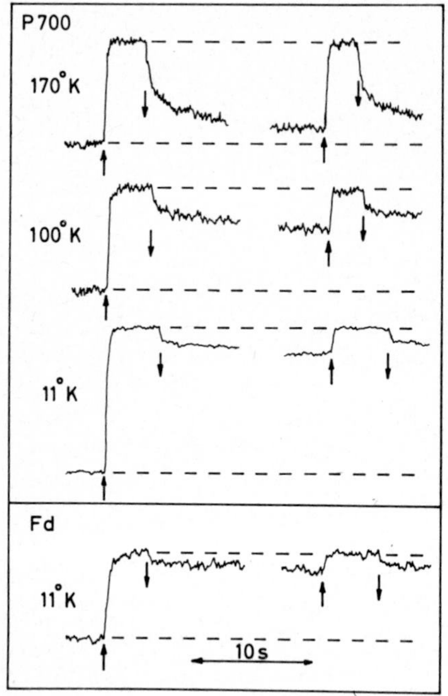

Fig. 2. Time courses of light-induced changes of the low field maxima of the EPR signals at g = 2.00 (P700) and 1.94 (Fd) at the temperatures indicated, with spinach chloroplasts frozen in the dark. White light on at upward, off at downward pointing arrows. Left hand curves: changes induced by the first illumination; right hand curves: changes after several 5 s light - 1 min dark cycles. Instrument settings: power: 2 mW (P700) and 10 mW (Fd); modulation amplitude: 5 G (P700) and 10 G (Fd); frequency: 9.080 GHz.

also 20 % of the available ferredoxin is photoreduced reversibly at 11 °K.

Furthermore, as can be seen in fig. 3, the light-on kinetics of the photoconversion of P700 and ferredoxin are also the same with spinach chloroplasts, which were frozen in the dark and then illuminated at 11 °K with weak (2 $mW.cm^{-2}$) and subsequently with strong actinic illumination (100 $mW.cm^{-2}$).

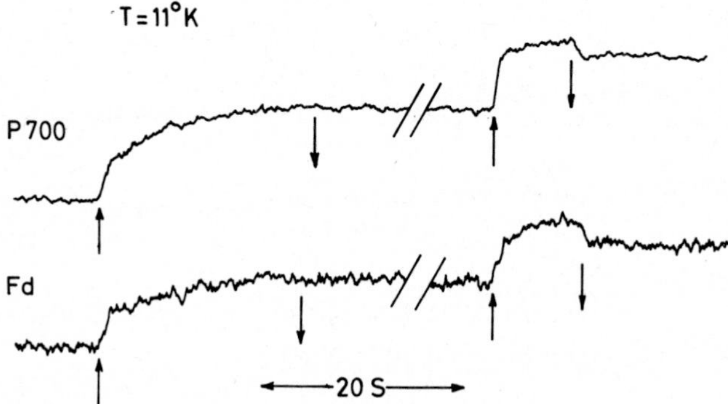

Fig. 3. Time courses at 11 °K of the light-induced changes of P700 and ferredoxin measured as in fig. 2. Left hand curves: changes at the first illumination period (I = 2 mW/cm^2). Right hand curves: changes at the second illumination period (I = 100 mW/cm^2) after a dark period of 1 min.

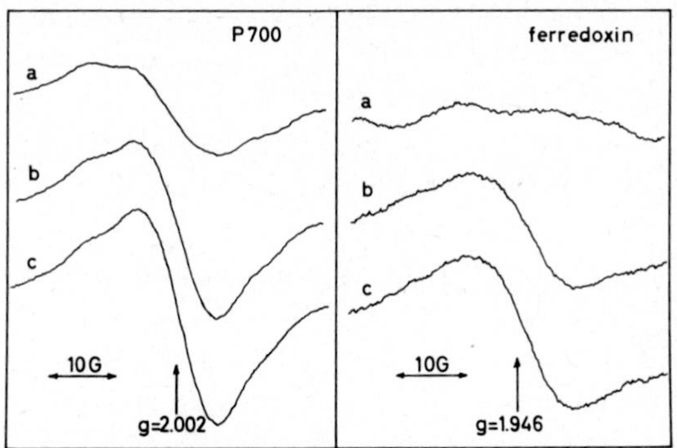

Fig. 4. EPR spectra at 11 ^{o}K of spinach chloroplasts (2 mg chlorophyll/ml) near g = 2.00 and 1.94. a, sample frozen in the dark; b, sample frozen in the dark, illuminated for about 30 s with white light at 77 ^{o}K and subsequently kept in the dark for 3 min at 77 ^{o}K; c, sample b during subsequent illumination with white light at 11 ^{o}K. Instrument settings as in fig. 2 at 11 ^{o}K. The right hand spectra (ferredoxin) were recorded with four times increased sensitivity.

Fig. 4 shows EPR spectra measured at 11 ^{o}K of the signals of $P^{+}700$ and reduced ferredoxin (near g = 1.94) after illumination at 77 ^{o}K (b) and during illumination at 11 ^{o}K (c) with the same sample. Spectra a are recorded with samples, which were frozen in the dark. If we take the signal height in spectra c (minus a) to indicate 100 % photoconversion, it is clear from fig. 4 that at 77 ^{o}K about 70 % of the total amount of P700 is irreversibly photooxidized, in agreement with the optical measurements and with experiments as in fig. 2. Furthermore, fig. 4 indicates that at 77 ^{o}K also about 70 % of the ferredoxin is photoreduced. Spectra b were obtained with a sample that was kept in the dark for three minutes at 77 ^{o}K after illumination. More P700 and ferredoxin were found to be converted irreversibly at 11 ^{o}K if the samples were cooled immediately after illumination at 77 ^{o}K. With different darktimes after illumination at 77 ^{o}K the relative amounts of $P^{+}700$ and reduced ferredoxin were always equal in the same sample. This is in agreement with the occurrence of a slow phase at the decay in the dark at 77 ^{o}K of both $P^{+}700$ (see also fig. 2) and of reduced ferredoxin.

Conclusively, there was always found a stoichiometric relation between the amounts of $P^{+}700$ and reduced ferredoxin, both in the light and in the dark. This indicates that ferredoxin is the only electron acceptor and that secondary reactions do not occur at temperatures below 150 ^{o}K. Apparently, in some reaction centers $P^{+}700$ and reduced ferredoxin are able to react back with each other, in others they are not. Measurements of the kinetics of the absorbance change at 703 nm and of the EPR signal at g = 2.00 indicated that the rate of the backreaction is independent of the duration of the preceding illumination and independent of the temperature between 10 and 150 ^{o}K (cf. fig. 2).

Especially in view of the latter observations, we have no ready explanation for the temperature dependence of the partial irreversibility of the photoconversion. One might speculate that this temperature dependence is due to the structure of ferredoxin. The ferredoxin molecule contains two (high spin) iron atoms at slightly inequivalent acceptor sites[20]. The iron atoms are antiferromagnetically coupled, causing a temperature dependence of the magnetic susceptibility[20], which may be related to the temperature dependence of the reversibility of the primary reaction. Since it is known[20] that the EPR signal of ferredoxin may reflect both iron atoms of the molecule, it might be postulated that a different iron atom is reduced in the reaction centers in which a backreaction between P^+700 and reduced ferredoxin occurs than in those that are photoconverted irreversibly.

4.2 Photosystem 2 reactions

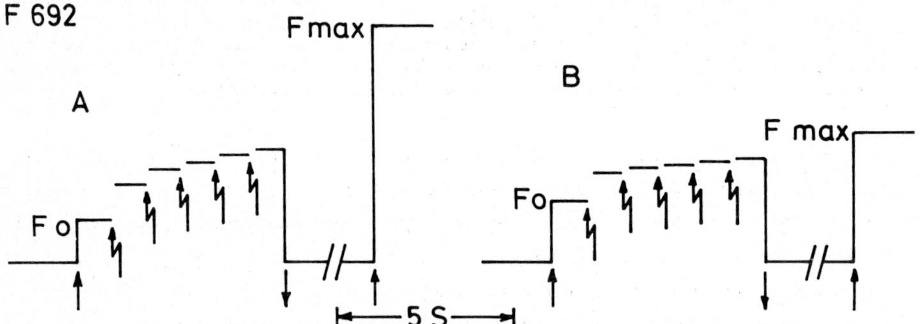

Fig. 5. Time course of the fluorescence yield (692 nm) with (B) and without (A) potassium ferricyanide (0.17 M) at 77 °K. The amplification of curve B is 2 times that of curve A. Non-actinic fluorescence measuring light (460 nm) on at upward, off at downward pointing arrows. A saturating green flash (8 µs; 500 - 600 nm) was given at every flash symbol. At the intersection, the sample was irradiated with blue light (420 - 500 nm, 70 $mW.cm^{-2}$) for 1 min. Chlorophyll concentration: 0.5 mg/ml.

It has been reported previously[11,12] that at 77 °K a short saturating flash caused an increase in the yield of the fluorescence of pigment system 2, which was much smaller than that caused by either continuous light at 77 °K or by a flash at room temperature. Models have been proposed[11,12] to explain this relative inefficiency of short flashes by a backreaction between the primary reactants P680 and Q. Fig. 5 (the amplification of curve B is 2 times that of curve A) shows that also in the presence of potassium ferricyanide short saturating flashes are relatively inefficient to induce an irreversible charge separation in photosystem 2 at 77 °K. The first flash, however, is much more efficient than the subsequent flashes to change the fluorescence yield. Therefore, in terms of models involving a backreaction between P^+680 and Q^-, one has to assume that also in the presence of ferricyanide secondary donors are available to reduce P^+680. Firstly, part of the reaction centers contain a donor which is converted with high efficiency at the first flash. This donor, apparently, is not a quencher of the fluorescence of pigment system 2. Secondly, comparison of A and B of fig. 5 indicates that the centers contain a donor, which is less efficient in reducing P^+680 than cytochrome $\underline{b}_{559}$, the secondary donor under conditions of fig. 5A (ref. 21). Furthermore, the maximal fluorescence yield was obtained only by strong continuous illumination: 15 saturating flashes induced an increase of the fluorescence yield, which was 70 % of that induced by continuous light. Similarly, Lozier et al.[4] have reported that about 30 % less C-550 was photoconverted irreversibly by ten saturating flashes than by continuous light. They suggested that this phenomenon may be explained by assuming that in about 30 % of the reaction centers the reaction between P^+680 and the secondary donor is slow compared with the backreaction.

Fig. 5 also shows that the fluorescence yield is decreased by the presence of ferricyanide both after flashes and after continuous light. In the presence of ferricyanide one of the ultimate products of the light-induced reactions of photosystem 2 at 77 °K is a quencher of the fluorescence of pigment system 2 (ref. 22).

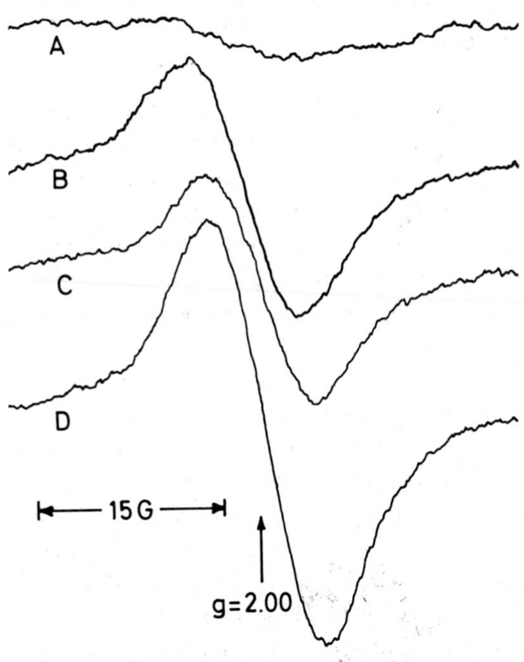

Fig. 6. EPR spectra at 110 °K with spinach chloroplasts (0.5 mg chlorophyll/ml) in the presence of glycerol (final concentration 55 % v/v). A, sample frozen in the dark; B, sample A during illumination; C, sample frozen in the dark in the presence of potassium ferricyanide (0.17 M); D, sample C after illumination for 7 s. Instrument settings as in fig. 2 for P700.

Fig. 6 shows EPR spectra near g = 2.00 with spinach chloroplasts, frozen in the dark and subsequently illuminated at 110 °K. Without ferricyanide (A and B) the spectra are similar to those at 11 °K (fig. 4) of P700. As discussed above, the signal height in spectrum B (minus A) indicates 100 % photooxidation of P700. Comparison of spectra B and C indicates that in the presence of ferricyanide P700 is completely oxidized in the dark. In agreement with this, EPR measurements at 15 °K

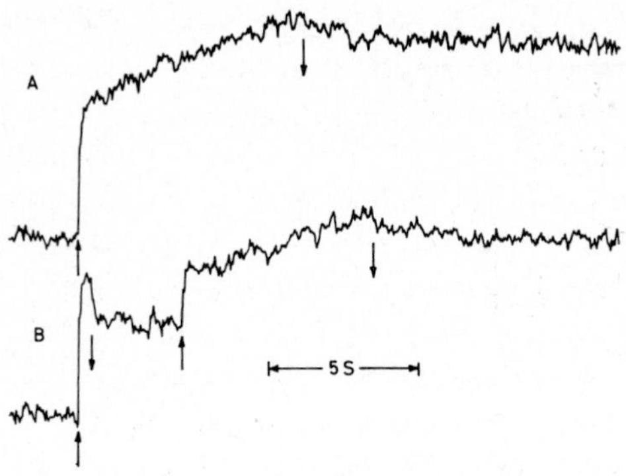

Fig. 7. Time course of the light-induced EPR signal at g = 2.00 in the presence of potassium ferricyanide (0.17 M) at 110 °K. Light on at upward, off at downward pointing arrows. Conditions as in fig. 6C.

indicated that in the presence of ferricyanide (0.17 M) photoreduction of ferredoxin did not occur (not shown). Spectra C and D show that at 110 °K in the presence of ferricyanide illumination causes also an increase in the extent of the signal near g = 2.00. The extent of the light-induced signal (spectrum D minus C) is about equal to that of the signal of $P^{+}700$. These results confirm the EPR measurements of Malkin and coworkers[13,14] and of Ke et al.[15] indicating an irreversible signal near g = 2.00 induced by illumination at low temperature in the presence of ferricyanide.

Fig. 7A shows the kinetics of the change of the EPR signal upon illumination at 110 °K in the presence of ferricyanide, measured with the magnetic field fixed at the maximum of the low field line. At the onset of irradiation a fast and a slow rise component can be distinguished. Only part (15 %) of the signal was found to be reversible after prolonged illumination (fig. 7A); the halftime of the decay was about three seconds. However, as can be seen in fig. 7B, about 30 % of the signal decayed rapidly ($\tau_{1/2} < 100$ ms) in the dark after a short (0.5 s) illumination period. If the sample was illuminated again after a dark period of three seconds (fig. 7B) the signal increased until the level obtained during the first, short illumination period, and subsequently more slowly as in fig. 7A. The decay in the dark after a light period of 6.5 s was also equal to that in fig. 7A.

The rapid decay of part of the signal after a short light period (fig. 7B) indicates that also at 110 °K a backreaction occurs at least in part of the reaction centers in the presence of ferricyanide. About 30 % of the signal decayed rapidly after illumination for periods between 400 and 700 ms. This indicates that the fraction of reversibly converted centers does not depend significantly on the duration of the illumination in this time region. Therefore, only 30 % of the reaction centers seem to be converted less efficiently to an irreversible state. Interestingly, the same conclusion was obtained from the fluorescence measurements (fig. 5).

In agreement with the assumption of a backreaction between P^+680 and Q^- the extent of the fast component of the light-on kinetics (fig. 7) was found to be strongly dependent on the intensity of illumination (not shown).

It is also clear from fig. 7B that after prolonged illumination the charge separation in the reaction centers is practically irreversible (cf. ref. 14). This suggests that also in the presence of ferricyanide P^+680 is reduced by a secondary electron donor at 110 ^oK. Furthermore fig. 7 indicates that the EPR signal at 110 ^oK of this secondary electron donor is identical to that of P^+680. Fig. 6 shows that the EPR signal is also similar to that of P^+700, which is known to be identical to that of an oxidized (dimeric) chlorophyll molecule[23]. Therefore, we suggest that in the presence of ferricyanide at 110 ^oK the secondary donor is a chlorophyll molecule.

Without ferricyanide the secondary electron donor at 77 ^oK is known[21] to be cytochrome $\underline{b}_{559}$. According to Murata et al.[11] the halftime of the reaction between cytochrome $\underline{b}_{559}$ and P^+680 is about 35 ms. The time response of our apparatus was too slow to permit accurate measurements of signals decaying with halftimes of less than 75 ms. At the onset of illumination at 110 ^oK, however, we could observe a spike, superimposed on the signal due to photooxidation of P700, and decaying during illumination with a halftime of less than 75 ms (not shown). This result is consistent with the hypothesis that at 110 ^oK cytochrome $\underline{b}_{559}$ is oxidized by the reaction center chlorophyll of photosystem 2.

As mentioned above the maximal fluorescence yield at 110 ^oK is decreased by the presence of ferricyanide. This may be explained from our measurements, assuming that the photooxidized chlorophyll molecules are quenchers of the fluorescence of pigment system 2.

Acknowledgements

We wish to thank Dr. J. Amesz for valuable discussions, Dr. B.F. van Gelder and Mr. R. Wever for helpful advice and for placing the EPR apparatus at the B.C.P. Jansen Institute, Amsterdam, at our disposal, and Mr. B. van Swol for technical assistance.

The investigation was supported in part by the Netherlands Foundation for Chemical Research (SON), financed by the Netherlands Organization for the Advancement of Pure Research (ZWO).

References

1. Vredenberg, W.J. and Duysens, L.N.M., 1965, Biochim. Biophys. Acta 94, 355.
2. Mayne, B.C. and Rubinstein, D., 1966, Nature 210, 734.
3. Yang, C.S. and Blumberg, W.E., 1972, Biochem. Biophys. Res. Comm. 46, 422.
4. Lozier, R.H. and Butler, W.L., 1974, Biochim. Biophys. Acta 333, 465.
5. Witt, H.T., Müller, A. and Rumberg, B., 1661, Nature 192, 967.
6. Androes, G.M., Singleton, M.F. and Calvin, M., 1962, Proc. Natl. Acad. Sci. U.S. 48, 1022.
7. Chance, B., Kihara, T., DeVault, D., Hildreth, W., Nishimura, M. and Hiyama, T., 1969, in: Progress in Photosynthesis Research, vol. III, ed. H. Metzner (H. Laupp Jr., Tübingen) p. 1321.
8. Cost, K., Bolton, J.R. and Frenkel, A.W., 1969, Photochem. Photobiol. 10, 251.
9. Witt, K., 1973, FEBS Letters 38, 112.
10. Malkin, R. and Bearden, A.J., 1971, Proc. Natl. Acad. Sci. U.S. 68, 16.
11. Butler, W.L., Visser, J.W.M. and Simons, H.L., 1973, Biochim. Biophys. Acta 325, 539.
12. Murata, N., Itoh, S. and Okada, M., 1973, Biochim. Biophys. Acta 325, 463.
13. Malkin, R. and Bearden, A.J., 1973, Proc. Natl. Acad. Sci. U.S. 70, 294.
14. Knaff, D.B. and Malkin, R., 1974, Biochim. Biophys. Acta 347, 395.
15. Ke, B., Sahu, S., Shaw, W. and Beinert, H., 1974, Biochim. Biophys. Acta 347, 36.
16. Visser, J.W.M., Amesz, J. and van Gelder, B.F., 1974, Biochim. Biophys. Acta 333, 279.
17. Amesz, J., Pulles, M.P.J., Visser, J.W.M. and Sibbing, F.A., 1972, Biochim. Biophys. Acta 275, 442.
18. Lozier, R.H. and Butler, W.L., 1972, FEBS Letters 26, 161.
19. Bennoun, P., 1970, Biochim. Biophys. Acta 216, 357.
20. Dunham, W.R., Palmer, G., Sands, R.H. and Bearden, A.J., 1971, Biochim. Biophys. Acta 253, 373.
21. Knaff, D.B. and Arnon, D.I., 1969, Proc. Natl. Acad. Sci. U.S. 63, 956.
22. Okayama, S. and Butler, W.L., 1972, Biochim. Biophys. Acta 267, 523.
23. Norris, J.R., Uphaus, R.A., Crespin, H.L. and Katz, J.J., 1971, Proc. Natl. Acad. Sci. U.S. 68, 625.

M. AVRON, *Proceedings of the Third International Congress on Photosynthesis*,
September 2-6, 1974, Weizmann Institute of Science, Rehovot, Israel
Elsevier Scientific Publishing Company, Amsterdam, The Netherlands, 1974

EXCITED STATES IN THE PHOTOCHEMISTRY OF BACTERIAL PHOTOSYNTHESIS

Roderick K. Clayton
Section of Genetics, Development and Physiology
Cornell University

and

William W. Parson
Department of Biochemistry
University of Washington

Reaction center (RC) preparations from Rhodopseudomonas spheroides strain R-26 show the following light-induced phenomena:

1. Optical absorbance changes[1] and an electron paramagnetic resonance (EPR) signal[2] indicating oxidation of the "P865" component of bacteriochlorophyll (BChl). The quantum efficiency for BChl oxidation is 1.02 ± 0.04 at room temperature[3]. At temperatures down to that of liquid helium the efficiency does not fall by a large factor[4]; it might be the same as at room temperature. The capacity for BChl oxidation is lost when the redox potential is brought below a mid-point of -0.05 v, presumably because the photochemical electron acceptor is reduced[5].

2. BChl fluorescence (maximum at 900 nm, corresponding to the 865 nm absorption band) with quantum yield 3.5×10^{-4} at room temperature in "open" RC's[6]. As the redox potential is lowered the fluorescence yield rises to about 10^{-3} in "closed" RC's[7]. Taking the intrinsic fluorescence lifetime to be 2×10^{-8} sec (from the area of the absorption band), the lifetime of the excited singlet state is $3.5 \times 10^{-4} \times 2 \times 10^{-8}$ or 7 psec in open RC's, and 20 psec in closed RC's. The fluorescence yield has not been measured at lower temperatures.

3. An EPR signal that suggests formation of a triplet state of BChl, observed below about 20°K in closed RC's but not in open RC's[8,9]. The signal decays with half-time about 5 μsec at 10°K; this decay time might reflect equilibration among spin states rather than quenching of the triplet. The quantum efficiency for formation of this EPR signal in closed RC's equals that for BChl oxidation in open RC's at the same temperature.

4. Two types of optical absorbance change (Δa) observed in closed RC's (at low redox potential), studied by ourselves in collaboration with R. Cogdell; unpublished. A "slow" change has the spectrum of triplet BChl[10] and a rise time less than 50 nsec at temperatures down to about 20°K. The decay half-time is

about 7 μsec at room temperature; it slows to about 100 μsec as the temperature is lowered to 100°K and remains so at temperatures down to about 20°K. The quantum efficiency for generation of this "slow" change is about 0.2 at room temperature. Judging from comparative measurements of signal amplitude versus exciting flash energy the efficiency rises, perhaps to a value near unity, as the temperature is lowered to about 20°K. Also at low redox potential, a "fast" Δa can be distinguished from the foregoing. The "fast" change has a rise time less than 5 nsec; the decay time is about 15 nsec (partly instrument-limited as measured) at room temperature and about 40 nsec at temperatures near 20°K. The spectrum of this change is similar but not identical to that of triplet BChl; the quantum efficiency for its formation appears to be about unity at both room and low temperatures.

We speculate that both the EPR signal and the "slow" Δa seen in RC's at low redox potential reflect a triplet-like state of BChl, and that the "fast" Δa indicates a precursor of this state. The lifetime of the "fast" Δa is far too great to correspond to the excited singlet state that gives rise to fluorescence. These metastable states may have the nature of charge-transfer states, with centers of positive and negative charge displaced relative to the neutral ground state of BChl. Observations thus far do not show whether these states, seen only in closed RC's, are intermediates in the photochemical process of open RC's.

From the high efficiency of BChl oxidation in open RC's at room temperature, we should expect a rise in fluorescence far greater than the threefold increase that is observed when the photochemistry is prevented by lowering the redox potential. This anomaly awaits clarification. If we assume that the singlet excited state of BChl is quenched by competing first-order processes, we can write, for open RC's,

$$-dN/dt = (k_f + k_d + k_p)N$$

where N is the number of molecules in the excited state, k_f is a rate constant for spontaneous radiative de-excitation (fluorescence), k_d is a constant for radiationless de-excitation, and k_p is a constant for the first step leading to BChl oxidation. In closed RC's the sum of k_d and k_p is replaced by a quenching constant k_q. We do not know to what extent the processes represented by k_d, k_p, and k_q involve the formation of triplet BChl, but we can recapitulate the following quantitative data, with values of k taken as reciprocals of characteristic lifetimes mentioned earlier, and quantum efficiencies denoted ϕ. The value of k_f is estimated from the area under the long wave absorption band.[6] Values of k_p, k_d, and k_q have been inferred from the efficiencies of fluorescence[6] and photochemistry[3].

Table 1

Rate constants and quantum efficiencies for light-induced processes in photochemical reaction centers from <u>Rps. spheroides</u>. Symbols are explained in the text.

	OPEN REACTION CENTERS	CLOSED REACTION CENTERS
Room temperature	$k_f = 5x10^7$ sec^{-1} $k_d << 10^{11}$ sec^{-1} $k_p = 1.4x10^{11}$ sec^{-1} ϕ_p (BChl oxid.) = 1.0	$k_f = 5x10^7$ sec^{-1} $k_q = 0.5x10^{11}$ sec^{-1} $k_p = 0$ $\phi_T = 0.2$ ("slow" Δa) ϕ ("fast" Δa) ≈ 1
Temperature $\lesssim 20^oK$	$k_f \approx 5x10^7$ sec^{-1} $\phi_p \approx 1$	$k_f \approx 5x10^7$ sec^{-1} $\phi_T \approx 1$ (triplet-like EPR signal and "slow" Δ a) ϕ ("fast" Δ a) ≈ 1

References

1. Clayton, R.K., 1962, Photochem. Photobiol. 1, 201.
2. McElroy, J.D., Feher, G., and Mauzerall, D.C., 1969, Biochim. Biophys. Acta 172, 180.
3. Wraight, C.A., and Clayton, R.K., 1974, Biochim. Biophys. Acta 333, 246.
4. Clayton, R.K., 1962, Photochem. Photobiol. 1, 305.
5. Dutton, P.L., Leigh, J.S., and Wraight, C.A., 1973, FEBS Letters 36, 169.
6. Zankel, K.L., Reed, D.W., and Clayton, R.K., 1968, Proc. Natl. Acad. Sci. U.S. 61, 1243.
7. Clayton, R.K., Fleming, H., and Szuts, E.Z., 1972, Biophys. J. 12, 46.
8. Wraight, C.A., Leigh, J.S., Dutton, P.L., and Clayton, R.K., 1974, Biochim. Biophys. Acta 333, 401.
9. Dutton, P.L., Leigh, J.S., and Seibert, M., 1972, Biochem. Biophys. Res. Commun. 46, 406.
10. Connolly, J.S., Gorman, D.S., and Seely, G.R., 1973, Ann. New York Acad. Sci. 206, 650.

M. AVRON, *Proceedings of the Third International Congress on Photosynthesis*, September 2-6, 1974, Weizmann Institute of Science, Rehovot, Israel
Elsevier Scientific Publishing Company, Amsterdam, The Netherlands, 1974

EPR DETECTABLE CONSTITUENTS IN WHOLE CELL RHODOSPIRILUM rubrum

G. A. Corker and S. A. Sharpe
IBM Thomas J. Watson Research Center
Yorktown Heights, New York 10598

1. Introduction

When dark-adapted, whole-cell Rhodospirillum rubrum is subjected to continuous illumination, one observes a complex kinetic response of the electron paramagnetic signal of the bacteriochlorophyll cation radical[1,2]. The simplest interpretation of the complex kinetics is that during the transition from dark-adapted state to a steady-state in the light, the reactions that control the flux of electrons into and out of the reaction center bacteriochlorophyll (P870) change because the components of the electron transport chain are oxidized and re-reduced or reduced and re-oxidized in dark reactions with significantly different rates. If this interpretation is correct, then the rates of these dark reactions vary over seven orders of magnitudes (ms to min lifetimes) and it should be possible, at least in principle, to observe by EPR the components involved in the dark reactions when controlled conditions of light and dark are used.

2. Methods

One liter cultures of R. rubrum were grown in modified Hunter's medium[3] or in this medium with $MnSO_4$ excluded. Following 24 hours of growth the bacteria from 5 liters were isolated and suspended in 10 to 20 ml of nutrient or buffer. The resulting suspension was equally distributed into 10 to 15 EPR tubes. The tubes were then subjected simultaneously to dark-light-dark periods which consisted of 45 min darkness, 45 min illumination and 45 min darkness. Tubes were removed periodically during this time and cooled to $77^{o}K$. The first tubes were cooled to $77^{o}K$ following the initial 45 minutes of darkness. These samples with predetermined light-dark histories at room temperature were then investigated by EPR at temperatures from $6^{o}K$ to $77^{o}K$.

3. Results

As shown in curves a of Figures 1 through 4 (also curve h of Fig. 2), eight distinct chemical species exhibit resonance absorption when 45 min dark adapted R. rubrum is examined at low temperatures by EPR. The approximate g-values of the signals are: (1)g=6.0, (2)g=4.3, (3)g=2.089, (4)g=1.86, (5)g=2.018, (6)g=2.044, (7)g=2.03,1.94 and 1.92, and (8)g=2.0,1.94 and 1.88. As also depicted in these four figures, illumination of dark-adapted bacteria causes changes in the signal amplitudes of some of the signals which are observed in the dark and also causes the appearance of other signals. When the bacteria are put into the dark following the illumination period, the amplitudes of some of the signals again undergo alterations. The signals which are observed in dark adapted samples and which are influenced by illumination of the bacteria at room temperature are: (1)g=6.0, (2)g=4.3, (3)g=2.018, (4)g=1.86, (5)g=2.0044 and (6)g=2.03, 1.94 and 1.92. The signals which appear during the illumination have g-values as follows: (1)g=9.4, 5.52 and 3.5,

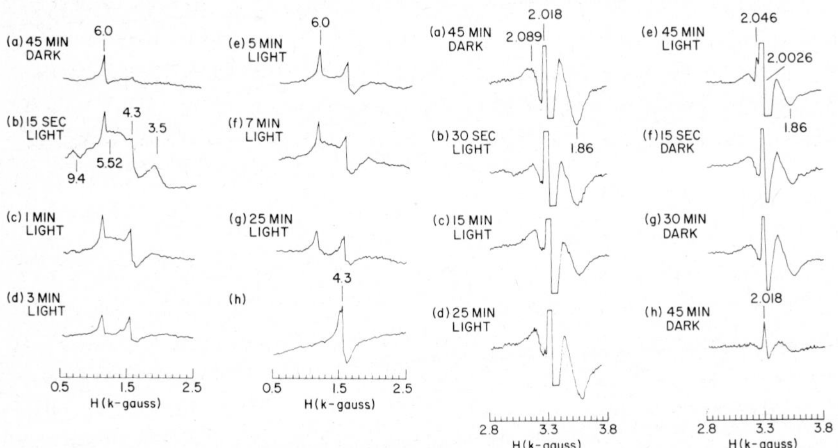

Fig. 1. EPR spectrum of dark-adapted whole-cell R. rubrum suspended in Mn free nutrient (a) and spectra showing changes in signals caused by room temperature illuminations for the times shown (b-g). Curve h shows spectrum of nutrient. Spectra obtained at 6°K.

Fig. 2. EPR spectrum of dark-adapted whole-cell R. rubrum suspended in phosphate buffer (a); the changes produced by illumination at room temperature for the times shown (b-e), and the changes which occur in the dark after illumination at room temperature (f-g). Spectra obtained at 6°K. The same sample was used to obtain curve h as used for curve a except with lower instrumental sensitivity.

(2)g=2.046 and 1.98 and (3)g=2.0026. We have grouped together in the above listings of g-values, those signals which always appear together and which probably are anisotropic g-values of the same chemical species. That the signal with a g-value of 2.0044 observed in dark-adapted bacteria is distinct from the bacteriochlorophyll signal is supported by the spectra shown in Figure 5 which were recorded at x-band (curves a through c) and at k-band (curves d and e)

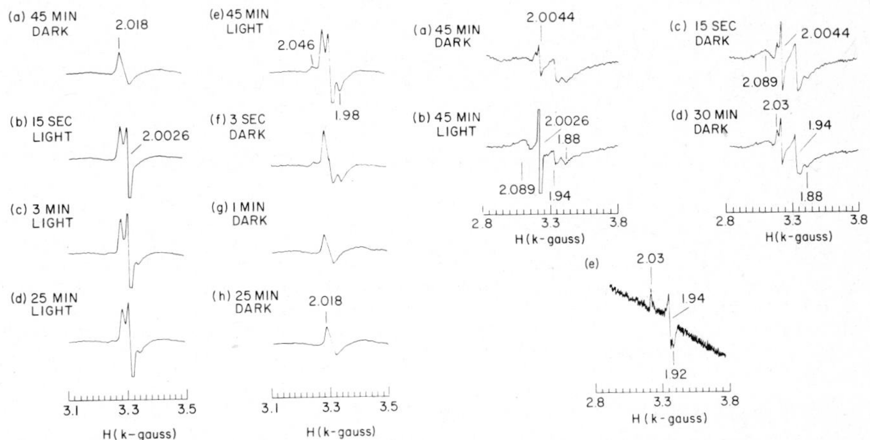

Fig. 3. EPR spectrum of dark-adapted whole-cell R. rubrum (a) and the changes which occur during illumination at room temperature (b-e) and during dark adaption after illumination (f-h). Spectra recorded at 6°K.

Fig. 4. EPR spectrum of dark-adapted whole-cell R. rubrum in phosphate buffer (a) and the changes produced in spectrum by illumination at room temperature (b) and by dark adaption after illumination (c-d). Spectra recorded at 38°K. Curve e was obtained at 77°K with 163mW of microwave power with a sample which was frozen, thawed and frozen, then illuminated for 10 minutes.

Because of their linewidths and field positions, most of the signals overlap one another. Thus, it is difficult to resolve them so as to obtain accurate plots of the variations in amplitudes with time of illumination. However, the variations in the amplitudes of some of the signals that are at least partially resolvable are shown in Figures 6 and 7. Figure 6 shows the relative signal amplitudes observed in bacterial suspensions that were illuminated for different lengths of time following 45 min of complete darkness. Figure 7 shows the amplitudes that are observed when the suspensions are placed in total darkness following 45 min illumination. As depicted in these figures, the amplitudes do not change monotonically but

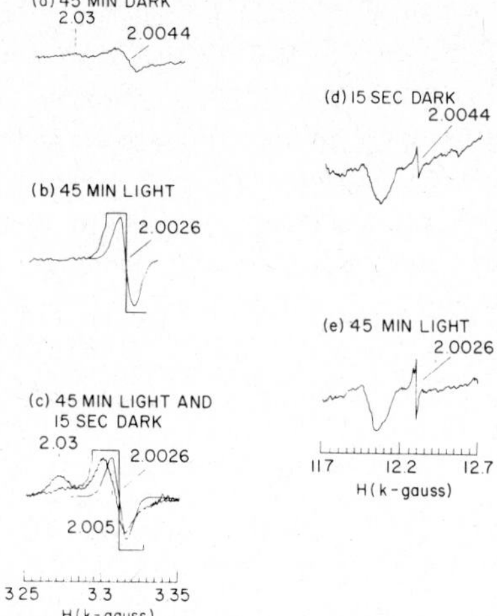

Fig. 5. Spectra recorded at x-band at 38°K showing a comparison of free radical signals observed in dark-adapted whole-cells (a) and in illuminated whole-cells (b). C contains superimposed spectra of the dark and the light signals. The spectra in curves d and e show K-band spectra of dark adapted and illuminated bacteria, respectively, recorded at 42°K.

exhibit complex time behavior. Included in Figures 6 and 7 are the photo-induced increases of the bacteriochlorophyll signal (curve b of Fig. 6 and curve a of Fig. 7) and of the g=2.0044 signal (lower curve in b of Fig. 7) when these suspensions were illuminated at 77°K.

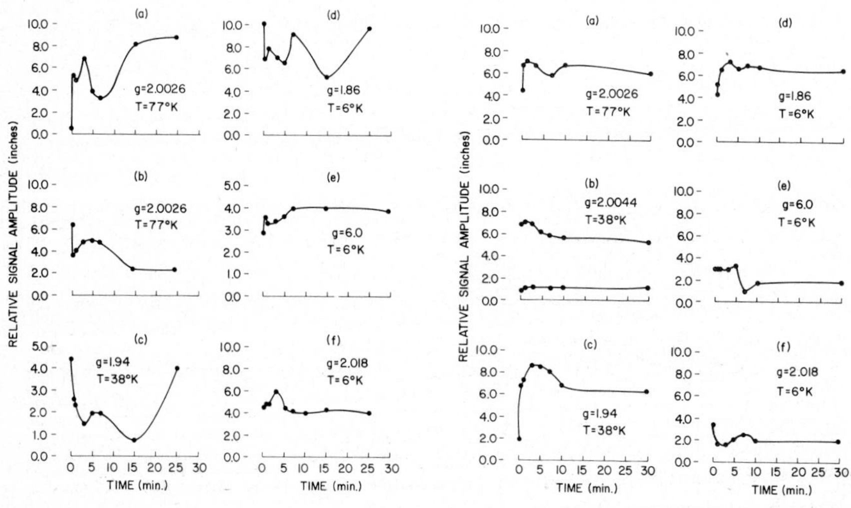

Fig. 6. The signal amplitudes of some EPR signals detected at the temperatures shown in whole-cell R. rubrum as a function of time of illumination at room temperature following 45 min in the dark. The curve in b shows the amount of $P870^+$ which can be photoproduced at 77°K in samples which had $P870^+$ "frozen in" to the heights shown in a.

Fig. 7. The signal amplitudes of some EPR signals detected in whole-cell R. rubrum as a function of dark time at room temperature following 45 min of darkness and 45 min of illumination. The plot in a represents the amount of P870+ which can be photoproduced in these samples at 77°K. The lower curve in b represents the non-reversible increase of the g=2.0044 signal which is produced by illumination of these samples at 77°K.

Although the amount of oxidized bacteriochlorophyll (as represented by the amplitude of the g=2.0026 signal) photoproduced at 77°K was reversibly reduced when the light was extinguished, the light-induced changes in the amplitudes of other signals at this temperature were not. Figure 8 shows non-reversible changes photo-induced at 77°K in the amplitudes of the g=2.0044, the g=6.0 and the g=1.86 signals. The amount of change of the g=6.0 signal is dependent upon the oxygen tension, greater change with decreasing oxygen tension.

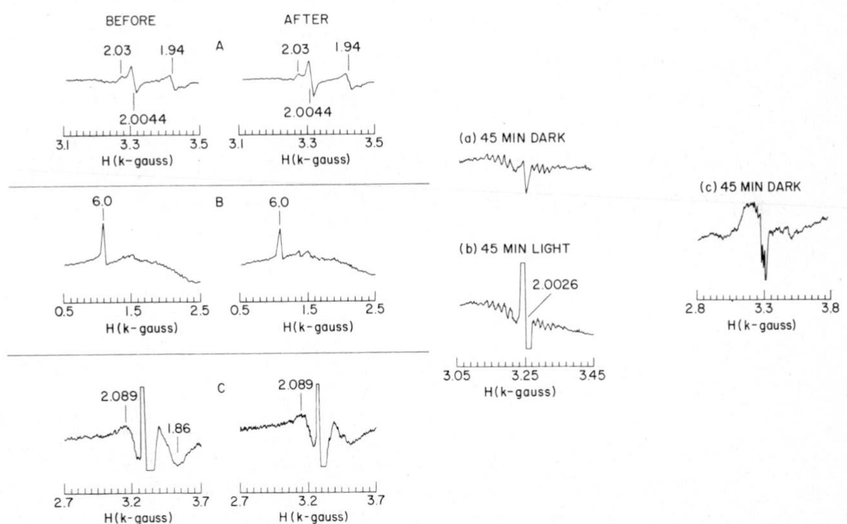

Fig. 8. Effects of illumination at 77°K upon amplitudes of the signals g=2.0044, g=6.0 and g=1.86 in whole-cell R. rubrum suspended in PO_4. Spectra in b and c recorded at 6°K; curve a at 38°K.

Fig. 9. EPR signals detected in R. rubrum chromatophores at 77°K (a and b) and at 38°K (e).

Most of the signals observed in whole cells are also detectable in chromatophores. The major differences between the observations with chromatophores compared to those with the whole cells are as follows: (1) the signals which resonate at g=2.03, 1.94 and 1.92 and at g=2.046 and 1.98 in the whole cells are not detected in chromatophores; (2) except for the lines at g=9.4, 5.52 and 3.5, the signals that are influenced by illumination of the chromatophores show either monophasic increases or decreases in time of illumination; (3) in whole cells, light causes an eventual decrease in the amplitude of the g=1.86 signal. In chromatophores, this signal increases in amplitude monotonically with time of illumination; (4) as shown in Figure 9, a multiline signal is detectable in chromatophores. This signal is not normally observed in whole cells.

The signals disappear at different times when whole-cell bacteria are heated. Figure 10 shows the variations in the amplitudes of some of the signals observed in bacterial suspensions whose temperatures were not controlled during the period of illumination. The temperatures in the EPR tubes rose almost exponentially from 22°C to 62°C with a halftime of approximately 2.5 min. The significant information contained in the figure is: (1) when the compound responsible for the g=1.86 signal is thermally destroyed, it is still possible to observe the signals with anisotropic g-values around g=1.94 observed at 38°K (Fig. 4); (2) when the g=6.0 and the g=1.86 signals are not detected, one cannot photo-induce the chlorophyll signal (g=2.0026) either at room temperature (curve a) or at 77°K (curve b).

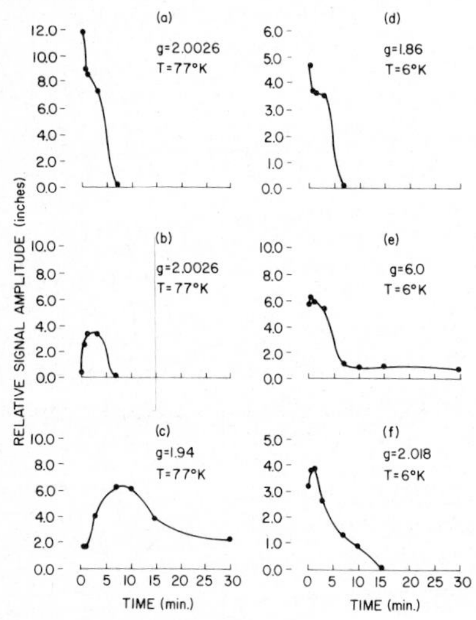

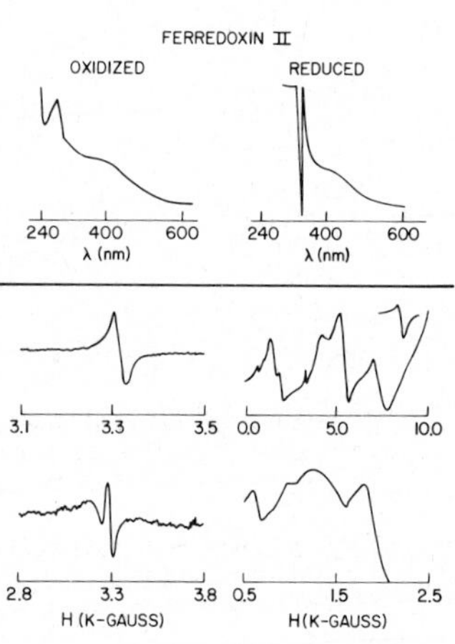

Fig. 10. The effects of heating and illumination upon the EPR signal amplitudes measured at the temperatures shown in the figure. The curve in b represents the amount of $P870^+$ which can be photo-produced at 77^oK in these samples. The temperature rises during illumination at room temperature from approximately 22^oC at time zero to about 60^oC at 7 min.

Fig. 11. Optical and EPR spectra of the oxidized and reduced forms of R. rubrum Ferredoxin II. EPR spectra recorded at 6^oK. The small signal inserted in spectra with 10 kilogauss sweep width was observed at 11.5 kilogauss with lower instrumental gain.

4. Discussion

The simplest interpretation of the data presented is that when dark adapted bacteria are illuminated, electron flow is initiated through dark redox reactions by the photochemical reaction involving the reaction center bacteriochlorophyll. The components involved in dark reactions that supply electron to $P870^+$ are oxidized; the constituents of the reactions that remove electrons from P870 are reduced; and, those involved in cyclic electron flow could be either oxidized or reduced. Thus, during the transition of the bacteria from the dark into the light, or vice versa, the amplitudes of the EPR signals change.

Although the data presented suggest that many of the EPR signals arise from components involved in the electron transport process in this bacterium, at present we cannot answer two significant questions raised by our results. That is, the identity of the compounds responsible for the signals and their relationship to one another and to P870 in the electron transport chain.

We are isolating components from R. rubrum in order to compare their EPR spectra with the spectra we observe with the whole cells. Shown in Figure 11 is the optical and the EPR spectra of both the oxidized and reduced form of an iron-sulfur protein which we have isolated according to a method of Shanmugan and Arnon[4], a protein they called Ferredoxin II. In the oxidized form, this protein exhibits a resonance with a low field peak at g=2.018 similar to the resonance we observe in whole cells with this g-value (see Figure 2, curve h and Fig. 3, curves a, g and h). In the reduced state, this protein exhibits a complex spectrum which extends from zero gauss to 12 kilogauss (see center curve on right of Figure 11). However, a portion of this spectrum resembles some of the signals we observe in the whole cells. Compare the low field portion of the spectrum shown in the lower curve to the right in Figure 11 to the signals shown in curve b of Figure 1. The signals in the whole cells with g-values of 9.4, 5.52 and 3.5 seem to arise from the reduced form of this protein.

The question of the relationship of these compounds in the transport chain, will be difficult to answer. However, the following observations suggest that the compounds responsible for the g=6.0, the g=1.86 and the g=2.0044 signals are intimately associated with the reaction center bacteriochlorophyll: (1)non-reversible changes are induced in these signals when dark adapted bacteria are illuminated at $77^{o}K$; (2)when bacteria are heated and the g=1.86 and g=6.0 signals are not detectable, one cannot photo-induced $P870^{+}$, and; (3)the room temperature time behavior of the 1.86 signal minics inversely the time behavior of the bacteriochlorophyll cation radical.

References

1. Corker, G. A. and Nicholson, W., 1970, Arch. Biochem. Biophys. 137, 75-83.

2. Corker, G. A. and Sharpe, S., 1974, Photochem. Photobiol. 19, 443-455.

3. Cohen-Bazire, G. W., Sistrom, R. and Stanier, R. V., 1957, J. Cellular Comp. Physiol. 49, 25-26.

4. Shanumugan, K. T. and D. I. Arnon, 1972, Biochem. Biophys. Acta. 256, 487-497.

M. AVRON, *Proceedings of the Third International Congress on Photosynthesis*, September 2-6, 1974, The Weizmann Institute of Science, Rehovot, Israel
Elsevier Scientific Publishing Company, Amsterdam, The Netherlands, 1974

THE ORIENTATION OF PHOTOSYNTHETIC PIGMENTS IN CHROMATOPHORE MEMBRANES AND IN PURIFIED REACTION CENTERS OF *RHODOPSEUDOMONAS SPHEROIDES*

Fred J. Penna, Dan W. Reed† and Bacon Ke
Charles F. Kettering Research Laboratory
Yellow Springs, Ohio 45387, U.S.A.

The reaction center of photosynthetic bacteria contains the specialized components which participate in the primary photochemical reaction. The primary photochemical event *in vivo* consists of transfer of excitation energy from the light-harvesting bacteriochlorophyll molecules to the reaction-center pigments and the resulting photooxidation of the reaction-center bacteriochlorophyll molecule, P870, which donates an electron to the associated primary electron acceptor, presumably an iron-sulfur protein molecule. Reaction centers in a purified form were first isolated from the blue-green mutant strain R-26 of *Rhodopseudomonas spheroides* by treatment with Triton X-100, a non-ionic detergent (1). Subsequently, purified reaction-center preparations have also been isolated from photosynthetic bacteria with other detergents (2-4). Extensive studies have been carried out with the *R. spheroides* reaction centers with respect to chemical composition (5), the identification of the free-radical EPR signal with the photooxidized P870 (6), the nature of the primary electron acceptor (7,8), and the electrostatic binding and rapid electron transfer between P870 and mammalian cytochrome *c* (9).

More recent studies of the molecular organization in the *R. spheroides* reaction centers revealed that they contain both bacteriochlorophyll *a* (BChl) and bacteriopheophytin *a* (Bpheo) in a molar ratio of 2:1, and most likely contain 2 Bpheo and 4 BChl molecules (10). In light of these pigment analyses, the absorption, circular-dichroism and fluorescence spectra of *R. spheroides* reaction centers in the oxidized and reduced states were investigated at 295 and 77°K. The spectral data were consistent with a single pigment complex in the reaction center containing six strongly interacting pigment molecules (11). Coupling between molecules of the pigment complex produces the rotational strength

†This manuscript was completed in the absence of Dr. Dan Reed who died in a plane crash after attending the 3rd International Congress on Photosynthesis in Israel (1974).

of the reaction center (12), and the strong coupling occurs between dipoles which do not have coplanar orientations. We report here some preliminary results on a further study of the molecular organization, especially with regard to the orientation of the pigment molecules in the *R. spheroides* reaction centers derived from linear-dichroism measurements.

The study of the orientation of photosynthetic pigments in chloroplasts by linear-dichroism measurements dates back more than three decades (13,14). With refined techniques in sample orientation and measuring instrumentation, Breton *et al.* (15,16) have recently detected a high degree of orientation of both the pigments and the structural proteins parallel to the plane of the membrane. From dichroism measurements, Morita and Miyazaki (17) showed that the bacteriochlorophyll molecules in "intact" cells of *Rhodopseudomonas palustris* and in air-dried films of lamellar fragments are largely orientated parallel to the lamellar membrane. More recently, Breton also measured pigment orientation in several photosynthetic bacteria (18).

We have measured linear dichroism with chromatophores and purified reaction centers (both prepared as in ref. 1) orientated on glass slides by the brush spreading method of Breton *et al.* (16). Linear dichroism was measured with the Kettering Circular Dichrometer (19) by placing an additional elasto-optic modulator operated as a steady quarter-wave retarder immediately behind the first modulator which was operated at 25 KHz.

The absorption and linear-dichroism spectra of *R. spheroides* chromatophores oriented on a glass slide by brush spreading are shown in Fig. 1 (both spectra were taken from the same sample). In the far-red region the broad linear-dichroism band (870 nm) corresponding to the major absorption band at 865 nm is positive, whereas the linear-dichroism band corresponding to the absorption band of BChl near 800 nm is negative. The linear-dichroism band corresponding to the absorption band of BChl at 592 nm is negative. The opposite signs of the 870 and 590 nm linear-dichroism bands are consistent with the notion that the far-red and the 590-nm bands represent the Q_y and Q_x transition moments which lie perpendicularly to each other in the plane of the porphyrin ring of the BChl molecule. The low absolute values of the measured dichroism indicate that the BChl molecules are either randomly oriented or inclined at an average angle of 35° with the membrane plane. In chromatophores, the orientations manifested by the linear-dichroism spectrum are

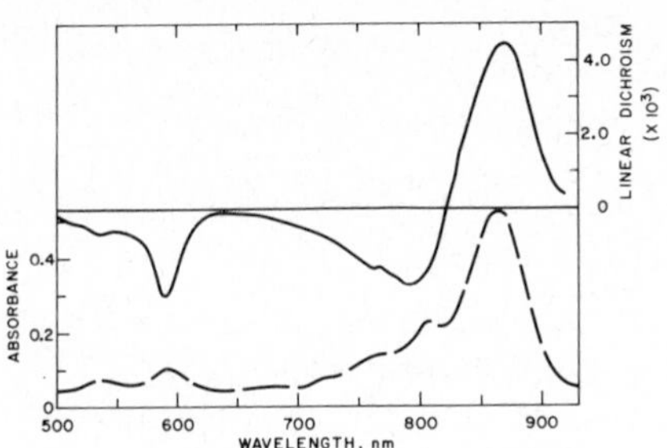

Figure 1. Absorption (bottom) and linear-dichroism (top) spectra of *R. spheroides* chromatophores at 295°K.

contributed predominantly by the bulk bacteriochlorophyll molecules; the contribution by the reaction-center pigments should be a very minute fraction.

In order to examine the pigment orientations in the reaction-center complex, linear dichroism was measured with purified reaction-center preparations similarly oriented on a glass slide. The absorption and linear-dichroism spectra measured at 295 and 77°K are shown in Fig. 2 (all spectra were taken from a single sample). The absorption spectrum of the oriented reaction-center preparation at 295°K appears similar to that in solution: three absorption bands at 865, 802 and 757 nm appear in the far-red region and two other maxima are at 597 and 535 nm. The linear-dichroism band corresponding to the Q_y transitions of BChl at 865 nm is positive, whereas that corresponding to the Q_y transition of Bpheo is negative. The linear-dichroism bands corresponding to the Q_x transitions of the 865-nm BChl and the Bpheo are exactly opposite to those of their Q_y transitions. The maximum of the longest-wavelength linear-dichroism band is also at 865 nm but a shoulder is

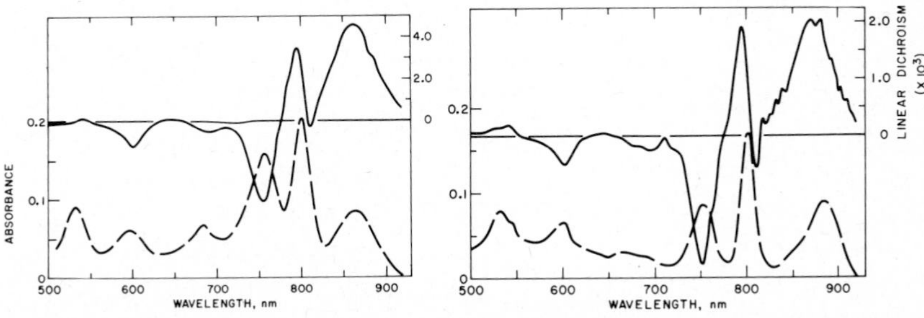

Figure 2. Absorption and linear dichroism spectra of purified reaction centers from *R. spheroides* R-26 at 275°K (left) and 77°K (right).

apparent near 883 nm. The linear-dichroism spectrum corresponding to the 802 nm absorption band consists of a negative band at 811 nm and a positive band at 796 nm.

At 77°K, the absorption maxima shift toward longer wavelengths and the absorption bands become narrower. The 865 nm absorption band shifts to 890 nm. The linear-dichroism band corresponding to the longest-wavelength absorption band apparently consists of a double band at 870 and 880 nm. Other linear-dichroism bands assume the same sign as those at 295°K, except that they are sharper. Both the absorption and linear-dichroism bands corresponding to the Q_x transition of Bpheo near 540 nm appear as double bands at 77°K.

The linear-dichroism spectra shown for the *R. spheroides* chromatophores and reaction-center preparations are true manifestations of the orientation of the BChl and Bpheo molecules. The type of sample materials and the fact that both positive and negative linear-dichroism bands corresponding to the X and Y transitions of the pigment molecules are observed ruled out any textural or form dichroism in the sample. The nature of the measuring technique used in the present work is also little affected by sample birefringence if any was present. In general, the linear-dichroism signals have the shape of the absorption bands for the transitions under consideration, but their magnitude and sign depend on the angle of orientation. The relatively low absolute values for the ($\Delta A/A$) term presumably arises from a low degree of orderliness in the sample system because of the small size of the reaction-center particles rather than an incomplete orientation of the pigment molecules within the particles.

Because of the low absolute values measured for our sample systems, we feel it may be premature to make a quantitative evaluation for the pigment arrangement in the reaction-center complex with the data available. However, some general conclusions may be drawn from data presented in Figs. 1 and 2. The apparent band splitting of the P870 linear-dichroism band at 77°K manifests a close interaction between two similar P870 BChl molecules. This is also consistent with previous circular-dichroism results (11). The positive and negative linear-dichroism bands associated with the absorption band at 800 nm suggests that the porphyrin planes of the two P800 BChl molecules are at an angle with each other. The Q_y transitions of Bpheo and P870 molecules also are not parallel.

Contribution No. 524 from the Charles F. Kettering Research Laboratory.

REFERENCES

1. Reed, D.W. and Clayton, R.K. (1968) Biochem. Biophys. Res. Commun., *30*, 471-475.
2. Jolchine, G., Reiss-Husson, R., and Kamen, M.D. (1969) Proc. Nat. Acad. Sci. USA, *64*, 650-653.
3. Thornber, J.P., Olson, J.M., Williams, D.M., and Clayton, M.L. (1969) Biochim. Biophys. Acta, *172*, 351-354.
4. Clayton, R.K. and Wang, R.T. (1971) Methods in Enzymology, *23*, 696-704.
5. Reed, D.W. (1969) J. Biol. Chem., *244*, 4936-4941.
6. Bolton, J.R., Clayton, R.K., and Reed, D.W. (1969) Photochem. Photobiol., *9*,209-218.
7. Reed, D.W., Zankel, K.L., and Clayton, R.K. (1969) Proc. Nat. Acad. Sci. USA, *63*, 42-46.
8. Dutton, P.L., Leigh, J.S., and Reed, D.W. (1973) Biochim. Biophys. Acta, *292*,654-664.
9. Ke, B., Chaney, T.H., and Reed, D.W. (1970) Biochim. Biophys. Acta, *216*, 373-383.
10. Reed, D.W. and Peters, G.A. (1972) J. Biol. Chem., *247*, 7148-7152.
11. Reed, D.W. and Ke, B. (1973) J. Biol. Chem. *248*, 3041-3045.
12. Sauer, K., Dratz, E.A., and Coyne, L. (1968) Proc. Nat. Acad. Sci., USA, *61*, 17-24.
13. Menke, W. (1943) Biochem. Zentralbl., *63*, 203-208.
14. Goedheer, J.C. (1957) Dissertation, State University, Utrecht.
15. Breton, J. and Roux, E. (1971) Biochem. Biophys. Res. Commun., *45*, 557-563.
16. Breton, J., Michel-Villaz, M., and Paillotin, G. (1973) Biochim. Biophys. Acta, *314*, 42-56.
17. Morita, S. and Miyazaki, T. (1971) Biochim. Biophys. Acta, *245*, 151-159.
18. Breton, J. (1974) Biochem. Biophys. Res. Commun., *59*, 1011-1017.
19. Breeze, R.H. and Ke, B. (1972) Anal. Biochem., *50*, 281-303.

M. AVRON, *Proceedings of the Third International Congress on Photosynthesis*,
September 2-6, 1974, Weizmann Institute of Science, Rehovot, Israel
Elsevier Scientific Publishing Company, Amsterdam, The Netherlands, 1974

ELECTRICAL POTENTIAL DIFFERENCE, pH GRADIENT AND PHOSPHORYLATION
ON THE RELATION BETWEEN THE TRANSMEMBRANE ELECTRICAL POTENTIAL DIFFERENCE, pH GRADIENT AND ATP FORMATION IN PHOTOSYNTHESIS

P. Graeber and H.T. Witt
Max-Volmer-Institut für Physikalische Chemie und Molekularbiologie
Technische Universität Berlin

1. Summary

Phosphorylation was measured simultaneously with the electrical potential difference $\Delta\varphi$ or with the pH gradient Δ pH across thylakoid membranes of spinach chloroplasts. Between $pH_{out}7$ and $pH_{out}9$ and Δ pH $\leq$ 2.7, the rate of ATP-formation is independent from the absolute proton concentration and depends only on the magnitude of Δ pH and $\Delta\varphi$. The relation between the rate of ATP-formation and Δ pH can be described by: $\lg \overline{\dot{ATP}} = b \cdot \Delta pH + const_2$. It is shown that the ΔH^+/ATP-ratio decreases strongly with increasing Δ pH as well as with increasing $\Delta\varphi$. It is suggested that the degree of coupling between proton flux and phosphorylation depends on Δ pH and on $\Delta\varphi$.

2. Introduction

In 1971 we reported on the relation between $\Delta\varphi$ and phosphorylation[1,2]:

1) The rate of electrical potential decay $\Delta\dot{\varphi}$ is identical with the rate of ATP-formation (at constant initial electrical potential difference).
2) The magnitude of the electrical potential difference $\Delta\varphi$ is proportional to the amount of generated ATP (at constant Δ pH).

3) The functional unit of the electrical events is identical with that of phosphorylation.

On the other hand relations between Δ pH and phosphorylation have been investigated under conditions where $\Delta\varphi$ was unknown or assumed to be zero[3,4,5].

In this work we report on the rate and amount of ATP-formation as a function of $\Delta\varphi$ as well as of Δ pH. The essential features of the experimental conditions are:

1) The electrical potential difference $\Delta\varphi$, Δ pH and proton flux ($\overline{\dot{H}^+}$) are known under all conditions.
2) $\Delta\varphi$ and Δ pH can be changed independently from each other and this at different pH_{out} values between 7 and 9.

To achieve these conditions we excite photosynthesis with periodical flashes of different intensities and frequencies ν. Periodical flash light was used instead of continuous light because flash induced electrical potential differences can be estimated more precisely in flash light than in continuous light.

3. Materials and Methods

Chloroplasts were freshly isolated from spinach grown in a BBC-phytocell or obtained from the local market using the method described by Winget[6]. Additionally 10 mM ascorbate had been added during grinding. The reaction volume was 1 ml and contained $2 \cdot 10^{-2}$m tricine adjusted to the applied pH_{out} with NaOH, 10^{-4}m benzylviologen, 10^{-2}m KCl, $5 \cdot 10^{-3}$m K_2HPO_4 containing ^{32}P with $1 - 3 \cdot 10^5$ counts/min, 10^{-2}m sucrose, $3 \cdot 10^{-4}$m ADP and chloroplasts giving a chlorophyll concentration of $2 \cdot 10^{-4}$m. For the Δ pH determination 9-aminoacridine was added to a final concentration of $5 \cdot 10^{-7}$m.

Electrical potential differences $\Delta\varphi$ were measured by the potential indicating absorption change ΔA at 515 nm. Since 1967 it was shown kinetically, spectroscopically and electrically that these absorption changes indicate electrical potential differences[7,8,9,10]. It was outlined that the indication of $\Delta\varphi$ by ΔA is (1) prompt, (2) linear and (3) transmembrane. (4) The relation between $\Delta\varphi$ and the absorption change ΔA is

$$\Delta\varphi = 50 \text{ mV} \cdot \Delta A / \Delta A_1 \qquad (1)$$

(ΔA_1 = absorption change in a saturating single turnover flash).

The proton flux $\overline{\dot{H}^+}$ was calculated according to eq.(2). Eq.(2) is the result of the following consideration. In a single turnover flash 2 protons are translocated across the thylakoid membrane[11] and an electrical potential difference of $\Delta\varphi_1$ = 50 mV is generated (see eq.(1). At arbitrary values of $\Delta\varphi$ the amount of the translocated protons is proportional to $\Delta\varphi$ [12]. Therefore, at any frequency and $\Delta\varphi$ the averaged proton flux per electron chain is $\overline{\dot{H}^+} = 2 \cdot \nu \cdot \Delta\varphi / \Delta\varphi_1$. From measurements of the O_2 yield per flash a number of 700 chlorophyll molecules per electron chain has been determined. Therefore, it results

$$\overline{\dot{H}^+} = \frac{2}{700} \cdot \nu \cdot \Delta\varphi / \Delta\varphi_1 = 2.8 \cdot 10^{-3} \cdot \nu \cdot \Delta\varphi/\Delta\varphi_1 \ \frac{\text{Mol } H^+}{\text{Mol Chl.s}} \qquad (2)$$

The pH_{in} value in the inner space of the thylakoid and $\Delta pH = pH_{in} - pH_{out}$ was measured by the light induced fluorescence quenching of 9-aminoacridine according to the method of Schuldiner et al.[13]. For the calculation of Δ pH a value of 10 l/Mol Chl was used for the inner space of the thylakoids. However, Rumberg et al.[14] have shown that the calculation of Δ pH outlined by Schuldiner et al.[13] is valid only under special conditions. A comparison with the methods of pH_{in} determination developed by Rumberg et al. (using the pH_{in} dependence of electron transport[15] or the pH dependent distribution of the amine imidazol[4,16]) shows that under the above described conditions the three methods give approximately the same result up to Δ pH = 2.7. Therefore, the following experiments are restricted to this limit.

Phosphorylation has been measured by ^{32}P as described previously[17]. The measurement of phosphorylation and the electrical potential difference or Δ pH was carried out simultaneously.

4. Results and Discussion

Fig.1 shows the rate of ATP-formation as a function of the pH_{in} value at three different pH_{out} values. The result seems to indicate that there exist a "threshold-value" of pH_{in} beyond which

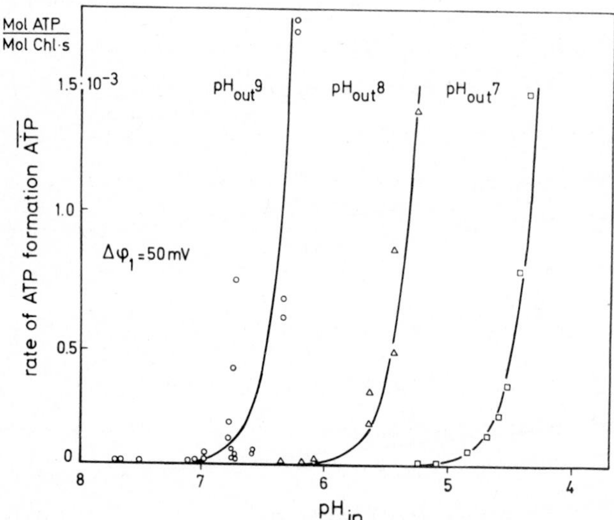

Fig.1. Rate of ATP-formation as a function of the pH_{in} value at different pH_{out} values between 7 - 9. The initial electrical potential difference is $\Delta\varphi_1$ = 50 mV.

practically no ATP is synthesized. Such threshold values have been reported by several authors[5,18] and have been interpreted as an indication of a critical Δ pH value necessary for ATP-formation. A similar "threshold-effect" has been found by Junge et al.[19] plotting the ATP yield versus the initial electrical potential difference. This has been interpreted[20] as an indication of a critical $\Delta\varphi$ value. However, if we plot lg $\overline{\dot{ATP}}$ versus Δ pH (see fig.2) no threshold can be seen and it seems that the reported "threshold" is only an apparent one due to the linear plot of an exponential function (see also below).

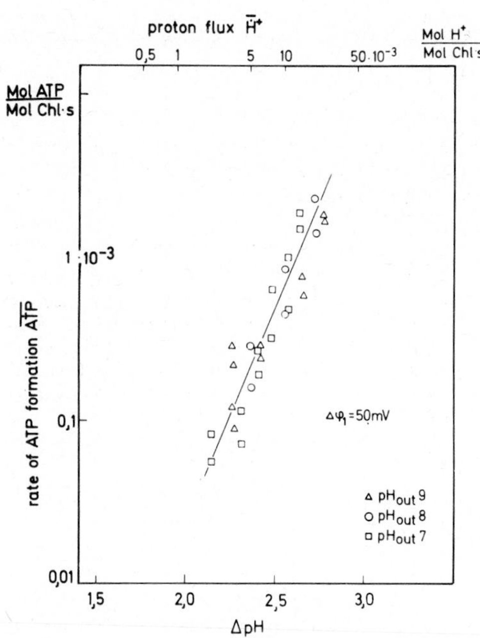

Fig.2. Logarithmic plot of the rate of ATP-formation versus Δ pH at different pH_{out} values. $\Delta\varphi_1$ =50 mV. The proton flux is depicted on the top.

The functional relation between $\overline{\dot{ATP}}$ and Δ pH is given by:

$$\lg \overline{\dot{ATP}} = b \cdot \Delta pH + const._2; \quad \Delta\varphi_1 = const. \tag{3}$$

Because the relation between the proton flux $\overline{\dot{H}^+}$ and Δ pH is given by [21]

$$\Delta pH = a \cdot \lg \overline{\dot{H}^+} + const._1 , \tag{4}$$

it follows from equation (3) and (4):

$$\lg \overline{\dot{ATP}} = c \cdot \lg \overline{\dot{H}^+} + const._3 \tag{5}$$

The proton flux $\overline{\dot{H}^+}$ is depicted on the top of fig.2. Additionally it can be seen that the rate of ATP-formation is independent from the absolute value of pH_{out} . This indicates that the rate of ATP-formation is not regulated by pH_{out} nor by pH_{in} but only by the difference between both, i.e. by the pH gradient.

The above described experiments have been carried out at a constant initial electrical potential of $\Delta\varphi_1$ = 50 mV. Fig. 3 shows the results at three different electrical potential differences (50 mV, 75 mV and 125 mV respectively).

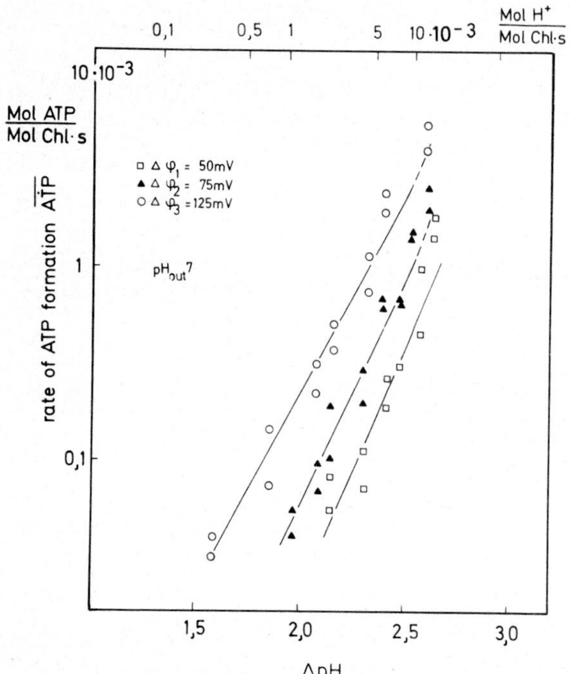

Fig.3. Logarithmic plot of the rate of ATP-formation versus Δ pH at different initial electrical potential differences between 50 - 125 mV. The proton flux is depicted on the top.

In all cases lg $\overline{\dot{ATP}}$ is proportional to Δ pH. However, the slopes as well as the absolute values are dependent on the electrical potential difference $\Delta\varphi$. At a constant Δ pH value, i.e. at a constant proton flux $\overline{\dot{H}^+}$, say Δ pH 2.2 or $\overline{\dot{H}^+} = 2.5 \cdot 10^{-3}$ $\frac{\text{Mol H}^+}{\text{Mol Chl s}}$ an increase of $\Delta\varphi$ from 50 mV to 125 mV gives a 5-10 fold increase of the rate of ATP-formation. At higher Δ pH values the stimulation is smaller.

If at a constant Δ pH, i.e. at a constant $\overline{\dot{H}^+}$ (see equation (4)), the ATP rate is increased by $\Delta\varphi$ this can only be possible when the yield of ATP-formation increases with $\Delta\varphi$. The yield is defined by the ratio of generated ATP per translocated protons: $\frac{ATP}{\Delta H^+}$. When the yield is increased due to the energetic properties of $\Delta\varphi$, the yield should be increased also by Δ pH. Fig.4 shows the dependence of the reciprocal yield, ΔH^+/ATP on Δ pH at different initial electrical potential differences. The corresponding proton flux (see eq.(4) is depicted on the top.

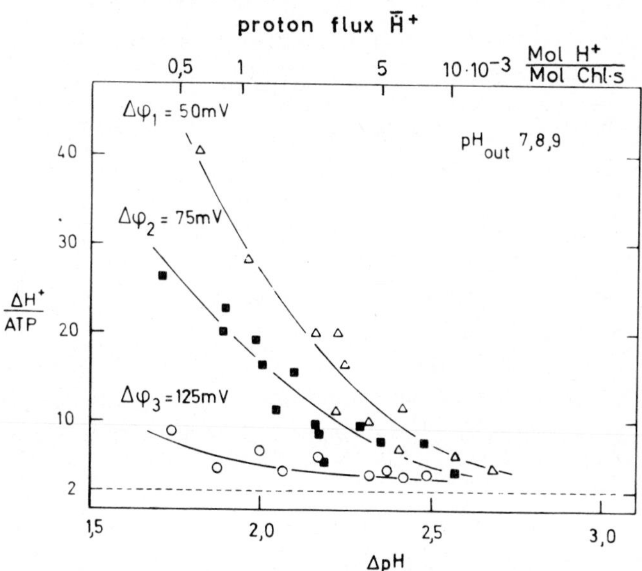

Fig.4. ΔH^+/ATP ratio as function of Δ pH at different initial electrical potential differences between 50 and 125 mV. Measurements have been carried out at pH_{out} 7, 8 and 9; no differences have been noticed. The dotted line corresponds to a ΔH^+/ATP ratio of two.

The yield was obtained by dividing the averaged proton flux $\overline{\dot{H}^+}$ by the rate of ATP-formation. $\overline{\dot{ATP}}$, using the data of fig.2 and fig.3. (The measured proton flux $\overline{\dot{H}^+}$ (see eq.(2)) is the total proton flux not separated in a basal and a phosphorylating one). It can be seen from fig.4 that for $\Delta\varphi_1$ = 50 mV the ΔH^+/ATP ratio decreases from 40 at ΔpH = 1.8 to 3 - 4 at ΔpH = 2.7. At $\Delta\varphi_3$ = 125 mV the ΔH^+/ATP ratio decreases from 6 at ΔpH = 1.8 to 3 - 4 at ΔpH = 2.7. This means, that at small $\Delta\varphi$ values the stimulation of the ATP yield by ΔpH is large whereas at higher $\Delta\varphi$ - values this stimulation is smaller. Furthermore, it can be seen that at a low ΔpH = 1.8 or $\dot{\overline{H^+}} \approx 0.6 \cdot 10^{-3} \frac{\text{Mol H}^+}{\text{Mol Chl s}}$ there exists a large stimulation of the ATP yield by increasing $\Delta\varphi$ from 50 mV to 125 mV, whereas at higher ΔpH = 2.5 or $\dot{\overline{H^+}} \approx 9 \cdot 10^{-3} \frac{\text{Mol H}^+}{\text{Mol Chl s}}$ only a small stimulation is observed. Similar stimulations were reported by Schuldiner et al.[5] by generating artificial diffusion potentials across thylakoid membranes. However, the magnitude of the induced electrical potential differences was not known.

According to our results the ΔH^+/ATP ratio depends on the energetic state of the membrane, i.e. it is supposed that the degree of coupling of the proton flux to ATP-synthesis depends on the pH gradient and the electrical potential difference. For instance, it can be suggested that at a constant ΔpH and, therefore, at constant proton flux the coupling is increased because an increasing part of the protons is focussed to the ATPase with increasing $\Delta\varphi$. The lowest measured ΔH^+/ATP ratio is 3 - 4.

In fig.5 the number of ATP molecules per flash and electron chain is depicted as a function of the initial electrical potential difference at different ΔpH values. The corresponding amount of translocated protons per flash and electron chain are shown on the top. The yield increases in a biphasic shape at ΔpH 2, suggesting a "threshold" at 50 mV. At ΔpH 2.7 the ATP yield increases, however, approximately linear with $\Delta\varphi$, and the "threshold" has disappeared. Obviously, the relation between ATP and $\Delta\varphi$ depends on ΔpH. The maximal slope at ΔpH 2 . 7 indicates again a value of 3 - 4 H^+ per ATP. These results explain the discrepancies between experiments which show a biphasic shape as reported by Junge et al.[19] and those with a linear shape as reported by Boeck and Witt[1,2].

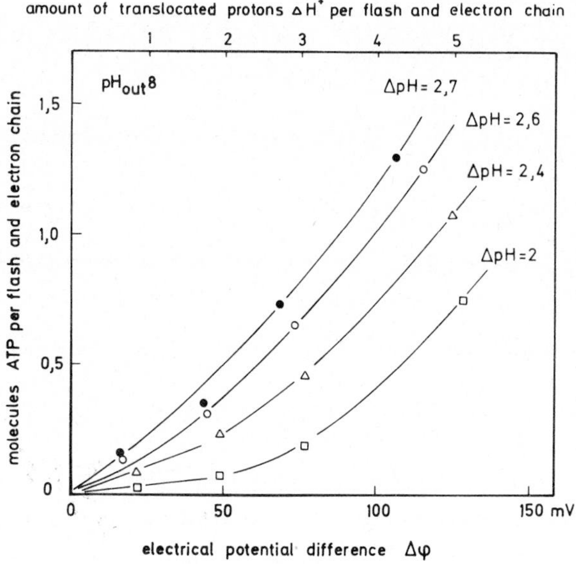

Fig.5. Molecules ATP per flash and electron chain as a function of the initial electrical potential difference at different Δ pH values between 2 and 2.7. pH_{out} =8. The amount of translocated protons per flash and electron chain is depicted on the top.

Summarising it can be said that an optimal ΔH^+/ATP ratio of ΔH^+/ ATP = 3 - 4 is observed. The extrapolation of the curves in fig.4 to high electrical potential differences or high Δ pH let us suggest that the ΔH^+/ATP ratio may reach even lower values. (The dotted line in fig.4 corresponds to a ΔH^+/ATP ratio of two). The results give evidence for a variable coupling between proton flux and ATP-formation and indicate that the degree of coupling depends on the pH gradient and the electrical potential difference.

It can be suggested that the ΔH^+/ATP ratio obtained by different authors reflects different degrees of coupling expecially by different electrical potential differences.

We thank Miss E. Schober for skilful technical assistance, Dr. C. Kötter from Schering AG. for supplying us with spinach grown in a phytocell, Prof. B. Rumberg for valuable discussions and for measuring the inner space of the thylakoids and Dr. G. Renger for measuring the number of chlorophyll molecules per electron chains.

References

1. Boeck,M. and Witt,H.T.(1972) Proc.Intern.Congr.Photosynthesis Res. Stresa 1971 (Forti,G., Avron,M. and Melandri,A. eds) pp.903-911, Junk N.V. Publishers, The Hague
2. Witt,H.T. (1974) Bioenergetics of Photosynthesis (Govindjee ed) pp.493-552, Academic Press, New York
3. Jagendorf,A.T. and Uribe,E.(1966) Proc.Nat.Acad.Sci.USA 55,170-177
4. Rumberg,B. and Schröder,H. (1974)Intern.Congr. Photobiol.6th Bochum 1972, in press
5. Schuldiner,S., Rottenberg,H. and Avron,M.(1973) Eur.J.Biochem.39, 455-462
6. Winget,G.D., Izawa,S. and Good,N.E.(1965)Biophys.Res.Commun.21, 438-443
7. Junge,W. and Witt,H.T.(1968) Z.Naturforsch.23b, 244-254
8. Emrich,H.M., Junge,W. and Witt,H.T.(1969)Z.Naturforsch.24b,144-146
9. Schmidt,S., Reich,R. and Witt,H.T.(1972)Proc.Intern.Congr.Photosynthesis Res.Stresa 1971 (Forti,G.,Avron,M.and Melandri,A.eds) pp.1087-1095, Junk N.V. Publishers, The Hague; Naturwiss. (1971) 58, 414-415
10. Witt,H.T. and Zickler,A. (1973) FEBS Letters 37, 307-310
11. Schliephake,W., Junge.,W. and Witt,H.T.(1968) Z.Naturforsch.23b, 1561-1578
12. Reinwald,E., Stiehl,H.H. and Rumberg,B.(1968)Z.Naturforsch.23, 1616-1617
13. Schuldiner,S.,Rottenberg,H.and Avron,M.(1972)Eur.J.Biochem.25,64-70
14. Rumberg,B., personal communication, preliminary results
15. Rumberg,B. and Siggel,U. (1969) Naturwiss.56, 130-132
16. Reinwald,E. (1970) Thesis, TU Berlin
17. Avron,M. (1960) BBA 40, 257-272
18. Schwarz,M. (1968) Nature (London) 219, 915-919
19. Junge,W.,Rumberg,B.and Schröder,H.(1970)Eur.J.Biochem.14,575-581
20. Junge,W. (1970) Eur.J.Biochem.14, 582-592
21. Gräber,P. and Witt,H.T. in preparation

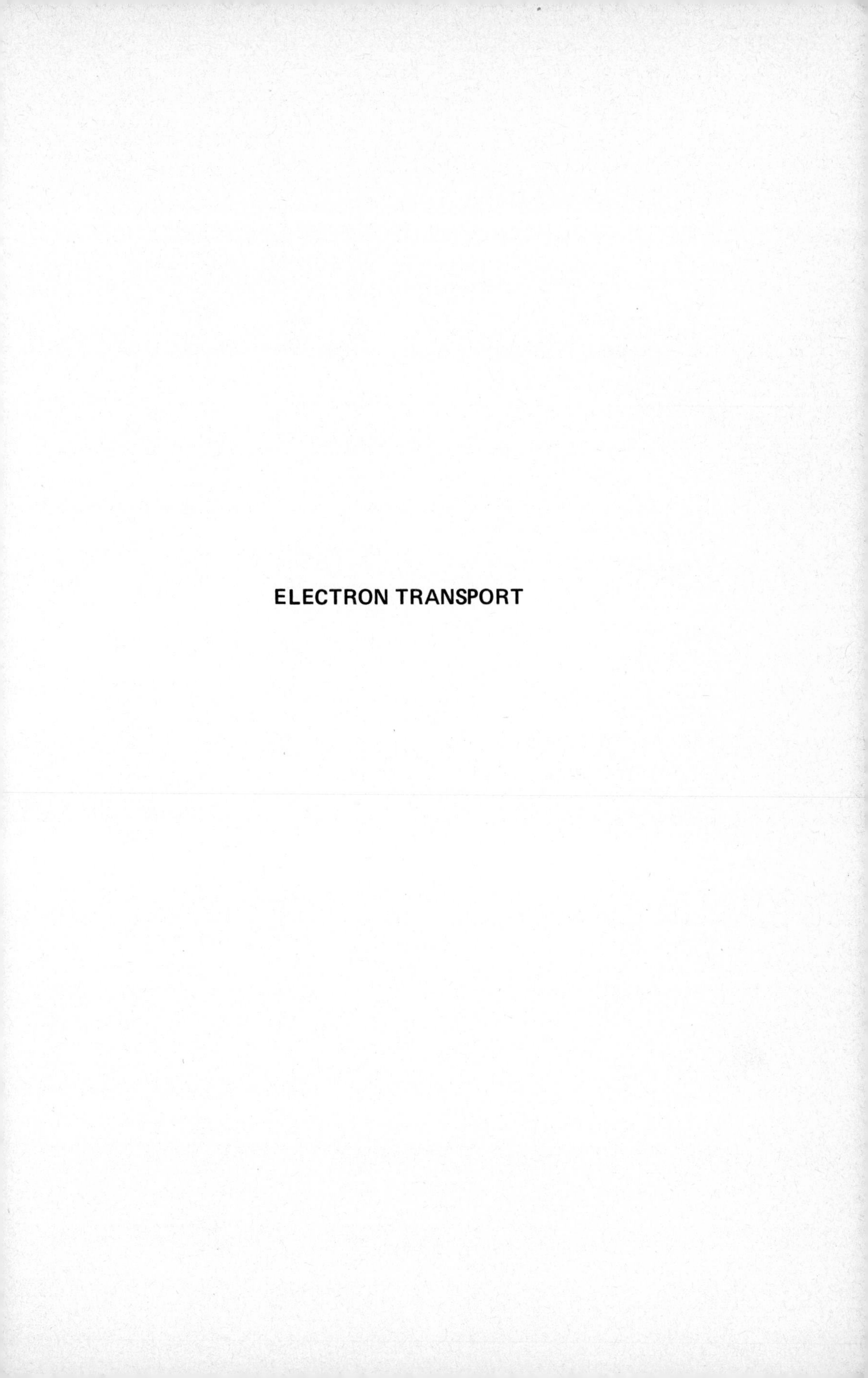

ELECTRON TRANSPORT

M. AVRON, *Proceedings of the Third International Congress on Photosynthesis*, September 2-6, 1974, Weizmann Institute of Science, Rehovot, Israel
Elsevier Scientific Publishing Company, Amsterdam, The Netherlands, 1974

ELECTRON FLOW ACROSS THE THYLAKOID MEMBRANE

Native and artificial energy conserving sites in photosynthetic electron flow.

A. Trebst

Lehrstuhl Biochemie der Pflanzen,
Ruhr-Universität Bochum, 463 Bochum

Recent studies on coupled ATP formation in non cyclic electron flow in chloroplasts led to the recognition of two energy conservation sites (Fig. 1). In addition to several reports on higher $^{P}/_{e_2}$ ratios than one in non cyclic photophosphorylation[1,2], inhibitor studies made it possible to functionally seperate two sites of ATP formation. Besides KCN (a plastocyanin inhibitor at high concentrations[3]) and polylysine[4] the plastoquinone antagonist DBMIB (dibromothymoquinone) proved to be the most suitable inhibitor to block electron flow between photosystem II and photosystem I.

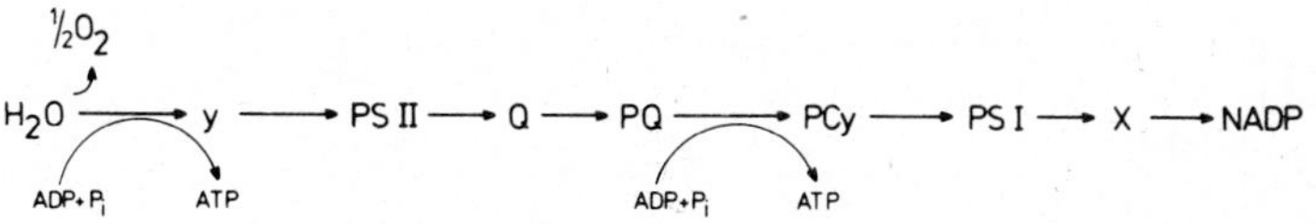

$$\frac{1}{2}O_2 \leftarrow H_2O \longrightarrow y \longrightarrow PS\,II \longrightarrow Q \longrightarrow PQ \longrightarrow PCy \longrightarrow PS\,I \longrightarrow X \longrightarrow NADP$$

(coupling: $ADP + P_i \rightarrow ATP$ at $H_2O \rightarrow y$ and at $PQ \rightarrow PCy$)

Fig. 1 Two coupling sites in the photosynthetic electron flow from water to NADP

1. DBMIB, an inhibitor between the two light reactions.

After we had introduced DBMIB in 1970[5] a review[6] at the congress at Stresa summarized the evidence till 1971. The results indicated that in the presence of DBMIB photoreductions by just photosystem II as well as photoreductions by photosystem I may be followed independently of each other. Of course an artificial electron acceptor for photosystem II or an artificial electron donor for photosystem I respectively had to be used. Since then a number of results have substantiated the view that DBMIB (at a concentration of 10^{-6}M) blocks the reoxidation of plastohydroquinone, but does not interfer with the reduction of plastoquinone[7-12].

2. A second ATP forming site connected with photosystem II.

The important recent development on energy conservation came from further studies of the properties of photoreductions by photosystem II (i.e. in the presence of DBMIB). It turned out that photoreductions by photosystem II are coupled to ATP formation with a P/e_2 ratio of about 0,5, i.e. about half the stoichiometry of non cyclic electron flow[7,13,14,15] (Tab. 1). The photoreduction by photosystem II (i.e. in the presence of DBMIB) of ferricyanide, mediated by lipophilic acceptors like phenylenediimine or benzoquinone (as introduced by Saha et al)[16] has very high rates of electron flow and coupled ATP formation. It does not include the coupling site between plastoquinone and cytochrome f, already recognized before[17-19], because this site has been made unoperational by DBMIB. Therefore a second energy conserving site had to exist.

Tab.1 ATP formation coupled to the photoreduction of ferricyanide by photosystem II (i.e. in the presence of DBMIB)

10^{-4} M mediator added	µmole/h/mg chlorophyll		P/e_2
	oxygen	ATP	
control without DBMIB	190	180	0.95
control with DBMIB	70	35	0.50
p-phenylenediamine	185	90	0.49
DAT = diaminotoluene	190	80	0.42
MMPD= 2-methyl-5-methoxy-p-phenylenediamine	220	120	0.54
TMPD= N-tetramethyl-p-phenylenediamine	70	25	0.36

3. Vectorial electron flow.

The final recognition of a second energy conserving site led to a new concept of non cyclic as well as cyclic photosynthetic electron flow[20,21]. Because it seemed that only the chemiosmotic hypothesis is able to explain the coupling in the area of photosystem II and the watersplitting reaction and, more important, uneven P/e_2 values. The easiest explanation for the coupling in photoreduction by photosystem II would be to assume that the proton, liberated in water photooxidation on the inside of the membrane, contributes to the pH gradient across the membrane, which in turn drives ATP formation. An uneven P/e_2 ratio results, when 2 protons/2 electrons are liberated inside per photosystem and 3 protons are required per one ATP formed, as has actually been measured[22,23] (see Skulashev[24] for a mechanism). This would result in a theoretical $P/_{e_2}$ value of 0.66 in a photoreduction by photosystem II and 1.33 in non cyclic photophosphorylation including both photosystems. Two energy conserving sites (proton translocating sites) therefore are not necessarily equivalent to two ATP coupling sites.

Vectorial electron flow in photosynthesis as required for a chemiosmotic coupling has been suggested by Mitchell and supported in particular by Witt and his colleagues[25,26,27] (Fig. 2). The present evidence for a sidedness of photosynthetic electron flow has recently been reviewed[20].

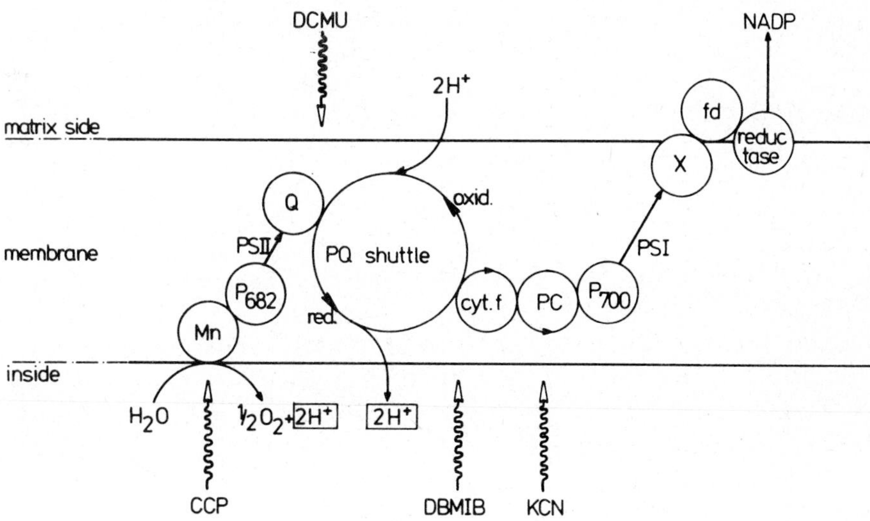

Fig. 2 - Non cyclic electron flow across the thylakoid membrane

4. Coupling of donor systems for photosystem I.

However, besides the recognition of the energy coupling in photosystem II with the help of DBMIB, also the coupling of photosystem I photoreductions in the presence of DBMIB posed a problem. Its explanation again led to still further support of a chemiosmotic coupling mechanism in photosynthesis. The problem is that photosystem I photoreductions and certain cyclic photophosphorylation systems are also not inhibited by DBMIB, nor is the ATP formation coupled to them[28,29] (Tab. 2). However, because DBMIB blocks the native ATP forming site, localized between plastoquinone and cytochrome f[17-19], still another energy conserving site must operate in such DBMIB insensitive coupled electron flow systems driven by photosystem I.

Tab.2 ATP formation coupled to the photoreduction of NADP by photosystem I (i.e. in the presence of DCMU and DBMIB)

donor added to 10^{-2} M ascorbate		µmole/h/mg chlorophyll NADPH	ATP	P/e_2
control without inhibitors		169	156	0.93
control with inhibitors		4	0.1	—
10^{-4}M	DCIP	97	42	0.44
	DAD	150	126	0.84
	TMPD	132	0.1	—
10^{-5}M	Indamine	120	130	1.08
	Tetramethyl-indamine	118	115	0.98
	Pentamethyl-indamine	120	0.1	—

This site cannot be localized between cytochrome f and NADP, because NADP reduction at the expense of certain artificial electron donors for photosystem I, like TMPD/ascorbate are not coupled to ATP formation[30,31] (also Tab. 2).

Two explain the coupling in certain photosystem I systems we have proposed therefore artificial energy conserving sites being able to replace native sites[21,29] (Fig. 3). An artificial energy conserving site is conceived to be a loop in the electron flow system consisting in a (native) electrogenic photosystem and an

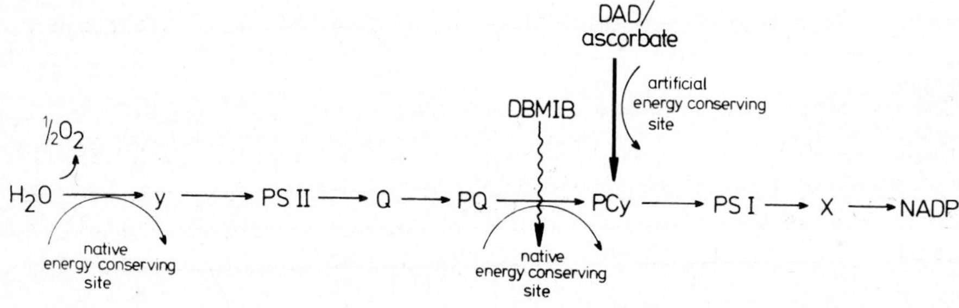

Fig.3 Artificial energy conserving site in a donor system for photosystem I

electron neutral proton transport by the (artificial) donor across the membrane. Our argument for such artificial energy conservation sites develops from the scheme for vectorial electron flow across the thylakoid membrane (Fig. 2). According to it the endogenous electron donor for photosystem I is oriented towards the inside of the membrane. Therefore the exogenous donor has to cross the membrane in order to get to the endogenous donor (plastocyanin). This is supported by the finding that only lipophilic donors for photoreductions by photosystem I or lipophilic cofactors of cyclic photophosphorylations are active as against hydrophilic redox carriers[32,33]. Furthermore, only if the redox carrier mediating electron flow through photosystem I is also a hydrogen carrying compound, i.e. is oxidized on the inside with a liberation of protons, then the system is coupled to ATP formation. If, however, an electron donor is used, which is oxidized to a radical without a liberation of a proton no ATP formation occurs. Examples for this are the comparison of the pair DAD vs. TMPD[30,33] and indamine vs. pentamethyl- indamine[34]. This is also shown in Tab. 2 and Fig. 4. Hauska and Prince have directly measured the proton uptake in such systems and showed that there is no pH gradient formed in the TMPD system[35].

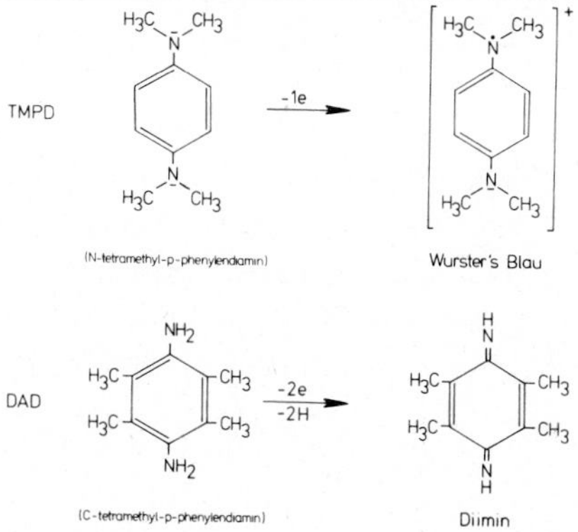

Fig. 4 - A proton is liberated when DAD is oxidized but not when TMPD is.

Tab. 3 DBMIB sensitivity of cyclic photophosphorylating systems (in the presence of 2×10^{-5}M DCMU)

Cofactor 10^{-4}M (added to a poised state)	µmole ATP/h/mg chlorophyll minus 5×10^{-6}M DBMIB	plus	inhibition
ferredoxin (3×10^{-5}M)	55	7	+
1,4 - naphthoquinone	39	<5	+
menadione	96	30	+
phenanthrenequinone	75	17	+
diaminodurene	107	75	-
methoxymethylphenylenediamine	63	57	-
dichlorophenolindophenol	31	22	-
indamine	192	228	-
dimethylindoaniline	324	324	-
phenazinemethosulfate	440	420	-
pyocyanine	280	255	-

5. Cyclic photophosphorylation.

In cyclic photophosphorylation also artificial energy conserving sites may be responsible for proton translocation and ATP formation[29]. Necessary again is that the cofactor is a hydrogen carrying redox compound, oxidized inside the membrane. Such systems are DBMIB insensitive (Tab. 3). This indicates that plastoquinone does not participate in the cycle. In most of such cyclic systems with cofactors like DAD, indamine or indoaniline (Tab. 3) plastocyanin seems to be included. But the PMS system is even KCN insensitive to a certain extent[3] which indicates that not even plastocyanin is participating in such a cyclic system. It appears then that the long known[36] PMS catalyzed cyclic system with the highest photosynthetic rates obtained in a cell free system[37] is also the most simple.

The minimum requirement for a cyclic photophosphorylation system by a chloroplast, i.e. for energy conservation in the photosynthetic membrane, is just an electrogenic photosystem and a proton translocation by the cofactor itself. This simplicity also explains why the PMS system is almost impossible to saturate by light[38]. Furthermore it becomes clear now that energy conservation (i.e. ATP formation) can be forced upon a part of the electron transport chain, which in vivo is not coupled (Fig. 5).

In DBMIB sensitive cyclic systems plastoquinone and therefore a native energy conserving site does participate (for example in the ferredoxin or naphthoquinone catalysed cyclic systems (Tab. 3). No particular lipophilicity is required, because these cofactors do not need to penetrate the membrane to the inside.

Fig. 5 Three possible cyclic electron flow systems

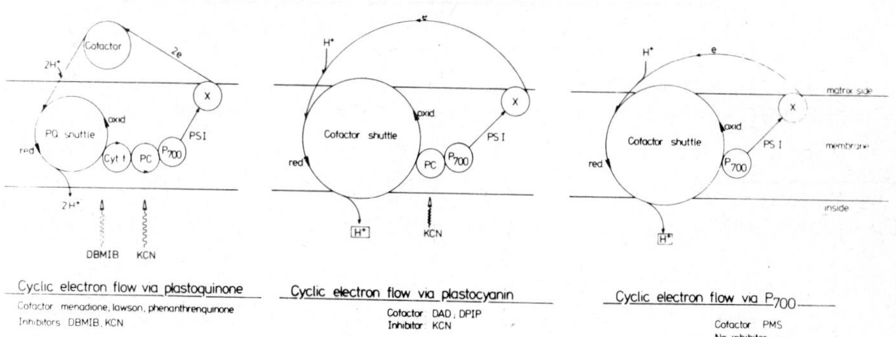

Cyclic electron flow via plastoquinone
Cofactor menadione, lawson, phenanthrenquinone
Inhibitors DBMIB, KCN

Cyclic electron flow via plastocyanin
Cofactor: DAD; DPIP
Inhibitor: KCN

Cyclic electron flow via P_{700}
Cofactor PMS
No inhibitor

The efficiency (or rather rates) of such systems are very high, as is well known for the PMS system[37]. Also the P/e_2 ratios in non cyclic photoreductions by photosystem I are high (Tab. 2). They are actually higher than expected because of overimposed cyclic electron flow, which shall not be discussed here. The point of this paper: i.e. artificial energy conservation and the minimal requirement of an electron flow system for coupling, is not to show new properties of such photoreductions or cyclic systems, but rather to clarify what part and kind of the thylakoid maschinery is involved. According to our concept the PMS system is then either the ideal system for studying the mechanism of the ATP forming system in photosynthesis because of the simplicity of the electron flow part or it should not be studied at all for that purpose by those not favoring the energy coupling mechanism implicated because of its artificiality.

6. Artificial electron donor systems for photosystem II.

The same approach (the use of electron donors, which may or may not carry hydrogen across the membrane) may be used also for artificial electron donor systems for photosystem II. Benzidine is such a well known donor[39]. It has been shown that NADP reduction at the expense of benzidine (water oxidation being blocked by tris treatment or NH_2OH) is coupled to ATP formation with a P/e_2 ratio of one, i.e. the same as in non cyclic flow from water[39,40]. As recently reported[41] and as Tab. 4 indicates the donor system tetramethyl-

-benzidine is coupled to a P/e_2 ratio of only 0.5 in NH_2OH treated chloroplasts. In both these donor systems the native energy conserving sites at the watersplitting reaction is inactivated by the hydroxylamine treatment, but the other native site at plastoquinone

Tab. 4 - NADP reduction and coupled photophosphorylation in electron donor systems for photosystem II in NH_2OH treated chloroplasts.

Additions to 0.5 µmoles donor and 3 mM ascorbate	µmoles NADPH	µmoles ATP	P/e_2
control without donorsystem	0.1	0.03	
benzidine	1.2	1.0	0.9
benzidine + $2\cdot10^{-5}$M DCMU	0.17		
tetramethylbenzidine	1.2	0.6	0.5
tetramethylbenzidine + $2\cdot10^{-5}$M DCMU	0.17		

is still operating. The proton translocation by this later site is supplemented by benzidine oxidation to yield a P/e_2 ratio of 1 but is not by tetramethyl-benzidine oxidation and therefore only a P/e_2 ratio of 0,5 is obtained with this donor[41] (Fig. 6).

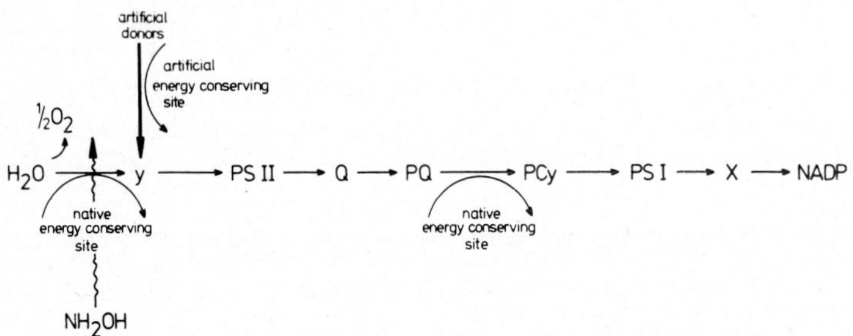

Fig.6 Artificial energy conserving site in a donor system for photosystem II supplementing a native energy conserving site depending on the hydrogen carrying property of an artificial donor

7. Summary

In conclusion then, both electrogenic photosystems are connected to a native proton translocating site, which complete the corresponding loop to yield an energy conserving site. These two native energy conserving sites are operating in non cyclic electron flow from water to NADP and together they yield ATP with a stoichiometry of probably 1.33 P per 2e. Only one site operates in photoreductions by photosystem II to yield a P/e_2 ratio of ∿0.6. Hall and colleagues put forward results to indicate a P/e_2 ratio of even 2 in non cyclic and 1 in photoreductions by photosystem II[1,42].

Both native proton translocation sites may be replaced by artificial proton translocation via an artificial donor system. Such artificial energy conservation is responsible for ATP formation in coupled donor systems for photosystem I and in some cyclic photophosphorylation systems. In such cyclic systems even a part of the electron flow system may be made coupled which is not in native electron flow. Also in electron donor systems for photosystem II an artificial proton translocation can be induced to replace the native reaction at the water splitting site.

References

1. Reeves, S.G. and Hall, D.O., 1973, Biochim. Biophys. Acta 314, 66.
2. West, K.R. and Wiskich, J.T., 1973, Biochim. Biophys. Acta 292, 197.
3. Ouitrakul, R. and Izawa, S., 1973, Biochim. Biophys. Acta 305, 105.
4. Ort, D.R., Izawa, S., Good, N.E. and Krogmann, D.W., 1973, Febs Letters 31, 119.
5. Trebst, A., Harth, E. and Draber, W., 1970, Z. Naturforsch. 25b, 1157.
6. Trebst, A., 1971, in: Proc. 2nd Int. Congr. Photosyn. Res. Stresa, vol. 1, eds. G. Forti, M. Avron and A. Melandri (Dr. W. Junk, N.V., The Hague) p. 399.
7. Trebst, A. and Reimer, S., 1973, Biochim. Biophys. Acta 305, 129.
8. Lozier, R.H. and Butler, W.L., 1972, Febs Letters 26, 161.
9. Haehnel, W., Berlin, personal communication
10. de Kouchkovsky, Y., Gif, personal communication
11. Bauer, R. and Wijnands, M.J.G., Z. für Naturforschg., in press
12. Amesz, J., 1973, Biochim. Biophys. Acta 301, 35.
13. Trebst, A. and Reimer, S., 1973, Biochim. Biophys. Acta 325, 546.

14. Izawa, S., Gould, J.M., Ort, D.R., Felker, P. and Good, N.E., 1973, Biochim. Biophys. Acta 305, 119.

15. Gimmler, H., 1973, Z. Pflanzenphysiol. 68, 289.

16. Saha, S., Ouitrakul, R., Izawa, S. and Good, N.E., 1971, J. Biol. Chem. 246, 3204.

17. Böhme, H. and Cramer, W.A., 1972, Biochemistry, 11, 1155.

18. Reinwald, E., Stiehl, H.H. and Rumberg, B., 1968, Z. Naturforsch. 23b, 1616.

19. Avron, M. and Chance, B., 1966, Brookhaven Symp. Biology 19,149.

20. Trebst, A., 1974, Ann. Rev. Plant Physiol. 25, 423.

21. Trebst, A. and Hauska, G., 1974, Naturwissenschaften 61, 308.

22. Junge, W., Rumberg, B. and Schröder, H., 1970, Eur. J. Biochem. 14, 575.

23. Schröder, H., Muhle, H. and Rumberg, B., 1971, in: Proc. 2nd Int. Congr. Photosyn. Res. Stresa, vol. 2, ed. G. Forti, M. Avron and A. Melandri (Dr. W. Junk, N.V., The Hague) p. 919.

24. Skulachev, V.P., 1971, Curr. Topics Bioenerg. 4, 127.

25. Witt, H.T., 1967, in: Nobel Symposium no. 5, ed. S. Claesson (Interscience) p. 261.

26. Witt, H.T., 1971, Quart. Rev. Biophys. 4, 365.

27. Junge, W. and Ausländer, W., 1974, Biochim. Biophys. Acta 333, 59.

28. Böhme, H., Reimer, S. and Trebst, A., 1971, Z. Naturforsch. 26b, 341.

29. Hauska, G., Reimer, S. and Trebst, A., 1974, Biochim. Biophys. Acta 357. 1.

30. Trebst, A. and Pistorius, E., 1965, Z. Naturforsch. 20b, 143.

31. Wessels, J.S.C., 1964, Biochim. Biophys. Acta 79, 640.

32. Hauska, G., 1972, Febs-Letters 28, 217.

33. Hauska, G., Trebst, A. and Draber, W., 1973, Biochim. Biophys. Acta 305, 632.

34. Oettmeier, W., Reimer, S. and Trebst, A., 1974, Plant Sci. Letters 2, 267.

35. Hauska, G. and Prince, R.C., 1974, Febs-Letters 41, 35.

36. Jagendorf, A.T. and Avron, M., 1958, J. Biol. Chem. 231, 277.

37. Avron, M., 1960, Biochim. Biophys. Acta 40, 257.

38. Avron, M. and Neumann, J., 1968, Ann. Rev. Plant Physiol. 19, 137.

39. Yamashita, T. and Butler, W.L., 1969, Plant Physiol. 44, 435.

40. Ort, D.R. and Izawa, S., 1973, Plant Physiol. 53, 595.

41. Harth, E., Oettmeier, W. and Trebst, A., 1974, Febs-Letters 43, 231.

42. Heathcote, P. and Hall, D.O., 1974, Biochem. Biophys. Res. Commun. 56, 767.

M. AVRON, *Proceedings of the Third International Congress on Photosynthesis*, September 2-6, 1974, Weizmann Institute of Science, Rehovot, Israel
Elsevier Scientific Publishing Company, Amsterdam, The Netherlands, 1974

ELECTRON TRANSPORT REACTIONS, ENERGY CONSERVATION REACTIONS AND PHOSPHORYLATION IN CHLOROPLASTS.

S. Izawa, D. R. Ort, J. M. Gould, and N. E. Good

Department of Botany and Plant Pathology, Michigan State University
East Lansing, Michigan 48824 (U. S. A.)

I. Location of the Sites of Phosphorylation

This part of the paper deals with the use of a variety of electron donors, electron acceptors, and specific electron transport inhibitors to isolate quite different oxidation-reduction reactions responsible for ATP formation in chloroplasts. Figure 1 summarizes the partial reactions thus investigated.

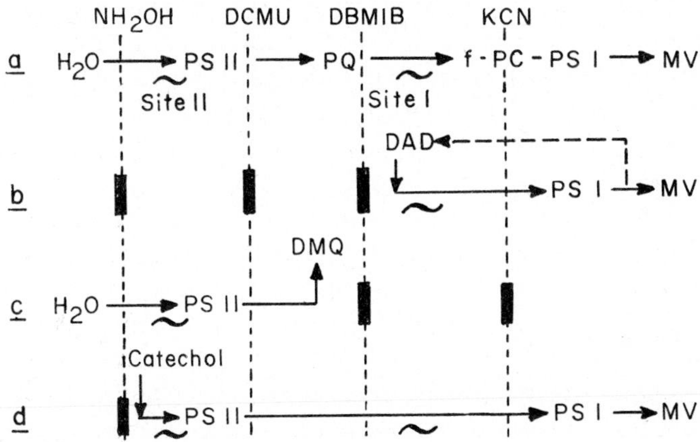

Figure 1. Chloroplast reactions currently available for the study of photosynthetic electron transport and phosphorylation

Figure 1a. The scheme shows the overall electron transport system, the probable positions of the transport blocks utilized, and the apparent location of the two sites of energy conservation. Let us first review the nature of the inhibitions indicated by the

Abbreviations used: DAD, diaminodurene; DBMIB, 2,5-dibromo-3-methyl-6-isopropyl-p-benzoquinone; $DCIPH_2$, reduced form of 2,6-dichlorophenolindophenol; DCMU, 3-(3,4-dichlorophenyl)-1,1-dimethylurea; DMQ, 2,5-dimethyl-p-benzoquinone; TMPD, N,N,N',N'-tetramethyl-p-phenylenediamine.

vertical bars across the transport chain. Hydroxylamine has long been known as an inhibitor of photosynthesis and the Hill reaction. More recently it has been recognized that this amine inactivates the water-oxidizing system (1-3) as do other amines including Tris (4). The great advantage of hydroxylamine-treatment is that it can be carried out in such a manner as to totally and specifically abolish water oxidation without adversely affecting the ability of the chloroplasts to phosphorylate ADP (5). Hydroxylamine-treated chloroplasts are highly active in the Photosystem II-dependent oxidation of those exogenous electron donors which can substitute for water. DCMU is the standard specific inhibitor of Photosystem II reactions. Dibromoisopropylbenzoquinone (DBMIB), introduced by Trebst and his associates (6), is a plastoquinone antagonist which seems to block the reoxidation of reduced plastoquinone (7,8) and thus prevent the transfer of electrons from Photosystem II to Photosystem I. Transfer of electrons between the two photosystems is also blocked by prior treatment of the chloroplasts with KCN (9). In this case the inhibition results from inactivation of plastocyanin (9,10) and some of the consequences of the block are quite different from the consequences of DBMIB inhibition (see below). The partial reactions of the electron transport system made available by the use of such inhibitors are shown in figure 1 (b,c and d).

Figure 1b. A number of exogenous electron donors such as diaminodurene (DAD) and reduced indophenol ($DCIPH_2$) can be oxidized by Photosystem I while $NADP^+$ or methylviologen (MV) is being reduced. These familiar reactions are insensitive to hydroxylamine, DCMU and DBMIB (7). In the past the electron transport has been difficult to measure accurately for two reasons. Superoxide formed by the aerobic oxidation of reduced methylviologen can spontaneously oxidize the electron donor with the consequence that the electron transport may be overestimated (11-14). Furthermore the tendency of the oxidized form of the donor to accept electrons, superceding the acceptor added, can result in a hidden cyclic electron flow with the consequence that the electron transport may be underestimated. This latter problem is particularly acute when the relatively inefficient $NADP^+$-ferredoxin acceptor system is used (15-17). Thus, at the present time the only reliable procedure for measuring the Photosystem I oxidations of exogenous electron donors involves the use of an efficient acceptor such as methylviologen under aerobic conditions, the use of ascorbate to prevent the accumulation of the oxidized form of the donor, and the use of large amounts of superoxide dismutase (13). When these precautions are taken the efficiency of the associated phosphorylation reaction can be measured with some precision. The ratio of ATP molecules formed to electron pairs transferred (P/e_2) measured in this manner is consistently about 0.6 regardless of the electron donor used (13,18 but see 19). We have called the site responsible for the phosphorylation "Site I" for historical reasons since this seems to be the site long known to control the rate of reduction of cytochrome *f* (20-22). As already implied, there are good reasons for believing that this is also the site involved in the so-called "cyclic phosphorylation" mediated by electron carriers such as DAD or pyocyanine. In any event, the reactions, cyclic or non-cyclic, are inhibited if plastocyanin is inactivated by KCN (9) or poly-L-lysine (23).

Figure 1c. Lipophilic strong oxidants such as dimethylquinone (DMQ), oxidized p-phenylendiamine (PD_{ox}) or oxidized diaminodurene (DAD_{ox}) can intercept electrons between the two photosystems (24). Thus the electron transport is totally independent of Photosystem I if DBMIB is present (25-27) or the chloroplasts have been treated with KCN (9) (see figure 2). The Photosystem II - dependent electron transport, which can be very rapid, is then coupled to phosphorylation with an efficiency (P/e_2) of about 0.4 (9,25-28). We have called the site of phosphorylation involved "Site II", both because it was the second identified and because it is associated with Photosystem II.

Figure 1d. As we have already indicated, hydroxylamine-treated chloroplasts can no longer obtain electrons from water but they can oxidize a number of exogenous substances by reactions which depend on Photosystem II and are therefore inhibited by DCMU. Thus electrons can be transported from ascorbate, benzidine (4), catechol (13), ferrocyanide, iodide, etc. (29), to methylviologen in the absence but not in the presence of DCMU. Again the electron transport is coupled to phosphorylation - but the efficiency of phosphorylation depends on the nature of the electron donor (29). With some donors Site I and Site II are operative but with other donors only Site I seems to be operative. This observation will be considered in more depth in the next section of the paper since it may be relevant to an understanding of the mechanism of phosphorylation at Site II.

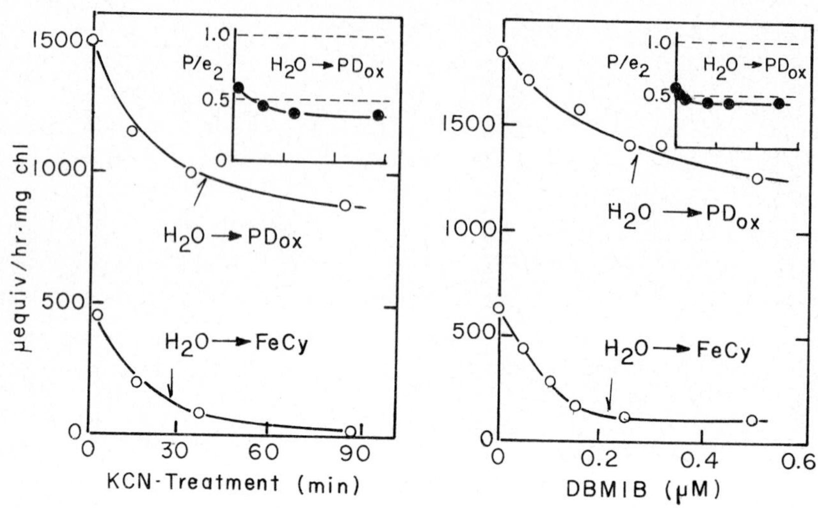

Figure 2. Reduction of a lipophilic strong oxidant with electrons from water. The reduction of oxidized p-phenylenediamine (PD_{OX}) has two components, a large component which depends only on Photosystem II and is therefore insensitive to KCN-treatments or DBMIB and a smaller component which is eliminated by those inhibitors. Note that the large Photosystem II component of the electron transport supports phosphorylation with the same efficiency (P/e_2) of about 0.4 regardless of which inhibitor is used to eliminate the Photosystem I contribution. Ferricyanide reduction, which involves both photosystems in these chloroplasts, is shown for comparison.

II. Characterization of the Sites of Phosphorylation

(a) Proton production and phosphorylation at Site II.

A great deal of evidence has accumulated which interrelates electron transport, the formation of hydrogen ion gradients, and ATP formation in chloroplasts. Much of this evidence supports the "chemiosmotic" explanation of phosphorylation proposed by Mitchell (30) and, indeed, some predictions of Mitchell's theory have been confirmed by experiments so beautifully that the theory has gained considerable popularity. Figure 3 illustrates the presumed workings of the chemiosmotic mechanism in chloroplasts when they are oxidizing water (upper) or exogenous donors (lower) (31; for a review see 32)

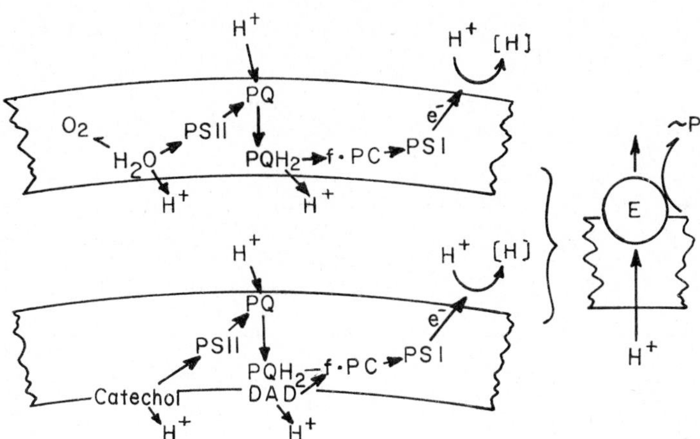

Figure 3. A conventional chemiosmotic interpretation of phosphorylation in chloroplasts. The upper figure shows a segment of a thylakoid membrane oxidizing water. The lower segment shows a segment oxidizing exogenous electron donors. The essential features of this model are: a) Photosystem I oxidations and Photosystem II oxidations both result in the accumulation of protons inside the thylakoid. b) These protons are extruded through the coupling factor in such a manner as to generate ATP.

According to the model hydrogen donors are oxidized on or near the inside of the membrane and as a result hydrogen ions are extruded into the inner space of the thylakoid. This is believed to happen when water is oxidized by Photosystem II (33-35) and again when some intermediate hydrogen carrier such as plastoquinone is oxidized by Photosystem I via cytochrome f (34). The inner protons then somehow generate ATP while diffusing out of the thylakoid via the coupling factor (30). The two proton-producing reactions thus constitute the two "sites" of phosphorylation described above.

The concept of internal proton production as an essential intermediate step in ATP formation has received very strong support from experiments with chloroplasts. As Hauska et al. (36) have pointed out, the fact that the oxidation of DAD by Photosystem I supports phosphorylation while the oxidation of TMPD does not may be a consequence of the fact that DAD oxidation results in hydrogen ion production whereas the removal of an electron from TMPD produces only the free radical. We have now shown that a similar correlation between proton release and ATP formation is observed when exogenous electron donors are oxidized by Photosystem II. Figure 4 compares the oxidation of catechol by hydroxylamine-treated chloroplasts with the oxidation of ferrocyanide by the same chloroplasts. When catechol is oxidized the efficiency of phosphorylation is almost

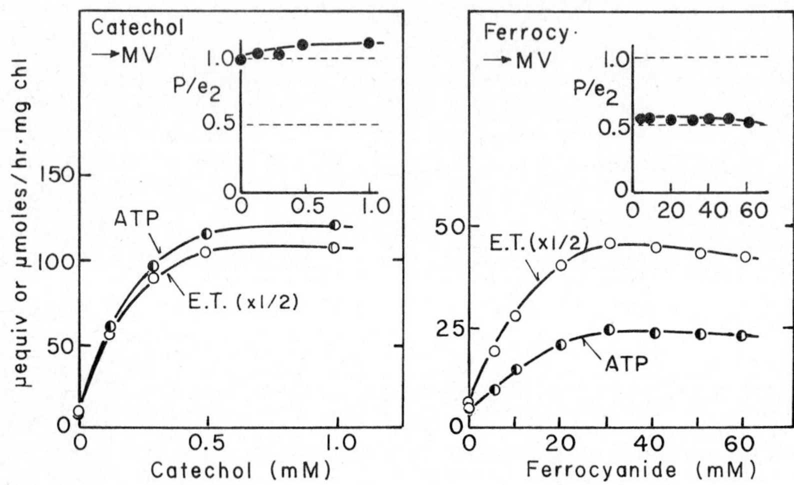

Figure 4. The oxidation of catechol and ferrocyanide by hydroxylamine-treated chloroplasts with methylviologen as electron acceptor. These reactions are inhibited by DCMU and therefore require Photosystem II. Note that the oxidation of catechol supports phosphorylation with an efficiency which implicates both Site I and Site II whereas the oxidation of ferrocyanide supports phosphorylation with the efficiency which is characteristic of the operation of Site I alone.

the same as when water is oxidized and therefore both Site I and Site II must be operative (13). On the other hand, when ferrocyanide is oxidized the efficiency of phosphorylation is lowered to about half although ferrocyanide at the concentrations used has no uncoupling effect (29). Indeed the phosphorylation associated with the transport of electrons from ferrocyanide to methylviologen has all of the properties of Site I phosphorylation and none of the properties of Site II phosphorylation (29; see also figure 6 below). Therefore it is reasonable to suppose that Site II is inoperative when ferrocyanide (pure electron donor) substitutes for water or for catechol (hydrogen donors). Incidentally, the fact that this reaction requires very high concentrations of ferrocyanide (29) tends to confirm the chemiosmotic notion that Photosystem II oxidations occur inside of the thylakoid membrane.

A similar correlation between proton production and the operation of Site II has been observed when other exogenous donors are oxidized by Photosystem II. On the basis of the efficiencies of phosphorylation, it would seem that oxidations of proton-producers such as hydroquinones and benzidine invariably result in energy conservation at Site II. In contrast, oxidations of stable metal complexes or of halide ions do not directly involve proton production and oxidations of such substances apparently do not support phosphorylation at Site II.

Table 1. Phosphorylation Efficiency as a Function of the Donor of Electrons to Photosystem II.

PS II donor	Conc.(mM)	E.T.*	P/e_2
(H_2O)	-	(410)	(1.15)
Catechol	0.5	204	1.10
p-Hydroquinone	0.5	200	0.97
p-Aminophenol	0.5	182	1.02
Benzidine	0.5	175	1.06
Dihydroxybiphenyl	1.0	76	1.09
I^-	20	200	0.46
Ferrocyanide	30	100	0.46
$Fe(dipyridyl)_3$**	1.5	70	0.52
$Mn(oxine)_2$**	0.5	145	0.61
$Mn(dipyridyl)_2$**	0.5	88	0.35

*Electron transport (uequiv/hr.mg chl)

**Secondary reactions associated with the oxidation of these metal complexes may result in large pH changes.

(b) Phosphorylation as a function of proton gradients and pH

As we have seen, an excellent case can be made for the involvement of protons in chloroplasts phosphorylation reactions. However the case for the involvement of trans-membrane proton concentration or activity gradients is not nearly as persuasive. In fact the simple version of the chemiosmotic mechanism depicted in figure 3 seems to be in conflict with a number of observations. Two major areas of conflict are illustrated by figure 5. This figure shows the time-course of phosphorylation at Site I and at Site II, both at pH 6.5 and pH 8.0.

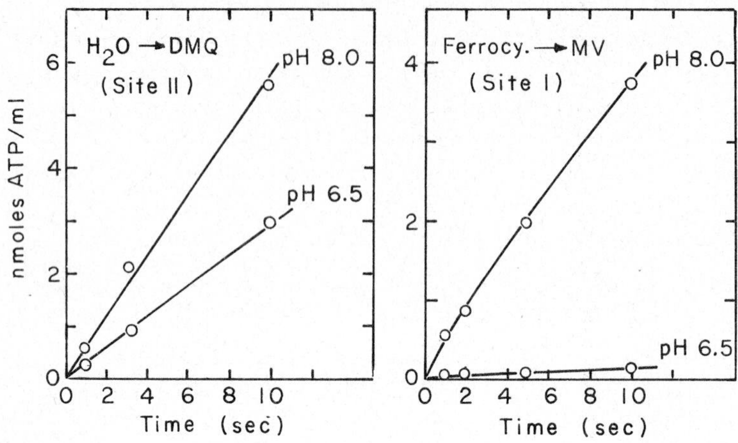

Figure 5. Phosphorylation as a function of the time of illumination.

Clearly phosphorylation is linear with time and therefore must have reached full efficiency within a small fraction of a second. Actually, there is reason to believe that phosphorylation starts very much sooner than this. The flash experiments of Boeck and Witt (37) suggest that phosphorylation occurs long before there is any appreciable difference between the concentrations of protons inside and outside the thylakoid. Therefore, if the escape of protons from the inner aqueous phase is to drive phosphorylation, the protons must be under the impetus of a very large, electron-transport-induced membrane potential. But the internal oxidation of ferrocyanide should produce the same charge separation and the same membrane potential as the oxidation of water or catechol and the oxidation of TMPD should produce the same membrane potential as the oxidation of DAD. Why then is there no phosphorylation by Site I or Site II when TMPD and ferrocyanide are oxidized if the requirement is a membrane potential rather than proton accumulation?

The second problem raised by figure 5 is equally puzzling from the chemiosmotic point of view. At pH 8 phosphorylation proceeds at a good rate whether driven by events at Site II or by events at Site I. At pH 6.5 or lower phosphorylation driven by Site II still goes on apace but phosphorylation driven by Site I stops. The precipitous drop in Site I phosphorylation at pH 6.5 is not primarily due to a decrease in electron transport. It is mostly due to a decrease in the efficiency with which electron transport supports phosphorylation. As figure 6 shows, the P/e_2 at Site I is very sensitive to changes in pH whereas the P/e_2 ratio at Site II is quite indifferent.

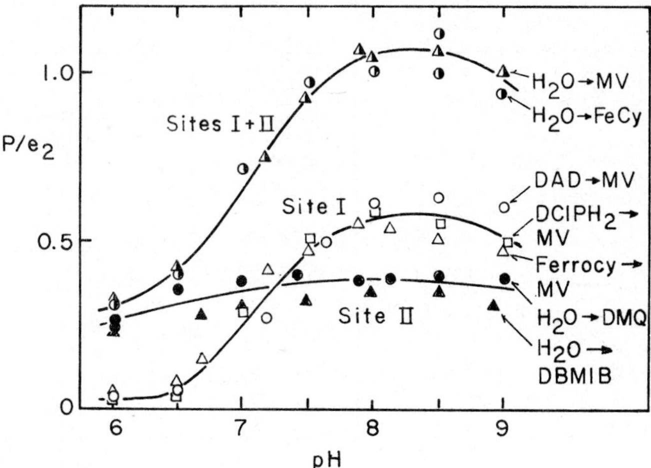

Figure 6. Photophosphorylation efficiency as a function of the external pH.

The sensitivity of Site I phosphorylation to low pH seems to be independent of the chemical nature of the electron donor used: ferrocyanide (anion of a strong acid); $DCIPH_2$ (anion of a weak acid); DAD (a weak base). This makes it extremely improbable that an identical uncoupling effect (due to the donors themselves) is responsible for the low pH inhibition. It is also important to point out that these pH dependence curves are not likely to be the trivial consequences of any other abnormal aspects of the reaction conditions employed. The sum of the curves for Site I and Site II phosphorylation efficiencies, regardless of the electron acceptors and electron donors used, is very close to the curve for the efficiency of the normal Hill reaction which uses both Site I and Site II (see also 38). Consequently it seems safe to conclude that the pH profiles for Site I and Site II, as measured by partial reactions and depicted in figure 6, approximate the true pH dependencies of those individual sites when they are operating in concert in the overall Hill reaction.

The conclusion that Site II is responsible for all of the phosphorylation below pH 6.5 is very difficult to reconcile with the chemiosmotic model. Proton uptake associated with Photosystem I reactions is very active, with a probable efficiency (H^+/e^- ratio) of about 1.0 regardless of pH. (See figure 7). Indeed the efficiency of proton accumulation seems independent of pH when the electron transport is through either site. How then can it be that below pH 6.5 the protons can be used for phosphorylation only if they are accumulated through the action of Site II? The implications of this

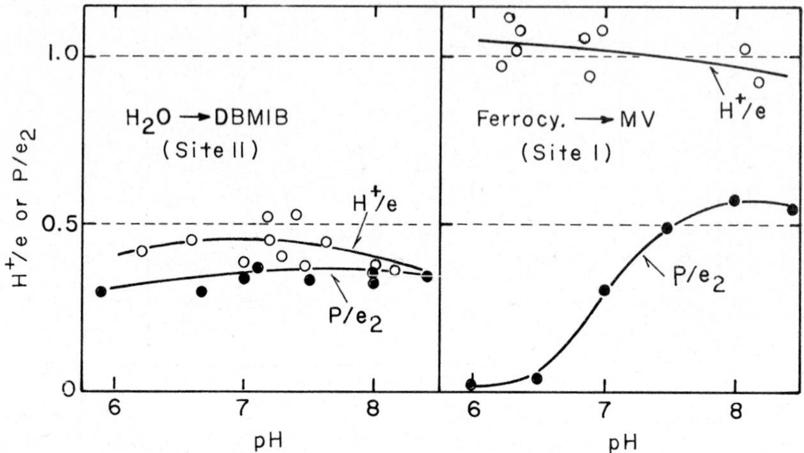

Figure 7. A comparison of the efficiencies of proton uptake and ATP synthesis at different pH's.

question are very important. If we accept the contention that accumulated protons are used to drive phosphorylation we are faced with a chemiosmotic conundrum: Utilization of the accumulated protons for phosphorylation seems to depend on how the protons are accumulated. Clearly this cannot be if both Site I and Site II deposit the protons destined to drive phosphorylation in a common pool, the inner space in the thylakoid.

III. A Proposed Refinement of the Chemiosmotic Theory

Considerations of the chemiosmotic theory of phosphorylation are usually predicated on this assumption that the gradients involved in the conservation of energy result from different concentrations of ions inside and outside the thylakoid. According to the model, there is only one energized state (the inside-outside gradient) and one ATP-synthesizing enzyme (the coupling factor) common to all phosphorylation sites. Furthermore, these "phosphorylation sites" must be thought of as the redox reactions which generate the critical trans-membrane proton gradient.

The concept of a single energized state in the form of a transmembrane hydrogen ion activity gradient is in accord with a great deal of evidence. When energization of the chloroplasts is separated in time from ATP synthesis, as in preillumination experiments (39) or in "acid-bath" experiments (40), the energized state almost certainly is an inside-outside gradient. There is also little doubt that the space inside the thylakoid becomes acid while electron transport is going on (41,42) and little doubt that concurrent ATP synthesis decreases (but does not abolish) the internal acidification (35, 43-45). Finally, as we have seen, it seems that proton production is required for phosphorylation. Such facts must be accomodated by any theory of phosphorylation and the chemiosmotic theory accomodates them admirably.

However, when it pictures the driving force of steady-rate phosphorylation strictly in terms of an inside-outside proton activity gradient, the chemiosmotic theory has corollaries which are not so obligingly consistent with observations. If the energized state of the chloroplast responsible for ATP synthesis consisted of a trans-membrane gradient, or indeed of any property of the membrane as a whole, site specificity with regard to the utilization of the conserved energy would be impossible. Thus there could be no site-specific inhibitions of photophosphorylation other than inhibitions of the redox reactions responsible for the gradient formation. Nevertheless, there are reasons for believing that the forbidden site specificities do exist. As we have shown above, low pH seems to produce an inhibition which is specific for Site I and this inhibition seems to have nothing to do with the formation of the common proton pool within the thylakoid; at low pH the proton gradient is formed but cannot be used for Site I phosphorylation. Similarly, very low concentrations of mercury partially inhibit Site I phosphorylation without inhibiting Site II phosphorylation at all (46). Again the inhibition seems to affect phosphorylation without affecting the earlier steps in the energy conservation.

For these reasons we are inclined to question the chemiosmotic theory as it is illustrated in figure 3. It may be that this simple model should be replaced by a modification which is basically similar to one already considered by Williams (47). Our version of this modified model is illustrated in figure 8. It differs from the original only in that the critical proton activity gradients are within the hydrophobic membrane rather than between the inner and outer aqueous phases. Being strictly local, a steep gradient could be formed almost instantaneously and the time-factor-proton-capacity problem alluded to in the discussion of figure 5 need not arise. Furthermore if Site I and Site II reside in different regions of the thylakoid – and the fact that Photosystem I can be physically separated from Photosystem II (48) implies that they do – each local gradient would presumably have to be utilized by a local coupling factor. In this connection it is of interest to note that the number of molecules of coupling factor in the chloroplast is of the same order of magnitude as the number of electron transport "chains" (49,50). These differently situated coupling factors could respond quite differently to pH or to mercury. Perhaps some critical region of Site II and its coupling factor is less accessible than the equivalent region of Site I to external hydrogen ions or added mercury.

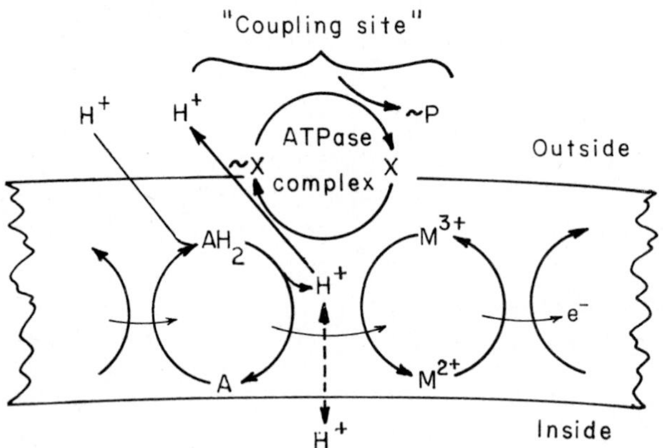

Figure 8. A modified model of the chemiosmotic mechanism which allows for site specificity in the utilization of the energized state for phosphorylation. In this model protons are produced and used within a hydrophobic region of the thylakoid membrane. During steady-state phosphorylation protons migrate to the outside through an ATP-synthesizing complex, the coupling factor, without having been in the inner aqueous phase of thylakoid. However protons can diffuse to the inner space from the local hydrophobic site of proton production-utilization and therefore the site may be in approximate equilibrium with the inner space. According to the model, each proton-producing oxidation center is of necessity associated with its own ATP-synthesizing coupling factor. Thus the coupling factors associated with Site I and Site II oxidation centers are differently situated and may respond in different ways to external conditions.

The fact that trans-membrane gradients <u>are</u> formed during electron transport (51) and the fact these gradients <u>are</u> used for ATP synthesis in the dark (39) can be accomodated by such a "microchemiosmotic" model if we retain the concept of an anisotropically organized membrane. The localized site of proton production and utilization need only be in communication with the inner space in order to permit massive net accumulations of protons. Presumably, such communication would also permit the utilization of inner space protons for phosphorylation as in "acid bath" or preillumination experiments. Thus the acidification of the inner space of the thylakoid, which has been taken to be an essential intermediate step in phosphorylation, may in fact represent a side-reaction which is only indirectly related to steady-state phosphorylation.

References

1. Joliot, A. (1966) Biochim. Biophys. Acta 126, 587.
2. Izawa, S., Heath, R.L. and Hind, G. (1969) Biochim. Biophys. Acta 180, 388.
3. Cheniae, G.M. and Martin, I.F. (1971) Plant Physiol. 47, 568.
4. Yamashita, T. and Butler, W.L. (1968) Plant Physiol. 44, 435.
5. Ort, D.R. and Izawa, S. (1973) Plant Physiol. 52, 595.
6. Trebst, A., Harth, E. and Draber, W. (1970) Z. Naturforsch 25b, 1157.
7. Lozier, R.H. and Butler, W.L. (1972) FEBS Letters 26, 161
8. Böhme, H., Reimer, S. and Trebst, A. (1971) Z. Naturforsch. 26b, 341.
9. Ouitrakul, R. and Izawa, S. (1973) Biochim. Biophys. Acta 305, 105
10. Izawa, S. Kraayenhof, R., Ruuge, E.K. and DeVault, D. (1973) Biochim. Biophys. Acta 314-328.
11. Elstner, E.F., Heupel, A. and Vaklinova, S. (1970) Z. Pflanzenphysiol. 62, 184.
12. Allen, J.F. and Hall, D.O. (1973) Biochem. Biophys. Res. Commun. 52, 856.
13. Ort, D.R. and Izawa, S. (1974) Plant Physiol. 53, 370.
14. Epel, B.L. and Neumann, J. (1973) Biochim. Biophys. Acta 325, 520.
15. Keister, D.L. (1963) J. Biol. Chem. 241, 3575.
16. Gromet-Elhanan, Z. and Avron, M. (1964) Biochemistry 3, 365.
17. Wessels. J.S.C. (1963) Proc. Roy Soc. London (1963) B157, 345.
18. Ort, D.R. submitted to Arch. Biochem. Biophys.
19. Goffer, J. and Neumann, J. (1973) FEBS Letters 36, 61.
20. Avron, M. and Chance B. (1966) Brookhaven Symp. Biol. 19, 149.
21. Kok, B., Joliot, P. and McGloin, M.P. (1969) Progress in Photosynthesis Research Vol. II, 1042
22. Böhme, H. and Cramer, W.A. (1972) Biochemistry 11, 1155.
23. Ort, D.R., Izawa, S., Good, N.E. and Krogmann, D.W. (1973) FEBS Letters 31, 119.
24. Saha, S., Ouitrakul, R., Izawa, S. and Good, N.E. (1971) J. Biol. Chem. 246, 3204.
25. Izawa, S., Gould, J.M., Ort, D.R., Felker, P. and Good, N.E. (1973) Biochim. Biophys. Acta 305, 119.
26. Trebst, A. and Reimer, S. (1973) Biochim. Biophys. Acta 305, 129.
27. Trebst, A. and Reimer, S. (1973) Z. Naturforsch. 28c, 710.
28. Gould, J.M. and Izawa, S. (1973) Eur. J. Biochem. 37, 185.
29. Izawa, S. and Ort, D.R. (1974) Biochim. Biophys, Acta 357, 127.
30. Mitchell, P. (1966) Biol. Rev. 41, 445.
31. Mitchell, P. (1966) in Regulation of Metabolic Processes in Mitochondria (Quaglierello et al. eds) Elsvier, p. 65.
32. Trebst, A. (1974) Ann. Rev. Plant Physiol. 25, 423.
33. Rumberg, B., Reinwald, E., Schroder, H. and Siggel, U. (1969) Progress in Photosynthesis Research Vol. III, 1974.

M. AVRON, *Proceedings of the Third International Congress on Photosynthesis*, September 2-6, 1974, Weizmann Institute of Science, Rehovot, Israel
Elsevier Scientific Publishing Company, Amsterdam, The Netherlands, 1974

NONCYCLIC PHOTOPHOSPHORYLATION IN PHOTOSYSTEM II AND PHOTOSYSTEM I : STOICHIOMETRY AND PHOTOSYNTHETIC CONTROL

P. Heathcote and D.O. Hall

Department of Plant Sciences, University of London King's College,
68 Half Moon Lane, London SE24 9JF

1. Introduction and Summary

The recent introduction of lipophilic ("Class III") electron acceptors[1] acting at PSII and the inhibitor DBMIB[2] (dibromothymoquinone) which acts after PSII has enabled an analysis of the photophosphorylation associated with PSII and PSI. This allows us to compare the two sites of energy conservation in PSII and PSI separately with the overall noncyclic photophosphorylation comprising PSII + I.

The stoichiometry ($P/_{2e}$) of ATP formed per pair of electrons transferred through PSII + I is considered to be two and the associated electron transport shows photosynthetic control by the phosphorylation mechanism, i.e. a state III/ state IV ratio of 4 to 6 (refs. 3 & 4). We have now shown that PSII itself has a $P/_{2e}$ approaching one and a photosynthetic control of about two[4,5]; in PSI a $P/_{2e}$ of one is observable as well as a photosynthetic control of about two.

The importance of establishing the $P/_{2e}$ stoichiometries has implications in the possible contribution of cyclic phosphorylation in the mechanism of CO_2 fixation where 1.5 ATP are required per $NADPH_2$ (per CO_2) and also in determining the number of protons transferred across the membrane per ATP formed, i.e. the H^+/ATP ratio.

Stoichiometries in noncyclic photophosphorylation (PSI + II)

H^+/e	$H^+/2e$	H^+/ATP	ATP/H^+	$ATP/_{2e}$
2	4	2	$^1/_2$	2.0
2	4	3	$^1/_3$	1.33
2	4	4	$^1/_4$	1.0

2. Materials and Methods

The techniques used were those of Heathcote and Hall[5] with extensive details in ref. 6.

3. Results

In Table I it is seen that the $P/_{2e}$ ratio measured as ADP/O in the oxygen electrode with photosynthetic control or as $ATP^{32}/_O$ using P_i^{32} uptake can approach two with PSI + II and approach one with PSII alone. The so-called Class III

TABLE I ADP/O and ATP^{32}/O with class I (PSI & II) and III (PSII) acceptors

	O_2 electrode expts. State III	+ NH_4	ADP/O	O_2 evolution (+ ADP + Pi)	overall ATP^{32}/O	computed PSI ATP^{32}/O
(1) FeCy (2mM)	43	82	1.77	70	1.90	-
" + DBMIB	6	6	-	7	0.94	-
(2)DAD_{ox}(0.1mM)	66	86	0.87	93	0.96)	1.28
+ DBMIB	54	18	0.65	63	0.62)	
(3)2,6DBQ (0.5mM)	58	66	0.87	74	0.96)	1.42
" + DBMIB	23	17	0.51	50	0.48)	
(4) PD_{ox} (0.1mM)	112	114	0.87	128	0.94)	
DBMIB	90	42	0.63	95	0.65)	1.25
(5) 2,5 DBQ (1mM)	41	50	0.91	88	0.92)	1.34
" + DBMIB	23	22	0.57	54	0.56)	

O_2 evolution in μmoles/mg. chlorophyll/hr. All class III acceptors contained 2mM FeCy in the reaction mixture. DBMIB was 1μM. Computed PSI ATP^{32}/O obtained by subtracting the DBMIB-insensitive component of electron transport from the overall ATP^{32}/O. ATP^{32} formation and ATP^{32}/O ratios were linear with time over 5 min. Dark controls were about 1% of FeCy rate.

TABLE II ADP/O and Photosynthetic Control in PSII

Electron acceptors present	Rates of electron transport (μmole O_2/mg Chl/hr) State II	State III	State IV	$+NH_4Cl$	Photosynthetic Control	ADP/O
FeCy	26	47	9	61	5.21	1.66
FeCy + DBMIB	5	5	5	3	-	-
FeCy + DAD	54	69	46	65	1.50	0.89
FeCy + DAD + DBMIB	43	50	33	23	1.51	0.65
FeCy + 2,6 DBQ	36	54	27	58	2.00	0.93
FeCy + 2,6 DBQ + DBMIB	24	33	23	28	1.44	0.69
FeCy + PD	48	71	47	61	1.51	0.76
FeCy +PD + DBMIB	39	57	38	29	1.50	0.54
FeCy + 2,5 DBQ	41	54	31	61	1.74	0.90
FeCy + 2,5 DBQ + DBMIB	27	32	20	24	1.60	0.70

FeCy = 3mM; DBMIB = 1μM; DAD = 1mM; 2,6-DBQ = 5mM; 2,5-DBQ = 1.0mM

Fig. 1. (see next page). Oxygen electrode traces showing two cycles of photosynthetic control by PSII in the presence of (a) 0.1mM DAD + 3mM Fe Cy; (b) 0.1mM DAD + 3mM Fe Cy + 1μM DBMIB; (c) 0.5mM 2,6-DBQ + 3mM Fe Cy ± 1μM DBMIB. The reaction mixture contained, in a total volume of 2ml, chloroplasts equivalent to 100μg chlorophyll, 50mM HEPES (pH 7.5), 100mM sorbitol, 5mM $MgCl_2$, 20mM NaCl, 2mM EDTA, 10mM K_2HPO_4. DAD, 2,6-DBQ, ADP, DBMIB, and NH_4Cl were injected as indicated. Photosynthetic control cycles could also be triggered by the addition of 10mM K_2HPO_4 to reaction mixtures containing 250 nmoles of ADP, once endogenous phosphate had been exhausted in the light. The temperature was maintained at 15°C. Illumination was with quartz iodine lamps through Cinemoid filters (5A) transmitting light between 540 and 740 nm; intensity was 9 x 10^4 ergs cm^{-2} sec^{-1} (measured with Spectroradiometer, Yellow Springs Instrument Co.) which was saturating for non-cyclic electron transport.

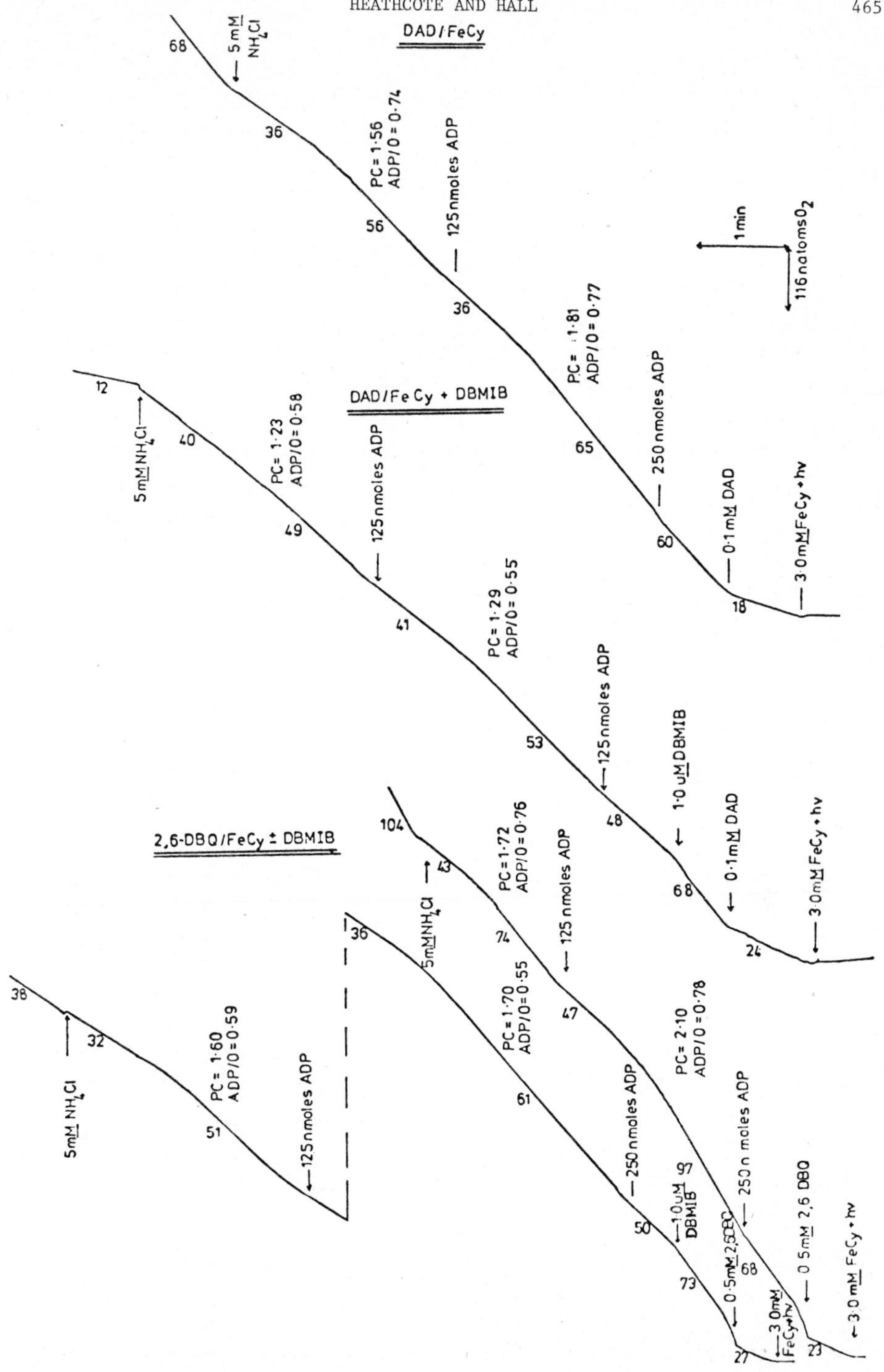
DAD/FeCy
5 mM NH4Cl
PC = 1·56
ADP/O = 0·74
125 nmoles ADP
PC = 1·81
ADP/O = 0·77
250 nmoles ADP
0·1 mM DAD
3·0 mM FeCy + hν
1 min
116 natoms O2
DAD/FeCy + DBMIB
5 mM NH4Cl
PC = 1·23
ADP/O = 0·58
125 nmoles ADP
PC = 1·29
ADP/O = 0·55
125 nmoles ADP
1·0 uM DBMIB
0·1 mM DAD
3·0 mM FeCy + hν
2,6-DBQ/FeCy ± DBMIB
5 mM NH4Cl
PC = 1·60
ADP/O = 0·59
125 nmoles ADP
5 mM NH4Cl
PC = 1·72
ADP/O = 0·76
125 nmoles ADP
PC = 1·70
ADP/O = 0·55
250 nmoles ADP
PC = 2·10
ADP/O = 0·78
250 n moles ADP
1·0 uM DBMIB
0·5 mM 2,6 DBQ
3·0 mM FeCy + hν
0·5 mM 2,6 DBQ
3·0 mM FeCy + hν

electron acceptors (DAD, diaminodurene; PD, phenylenediamine, 2,6- and 2,5-DBQ, dimethylbenzoquinone), kept in the oxidized state by the presence of 2mM ferricyanide (Fe Cy), accept electrons at PSII. Phosphorylation associated with PSII is relatively insensitive to DBMIB whereas electron transport to Fe Cy is inhibited by DBMIB. The extent of the photosynthetic control in PSII depends largely on the state of the isolated chloroplasts (Type A)[7] and the assay conditions; Table II shows values for PSII (Fe Cy is given for comparison) obtained from reasonably good spinach leaves. Obviously the control of electron transport is not nearly as good as the PSII + I systems. The inhibitory action of NH_4 ions with DAD and PD in the presence of DBMIB is quite evident[2].

Two cycles of photosynthetic control in PSII are shown in Figure 1 with DAD ± DBMIB and 2,6-DBQ_{ox} ± DBMIB. Thus control of electron transport by phosphorylation in PSII is capable of repetition and is relatively insensitive to DBMIB.

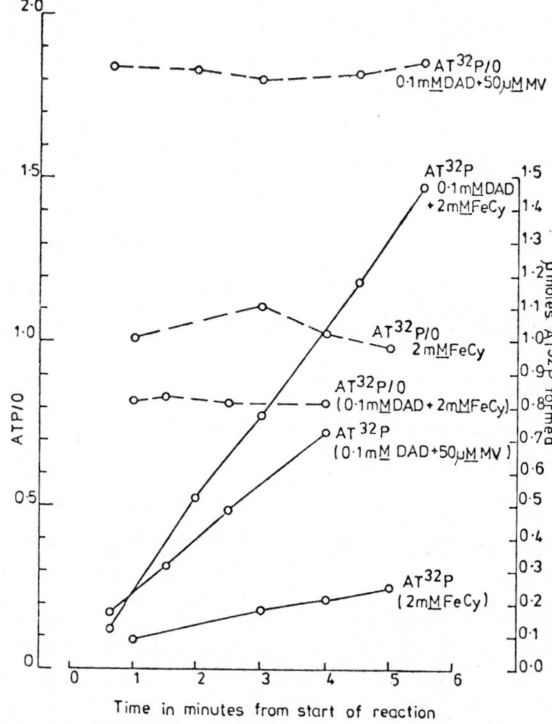

Fig. 2. Linearity of ATP^{32} formation and measured ATP^{32}/O ratios over a period of several minutes with three different electron acceptor systems. Reaction conditions as Fig. 1 except Fe Cy, 2mM; ADP, 1.25mM; MV, 50uM; and NaN_3, 3mM added in the presence of MV as a H_2O_2 trap. Illumination started and stopped reaction. Dark controls were about 1% of the Fe Cy rate.

In Figure 2 the phosphorylation associated with Fe Cy (PSII + I), DAD_{ox} (PSII) and methylviologen (MV; PSII + I) in the presence of DAD is shown to be linear over a five minute period. This is of experimental importance when measuring both $ADP/_0$ and $ATP^{32}/_0$ ratios.

Figure 3 shows that DBMIB itself can act as a phosphorylating Class III electron acceptor at concentrations above about 1μM when it is kept oxidized by Fe Cy[1]. This was confirmed by measuring the reduction of Fe Cy itself.

The action of ATP as an energy transfer inhibitor[8,4] is shown in Figure 4 for Fe Cy (PSII + I; top half) and for 2,6-DBQ_{ox} (PSII; bottom half). It is seen that ATP has a similar inhibitor action (most marked on the non-phosphorylating State II) in PSII as in the overall PSII + I. Further data is also provided in Table V.

The ability of DAD, but not PD or 2,6-DBQ, to bypass the action of DBMIB blocking electron transport between PSII and PSI is shown in Table III. The $P/_{2e}$ doubles when a PSII + I (H_2O ⟶ DAD/DBMIB ⟶ MV) system replaces a PSII (H_2O ⟶ DAD_{ox} ± DBMIB) system. Thus DAD can act as a phosphorylating PSII acceptor and donate electrons to PSI past a DBMIB block.

Phosphorylation is known to be associated with PSI non-cyclic electron transport but here (Figure 5) we show two cycles of photosynthetic control, albeit the

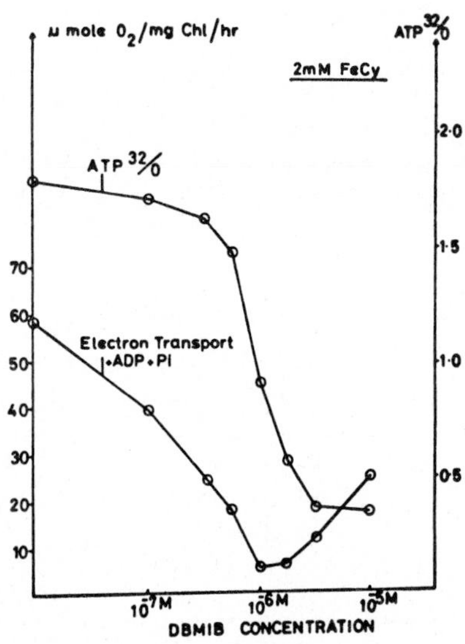

Fig. 3. Effect of DBMIB upon phosphorylating electron transport and ATP^{32} formation in presence of Fe Cy. Conditions as Fig. 2.

values are rather low; again the optimum conditions have not yet been established. The $P/_{2e}$ ratio associated with PSI alone is shown in Table IV where the action of ascorbate, superoxide dismutase and catalase are considered[9] in calculating these ratios; these values are compared with PSII + I (Fe Cy or MV) and also the PSI non-cyclic phosphorylation is shown to be insensitive to DBMIB. Table V shows that ATP also inhibits states II and III electron transport of PSI as it does in PSII.

4. Discussion

These results seem to confirm that there are indeed two separate sites of energy conservation in non-cyclic electron transport - one associated with PSII and the other with PSI. Both sites show $P/_{2e}$ ratios approaching one and show photosynthetic control. The control of electron transport by phosphorylation is

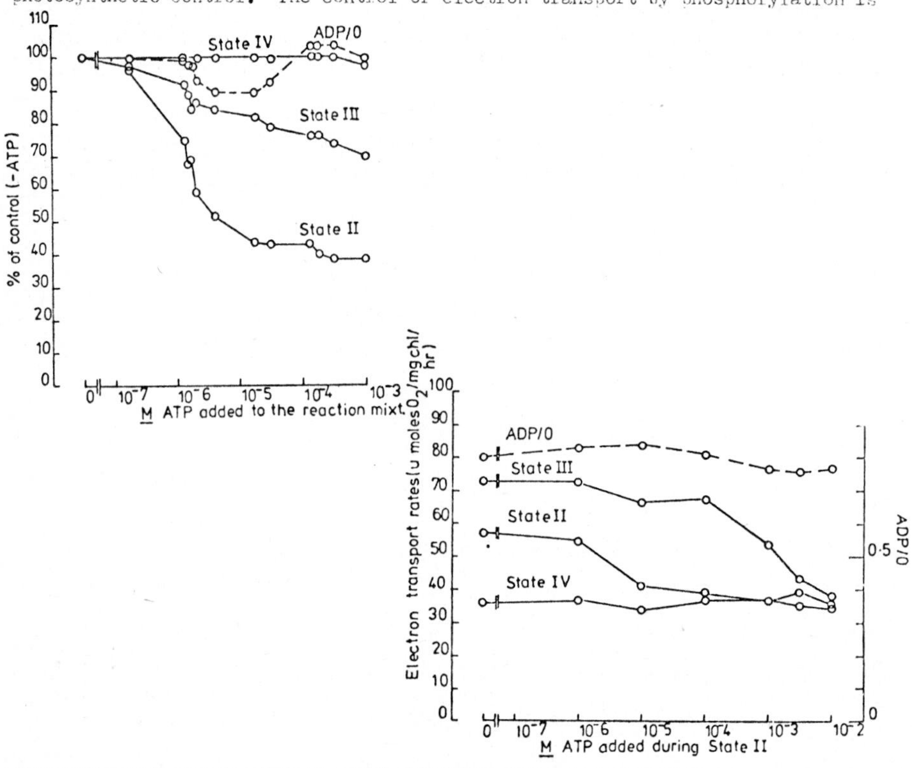

Fig. 4. Effect of ATP on electron transport and photophosphorylation in the presence of 2mM Fe Cy (top) and 0.5mM 2,6-DBQ + 3mM Fe Cy (bottom). Conditions otherwise as in Fig. 1. ATP injected in States II, III or IV had the same effect as addition to the reaction mixture from the start.

not as marked as when PSII and PSI operate in series to Class I electron acceptors at PSI, e.g. MV and Fe Cy. The reasons for this are as yet unknown but may reflect the tight control on the movement of electrons and protons across the membrane.

Work with isolated chloroplasts needs a careful control of the structural state of the chloroplasts in order to measure optimum coupling of electron transport, phosphorylation and carbon fixation. It is quite possible that conflicting values of the $P/_{2e}$ of PSII or PSI or PSII + I are reflections of the experimental conditions. In experiments with complete chloroplasts there is a

TABLE III ADP/O photosynthetic control and ATP^{32}/O with PSI + II and PSII systems

	O_2 electrode expt.			ATP^{32} formation expt.	
	State III	PC	ADP/O	O_2 exchange	ATP^{32}/O
H_2O — FeCy	+ 51	3.40	1.60	+ 65	1.68
H_2O — FeCy/DBMIB	+ 5	-	-	+ 6	0.69
H_2O — DAD_{ox}	+ 76	1.28	0.77	+ 92	0.85
H_2O — DAD_{ox}/DBMIB	+ 51	1.16	0.60	+ 52	0.54
H_2O — MV	- 41	3.71	1.64	- 51	1.74
H_2O — MV/DBMIB	- 1	-	-	- 2	0.33
H_2O — DAD — MV	-24	4.00	1.64	- 52	1.77
H_2O — DAD/DBMIB — MV	- 18	2.61	1.44	- 18	1.57
H_2O — PD — MV	- 25	1.66	1.49	*	*
H_2O — PD/DMIB — MV	- 5	-	-	*	*
H_2O —2,6 DBQ — MV	- 5	-	-	*	*
H_2O — DBQ/DBMIB — MV	0	-	-	*	*

Conditions as in Table I; MV = 50μM. * = not measured.

TABLE IV ADP/O and equivalent P/2e for PSI, PSII and PSI + II

	State III	$ADP/_0$	$P/_{2e}$	State III	$ATP^{32}/_0$	$P/_{2e}$
FeCy	+ 49	1.61	1.61	+ 75	1.71	1.71
FeCy + DBMIB				+ 7	0.85	0.85
FeCy + DAD	+116	0.79	0.79	+ 86	0.81	0.81
" " + DBMIB	+ 92	0.59	0.59	+ 56	0.51	0.51
FeCy + 2,6-DBQ	+ 97	0.79	0.79			
" " +DBMIB	+ 45	0.56	0.56			
MV	- 62	1.49	1.49	- 34	1.70	1.70
" + DBMIB				- 3	0.18	0.18
MV + Asc/DPIP/DCMU	- 61	0.29	*	- 64	0.31	*
" " +KCN	- 77	0.21	0.84	- 85	0.22	0.88
" " +SOD	- 39	0.43	0.86	- 41	0.43	0.86
" " " +DBMIB				- 40	0.40	0.80

* = Cannot be caluclated because of variable presence of SOD and catalase in chloroplasts.

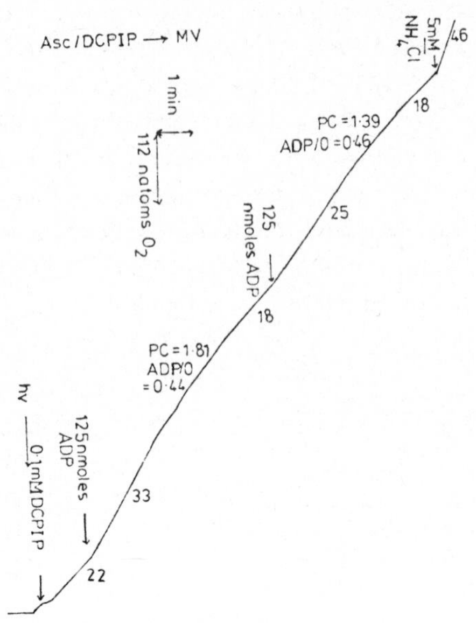

Fig. 5. Oxygen electrode trace showing two cycles of photosynthetic control by PSI in the presence of MV (50μM), Ascorbate (2mM), DCPIP (0.1mM), NaN_3 (3mM), bovine superoxide dismutase (400 IU) and DCMU (5μM).

TABLE V ATP inhibition of electron transport in PSI and PSII

Constituents of reaction mixture	rates of electron transport μmole O_2 exchanged/mg. chl/hr.				Photo-synthetic control	ADP/O
	State II	State III	State IV	NH_4Cl		
DAD_{ox}	71	85	50	93	1.71	0.78
" + ATP	46	68	44	85	1.55	0.72
DAD_{ox} + DBMIB	43	51	36	23	1.42	0.62
" " + ATP	31	36	28	18	1.29	0.67
2,6-DBQ_{ox}	57	73	36	86	2.03	0.80
" + ATP	37	54	37	79	1.46	0.77
2,6 DBQ_{ox} + DBMIB	50	61	36	38	1.70	0.55
" " + ATP	39	54	34	46	1.60	0.55
An/DPIP/DCMU/MV/SOD	23	32	21	63	1.52	0.39
" + ATP	19	25	21	56	1.19	0.38

DAD (0.1mM) and 2,6-DBQ (0.5mM contain 3mM FeCy ; DBMIB = 1μM ; DPIP = 0.1mM ; Ascorbate = 2mM ; DCMU = 5μM ; MV = 50μM ; SOD = 400 units ; ATP = 1mM question of the possibility of flexible coupling [10] between phosphorylation and electron transport which may reflect various control mechanisms in the chloroplast with all its pools of donors and acceptors and its various compartments.

Reports of work with algae are very relevant to our discussion. Gimmler[11] has shown with whole cells of Dunaliella that there is photophosphorylation associated with DBMIB - insensitive conformational changes and that fluorescence of PSII is under the energy control of its coupling site. Kylin[12] has shown in Scenedesmus by titration with DCMU and desaspidin that there are two sites of phosphorylation in non-cyclic electron transport. Careful measurement of the quantum requirement of photosynthesis in synchronous cultures of Scenedesmus by Senger[13] gave an average value of 7.99 - this figure of 8 certainly supports the notion that the $P/_{2e}$ of non-cyclic photophosphorylation is 2 and that under optimum conditions cyclic phosphorylation need not contribute to the process of CO_2 fixation to the level of sugar phosphates.

References

1. Saha, S., Ouitrakul, R., Izawa, S. and Good, N.E., 1971, J. Biol. Chem., 246, 3204; Izawa, S., Gould, J.M., Ort, D.R., Felker, P. and Good, N.E., 1973, Biochim. Biophys. Acta, 305, 119; Gould, J.M. and Izawa, S., 1973, Biochim. Biophys. Acta, 314, 211; Ort, D.R. and Izawa, S., 1973, Plant Physiol., 52, 595; Izawa, S. and Ort, D.R., 1974, Biochim. Biophys. Acta, 357, 127.
2. Trebst, A., Harth, E. and Draber, W., 1970. Z. Naturforsch., 25b, 1157; Trebst, A., 1974, Ann. Rev. Plant Physiol., 25, 423.
3. West, K.R. and Wiskich, J., 1968, Biochem. J., 109, 527; West, K.R. and Wiskich, J., 1973, Biochim. Biophys. Acta, 292, 197.
4. Hall, D.O., Reeves, S.G. and Baltscheffsky, H., 1971, Biochem. Biophys. Res. Comm., 43, 359; Reeves, S.G. and Hall, D.O., 1973, Biochim. Biophys. Acta, 314, 66.
5. Heathcote, P. and Hall, D.O., 1974, Biochem. Biophys. Res. Comm., 56, 767.
6. Heathcote, P., 1974. Ph.D. Thesis, University of London King's College.
7. Hall, D.O. 1972, Nature New Biol., 235, 1256.
8. Shavit, N. and Herscovici, A., 1970, FEBS Lett., 11, 125.
9. Allen, J.F. and Hall, D.O., 1973, Biochem. Biophys. Res. Comm., 52, 856; Allen, J.F. and Hall, D.O., 1974, Biochem. Biophys. Res. Comm., 58, 579; Lumsden, J. and Hall, D.O., 1974, Biochem. Biophys. Res. Comm., 58, 35.
10. Heber, U., 1973, Biochim. Biophys. Acta, 305, 140; Heber, U., 1973, Abstr. Br. Photobiol. Soc. Mtg., October.
11. Gimmler, H., 1973, Z. Pflanzenphysiol., 68, 289 and 385.
12. Kylin, A., Sundberg, A. and Tillberg, J.E., 1972, Physiol. Plant., 27, 376; Kylin, A. and Okkeh, A., 1974, Physiol.,Plant. 30, 58.
13. Senger, H., 1971, in: Proc. II Intl. Congr. Photosynthesis Research, vol.1, eds. G. Forti, M. Avron, and A. Melandri (Dr. W. Junk, N.V., The Hague) p.723.

M. AVRON, *Proceedings of the Third International Congress on Photosynthesis*,
September 2-6, 1974, Weizmann Institute of Science, Rehovot, Israel
Elsevier Scientific Publishing Company, Amsterdam, The Netherlands, 1974

STUDIES ON SUBCHLOROPLAST PARTICLES

Similarity of Grana and Stroma Photosystem I and the Protein Composition of Photosystem I and Photosystem II Particles

J.S.C. Wessels and M.T. Borchert
Philips Research Laboratories, Eindhoven (Netherlands)

Summary

No significant difference was observed between grana and stroma photosystem I particles either in chlorophyll a/b ratio or in P700 content. The lower chlorophyll a/b and P700/chlorophyll ratios, previously found in grana photosystem I particles, may be due to contamination by the accessory complex of photosystem II.

The protein composition of chloroplast membranes, as resolved by sodium dodecylsulphate-polyacrylamide gel electrophoresis, was compared with that of subchloroplast particles. Distinctly different protein patterns have been obtained for photosystem I particles, the reaction-center complex of photosystem II and the accessory complex.

Introduction

It has been well established that mechanical treatment of chloroplasts causes the release of stroma lamellae which have only photosystem I (PS I) activity, leaving a grana fraction containing both PS I and PS II [1-3]. The initial action of digitonin on spinach chloroplasts is similar to that of the mechanical methods: the stroma lamellae form round vesicles between the grana stacks and separate from them [4-6]. The stroma vesicles have only PS I activity and can be further disrupted by the detergent to yield PS I particles [6]. Upon prolonged treatment with digitonin, however, the grana lamellae are also broken and can be fractionated into PS I and PS II fractions [6-10].

These results imply that there may be two kinds of PS I in spinach chloroplasts. One of these is located in the stroma lamellae, and the other type of PS I, which may be closely associated with PS II, is located in the grana region. However, the question of whether the two PS I particles are nearly the same or significantly different has not been settled yet.

On the basis of the P700 contents of whole chloroplasts and PS I fragments released from stroma lamellae, Sane et al. [2] suggested that PS I in the stroma may have a smaller photosynthetic unit size than photosystem I in the grana. Previously we have reported [6] that the two kinds of PS I particles are very similar except for the chlorophyll a/b and P700/chlorophyll ratios, which were shown to be somewhat lower for the grana PS I particles. Similar results were obtained by Arntzen et al. [7]. In addition, these authors found that the grana PS I fraction, but not the stroma PS I fraction, recombined with the grana PS II fraction to reconstitute electron transport activity from diphenylcarbazide to $NADP^+$. It was suggested that stroma PS I may be deficient in a factor which is necessary to properly link these particles with PS II. Gasanov and French [9] showed that the short-wavelength forms of chlorophyll a and also chlorophyll b-650 nm in the PS I fraction from grana lamellae comprise relatively more total area than these same forms in the stroma PS I fraction. Gasanov and Govindjee [11] reported that the fluorescence properties of PS I from grana and stroma are significantly different. The ratio of the fluorescence intensity at 685 nm to 735 nm in PS I from grana was about twice that in PS I from stroma. According to Boardman [8], on the other hand, there are no significant differences between PS I particles obtained from stroma and grana membranes either in chlorophyll a/b ratio or in P700 content.

Recently we have reported [12] that by density-gradient centrifugation of the 100 000 x g supernatant of spinach chloroplasts treated with a high concentration of digitonin three kinds of chlorophyll-containing particles can be obtained: PS I particles (F_I), particles containing the reaction center of PS II (F_{II}) and particles containing the accessory complex of PS II (F_{III}) (Fig. 1). Table I summarizes some properties of these particles after purification by means of chromatography on a DEAE-cellulose column.

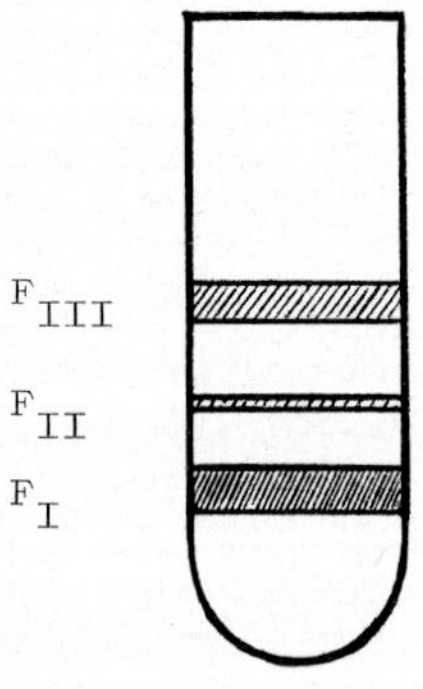

Fig. 1.

Location of the three green bands in a gradient tube after density-gradient centrifugation of the 100 000 x g supernatant of digitonin-treated chloroplasts.

Table I

Some properties of purified subchloroplast particles

Fraction	Abs.max. (nm)	Chl a/b	Chl/P700	PS-I activity (Asc.-DCIP→$NADP^+$)	PS-II activity (DPC→DCIP)
F_I	680	7.5-10	110-130	1700-2000	0
F_{II}	674-675	7-9	0	0	200-300
F_{III}	674-675	1.3-1.5	0	0	0

DCIP, 2,6 - dichlorophenol-indophenol; DPC, 1,5 - diphenylcarbazide
The activities are expressed in µmoles/mg chlorophyll per hour.

Since during the isolation of grana PS I, contrary to that of stroma PS I, a lot of the accessory complex F_{III} is released too, it is conceivable that the lower chlorophyll a/b ratio and P700 content, previously found for grana PS I particles, may be due to some contamination by F_{III}. The results reported here support this view, and show that essentially there is no significant difference between PS I particles obtained from stroma and grana lamellae.

Methods

Chlorophylls a and b were measured spectrophotometrically by the method of Arnon [13]. When the chlorophyll a/b ratio was higher than 7, the spectrofluorimetric method of Boardman and Thorne [14] was used.

PS I and PS II activities were determined as described previously [12]. PS I activity was assayed with ascorbate - dichlorophenol - indophenol (DCIP) as electron donor and $NADP^+$ as electron acceptor, PS II activity with 1,5 - diphenylcarbazide (DPC) as donor and DCIP as acceptor. In reconstitution studies, $NADP^+$ reduction was measured with DPC as electron donor. Ascorbate and DCIP were then replaced by 3 µmoles of DPC, and the absorbance increase was multiplied by 0.66 in order to correct for the oxidation of DPC to diphenylcarbodiazone [7].

Absorption spectra were recorded with a Cary, model 14R, spectrophotometer, fluorescence spectra with a Perkin-Elmer, type MPF-2A, fluorescence spectrophotometer. Fluorescence measurements at low temperature were carried out in 60% glycerol.

P700 was determined from the light-induced absorbance change at

698 nm and from the oxidized minus reduced difference spectrum as described by Yamamoto and Vernon [15], using an Aminco - Chance dual-wavelength spectrophotometer, type 4 - 8450, fitted with a side-illumination attachment.

The isolation of subchloroplast particles from spinach was carried out as described previously [12], using a zonal rotor to enhance both the yield and the purity of the particles.

Sodium dodecylsulphate - polyacrylamide gel electrophoresis was performed as described by Hoober [16] except that the proteins were solubilized by heating in a boiling water bath for 90 s [17]. Usually, about 50 µg protein was applied to each gel. The gels were stained with Coomassie brilliant blue and destained according to the method of Weber and Osborn [18]. Scanning of the gels was carried out with a Joyce-Loebl densitometer at a wavelength of 530 - 550 nm. Molecular weights were estimated from a calibration curve obtained with standard proteins.

Results and Discussion

PS I particles were isolated by density-gradient centrifugation of a 100 000 x g supernatant of digitonin-treated chloroplasts in an MSE, model B XV, zonal rotor [12]. When chloroplasts are treated with 1.3% digitonin, density-gradient centrifugation of the 100 000 x g supernatant produces mainly F_{I}, with some F_{III} and only a little bit of F_{II} [4,6,12]. Though some contamination by grana PS I particles seems probable, we may assume that most of the PS I particles prepared in this way are derived from the stroma lamellae. When the 30 000 x g sediment of the digitonin fragmentation, which is relatively rich in the photosystems derived from the grana lamellae, is treated with digitonin in the presence of 0.35 M NaCl, a 100 000 x g supernatant is obtained which produces much more F_{III} and F_{II} [12]. In this case most of the PS I particles isolated by density-gradient centrifugation may be assumed to be derived from the grana lamellae.

The particles enriched in stroma PS I and those enriched in grana PS I were passed through a Sephadex G-25 column, and then adsorbed on a DEAE-cellulose column, equilibrated with 0.01 M Tris buffer (pH 7.0) containing 0.2% digitonin [12,19]. Elution was achieved by gradually increasing the molarity of the buffer. PS I and PS II activities of all fractions eluted from the column were determined as well as the absorption spectrum and the chlorophyll a/b ratio. Similarly, we have studied these properties of fractions eluted from the column during the purification of F_{II} and F_{III} particles. A survey of the results is represented in Tables II - V. In order to avoid contamination by F_{II} as much as possible, the purification of F_{III} represented in Table V was carried out with a 100 000 x g supernatant enriched in stroma PS I.

Table II

Some properties of fractions eluted from a DEAE-cellulose column during the purification of F_I enriched in stroma PS I

Fraction eluted with:	Abs.max. (nm)	µg Chl	Chl a/b	PS-I activ. (ASC-DCIP→$NADP^+$)	PS-II activ. (DPC→DCIP)
0.05 M Tris	680	2100	8.7	1900	0
0.075 M Tris	679.5	700	7.6	1800	0
0.1 M Tris	678	330	6.3	1200	60
0.1 M Tris +0.1 M NaCl	676.5	180	7.2	700	130

Table III

Some properties of fractions eluted from a DEAE-cellulose column during the purification of F_I enriched in grana PS I.

Fraction eluted with:	Abs.max. (nm)	µg Chl	Chl a/b	PS-I activ. (ASC-DCIP→$NADP^+$)	PS-II activ. (DPC→DCIP)
0.05 M Tris	680	1260	8.5	1780	0
0.075 M Tris	678	550	5.2	970	20
0.1 M Tris	676	700	4.6	450	90
0.1 M Tris +0.1 M NaCl	675	210	7.1	150	220

Table IV

Some properties of fractions eluted from a DEAE-cellulose column during the purification of F_{II}.

Fraction eluted with:	Abs.max. (nm)	µg Chl	Chl a/b	PS-I activ. (ASC-DCIP→$NADP^+$)	PS-II activ. (DPC→DCIP)
0.05 M Tris	677	210	6.3	1280	0
0.075 M Tris	675	1700	3.8	120	80
0.1 M Tris	674	260	5.1	30	150
0.1 M Tris +0.1 M NaCl	674	520	8.1	0	270

Table V

Some properties of fractions eluted from a DEAE-cellulose column during the purification of F_{III}.

Fraction eluted with:	Abs.max. (nm)	µg Chl	Chl a/b	PS-I activ. (ASC-DCIP→$NADP^+$)	PS-II activ. (DPC→DCIP)
0.05 M Tris	675	850	1.4	50	0
0.075 M Tris	675	260	1.5	0	8
0.1 M Tris	675	130	3.0	0	50
0.1 M Tris +0.1 M NaCl	675	30	4.3	0	110

Table VI

Chromatography of a fraction containing both PS I and PS II activity *) on a DEAE-cellulose column

Fraction eluted with:	Abs.max. (nm)	µg Chl	Chl a/b	PS-I activ. (ASC-DCIP→$NADP^+$)	PS-II activ. (DPC→DCIP)
0.075 M Tris	679.5	230	8.2	1020	0
0.1 M Tris	676	90	3.4	220	70
0.1 M Tris +0.1 M NaCl	675	140	7.6	30	120
Fr. before 2nd DEAE	677	700	5.0	650	55

*) Fractions eluted at 0.075 and 0.1 M Tris buffer in the procedure summarized in Table III were combined and applied to the column in 0.075 M Tris.

Table VII

Chromatography of a fraction mainly consisting of the reaction-center complex and the accessory complex of PS II *) on a DEAE-cellulose column.

Fraction eluted with:	Abs.max. (nm)	µg Chl	Chl a/b	PS-I activ. (ASC-DCIP→$NADP^+$)	PS-II activ. (DPC→DCIP)
0.075 M Tris	675	250	2.0	110	0
0.1 M Tris	675	150	9.8	20	140
0.1 M Tris +0.1 M NaCl	674	70	5.9	0	90
Fr. before 2nd DEAE	675	600	3.8	90	80

*) Fraction eluted at 0.075 M Tris buffer in the procedure summarized in Table IV was applied to the column in 0.075 M Tris.

It is seen that in addition to pure PS I and PS II particles (e.g. Table II, 0.05 M Tris fraction and Table IV, 0.1 M Tris + 0.1 M NaCl fraction, respectively) there are several fractions that have both activities. It proved to be impossible to separate the PS I and PS II activities by subjecting these fractions a second time to the chromatographic procedure described above. However, when the fractions were diluted and applied to the DEAE-cellulose column at a concentration of 0.075 M Tris buffer, a reasonable separation could be achieved between PS I and PS II particles. This result suggests that the particles may be able to associate with each other. Table VI shows that the fractions eluted with 0.075 M Tris and 0.1 M Tris + 0.1 M NaCl correspond to almost pure PS I and PS II particles, respectively, as regard to their photochemical activities and chlorophyll a/b ratios. The fraction eluted with 0.1 M Tris, on the other hand, is characterized by low photochemical activities and a low chlorophyll a/b ratio, indicating that this fraction is enriched in the accessory complex F_{III}. Similarly we could accomplish a good separation in a fraction mainly consisting of the reaction center complex of PS II and the accessory complex. (Table IV, 0.075 M Tris fraction). The photochemical activities and the chlorophyll a/b ratios of the separated fractions are given in Table VII.

When we now compare the PS I particles obtained from an F_I enriched in stroma PS I (Table II) with those obtained from an F_I enriched in grana PS I (Table III) and those obtained from an associated particle (Table VI), it is evident that there is no significant difference between these particles in the chlorophyll a/b ratio. Neither could we find any significant difference in the P700 content; the value of the chlorophyll/P700 ratio in the PS I particles varied between 110 and 130. The chlorophyll a/b and chlorophyll/P700 ratios are also similar to those previously determined for PS I particles isolated from stroma vesicles [6].

Computer-assisted curve analyses of the absorption spectra of the differently prepared PS I particles, performed in our laboratory by Dr. O. Elgersma, did not reveal any significant difference in the relative proportions of the different chlorophyll forms.

From these experiments we may conclude that the lower chlorophyll a/b and P700/chlorophyll ratios previously found in grana PS I particles were due to contamination by the accessory complex F_{III}. The difference in the relative amounts of fluorescence emitted at 685 nm and 735 nm in PS I from grana and stroma, recently reported by Gasanov and Govindjee [11], could also be due to the presence of accessory complex in their grana preparation. It is well-known that at liquid nitrogen temperature the ratio of the 735 nm band to the 685 nm band is much higher in PS I than in PS II. The fluorescence spectra of PS I particles and accessory complex F_{III} are shown in Fig. 2. It is apparent from this figure that contamination of PS I by F_{III} will give rise to an increase in the ratio of the 685 nm band to the 735 nm band. In fact we found that the ratio of the fluorescence intensities at 685 and 735 nm was higher as the PS I fraction had a lower chlorophyll a/b ratio.

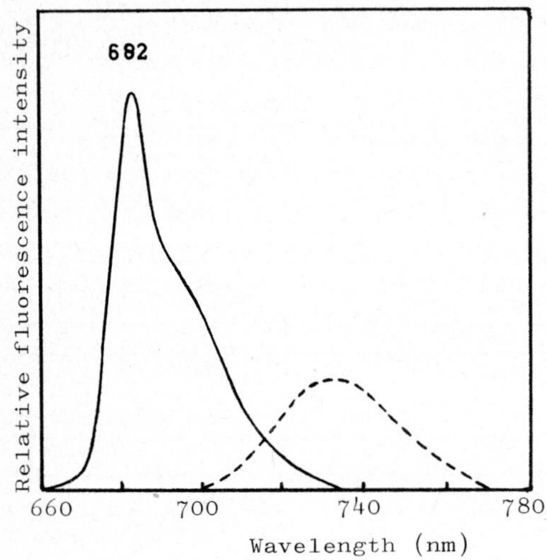

Fig. 2.

Fluorescence emission spectrum of PS I particles (....) and accessory complex F_{III} (——) at liquid nitrogen temperature. Chlorophyll concentration 6.2 and 2.2 µg/ml, respectively. Wavelength for fluorescence excitation 440 nm.

In order to investigate whether there could be a difference in protein composition between stroma and grana PS I particles, we subjected the particles to sodium dodecylsulphate (SDS)-polyacrylamide gel electrophoresis. Fig. 3A shows the densitometer tracing of an SDS-gel electrophoresis pattern obtained from washed, EDTA-treated chloroplast membranes. The main components are found in the 20-30 and 60 kilodalton ranges, confirming results previously obtained by Remy [20], Levine et al. [21] and Klein and Vernon [22]. Levine et al. [21] have shown that the chloroplast membrane polypeptides with molecular weight in the 60 kilodalton range are associated with a digitonin fraction enriched in PS I (144 000 x g fraction). Polypeptides in the 20-30 kilodalton range, labelled II a, b and c, were found to predominate in a fraction enriched in PS II (10 000 x g fraction).

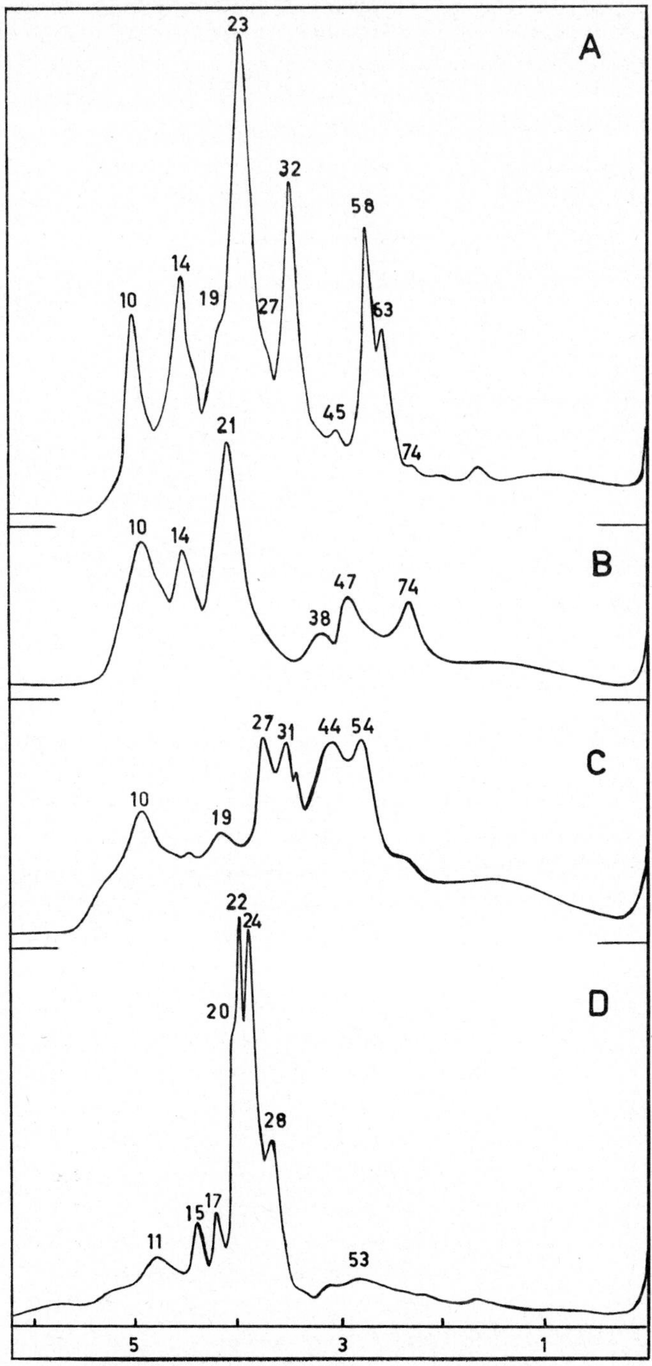

Fig. 3.

Densitometer tracings of polyacrylamide gels following electrophoresis of SDS-solubilized proteins of: spinach chloroplast lamellae (A), stroma and grana PS-I particles (B), the reaction-center complex of PS-II, F_{II} (C), and the accessory complex F_{III} (D). The numbers indicate molecular weight in kilodalton. Ordinate and abscissa represent absorbance in arbitrary units, and migration to the anode in cm, respectively.

The SDS protein pattern of PS I particles obtained from stroma and grana lamellae is represented in Fig. 3B. In both systems the main component is found at 21 kilodalton. Other proteins are distinctly observed at 10, 14, 47 and 74 kilodalton. No difference was found in the protein composition of the differently prepared PS I particles, supporting our conclusion that stroma and grana PS I are identical.

Recently, Anderson and Levine [23] reported that polypeptides from the extensively unstacked membranes of chloroplasts from the chlorophyll-deficient mutant strains of barley and pea as well as those obtained from the agranal bundle sheath cell chloroplasts of maize are deficient in the polypeptides II b and c. Since both the barley and pea mutant chloroplasts and the maize bundle sheath chloroplasts possess PS II activities, it was proposed that these polypeptides are required for membrane stacking in higher plant chloroplasts. The question of whether the polypeptide II a may be required for PS II activity was not answered by these authors because fractionation of the photosystems of the mutants or the agranal bundle sheath chloroplasts of maize was not successful. It was of interest, therefore, to study the SDS-gel electrophoresis pattern of both the reaction-center complex of PS II (F_{II}) and the accessory complex F_{III}.

Fig. 3C shows that in the reaction-center complex of PS II prominent bands are actually observed in the region characteristic of polypeptide II a (27 and 31 kilodalton). The 22 and 24 peaks, on the other hand, which predominate in F_{III} (Fig. 3D) and may be identical with polypeptides II b and c, are missing in F_{II}. These results suggest that polypeptide II a is associated with PS II activity, and that the accessory complex F_{III} is required for stacking.

Thornber and Highkin [24] have recently shown that PS II chlorophyll-protein, one of the two major chlorophyll-protein complexes in photosynthetic membranes of higher plants and green algae, is not present in a mutant of barley lacking chlorophyll b. As this mutant can live photoautotrophically a more appropriate name, light-harvesting chlorophyll a/b - protein, was proposed for this pigment - protein complex. They further reasoned that the complex exerts a strong influence on the organization of photosynthetic lamellae in higher plant chloroplasts by maintaining lamellae in contact with each other. Upon subjecting F_{III} to SDS-polyacrylamide gel electrophoresis by methods described by Thornber [24,25] we obtained two chlorophyll-containing zones, a fast migrating zone containing free pigment, and a zone of intermediate mobility characteristic of the PS II chlorophyll-protein. This result indicates that the accessory complex F_{III} may be identical with the light-harvesting chlorophyll a/b - protein described by Thornber and Highkin, and is consistent with the view that this complex is required for membrane stacking in chloroplasts.

Attempts to reconstitute electron transport activity from DPC to $NADP^+$ by recombining highly purified PS I particles and the reaction-center complex of PS II were not yet successful. Recombination was carried out under various conditions, among other things, by using methods described by Arntzen et al. [7] and Ke and Shaw [26]. Partial reconstitution, up to 50 µmoles $NADP^+$ reduced/mg chlorophyll per hour, could be obtained, however, with impure preparations. It is possible that purification of the digitonin particles involves the loss of a factor that is necessary for reconstitution of the photochemical activity. It should be pointed out that Ke and Shaw [26] reported that certain batches of their Triton PS I particles, for unknown reasons, would occasionally not reconstitute. Furthermore, the SDS protein patterns of digitonin subchloroplast particles, described in this paper, seem to be somewhat different from those of Triton particles [22].

References

1. Jacobi, G. and Lehmann, H. (1969) in Progress in Photosynthesis Research (Metzner, H. ed.), Vol. I, p. 159, Laupp, Tübingen.
2. Sane, P.V., Goodchild, D.J. and Park, R.B. (1970) Biochim. Biophys. Acta 216, 162.
3. Arntzen, C.J., Dilley, R.A. and Neumann, J. (1971) Biochim. Biophys. Acta 245, 409.
4. Wessels, J.S.C. and van Leeuwen, M.J.F. (1971) in Energy Transduction in Respiration and Photosynthesis (Quagliariello, E., Papa, S. and Rossi, C.S., eds.), p. 537, Adriatica Editrice, Bari.
5. Goodchild, D.J. and Park, R.B. (1971) Biochim. Biophys. Acta 226, 393.
6. Wessels, J.S.C. and Voorn, G. (1972) in Proc. 2nd Int. Congress on Photosynthesis Research (Forti, G., Avron, M. and Melandri, A., eds.), Vol. I, p. 833, Dr. W. Junk, The Hague.
7. Arntzen, C.J., Dilley, R.A., Peters, G.A. and Shaw, E.R. (1972) Biochim. Biophys. Acta 256, 85.
8. Boardman, N.K. (1972) Biochim. Biophys. Acta 283, 469.
9. Gasanov, R.A. and French, C.S. (1973) Proc. Natl. Acad. Sci. U.S. 70, 2082.

10. Brown, J.S. and Gasanov, R.A. (1974) Photochem. Photobiol. 19, 139.

11. Gasanov, R.A. and Govindjee (1974) Z. Pflanzenphysiol. 72, 193.

12. Wessels, J.S.C., van Alphen-van Waveren, O. and Voorn, G. (1973) Biochim. Biophys. Acta 292, 741.

13. Arnon, D.I. (1949) Plant Physiol. 24, 1.

14. Boardman, N.K. and Thorne, S.W. (1971) Biochim. Biophys. Acta 253, 222.

15. Yamamoto, H.Y. and Vernon, L.P. (1969) Biochemistry 8, 4131.

16. Hoober, J.K. (1970) J. Biol. Chem. 245, 4237.

17. Anderson, J.M. and Levine, R.P. (1974) Biochim. Biophys. Acta 333, 378.

18. Weber, K. and Osborn, M. (1969) J. Biol. Chem. 244, 4406.

19. Wessels, J.S.C. (1968) Biochim. Biophys. Acta 153, 497.

20. Remy, R. (1971) FEBS Letters 13, 313.

21. Levine, R.P., Burton, W.G. and Duram, H.A. (1972) Nature New Biol. 237, 176.

22. Klein, S.M. and Vernon, L.P. (1974) Photochem. Photobiol. 14, 343.

23. Anderson, J.M. and Levine, R.P. (1974) Biochim. Biophys. Acta 333, 378.

24. Thornber, J.P. and Highkin, H.R. (1974) Europ. J. Biochem. 41, 109.

25. Thornber, J.P. (1970) Biochemistry 9, 2688.

26. Ke, B. and Shaw, E.R. (1972) Biochim. Biophys. Acta 275, 192.

M. AVRON, *Proceedings of the Third International Congress on Photosynthesis*, September 2-6, 1974, The Weizmann Institute of Science, Rehovot, Israel
Elsevier Scientific Publishing Company, Amsterdam, The Netherlands, 1974

ELECTRON TRANSPORT IN PHOTOSYSTEM II

Bessel Kok, Richard Radmer and Charles F. Fowler

Martin Marietta Laboratories (RIAS)
1450 South Rolling Road
Baltimore, Maryland 21227

Equal abundance of System I and System II traps

In the experiment shown in Fig. 1 we measured flash yields of viologen reduction. Isolated spinach chloroplast were deposited on the bare polarograph electrode using positive polarization (1). As indicated in the figure, the flow medium contained 50 mM ammonium chloride, an agent which progressively inhibits Photosystem II (2). A sequence of flashes (10 per sec) was given after a 2-second pre-illumination with strong light, which presumably reduces the pool of electron carriers between the photoacts. Consequently, upon termination of the white light, P700 returns to its fully reduced state (3) so that the height of the first flash reflects the maximum possible flash yield of viologen (Y_{max}). Figure 1 shows that after about 6 flashes the flash yield begins to decline because the pool becomes depleted and after about 20 flashes a steady state yield is attained, Y_{ss} being about 75% of Y_{max}. Since the only source of electrons for System I in the steady state is System II this Y_{ss} value indicates that the System II traps in the chloroplast suspension reduce 75% of the System I traps. Since we know that 10% of the photoevents in System II are ineffective ("misses"(4)), the ratio System II/System I trap must be at least 0.83. We have observed even better ratios and conclude that in good material the number of System II traps is about equal to the number of System I traps.

Figure 1 (left) further shows that the high concentration of ammonium in the flow medium causes a progressive decline of the steady state flash yield of viologen (i.e., a decrease of the number of active System II traps). Control experiments using a concentration polarograph and continuous light showed a very similar loss of the rate.

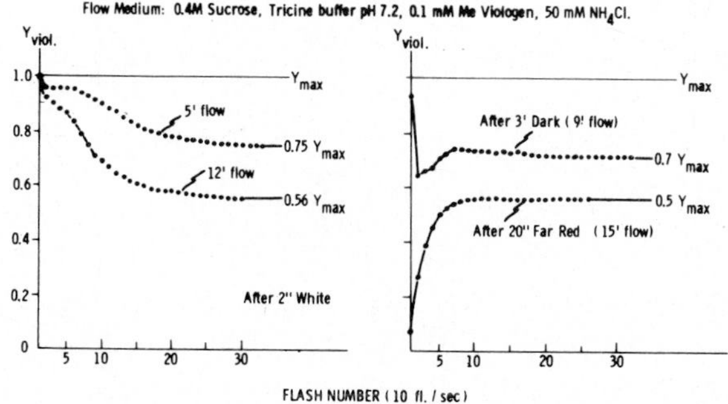

Fig. 1

Aging of the chloroplasts has a similar albeit slower effect as the high concentration of ammonium ion used in Expt. Fig. 1. In the experiments reported below, we used the ratio Y_{ss}/Y_{max} as an index of the ratio of active System II/System I traps.

Lost equivalents between Systems I and II

Figure 1 (right) shows two additional observations. When a flash sequence was given after a 3-minute dark period the first yield is nearly maximal, which shows that P700 had become reduced in the dark. The second flash, however, is low compared to Y_{ss} which is only gradually approached after about 5 flashes. Initially, not all reduced equivalents produced by System II by the first (second, etc) flash reach P700. We explain the deficit by assuming that the "lost equivalents" instead reduce another intermediate ("C") which reacts slowly with $P700^{+}$, if at all. (An alternate explanation, that electrons from X^{-} initially reduce a side pool instead of viologen, seems untenable. In this case the first and not the second flash would be the lowest).

The same deficit pool is also seen in the last of the four experiments in Fig. 1 (right). In this case, the flash sequence was preceded by an illumination with 720 nm light. This light was shut off ~0.1" before the first flash was given. As expected, this left P700 largely oxidized so that the yield of the first flash was quite low. Surprisingly, however, also the next four flashes are (to a decreasing extent) lower than Y_{ss}.

There is no reason to assume that System II is non-functional after far red light -- to the contrary all Q's are in the oxidized (active) state. We expect the first flash to reduce all Q's in the system and these to in turn reduce all P's; i.e., Y_2 to be maximal. Therefore, we again conclude that one or more reducing equivalents made by System II, do not go to P700 and instead reduce an intermediate "C" located on a side path.

The latter experiment is essentially repeated in Fig. 2, using fresh and aged chloroplasts. The fresh chloroplasts show much

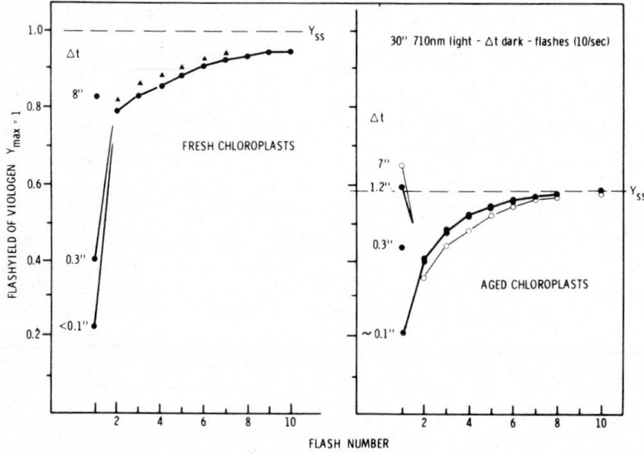

Fig. 2

higher values of $Y_2 \longrightarrow Y_{ss}$ but still a clear initial deficit, which is about equal to that seen in the aged chloroplasts. With only 0.1" separation between the far red preillumination and the flash sequence, Y_1 is low -- as expected. However, if a dark time (Δt) is given between the two illuminations, we notice a) a pronounced increase of Y_1 (i.e., a reduction of P700)with a time constant of about a second and b) a simultaneous increase of the deficit pool (e.g. closed vs open circles Fig. 2, right). We interpret this observation as a slow (~1 sec) reduction of $P700^+$ in the dark by "C" which apparently is being reduced even in far red illumination. This loss of electrons from C then is made up again in the subsequent flash sequence.

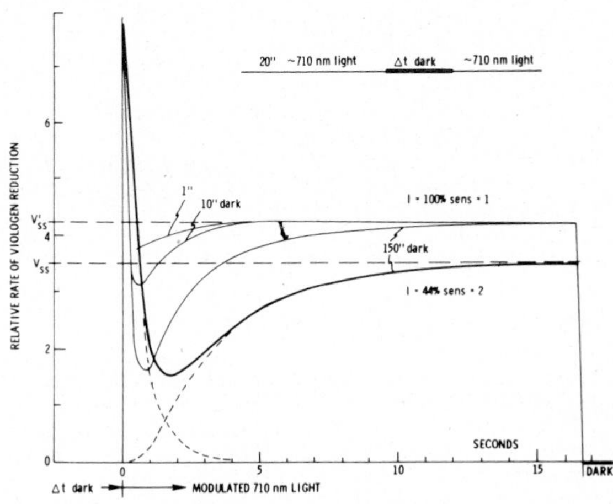

Fig. 3

Figure 3 shows essentially the same phenomenon: In this case we measured the modulated rate of viologen reduction using fresh chloroplasts and a modulated 710 nm beam (which, like a flash in aged chloroplasts sensitizes mainly System I). The light was given for 20" to attain a steady state rate, then simply switched off and switched on again after Δt seconds dark. With increasing dark time an initial spike (a higher rate of viologen reduction or more reduced P700) develops, followed by a dip. The time course of the rate (drawn as a heavy line) can be analyzed (dashed curves) as being composed of (1) a first order depletion of the amount of P700 which was reduced in dark and (2) the filling of a deficit pool (which was presumably oxidized by $P700^+$ in the dark period).

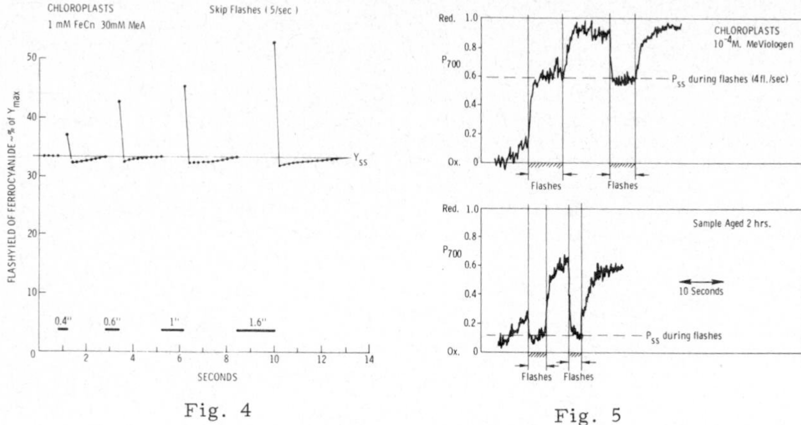

Fig. 4 Fig. 5

Figure 4 shows an equally simple experiment with a similar result. Here we gave a steady sequence of flashes (5/sec) and skipped 1, 2, 3, etc flashes in the sequence. The first flash following an interruption appears to be increased (half time of the order of a second) and the next few flashes show a deficit. Note that in this experiment ferricyanide rather than (actually in addition to) methylviologen was used as an electron acceptor; this did not essentially change the phenomena.

Figure 5 illustrates that (presumably) the same phenomenon can also be observed by monitoring P700 spectroscopically instead of via viologen reduction. In the first experiment (top), fresh chloroplasts were used. After P700 was brought in its oxidized state by a 720 nm light, a sequence of ~40 flashes was given. During this sequence P700 attained a steady state in which it was ~60% reduced (indicating that System II was rather deficient with respect to System I). Following the flashes, we notice a slow return to the fully reduced state with a half time of about a second. A second flash sequence brings the pigment to the same steady state level of reduction and we again notice the slow reduction after the last flash. The same experiment was repeated after the chloroplasts had aged ~2 hours at room temperature (Fig. 5, bottom). The flashes now caused ~90% oxidation of P700--indicating that the number of active System II traps had become quite low. Following the last flash, we observe the reduction of about 50% of P in a few seconds. Apparently, in both cases the flashes produced an amount of reductant "C" which, because of its slow reaction with P^+ cannot be identified with any member of the normal $Q \longrightarrow P$ chain.

We have also performed experiments similar to those shown in Fig. 5 using continuous light instead of flash sequences. The slow post-illumination reduction of ~50% P can only be seen with long wave light (or 650 light and aged chloroplasts) which does not leave any rapidly reacting electrons in the $Q \rightarrow A_2 \rightarrow A_1 \rightarrow P$ chain.

A striking feature of these experiments is that even 710 nm light which yields very few System II turnovers relative to System I (not unlike the effect of flashes in aged chloroplasts in Fig. 5) can reduce "C" provided it has high enough intensity. Thus a strong 710 light can be more effective than a weak 690 nm beam.

Interpretation

Our interpretation of these data is shown in Fig. 6. We assume a secondary System II reductant "C" which is rapidly reduced by primary reductant Q^-. It thus competes successfully with A_2, the normal electron acceptor of Q^-. On the basis of the above data, we estimate that the time constants for the two steps are about equal (~200 μsec (3, 7, 8)). Additional support for the direct connection $Q \rightleftharpoons C$ in the scheme of Fig. 6 rests on:

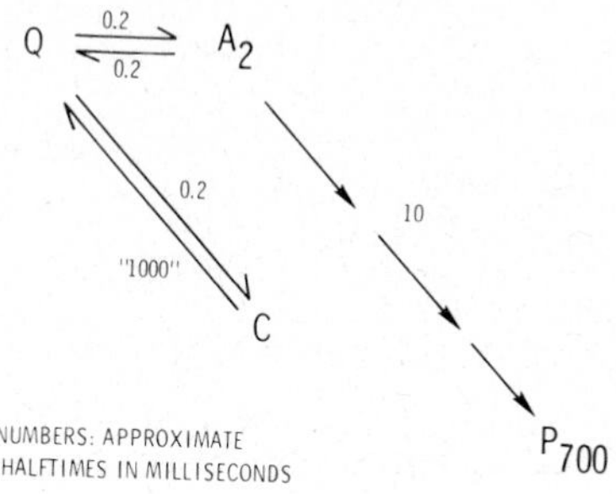

Fig. 6

(a) Observations with Hg^{++} poisoned chloroplasts: Radmer and Kok (9) showed that in this material the Q-P chain appeared to be blocked after A_2. Starting from the oxidized state System II can reduce ~4 "low K" electron acceptors (A_2) and one "high K" acceptor--this is presumably "C".

(b) Observations in very high light (Morin(10)) and with DCMU poisoned chloroplasts (Doschek and Kok(11)) in which the chain appears to be blocked after Q. These indicate that, starting from the oxidized state System II can very rapidly reduce two electron acceptors (within $<$0.5 msec).

In contrast, the back reaction in which C^- reduces Q is quite slow. It can only be seen indirectly when the electron goes from C^- through the Q-A chain to P^+700 (in about a second). We thus arrive at a high equilibrium constant (~10^4) i.e., a large difference between the midpoint potentials of Q and C (200-300 mV). We tentatively assume that the concentration of C equals that of Q, i.e., one per reaction chain. The rapid forward reactions and slow back reaction between Q and C readily explain the peculiar color and intensity dependence of the redox state of C: as long as a light beam, of any wavelength, provides more than one photon per second to the System II centers, C will become reduced.

Different reaction chains presumably communicate via the quinone pool. (12) Thus, in strong light the good System II units can reduce the C^-'s connected to "bad" System II units. Consequently, following an illumination, C^- can reduce a considerable fraction of P700 in the system.

Is "C" identical with cytochrome b-559 ?

The peculiar color and intensity dependence of "C" reminded us of the behavior of cyt b-559 as described in some reports, and seen in some preliminary experiments in this laboratory. Unfortunately the literature concerning this cytochrome is confusing (re 13) and not unlike the holy scriptures, seems to offer a chapter and verse for every occasion or theory. Ikegami and Katoh (14) estimated a midpoint potential of 360 mV for the "second reductant" seen in the fluorescence rise curve in the presence of DCMU. They suspected, but could not confirm that cyt b-559 was involved.

Space does not allow elaboration of these matters or of the fact that our own data (exemplified in Figs. 2-5) show additional complexities and uncertainties. There probably is more than one place where electrons can "hang up" between the photoacts (e.g. Bouges-Bocquet(15), Fowler, unpublished). Moreover, the polarograph electrode with positive (or negative) polarization acts as an electron sink (or source) which can add or subtract to the pools. However, the main hypothesis we want to present in this paper (Fig. 10) does not necessarily require homogeneity of "C" and for the moment the notation C559 aids simplicity.

Dark equilibration of the S states with O_2

One of the cornerstones of the O_2 evolution model of Kok et al(4) is the assumption that the S_1 state is stable in the dark. The main support for this assumption is the observation that the ratio S_1/S_0 attained after deactivation can be changed by giving e.g., one or three preflashes. We have always been uncomfortable with the assumption of two stable states. Joliot et al (16) observed that the setting of the states by preflashes was not permanent; in long enough dark periods the system tends to revert to the "normal" ratio S_1/S_0=3. Bouges-Bocquet(8) went one step further and showed that in the presence of ferricyanide all trapping centers attained the S_1 state ($S_1/S_0=\infty$), while reduced indophenol induced a low ratio $S_1/S_0(\lesssim 1)$. She concluded that the midpoint potential of the reaction $S_1+e \rightleftharpoons S_0$ is quite low. We consider this interpretation, although not impossible, unsatisfactory from an energetic viewpoint.

We have confirmed and extended these same observations, and encountered a great deal of variability and the usual number of complicating phenomena. For the moment, we restrict ourselves to the most perturbing aspects:

(a) How can the S_1 and S_0 states equilibrate in the dark, while on the other hand we know that the trapping centers operate entirely independent of each other?

(b) How can a relatively low potential agent like ferricyanide or benzoquinone produce S_1 states?

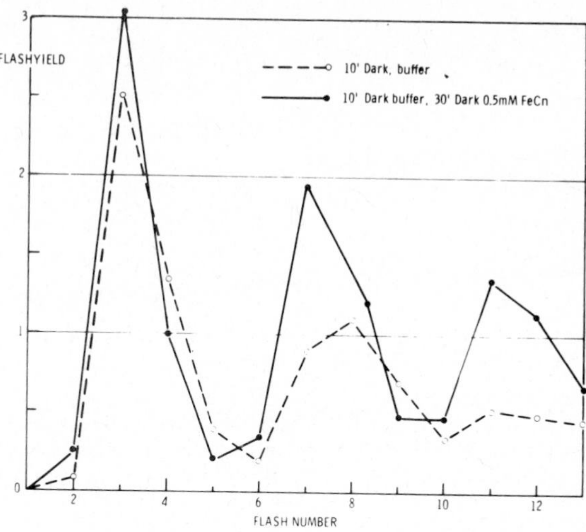

Fig. 7

A strong argument that this "oxidation" is only an indirect effect is given by the results shown in Fig. 7. Two parallel samples of chloroplasts were incubated for 10' in regular buffer medium. This caused complete deactivation as shown by the oscillation of the O_2 flash yield measured in one sample (open circles connected by dashed line). The other sample was subsequently exposed for 30' to 0.5 mM ferricyanide and also measured in this solution (dots, solid line in Fig. 7). Note that not only Y_3 but also Y_2 increased in dark. Thus we either have to assume either a) ferricyanide can oxidize the system to the S_2 level -- a very unattractive thought indeed, or b) a stronger oxidant is involved; the obvious candidate is molecular O_2 which has ready access to the "O_2 evolution box." Therefore we assume that not only S_4 but all S states equilibrate with O_2, and that the "normal" ratio S_1/S_0 which is slowly attained in dark reflects the reactions $S_0+O_2 \longrightarrow S_1$ and $S_1+O_2 \longrightarrow S_2$ balanced by the simultaneously occurring reductions which we call "deactivation." Scheme Fig. 10, discussed below, incorporates these oxidations and at the same time accounts for the various effects of low potential agents like ferricyanide and ascorbate.

Deactivation

The rate of deactivation -- the loss of O_2 precursor states in the dark -- varies widely with conditions. The half life of S_3 can be as brief as 0.6 sec and as long as 5 minutes (see Figs. 8 and 9) -- a 400 fold variation. Drs. Joliot and Forbush as long as seven years ago were aware that there must be two paths of deactivation: a "fast" one observed after bright light and a slow one prevailing after weak light. The analysis by Radmer and Kok(6) (illustrated in Fig. 8) showed that the fast (0.6 sec) path is predominant when primary

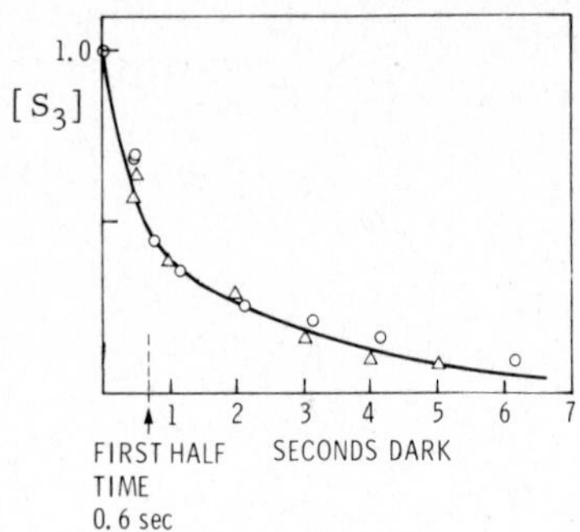

Fig. 8

acceptor Q (and thus its associated pool A_2) are reduced, which is the case in strong light or in the absence of an electron acceptor. This path becomes inoperative when the Q A_2 complex is oxidized, for instance, in weak long wave light. The analysis of Forbush et al(5) concerned exclusively the second slow deactivation path. Forbush postulated the existence of a reductant "R" which had a concentration of about one per trapping center and was generated by a single System II excitation.

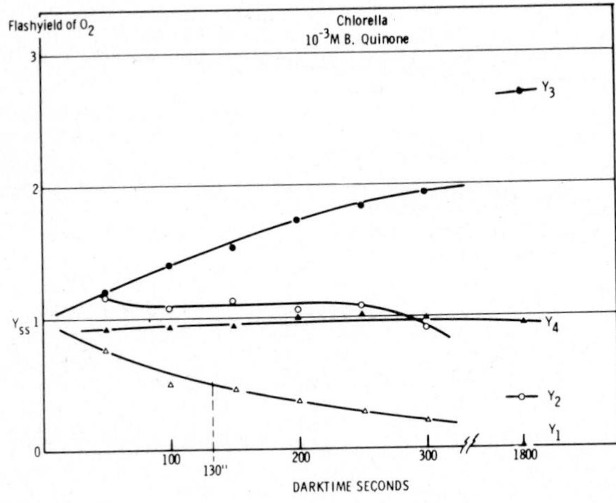

Fig. 9

The intermediate "C" described above, which we have tentatively identified with cyt b-559, might be identical with "R", since: (a) "C" is rapidly reduced by the photoreductant of System II (see above); (b) in isolated chloroplasts in dark cyt b-559 seems to occur ~70% in the reduced form (17, 18); and (c) Knaff and Arnon (19) observed that cyt b-559 is photooxidized at liquid nitrogen temperature, an observation which has been confirmed in several laboratories. This implies an intimate association with the primary photooxidant of System II, presumably "P680" first observed by Döring et al(20). We now assume that this path also operates at room temperature and is involved in the "slow deactivation."

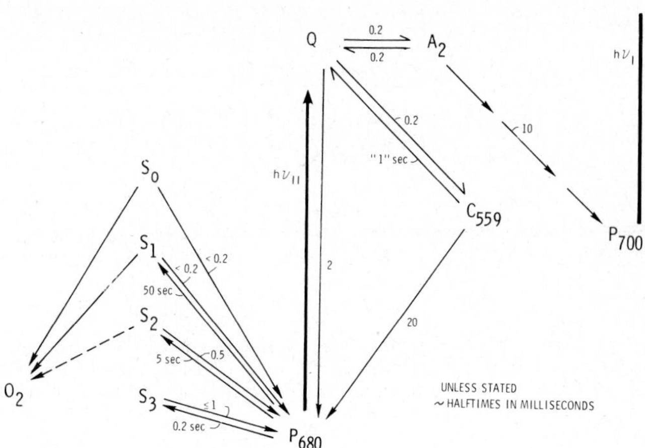

Fig. 10

A more complete electron flow scheme

The scheme of Fig. 10 extends that of Fig. 6 so as to include:

(1) The primary photoact of System II, in which we assume that the role of P680 is analogous to that of P700 in System I. In addition we assume a 2-msec back reaction: $Q+P680 \underset{2msec}{\overset{h\nu}{\rightleftharpoons}} Q^{-}+P680^{+}$. This backreaction causes the "fast deactivation" discussed above.

(2) The backreaction $P680^{+} + C559^{-} \xrightarrow{20\ msec} P680 + C559$. We assume that the oxidation of $C559^{-}$ by $P680^{+}$ is about ten times slower than the oxidation of Q^{-}. In this way the scheme accommodates the "slow deactivation" discussed above.

(3) Forward and backreactions between P680 and the four S states.

(4) Reactions between S_0, S_1 and S_2 and molecular O_2 (state S_4 and O_2 evolution are not shown).

The precision of the rate constants assigned to the various steps in Fig. 10 varies. Some rest on direct observations (4, 5, 7, 8), others are minimum values, while the remaining ones are chosen for overall best fit (e.g. P680 ⟶ Q^-, P680 ⟶ C^-). Awaiting more detailed analyses (which are presently being undertaken) we should concentrate on the merits of the framework rather than the detail. A few aspects should be further clarified or stressed.

(1) The "misses" of Photosystem II (5) arise from a competition between the S enzyme and Q for the "hole" in P680. For instance, the electron transfer $S_0 \longrightarrow P680^+$ takes place in $\leqslant 0.2$ msec and with 10% misses. Thus the competing electron transfer $Q^- \longrightarrow P680^+$ should be 10x slower and have a reaction time of 2 msec. Since the latter transfer is presumably independent of the state of the S system, the % misses in each step $S_n \longrightarrow S_{n+1}$ should reflect the respective electron transfer times.

(2) On first sight it seems a bit peculiar that, despite the rapid time constants of 2 and 20 msec, deactivation requires seconds (fast path to Q) to minutes (slow path to C559). This becomes clear if one considers that the electron transfer from $Q^- \longrightarrow S_n$ (and $C^- \longrightarrow S_n$) involves three states of the system:

$$Q^-P^-S \underset{k_{-1}}{\overset{k_1}{\rightleftarrows}} Q^-P\,S^- \xrightarrow{k_2} Q\,P^-S^-$$

In this sequence the rate of formation of $Q\,P^-S^-$ from Q^-P^-S will be $k_{obs} \simeq k_1k_2/(k_{-1}+k_2)$.

For example, the observed rate of rapid S_3 deactivation (k_{obs}) is ~1 sec (Fig. 8).

We took $k_{-1}= 2000\ sec^{-1}$ (the observed relaxation time for the transition S_2+$P680^+$ ⟶ S_3+ P680 (3, 8)) and $k_2=500\ sec^{-1}$ (assumed on the basis of ~10% misses). We then computed $k_1= 5\ sec^{-1}$ for the reaction S_3+ P680 ⟶ S_2+ $P680^+$.

The rate of this backreaction (the essence of deactivation) is different for the different S states. The overall process is slowed down because the hole in $P680^+$ (state Q^- PS^-) will tend to revert to S rather than to Q (four out of five times in the above example).

(3) The same reasoning holds for the slow deactivation path from C559, the halftimes being ~10x longer. Under conditions that Q is oxidized the rate of deactivation will be proportional to the degree of reduction of C559, the reason why the process can be retarded by relatively mild oxidants. For instance, benzoquinone in high concentrations can yield a 5 min half life of S_3 which would imply that C559 is 97% oxidized.

(4) The equilibrium distribution of S_0, S_1 and S_2 in dark is a balance between oxidation by O_2 and reduction by C^- (ultimately by endogenous reductants in the "soup" which reduce C^-).

(5) The scheme yields a rough estimate of the midpoint potentials of the various S states with respect to P680. Using the numbers in Fig. 10, we estimate: $S_2/S_3 K = 400$, $\Delta E = 160$ mV; S_1/S_2 $K = 2.5 \times 10^4$, $\Delta E = 260$ mV; S_0/S_1 $K = 2.5 \times 10^5$, $\Delta E = 320$ mV. For energetic reasons we must assume that the midpoint potential of the $P680/P680^+$ is more oxidized than 820 mV. Consequently, our estimate implies that the S_0/S_1 couple has a potential more oxidized than 500 mV.

(6) One attractive feature of the scheme is that the "O_2 box" communicates exclusively with water, O_2 and P680.

(7) The hypothesis that deactivation occurs exclusively via $P680^+$, presumably oxidized chlorophyll, may help explain the emission of delayed light by System II.

Acknowledgements

This work was supported in part by the Atomic Energy Commission, Contract AT(11-1)-3326, and the National Science Foundation, Contract NSF-C705.

References

1. P. Joliot and A. Joliot (1968), Biochem. Biophys. Acta. 153, 625.
2. S. Izawa, R. L. Heath and G. Hind (1969), Biochem. Biophys. Acta. 180, 388.
3. B. Kok, P. Joliot and M. P. McGloin (1969), p. 1042, Vol. II in: Progress in Photosynthesis Research. H. Metzner, Ed., Tübingen.
4. B. Kok, B. Forbush and M. P. McGloin (1970), Photochem. Photobiol. 11, 457.
5. B. Forbush, B. Kok and M. P. McGloin (1971), Photochem. Photobiol. 14, 307.
6. R. Radmer and B. Kok (1973), Biochem. Biophys. Acta. 314, 28.
7. K. L. Zankel (1973), Biochem. Biophys. Acta. 325, 138.
8. B. Bouges-Bocquet (1973), Biochem. Biophys. Acta. 292, 772.
9. R. Radmer and B. Kok (1974), Biochem. Biophys. Acta., in press.

10. P. Morin (1964), J. Chim. Phys. 61, 674.

11. W. W. Doschek and B. Kok (1972), Biophys. J. 12, 832.

12. U. Siggel, G. Renger, H. H. Stiehl and B. Rumberg (1972), Biochem. Biophys. Acta. 256, 328.

13. R. Radmer and B. Kok (1975), Ann. Rev. Bioch., in press.

14. I. Ikegami and S. Katoh (1973), Plant and Cell Physiol. 14, 829.

15. B. Bouges-Bocquet (1973), Biochem. Biophys. Acta. 314, 250.

16. P. Joliot, A. Joliot, B. Bouges and G. Barbieri (1971), Photochem. Photobiol. 14, 287.

17. W. A. Cramer and W. L. Butler (1967), Biochem. Biophys. Acta. 143, 332.

18. F. A. McEvoy and W. S. Lynn (1972), Arch. Bioch. Bioph. 150, 624.

19. D. B. Knaff and D. I. Arnon (1969), Proc. Nat. Ac. Sc. 63, 956.

20. G. Döring, H. H. Stiehl and H. T. Witt (1967), Z. f. Naturf. 22, 639.

M. AVRON, *Proceedings of the Third International Congress on Photosynthesis*, September 2-6, 1974, Weizmann Institute of Science, Rehovot, Israel
Elsevier Scientific Publishing Company, Amsterdam, The Netherlands, 1974

STUDIES ON THE OXYGEN EVOLVING SYSTEM IN FLASHED LEAVES

Reto J. STRASSER*

Department of Botany, Laboratory of Photobiology
University of Liege
4000 Sart Tilman Liege (Belgium)

I. INTRODUCTION

Generally, a functioning PS II in vivo can be written as a D.P.A-system, where P represents the active centre, A the primary and the following electron acceptors of P and D the whole electron transport chain between water and the active centre of PS II. We were working with a biological material which shows some deficiencies on the water splitting side of PS II. For such a deficient system we write R instead of D. We could show that, in this system, P as well as A are present as usual. We therefore can write for our system the term R P.A . The plants (mostly beans, oats, barly, lemna etc.) were grown under a flash regime in which they were illuminated every 15 min by a 1 ms strong polychromatic flash during 5 to 7 days to get so-called flashed leaves (FL). Thylakoids from such leaves we call flashed thylakoids (FT). At the beginning of the first continuous illumination of flashed leaves with photosynthetic light, they show no oxygen evolution and no variable fluorescence.

After some 3 to 6 minutes of illumination with photosynthetic actinic light flashed leaves show a strong oxygen evolution. This phenomenon we call the induction of PS II-activity in flashed leaves. Correlated induction phenomena can be observed by fluorescence, delayed light emission and absorption, as well as by electron microscopic measurements. The induction mechanism therefore transforms flashed leaves, FL, (containing a PS II in a R P.A-form) into flashed induced leaves, FLI, (containing a PS II in a complete D.P.A-form). This reaction needs photosynthetic light. If flashed leaves were illuminated with very weak or even with weak green light, they become sensitized. Sensitized leaves show on illumination with actinic light an oxygen outburst as well as variable fluorescence. We have arguments to belief that flashed sensitized leaves (FLS) contain a biochemicaly complete donor system D which is at the moment after sensitization somehow not "connected" to the active centre. We symbolyze such a sensitized system with the term D P.A . The data have been reviewed elsewhere[1] and are summarized in table 1.

---

*present address: University of California, Department of Biology, La Jolla San Diego, California 92037, USA

Table 1 Characterization of three different states during the induction of PS II-activity in flashed leaves

FLASHED LEAF →	SENSITIZED LEAF →	INDUCED LEAF
No capacity for O_2 evolution	Capacity for O_2 outburst	Capacity for O_2 outburst
No capacity for variable fluorescence	Capacity for variable fluorescence	Capacity for variable fluorescence
No capacity for variable DF	No or little capacity for variable DF	high capacity for variable DF
Level of DF high	Level of DF high	Level of DF low
Abs 505 nm low	Abs 505 nm low	Abs 505 nm high
Stroma thylakoids are unfused	Stroma thylakoids are unfused	Stroma thylakoids are mostly fused
R P.A →	D P.A →	D.P.A

DF: modulated o.3 ms delayed light emission

To see the correlations of polarographic, fluorescence, luminescence, and absorption measurements in such a sensitive system, simultaneous measurements of several parameters on the same sample (leaf or thylakoid preparations) are neccessary. For this reason a special technical system with multibranched fibre optics and a light guiding oxygen electrode has been constructed[1,2] which is shown schematically in fig.1. For the in vitro experiments it was therefore possible to measure the oxygen evolution as well as the reduction of an artificial electron acceptor at the same time.

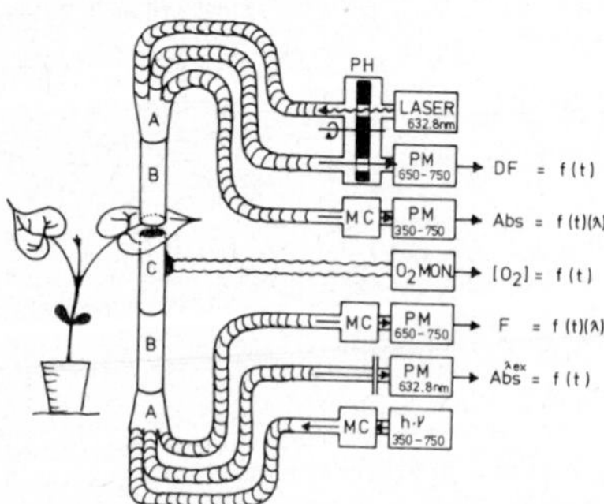

Fig. 1. A technical system which allows the simultaneous measurements of delayed light emission (DF), absorption (Abs), oxygen (O_2), fluorescence (F).
PH: Phosphoroscope, MC: Monochromator or filter, PM: Photomultiplier, A: Three arm flexible fibre optics, B: Glas or quarz rod for lightmixing, C: Light-guiding oxygen electrode. (All optical parts available by Jenaer Glaswerk SCHOTT & Gen. Mainz, W-Germany, Ref.Nr. 2.1304-0)

II. TERMINOLOGY

We distinguish two different groups of mechanisms in PS II.

1. The induction mechanism: The mechanism which changes a biological system which never was involved in photosynthetic work before (R P.A), into an active D.P.A-form. The induction process can be divided in two kinds of light dependent reactions (fig. 2).

1.1 The photosensitization: which changes the R P.A-form (e.g. flashed leaves FL) into a D P.A-form (e.g. flashed sensitized leaves FLS). It is a low energy multiquantum process with one or more dark steps.

1.2 The photoengagement: which changes the D P.A-form (e.g. flashed sensitized leaves FLS) into a D.P.A-form (e.g. flashed induced leaves FLI). It depends on photosynthetic actinic light and "connects" for the first time the oxygen evolving system to the active centre of PS II.

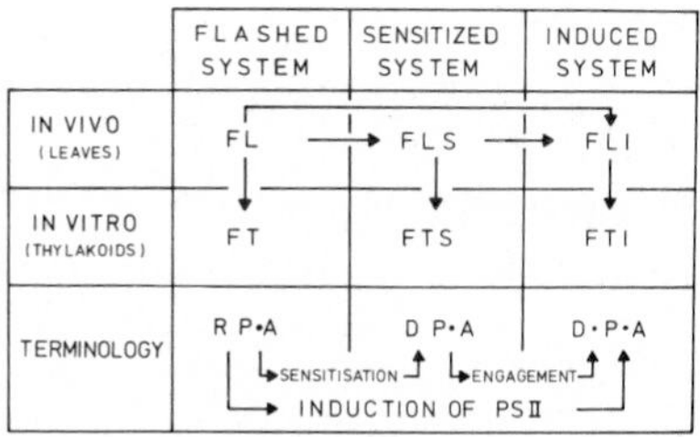

Fig. 2. Comparison of the different biological systems with the corresponding terms.

2. The action mechanism: The mechanism which is responsible for the photosynthetic activities of the photosynthetic units in the intact D.P.A-form.

2.1 The photoequilibration brings the substates $S_0S_1S_2S_3S_4$ of the D.P.A-cycle from a steady state in the dark into a steady state in the light.

2.2 The photoactivation (used in the same way as Joliot et al.) is the beginning of the photoequilibration reactions.

III. RECENT RESULTS ABOUT THE INDUCTION OF PS II ACTIVITY

Until now the induction and the sensitization phenomena could only be observed in whole leaves but never in thylakoid preparations. Flashed or flashed induced thylakoids can therefore be analyzed by photosynthetic electron transport tests. The typical results of such tests are given in table 2 for normal, flashed, or flashed induced thylakoids, without or after tris, KCl, or NH_2OH washing.

Table 2 Photosynthetic electron transport tests of different thylakoid preparations

treatment	test	normal thylakoids NT	flashed thylakoids FT	induced thylakoids FTI
	$H_2O \rightarrow EA$	+	-	+
	$BZ \rightarrow EA$	+	-	+
	$Mn^{2+} \rightarrow EA$	+	+	+
	$DPC \rightarrow EA$	+	+	+
after tris treatment[3]	$H_2O \rightarrow EA$	-	-	-
	$BZ \rightarrow EA$	+	-	+
	$Mn^{2+} \rightarrow EA$	+	+	+
	$DPC \rightarrow EA$	+	+	+
and after KCl treatment[4]	$H_2O \rightarrow EA$	-	-	-
	$BZ \rightarrow EA$	-	-	-
	$Mn^{2+} \rightarrow EA$	+	+	+
	$DPC \rightarrow EA$	+	+	+
and after NH_2OH treatment[5]	$H_2O \rightarrow EA$	-	-	-
	$BZ \rightarrow EA$	-	-	-
	$Mn^{2+} \rightarrow EA$	-	-	-
	$DPC \rightarrow EA$	+	+	+

EA: Electron acceptor such as dichlorophenolindophenol or ferricyanide
BZ: Benzidine, DPC: Diphenylcarbazide. All + reactions were DCMU sensitive

Based on the conventional linear view of an electron transport chain as indicated in fig. 3 we can distinguish at least four different sites for exogeneous electron-donors which bring electrons to PS II.

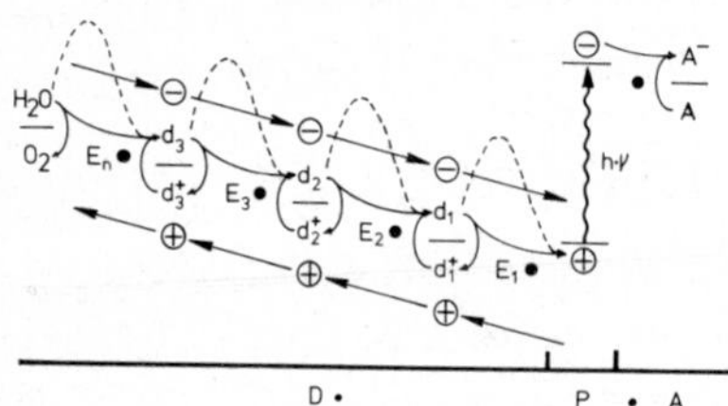

Fig. 3 Linear presentation of the electron transport chain from water to the active centre of PS II. d/d^+ :electron carrier, E: any kind of catalyst which catalyzes the reduction of d^+
The point · always indicates the reducing side of a redox-couple.

With the functional tests used, no difference between the behaviour of normal green (NT) and flashed induced (FTI) thylakoid preparations could be found. That means flashed untreated thylakoids show the same photosynthetic activities as normal thylakoids after tris and KCl treatment. They therefore are still able to photooxidize Mn^{2+}, DPC, NH_2OH but not benzidine or water (fig. 4).

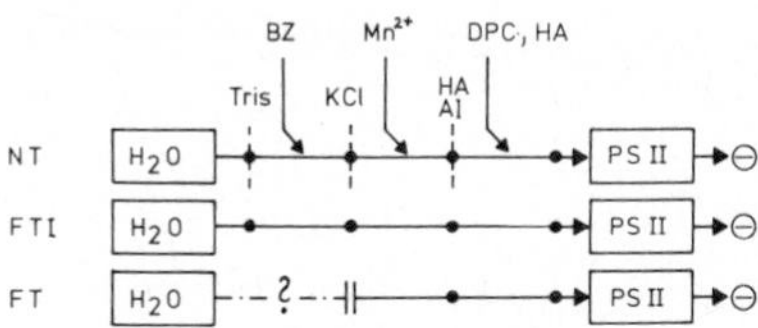

Fig. 4 Sites of electron donation of benzidine BZ, manganese Mn^{2+}, diphenylcarbazide DPC, hydroxylamine HA, and sites of inhibition of tris-, KCl-, or hydroxylamine- washing or after ageing AI. NT: normal thylakoids, FT: flashed thylakoids, FTI: flashed induced thylakoids, −//− deficient site in flashed leaves.

Considering such a chain of several electron carriers, d_i, combined with catalysts E_i, (any kind of catalyst e.g. enzymes, special membrane conformations etc.) it is possible to reconstitute an intact oxygen evolving system by interaction of two types of thylakoids which are not able to produce oxygen alone. For these experiments we used i) normal tris-washed green thylakoids NTTris and ii) flashed thylakoids FT. Mixing such preparations (about 1 mg Chl per ml) for about 3 min without any additions and diluting afterwards with the reaction medium results in a mixture which has the photosynthetic activity shown in fig. 5.

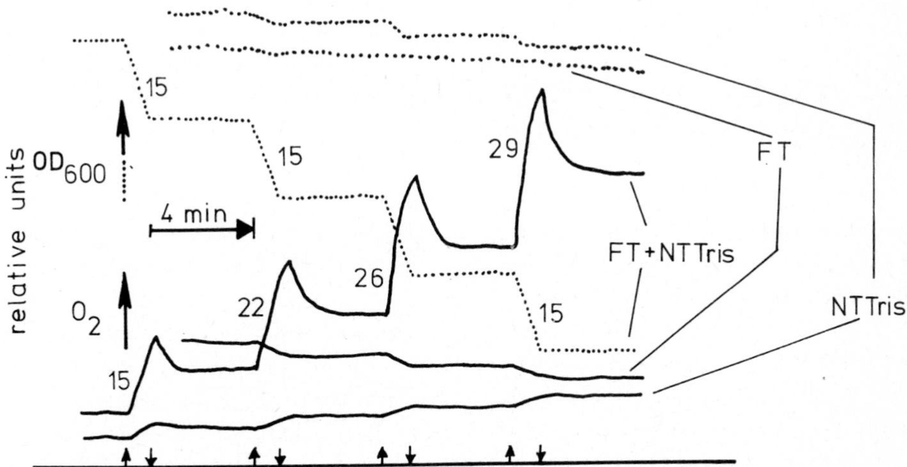

Fig. 5 Reconstitution of Hill-activity (O_2-production and reduction of DCPIP) by mixing tris-washed normal green thylakoids (NTTris) with flashed thylakoids (FT) of equal amount of chlorophyll a . All curves were recorded under the same technical conditions and amplifications.

Today we can't say if the tris washed or the flashed or both types of thylakoids were reconstituted. Fig. 6 shows schematically a way to explain the measured data.

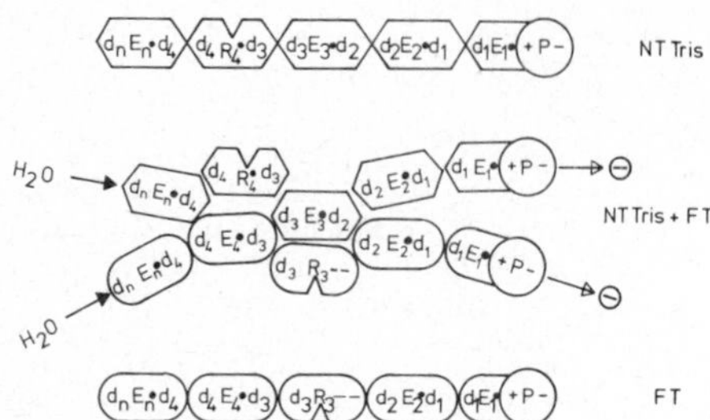

Fig. 6 Schematic representation of the reconstitution of an intact electron transport chain between water and the active centre of PS II. R instead of E indicates a defective step.

We have to emphazise here that the reconstitution only occurs, when each type of thylakoids shows the activities as indicated in tab. 2. These experiments provide further support for the sugestion that the induction of PS II is correlated to an aggregation process[6] which can also be seen as a fusion of primary thylakoids or by absorption changes in the carothenoid region[1,7].

IV. DISCUSSION

As a working hypothesis we propose that flashed thylakoids are unable to rereduce the photooxidized manganese. During the induction period a manganese reducing system may be created so that the photooxidized manganese reacts with the reducing system and is incorporated into one or more Mn-protein-complex(es), located at the defective site in the electron transport chain. This hypothesis would be in agreement with data of other authors[5] reporting an incorporation of free Mn^{2+} during the induction period of PS II in heterotrophically dark grown Chlorella. Fig. 7 represents the observed facts in a hypothetical scheme of PS II and indicates where the induction reactions (R to E) may occur.

The fluorescence and delayed light emisson properties of flashed leaves before and after induction (not shown in this report) can also be explaned by this scheme.

We see the complete PS II as a kind of biphasic system with a lipophilic phase (pigmented active centres and immediately adjacent carriers, which also can be involved in delayed light emission) and a hydrophilic phase. The induction may reflect the "connection" of such two phases in the thylakoid membrane of flashed leaves.

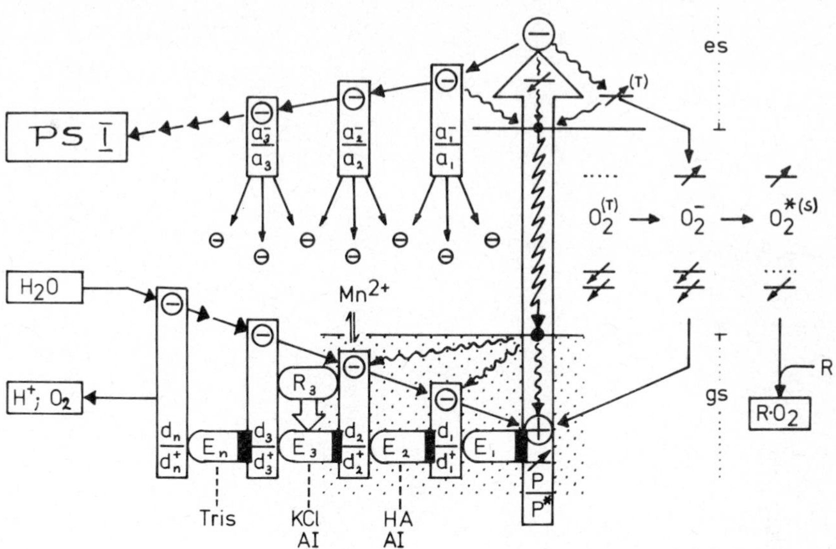

Fig. 7 Hypothetical presentation of the electron transport chain of PS II ⟶ : redox reactions, ⁓⁓➝: radiationless deexcitations, ⁓⁓➝: light emission as fluorescence or delayed light emission, ⇨ : excitation by photosynthetic actinic light, ⊏ : loose binding side of the electron donor d_i to the catalytic system E_i , ⊐▪ : firm binding between the catalytic system E_i and the electron acceptor d^+_{i-1} , ∴∴∴ : hydrophobic phase (may be cristal-like). gs : ground states of P, es : excited states of P^*, ---- : sites of inhibition by tris, KCl, hydroxylamine or ageing. The formation of excited molecular oxygen is a proposition to explain the light dependent oxygen uptake by flashed thylakoids (see also ref. 8).

References:

1. Strasser, R. J., 1973, Arch. Int. Physiol. Biochim. 81, 935-955
2. 1974, Experientia, 30, 320
3. Yamashita, T. & Butler, W. L., 1968, Plant Physiol. 43, 1978-1986
4. Asada, K. & Takahashi, M., 1971 in: Photosynthesis and Photorespiration, Hatch, M. D., Osmond, C. B., Slatyer, R. O. (eds), Wiley Interscience, New York, pp. 387-393
5. Cheniae, G. H. & Martin, I. F., 1973, Photochem. Photobiol. 17, 441-459
6. Ninnemann, H., & Strasser, R. J., 1974, these proceedings
7. Coumans, M., & Strasser, R. J., 1972, VI. Int.Congr.on Photobiol. Abstr. 244
8. Remy, R., 1973, Photochem. Photobiol. 18, 409-416

This work has been suported by EMBO (European Molecular Biology Organization) and by the Schweizerischer Nationalfonds.

M. AVRON, *Proceedings of the Third International Congress on Photosynthesis*, September 2-6, 1974, Weizmann Institute of Science, Rehovot, Israel
Elsevier Scientific Publishing Company, Amsterdam, The Netherlands, 1974

INDICATIONS FOR THE EXISTENCE OF A SPECIAL ELECTRON ACCEPTOR OF PHOTOSYSTEM II IN PHOTOHETEROTROPHICALLY CULTIVATED CHLAMYDOBOTRYS STELLATA NOT BELONGING TO THE NORMAL NON-CYCLIC ELECTRON TRANSPORT

D. MENDE and W. WIESSNER

Pflanzenphysiologisches Institut, Abt. Exp. Phykologie, Göttingen, Germany

1. Introduction

It has been shown previously that autotrophically cultivated Chlamydobotrys stellata behaves photosynthetically like all other green algae and higher plants[16], but that photoheterotrophic growth for which only photosystem I activity is required[15], leads to an almost grana-free chloroplast structure[17,18], to a diminished capacity for photosynthetic CO_2-fixation[13,14], to a reduced electron transport between the photosystems II and I and to a low photosystem II fluorescence[19,20]. The appearance of these characteristic features of autotrophic and of photoheterotrophic Chl. stellata can be followed easily, when autotrophic cells are transferred to photoheterotrophic growth conditions and vice versa. It requires not more than 4 to 6 hours. These observations lead to the question, how the differences in photosynthetic activity can be brought about by the actual carbon source during growth, in particular how the activity of the photosystem II dependent electron transport is regulated by the in vivo requirements.
In the present study we analyzed the behaviour of fluorescence and of delayed light emitted by Chl. stellata cultivated under different growth conditions. The results are discussed with respect to the mechanisms by which the photosystem II dependent electron transport might be regulated in vivo.

2. Material and methods

Chlamydobotrys stellata, strain 10-1e from the Sammlung von Algenkulturen des Pflanzenphysiologischen Instituts der Universität Göttingen, was cultivated either autotrophically (light + CO_2) or photoheterotrophically (light + acetate) as described by Wiessner[16]. Continous illumination, light intensity 9000 lux, temperature 30 °C. All measurements were done with whole cells. Fluorescence and delayed light emission were measured in the apparatus shown in fig. 1. The wavelength of the excitation light was 470 nm. The photomultiplier (RCA 4463) was blocked with a 679 nm interference filter (Schott and Gen., Mainz), in order to measure only the light emitted by the algae. The signals were stored in a transient recorder (biomation 802) and recorded with a Servogor S recorder. The decay half time of the flash, by which the delayed light was excited, was ≤ 10 μsec.

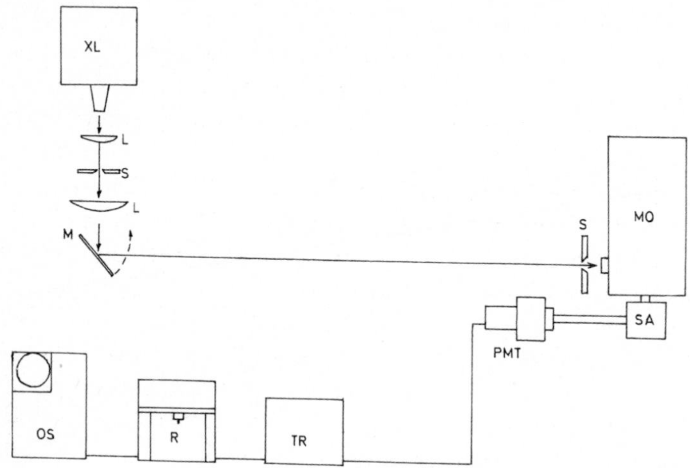

Fig. 1 : Schematic diagram of the fluorescence and delayed light measuring apparatus. XL = xenon lamp, L = lens, S = slit, M = mirror, MO = monochromator, SA = sample, PMT = photomultiplier, TR = transient recorder, R = recorder, OS = oscilloscope

3. Results

3.1. Fluorescence emission of Chl. stellata adapting from autotrophy to photoheterotrophy

It is generally accepted that the redox state of an intermediate of the photosynthetic electron transport, Q, determines the fluorescence intensity at room temperature, which therefore reflects the activity of the non-cyclic electron transport. The inhibitor

3-(3,4-dichlorophenyl)-1,1-dimethylurea (DCMU) normally increases the fluorescence because it blocks the oxidation of Q be photosystem I. The increase in fluorescence emitted by DCMU poisoned cells therefore can be considered to be a relative measure for the normal rate of non-cyclic electron flow. In order to follow the development of the disconnection of the e^--transport between the two photosystems, we therefore measured the fluorescence emitted by this alga at different stages of adaptation in the presence of DCMU and in non-poisoned cells. The results showed (fig. 2) that the fluorescence from previously autotrophic cells emitted 15 min after the change to photoheterotrophic growth conditions (phase 1) is as high as the one of DCMU poisoned cells (5.10^{-6} M DCMU). Another 15 min later the fluorescence has dropped and reaches soon a level 40 % below its initial rate (phase 2). DCMU increases only the phase 2 but not the phase 1 fluorescence [19].

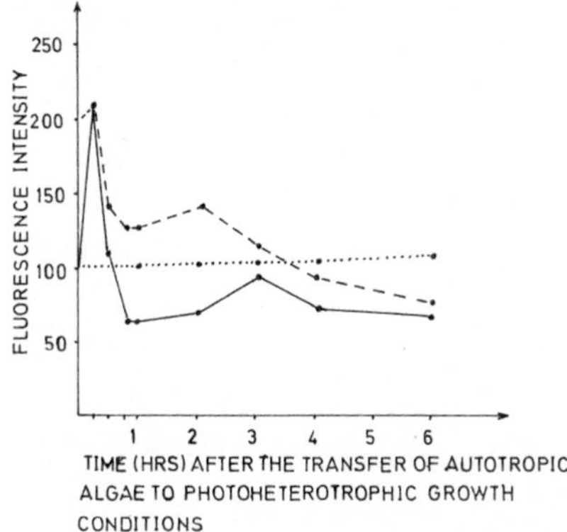

Fig. 2 : Fluorescence emitted from Chl. stellata adapting from autotrophy to photoheterotrophy. ...... autotrophic cells, ——— previously autotrophic cells adapting to photoheterotrophy. ------ fluorescence emitted in presence of DCMU

It follows that the disconnection of the photosynthetic e^--flow between the two photosystems starts with its complete blockage. Afterwards the e^--flow partially recovers, its efficiency, however, drops significantly.

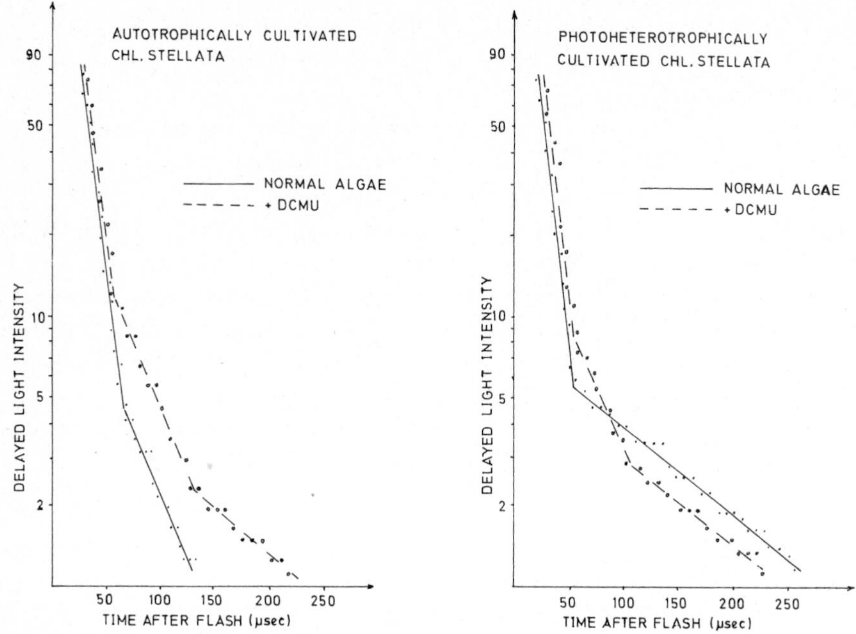

Fig. 3 and 4 : Log of delayed fluorescence of autotrophically and of photoheterotrophically cultivated Chl. stellata in the absence and in the presence of DCMU as a function of time

The first delayed light component has a half time of about 10 μsec, the second one of about 35 μsec and the third of 190 μsec. (The latter one has not been plotted in fig. 3.) In photoheterotrophic cells (fig. 4) the 35 μsec and the 190 μsec components are missing. Instead of them a new 60 μsec component shows up. The addition of DCMU has no influence on the 10 μsec component, but it leads to the appearance of the 35 μsec component also in the photoheterotrophic cells and of the 60 μsec component in the autotrophic organisms. The 190 μsec component of the autotrophic alga is completely abolished by this inhibitor.

Taking for granted that each component of the observed delayed light arises from a back reaction between a special redox compound reduced and one oxidized by photosystem II, then four different redox compounds on the reducing side of PS II can take part in such back reactions, and the redox potential of the component with

3.2. Delayed light emission of Chl. stellata

Next we tried to find a common explanation for both, the reduction of the electron flow between the two photosystems and the maintenance of a fairly oxidized state of Q at the same time. Both do not, at least not during the first 4 hours of the adaptation process from autotrophy to photoheterotrophy depend on a reduced electron flow on the oxidizing side of PS II (Niemeyer and Wiessner in preparation). In continuation of an hypothesis of Wiessner and Fork [19], we therefore looked for the existence of an electron acceptor on the reducing side of PS II, which should have the following properties : 1. It should not operate in the autotrophic cell. 2. It should be activated to accept electrons from PS II during the adaptation from autotrophy to photoheterotrophy. 3. It should replace a normal intermediate of the non-cyclic electron transport system located before the DCMU block. The best test of this hypothesis -- at least under our experimental limitations -- was the analysis of the delayed fluorescence emitted from autotrophic and from photoheterotrophic cells as well as at different stages of the adaptation process. We choose this test because it has been demonstrated that back reactions within PS II, involving perhaps different components of the electron transport system, participate in the delayed light emission[1,4,7,8,11,12,21].
An analysis of the delayed light decay in the time range between 10 and 250 μsec reveals that at least 3 different reactions are responsible for the emission from autotrophic Chl. stellata (table 1, fig. 3) [21].

Table 1

Delayed light components of autotrophic and of photoheterotrophic Chlamydobotrys stellata

Growth conditions of the algae		Decay half time of the components 10 μsec	35 μsec	60 μsec	190 μsec
autotrophic	normal	+	+	-	+
	+ DCMU	+	+	+	-
photoheterotrophic	normal	+	-	+	-
	+ DCMU	+	+	+	-

+ = present, - non present

the shortest half life of decay should be next to P680*. Having this in mind we probably can attribute the 10 µsec component to the primary acceptor of PS II. (However, we were not able to measure decay times shorter than 10 µsec.) This component and the ones next in line, the 35 µsec and the 60 µsec component, must be located in the electron transport chain from PS II to PS I before the DCMU block. Because the third component, 190 µsec, is sensitive to poisoning with DCMU, it belongs to the transport chain behind the DCMU block and also behind the point of disconnection of the electron flow in photoheterotrophic algae.

After we had thus established the presence of differences in the delayed light components between the autotrophic and the photoheterotrophic Chl. stellata, we followed the behaviour of the 35 µsec and the 60 µsec component during the period of adaptation of this alga from autotrophy to photoheterotrophy. The data presented in fig. 5 demonstrate that the 35 µsec component has disappeared after the 3 hour of adaptation and is replaced by the 60 µsec component. This coincidence might reflect a direct relationship between these two components.

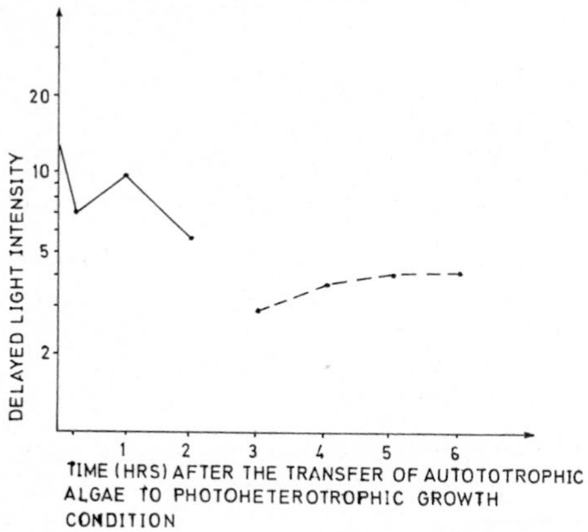

Fig. 5 : Relative intensity of the 35 µsec (————) and the 60 µsec (------) component of delayed light as emitted from Chl. stellata adapting from autotrophy to photoheterotrophy

4. Final conclusions and hypothesis

From the work reported here we have come to the following explanation of the mechanism which regulates the efficiency of the photosystem II dependent electron transport in Chlamydobotrys stellata (fig. 6).

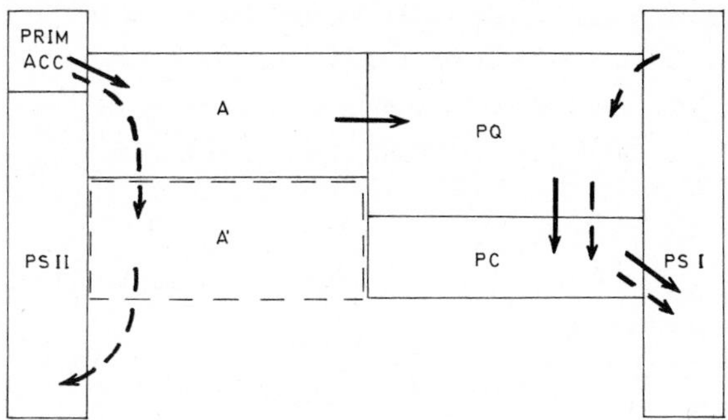

Fig. 6 : Schematic representation of the possible mechanism of the photosynthetic electron flow in the thylakoidmembranes of autotrophically (———) and photoheterotrophically (-----) cultivated Chl. stellata.
PS II + Prim.Acc. = photosystem II with its primary electron acceptor, A component of the electron transport chain in its low potential form. A' = same as A, but in a high potential form. PQ = plastoquinone, PC = plastocyanine, PS I = photosystem I

The initiating step for the decrease in the apparent photosystem II efficiency is the absence of CO_2 as the final electron acceptor of photosynthesis. In the light this leads soon to a highly reduced state of all redox compounds of the electron transport chain, because photosystem II continues to operate, but photosystem I has ceased to oxidize the PS II and PS I connecting electron transport chain. This state is reflected by the incease in fluorescence during phase 1 of the adaptation period from autotrophy to photoheterotrophy. As a next step it brings about a change of the redox potential of one special component of the transport chain (the

35 μsec component of delayed light) to a more positive value (60 μsec component). This component from now is unable to reduce photosystem I. In a cyclic way it recycles photosystem II activated electrons at high speed and efficiency. This renders possible the decrease in fluorescence and of the electron transport between the two photosystems. At the same time it facilitates a high efficient cyclic photosystem I without interference by the photosystem II dependent electron transport. Even though we do not yet know the actual nature of this postulated redox compound, we like to speculate that it is of protinous nature and possibly identical with cytochrome b_{-559} in its low and high potential form [2,3,5,6,10].

This work was supported by the Deutsche Forschungsgemeinschaft.

References

1. Amesz, J. and Kraan, G.P.B., 1971, Umschau, Biophysikalisches Inst. der Universität Leiden 19, 715
2. Boardmann, N.K., Anderson, J.M. and Hiller, R.G., 1971, Biochim. Biophys. Acta 234, 126
3. Cramer, W.A. and Böhme, H., 1972, Biochim. Biophys. Acta 256, 358
4. Duysens, L.N.M. and Sweers, H.E., 1963, in :Studies in Microalgae and Photosynthetic Bacteria, eds. Jap. Soc. plant physiol. (The University of Tokyo Press, Tokyo) p. 353
5. Erixon, K. and Butler, W.L., 1971, Biochim. Biophys. Acta 234, 381
6. Floyd, A., Chance, B. and Devault, D., 1971, Biochim. Biophys. Acta 226, 103
7. Itoh, S., Katoh, S. and Takamiya, A., 1971, Biochim. Biophys. Acta 245, 121
8. Jursinic, P. and Govindjee, 1972, Photochem. Photobiol. 15, 331
9. Kautsky, H., Appel, W. and Amman, H., 1960, Biochem. Z. 332, 277
10. Ke, B., Vernon, L.P. and Chaney, T.H., 1972, Biochim. Biophys. Acta 256, 345
11. Lavorel, J., 1969, in : Progress in Photosynthesis Research, vol. 2, ed. H. Metzner (H. Laupp Jr., Tübingen) p. 883

12. Mayne, B.C., 1967, Photochem. Photobiol. 6, 189
13. Merrett, M., 1969, Arch. Mikrobiol. 65, 1
14. Wiessner, W., 1963, Arch. Mikrobiol. 45, 33
15. Wiessner, W., 1965, Nature (London) 205, 56
16. Wiessner, W., 1968, Planta (Berl.) 79, 92
17. Wiessner, W. and Amelunxen, F., 1969, Arch. Mikrobiol. 66, 14
18. Wiessner, W. and Amelunxen, F., 1969, Arch. Mikrobiol. 67, 357
19. Wiessner, W. and Fork, D.C.,1970, Carnegie Inst. Year Book 69, 695
20. Wiessner, W. and French, C.S., 1970, Planta (Berl.) 94, 78
21. Zankel, K.L., 1971, Biochim. Biophys. Acta 245, 373

M. AVRON, *Proceedings of the Third International Congress on Photosynthesis*,
September 2-6, 1974, Weizmann Institute of Science, Rehovot, Israel
Elsevier Scientific Publishing Company, Amsterdam, The Netherlands, 1974

PHOTO-OXIDATION OF THE PLASTOHYDROQUINONE-9 POOL IN PLASTOGLOBULI DURING ONSET OF PHOTOSYNTHESIS

K.H.Grumbach and H.K.Lichtenthaler
Botanical Institute (Plant Physiology) University of Karlsruhe
D-75 Karlsruhe, FRG

Summary

Photooxidation of plastohydroquinone-9 to plastoquinone-9 was investigated in etiolated and green plants under different light treatments.

1) Plastohydroquinone-9 is mainly located in the osmiophilic plastoglobuli of the plastid stroma while the oxidized form (plastoquinone-9) is mainly associated with the photochemical active thylakoids.
2) In etiolated Hordeum seedlings, which contain many plastoglobuli, we observed a partial photooxidation of the plastohydroquinone-pool in flash light and in white, red and blue light.
3) Photooxidation of plastohydroquinone-9 was also detected in green leaves of Ficus and Fagus which are known to contain large plastoglobuli and high excess amounts of plastohydroquinone. The oxidation is hardly seen in young growing leaves.
4) Darkness, in turn, results in a partial reduction of the oxidized plastoquinone - pool in the plastoglobuli to plastohydroquinone-9.
5) The possible correlation of the oxidation-reduction-change in the plastoglobulus with invers changes in the redox-degree of carotenoids in the thylakoid membrane is discussed.

1. Introduction

The lipophilic benzoquinone plastoquinone-9 was detected in all photosynthetic organisms, except in bacteria, as one obligatory component of non cyclic electron transport close to photosystem II[1,2]. In this flow plastoquinone-9 is suggested to act in a large pool as the terminal acceptor of photosystem II[3,4], which is oxidized by photosystem I[5]. This membrane bound pool between the two light reactions possesses a high quantum efficiency for photoreduction and photooxidation.

Most chloroplasts, in particular those of fully developed and older green leaves contain high excess amounts of plastoquinone-9, which is deposited outside the membrane in the osmiophilic plastoglobuli mainly in the reduced form as plastohydroquinone-9[6,7,8,9]. Plastoglobuli are present in all stages of plastid development and degeneration as a structural component. In the plastid stroma they are arranged solitary in a big size or in groups of small sizes close to the grana and stroma thylakoids[7,8,9].

In this paper we have investigated the possible role of plastohydroquinone-9 accumulated in the plastoglobuli as a proton and electron donor to the photosynthetic thylakoid membrane.

2. Materials and Methods

Barley seedlings were grown for 6 or 7 days in continuous darkness or in white light (18000 lux, Osram fluora lamps 55W; 23 ± 1°C; 50 % relative humidity). Thereafter the etiolated plants were exposed to various light treatments (flash light: Mecablitz; red light: Philips TL 40 W/15 lamps, plexiglass red filter Fa. Röhm, no. 555, 1000 erg/cm^2 x sec; blue light: Philips TL 40 W/18 lamps, plexiglass blue filter Fa. Röhm, no. 627, 1500 erg/cm^2 x sec; strong white light: Philips Atralux lamps 150 W). The leaves of intact plants from Hordeum vulgare, Ficus elastica and Fagus silvatica were exposed to prolonged darkness or darkness with subsequent re-illumination. The plant material was homogenized and extracted at 5°C in an aceton-petrol-ether mixture with an Ultra Turrax Homogenisator. Lipoquinones and pigments were separated by thin layer chromatography and then quantitatively estimated photometrically[10,11,12].

3. Results and Discussion

During the thylakoid synthesis in young developing leaves, chloroplasts contain almost no plastoglobuli and only little excess plastoquinones[7]. With increasing age of leaves chloroplasts accumulate high amounts of excess plastoquinone-9 which is not needed for photosynthesis. This excess amount is deposited mainly as plastohydroquinone-9 in the osmiophilic plastoglobuli of the plastid stroma[7,8,9]. The plastoquinone of plastoglobuli is thus present to a larger degree in the reduced form than in the chloroplasts from which they were isolated (table 1). The thylakoid fraction of the same chloroplasts contain plastoquinone mostly in the oxidized form. Since

Table 1

Comparison of the percentage amount of total plastoquinone-9, which is present in its reduced form as plastohydroquinone-9 in whole chloroplasts and in the thylakoid and plastoglobuli fractions.

	chloroplasts	thylakoids	plastoglobuli
Spinacia	52 %	26 %	61 %
Fagus	81 %	19 %	88 %
Ficus	65 %	16 %	81 %

the thylakoid fraction still contained some single plastoglobuli, the percentage of reduced plastoquinone in the thylakoids is certainly even lower than given in table 1.

Etiolated barley seedlings contain numerous plastoglobuli arround the prolamellar body which are used up during light induced thylakoid formation [13]. It had been shown before, that the plastohydroquinone-9 pool of etioplasts is partially photooxidized to plastoquinone-9 upon illumination [10]. We repeated this experiment with short time illumination by flashes. Two flashes (with a subsequent dark phase of 15 min.) are sufficient to transform about one third of the plastohydroquinone-9 to its oxidized form (table 2). The percentage of plastohydroquinone decreases correspondingly from 77 to 49 %. Further flashes do not change this redox-level. The two flashes induced some chlorophyll formation from protochlorophyllide (or protochlorophyll). It had been postulated that plastohydroquinone may serve as hydrogen donor for the reduction of protochlorophylide [14]. Our data show that there is about 5 times more plastohydroquinone oxidized than would be needed for protochlorophyllide reduction (table 2). This indicates that the hydrogen (or electrons + protons) of plastohydroquinones-9 are used for other processes too.

The photooxidation of plastohydroquinone in etioplasts is not dependant on white light. It also occurs in red and blue light (table 3). In all cases we find the decrease of the plastohydroquinone content paralleled by an almost aequivalent increase in the level of plastoquinone.

A partial photooxidation of plastohydroquinone to plastoquinone also occurse in green leaves, provided that the chloroplasts contain excess plastoquinone in the reduced form in plastoglobuli. The latter has been shown before in the case of green leaves of Ficus and green

Table 2
Chlorophyll a formation and changes in the redox level of plastoquinone-9 in 6 days old etiolated Hordeum seedlings (type Asse) in flash light (10^{-3} sec + 15 min darkness) (μg/100 shoots).

	6 d dark	no. of flashes 2	10	30
Chlorophyll a		14	19	25
Plastoquinone - 9	64 ⟹	141	144	180
Plastohydroquinone - 9	210 ⟹	137	140	140
% Plastohydroquinone - 9	77	49	49	44

Table 3
% plastohydroquinone-9 in 7 days old etiolated barley seedlings and upon illumination.

7 d dark	light	
52 ⟶	21 :	4 h white light (2000 lux Osram HNW 202)
47 ⟶	30 :	6 h white light (1800 lux Osram fluora)
58 ⟶	46 :	6 h red light (1000 erg/cm^2·sec)
32 ⟶	22 :	6 h blue light (1500 erg/cm^2·sec)

sun leaves of Fagus[7,15,16]. Green Ficus leaves which were kept in darkness for 16 hours and then illuminated with strong light, show this partial photooxidation of plastohydroquinone (table 4). When green leaves were brought from light to darkness we observed a partial reduction of plastoquinone to plastohydroquinone (table 5). Re-illumination results again in a partial photooxidation of the plastohydroquinone-pool. This change is fully obtained within 15 min (and possibly much earlier) and is not altered by further light or darkness (table 5). Since the plastohydroquinone present in chloroplasts is almost completely bound to the plastoglobuli[7,8,9,16], we conclude that the partial oxidation of plastohydroquinone to its quinone in light and the subsequent reduction in darkness proceeds in the plastoglobuli.

$$PQ\text{-}9 \cdot H_2 \underset{\text{dark}}{\overset{\text{light}}{\rightleftharpoons}} PQ\text{-}9 + 2\,[H]$$

We assume that the hydrogen (or electrons + protons) from the plastohydroquinone-pool are donated in the light to the thylakoid membrane and get retransformed to the plastoglobulus in the dark. This reaction only occurs when plastoglobuli with excess plastohydroquinone are present. According to this hypothesis the plastohydroquinone-pool in plastoglobuli if present would contribute to the building up of a membrane potential in the light during onset of photosynthesis.

Table 4
Photooxidation of plastohydroquinone-9 in darkened green leaves of Ficus elastica upon illumination (μg/100 cm² leaf area).

	16 h darkness		+ 2 min 30 000 lux
Chlorophyll a	3985		3850
Chlorophyll b	1370		1260
Plastoquinone-9	312	←	465
Plastohydroquinone-9	690	→	554
% Plastoquinone-9	31		46
% Plastohydroquinone-9	69		54

In very young chloroplasts e.g. from pale green Ficus leaves which contain almost no plastoglobuli and almost no extrathylakoidal plastohydroquinone-pool, we could not detect this effect. In young fully green Ficus leaves which contain some plastohydroquinone we found an increase of the plastohydroquinone-pool in darkness with a concommittant decrease in the plastoquinone-9-content. Parallel to this we found an inverse change in the redox level of the membrane bound carotenoids with a particular decrease of the ß-carotene level and an increase in the lutein, antheraxanthin and neoxanthin level (table 7). Small inverse changes are seen, too, in violaxanthin and zeaxanthin content. This could indicate that the light driven xanthophyll cycle[17] might operate in Ficus leaves. In older green Ficus leaves we could, however, not clearly correlate the photooxidation of the plastohydroquinone-9-pool with the photoreduction of violaxanthin via antheraxanthin to zeaxanthin. This has been proved with sun leaves of Fagus

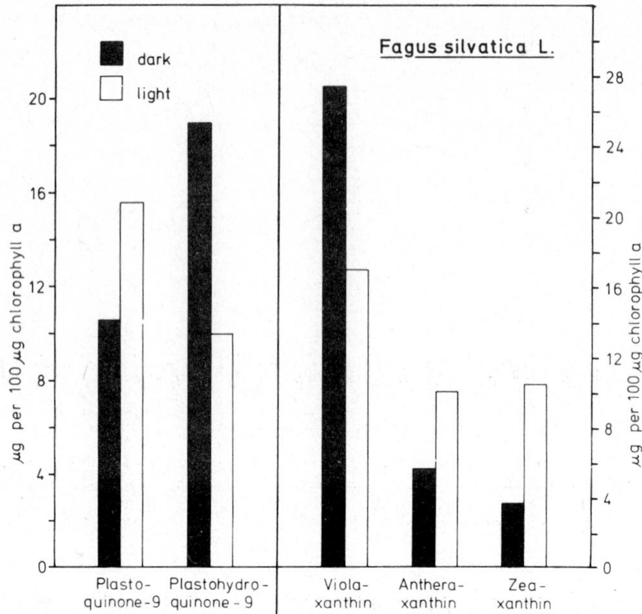

Fig.1. Changes in the redox level of plastoquinone-9 and of carotenoids in green darkened sun leaves of Fagus silvatica upon illumination.

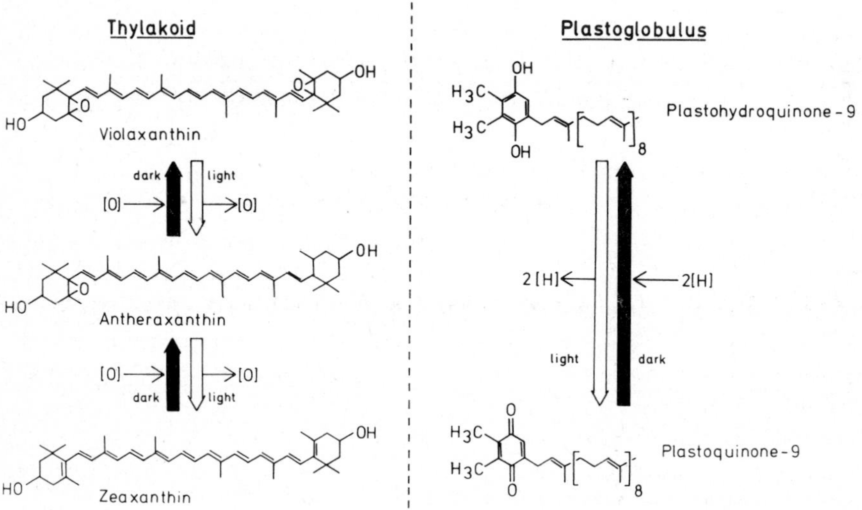

Fig.2. Possible correlation of plastohydroquinone-9 photooxidation in the plastoglobulus with photoreduction of violaxanthin in the thylakoid membrane.

Table 5
Changes in the redox level (% values) of plastoquinone-9 in older leaves of Ficus elastica.

	continuous white light		darkness 15'	60'		light 15'	60'
Plastoquinone-9	16	→	9	8	→	25	21
Plastohydroquinone-9	84	→	91	92	→	75	79

Table 6
Lipoquinone and pigment content of young green and older dark green Ficus leaves (µg per 100 cm^2 leaf area).

	young	older
Chlorophyll a	3456	5635
Chlorophyll b	1189	2160
Carotenoids	594	1056
Plastoquinone-9	422	335
Plastohydroquinone-9	58	1995
	480	2330

silvatica. Green fully functional Fagus leaves which were kept in darkness for 16 hours show a photooxidation of the plastohydroquinone-9-pool and parallel, in the membrane, a nearly stoichoimetrically transformation of violaxanthin to anthera- and zeaxanthin (fig.1). From this it looks like that these two redox changes may be correlated. We thus propose the following hypothetically sheme (fig.2). With the onset of photosynthesis in darkened green Fagus leaves a photooxidation of plastohydroquinone-9 located in plastoglobuli takes place. The resulting hydrogen or eventually free electrons and protons are given to the membrane. There they help to build up a membrane potential and can act in photosynthetic electron transport, enzyme activation or photoreduction of carotenoids inside the thylakoid membrane (e.g. photoreduction of violaxanthin to anthera- and zeaxanthin).

Table 7
Changes in the redox level of plastoquinone-9 and carotenoids in young green leaves of Ficus elastica in light and subsequent darkness (μg per 100 cm^2 leaf area).

	daylight	+ 2 h strong light (100 000 lux)	+ 2 h darkness
Chlorophyll a	3456	3796	3890
Chlorophyll b	1189	1263	1260
Plastoquinone-9	422	428	340
Plastohydroquinone-9	58	76	136
% Plastoquinone-9	88	85	69
% Plastohydroquinone-9	12	15	31
ß-Carotene	224	257	222
Lutein	250	262	290
Violaxanthin	39	33	38
Antheraxanthin	9	11	25
Zeaxanthin	7	37	34
Neoxanthin	65	65	96
Carotenoids	594	665	705

In darkness the membrane potential is depleted which results in a plastoquinone reduction in plastoglobuli and gives rise to the oxidation of carotenoids such as the well known dark back-reaction of zeaxanthin to violaxanthin. This hypothesis will be proved by further investigations.

This paper was sponsored by the Deutsche Forschungsgemeinschaft.

References

1. Fuller, R.C., Smillie, R.M., Rigopoulos, N. and Youn, A.V., 1961, Arch. Biochem. Biophys. 95, 197.
2. Lichtenthaler, H.K., 1968, Planta (Berl.) 81, 140.
3. Stiehl, H. and Witt, T., 1969, Z. Naturforsch. 24b, 1588.
4. Trebst, A., Proceedings of the IInd International Congress on Photosynthesis Research, p. 399, D.W. Junk, N.V. Publishers, Den Haag 1972.
5. Amesz, J., van der Engh, G.J. and Visser, J.W.M., Proceedings of the IInd International Congress on Photosynthesis Research, p.419, D.W. Junk, N.V. Publishers, Den Haag 1972.
6. Lichtenthaler, H.K., 1970, Septième Congrés International de microscopie électronique, Grenoble.
7. Lichtenthaler, H.K. and Weinert, H., 1970, Z. Naturforsch. 25b, 619.
8. Lichtenthaler, H.K. and Sprey, B., 1966, Z. Naturforsch. 21b, 690.
9. Lichtenthaler, H.K., 1969, Protoplasma 68, 65 and 315.
10. Lichtenthaler, H.K., 1969, Biochim. Biophys. Acta 184, 164.
11. Ziegler, R. and Egle, K., 1965, Beitr. Biol. Pflanzen 41, 11.
12. Hager, A. and Meyer-Bertenrath, T., 1967, Planta (Berl.) 76, 149.
13. Sprey, B. and Lichtenthaler, H.K., 1966, Z. Naturforsch. 21b, 697.
14. Oku, T. and Tomita, G., 1970, Photosynthetica 4 (4), 295.
15. Lichtenthaler, H.K., Progress in Photosynth. Res. Vol. I 304, 1969.
16. Lichtenthaler, H.K., 1971, Z. Naturforsch. 26b, 832.

M. AVRON, *Proceedings of the Third International Congress on Photosynthesis*, September 2-6, 1974, Weizmann Institute of Science, Rehovot, Israel
Elsevier Scientific Publishing Company, Amsterdam, The Netherlands, 1974

LOCALIZATION AND FUNCTION OF CYTOCHROME 552 IN EUGLENA GRACILIS

G.F.Wildner and G.A.Hauska

Lehrstuhl Biochemie der Pflanzen, Ruhr-Universität Bochum, Germany

SUMMARY

Euglena chloroplasts isolated by Yeda press treatment retain up to 50% of cytochrome 552. It can be fully liberated by sonication showing that it is not tightly bound to the membrane. The rates of electron flow from water to $NADP^+$ and of cyclic photophosphorylation mediated by phenazine methosulfate correlate with the content of endogenous cytochrome 552 in the thylakoids. $NADP^+$ reduction but not cyclic photophosphorylation could be restored by supplementation of cytochrome 552. Sonication and detergent action in addition caused an exposure of the originally burried oxidation sites, now able to react with external cytochrome 552. Trapping the cytochrome in the thylakoid vesicles during sonication stimulated their phosphorylating capacity.

Monospecific antibodies against cytochrome 552 did not agglutinate chloroplasts and did not inhibit photosynthetic electron flow from water to $NADP^+$ and cyclic photophosphorylation showing that the native site of cytochrome 552 is inside the thylakoids.

INTRODUCTION

Chloroplasts from Euglena gracilis contain a c-type cytochrome with an α-band at 552 nm [1]. In contrast to cytochrome f from spinach it is a soluble protein and is easily lost during cell disruption[2] and only part of it remains in the chloroplast fraction. Since only class II chloroplasts are obtained even by the employment of mild cell breaking conditions, the question remains if cytochrome 552 is part of the stroma or is loosely bound to the thylakoid lamellae. Katoh and San Pietro[3] isolated Euglena chloroplasts which showed no $NADP^+$ reduction except after addition of cytochrome 552.

Concerned about the general topography of photosystem I in the chloroplast membrane (see rf. 4 for a recent review) it should be assumed that cytochrome 552 is indeed localized in the thylakoids and should be oxidized rather on the inside of the membrane. Similarly cytochrome c_2 in Rhodopseudomonas capsulata[5], and cytochrome f and plastocyanin in spinach[6] seem to react at the oxidation site of the reaction center on the inner side of the membrane.

In this communication evidence for an analogue location of cytochrome 552 in Euglena thylakoids is presented by the results of reconstitution experiments and the use of monospecific antibodies. In addition we could correlate the varying amount of cytochrome 552 retained in chloroplasts isolated by different pressure cell methods with photosynthetic activities.

METHODS AND MATERIALS

Cultivation of Euglena cells and the different methods for isolation of chloroplasts using the Yeda press (Yeda Res. & Dev. Co. Rehovot, Israel) or the Ribi Cell fractionator (Sorvall Co., Newtown, USA) as well as the preparation of the immune serum against cytochrome 552 will be described extensively[7,8] elsewhere.

Cytochrome 552 was purified according to the procedure by Perini et al[2] and spinach plastocyanin was prepared according to the method of Anderson and McCarty[9].

Photooxidation of cytochrome 552 and difference spectra were recorded in an Aminco DW-2 UV/VIS spectrophotometer, the details of the procedure and the assay conditions for $NADP^+$ reduction will be reported[7,8].

Cyclic photophosphorylation was measured according to McCarty and Racker[10].

RESULTS

By the use of the French press as a cell breaking device only class II chloroplasts could be obtained which lacked CO_2 fixation completely. In the course of the isolation of the chloroplasts a large amount of cytochrome 552 was lost but the extent was depending on the isolation procedure. Table I shows that using the Ribi cell fractionator (Method A,6 000 psi) on a chlorophyll basis. 30% of the cytochrome was found in chloroplasts , the rest was found in the supernatant. By the application of lower pressure conditions (Method B,650 psi) and the use of the Yeda press the content of the cytochrome 552 in the chloroplast fraction was significantly higher, up to 50% was retained in the chloroplast thylakoids. The content of the chloroplasts on cytochrome 558 and cytochrome 563 which are tightly membrane bound, were unchanged during the isolation procedure.

Table I

Correlation of cytochrome 552 content with $NADP^+$ reduction and cyclic phosphorylation in Euglena chloroplasts isolated after different pressure cell treatment.

552 stands for cytochrome 552, the rates for $NADP^+$ reduction and ATP formation in cyclic photophosphorylation are given in µmoles NADP, or ATP formed per mg chlorophyll and hr. The assay mixture for $NADP^+$ reduction contained: 50 mM HEPES-NaOH pH 6.5, 50 mM NaCl, 5 mM $MgCl_2$, 0.25 mM $NADP^+$, 10 µM ferredoxin and chloroplasts equivalent to 10 µg chlorophyll in 1 ml. The reaction mixture for cyclic photophosphorylation contained: 50 mM Tricine-NaOH, pH 8.0, 50 mM NaCl, 5 mM $MgCl_2$, 5 mM ADP, 2 mM phosphate containing 10^6 cpm ^{32}P, 0.05 mM PMS, 5 mM ascorbate, 0.02 mM DCMU, 1 mg defatted bovine serum albumine and chloroplasts equivalent to 30 µg chlorophyll in 1 ml. The recommended addition of hexokinase and glucose[11] did not increase the observed rates.

	Addition of 552	Retained 552 in %	$NADP^+$ reduction	cyclic photophosphorylation
METHOD A	-	30	5	25
	1.7 µM	-	45	25
	6.8 µM	-	87	22
METHOD B	-	50	29	83
	1.7 µM	-	56	78
	6.8 µM	-	100	75

The rates of $NADP^+$ reduction and cyclic photophosphorylation given in Table I indicate a correlation between the cytochrome content and its function in electron transport, and PMS-mediated cyclic photophosphorylation. In both chloroplast preparations the addition of external cytochrome 552 stimulated the rate of $NADP^+$ reduction but had no effect on cyclic photophosphorylation. Cyclic photophosphorylation could only be enhanced by trapping cytochrome 552 inside the thylakoid vesicles during sonication (Table II), analogous to plastocyanin and spinach chloroplasts[6]. These data indicate that only internal cytochrome 552 participates in coupled electron flow mediated by PMS around photosystem I. Spinach plasto-

Table II

Protection of cyclic photophosphorylation in Euglena chloroplasts during sonication by the presence of cytochrome 552 or plastocyanin. The incubation mixture was the same as described in the legend of Table I. 552 stands for cytochrome 552 and PC for spinach plastocyanin. The rate of ATP formation is given in μmoles formed per mg chlorophyll and hour. Sonication was perfermed as described[6].

	Additions		Cyclic Photo-phosphorylation
	during sonication	during assay	
Chloroplasts	-	-	35
	-	2 μM 552	33
	-	2 μM PC	36
Sonicated Chloroplasts	-	-	5
	-	2 μM 552	5
	-	2 μM PC	5
	20 μM 552	-	11
	70 μM 552	-	35
	140 μM 552	-	42
	10 μM PC	-	9
	20 μM PC	-	18
	40 μM PC	-	40

cyanin could replace cytochrome 552. Stimulation of $NADP^+$reduction by external cytochrome 552 implies that there are oxidation sites also on the outer surface of the membrane, but this could be explained by disorientation of photosystem I caused by mechanical treatment. This consideration was substantiated by the results of Fig. 1, which show the photooxidation of externally added cytochrome 552 and spinach plastocyanin by chloroplasts at different stages of fragmentation. Sonication of the chloroplasts and treatment with cholate increases the photooxidation of cytochrome 552 or of spinach plastocyanin drastically.

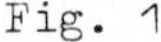

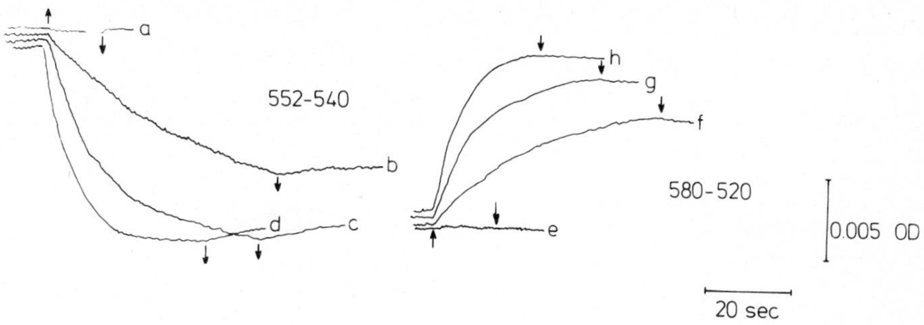

Fig. 1. Photooxidation of externally added cytochrome 552 and spinach plastocyanin by Euglena chloroplasts at different degrees of fragmentation.

The assay mixture of 3 ml contained: 50 mM Tricine-NaOH, pH 8.0, 50 mM NaCl, 5 mM $MgCl_2$, 2 x 10^{-5}M DCMU, 10^{-4}M anthraquinone-2-sulfonate, chloroplasts equivalent to 0.02 mg chlorophyll. Cytochrome 552 was added to 0.7 µM in the cases b,c,d, and spinach plastocyanin was added to 1.4 µM in the cases f,g and h. Traces a and e correspond to the chloroplast preparation (Method B) without addition, traces b and f show the change after the addition of cytochrome 552 and plastocyanin, respectively traces c and g show the corresponding change in chloroplasts sonicated in the assay medium four times for 15 sec; traces d and h depict the changes after addition of cholate to 1% before the assay. Upward arrows reflect "light on", downward arrows" light off".

A further method to investigate the localization of cytochrome 552 was the application of monospecific precipitating antibodies. The immunoglobulins did not agglutinate chloroplasts neither directly nor in the indirect agglutination test[12,13], indicating that the remaining cytochrome 552 in the chloroplast lamellae was inaccesible to the antibody. Accordingly, the immune serum did not inhibit photooxidation of the endogenous cytochrome 552 and no effect on the $NADP^+$reduction or cyclic photophosphorylation could be observed. In the presence of 0.2% cholate, however, cytochrome 552 in the chloroplasts is now accesible to the antibody as shown by the block of its photooxidation (Fig. 2).

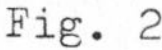

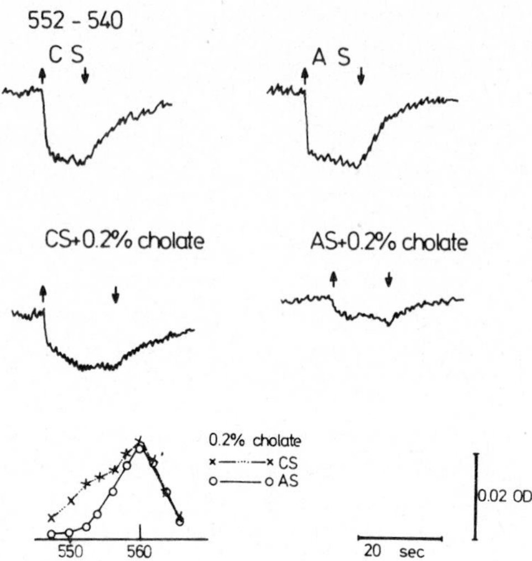

Fig. 2. Effect of immunoglobulins on photooxidation of endogenous cytochrome 552 in Euglena chloroplasts.

The assay mixture contained: 50 mM Tricine-NaOH, pH 8.0, 50 mM NaCl, 5 mM $MgCl_2$, 0.1 mM anthraquinone-2-sulfonate, 0.03 mM ascorbate and Euglena chloroplasts equivalent to 0.1 mg chlorophyll in 3 ml volume and 0.1 ml gamma-globulin fraction (50 mg/ml) prepared either from control (CS) or immune serum (AS) were present. The upward arrows indicate "light on" and the downward arrows "light off".

It was ascertained by immunological cross reactions that the endogenous cytochrome 552, which could be liberated by sonication was identical with the cytochrome 552 already liberated during the isolation of the chloroplasts. The immune serum did block the photooxidation of external cytochrome 552, but did not react with cytochrome 552 trapped in the thylakoids during sonication.

DISCUSSION

The inaccessibility of cytochrome 552 in the chloroplast membrane to antibodies on one hand, and the oxidation of external cytochrome 552 on the other creates a discrepancy which calls for an explanation. Since similar situations have been met in the cases of mammalian cytochrome c and the mitochondrial membrane[14], cytochrome c_2 and the membrane of photosynthetic bacteria[5] and plastocyanin and the membrane of spinach chloroplasts[6], this double reaction mode of such peripheral[15] membrane proteins might reflect a general phenomenon. The question arises whether these two reaction sites are resembling electron flow in vivo, or whether one of them is introduced as an isolation artefact. The latter possibility is more likely[14,5,6] and seems to gain support also by the data presented here. This is summarized in the scheme, which depicts the orientation of electron transport in the membrane of Euglena chloroplasts according to Mitchells loops[16]. We think that the physiological

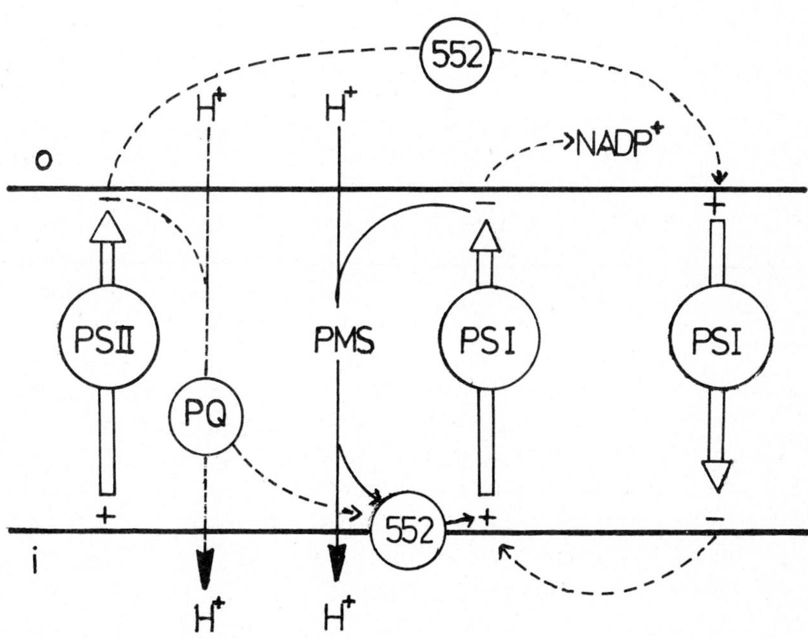

location of cytochrome 552 is the one not accessible to antibodies, because there it functions in electron transport coupled to energy conservation. The external oxidation site we consider as an artefact because further fragmentation increases this reaction. Such artificial, external oxidation sites could reflect several modes changing the accessibility to photosystem I. The scheme depicts a tumbled disorientation in the membrane, but non-vesicular structures or vesicle populations with different surfaces exposed, would equally account for the observation. In an intact chloroplast, stripped by the outer membrane, no oxidation of external cytochrome 552 should occur, resembling the situation with chloroplasts from spinach and plastocyanin[17]. Unfortunately we did not succeed in preparing such intact chloroplasts from Euglena.

Following the scheme, short circuits of energy conserving electron flow, either by draining of electrons, or by providing for an "anti-loop" of PMS around the disoriented photosystem I should occur, and could account for the low efficiency of energy conservation in isolated Euglena chloroplasts. However, external cytochrome 552, which according to this view should increase these short circuits, did not inhibit cyclic phosphorylation with PMS. This suggests t at there is no connection between the two populations of photosystem I.

As with plastocyanin in spinach chloroplasts[6] we have to consider three possibilities for the location of internal cytochrome 552. It either is located on the inner, aqeous surface, or within the membrane core, or even on the outer surface without exposure of its antigenic determinants. We favor the first possibility because of its analogy to the location of cytochrome c in submitochondrial vesicles[14] and of cytochrome c_2 in chromatophores[5]. It is also supported by the fact that cytochrome 552 is a very hydrophilic protein, but needs perturbance of the membrane for liberation.

Sonication in the presence of cytochrome 552 prevents the liberation of internal cytochrome 552 and preserves cyclic photophosphorylation, but surprisingly we could not observe simultaneous preservation of NADP-reduction by water. Sonication thus interrupts electron flow from photosystem II to internal cytochrome 552 in an irreversible way, and at the same time allows external cytochrome 552 to bridge between the two photosystems. Also French press treatment seems to cause such changes which would explain why chloroplast preparations retaining substantial amounts of cytochrome 552 are almost totally inactive in NADP-reduction (s.Table 1).

Remembering, that in an isolated system we naturally face some artefacts, we can state in summary that cytochrome 552 has to be located inaccessible to antibodies, probably on the inner surface of the thylakoid membrane, in order to participate in coupled electron transport. The topography of photosystem I in Euglena, therefore, fits into a general concept for the topography of energy transducing biomembranes[5], as postulated by Mitchell[16].

ACKNOWLEDGEMENTS

The authors are indebted to Prof. Dr. A. Trebst for encourangement and interest and Dr. R. Berzborn for his advice in immunological techniques. The work was supported by Deutsche Forschungsgemeinschaft.

REFERENCES

1. Davenport, H.E. and Hill, R., 1954, Proc. Roy. Soc. (London) B 139, 327
2. Perini, F., Schiff, J.A. and Kamen, M.D., 1964, Biochim. Biophys. Acta 88, 74
3. Katoh, S. and San Pietro, A., 1967, Arch. Biochem. Biophys. 118, 488
4. Trebst, A., 1974, Ann. Rev. Plant Physiol. 25, 423
5. Prince, R.C., Hauska, G.A., Crofts, A.R., Melandri, A. and Melandri, B.A., abstract for this congress.
6. Hauska, G., McCarty, R.E., Berzborn, R. and Racker, E., 1971, J. Biol. Chem. 246, 3524
7. Wildner, G.F. and Hauska, G.A., 1974, Arch. Biochem. Biophys., in press
8. Wildner, G.F. and Hauska, G.A., 1974, Arch. Biochem. Biophys., in press
9. Anderson, M.M. and McCarty, R.E., 1969, Biochem. Biophys.Acta 189, 193
10. McCarty, R.E. and Racker, E., 1967, J. Biol. Chem. 242, 3435
11. Kahn, J.S., 1966, Biochem. Biophys. Res. Commun. 24, 329
12. Berzborn, R., 1968, Z. Naturforschg. 23b, 1096

13. Uhlenbruck, G. and Prokop, O., 1967, Dtsch. med. Wochenschriften 92, 940
14. Racker, E., Burstein, C., Loyter, A. and Christiansen, P.O., 1970, in Electron transport and energy conservation, Tager, J. M., Papa, S., Quagliarello, E. and Slater, E.C., eds, Adriatica Editrice, Bari, p. 235
15. Singer, S. J. and Nicholson, G.L., 1972, Science 175, 720
16. Mitchell, P., 1966, chemiosmotic coupling in oxidative and photosynthetic phosphorylation, Glynn Res. Ltd., Bodmin, Cornvall, England
17. Sane, P.V. and Hauska, G., 1972, Z. Naturf. 27b, 932

M. AVRON, *Proceedings of the Third International Congress on Photosynthesis*, September 2-6, 1974, Weizmann Institute of Science, Rehovot, Israel
Elsevier Scientific Publishing Company, Amsterdam, The Netherlands, 1974

ON THE PATHWAY AND MECHANISM OF THE CYTOCHROME b-559 PHOTOREACTIONS

P. Horton, H. Böhme*, and W. A. Cramer

Department of Biological Sciences
Purdue University
West Lafayette, Indiana 47907 U.S.A.

*Present address: University of Konstanz, Fachbereich Biologie, D-7750 Konstanz, Giessberg, Postfach 733, Germany

Introduction

There is a component or a state of cytochrome b-559 which has a high potential ($E_m \gtrsim 0.3$ V), as measured in the dark[1]. This component has an anomalous position in the two pigment system framework of green plant photosynthesis. The problem is that this cytochrome seems to be structurally linked to PS II as inferred from its oxidation by system II at 77°K (e.g., ref. 2-4), but its redox potential in the dark places it midway between the water oxidation pathway and the mild reductant generated in PS II. It is also anomalous because in chloroplasts it has been difficult to experimentally detect significant light-induced absorbance changes of this cytochrome without perturbing the photosynthetic system in a nonphysiological manner. The treatments which have been used are (a) addition at room temperature of particular uncouplers or membrane reactive reagents such as the carbonyl cyanide phenylhydrazone compounds[5], and (b) freezing to liquid nitrogen temperature[2-4] or washing chloroplasts with high concentrations of Tris buffer[2]. With treatments (b) the water oxidation reactions are blocked and the cytochrome b-559 is preferentially oxidized by PS II. It seems to be possible to gain much information about the PS II reaction center photochemistry through study of the low temperature photoreactions of this cytochrome[2-4]. In the present work, however, we shall concentrate on the pathway of cytochrome b-559 oxidation and reduction in chloroplasts at physiological temperatures.

Hypotheses for pathways involving cytochrome b-559

There are two proposals currently in the literature for the functioning pathway of cytochrome b-559: (1) The suggestion that this cytochrome can function in the main chain, despite its high redox potential in the dark, is based (a) on studies of the cytochrome b-559 photoreactions in wild type and mutant strains of the alga *C. reinhardi*[6], and (b) on chloroplast photoreactions of this cytochrome in the presence of low or intermediate concentrations of FCCP[2,7,8]. In both of these cases the cytochrome is oxidized and reduced, respectively, by moderate to high intensities of far-red and red actinic light in a pattern very similar to other components in the main chain such as cytochrome _f_ or plastoquinone.[5] That part of the cytochrome _b_-559, which is initally high potential in the dark, can be oxidized by PS I in the presence of FCCP is shown in an experiment (fig. 1) where the far-red light intensity at 732 nm is

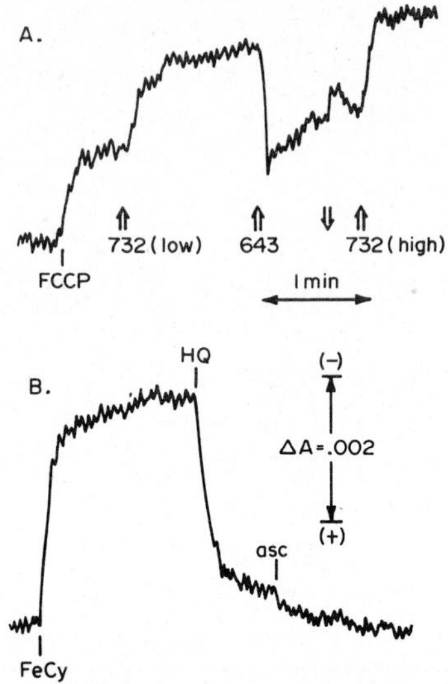

Fig. 1. Oxidation of cyt _b_-559 by far-red light (A) compared to chemical oxidation (B). Intensities (erg/cm^2-sec): 732 nm (low), 6.1 x 10^3; 732 nm (high), 6.3 x 10^4; 643 nm, 6.4 x 10^4. (A) 1 μM FCCP. (B) 250 μM FeCy, 1 mM HQ.

deliberately set at 1/10 that of the red actinic light, so that there can be no doubt that system I drives the photooxidation. It should be noted that the increase in photooxidation of _b_-559 occurs at FCCP concentrations which do not detectably inhibit the steady-state level of oxygen evolution with methyl viologen[9] or diaminodurine-ferricyanide (unpublished data) as acceptor, although it is possible that individual steps in oxygen evolution, which would not be rate-limiting for the overall process, are inhibited[10]. FCCP may also inhibit PS II more effectively at lower actinic light intensities, which together with interference by dark reactions, tends to complicate measurements of quantum requirements for the light reactions of the cytochrome.

Another proposal (2) for cytochrome _b_-559 function, which is also partly based on experiments with FCCP, is that it functions in a cycle about photosystem II[11]. Because it is uncertain that such a cyclic pathway is significant at physiological light intensities and temperatures we choose to concentrate on proposal (1), i.e., that cytochrome _b_-559 can transfer electrons in the main chain near PS II. We will discuss the pathway for the light reactions in the presence of FCCP to some extent, but will base our conclusions on light reactions for b-559 measured in the absence of FCCP under experimental conditions that are close to physiological for the *in situ* system.

What is the pathway for b-559 oxidation by PS I in the presence of FCCP? Since the cytochrome is initially high potential in the dark, electron transfer to PS I must bypass plastoquinone and the bulk of the main chain unless the cytochrome can assume a significantly lower potential in the light. In this context, it will be assumed here that the redox properties of membrane bound cytochromes in general are not immutable, but can be modified through alterations in the membrane environment or changes in differential ligand binding to oxidized and reduced states. As an example of this phenomenon in the chloroplast system, we have shown by titration that actinic light in the presence of FCCP or NH_4Cl can cause a negatively directed shift of 100-150 mV in the midpoint potential of part of the chloroplast cytochrome b_6 (fig. 2)[12]. It appears that the pathway

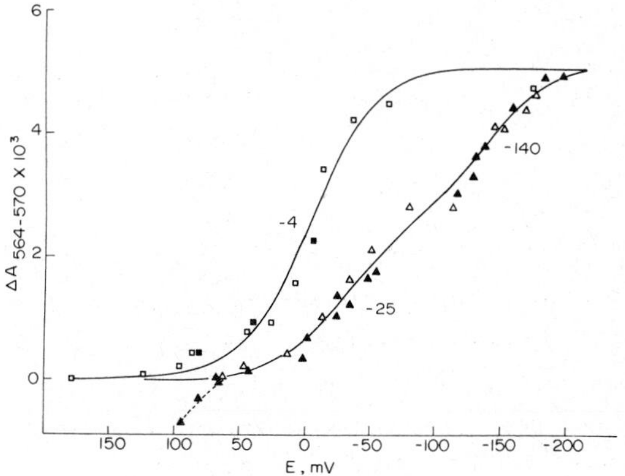

Fig. 2. The effect of FCCP (10 μM) on the redox titration of cytochrome b_6[12].

for oxidation of cytochrome b-559 in the presence of FCCP does include plastoquinone, since the oxidation is inhibited by the quinone analog DBMIB, and the inhibition is reversed by added plastoquinone[8]. The b-559 component oxidized by plastoquinone must then transfer electrons in the light at a much lower potential than measured in the dark. Thus, we propose that in the presence of FCCP, light causes a *transient* conversion of cytochrome b-559 to a low potential form which is oxidized by plastoquinone[7,8].

The effect of FCCP in stimulating photooxidation of cytochrome b-559 by PS I, is not shared by other uncouplers of photophosphorylation such as NH_4Cl or atebrin[5], nor uncouplers such as gramicidin which are very potent in their ability to increase proton conductance across membranes (fig. 3). Other agents

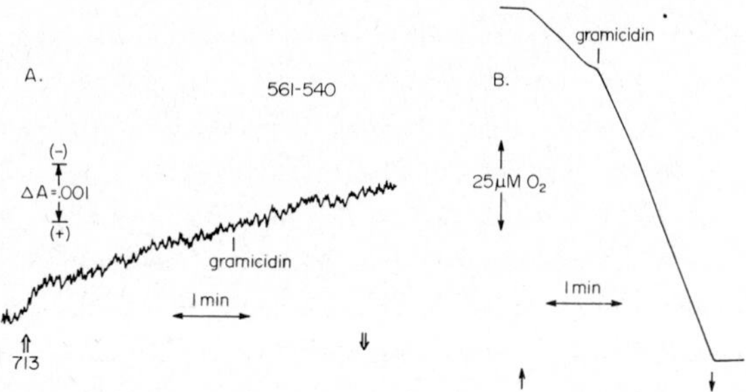

Fig. 3. Effect of gramicidin on cyt $\underline{b}_{559}$ oxidation (A) and oxygen evolution (B). 713 nm light intensity, 6.5 x 10^4 erg/cm^2-sec. Oxygen uptake measured with methyl viologen (.1 mM) and azide (.2 mM). Chl. conc., 80 (A) and 20 (B) μg/ml. Gramicidin, 10 μg/ml.

such as antimycin A and desaspidin have been shown to stimulate photooxidation of b-559[13], and we have found that the uncouplers S-13, 1799, as well as TTFB will also act in this way, although the red light reversibility is marginal with the latter two compounds (unpublished data). The b-559 photoreactions in the presence of FCCP are then somewhat unique in terms of uncoupler effects. In order to seriously discuss any model for the pathway of a functioning cytochrome b-559 it is necessary to demonstrate photoreactions of this cytochrome under more physiological conditions.

Experimental problems

Why is a light-induced oxidation of cytochrome b-559 not generally observed with chloroplasts in the absence of an FCCP-like reagent? There are several experimental problems involved in studying the turnover of cytochrome b-559: (1) If the cytochrome b-559 is initially reduced, significant net oxidation by system I may not occur if the rate constants for the oxidative steps are much smaller than those for reduction. The oxidative pathway to PS I has at least one slow step which is rate limiting for non-cyclic electron transport with a characteristic transfer time of about 20 msec. The rate constant for reduction is about 40 times faster according to the kinetic data of Witt and coworkers[14]. The DBMIB inhibition of b-559 photooxidation[7,8] implies that cytochrome b-559 transfers electrons to the plastoquinone pool. If b-559 functions in the main chain, there might then be two relatively slow steps on the system I side of b-559. The existence of more than one step would be required to explain why uncouplers like NH_4Cl do not stimulate oxidation of b-559. The experimental conditions for detecting b-559 oxidation under physiological conditions could then perhaps be improved (a) by the use of lower intensities and/or longer wavelengths of far-red illumination in order to decrease excitation of PS II, and (b) the use of compounds which might bypass slow steps on the system I side of b-559.

Other experimental problems which may obscure light induced b-559 oxidation include (2) the photoreduction of cytochrome b_6 by PS I, which occurs at high actinic light intensities[15]. This b_6 absorbance change, which would be oppositely directed to that of b-559 oxidation, would lead to a decreased absorbance change at 559 nm. One might try to counteract the b_6 reduction by again using low intensities of far-red light and an autoxidizable electron acceptor like methyl viologen to prevent electrons from reducing b_6. (3) One can ask the question as to whether there are any other conditions in the spectrophotometer which make the measurement conditions dissimilar to those used in other assays of chloroplast activity (e.g., oxygen electrode assays). One such condition is in fact the actinic light intensity, which at 6 x 10^4 ergs/cm^2-sec, is about 5-10 times lower as it strikes the cuvette than the intensities commonly used in oxygen electrode measurements. The effective light intensity inside the cuvette is of course even lower because the chlorophyll concentration in the cuvette at 50-80 μg/ml is about 3-4 times larger than that generally used in oxygen measurements.

Results and discussion

We have found more physiological conditions for observing cytochrome b-559 turnover in chloroplasts typical of a component in the main chain: (1) In the presence of low concentrations of PMS (~ 0.5 μM) (fig. 4A) and (2) after a pre-illumination regime in which the chloroplasts are exposed to a more intense red light (~ 2 x 10^6 ergs/cm^2-sec., no interference filter, PMT turned off), the cytochrome b-559 is oxidized by relatively low intensities of far-red light (732 nm) and reduced by red light, as shown in fig. 4B. As a reference, the oxidative change caused by 500 μM ferricyanide is shown in fig. 4C.

In light of the above discussion on experimental problems we note that the far-red response is appreciably greater when 732 nm far-red light of relatively low intensity is used compared to 713 nm at the normal actinic intensity (fig. 5).

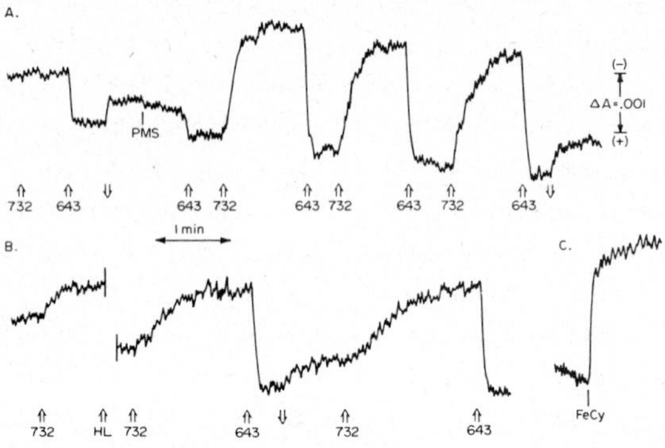

Fig. 4. Light-induced absorbance changes at 560 nm (A) after addition of 1 μM PMS and (B) after exposure to 2 min high intensity (HL) red illumination. Light intensities (erg/cm^2-sec): HL, 2 x 10^6; 732 nm (A), 643 nm (A), and 643 nm (B), 6.3 x 10^4; 732 nm (B), 2.4 x 10^3. Chl. conc., 80 μg/ml. (C) Oxidation by 0.5 mM FeCy. Reference, 570 nm.

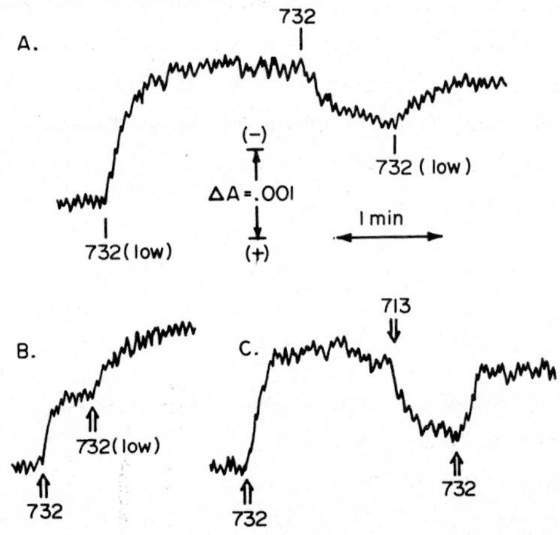

Fig. 5. Oxidation of cyt $\underline{b}_{559}$ (560/570) by different intensities and wavelengths of far-red light; A and C (1 μM PMS), B (pre-treatment with 2 min high light). Light intensities (erg/cm^2-sec): 732 and 713 nm, 6.3 x 10^4; 732 (low), 6.3 x 10^3 (A), and 2.4 x 10^3 (B). Chl. conc., 80 μg/ml.

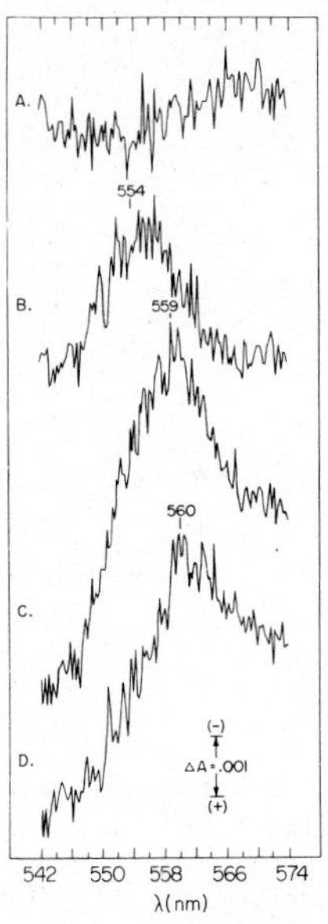

Fig. 6. Difference spectra for absorbance changes of fig. 4A, measured as in ref. 16. (A) dark (PMS) minus dark. (B) Far-red (low) minus dark. (C) Far-red (PMS) minus dark. (D) C minus B. Time const, 0.2 sec.

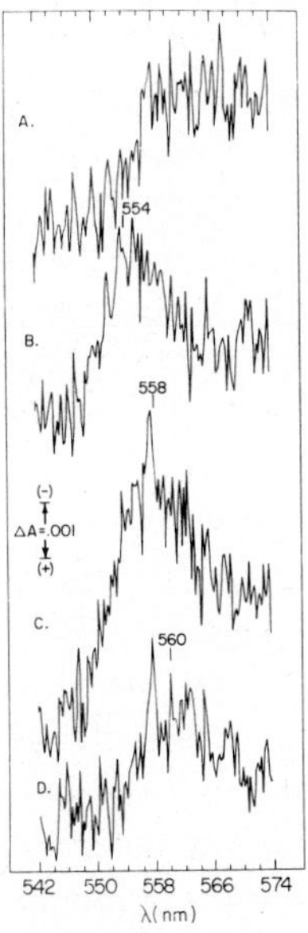

Fig. 7. Difference spectra for absorbance changes of fig. 4B. (A) (Dark after HL) minus dark; (B) Far-red (low) minus dark; (C) Far-red (low) after HL minus dark; (D) C minus B.

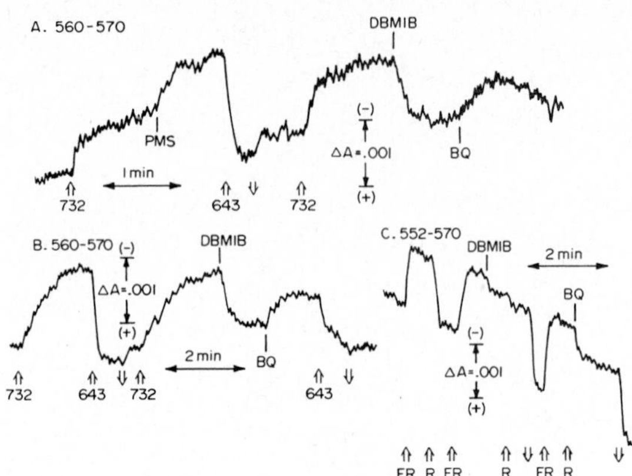

Fig. 8. Inhibition of cyt $\underline{b}_{559}$ oxidation by DBMIB. (A) With 0.5 μM PMS, 3 μM DBMIB. (B) After "light activation," with 2 μM DBMIB. 643 nm light intensity, 2.4 x 10^3 erg/cm^2-sec. (C) DBMIB (2 μM) inhibition of red light reduction of cyt $\underline{f}$. 732 and 643 nm light intensities, 1.3 and 1.6 x 10^4 erg/cm^2-sec., except where noted. Chl. and BQ conc., 75 μg/ml and 20 μM.

We do not know at present whether this is because the low intensity 732 nm light results in a slower rate of photoreduction of $\underline{b}$-559 by PS II and/or a slower rate of $\underline{b}_6$ reduction by PS I. The stimulation by PMS of $\underline{b}$-559 oxidation occurs only at low concentrations of PMS and seems to be inhibited above 5 μM PMS. Computer-corrected single-beam spectra taken at room temperatures for the far-red light-induced absorbance changes (a) in the presence of PMS (fig. 6) and (b) after "light activation" (fig. 7), show that the component which is principally photooxidized by far-red light under these conditions is cytochrome $\underline{b}$-559 (figs. 6C,D and 7C,D). It can be seen in addition that the photooxidation under both of these conditions is inhibited by DBMIB (2-3 μM) and reversed by the addition of benzoquinone (20 μM), as shown in figs. 8A and B.

The "light activation" experiment described here is done under conditions which are physiological for the _in situ_ system, in that no additions other than a Hill oxidant are made. The results indicate that a large part of the cytochrome $\underline{b}$-559 can be reversibly oxidized by low intensities of far-red light through a pathway that includes plastoquinone. We would conclude that it is a relatively low potential species of $\underline{b}$-559 which is oxidized. Thus, we think that there is now reason to believe that cytochrome $\underline{b}$-559 undergoes a light-activated negatively directed potential change and possibly a conformational change which allows it to function in the main chain near PS II. In this context it should be noted that cytochrome $\underline{f}$ and/or the thylakoid membrane undergo conformational changes after actinic illumination which lead to an increased accessibility to the charged oxidant ferricyanide[16]. A model for the photosynthetic electron transport chain which includes the possibility of cytochrome b-559 functioning in the main chain is shown in fig. 9.

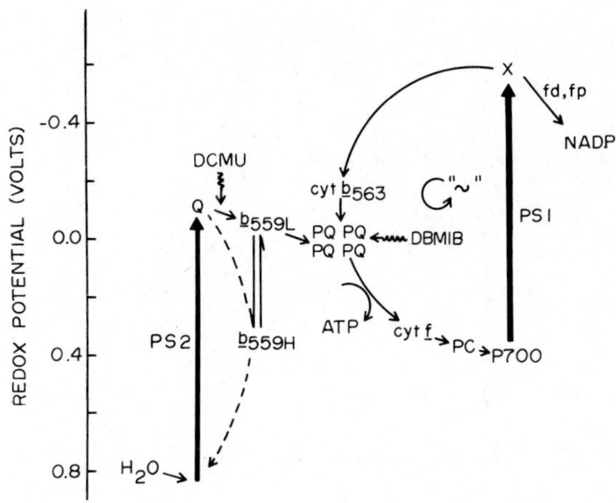

Fig. 9. Formulation of the Z scheme for photosynthetic electron transport including the possibility of two redox states for cytochrome b-559.

In the model of fig. 9 a mechanism is needed for the potential change of cytochrome b-559. The special effects of FCCP-like reagents, PMS and "light activation" must then be examined as to whether they exert their special effect on b-559 by facilitating this mechanism, or by altering the rate constants for oxidation and reduction. The simplest proposal for the mechanism of action of PMS would be that it creates a bypass around a slow step in the pathway of oxidation of low potential b-559. The potential of PMS (E_{m7} = + .08V) is appropriate for such a bypass, and there is evidence in the literature that PMS can bypass the slow step in the oxidation of cytochrome b_6 by photosystem I[15]. It is of interest with regard to the mechanism of action of the CCP compounds that the pool of electron carriers between the two photosystems appears to be largely oxidized after red actinic illumination in the presence of low concentrations of CCCP[17]. Regarding the mechanism of b-559 conversion to a low potential form, one might ask what chemical changes occur in the light? An obvious chemical change involves proton movement into and possibly across the chloroplast membrane. Does proton movement or formation of transmembrane or local charge gradients have any relation to conformational and redox changes of the cytochrome b-559? If proton uptake would cause a negatively directed change in E_m, a dark acid transition of the chloroplasts would be expected to form a lower potential species of b-559 which might be reflected in cytochrome oxidation. An acid-base

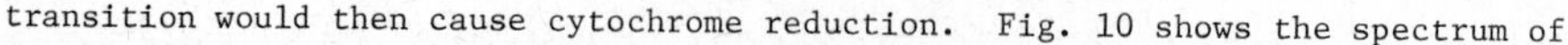

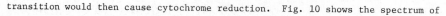

transition would then cause cytochrome reduction. Fig. 10 shows the spectrum of

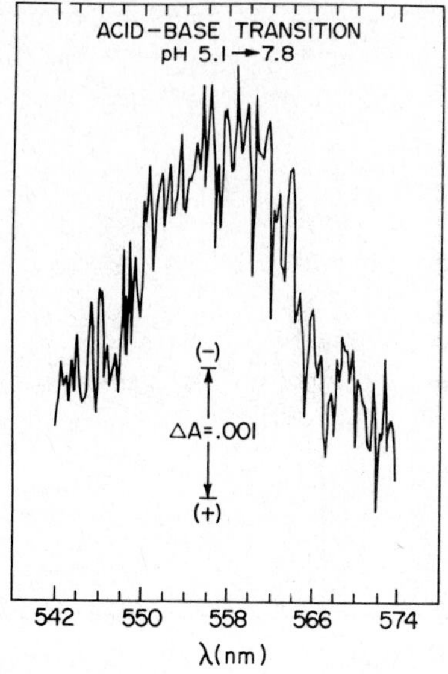

Fig. 10. The effect of an acid-base change on the cytochrome oxidation state. Chloroplasts (80 μg Chl/ml) were incubated in 2 mM $MgCl_2$, 10 mM NaCl and exposed to four cycles of acid (pH ~ 5 with succinate) and base (pH 7.5-8 with NaOH). Spectra were recorded as in fig. 6 for the 4th cycle before and after addition of alkali and the difference spectrum, pH 5.1 minus pH 7.76, was computed on-line.

the absorbance change caused by addition of alkali to chloroplasts at pH 5.1. The acid-base transition to pH 7.8 causes a reduction of cytochrome b-559 and possibly cytochrome f. The chemical acid-base transition experiments are still in a preliminary state but may provide some insight on the mechanism of light-induced redox and conformational changes of cytochrome b-559.

Acknowledgment.

We would like to thank J. J. Donnell for her participation in some of these experiments, Dr. R. Dilley for helpful discussions, and Mrs. Mona Imler for help with the manuscript. This research was supported by NSF grant GB-34169X and NIGMS Career Development Award I K04 GM 29735.

References

1. Bendall, D. S., 1968, Biochem. J., 109, 46p.
2. Knaff, D. B., and Arnon, D. I., 1969, Proc. Natl. Acad. Sci. U.S. 63, 956.
3. Butler, W. L., 1972, Proc. Natl. Acad. Sci., U.S., 69, 3420.
4. Vermeglio, A., and Mathis, P. 1973, Biochim. Biophys. Acta, 314, 57.

5. Cramer, W. A., and Butler, W. L., 1967, Biochim. Biophys. Acta, 143, 332.
6. Levine, R. P., and Gorman, D. S., 1966, Plant Physiol., 41, 1293.
7. Cramer, W. A., Fan, H. N., and Böhme, H., 1971, J. Bioenerg., 2, 289.
8. Böhme, H., and Cramer, W. A., 1971, FEBS Lett., 15, 349.
9. Cramer, W. A., and Böhme, H., 1972, Biochim. Biophys. Acta, 256, 358.
10. Renger, G., 1972, Physiol. Veg., 10, 329.
11. Ben-Hayyim, G., 1974, Eur. J. Biochem., 41, 191.
12. Böhme, H., and Cramer, W. A., 1973, Biochem. Biophys. Acta, 325, 1275.
13. Hind, G., 1968, Photochem. Photobiol, 7, 369.
14. Witt, H. T., 1971, Quart. Rev. Biophys., 4, 365.
15. Böhme, H., and Cramer, W. A., 1972, Biochim. Biophys. Acta, 283, 302.
16. Horton, P., and Cramer, W. A., 1974, Biochim. Biophys. Acta., in press.
17. Marsho, T. V., and Kok, B., 1974, Biochim. Biophys. Acta, 333, 353.

M. AVRON, *Proceedings of the Third International Congress on Photosynthesis*,
September 2-6, 1974, Weizmann Institute of Science, Rehovot, Israel
Elsevier Scientific Publishing Company, Amsterdam, The Netherlands, 1974

PHOTOOXIDATION OF CYTOCHROME *b*-559 IN CHLOROPLAST FRAGMENTS OF NON-PHOTOSYNTHETIC MUTANTS OF *CHLAMYDOMONAS REINHARDTI*

Jacques Garnier and Jeannine Maroc

Laboratoire de Photosynthèse, C.N.R.S.
91190 Gif-sur-Yvette, France

1. Summary

Photoinduced absorbance changes related to cytochrome *b*-559, in chloroplast fragments of wild type and non-photosynthetic mutants of *Chlamydomonas reinhardti*, are investigated. Three mutant strains were used : *Fl* 5, which lacks P700 ; *Fl* 9 and *Fl* 15, which are deficient in cytochromes *c*-553 and *b*-563. In presence of antimycin 0.05 mM or of FCCP 0.01 mM, two kinds of cytochrome *b*-559 oxidation were observed : (1) a System II-dependent photooxidation, inhibited by CMU, occurring in all the strains, and (2) a CMU-insensitive, System I-dependent photooxidation, occurring in *Fl* 9, *Fl* 15 and wild type, but not in *Fl* 5. In these two kinds of photooxidation, the hydroquinone-reducible form of cytochrome *b*-559 was mainly involved. These data suggest that the high-potential form of cytochrome *b*-559 and cytochrome *c*-553 are involved in different electron transport chain.

2. Introduction

The cytochrome *b*-559 of higher plants chloroplasts exists in various forms having different oxidation-reduction potentials[1,2]. The high-potential (HP) form, reducible by hydroquinone, appears to be the main form in fresh chloroplasts in absence of any inhibitor[3]. But this form is unstable and various treatments transform it in low-potential (LP) form(s) reducible by ascorbate[1,4]. Another form, reducible only by dithionite, exists also in fresh chloroplasts[2].

When chloroplasts are illuminated in presence of various inhibitors of the electron flow (DCMU, antimycin, CCCP, FCCP), a photooxidation of the cytochrome *b*-559 occurs. Depending on the cases, this photooxidation can be attributed to photosystem I activity[4-7] or to photosystem II activity [8-11]. In addition, many of

---

Abbreviations : CCCP, carbonylcyanide *m*-chlorophenylhydrazone ; CMU, 3-*p*-chlorophenyl-1,1-dimethylurea ; DCMU, 3-(3,4-dichlorophenyl)-1,1-dimethylurea ; FCCP, carbonylcyanide *p*-trifluoromethoxyphenylhydrazone ; HP, high-potential ; LP, low-potential ; P700, chlorophyll *a* holochrome, active pigment in photosystem I.

these inhibitors (antimycin, CCCP, FCCP) can transform the cytochrome *b*-559 HP form into the LP form[4]. A System II-dependent cytochrome *b*-559 HP photooxidation, occurring at -196°C in absence of any inhibitor, has been observed also by different authors[3,12].

Studying non-photosynthetic mutants and the wild type of *Chlamydomonas reinhardti*, Levine and Gorman[13] showed a photoreduction of the cytochrome *b*-559 by System II and its photooxidation by System I in chloroplast fragments preparations. Afterwards, Epel et al.[14,15], carrying out a spectroscopic analysis of chloroplast fragments of several strains of low-fluorescent and high-fluorescent mutants, distinguished two pools of ascorbate-reducible cytochrome *b*-559 and a dithionite-reducible form.

In this paper, we report some investigations on the effects of antimycin, FCCP and CMU on photoinduced absorbance changes of the cytochrome *b*-559 in chloroplast fragments of the wild type and non-photosynthetic mutants of *C. reinhardti*. Three mutant strains were used : *Fl* 5 which lacks the active pigment of System I, P700[16] ; *Fl* 9 and *Fl* 15 which are deficient in bound cytochrome *c*-553 and in cytochrome *b*-563[16,17].

3. Materials and methods

The three mutants *Fl* 5, *Fl* 9 and *Fl* 15, isolated in our laboratory, and the wild type of *C. reinhardti* were previously described[16-18]. Algae were grown axenically, under low light intensity, in Tris-acetate-phosphate medium[19].

For chloroplast fragments preparation, the cells were suspended in phosphate buffer (pH 7.5) and were disrupted by treatment in a gas pressure device ("Yeda Press", Yeda Scientific Instruments)[20], at 4°C ; the chloroplast fragments were separated by centrifugation at 480 g for 6 min, then at 30 000 g for 15 min. Chlorophyll was extracted and determined in 80 % acetone[21,22].

The measurements of the cytochrome *b*-559 absorbance changes were carried out in an Aminco-Chance dual-wavelength spectrophotometer. The actinic light was provided by a side xenon lamp and two combinations of filters were employed : Balzers interference filter K6 + Schott cut-off filter RG2 + anticaloric filter Balzers Calflex B_1K_1 (half-band width : 630-680 nm, maximum at 650 nm) or the same anticaloric filter + Balzers interference filter K7 (half-band width : 680-750 nm, maximum at 711 nm). With these both sets, the light intensity was $2 \cdot 10^4$ $ergs \cdot cm^{-2} \cdot sec^{-1}$. Because the transmission spectra of these interference filters were broad in their bases, the illuminations given by these two combinations were not quite selective for each photosystem. The photomultiplier was protected by a Corning 4-76 glass filter. An extinction coefficient of 20 $cm^2 \cdot mole^{-1}$ for the cytochrome *b*-559[23] was used.

4. Results

Fig. 1 shows absorption difference spectra : light *minus* dark, at liquid nitrogen temperature, of chloroplast fragments of the wild type and of the three mutants. Two negative peaks, at 547 nm and 557 nm, appear on each of these spectra. These characteristic peaks indicate respectively the photoreduction of the electron carrier C-550 and the photooxidation of the cytochrome *b*-559 by System II[12-14].

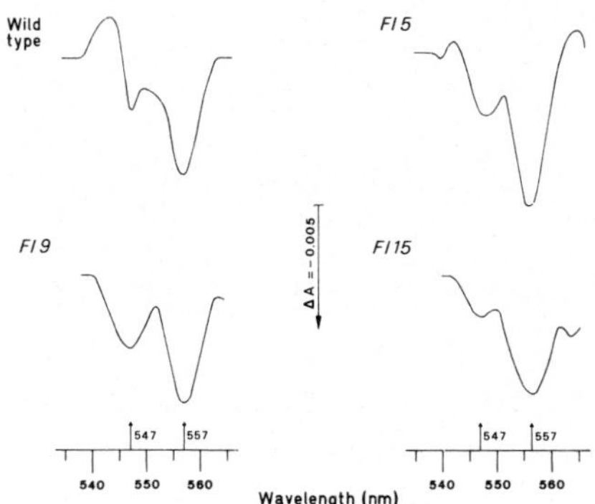

Fig. 1. Difference spectra : light *minus* dark, at - 190°C, of chloroplast fragments of the wild type and the mutants *Fl* 5, *Fl* 9, *Fl* 15 of *C. reinhardti*. Chloroplast fragments in phosphate buffer 0.01 M (pH 7.5) in concentrations corresponding to 250 (Wild type, *Fl* 9), 550 (*Fl* 5) and 200 (*Fl* 15) μg chlorophyll/ml. Reducing agent prior to freezing : Na ascorbate. Actinic light : 630-680 nm, $2 \cdot 10^4$ $ergs \cdot cm^{-2} \cdot sec^{-1}$.

The main results concerning the redox changes of cytochrome *b*-559 at room temperature are given in Table 1 and are illustrated in Figs. 2 to 6.

Figs. 2 and 3 show that photoinduced absorbance increases at 559 nm, indicating cytochrome *b*-559 photoreduction, were observed in chloroplast fragments of the four strains, in absence of any inhibitor. This cytochrome *b*-559 photoreduction occurred in the mutant *Fl* 5 devoid of P700, it was somewhat more important in 630-680 nm light than in 680-750 nm light, and it was inhibited by CMU : it is a System II reaction which is similar to the System II-driven cytochrome *b*-559 photoreduction pointed out by Levine and Gorman[13] in chloroplast fragments of the wild type and three mutants of *C. reinhardti*. In the case of chloroplast fragments of *Fl* 9, often, the cytochrome *b*-559 was already reduced in the dark and it was necessary, before measuring its photoreduction, to add $K_3Fe(CN)_6$ in the preparations. The

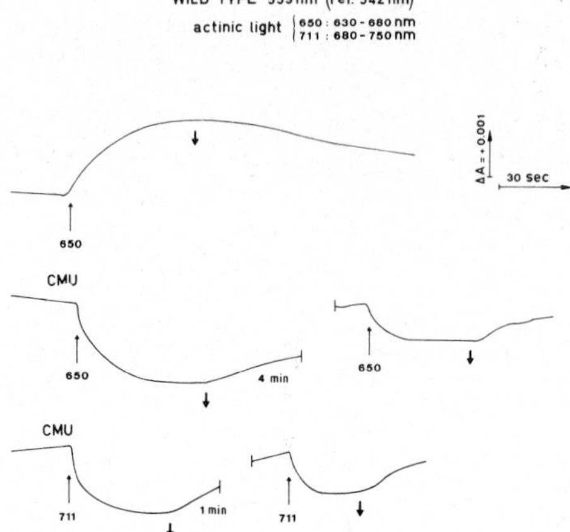

Fig. 2. Photoreduction of the cytochrome *b*-559 and its photooxidation in presence of CMU in chloroplast fragments of the wild type of *C. reinhardti*. Arrows up : light on ; arrows down : light off. For the experimental conditions, see the legend of Table 1.

photoreduction of cytochrome *b*-559 did not occur in presence of hydroquinone (Table 1) ; this fact indicates that it is a high-potential form of cytochrome *b*-559 which was photoreduced. When CMU was added, the photoreduction was inhibited and a System I-dependent photooxidation of the cytochrome *b*-559 was observed in chloroplast fragments of the wild type, *Fl* 9 (with no $K_3Fe(CN)_6$ in reaction mixture) and *Fl* 15, but not in chloroplast fragments of *Fl* 5 (Figs. 2,3).

Figs. 4, 5, 6 show the effects of antimycin $5 \cdot 10^{-5}$ M and of FCCP $9.6 \cdot 10^{-6}$ M on the cytochrome *b*-559 in chloroplast fragments of the wild type and of the mutants *Fl* 5 and *Fl* 9. Chloroplast fragments of *Fl* 15 behaved similarly to those of *Fl* 9. In these figures, decreases in absorbance at 559 nm indicate that, in presence of antimycin or FCCP, the cytochrome *b*-559 was photooxidized in chloroplast fragments of all the strains, provided that hydroquinone or ascorbate had been previously added in order to reduce this cytochrome in the dark. These photooxidations could occur again several times, after few minutes periods of darkness. In all the strains, the photooxidation with FCCP were clearly more important in presence of ascorbate than in presence of hydroquinone (Table 1). With antimycin, the differences were smaller.

In chloroplast fragments of the mutant *Fl* 5, the cytochrome *b*-559 photooxida-

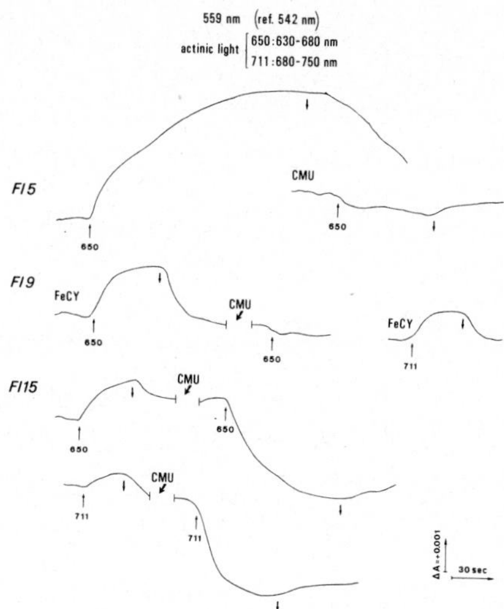

Fig. 3. Photoreduction of the cytochrome *b*-559 and its photooxidation in presence of CMU in chloroplast fragments of the mutants *Fl* 5, *Fl* 9, *Fl* 15 of *C. reinhardti*. Arrows up : light on ; arrows down : light off. FeCY : $K_3Fe(CN)_6$ $1.2 \cdot 10^{-3}$ M. For the other experimental conditions, see the legend of Table 1.

tion was fully inhibited by CMU $1 \cdot 10^{-4}$ M (Fig. 5). But it was not the case for the mutants *Fl* 9 and *Fl* 15 and for the wild type. Indeed, in chloroplast fragments of these three strains, a large part of the cytochrome *b*-559 was still photooxidized in presence of antimycin and CMU, as well as in presence of FCCP and CMU. This CMU-insensitive photooxidation was particularly important in the cases of the mutants *Fl* 9 and *Fl* 15 (Table 1). Fig. 6 shows that, for *Fl* 9, it was somewhat greater in 680-750 nm light than in 630-680 nm light.

5. Discussion

The different activities of the mutants allow to distinguish two kinds of cytochrome *b*-559 photooxidation occurring in presence of antimycin or FCCP :

-(1) A photosystem II-dependent photooxidation, inhibited by CMU, which was observed in all the strains, including the mutant *Fl* 5 devoid of P700. This photooxidation is comparable to the System II-driven photooxidations occurring in chloroplasts of higher plants treated by CCCP or FCCP, which were described by Hiller et al.[8], Ben-Hayyim[9,11], Satoh and Katoh[24], Anderson et al.[10]. It proceeds through

Table 1

Cytochrome *b*- 559 photooxidation in chloroplast fragments of the wild type and the mutants *Fl* 5, *Fl* 9, *Fl* 15 of *C. reinhardti*

Additions	Cytochrome *b*-559 photooxidized (μmoles/mmole chlorophyll)			
	Wild type	Mutants *Fl* 5	*Fl* 9	*Fl* 15
None	- 0.71 (*)	- 1.76	- 0.28	- 0.46
HQ	0.26	0.18	n.m.	0.27
CMU	0.71	0.00	1.39	1.22
ANT + HQ	0.62	0.36	1.31	0.85
ANT + HQ + CMU	0.45	0.00	0.83	n.m.
ANT + ASC	0.80	0.59	1.42	1.15
ANT + ASC + CMU	0.32	0.00	n.m.	0.95
FCCP + HQ	0.71	0.66	1.30	0.87
FCCP + HQ + CMU	0.36	0.00	1.11	n.m.
FCCP + ASC	1.47	0.81	2.07	1.57
FCCP + ASC + CMU	0.36	0.00	1.30	1.07

(*) The minus sign (-) indicates a photoreduction, in place of a photooxidation. Chloroplast fragments (100 μg chlorophyll/ml) in phosphate buffer 0.01 M, KCl 0.02 M, $MgCl_2$ $2.5 \cdot 10^{-3}$ M (pH 7.5). Actinic light : for the three upper lines : 630-680 nm, for the following lines : 680-750 nm, $2 \cdot 10^4$ ergs$\cdot$cm$^{-2}\cdot$sec^{-1}. ANT : antimycin $5 \cdot 10^{-5}$ M, ASC : Na ascorbate $1.2 \cdot 10^{-3}$M, CMU : $1.2 \cdot 10^{-4}$ M, FCCP : $9.6 \cdot 10^{-6}$ M, HQ : hydroquinone $1.2 \cdot 10^{-3}$ M, n.m. : non measured.

a cyclic electron transfer around the System II (see scheme, Fig. 7) involving essentially a hydroquinone-reducible HP form of the cytochrome *b*-559[8,24].

-(2) A CMU-insensitive photooxidation which occurs in *Fl* 9, *Fl* 15 and the wild type, but not in *Fl* 5. This photooxidation appears to be System I-dependent ; it occurs also in presence of CMU alone without antimycin or FCCP (Figs. 2,3).

Different authors had described System I-dependent photooxidations of the cytochrome *b*-559, induced by CCCP[5,6], FCCP[4,7] or antimycin [4,6] in chloroplasts of higher plants. In all these photooxidations, it is a LP form (less than 100 mV) of cytochrome *b*-559 which would be oxidized by System I through the classical electron

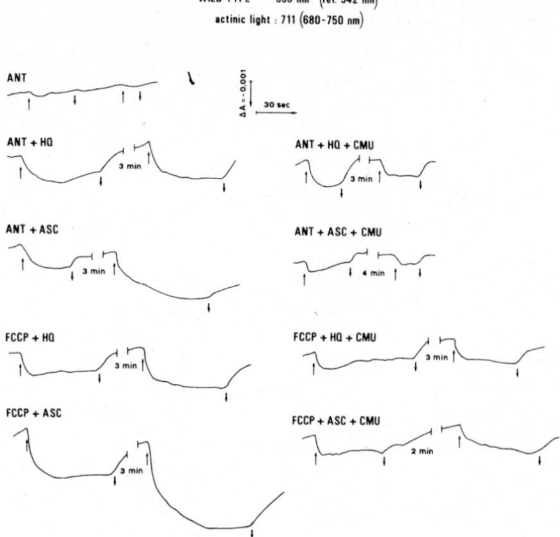

Fig. 4. Photooxidation of the cytochrome *b*-559 in presence of antimycin and FCCP in chloroplast fragments of the wild type of *C. reinhardti*. Arrows up : light on ; arrows down : light off. For the experimental conditions and for the abbreviations, see the legend of Table 1.

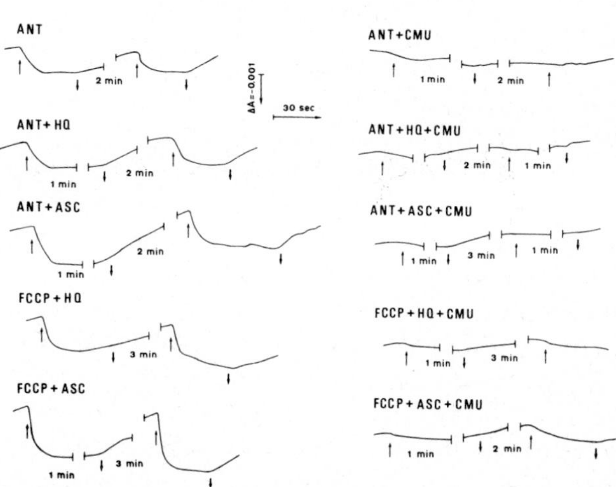

Fig. 5. Photooxidation of the cytochrome *b*-559 in presence of antimycin and FCCP in chloroplast fragments of the mutant *Fl* 5 of *C. reinhardti*. Arrows up : light on ; arrows down : light off. For the experimental conditions and for the abbreviations, see the legend of Table 1.

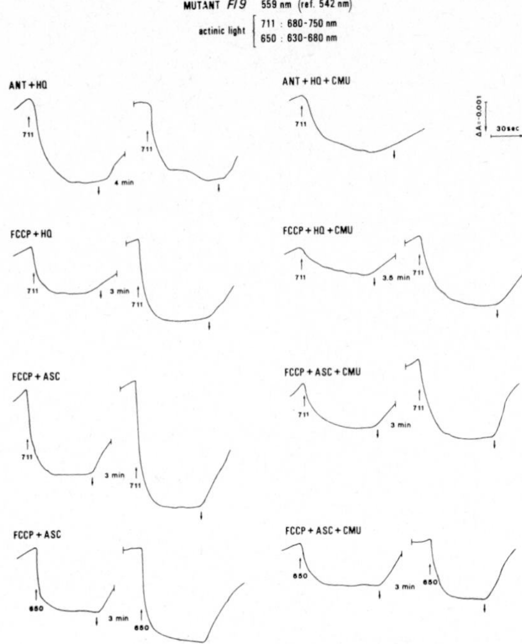

Fig. 6. Photooxidation of the cytochrome *b*-559 in presence of antimycin and FCCP in chloroplast fragments of the mutant *Fl* 9 of *C. reinhardti*. Arrows up : light on ; arrows down : light off. For the experimental conditions and for the abbreviations, see the legend of Table 1.

transport chain, including cytochrome *f* and plastocyanin. With chloroplast fragments of *C. reinhardti* (obtained by sonic disruption of cells), Levine and Gorman[13] observed a System I-type photooxidation of cytochrome *b*-559 occurring in the wild type, in absence or in presence of DCMU. This kind of photooxidation did not occur in chloroplast fragments of the mutants *ac*-206 and *ac*-208, which lack respectively cytochrome *c*-553 and plastocyanin.

But the System I-dependent photooxidation of cytochrome *b*-559 observed in our experiments is quite different of those mentionned above : (1) it involves mainly a hydroquinone-reducible HP form of cytochrome *b*-559, and (2) it is more important in the mutants *Fl* 9 and *Fl* 15, which are deficient in bound cytochrome *c*-553, than in the wild type. Therefore, the present data suggest that the HP form of the cytochrome *b*-559 and the cytochrome *c*-553 are involved in different electron transport chains. In presence of antimycin of FCCP, which inhibit the electron flow from water to System II, the cytochrome *b*-559 HP is photooxidized both by the System II and by the System I. In presence of CMU, the System II-driven photooxidation is inhibited and the cytochrome *b*-559 is photooxidized by System I alone. In this lat-

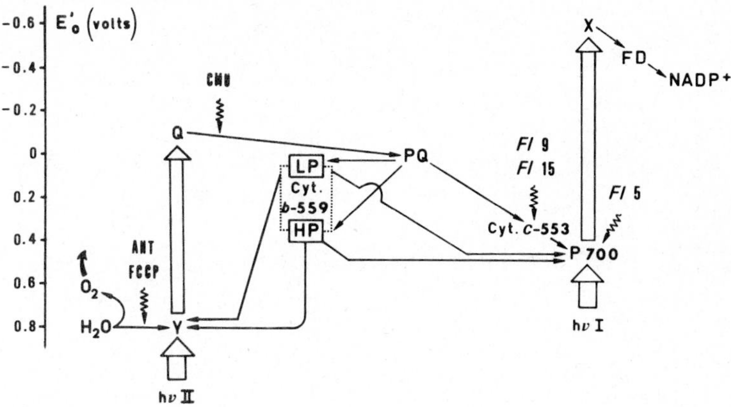

Fig. 7. Simplified scheme for photosynthetic electron transport, showing the possible pathways for the photoreduction and the photooxidation of the cytochrome *b*-559. Zigzag arrows indicate the sites of the inhibitors or the mutants deficiencies. ANT : antimycin ; Cyt. : cytochrome ; HP, LP : high, low potential ; FD : ferredoxin ; PQ : plastoquinone ; X : primary acceptor of the photoreaction I ; Y, Q : primary donor and acceptor of the photoreaction II.

ter case, a competition between the cytochrome *c*-553 and the cytochrome *b*-559 HP would occur, for electron donating to System I. Such a competition can explain the fact that the cytochrome *b*-559 photooxidation appears more important in the mutants deficient in cytochrome *c*-553 than in the wild type. These interpretations are illustrated in the scheme of Fig. 7. The occurrence of this cytochrome *b*-559 HP photooxidation by System I is strongly depending on the state of the chloroplast fragments : indeed, we have observed elsewhere[25] that this kind of photooxidation did not occur in chloroplast fragments of the mutant *Fl* 9 prepared by ultrasonic disintegration of cells, and that, in whole cells, it appeared only if *p*-benzoquinone had been added.

In the both kinds of photooxidation, we measured, the hydroquinone-reducible cytochrome *b*-559 HP was mainly involved. But in several cases, especially with FCCP in absence of CMU, the photooxidations were more important in presence of ascorbate than in presence of hydroquinone ; thus, FCCP appeared more efficient than antimycin in transforming the hydroquinone-reducible HP form into ascorbate-reducible LP form. Therefore, it is difficult to know exactly the potential of the cytochrome *b*-559 which was involved in each case. Also the question, whether the cytochromes *b*-559, which are photooxidized respectively by System II and by System I, belong to the same pool or to different pools, remains unanswered.

References

1. Wada, K. and Arnon, D.I., 1971, Proc. Nat. Acad. Sci. USA, 68, 3064.
2. Erixon, K., Lozier, R. and Butler, W.L., 1972, Biochim. Biophys. Acta, 267, 375.
3. Boardman, N.K., Anderson, J.M. and Hiller, R.G., 1971, Biochim. Biophys. Acta, 234, 126.
4. Cramer, W.A., Fan, H.N. and Böhme, H., 1971, Bioenergetics, 2, 289.
5. Cramer, W.A. and Butler, W.L., 1967, Biochim. Biophys. Acta, 143, 332.
6. Hind, G., 1968, Photochem. Photobiol., 7, 369.
7. Böhme, H. and Cramer, W.A., 1971, FEBS Letters, 15, 349.
8. Hiller, R.G., Anderson, J.M. and Boardman, N.K., 1971, Biochim. Biophys. Acta, 245, 439.
9. Ben-Hayyim, G., 1972, FEBS Letters, 28, 145.
10. Anderson, J.M., Than-Nyunt and Boardman, N.K., 1973, Arch. Biochem. Biophys., 155, 436.
11. Ben-Hayyim, G., 1974, Eur. J. Biochem., 41, 191.
12. Vermeglio, A. and Mathis, P., 1973, Biochim. Biophys. Acta, 292, 763.
13. Levine, R.P. and Gorman, D.S., 1966, Plant Physiol., 41, 1293.
14. Epel, B.L., Butler, W.L. and Levine, R.P., 1972, Biochim. Biophys. Acta, 275, 395.
15. Epel, B.L. and Butler, W.L., 1972, Biophys. J., 12, 922.
16. Maroc, J. and Garnier, J., 1973, Biochim. Biophys. Acta, 292, 477.
17. Garnier, J. and Maroc, J., 1970, Biochim. Biophys. Acta, 205, 205.
18. Garnier, J. and Maroc, J., 1972, Biochim. Biophys. Acta, 283, 100.
19. Gorman, D.S. and Levine, R.P., 1965, Proc. Nat. Acad. Sci. USA, 54, 1665.
20. Shneyour, A. and Avron, M., 1970, FEBS Letters, 8, 164.
21. MacKinney, G., 1941, J. Biol. Chem., 140, 315.
22. Arnon, D.I., 1949, Plan Physiol., 24, 1.
23. Lundegårdh, H., 1962, Physiol. Plantarum, 15, 390.
24. Satoh, K. and Katoh, S., 1972, Plant Cell Physiol., 13, 807.
25. Maroc, J. and Garnier, J., in preparation.

M. AVRON, *Proceedings of the Third International Congress on Photosynthesis*, September 2-6, 1974, Weizmann Institute of Science, Rehovot, Israel
Elsevier Scientific Publishing Company, Amsterdam, The Netherlands, 1974

LIGHT-INDUCED ABSORBANCE CHANGES OF PLASTOCYANIN IN SITU AND ITS FUNCTIONAL ROLE IN CHLOROPLASTS

Wolfgang Haehnel

Max-Volmer-Institut für Physikalische Chemie und Molekularbiologie, Technische Universität Berlin

1. Summary

Absorbance changes induced by excitation with different groups of saturating flashes after preillumination with System I light were measured in spinach chloroplasts.

1. The difference spectrum of plastocyanin (Pc) in situ was separated with respect to the superimposing absorbance changes of Pc and chlorophyll a_I (Chl-a_I).
2. The oxidation and reduction kinetics of Pc were observed without superposition with other absorbance changes at 584 nm. Simultaneously the oxidation kinetics of the reduced plastoquinone (PQ) pool and the reduction kinetics of cytochrome f (Cyt-f) and Chl-a_I were studied.
3. A comparison of the electron equivalents yielded a quantitative electron transfer from the reduced PQ pool to Cyt-f^+, Pc^+ and Chl-a_I^+. Relative concentrations of 0.45:1:1 were determined for Cyt-f, Pc and Chl-a_I, respectively.
4. The kinetics of these components gave evidence for a function of Cyt-f in parallel to Pc. In addition, they indicated that most of the linear electron transport passes Pc.

2. Introduction

Plastocyanin, a protein containing two atoms of copper per molecule, was discovered and characterized by Katoh[1,2,3]. It was shown to function in photosynthetic electron transport between the two light reactions by many lines of investigation. However, the sequential relationship between plastocyanin (Pc), cytochrome f (Cyt-f) and chlorophyll a_I (Chl-a_I) is the subject of controversial hypotheses. The interpretation of the effects of externally added Pc on photochemical reactions in chloroplast fragments suffers from the result that Pc in situ is located on the inner side of the thylacoid membrane[4]. Experiments with chloroplasts incubated with the inhibitors

$HgCl_2$[5] and KCN[6] gave evidence for the view that Pc mediates the electron flow between Cyt-f and Chl-a_I. Recently light-induced changes of the EPR signal of Pc in situ were detected[7,8]. The results are consistent with the role of Pc as the electron carrier between the two light reactions. More direct evidence would be expected by the kinetics of the light-induced absorbance changes of Pc in situ. The only spectrophotometric measurements ascribed to Pc are reported for whole cells of a marine alga[9]. However, the inhibition of these absorbance changes by salicylaldoxime[10] is not understood because this copper chelating reagent does not react with Pc[11]. In this contribution a method to separate the absorbance changes of Pc in spinach chloroplasts is reported with special regard to superimposing absorbance changes.

3. Materials and Methods

Spinach chloroplasts were prepared according to ref. 12. In order to prepare broken chloroplasts these chloroplasts were osmotically shocked in 1 mM MES buffer at pH 6.4 for 20 min. The reaction mixture contained these chloroplasts at a concentration of 10 μM chlorophyll, 20 mM Tricine buffer at pH 7.2, 20 mM KCl, 1 mM $MgCl_2$, 10 μM benzylviologen as electron acceptor and 0.5 μM gramicidin D. Gramicidin D was added for uncoupling as well as to avoid the superposition of the investigated absorbance changes with the electrochroic absorbance change[13]. Fig. 1 illustrates the excitation conditions and the corresponding situation in the electron transport chain. Preillumination with System I light for 4.5 sec (720 nm, $\Delta\lambda$=15 nm, $1.3\cdot10^4$ ergs $cm^{-2}s^{-1}$) was used to oxidize the electron carriers between the light reactions of Chl-a_{II} and Chl-a_I. Light reaction II produces electrons with a half-life of 0.6 ms [14]. Therefore each of the saturating flashes (20 μs) spaced at 1.6 ms should produce nearly one electron. Because of the rate-limiting oxidation of plastohydroquinone (PQH_2) with a half-life of 20 ms, most of these electrons accumulate in the pool of PQ and the electron acceptors of PQ remain oxidized[15]. The amount of the accumulated electrons depends on the number of the flashes. After the last flash one should observe the slow reoxidation of PQH_2 via the rate-limiting step and simultaneously the reduction of its electron acceptors of Photosystem I. These reactions were investigated. Every reduced state of the electron carriers of Photosystem I can be realized by this method of "titration by flashes". To avoid any reoxidation the

System I light was switched off during the measurement of the absorbance changes.

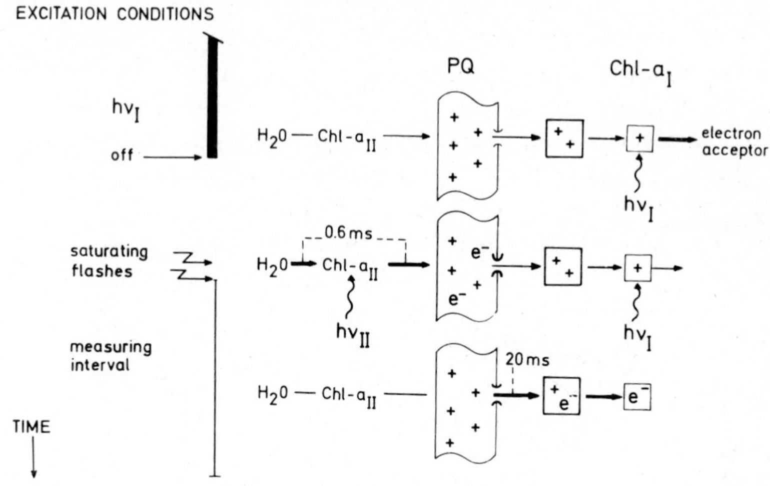

Fig. 1. Excitation conditions. Chloroplasts were periodically excited with System I light and during a following dark period of 0.5 sec with different numbers of saturating flashes spaced at 1.6 ms. The absorbance changes after the last flash were stored in an averager during every measuring interval. On the right-hand side the reactions in the linear electron transport and the redox states of the pool of PQ and its acceptors with an electron capacity of three[15] are symbolized.

For registration a repetitive flash photometer with double beams similar to that described in ref. 16 was used with an electrical band width of 5 kHz. The signals induced by flashes of blue (Schott filter BG 23, 6 mm) or red light (Schott filter RG 610, 3 mm), respectively, were stored in the averager (Fabri-Tek 1072) with a frequency of 0.2 Hz. The content of the sample cuvette was changed after every 32 signals. The optical path length of the cuvette was 20 mm, the temperature of the sample 22°C. The influence of light-induced scattering changes on the absorbance changes shown in Fig. 4, 5, 6 was minimized by placing an opal glass on the cuvette and a light pipe above the photomultiplier.

4. Results

Indirect information about the reaction of Pc is expected from measurements of the absorbance changes of PQ, Cyt-f and Chl-a_I. Fig. 2 presents the simultaneous transients during the oxidation of PQH_2 and the reduction of Cyt-f^+ and Chl-a_I^+ after the last flash as a function of the number of the preceding flashes. As mentioned above, PQH_2 is oxidized in the rate-limiting time of 20 ms. The half-life of the reduction kinetics of Cyt-f^+ shows an approx. constant value of 17-20 ms while the first half-life of the reduction kinetics of Chl-a_I^+ decreases with increasing flash numbers from 13 ms to 4.3 ms. The amplitude of the PQH_2 oxidation increases to only three-times the amplitude after a single flash (see Fig. 2 D) although the amplitude of the preceding reduction of the PQ pool becomes greater (not shown). This is an important effect attributed to the pool size

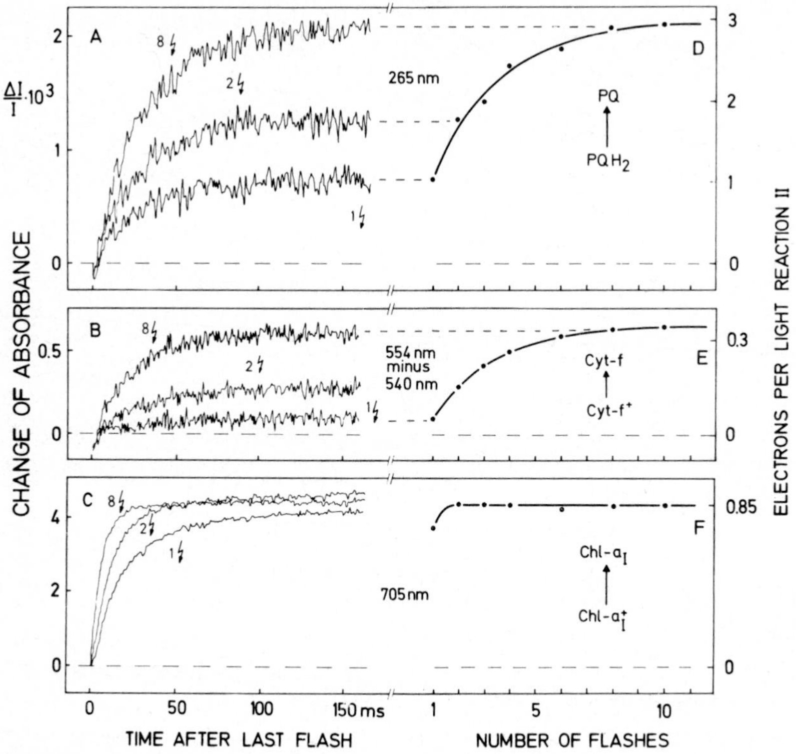

Fig. 2. Absorbance changes after the last flash of a flash group as a function of the number of the flashes. A and D, absorbance changes at 265 nm ($\Delta\lambda$ = 2.5 nm). B and E, differences of absorbance changes at 554 nm minus 540 nm ($\Delta\lambda$ = 2 nm). C and F, absorbance changes at 705 nm ($\Delta\lambda$ = 3 nm). Left: time courses after one, two, and eight flashes. Right: extent of the absorbance changes. Every trace is the average of 64 signals.

of the oxidized electron carriers of Photosystem I[15]. The reduction of Cyt-f^+ shows a small amplitude of about 15% after one flash and increases to its maximal value after ten flashes. The amount of Chl-a_I^+ being reduced after one flash yields about 91% of the maximal amount found after two or more flashes. The almost complete reduction of Chl-a_I^+ after one flash suggests an equilibrium between Chl-a_I and its primary electron donor well towards Chl-a_I. To compare the electrons released by PQH_2 and accepted by Cyt-f^+ and Chl-a_I^+, the amplitudes of the absorbance changes in Fig. 2 were converted into electron equivalents related to one light reaction II. It was assumed that PQ accepts all the electrons from light reaction II. Therefore the amplitude of PQ after one flash is proportional to the light reactions II[15]. For Cyt-f the difference extinction coefficient $\Delta\varepsilon_{554-540} = 2.2 \cdot 10^4\ M^{-1}cm^{-1}$ (ref. 17) was used and for Chl-a_I $\Delta\varepsilon_{705} = 6.4 \cdot 10^4\ M^{-1}cm^{-1}$ (ref. 18). The concentrations of Cyt-f and Chl-a_I were related to the number of light reactions II determined by the oxygen yield per flash. The preparation used in Fig. 2 had a ratio of 530 chlorophylls per light reaction II (measurement by Dr G. Renger). The resulting electron equivalents are indicated on the right-hand side of Fig. 2. The electron equivalents of 0.85 of Chl-a_I are different from one as previously assumed[19]. This value has also been deduced from comparing the kinetics of PQ and Chl-a_I. Details will be reported in a subsequent paper. Cyt-f was found to be engaged in linear electron transport with an electron capacity of only 0.4 per light reaction II.

The difference between the electron equivalents per light reaction II released from PQH_2 and the electron equivalents accepted by Cyt-f^+ and Chl-a_I^+ is presented in Fig. 3. This calculated difference should correspond to the electrons accepted by other electron carriers between PQ and Chl-a_I. The only known is Pc. According to the kinetics in Fig. 3, left, one can expect reduction kinetics of Pc to start with a lag and to be followed by a rise with a half-life of about 20 ms. Since a reduction of Pc causes a bleaching of this protein negative absorbance changes must be expected. The maximal amplitude should correspond to about 1.75 electrons per light reaction II.

The low extinction coefficient of Pc, the width of its absorbance band at 597 nm [3], the superposition with other absorbance changes, and light-induced scattering changes complicated the identification

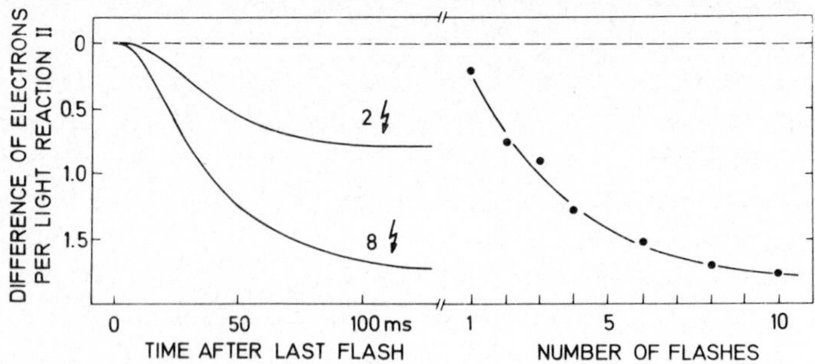

Fig. 3. Difference between the electron equivalents released from PQH_2 and the electron equivalents accepted by Cyt-f^+ and Chl-a_I^+ related to one light reaction II. The calculation is based on the absorbance changes in Fig. 2 and the assumptions defined in the text. Left: time course after two and eight flashes. Right: amount as a function of the number of the flashes.

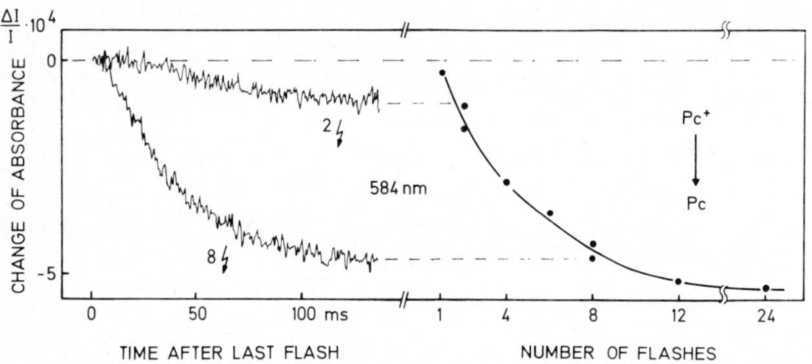

Fig. 4. Absorbance changes at 584 nm ($\Delta\lambda$ = 5 nm) after the last flash of a flash group as a function of the number of the flashes. Left: time course after two and eight flashes. Right: extent of the absorbance changes. Every trace is the average of 512 signals.

of corresponding absorbance changes of Pc. However, at about 584 nm no superimposing absorbance changes should occur because the difference spectrum of Chl-a_I crosses zero[18,20] and the absorbance changes of the cytochromes are negligible. Fig. 4 shows the absorbance changes after the last flash at 584 nm as a function of the number of the preceding flashes. The kinetics after two and eight flashes as well as the amplitude of the absorbance change as a function of the flash number show a good agreement with the calculated difference of electron equivalents in Fig. 3. The time course in Fig. 4, left, shows a lag of about 15 ms in the beginning after two flashes and of about 5 ms after eight flashes. These lags are followed by reduction kinetics of Pc^+ with half-lives of 18-20 ms. The equilibrium constant of the electron transfer from Pc to Chl-a_I is estimated from their reduction kinetics after eight flashes to be about K ≈ 20. The amplitude shown in Fig. 4, right, is about zero after one flash and increases to the maximal value after twelve flashes.

This dependence made it possible to separate the light-induced difference spectrum of Pc from the interfering difference spectrum of Chl-a_I. The difference spectra of the absorbance changes after

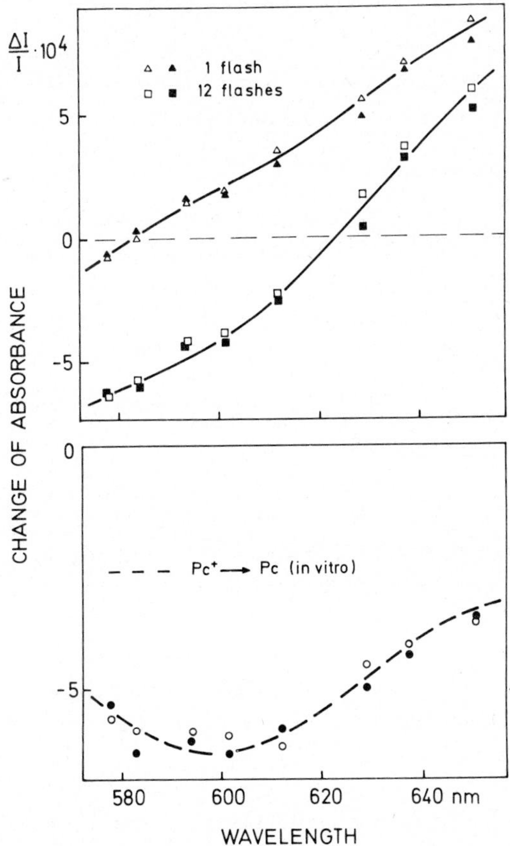

Fig. 5. Above: absorbance changes as a function of the wavelength after one flash (triangles) and a group of twelve flashes (squares). Below: difference between the absorbance changes after a group of twelve flashes and the absorbance changes after one flash multiplied by 1.1 as a function of the wavelength. The difference is calculated from the amplitudes shown above. The dashed line traces the reduced minus oxidized difference spectrum of Pc in vitro reported by Katoh[3].

one flash and after twelve flashes are presented in Fig. 5, above. The difference spectrum after one flash shows a good agreement with the difference spectrum of the reduction of $Chl\text{-}a_I^+$ [18,20]. The absorbance changes after twelve flashes should be due to the reduction of Pc^+ in addition to the reduction of $Chl\text{-}a_I^+$. Related to the amplitude after one flash the amplitude of $Chl\text{-}a_I$ after twelve flashes increases by a factor of 1.1 (see Fig. 2 F). Therefore the difference between the absorbance changes after twelve flashes $(\frac{\Delta I}{I})_{12ϟ}$ and those after one flash $(\frac{\Delta I}{I})_{1ϟ}$ multiplied by the factor 1.1 is expected to be due to the absorbance changes of Pc $(\frac{\Delta I}{I})_{Pc}$:

$$(\frac{\Delta I}{I})_{Pc} = (\frac{\Delta I}{I})_{12ϟ} - 1.1 \cdot (\frac{\Delta I}{I})_{1ϟ}$$

In this way the amplitudes in Fig. 5, below, were calculated. They show a good agreement with the in vitro difference spectrum of the reduction of Pc^+ [3]. At wavelengths shorter than 570 nm the superposition with absorbance changes of Cyt-f was observed.

From the amplitude of the separated difference spectrum at 597 nm and the extinction coefficient of Pc related to a gram atom copper, $\Delta\varepsilon_{597} = 4.9 \cdot 10^3\ M^{-1} cm^{-1}$ (ref. 2), it was estimated that Pc^+ accepts about 1.6 electrons per light reaction II. This is a first approximation. Because of the equilibrium between Pc and $Chl\text{-}a_I$ some Pc^+ must be reduced even after one flash. According to the reduced amounts of Cyt-f and $Chl\text{-}a_I$ after one flash (see Fig. 2, E and F) Pc accepted about 0.17 electrons per light reaction II. The corresponding small absorbance changes have been subtracted together with the absorbance changes of $Chl\text{-}a_I$. Therefore Pc^+ should accept about 1.77 electrons per light reaction II. 1.75 electrons per light reaction II were estimated to be accepted by an electron carrier of Photosystem I different from Cyt-f and $Chl\text{-}a_I$ (see above). This agreement suggests that the electrons released from PQH_2 are quantitatively accepted by Pc^+, $Cyt\text{-}f^+$ and $Chl\text{-}a_I^+$. Any other electron carrier between PQ and $Chl\text{-}a_I$ can be excluded (the maximal error is about $\pm$ 0.3 electrons per light reaction II).

By the method of "titration by flashes" definite redox states of the electron carriers between the light reactions were realized. Fig. 6 shows the time courses of the reoxidation of Pc, if weak System I light is subsequently presented. After four flashes Pc^+ was partially reduced (cp. Fig. 4, right). The reoxidation of Pc immediately started after switching on the System I light. After

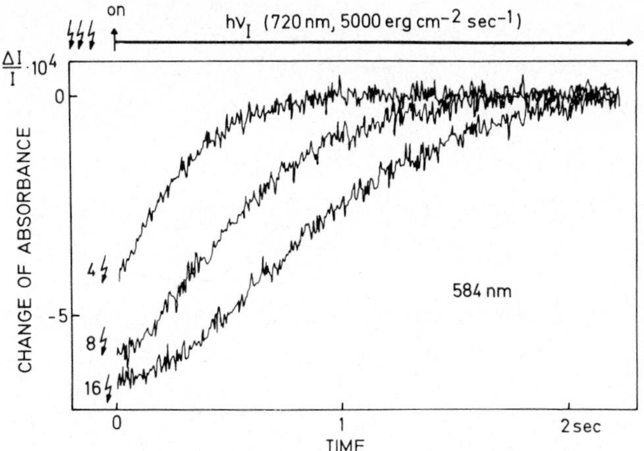

Fig. 6. Absorbance changes at 584 nm as a function of time induced by System I light after previous excitation with a group of four, eight, and sixteen flashes. The System I light with an intensity of $5 \cdot 10^3$ ergs $cm^{-2}sec^{-1}$ was switched on 0.4 sec after the flash groups. 64 signals have been averaged for every trace with a repetition rate of 0.16 Hz.

sixteen flashes Pc^+ was fully reduced. The following reoxidation of Pc is preceded by an initial delay. As reported above after four flashes the accumulated PQH_2 was fully oxidized but after sixteen flashes the oxidation of PQH_2 was limited by the pool size of its acceptors. Obviously, the remaining PQH_2 after sixteen flashes initially keeps Pc in the reduced state. Similar lags have been observed in the oxidation of Cyt-f and Chl-a_I by System I light after preillumination with strong white light[21].

5. Discussion

The estimate of the electron equivalents accepted by the active electron carriers of Photosystem I is supported by the quantitative agreement with the electron equivalents released by PQH_2. Related to Chl-a_I, the relative concentrations of Cyt-f, Pc and Chl-a_I are 0.45:1:1, respectively (one molecule of Pc contains two atoms of copper[2]). The ratio of Pc to Chl-a_I of one suggests a structural coupling of both electron carriers. The electron equivalents being accepted by Pc^+ exceed those accepted by Cyt-f^+ by the factor of 1.75:0.4 = 4.4. An approximately equal ratio of 4.7 copper atoms of Pc per one molecule of Cyt-f has been determined by chemical methods in spinach chloroplasts[22]. Both corresponding values differ from the ratio of 2:1 which is implicated by the model of a simple linear arrangement of these molecules as well as from the ratio of 1:1 as suggested by Marsho and Kok for the electron equivalents of Cyt-f and an unknown electron donor of Chl-a_I[21].

The absolute concentration of Pc estimated from the absorbance changes to be 1:700 chlorophylls is smaller than the chemical determined concentration of about 1:400 [22,4] or 1:600 chlorophylls[2]. This difference may be caused by isolated Photosystems I possibly located in the stroma lamellae[23]. Components of such Photosystems I would not be detected because only absorbance changes of electron carriers coupled to the linear electron transport are induced by flashes after preillumination with System I light.

The equilibrium constant of $K \approx 20$ estimated for the oxidation-reduction reaction of Pc and Chl-a_I is consistent with the midpoint potentials of Pc, $E_o' = 0.37$ V [3], and of Chl-a_I, $E_o' = 0.45$ V [24]. The consequence of this equilibrium is the lag of the reduction of Pc^+ although Chl-a_I^+ is reduced to a high extent at the same time. This lag and the half-life of the reduction of Chl-a_I^+ decrease at increasing flash numbers because the electron transfer via the rate-limiting step becomes faster at increasing concentrations of PQH_2[15]. The time course of the reduction of Pc^+ indicates an electron transfer from PQH_2 via Pc^+ to Chl-a_I. In contrast to Pc^+ the reduction of Cyt-f^+ does not start with a lag (see Fig. 2 B and ref. 19). Therefore the electron transfer from PQH_2 to Cyt-f^+ must run via a side path in parallel to the electron path via Pc^+ to Chl-a_I^+. The transfer of electrons from Cyt-f to Chl-a_I^+ seems not to be possible in the dark.

The oxidation kinetics of Pc and Cyt-f should correspond to the fast reduction kinetics of Chl-a_I^+ with half-lives of 20 μs and 200 μs [25,26]. First measurements revealed an oxidation of Pc with 200 μs and the oxidation of Cyt-f to be faster. These reaction times exclude the frequently assumed sequence Cyt-f $\rightarrow$ Pc $\rightarrow$ Chl-a_I but support the function of Pc in parallel to Cyt-f.

Acknowledgements

I thank Irma Jürgens for her valuable technical assistance and Waltraud Falkenstein for the graphs.

6. References

1. Katoh, S., 1960, Nature 186, 533
2. Katoh, S., Suga, I., Shiratori, I. and Takamiya, A., 1961, Arch. Biochem. Biophys. 94, 136
3. Katoh, S., Shiratori, I. and Takamiya, A., 1962, J. Biochem. Tokyo 51, 32
4. Hauska, G.A., McCarty, R.E., Berzborn, R.J. and Racker, E., 1971 J. Biol. Chem. 246, 3524
5. Kimimura, M. and Katoh, S., 1972, Biochim. Biophys. Acta 283, 279
6. Izawa, S., Kraayenhof, R., Ruuge, E.K. and DeVault, D., 1973, Biochim. Biophys. Acta 314, 328
7. Malkin, R. and Bearden, A.J., 1973, Biochim. Biophys. Acta 292, 169
8. Visser, J.W.M., Amesz, J. and van Gelder, B.F., 1974, Biochim. Biophys. Acta 333, 279
9. De Kouchkovsky, Y. and Fork, D.C., 1964, Proc. Natl. Acad. Sci. U.S. 52, 232
10. Fork, D.C. and Urbach, W., 1965, Proc. Natl. Acad. Sci. U.S. 53, 1307
11. Katoh, S. and San Pietro, A., 1966, Biochem. Biophys. Res. Commun 24, 903
12. Winget, G.D., Izawa, S. and Good, N.E., 1965, Biochem. Biophys. Res. Commun. 21, 438
13. Junge, W. and Witt, H.T., 1968, Z. Naturforsch. 23 b, 244
14. Vater, J., Renger, G., Stiehl, H.H. and Witt, H.T., 1968, Naturwissenschaften 55, 220
15. Stiehl, H.H. and Witt, H.T., 1969, Z. Naturforsch. 24 b, 1588
16. Döring, G., Stiehl, H.H. and Witt, H.T., 1967, Z. Naturforsch. 22 b, 639
17. Forti, G., Bertolè, M.L. and Zanetti, G., 1965, Biochim. Biophys. Acta 109, 33
18. Hiama, T. and Ke, B., 1972, Biochim. Biophys. Acta 267, 160
19. Haehnel, W., 1973, Biochim. Biophys. Acta 305, 618
20. Witt, H.T., 1971, Quart. Rev. Biophys. 4, 365
21. Marsho, T.V. and Kok, B., 1970, Biochim. Biophys. Acta 223, 240
22. Plesničar, M. and Bendall, D.S., 1970, Biochim. Biophys. Acta 216, 192
23. Arntzen, C.J., Dilley, R.A., Peters, G.A. and Shaw, E.R., 1972, Biochim. Biophys. Acta 256, 85

24. Kok. B., 1961, Biochim. Biophys. Acta 48, 527

25. Haehnel, W., Döring, G. and Witt, H.T., 1971, Z. Naturforsch. 26 b, 1171

26. Haehnel, W. and Witt, H.T., 1972, 2nd Int. Congr. Photosynth. Res., Stresa (Forti, G., Avron, M. and Melandri, A., eds.), Vol. 1, p. 469, Dr. W. Junk N.V. Publishers, The Hague

M. AVRON, *Proceedings of the Third International Congress on Photosynthesis*, September 2-6, 1974, Weizmann Institute of Science, Rehovot, Israel
Elsevier Scientific Publishing Company, Amsterdam, The Netherlands, 1974

ISOLATION AND CHARACTERIZATION OF A FACTOR WHICH RESTORES THE HILL REACTION FROM PHORMIDIUM LURIDUM

E. Tel-Or and M. Avron

Biochemistry Department, Weizmann Institute of Science
Rehovot, Israel

1. Introduction

Photosynthetic activity of cell free preparations from blue green algae is dependent on the structural integrity of the particles. The components of both electron transport and photophosphorylation activities are located in the thylakoids, which seem to be much more fragile than the thylakoids of the higher plant chloroplasts[4,5]. Membrane fragments prepared by mechanical rupture of blue green algal cells did not possess all the photochemical activities associated with higher plant-chloroplasts. Lysozyme digestion of the bacterial type cell wall of the blue green algae followed by osmotic rupture, was found to produce preparations which retain the complete photosynthetic capacity, when compared to higher plant chloroplasts[4,5].

The Hill activity of higher plant chloroplasts is a stable system which withstands rough mechanical or osmotic procedures. Blue green algal membrane fragments were found to lose Hill activity when prepared in hypotonic medium. The presence of high molecular weight solutes was required to maintain photosystem 2 activity[8,9].

Many studies including in vivo studies with Anacystis, indicated the role of endogeneous manganese in the oxygen evolving apparatus, probably bound in a catalytic site[6, 7, 10].

In this communication the isolation and characterization of a Hill factor from Phormidium luridum are described. Hill activity of the depleted particles was restored by the purified Hill factor, which contains manganese. The site of its activity was localized, to be within the oxygen evolving apparatus.

2. Methods

Algae: Phormidium luridum: var. olivaceae was obtained from the culture collection at Indiana University. The cells were grown on medium C of Kratz and Myers[12], flushed with air supplemented with 5% CO_2. Cells were harvested during the late logarithmic phase of growth by centrifugation at 3000g for 10 minutes.

Spheroplast preparation: Cells were washed with mannitol-buffer solution (0.5 M mannitol + 30 mM phosphate or tricine-maleate, pH 7.5) and suspended in the same solution containing lysozyme at a final concentration of 0.05% (W/V). The cells were incubated for 30 minutes at 37°C, illuminated with 500 foot candles of white light (which results in similar preparations as those obtained after 150 min. incubation in the dark). The suspension was passed through glass wool and the spheroplasts were collected by centrifugation at 1000 g for 10 min. The pellet was resuspended in a minimal volume of mannitol-buffer.

Spheroplasts fragments: Spheroplasts were washed hypotonically by suspension in 1:10 dilution (v/v) with 30 mM phosphate or tricine-maleate, pH 7.5. The suspension was passed once for homogenous breakage in a YEDA Press cell [14] at 750 PSI and centrifuged at 5000 g for 10 min. The pellet was suspended in a minimal volume of mannitol-buffer solution. Chlorophyll a content was estimated spectrophotometrically after extraction with 80% acetone, according to Mackinney (1941).

Measuring techniques: Ferricyanide reduction was measured in a Cary 14 spectrophotometer at 420 nm with a scatter attachment. The photomultiplier was protected by a Corning C.S. 4-96 filter plus a Wratten No. 34 filter. Samples were illuminated with a 500 W slide projector lamp filtered through a Baird-Atomic interference filter transmitting between 600-720 nm. ATP formation was measured at 20°C with 140,000 lux white light. Ferrocyanide produced was measured chemically[2] and ATP[32] according to Avron [1]. Oxygen evolution and consumption was measured with a Clark type oxygen electrode. Spectra were measured with a Cary 16 spectrophotometer. Manganese was measured with an Atomic Absorption Spectrophotometer Model 153 of Instrumentation Laboratory, Inc. The atomic emission of manganese was determined at 403 nm.

The preparation of Hill factor from Phormidium luridum cells: The procedure was carried out at room temperature.

1. Harvested cells were washed with 10 mM phosphate buffer pH 7.5, centrifuged and lyophilized. 15 g of lyophilized material was suspended in 50 ml of 10 mM phosphate buffer pH 7.5, mixed for 10 minutes and centrifuged at 5000g for 10 minutes. The supernatant was stored and the pellet was resuspended with phosphate buffer, homogenized, mixed for 10 minutes and centrifuged. Both supernatants were pooled, boiled for 10 minutes and centrifuged at 5000g for 10 minutes.

2. The supernatant was filtered through a Diaflo ultrafiltration membrane type UM-2 with nitrogen pressure of 50 PSI.

3. The filtrate was passed through a 2 x 20 cm DEAE-cellulose column equilibrated with 10 mM phosphate buffer, pH 7.5, and washed with the same buffer. The material absorbing at 250 nm which appeared in the front of the eluate was collected, lyophilized, suspended in 3 ml of water and loaded on a 2 x 200 cm sephadex G-10 column. The column was washed with water and samples were checked for absorption at 250 nm, Hill

activity and manganese content.

4. The active material was collected and passed through 2 x 20 cm Amberlite IRA-400 anion exchanger column in its OH^- form. The active material was eluted with water and appeared in the front of the column. It was collected and loaded on a 2 x 20 cm Amberlite IR-120 cation exchanger in its H^+ form. The column was washed with water and the active material eluted with 1-4 N HCl. The material absorbing in 250 nm was lyophilized to eliminate HCl, and the activity of the neutralized material measured.

3. Results

Isolation: Intact spheroplasts, lysed in the reaction mixture, showed good rates of electron transport which were stimulated by phosphorylation conditions. Isotonic washes of the spheroplasts decreased both the rate of electron transport and photophosphorylation. Hypotonic washes caused more pronounced decreases in activity (Table 1). One wash decreased the rate of ferricyanide reduction by two thirds, that of the coupled photophosphorylation by more than 95%, and that of PMS mediated photophosphorylation by 85%.

Table 1

Effect of hypotonic washes on photosynthetic activity of the spheroplasts

Spheroplasts	Electron Acceptor or Mediator	Ferricyanide Reduction μmoles x mg chl^{-1} x h^{-1}	Photophosphorylation μmoles ATP x mg chl^{-1} x h^{-1}	P/2e
Intact	Ferricyanide	1025	460	0.9
Once washed	Ferricyanide	333	16	0.1
Twice washed	Ferricyanide	104	3	0.06
Intact	PMS	--	770	--
Once washed	PMS	--	126	--
Twice washed	PMS	--	5	--

Spheroplasts were washed by dilution 1:30 (v/v) with 50 mM tricine-maleate pH 7.5, homogenized, and centrifuged at 5000 g for 10 minutes. The pellet was resuspended in minimal volume of 0.5 M mannitol + 30 mM tricine maleate pH 7.5. The second wash was identical to the first.

The reaction mixture contained in a volume of 3 ml in mM: Tricine-maleate pH 7.5 - 25, $MgCl_2$-20, ADP-3.3, phosphate-3.3 (containing 6 x 10^6 CPM P^{32}) ferricyanide-1 or PMS-0.03, ascorbate-6.6 and DCMU- 1.5 x 10^{-3}. Intact spheroplasts, once and twice washed contained 56,56 and 42 μg chlorophyll, respectively.

Since the Hill reaction is a very stable membrane bound system in higher plant chloroplasts, it was of interest to see whether some component required for Hill activity was released and may be present in the supernatant of the washed membrane fragments.

The supernatant of the washed spheroplasts was indeed found to be active in restoring Hill activity to washed fragments (Fig. 1). Several water soluble components were present in the supernatant, including phycocyanin. It was found that boiling for ten minutes and ultrafiltration through a Diaflo membrane which passes components of molecular weight lower than 1000 daltons did not affect the Hill activity (Fig. 1), but removed essentially all the phycocyanin. The active component was clearly a heat-stable, low molecular weight material.

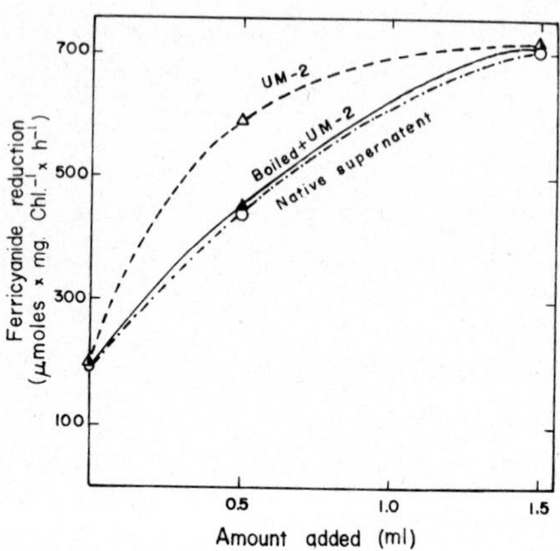

Fig. 1. Effect of native, boiled and filtered Hill factor on ferricyanide photoreduction by washed spheroplasts:

Ferricyanide photoreduction was measured spectrophotometrically in a reaction mixture containing in mM: KCl-100; ferricyanide-1 tricine-maleate, pH 7.5-25; spheroplasts once washed by dilution of 1/10 (v/v) in 0.1 M mannitol + 30 mM tricine-maleate pH 7.5, containing 30μg chlorophyll, and the amount of Hill factor indicated.

Several controls were run to ensure that the effect was not simulated by uncouplers, and that the factor had little to no stimulation activity in non-washed spheroplasts.

The active component was not adsorbed on Dowex-l or DEAE-cellulose anion exchangers, but could be purified by passing it through such a DEAE cellulose column. The spectra of the Hill factor before (crude) and after the DEAE-cellulose step are shown in Fig. 2. The crude material absorbed in the UV region with a maximum around 254 nm, and was accompanied by a fluorescent component. During fractionation on DEAE-cellulose two major peaks were observed: an inactive fluorescent material (2nd peak) and the active material (1st peak) which appeared in the front of the column and had one absorption peak at 254 nm.

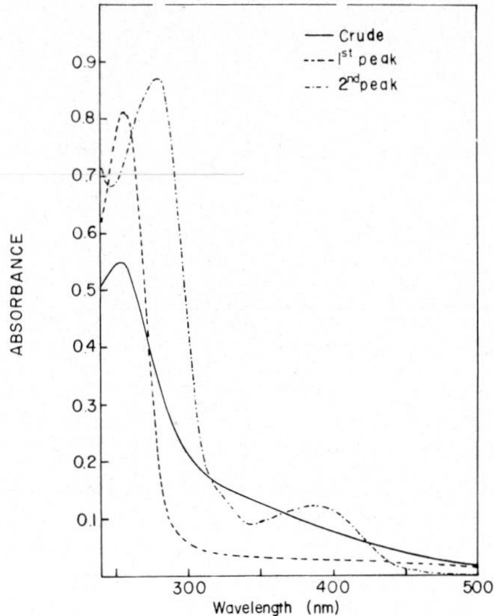

Fig. 2. Absorption spectra of the Hill factor before and after separation on DEAE-cellulose. Crude-before passing through column. 1st peak-fraction active as Hill factor. 2nd peak-fraction inactive as Hill factor.

The active hill factor obtained from the DEAE-cellulose column was further fractionated on a sephadex G-10 column. The material which absorbed at 254 nm, appeared in two peaks and both were active in enhancing the Hill reaction of washed membranes. Atomic absorption studies indicated that manganese was present and its content in the fractions roughly paralleled the activity and the absorption at 254 nm. (Fig. 3).

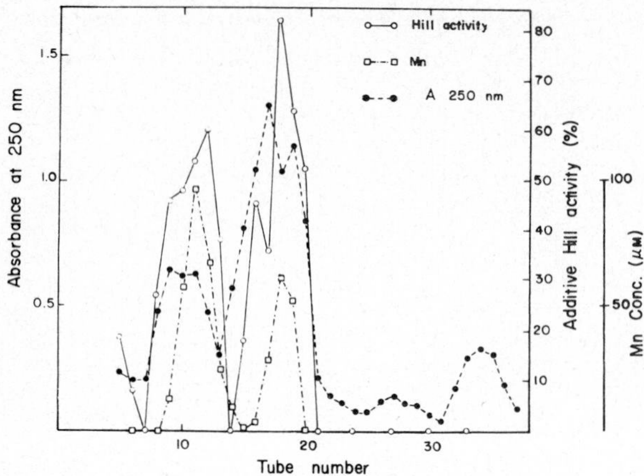

Fig. 3. Fractionation of the Hill factor on Sephadex G-10.

Oxygen evolution was measured in a reaction mixture similar to that described under Fig. 1 with spheroplast fragments containing 45μg chlorophyll.

Addition of 1-2 mM EDTA did not abolish the stimulating effect of the Hill factor, and therefore it seems likely that if necessary for activity, the metal is present in a bound form. A similar fractionation on a Biogel P-2 column, gave essentially identical results regarding the correlation between activity, absorption and manganese content. Comparison with substances of known molecular weight indicate that the molecular weight of the active component was lower than 1000 daltons.

The active component was adsorbed on the cation exchange Amberlite IR-120 (H^+ form) and only high concentrations of hydrochloric acid eluted the active component (Fig. 4).

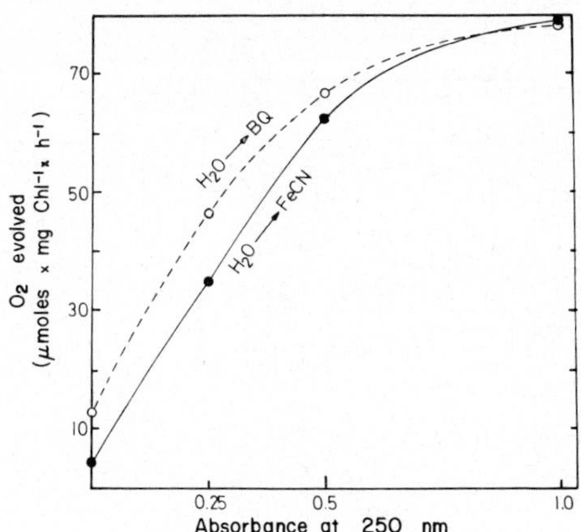

Fig. 4. Activity of the Hill factor after fractionation on Amberlite IR-120(H^+).

Reaction mixture contained in a volume of 2.5 ml in mM: phosphate, pH 7.5, 30; and either ferricyanide, 1, or benzoquinone, 2.5; spheroplast fragments containing 80µg chlorophyll, and the amount of Hill factor indicated (as absorbance at 250 nm).

Site of Action: As can be seen in Fig. 5, the factor had no effect on the photosystem I sensitized electron transfer from ascorbate + dichlorophenolindophenol to diquat in the presence of DCMU. Its site of action must, therefore, be in the vicinity of photosystem 2.

Table 2 illustrates that the factor did not stimulate the Hill reaction (measured as oxygen uptake) when ascorbate and hydroquinone were added to the reaction mixture. Ascorbate and hydroquinone were previously shown to be an excellent electron donating couple of photosystem II [18,19,3], and this reaction was indeed completely sensitive to DCMU in the spheroplast fragments. It seems likely therefore that the site of action of the Hill factor is between the site of water splitting and that of electron donation by ascorbate + hydroquinone.

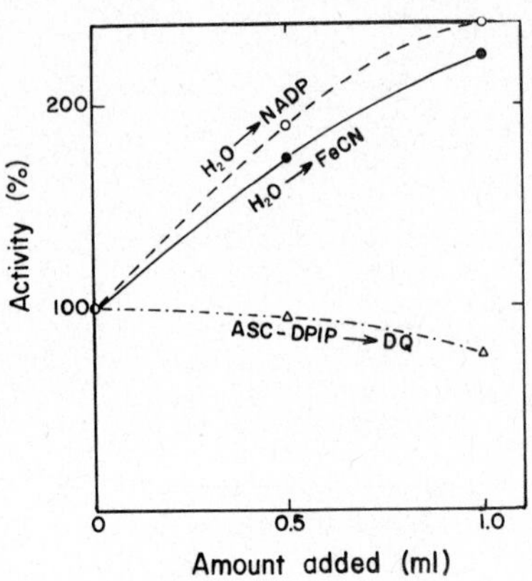

Fig. 5. The effect of the Hill factor on photosystem 1 and 2 sensitized reactions.

Reaction mixture for NADP photoreduction contained in a total volume of 2.5 ml in mM: phosphate pH 7.5, 30; NADP, 1; 150 μg Swiss Chard ferredoxin and a saturating amount of ferredoxin-NADP-reductase. For ferricyanide reduction: phosphate pH 7.5, 30; ferricyanide 1. For DQ photoreduction: phosphate, pH 7.5, 30; ascorbate, 10; DPIP, 0.05, DQ, 0.05; DCMU, 0.01; NaN_3, 0.5. In addition all reaction mixtures contained spheroplast fragments containing 56 μg chlorophyll and the amount of Hill factor (after the sephadex step) indicated. Oxygen evolution or consumption were measured and the rates for 100% activity were 15 and 27 μmoles oxygen evolved x mg chl^{-1} x h^{-1} with NADP and ferricyanide, respectively, and 420 μmoles oxygen consumed x mg chl^{-1}x h^{-1} with DQ

TABLE 2

Effect of the Hill factor on reactions sensitized by photosystem 2

Preparation	$H_2O \rightarrow$ Ferricyanide	$H_2O \rightarrow$ Benzoquinone	Ascorbate + Hydroquinone $\rightarrow O_2$
		(μmoles O_2 x mg chl^{-1} x h^{-1})	
Spheroplasts	108	175	198
Washed spheroplasts	23	55	246
Washed spheroplasts + Hill factor	51	110	164

The reaction mixture contained in a volume of 2.5 ml in mM: phosphate pH 7.5, 30; ferricyanide, 1, or benzoquinone, 0.625; or hydroquinone, 2.5, + ascorbate, 10, + NaN_3, 0.5; spheroplasts or spheroplasts fragments containing 50 μg chlorophyll and the amount of Hill factor (after the Sephadex step) indicated.

4. Discussion

Several investigators have previously looked into the extreme lability of the Hill reaction of cell free preparations from blue green algae[8,9,11,17]. The inhibition of the system, could be prevented by including in the isolation media components like polyethylene glycol, sucrose, carbowax and mannitol. Reconstitution of the Hill activity of depleted preparations was a much more difficult task. Frederick and Jagendorf[8] isolated and partially purified a high molecular weight acid labile component which restored the photoreduction of ferricyanide. We report here the isolation of a low molecular weight Hill factor which is obviously different from that just mentioned.

The Hill factor is released by a relatively gentle procedure from the spheroplasts, which may be responsible for the relative ease of reconstitution. The properties of the Hill factor were different from those of most known electron transport components. It is a low molecular weight compound with absorption in the UV region . The presence of bound manganese in the active component may clearly relate to the large body of evidence which implicates manganese as a component of the oxygen evolving complex[6], which fits with the site of action indicated by this study.

The site of action of the Hill factor could clearly be defined to an electron transfer step in the vicinity of photosystem 2, and most probably between the site of water decomposition and that for the donation of electrons by ascorbate and hydroquinone. A similar loss of Hill activity in blue-green algae which was restored by the addition of electron donors like Mn^{++}, diphenylcarbozide or hydroxylamine, was reported[15,16].

References

1. Avron, M., 1961, Anal. Biochem. 2, 535.
2. Avron, M. and N. Shavit, 1963, Anal. Biochem. 6, 549.
3. Ben-Hayyim, G. and M. Avron, 1970, Eur. J. Biochem. 15, 155.
4. Biggins, J., 1967a, Plant Physiol. 42, 1442.
5. Biggins, J., 1967b, Plant Physiol. 42, 1447.
6. Cheniae, G.M., 1970, Ann. Rev. Plant Physiol. 21, 467.
7. Cheniae, G.M. and I. F. Martin, 1971, Biochim. Biophys. Acta 253, 167.
8. Frederick, W. W. and A. T. Jagendorf, 1964, Arch. Biochem. Biophys. 104, 39.
9. Fujita, Y. and R. Suzuki, 1971, Plant Cell Physiol. 12, 641.
10. Gerhardt, B., 1966, Ber. Dent. Bot. Ges. 79, 63.
11. Jansz, E. R. and F. I. Maclean, 1972, Can. J. Microbiol. 18, 1727.
12. Kratz, W. A. and J. Myers, 1955, Am. J. Botany 42, 282.
13. Mackinney, G., 1941, J. Biol. Chem. 140, 315.

14. Shneyour, A. and M. Avron, 1970, FEBS Letters 8, 164.

15. Sofrova, D., V. Slechta and S. Leblova, 1974, Photosynthetica 8, 34.

16. Suzuki, R. and Y. Fujita, 1972, Plant Cell Physiol. 13, 427.

17. Ward, B. and J. Myers, 1972, Plant Physiol. 50, 547.

18. Yamashita, T. and W. L. Butler, 1968, Plant Physiol. 43, 1978.

19. Yamashita, T. and W. L. Butler, 1968, Plant Physiol. 43, 2027.

M. AVRON, *Proceedings of the Third International Congress on Photosynthesis*,
September 2-6, 1974, Weizmann Institute of Science, Rehovot, Israel
Elsevier Scientific Publishing Company, Amsterdam, The Netherlands, 1974

ELECTRON ACCEPTORS OF PHOTOSYSTEM II

B. Bouges-Bocquet
Institut de Biologie Physico-Chimique
13, rue Pierre et Marie Curie
75231 Paris Cedex 05, France

1. Introduction

In the photosynthetic system, the two photochemical reactions in series transfer electrons from water, through a chain of electron carriers, to a terminal electron acceptor (see review[1] or [2]). The primary electron acceptor Q, for System II, while not chemically identified, has been characterized by its fluorescence quenching properties[3]. Q may be reduced a short flash[4], so, it seems likely that Q is a one electron acceptor. On the other hand, the plastoquinone pool of electron carriers located between the two photosystems[5,6] are two electron carriers.

We have tried to elucidate the nature of the transfer between Q and the pool of plastoquinones.

2. Electrons transferred from System II

In this paragraph, we shall summarize rapidly reference[7].

We studied the number of electrons, U_n, transferred by the $n^{\underline{th}}$ flash of a sequence from System II to the pool, and available to System I. We measured via methylviologen reduction by System I :

- N_n, the number of electrons remaining in the pool after n flashes and available for transfer to System I.
- MV_n, the number of electrons transferred by the $n^{\underline{th}}$ flash through System I.

The difference $N_n - N_{n-1}$ equals the number of electrons (U_n) transferred by the $n^{\underline{th}}$ flash to the pool and available for System I minus the number of electrons (MV_n) transferred by the $n^{\underline{th}}$ flash through System I.

$$N_n - N_{n-1} = U_n - MV_n$$

$$\boxed{\text{System II}} \xrightarrow{U_n} \boxed{\text{Pool} \mid \text{System I}} \xrightarrow{MV_n} \qquad \text{Diagram 1}$$

U_n was calculated from the experimental values of the other parameters (Fig.1). Sequence U_n exhibits oscillations with period 2.

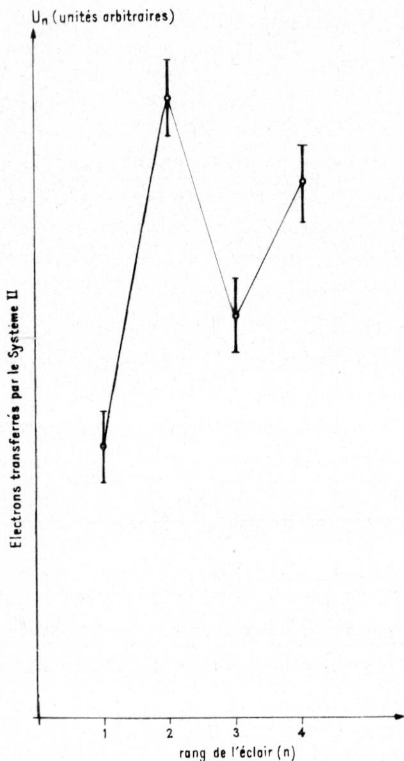

Fig. 1. Numbers U_n of electrons transferred from System II to the pool available for System I, by the nth flash of a series, for spinach chloroplasts in presence of 10^{-4} M methylviologen. The time between the flashes was 1 s.

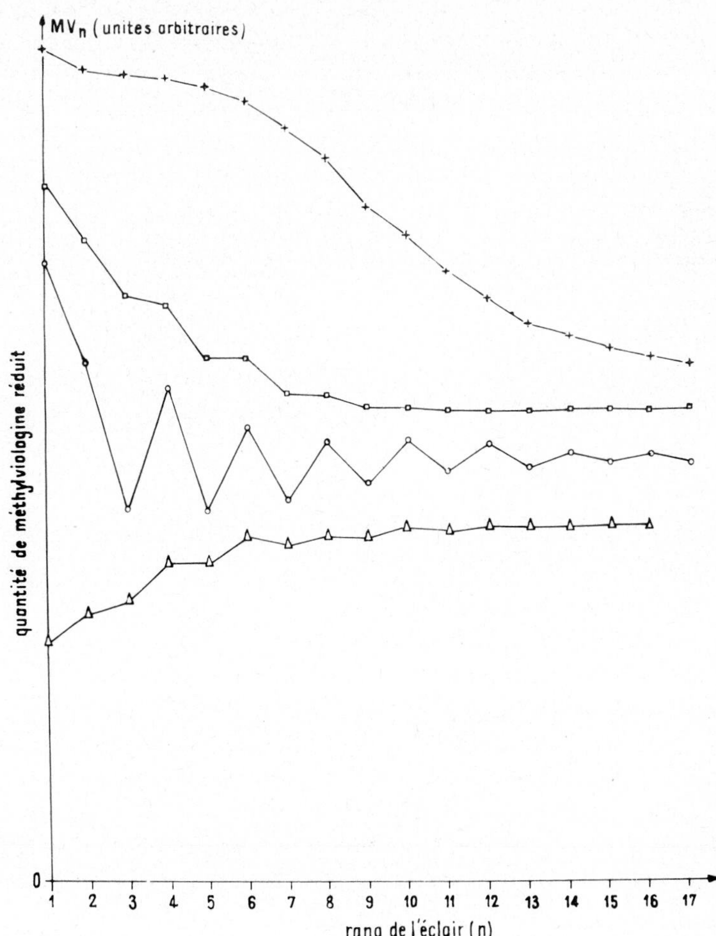

Fig. 2. Quantities MV_n of methylviologen reduced by the nth flash or a series for spinach chloroplasts in presence of DCMU 5.10^{-7} M (inhibition for oxygen : 50%) and 10^{-4} M methylviologen. The time between two flashes was 0.300 s. The series of flashes were given as follows : △ 4 s dark after 20 s far red light ; O , 2 mn 20 s dark after 20 s far red light ; □ , 10 mn dark after 20 s far red light ; + , 4 s dark after 10 s bright white light.

We could also observe damped oscillations with period 2 for MV_n (Fig.2) - when System II is limiting (on chloroplasts partially inhibited by DCMU, or aged chloroplasts, i.e. chloroplasts where there are fewer System II than System I centers connected to the pool) - and when the pool between the two photosystems is largely oxidized (i.e. after far red light).

Thus it is reasonnable to suppose that the MV_n oscillations, observed on Figure 2, originate in the electrons coming from System II.

We proposed the following model to explain this periodicity of 2:
- it is possible to store one minus charge on the secondary acceptor B of photosystem II.
- B^- is stable for at least 1s (time between two flashes).
- electron exchange between the pool and QB occurs two at a time.

This may be represented by one of the two reactions :

$$QB^{2-} + PQ \rightleftarrows QB + PQ^{2-} \qquad \text{Reaction 1}$$

or

$$Q^-B^- + PQ \rightleftarrows QB + PQ^{2-} \qquad \text{Reaction 2}$$

These assumptions are in contradiction with the model of Stiehl and Witt[8]. But they have been confirmed, by completely different types of experiments by Velthuys and Amesz[9].

Two types of reactions may occur between this pair of electrons and System I donors.

a) the two electrons are transferred to the same System I centers. The System I centers react only once during a short flash ; two flashes are necessary for the transfer of the two electrons to the methylviologen. This represents on the donor side of System I, the same model as for acceptors of System II. But this type of reaction produces no oscillations for MV_n when System II is limiting (see diagram 1).

b) the two electrons may be transferred one by one to two different System I centers ; when System II is limiting, and after intermediary pools have been emptied, the number of oxidized System I centers before the nth flash, equals the number of electrons transferred from System II by the $(n-1)^{th}$ flash. According to diagram 1, we then obtain :

$$MV_n = U_{n-1}$$

Both sequence U_n (Fig.1) and MV_n (Fig.2) oscillate with period 2. But an inversion of the oscillations is observed between the two figures : in effect, in both cases, the minima are obtained for odd values of n.

A difference between Figure 1 and 2 is the presence of DCMU for Fig.2. Diner [10] obtained, using chloroplasts partly inhibited by DCMU, evidence for a supplementary System II acceptor, which does not transfer its electron to the pool. The supplementary acceptor could explain an inversion of the oscillations at high DCMU concentration.

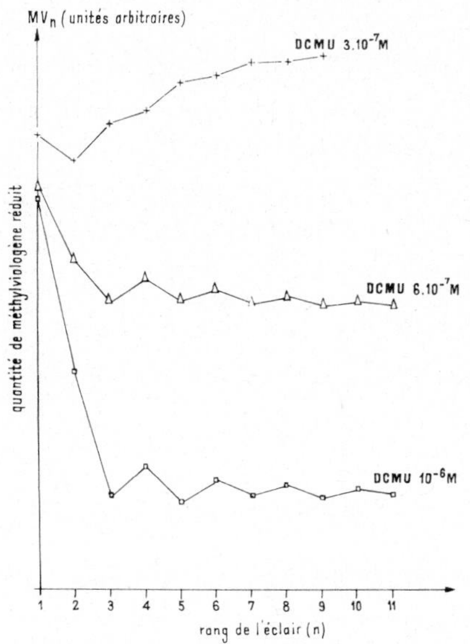

Fig.3. Quantities MV_n of methylviologen reduced by the nth flash of a series for spinach chloroplasts in presence of DCMU and 10^{-4}M methylviologen. The concentration of DCMU is indicated on each curve. The series of flashes was given 2mn 30s dark after 20 s far red light.

In support of this assumption, at lower concentration of DCMU (Fig.3), we could observe a sequence MV_n the minima of which occured for even values of n.

We then compared U_1/U_2 and MV_2/MV_3 in presence of DCMU 5.10^{-7}M. The results : U_1/U_2= 1.4 and MV_2/MV_3 = 1.5 are in favor of the relation $MV_n = U_{n-1}$ and reaction mechanism b.

Eventhough, because of the experimental uncertainties, it can't be completely excluded that some centers react according to a (a) - type mechanism.

3. Electrons transferred after the first flash in absence of DCMU.

Figure 1 indicated that some electrons are transferred after the first flash, on dark adapted material. This may be interpreted in 3 ways :

1- A certain number of System II centers may transfer their electron one by one to the acceptors oxidizable by System I. This assumption would introduce supplementary damping for U_n.

A striking point is that everytime sequence MV_n exhibits oscillations, the damping of the oscillations is 0.12, i.e. equal to the damping of oxygen oscillations[11]. To interprete the oxygen damping, Kok et al[11] proposed that during each short saturating flash of a sequence, 12% of the System II centers are photochemically inactive. Thus, the damping for U_n ought to at least equal 12%.

- Damping for $U_n \geq$ Damping for oxygen
- Damping for $MV_n \geq$ Damping for U_n
- Damping for $MV_n =$ Damping for oxygen

=> Damping for U_n = Damping for oxygen

Sequence U_n presents no supplementary damping. Assumption 1 is unlikely. Electrons exchange between the pool and the System II acceptors occurs two at a time for all the System II chains.

Two others assumptions are possible :

2- some B^- is stable in the dark.

3- electron transfers are different after the first flash, and after the following flashes. This type of assumption was proposed by Zankel[12].

4. Equilibirum constants

A) Rates of electron transfers.

Just after a photochemical reaction a center is in photochemically inactive state. The reactions leading from this inactive state to a photochemically active one are called turnover reactions and reveal dark limiting steps. For System II, these limiting steps have been studied by Kok et al[11], then by Bouges-Bocquet[13,14].

According to recent work of Diner[10], the turnover reveals electron transfers (half-time about 200µs to 400µs), on acceptor side of System II.

This rapid transfers are incompatible with reaction : ($Q^-B^- + PQ \rightarrow QB + PQ^{2-}$), which, according to Joliot[5] ought be slow (half-time around 20ms). So we could show[15] that reaction 1 was to be prefered to reaction 2.

Q being the primary acceptor of System II, Q^- is the photochemically inactive form of the acceptor, and the electron transfer on acceptor side of System II may be represented by the following scheme :

$$\begin{array}{c} PQ^{2-} \nwarrow \nearrow PQ \\ QB \overset{h\nu}{\rightarrow} Q^-B \rightleftarrows QB^- \overset{h\nu}{\rightarrow} Q^-B^- \rightleftarrows QB^{2-} \overset{h\nu}{\rightarrow} Q^-B^{2-} \\ PQ^{2} \swarrow \searrow PQ \end{array} \qquad \text{Diagram 2}$$

The observed half-time of electron transfers form Q^- to B or B^- are 200μs to 400μs[10].

This scheme leads to an heterogeneity of the secondary electrons acceptors. This is in agreement with the results of Joliot[15], Malkin[16] and recently of Radmer and Kok[17]. A_1 would represent Q + B, that is 3/10 of the total amount $A_1 + A_2$. This is compatible with the experimental evaluations.

The electron transfer rate from B^{2-} to PQ is the rate measured by Joliot[5] for transfers between acceptors A_1 and acceptors A_2 : it is about $15s^{-1}$ at 20°C.

B) Deactivation kinetics

The deactivation of oxygen precursors S_2 and S_3 formed on dark adapted chloroplasts by one or two flashes[11] have been studied in ref[18, 19, 20]. In[21], we showed that for the deactivation of the S states, it is reasonnable to assume that the deactivation reactions are the exact inverse mechanisms of the photochemical reactions and electron transfers.

After dark adaptation, most acceptors are in QB state (see the first paragraph), and most donors are in S_1 state (see ref[11]).

After one flash, the acceptors are in QB^- state and the donors in S_2 state. The deactivation of S_2 after one flash may be written:

$$S_2 \text{ Chl } QB^- \underset{ki}{\overset{kd}{\rightleftarrows}} S_2 \text{ Chl } Q^-B \xrightarrow{kr} S_1 \text{ Chl } QB$$

This simple mechanism could explain why S_2 deactivation after one flash is almost exponential[19]. Let us call k_2 the rate of observed deactivation :

$$k_2 = \frac{ki}{kd} \times k_r$$

(for more details on the computation, see ref[15]).

For Chlorella, the half-time for S_2 deactivation is 10s. The half-time of the reaction S_2 Chl $Q^-B \rightarrow S_1$ Chl QB was estimated to be 0.5s[15] So the equilibirum constant between (QQ^-) and (BB^-) is :

$$K = \frac{ki}{kd} = \frac{k_2}{k_r} \sim 20$$

C) Apparent equilibrium constant between Q and pool PQ

Joliot,A.[22] observed an equilibirum constant between Q and pool PQ which seemed to be 1 in the light. But several authors[23,16], found a complex relation between Q and PQ, the equilibrium constant variing between 1 and 100 in the dark. Several models were proposed to account for this complexity (Malkin[16], Joliot and Joliot[24] and Lavergne[25].

From diagram 2, we computed the relations between Q and PQ. In order to simplify the computation, we admitted that B is an ordinary plastoquinone molecule, fixed to the center, and that its equilibrium constant with the pool PQ is 1.

To account for an equilibrium constant equal to 1 between Q and PQ in the light[22], we must assume an equilibrium between $(Q\ Q^-)$ and $(B^-\ B^{2-})$ close to 0.5 (for details, see ref[15]). In these conditions, in the dark, the equilibrium constant varies between 0.5 and 20.

D) Action of a strong reductant

These equilibrium constants explain the action of dithionite observed by Velthuys and Amesz[9] : a strong reductant (dithionite) is introduced in a chloroplast suspension after 1, 2 or n flashes. After an even number of flashes, few centers go to Q^-, while after an odd number of flashes, almost all centers go to Q^-.

According to diagram 2, if all centers are in state QB before addition of dithionite, 66% will be in state Q^-B^- and 33% in state QB^{2-} after reduction via PQ. On the other hand, if all centers were in state QB^- before addition of dithionite, all centers would be in state Q^-B^{2-} after reduction. Thus, after reduction, the minimum amounts of Q^- will be observed for even number of preilluminating flashes (original state : QB) and maximum amounts will be observed after odd number (original state : QB^-).

We also found, without introducing any heterogeneity of the centers, that, in the dark, dithionite reduces only a part of the centers. This is in agreement with the experiments of Joliot and

Joliot[24].

E) Fluorescence recovery after n flashes

From diagram 2 and the equilibrium constants assumed in the preceeding paragraph, we may compute the recovery kinetics of fluorescence (disappearance of Q^-) when some + charge is present.

Three phases are experimentaly observed[19]. Their half-times are respectively 300μs, 100ms, and 2s (after two flashes) or 10s (after one flash). Diagram accounts well for these three phases when the following rates are chosen for the reactions

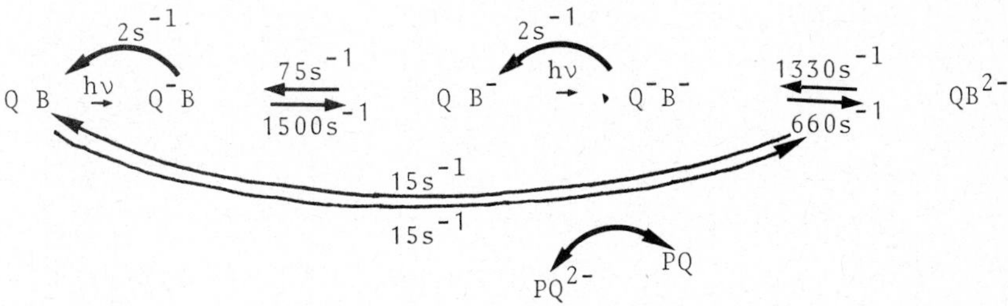

(These rates are just order of magnitude of the real rates).

The computed rate of the rapid phase is independent of $[PQ^{2-}]$ the rate of the intermediate phase is slightly dependent on $[PQ^{2-}]$, the rate of the slow phase is linearly dependent of $[PQ^{2-}]$ for Q^-B^-. This is in agreement with Lavergne's observations [25].

For these conditions, the amplitude of the intermediate phase would be 15% of the total after one flash, 42% after two flashes, 28% after three flashes. This fits qualitatively with the experimental results of Joliot et al[19].

Thus, the low equilibrium constant between ($Q\ Q^-$) and ($B\ B^{2-}$) may account for the strong amplitude of the 100ms phase observed after two flashes.

References

1. Witt, H.T., 1971, Q. Res. Biophys. 4, 365.
2. Amesz, J., 1973, Biochim. Biophys. Acta. 301, 35.
3. Duysens, L.N.M., 1963, Natl. Acad. Sci. Natl. Res. Counsil. Publ. 1145, 117.
4. Morin, P., 1964, J. Chim. Phys. 61, 674.
5. Joliot, P., 1965, Biochim. Biophys. Acta. 102, 116.
6. Rumberg, B., Schmidt-Mende, P., Weikard, J. and Witt, H.T., 1963 Natl. Acad. Sci. Natl. Res. Council Publ. 1145, 18.

7. Bouges-Bocquet, B., 1973, Biochim. Biophys. Acta. 314, 250.
8. Stiehl, H.H. and Witt, H.T., 1969, Z. Naturforsch 24b, 1588.
9. Velthuys, B.R. and Amesz, J., 1974, Biochim. Biophys. Acta. 333, 85.
10. Diner, B., 1974, to be published
11. Kok, B., Forbush, B. and McGloin, M., 1970, Photochem. Photobiol. 11, 457.
12. Zankel, K., 1973, Biochim. Biophys. Acta. 325, 138.
13. Bouges, B., 1972, Biochim. Biophys. Acta. 256, 381.
14. Bouges-Bocquet, B., 1973, Biochim. Biophys. Acta. 292, 772.
15. Bouges-Bocquet, B., 1974, Thèse de Doctorat d'Etat. Paris.
16. Malkin, S., 1971, Biochim. Biophys. Acta. 234, 415.
17. Radmer, R. and Kok, B., 1973, Biochim. Biophys. Acta. 314, 28.
18. Joliot, P., Barbieri, G. and Chabaud, R., 1969, Photochem. Photobiol. 10, 309.
19. Joliot, P., Joliot, A., Bouges, B. and Barbieri, G., 1971, Photochem. Photobiol. 14, 307.
20. Forbush, B., Kok, B. and McGloin, M., 1971, Photochem. Photobiol. 14, 307.
21. Bouges-Bocquet, B., Bennoun, P. and Taboury, J., 1973, Biochim. Biophys. Acta. 325, 247.
22. Joliot, A., 1965, Physiol. Veg. 3 (4), 329.
23. Forbush, B. and Kok, B., 1968, Biochim. Biophys. Acta. 162, 243.
24. Joliot, P., and Joliot, A., 1971, in Proceedings of IInd International Congress on Photosynthesis Research (ed. W. Junk), 569.
25. Lavergne, J., 1973, Thèse de Doctorat de 3e Cyle. Orsay.

M. AVRON, *Proceedings of the Third International Congress on Photosynthesis*,
September 2-6, 1974, Weizmann Institute of Science, Rehovot, Israel
Elsevier Scientific Publishing Company, Amsterdam, The Netherlands, 1974

DEPENDENCE OF THE TURNOVER AND DEACTIVATION REACTIONS OF PHOTOSYSTEM II ON THE REDOX STATE OF THE POOL A VARIED UNDER ANAEROBIC CONDITIONS

Bruce DINER
Institut de Biologie Physico-Chimique
13, rue Pierre et Marie Curie
75231 Paris Cedex 05 France

Introduction

Chlorella cells placed under an anaerobic atmosphere undergo an immediate reduction of the A pool. The redox state is determined by a balance established between reduction of the pool by an endogenous reductant and pool reoxidation by molecular oxygen[1]. This balance provides a stable means of regulating the redox state of the pool for periods as long as two hours. This technique was used for the measurement of steady-state turnover in Photosystem II by repetitive double flash excitation of Chlorella[2]. These experiments demonstrated that the turnover time slowed as the pool became reduced. Forbush and Kok[3] earlier demonstrated, in chloroplasts, that the fluorescence relaxation following each of a series of short flashes also slowed as the pool became reduced.

In the experiments to be presently described, we again resorted to the anaerobic technique for varying the redox state of the pool. However, in these experiments, an effort was made to specifically measure the $S_1' \rightarrow S_2$ and $S'_2 \rightarrow S_3$* turnover as the background oxygen concentration was varied and to correlate the kinetics observed to the redox state of the pool, measured by fluorescence induction.

The description of the equilibrium relation between primary acceptor, Q, and the pool, A, has been attempted by numerous authors[3-6], but was complicated by an apparent dependence of the equilibrium on whether the measurement was made under continuous light or darkness[4]. The models, on which the estimations of equilibrium were based, assumed a one-electron interaction between Q and

*The notation $S'_n \rightarrow S_{n+1}$, used in this paper, refers to the dark transition between intermediate state S_n', formed by flash excitation of a center in state S_n, and the successive more oxidized state, S_{n+1}.[9] This dark transition includes reactions on both the donor and acceptor sides of Photosystem II.

the pool. The recent work of Bouges-Bocquet[7] and Velthuys and Amesz[8] has demonstrated the existence of an intermediate acceptor B, between Q and A, which interacts with Q by a one-electron transfer and with the pool by a two-electron transfer. Thus, an understanding of the dependence of Q on the pool redox state requires a study of the Q to B and of the B to A equilibria.

The results presented here permit an estimation of the equilibrium constant relating B and the plastoquinone pool, A, as well as the rate constant for the two-electron transfer from B to A.

Materials and Methods

Chlorella pyrenoidosa were grown on Knop medium containing Arnon's trace elements A_6 and B_6, Prior to use, cells were suspended in 0.1M Naphosphate pH 7.0, containing 0.1M KCl at a chlorophyll concentration of about 300 ug/ml.

Chloroplasts were prepared from market spinach according to the method of Avron and stored at -70° in 0.05M Tris-HCl buffer pH 7.8, containing 0.01 M NaCl, 0.4 M sucrose and 5% dimethylsulfoxide (Me_2SO_4). When used for oxygen measurements, the chloroplasts were diluted to a chlorophyll concentration of 300ug/ml. using the same buffer as above, containing, in addition, 0.1 M KCl but without Me_2SO_4.

Oxygen was detected using a polarograph similar to that described by Diner and Mauzerall[1], which permitted equilibration of cells or chloroplasts with gases of varying oxygen content (ranging from 5 ppm O_2 to air). The 5 ppm O_2 gas was purified argon. All other mixtures (prepared by L'Air Liquide) were of oxygen in nitrogen. These gases were humidified by passage through a water bubbler. The gas flow rate was 40 ml./m.

Flash illumination was provided by Xenon flash lamps (General Radio, stroboslave, Model 1539-A) with a half-height of 4μs. All flashes used in the experiments, to be described, were saturating.

For measurements of fluorescence induction, cells or chloroplasts were illuminated on the platinum electrode, during the course of an oxygen experiment. Continuous light was filtered through one BG 38 (Schott) and two 4-96 (Corning) blue filters. The illumination intensity corresponded to 40 photons/sec-center. A photomultiplier (Radiotechnique XP 1002), placed close to the polarograph, detected the emitted fluorescence through one 2-64

(Corning) and two Rubalith red blocking filters.

Results

Chlorella cells were equilibrated for 30m with argon containing 5 ppm O_2. These cells, dark adapted for 4 m, were exposed to a sequence of saturating flashes spaced 320 ms apart (Fig. 1). The maximum oxygen yield was observed on the fifth flash, while the third amounted to only 50% of the fifth. A small secondary maximum was observed on the eighth flash. Air was then admitted to the cuvette and the cells equilibrated with 21% O_2. A flash sequence similar to the first was given (Fig. 1), after 4 m dark adaptation, and shows the normal flash pattern observed for Chlorella under aerobic conditions.

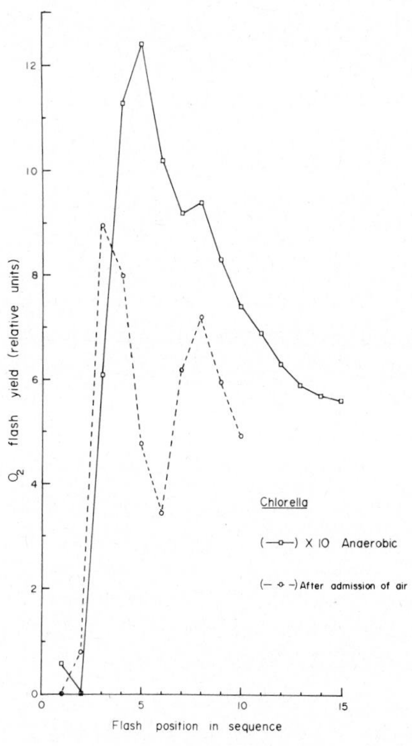

Fig. 1. Oxygen yield per flash for a sequence of flashes given to Chlorella cells under aerobic and anaerobic conditions. Chlorella cells were equilibrated with an atmosphere of 5 ppm O_2 in argon for 30 m. After 4 m dark adaptation, a series of saturating flashes were given spaced 320 ms apart (□). Air was then admitted and the cells adapted to the 21% O_2 atmosphere. After 4 m dark adaptation, a second sequence was given identical to the first (O). The flash yields of the anaerobic curve are multiplied by 10.

The anaerobic sequence shows a level of "misses"[9] appreciably elevated relative to the aerobic case, nevertheless, vestiges of the aerobic sequence remain, as evidenced by the secondary maximum on the eighth flash. This secondary maximum was always observed in anaerobic sequences. Considering that the third flash is more than ten times inhibited relative to the aerobic maximum, it is surprising that an oscillating sequence is observed at all. The third flash yield was found to be roughly parallel to the pool redox state.

The reduction of the pool under anaerobic conditions should transfer the rate limiting step for the turnover to secondary reactions on the acceptor side of Photosystem II. To study these reactions and their dependence on the redox state of the pool, turnover measurements were made at various background oxygen concentrations.

Chlorella cells were equilibrated for 30 minutes with gases containing various oxygen concentrations. The cells were then flashed in the same way preceding each kinetic time point. This treatment consisted of exposing the cells to 10 flashes during which a measurement was made, followed by a 5 m dark adaptation. This cycle was then repeated for a new time point. Turnover $S_1' \rightarrow S_2$ (Fig.2, upper) was measured by varying the time between the 1st and 2nd flashes and detecting oxygen on the 3rd flash, 320 ms after the 1st. $S'_2 \rightarrow S_3$ (Fig.2, lower) was measured by varying the time between the 2nd and 3rd flashes and detecting 0_2 on the 3rd. The 1st and 2nd flashes were separated by 320 ms. The remaining 7 flashes in both turnover experiments were spaced 320 ms apart.

Under the most anaerobic conditions (5 ppm 0_2) a radical difference is observed between $S_1' \rightarrow S_2$ and $S_2' \rightarrow S_3$. $S_1' \rightarrow S_2$ shows an initial sigmoidal phase composed of a 0.5 ms ($t_{1/2}$) component and a second 15 ms ($t_{1/2}$) component, followed by a much slower phase with a half rise time of 50 ms. $S_2' \rightarrow S_3$ shows much faster sigmoidal kinetics with a half-rise of 1.3 ms. On raising the 0_2 concentration to 1675 ppm 0_2, the ultra slow phase of $S_1' \rightarrow S_2$ diminishes while the sigmoid remains. At 1675 ppm 0_2, $S_2' \rightarrow S_3$ is practically identical to the aerobic curve showing a half rise of 0.5 ms, but with a slightly more pronounced sigmoid.

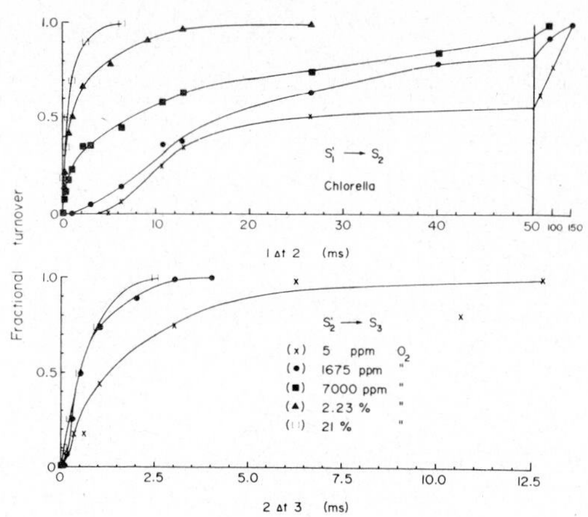

Fig.2. $S_1' \rightarrow S_2$ (upper) and $S_2' \rightarrow S_3$ (lower) turnover for Chlorella cells equilibrated with the indicated atmospheres for > 30 m. A series of 10 flashes were given, during which a time point was measured. Five minutes dark adaptation followed, prior to the measurement of the next time point. The time was varied between the 1st and 2nd ($S_1' \rightarrow S_2$) or between the 2nd and 3rd flashes ($S_2' \rightarrow S_3$) and oxygen detected on the 3rd flash. The time between all flashes (a total of 10) other than the closely spaced pair was 320 ms. The fraction of centers having recovered following the 1st or 2nd flash is expressed on the ordinate. Note the different time scales on the upper and lower figures.

While $S_2' \rightarrow S_3$ changes little as the O_2 concentration is further increased, $S_1' \rightarrow S_2$ continues to evolve. The ultra slow 50 ms phase of $S_1' \rightarrow S_2$ practically disappears at 7000 ppm O_2 while a fast rise ($t_{1/2} \simeq$ 0.5ms) is observed along with what remains of the 15 ms component. At 2.23% O_2 about 75% of the centers show the 0.5 ms phase, the rest recovering in 15 ms. In air only the fast (0.5 ms) component remains.

The redox state of the A pool has been determined by measuring the area defined by the fluorescence induction curve and the F_{max} level. The validity of this measurement is based on the demonstration by Delosme et al[10] of the linear relation existent between the Photosystem II electron transfer rate and the fluorescence yield. This measurement is a difficult one to make in algae because of the turnover of Photosystem I, despite its partial inhibition by anaerobic conditions[11]. This problem is avoided in chloroplasts in that Photosystem I is unable to turnover in the absence of an electron acceptor. Therefore, the turnover experiments were repeated using chloroplasts and were correlated with the pool redox state.

In algal cells, the pool becomes reduced in the dark under anaerobic conditions, because of a balance between pool reduction by an endogeneous reductant and its oxidation by molecular oxygen. Chloroplasts, when exposed to anaerobic conditions, in the dark do not undergo pool reduction, indicating an absence of the reductant present in algal cells. Instead, for chloroplasts, the pool was reduced by preillumination, with the anaerobic conditions greatly slowing its reoxidation.

Chloroplasts were equilibrated for 30 minutes with an anaerobic atmosphere. By subjecting the chloroplasts to a cycle consisting of 19 flashes (320 ms apart) followed by 7 m dark at 5 ppm O_2, the pool would reoxidize only slightly to a state, 90-95% reduced. By varying the background oxygen but maintaining the pretreatment constant a whole range of pool oxidation states could be obtained.

Using this technique, a family of $S_1' \rightarrow S_2$ (Fig.3, upper) and $S_2' \rightarrow S_3$ (Fig.3, lower left) curves were measured at varying oxygen concentrations and thus at varying redox states of the pool. The latter were measured by the area above the fluorescence induction curve (not shown) and compared to that obtained under aerobic conditions in the presence and absence of 10^{-4}M DCMU. The aerobic fluorescence induction curve showed a pool size of 20 equivalents, based on the area above the DCMU curve as equal to one equivalent.

The $S_1' \rightarrow S_2$ curves show an evolution similar to that observed for algal cells, but with kinetic components somewhat less clearly defined than for Chlorella. A sigmoidal curve is observed, at 5 ppm O_2, composed of a 0.5 -1 ms and a 12 ms component, but lacking the 50 ms phase. As for algal cells, an increase in the background oxygen concentration increases the fraction of centers which recover with a 0.5 -1ms half-time.

The fraction of centers, recovering with a $t_{1/2}$ of 0.5 ms is plotted (as $\frac{B_{ox}}{B_{tot}}$, for reasons to be explained later) as a function of the fraction of the pool oxidized, $\frac{A_{ox}}{A_{tot}}$ (Fig.3, lower right). The fact that the experimental points fall on a straight line means that the fraction of centers recovering in 0.5 ms is directly proportional to the fraction of the pool oxidized.

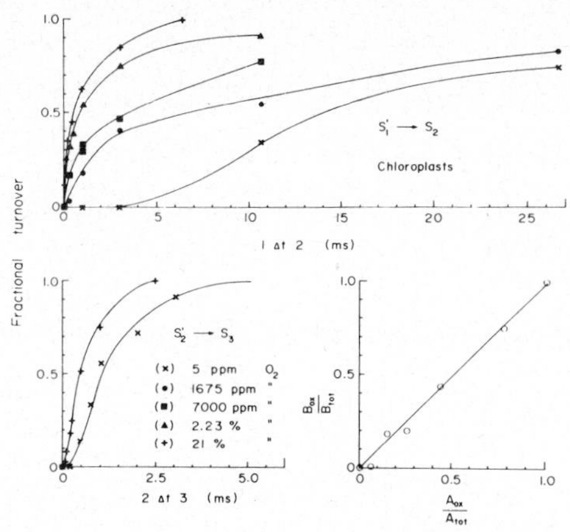

Fig. 3. $S_1' \rightarrow S_2$ (upper) and $S_2' \rightarrow S_3$ (lower left) turnover for chloroplasts equilibrated with the indicated atmospheres > 30m. The turnover times were measured in the same way as for Fig.2. except that the pretreatment was different prior to the measuring of the kinetic points.This consisted of giving a series of 19 flashes followed by a 7 m period of dark adaptation. Note the difference in time scales.

Lower right: Fraction of centers recovering ($S_1' \rightarrow S_2$) with 0.5 ms half-time as a function of the redox state of pool A. For reasons explained in the text, the ordinate is equivalent to the redox state of secondary acceptor, B. The ordinate values are taken from the curves shown (upper) plus other experiments not shown. The abscissa values were obtained by measuring the area defined by the fluorescence induction curve and F_{max} during the turnover experiments and with the same light-dark pretreatment as for the kinetic points.

$S_2' \rightarrow S_3$ in chloroplasts is also similar to that observed in Chlorella, in that only a small acceleration of the turnover is observed in going from 5 ppm O_2 ($t_{1/2} \simeq 1$ ms) to air ($t_{1/2} \simeq 0.45$ms). The turnover results, in general, indicate that for $S_1' \rightarrow S_2$ the turnover rate is a function of the pool redox state, whereas for $S_2' \rightarrow S_3$, the rate is largely independent of the pool.

Another way of measuring the interaction of the Photosystem II centers with the pool is by a study of the deactivation of the S states. Conditions which increase the concentration of Q^- and $A^=$ would be expected to increase the rate of deactivation if these were substrates for the deactivation reaction.

These measurements were performed on Chlorella cells at an O_2 concentration of 10 ppm. Deactivation of S_2 was measured by exposing cells, dark adapted for 5 m, to a single flash followed a variable time later by a flash pair given 160 ms apart. For deactivation of S_3, cells, dark adapted for 5 m, were exposed to a flash pair, given 160 ms apart, followed a variable time later by a 3rd flash. Oxygen was detected on the 3rd flash in both cases. A similar procedure was used for deactivation measured under aerobic conditions, except that cells were dark adapted for 4 m instead of 5.

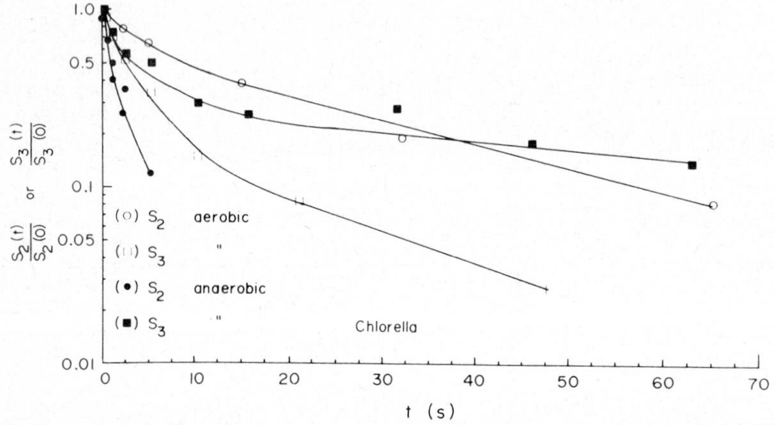

Fig. 4. Deactivation of S_2 and S_3 at 10 ppm O_2 and under aerobic conditions for Chlorella. For the anaerobic conditions S_2 (●) was measured by giving a single flash to cells, dark adapted for 5m, followed a variable time later by a flash pair given 160 ms apart. For S_3 (■) a flash pair was given 160 ms apart followed a variable time later by a third flash. Oxygen was detected on the 3rd flash in both cases. An identical procedure was used for aerobic cells except that dark adaptation was 4 m instead of 5.

Deactivation of S_2 is considerably accelerated by anaerobic conditions, showing a half-decay time of 0.8 s as opposed to 10 s under aerobic conditions. Deactivation of S_3 initially shows almost the same kinetics under both conditions, with a half-decay time of 3.5 s under anaerobic and 2.5 s under aerobic atmospheres. The S_3 curves continue to diverge at longer times with the anaerobic curve decaying more slowly than the aerobic. Thus an anomaly, similar to that observed for the turnover, is observed for deactivation. The decay of S_2 is greatly accelerated, probably by the increased concentration of $A^=$ and Q^-, while S_3, apparently independent of the highly reduced pool, decays even more slowly than under aerobic conditions.

Discussion

Turnover : Bouges-Bocquet[7] and Velthuys and Amesz[8] have demonstrated that the first electron transferred from Photosystem II is stored on a secondary acceptor and is not accessible to the pool. Upon transfer of a second electron, the now accumulated electron pair is transferred to the pool. The latter authors in experiments involving dithionite reduction of the pool found that a) the primary plus secondary acceptors comprised a pool consisting of 3 equivalents and b) that reverse transfer from the large A pool to the 3-equivalent pool also occured by means of electron pairs.

These results provide a basis for the interpretation of the results obtained here. Under the most anaerobic conditions, for Chlorella, the turnover $S_1' \rightarrow S_2$ is composed of a sigmoid with 0.5 ms and 15 ms components, followed by an appreciably slower phase with a 50 ms half time. As the O_2 concentration is raised to 1675 ppm O_2, the latter phase largely disappears but the sigmoid remains. The sigmoidal kinetics are indicative of two reactions in series, both of which are required for the regeneration of a Photosystem II center. Furthermore, Joliot by measuring oxygen emitted by flashes of varying length[12], observed the presence of a small and large pool, with a rate limiting step of roughly 20 ms (per 2 equivalents), linking the two. This ensemble of evidence leads to the following description of Photosystem II following the first flash, under anaerobic conditions :

$$S_1QB^{=} \xrightarrow{h\nu} S_2Q^-B^{=} \xleftrightarrow[15ms]{A \;\rightarrow\; A^{=}} S_2Q^-B \underset{0.5ms}{\leftrightarrow} S_2QB^-$$

Photosystem II centers, detected on the 3rd flash (i.e. state S_1 in the dark) exist almost exclusively in the $QB^{=}$ configuration. Upon photoexcitation, these centers form $Q^-B^{=}$ and are unable to undergo further charge separation until Q is reoxidized. This reoxidation requires a two-step process, oxidation of $B^{=}$ by A in 15 ms followed by oxidation of Q^- by B in 0.5 ms. At 5 ppm O_2 (Chlorella) the slow 50 ms phase might be the diffusion time within the heavily reduced pool for an A to diffuse to a $B^{=}$, initially adjacent to an $A^{=}$.

The 2nd and 3rd flashes induce the following events :

$$S_2QB^- \xrightarrow{h\nu} S_3Q^-B^- \underset{0.5ms}{\leftrightarrow} S_3QB^= \xrightarrow{h\nu} S_0Q^-B^= + O_2$$

The $S_2' \rightarrow S_3$ curves at 1675 ppm O_2 (Chlorella) show a rapid turnover, little effected by the heavily reduced pool. This is because the reaction $S_3Q^-B^- \rightarrow S_3QB^=$ is independent of the pool and is identical to the reaction that occurs under aerobic conditions. Furthermore these results argue against the existence of S_1QB^-, in the dark, because such centers would, after the first flash, be in configuration $S_2QB^=$ which would show a slow $S_2' \rightarrow S_3$ turnover.

At 5 and 1675 ppm O_2 almost no centers exist in the form S_1QB, in the dark, as these centers would show a rapid turnover with a $t_{1/2}$ of 0.5 ms. As the O_2 concentration is increased, however, the fraction of centers in state S_1QB in the dark increases, as evidenced by the appearance of the 0.5 ms phase at 7000 ppm O_2 and above for Chlorella. Thus, Photosystem II centers in state S_1, in the dark and detected on the 3rd flash, exist only in forms S_1QB or $S_1QB^=$, the proportion of each depending on the O_2 concentration.

A study of the dependence of this proportion on the redox state of the pool was carried out in spinach chloroplasts which show anaerobic turnover kinetics similar to those described in Chlorella. Thus, each species of the equilibrium $S_1QB^= + A \leftrightarrow S_1QB + A^=$ could be evaluated. $\frac{S_1QB}{S_1QB + S_1QB^=}$ was determined by measuring the fraction of centers, detected on the 3rd flash, which showed a $t_{1/2}$ of 0.5 ms for the $S_1' \rightarrow S_2$ turnover. In Fig. 3, the shorthand, $\frac{B_{ox}}{B_{tot}}$ stands for this ratio. $\frac{A}{A+A^=}$ $\left(\frac{A_{ox}}{A_{tot}}\right)$ was determined by measuring the area defined by the fluorescence induction curve and F_{max}. A_{tot} was taken as the area above the aerobic fluorescence induction curve. A plot of $\frac{B_{ox}}{B_{tot}}$ versus $\frac{A_{ox}}{A_{tot}}$ gives a straight line which means that B and A are related by an equilibrium constant equal to 1 (at least for centers in state S_1). Thus B is very likely chemically equivalent to the members of pool A. The fact that B is able to store one electron [1, 8] between Q and A means that B is not interchangeable with A. Thus B is likely to be a plastoquinone fixed to a Photosystem II center.

An equilibrium constant of 1 between B and A means that for 90-95% pool reduction, 5-10% of the S_1 centers should be in configuration S_1QB. Following the first flash, these centers are converted to S_2Q^-B which would be expected to turnover rapidly. It is necessary to assume with Velthuys and Amesz[8] that these centers equilibrate with the pool and that 90-95% are in form $S_2Q^-B^=$ and are thus "missed" on the 2nd flash. Therefore the fraction of centers in S_1QB, in the dark, and detectable on the 3rd flash would represent less than 10% of the $S_1' \rightarrow S_2$ turnover curve. Turning the argument around this is another way of saying that equilibirum $S_2Q^-B^= + A \longleftrightarrow S_2QB^- + A^=$ ($Q^- + B \longleftrightarrow Q + B^-$ assumed to have a large K, see below and refs. 13,15) has an equilibrium constant ≤ 1. The deactivation results also support this conclusion.

Deactivation :

The deactivation experiments (Fig. 4) were performed with Chlorella under a 10 ppm O_2 atmosphere. Almost all S_1 centers, detected on the 3rd flash were in configuration $S_1QB^=$ in the dark. Following the 1st flash, these centers were transformed to $S_2Q^-B^=$ which, via the pool, is in equilibrium with S_2QB^-. Under aerobic conditions, centers detected on the 3rd flash, are primarily in form S_1QB before the 1st flash and in S_2QB^- afterwards . A comparison of the aerobic and anaerobic S_2 deactivation curves shows a 10-12 fold factor between the rates of deactivation ($t_{1/2} \simeq 0.8$ s,anaerobic and 10 s, aerobic).

Bennoun[14] has shown that, following a saturating illumination of Chlorella in 20μM DCMU, reoxidation of Q^- and thus deactivation of S_2 occurs with a $t_{1/2}$ of 0.75 s. The fact that anaerobic deactivation of S_2 occurs at the same rate ($t_{1/2} \simeq 0.8$ s), suggests that here too Q^- is the substrate for deactivation and that most centers remain in configuration $S_2Q^-B^=$ following the 1st flash. The fact that the pool is probably 90-95% reduced (if an extrapolation from chloroplasts to Chlorella is justified) means that $S_2Q^-B^= + A \leftrightarrow S_2QB^- + A^=$ also has an equilibrium constant of roughly 1. As Q^- is probably also a substrate for deactivation of S_2 under aerobic conditions than $\frac{[Q^-]\text{anaerobic}}{[Q^-]\text{aerobic}} = 10$ to 12 (rate proportional to kQ^-) and for reaction $Q^- + B \longleftrightarrow Q + B^-$ $K \simeq 10$ to 12. A similar conclusion for the latter equilibrium was reached by Bouges-Bocquet[13, 15] ($K \simeq 20$).

The deactivation of S_3, unlike that of S_2, is not at all accelerated by anaerobic conditions and is even slowed relative to the aerobic case. Following 2 flashes, under anaerobic conditions, most centers, detectable on the 3rd flash, should be in state $S_3QB^= \leftrightarrow S_3Q^-B^-$. Under aerobic conditions, two flashes produce $S_3QB^=$ which transfers an electron pair to the fully oxidized pool leaving most centers in configuration S_3QB. If an equilibrium is assumed between $S_3Q^-B^-$ and $S_3QB^=$(even if weighted toward the latter) then under anaerobic conditions the concentration of Q^- should be higher than under aerobic conditions and the deactivation more rapid. Because the opposite is observed, it is necessary that either a) the reaction $Q^-B^- \rightarrow QB^=$ is irreversible or b) if this reaction is reversible that neither Q^- nor $A^=$ are substrates for S_3 deactivation. In either case neither Q^- nor A^- are substrates.

The deactivation results for S_3 are consistent with an observation of Lemasson and Barbieri[16], for chloroplasts, in which reduction of the pool by strong white light did not accelerate the rate of S_3 deactivation over that observed for the oxidized pool following 2 flashes.

That S_2 is sensitive to the redox state of Q and A while S_3 is not means that the substrates for deactivation are different in the two cases and that S_3 is somehow protected from Q^- and $A^=$. Why S_3 deactivation is slowed by anaerobic conditions, however remains unclear.

Acknowledgement:

The author is a Helen Hay Whitney Research Fellow and gratefully acknowledges the support of the Foundation.

References:

1. Diner, B. and Mauzerall, D., 1973, Biochim. Biophys. Acta. 305, 329.
2. Diner, B. and Mauzerall, D., 1973, Biochim. Biophys. Acta. 305, 353.
3. Forbush, B, and Kok, B., 1968, Biochim. Biophys. Acta. 162, 243.
4. Malkin, S., 1971, Biochim. Biophys. Acta. 234, 415.
5. Velthuys, B.R. and Amesz, J., 1973, Biochim. Biophys. Acta. 325, 126.
6. Joliot, A., 1968, Physiol. Veg. 6, 235.
7. Bouges-Bocquet, B., 1973, Biochim. Biophys. Acta. 314, 250.

8. Velthuys, B.R. and Amesz, J., 1974, Biochim. Biophys. Acta. 333, 85.

9. Kok, B., Forbush, B. and McGloin, M., 1971, Photochem. Photobiol. 11, 457.

10. Delosme, R., Joliot, P. and Lavorel, J., 1959, C.R. Acad. Sci. Paris, 249, 1409.

11. Joliot, P., 1961, J. Chim. Phys. 58, 583.

12. Joliot, P., 1965, Biochim. Biophys. Acta. 102, 116.

13. Bouges-Bocquet, B. - these proceedings.

14. Bennoun, P., 1970, Biochim. Biophys. Acta. 216, 357.

15. Bouges-Bocquet, B., 1974, Thèse de Doctorat d'Etat, Paris.

16. Lemasson, C. and Barbieri , G., 1971, Proc. 2nd Int. Congress on Photosynthesis, Stresa, 753.

M. AVRON, *Proceedings of the Third International Congress on Photosynthesis*, September 2-6, 1974, Weizmann Institute of Science, Rehovot, Israel
Elsevier Scientific Publishing Company, Amsterdam, The Netherlands, 1974

STIMULATION BY ADDED LIPIDS OF PHOTOPHOSPHORYLATION ASSOCIATED WITH PHOTOSYSTEM I.

D. M. HAWCROFT & J. FRIEND
Department of Plant Biology, University of Hull, England.

1. Summary

Addition of small amounts of C_{18} fatty acids or carotenoids to reaction mixtures stimulates PMS-catalyzed photophosphorylation by isolated broad bean chloroplasts. In addition these lipids stimulate photoreduction reactions associated with photosystem I.

Since the carotenoids and fatty acids have different effects on non-cyclic electron transport and the associated phosphorylation, these results support the hypothesis that there are separate sites for cyclic and non-cyclic photophosphorylation, and also indicate that the molecular environments of the two sites are different.

2. Introduction

Addition of carotenoids to isolated chloroplasts can either stimulate or inhibit the photoreduction of indophenol and ferricyanide and non-cyclic photophosphorylation. The observed stimulatory or inhibitory effects are dependent upon the concentration of added carotenoid and are due to the carotenoids acting both as uncouplers and energy transfer inhibitors[1]. The carotenoids are thus acting in a similar manner to unsaturated fatty acids originally investigated by McCarty and Jagendorf[2] and more recently by Siegenthaler[3]. However the concentration at which these two types of lipid are effective is quite different; the maximum effect of linolenic acid is at a concentration 40 times greater than that of violaxanthin when measured as stimulation of indophenol reduction under non-phosphorylating conditions. The present results show that added lipids stimulate both photophosphorylation and photoreduction reactions associated with photosystem I.

3. Materials and Methods

Broad bean plants (Vicia faba L. var Aquadulce) were grown in the glasshouse. Chloroplasts were isolated in the sodium chloride, phosphate, EDTA medium of Good et al[4] and resuspended in 0.4M sucrose, 0.03M Tricine pH 7.3.

For measurement of PMS-mediated photophosphorylation reaction mixtures contained, in 3 ml, 50 µmoles 0.5M tricine/NaOH pH 7.8, 60 µmoles NaCl, 12 µmoles $MgCl_2$ and 0.1 µmoles PMS and chloroplasts containing 45 µg chlorophyll. Photophosphorylation was measured as the disappearance of inorganic phosphate by the method of Hill and Walker[5].

Methyl viologen reduction was measured spectrophotometrically under anaerobic conditions in a reaction mixture containing tricine/NaOH, NaCl, $MgCl_2$ as above and in addition 100 µmoles methyl viologen, 10 µmoles cysteine HCl pH 7.8 and 71.2 nmoles dichlorophenol indophenol and chloroplasts containing 49 µg chlorophyll. For measurement of NADP reduction, in addition to tricine/NaOH, NaCl and $MgCl_2$, the reaction mixture contained 1.6 µmoles NADP, 8 µmoles ADP, excess ferredoxin, 12 µmoles phosphate with ^{32}P to give 3 x 10^6 c.p.m., 55 µmoles ascorbate, 71.2 nmoles dichlorophenol indophenol and 60 nmoles DCMU as indicated and chloroplasts equivalent to 41 µg chlorophyll. NADP reduction was measured spectrophotometrically and ATP by the method of Avron[6].

Lutein and violaxanthin were isolated from sugar beet leaves, and diadinoxanthin[7] from _Euglena gracilis_ strain Z by standard procedures[8].

Carotenoids and fatty acids were added to reaction mixtures dissolved in 0.05 ml spectroscopically pure ethanol; the same amount of ethanol was added to the controls. Detergents were added in distilled water.

4. Results

The effects of the addition of lipids to reaction mixtures on PMS-mediated photophosphorylation are seen in figs.1, 2 and 3 when it can be seen that at the appropriate concentrations, it is possible to observe stimulation by both ionic and non-ionic detergents (fig.1), by C_{18} fatty acids (fig.2) and by carotenoids (fig.3). It should be noted that there is at the most a 2-fold difference between the optimum concentrations of C_{18} fatty acids and carotenoids whereas the optimum concentration for the two detergents is about 10 to 20-fold higher.

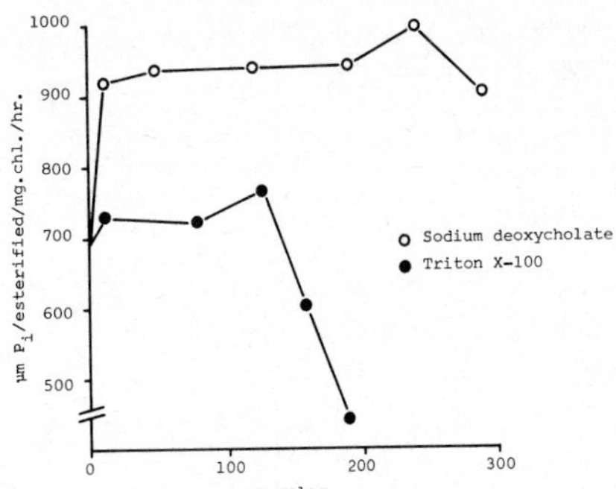

Fig.1 Effects of detergents on PMS-mediated photophosphorylation

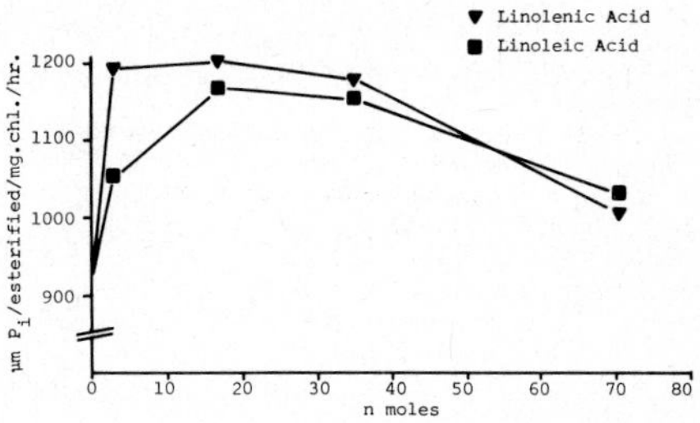

Fig.2 Effects of C_{18} fatty acids on PMS-mediated photophosphorylation.

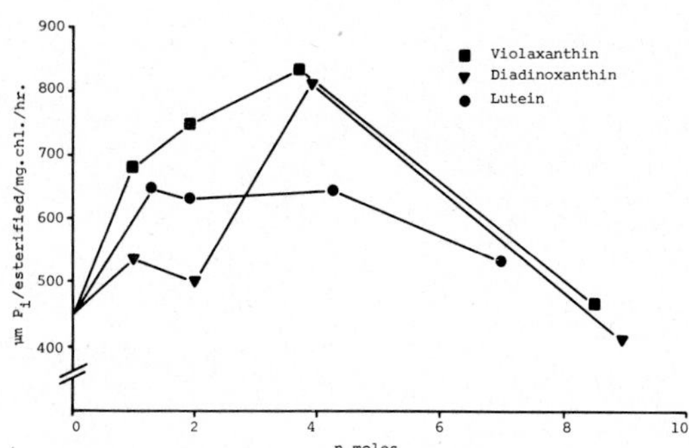

Fig.3 Effects of carotenoids on PMS-mediated photophosphorylation

As with the interaction of added lipids with photosystem II reactions, the stimulatory effects on PMS-mediated photophosphorylation vary with different chloroplast preparations; the stimulation given by C_{18} unsaturated fatty acids ranges between 20 and 50% and by carotenoids between 40 and 90%. Linolenic acid has its maximum stimulatory effect at a concentration between 2-4 fold as great as that of violaxanthin. In addition it seems that only unsaturated C_{18} fatty acids will stimulate PMS-mediated photophosphorylation. There is little difference in the activities of the three carotenoids which mainly differ structurally in the number of epoxide groups. However linoleyl alcohol and methyl linoleate have their maximum effect at half the concentration of linoleic acid, from which it is concluded that an ionizable carboxyl group extends the range of stimulatory concentrations.

The effects of linolenic acid on electron transport associated with photosystem I have also been examined. From the results in table I it will be seen that methyl viologen reduction from $DPIPH_2$ is stimulated both in the absence and presence of uncoupling levels of $(NH_4)_2SO_4$; similar results were obtained when NADP reduction was measured.

TABLE I.

Effects of linolenic acid on photoreduction of methyl viologen using $DPIPH_2$ as electron donor expressed as a percentage of the control value

Linolenic acid addition	$- (NH_4)_2SO_4$	$+ (NH_4)_2SO_4$ (1.0 mM)
None (control)	100	260
12 μM	133	305
24 μM	289	500

Linolenic acid also stimulates NADP reduction using $DPIPH_2$ (Table II) and gave further stimulation in the presence of $(NH_4)_2SO_4$. Similar stimulatory effects, but of a smaller magnitude were found when electrons were donated from water; the smaller effects are probably the net results of an inhibition of photosystem II and a stimulation of photosystem I activities.

TABLE II

Effects of linolenic acid on photoreduction of NADP and ATP formation expressed as a percentage of control value

	(a) $DPIPH_2$ as electron donor		(b) Water as electron donor	
Additions	$NADPH_2$	ATP	$NADPH_2$	ATP
None (control)	100	100	100	100
+ linolenic acid (12μM)	128	135	117	114
+ $(NH_4)_2SO_4$ (1mM)	152	3.4	128	2.1
+ linolenic acid (12μM) and $(NH_4)_2SO_4$ (1mM)	165	3.2	138	1.4

5. Discussion

It is clear that carotenoids and C_{18} fatty acids have remarkably different effects on the biochemical reactions associated with photosystems I and II.

These lipids stimulate both photoreduction and photophosphorylation associated with photosystem I whereas at similar concentrations they cause uncoupling of photosystem II. There is evidence for at least two sites of ATP formation associated with photosynthetic electron transfer[9,10,11]. According to the scheme of Neumann _et al_[9], the photosystem II phosphorylating site would be site B and system I phosphorylation would occur at either site A or on the "cyclic" site[10]. On the other hand if the scheme of Bradeen & Winget[11] were used to explain the reactions, presumably the phosphorylation reactions would occur at the two parallel sites between PQ and PC.

Whichever scheme is acceptable it seems obvious that the phosphorylation sites associated with the two light reactions respond differently to added lipids possibly because they are in different physical environments.

The present results also suggest that re-activation experiments, such as those in which it has been found that the loss of photosystem I activity from lyophilized chloroplasts extracted with heptane can be restored by addition of lipids[12], should be interpreted with a great deal of caution, especially in view of our finding that on increasing the concentration of added lipids, the stimulatory effects are succeeded by inhibitory ones.

6. References

1. Friend, J. & Hawcroft, D.M. 1967, Biochem. J. 104, 60P.
2. McCarty, R.E. & Jagendorf, A.T. 1965, Plant Physiol. 40, 725.
3. Siegenthaler, P.A. 1973, Biochim. Biophys. Acta, 302, 153.
4. Good, N.E., Winget, G.D., Winter, W., Connolly, T.N., Izawa, S. & Singh, R.M.M. 1966, Biochemistry 5, 466.
5. Hill, R. & Walker, D.A. 1959, Plant Physiol. 34, 240.
6. Avron, M. 1960, Biochim. Biophys. Acta. 40, 257.
7. Goodwin, T.W. 1971, in "Aspects of Terpenoid Chemistry & Biochemistry" p315, Ed. T.W.Goodwin, Academic Press, London and New York.

8. Davies, B.H. 1965, in "Chemistry & Biochemistry of Plant Pigments", p489, Ed. T.W.Goodwin, Academic Press, London and New York.
9. Neumann, J., Arntzen, C.J. & Dilley, R.A. 1971, Biochemistry 10, 866.
10. Avron, M. & Neumann, J. 1968, Ann. Rev. Plant Physiol.19, 137.
11. Bradeen, D.A. & Winget, G.D. 1974, Biochim. Biophys. Acta, 333, 331.
12. Brand, J., Krogmann, D.W. & Crane, F.L. 1971, Plant Physiol. 47, 135.

Acknowledgment.

D.M.H. was in receipt of a Science Research Council studentship. Present address of D.M.H. is Lanchester Polytechnic, Priory St. Coventry.

M. AVRON, *Proceedings of the Third International Congress on Photosynthesis*,
September 2-6, 1974, Weizmann Institute of Science, Rehovot, Israel
Elsevier Scientific Publishing Company, Amsterdam. The Netherlands, 1974

REACTION CENTER P700 FROM CHLOROPLASTS

Nathan Nelson and Carmela Bengis

Department of Biology

Technion-Israel Institute of Technology, Haifa, Israel

1. Summary

A reaction center from chloroplasts active in NADP photoreduction by ascorbate, was purified by means of detergent treatment, differential centrifugation, column chromatography and sucrose gradient. Ferredoxin, feredoxin-NADP-reductase and plastocyanin were required for the reaction. The preparation contains five classes of polypeptide chains with apparent molecular weights of 70,000, 25,000, 20,000, 18,000 and 16,000 as determined by gel electrophoresis in sodium dodecyl sulfate. The NADP photoreduction correlated to the amount of 70,000 molecular weight polypeptide and to the presence of 20,000, 18,000 and 16,000 polypeptides. About 4 non-heme iron atoms per one P700 were detected in the preparation. Antibody, prepared against the active reaction center, inhibited NADP photoreduction catalyzed by the purified reaction center as well as by isolated chloroplasts. The antibody had no effect on any of the other reactions tested including methyl viologen photoreduction. Treatment with 0.5% sodium dodecyl sulfate abolished the NADP photoreduction activity and released the low molecular weight subunits, which were removed by sucrose gradient from the high molecular weight one. The P700 signal is associated with the 70,000 molecular weight polypeptide. The antibody interacted on immunodiffusion plates with any subchloroplasts preparation with P700 signal including the purified 70,000 molecular weight polypeptide. It is concluded that both the primary oxidation and the primary reduction in photosystem I are associated with the 70,000 molecular weight polypeptide.

2. Introduction

Reaction center particles from photosynthetic bacteria have been isolated and purified extensively (1-8). The preparation consists of three polypeptides with molecular weights of 27,000, 22,000 and 19,000, none heme iron and four molecules of bacteriochlorophyll per reaction center (3,4). The polypeptide composition of chloroplasts reaction center is not known. Takamiya (9) reported on purification of chlorophyll-protein complexes from several sources with molecular weight of about 70,000. Thornber (10) purified a chlorophyll a-protein complex with molecular weight of about 150,000 with no photochemical activity. Later on, using a modified procedure, one P700 per 80 chlorophyll a molecules was found in the preparation (11), the presence of cytochromes b_6 and f in the latter was the only observed difference (12). All of the above mentioned preparations were not active in NADP photoreduction.

Polypeptide composition of chloroplasts membranes and photosystems I and II enriched particles, was analysed recently in several laboratories (13-19). The preparations were active in NADP photoreduction but multiple bands were observed on SDS gels. The participation of certain polypeptides in photosystems I and II was deduced from their relative intensity on the gels.

It is the purpose of this communication to describe a preparation of a reaction center from chloroplasts composed of 5-6 polypeptides and active in NADP photoreduction, as well as a preparation of a single polypeptide having P700 signal.

3. Methods

Chloroplasts from Swiss chard leaves and photosystem I particles depleted of cytochromes b_6 and f were prepared in a fashion described previously (20). Ferredoxin (21), ferredoxin - NADP-reductase (21) and plastocyanine (22) were prepared as described. The Ouchterlony immunodiffusion reaction (23), NADP photoreduction (21), protein concentration (24), chlorophyll concentration (25), non-heme iron (26) and pigments content (27) were assayed by published procedures. Gel electrophoresis in the presence of SDS was performed as described by Weber and Osborn (28). The gels were fixed, stained, destained and scanned as previously described (29). Antibody against the active reaction center was obtained by mixing 1ml (0.5mg protein) with 1.5ml of complete Freund's adjuvant on a vortex and immediate injection to rabbits as previously described (29). Absorption spectra were recorded by Cary 118C and light induced P700 signal at 430nm by Aminco-Chance dual wavelength spectrophotometer.

4. Results

Preparation of a reaction center active in NADP photoreduction:

Photosystem I particles which were prepared in a large quantity (20) were kept frozen at -70°, for as long as six months, in a medium containing 0.4M sucrose, 0.01M NaCl, 0.01M Tricine-NaOH (pH8) and at chlorophyll concentration of 3mg per ml. A portion of 30ml was thawed and Triton X-100 was added as 20% solution to give a final concentration of 4%. After incubation over night at 4° the solution was applied on DEAE-cellulose column (Whatman DE 11) 2 x 25cm, which was equilibrated with solution containing 50mM Tris-Cl (pH8) and 0.2% Triton X-100.

The column was washed with 100ml of the same buffer. About 50% of the total chlorophyll and 15% of the protein were washed out of the column and most of it appeared on SDS gel in a position of about 23,000 dalton, which is probably general chlorophyll-protein complex. The reaction center was eluted with a linear NaCl gradient from 0 to 400mM (200ml in each chamber) in the same buffer.

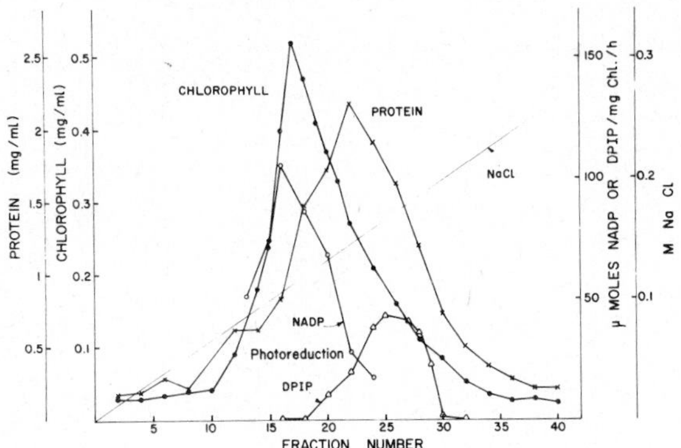

Fig. 1. Elution pattern of the reaction center from DEAE-cellulose column. Fractions of 6,6ml were collected and assayed. For NADP photoreduction the reaction mixture contained, in a final volume of 1ml, 20μ moles of MES-Tricine (pH8), 40μ moles of NaCl, 0,5μ moles of NADP, 3n moles of ferredoxin, 2,5n moles of plastocyanin, 0,05n moles of ferredoxin-NADP-reductase, 0,1% Triton X-100 and particles equivalent to 10-15μg of chlorophyll. The reaction mixture was illuminated, through a red filter (Corning 2403), with light intensity of 4,8 x 10^5 ergs per cm^2 per sec. The absorbance changes at 350nm were recorded by Cary 118C in which a blue filter (Corning 9782) was placed in front of the phototube. For 2,6-dichlorophenol indophenol (DPIP) photoreduction the reaction mixture contained, in a final volume of 1ml, 50μ moles of MES-Tricine (pH6,5), 30n moles of DPIP, 2 μmoles of diphenyl carbazide and particles equivalent to 10-20μg of chlorophyll. The reaction mixture was illuminated as for NADP photoreduction and recorded, with the same set of filters, at 600nm.

Fig. 1. shows the elution profile of a DEAE cellulose column. Fractions with NADP photoreduction activity were separated completely from a residual photosystem II activity. The presence of 0.2% Triton X-100 in the elution buffer was necessary to obtain good separation. With lower or high Triton concentrations poor separations were obtained and in the presence of 1% Triton NADP photoreduction activity could not be recovered. Fig. 2 demonstrates scanes and photographs of SDS gels of two peak fractions of NADP and DPIP photoreduction. Protein bands at 70, 20, 18 and 16 kilodalton are evident in photosystem I while in fractions with photosystem II activity they almost completely disappeared and distinct bands at 44 and 29 kilodalton appeared. The NADP photoreduction activity correlates with the density of the 70 kilodalton peptide in several fractions and various preparations.

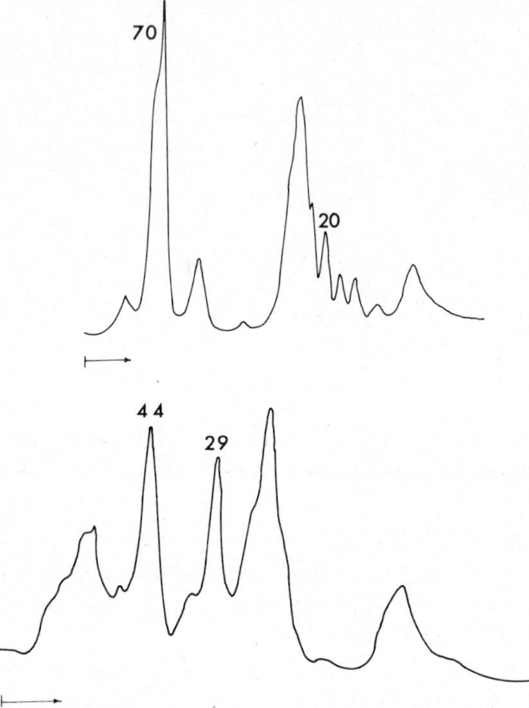

Fig. 2. The SDS gel electrophoresis pattern of fractions 16 (top) and 25 from the DEAE-cellulose column. Experimental conditions were as described in the methods. Fifty µl of each fraction were applied on the gels.

The most active fractions (15-19) were applied (0.5ml to each tube) on sucrose gradient 5-25% in 50mM Tris-Cl (pH8) and 0.2% Triton X-100 and run for 15 hours in SW-50 Spinco rotor at 35,000rpm. Fractions of 0.5ml were collected and the active fractions(the lower green band) were kept at 0° in the dark up to three days. Freezing of the fraction or increasing the Triton concentration during the gradient to 1% caused a complete loss of the NADP photoreduction activity.

Properties of the reaction center:

The NADP photoreduction activity was dependent on addition of ferredoxin, ferredoxin NADP reductase and plastocyanine. The specific activity at pH8 and under the conditions described in Fig. 1, increased about 8 folds on chlorophyll basis and about 10 folds on protein basis. Fig. 3 depicted the influence of the pH on NADP photoreduction by the reaction center. Maximum at pH7 was obtained, however different amounts of the soluble electron carriers were needed to give maximal rate at the given pH. The absorption spectrum of the reaction center given in Fig. 4 shows a maximum at 673nm.

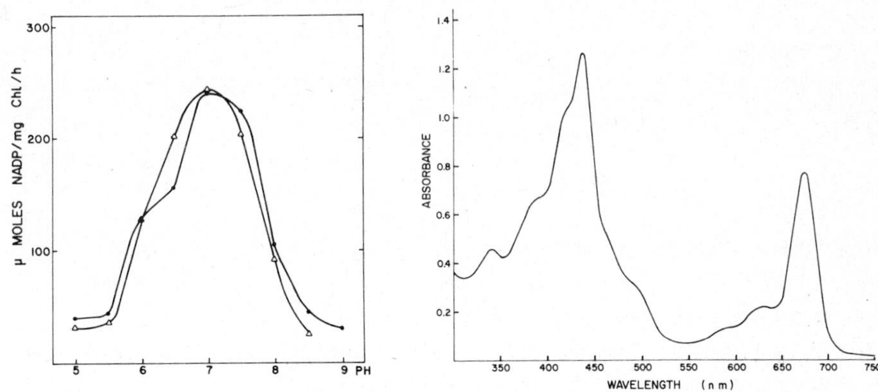

Fig. 3. (left) Photoreduction of NADP by the reaction center in various pH. Conditions as described in the legend of Fig. 1. Purified reaction center equivalent to 8μg chlorophyll was used. Δ; 3n moles of DPIP were added to the reaction mixture, O; The amounts of ferredoxin, ferredoxin-NADP-reductase and plastocyanin were doubled.

Fig. 4. (right) Absorption spectrum of the purified reaction center. The spectrum was recorded on a Cary 118C spectrophotometer using 1-cm quartz cuvettes.

The reaction center is almost free of chlorophyll b and the ratio of chlorophyll _a_ to chlorophyll _b_ is higher than 40. Neoxantin lutein and violaxantin are missing and the amount of β carotene is larger in comparison with chloroplasts. One P700 per 80-100 chlorophyll molecules was found in the preparation assuming 64,000 as a molar extinction coefficient at 700nm (30). Cytochromes could not be detected in the reaction center neither by spectrophotometry nor by antibody against cytochrome f (31). The preparation contained 3,5n moles of non-heme irons per 100n moles of chlorophyll. On SDS gels the purified reaction center dissociates to five distinct bands of molecular weights of 70,000, 25,000, 18,000 and 16,000 (fig. 5). Additional more diffused band was detected at the position of 8,000 daltons.

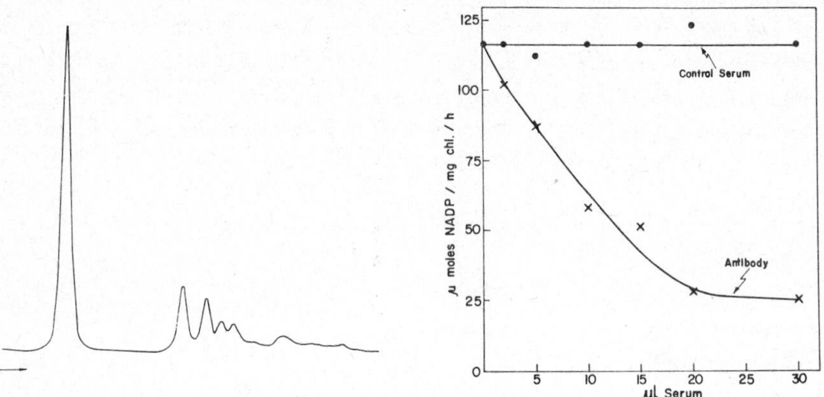

Fig. 5. (left) The SDS gel electrophoresis pattern of the purified reaction center. Experimental conditions were as described in the methods. Phosphorylase-a (M.W. 94,000), bovine albumin (M.W. 68,000), pyruvate kinase (M.W. 57,000), ovalbumin (M.W. 43,000), pepsin (M.W. 35,000) and α-chymotrypsin and its subunits (M.W. 25,000, 13,000 and 11,000) were used as standards.

Fig. 6. (right) Effect of antibody against purified reaction center on NADP photoreduction by lettuce chloroplasts. The reaction mixture contained in a final volume of 1ml: 50μ moles of Tricine-NaOH (pH8), 50μ moles of NaCl, 5μ moles of NH_4Cl, 0.5μ moles of NADP, 0,003μ moles of ferredoxin, chloroplasts equivalent to 17μg of chlorophyll and the volume of antibody indicated in the figure. After 10 min. of incubation at room temperature it was illuminated by white light ($1,5.10^6$ ergs per cm^2 per sec) for 1 min.

Antibody against the active reaction center, inhibited NADP photoreduction not only by the purified reaction center but also by chloroplasts from lettuce leaves (fig. 6). Photophosphorylation in the presence of NADP was inhibited accordingly to the inhibitions of NADP photoreduction. None of the other reactions tested were inhibited including photoreduction of methyl viologen. On immunodiffusion plates precipitation arcs without any spare were obtained with any one of the subchloroplasts preparations capable of NADP photoreduction (fig. 7). No positive reaction could be detected using the purified reaction center and antibodies against CF_1 and its five individual subunits (29), cytochrome f (31), plastocyanine (32) and ferredoxin NADP reductase (21).

Purification of the polypeptide associated with P700 (P700 reaction center):

The purified reaction center was incubated with 0.5% SDS for 20 min. at 0° and diluted with equal volume of distilled water. The SDS treatment abolished the NADP photoreduction and shifted the P700 peak to 697nm. Portions of 0.5ml were applied on sucrose gradient in identical conditions as in the last step of purification of the active reaction center. Two green bands were formed and collected separately. The absorption spectrum of the lower blueish-green band, given in Fig. 8, shows marked decrease in the β-carotene region in comparison with the active reaction center. Analysis of this preparation revealed that 80% of the β-carotene was removed by the treatment. The preparation contained one P700 per 45 chlorophyll a molecules. The absorption spectrum of the higher yellowish-green band, given at fig. 9, shows increase in the β- carotene region and indeed this preparation is enriched with it in comparison with the active reaction center. No P700 signal was detected in this fraction. Fig. 10 shows the SDS gel pattern of the blueish-green band. Nearly a single band with molecular weight of 70,000 was obtained. The low molecular weight subunits were found in the yellowish-green band from the sucrose gradient.

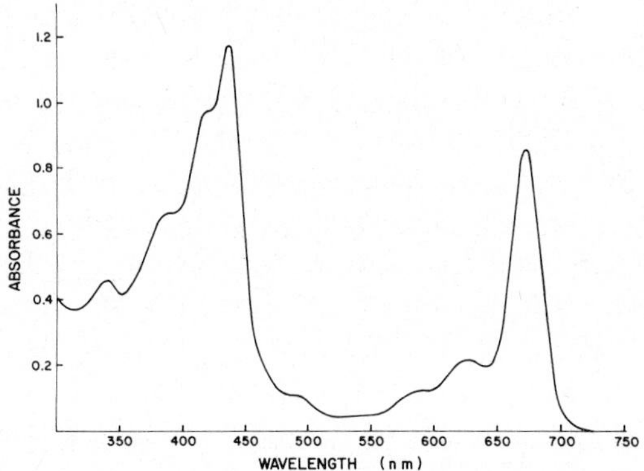

Fig. 8. Absorption spectrum of the P700 reaction center.

The P700 reaction center interacted, on immunodiffusion plates, with the antibody obtained against the active reaction center, while the fraction containing the low molecular weight subunits did not interact (fig. 11). The arc, obtained with the active reaction center, combined with the arc obtained with the P700 reaction center with no spare between them.

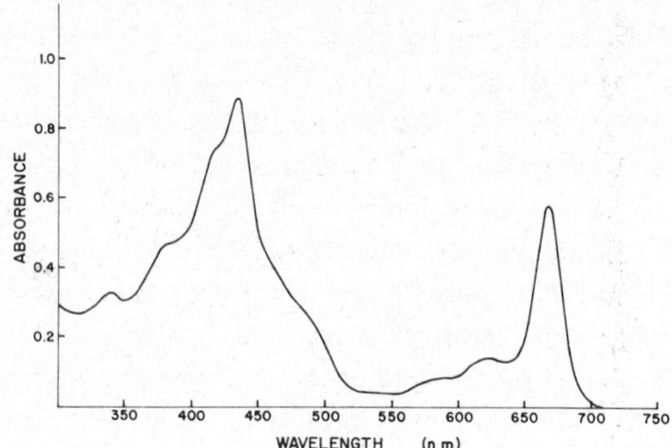

Fig. 9. Absorption spectrum of the yellowish-green fraction from sucrose gradient.

Fig. 7. (top) Interaction between the reaction center and its antibody in immunodiffusion experiments. 1) 5μl serum from rabbit I injected with the reaction center. 2) 5μl serum from rabbit II injected with the reaction center. 3) Control serum. 4) Anti γ subunit of CF_1. a) Purified reaction center (0.6μg chlorophyll). b) Active fraction after DEAE-cellulose column (2μg chlorophyll). c) Photosystem I particles (12μg chlorophyll). d) Inactive fraction from the sucrose gradient (0.3μg chlorophyll). e) Lettuce chloroplasts dissolved in 4% Triton (6μg chlorophyll). f) Chloroplasts pigments extracted by 80% aceton and dissolved in 4% Triton (10μg chlorophyll). The immunodiffusion plates contained 0.5% Triton X-100.

Fig. 10. (mid.) The SDS gel of the purified P700 reaction center (lower gel) and SDS gel of the reaction center active in NADP photoreduction.

Fig. 11. (bottom) Interaction between the antibody, active reaction center and purified P700 reaction center. 1) 10μl antibody from rabbit I. 2) 10μl antibody from rabbit II. a) Purified active reaction center (1μg chlorophyll). b) Purified P700 reaction center (bluish-green fraction after SDS treatment - 70 Kilodalton subunit 0.6μg chlorophyll). c) Yellowish-green fraction after SDS treatment (low molecular weight subunits 0.5μg chlorophyll).

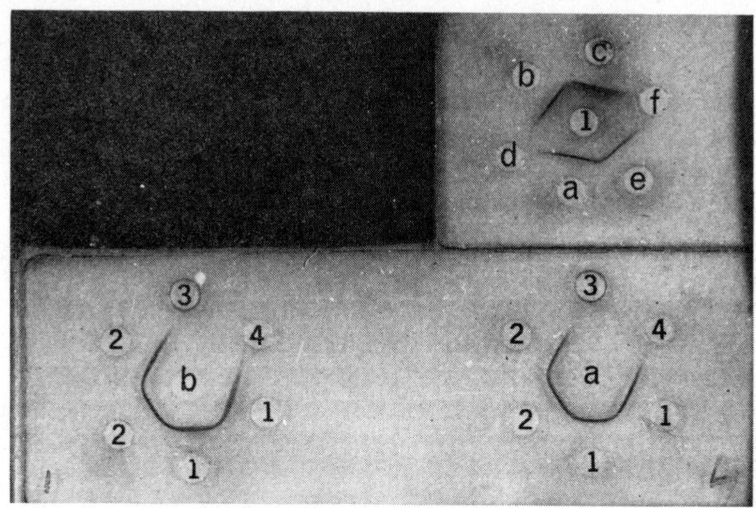
c
b
f
1
d
a
e
3
2
4
b
1
2
1
3
2
4
a
2
1
1

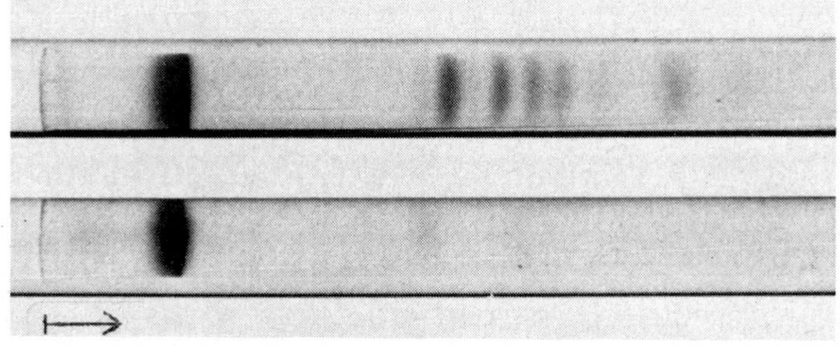

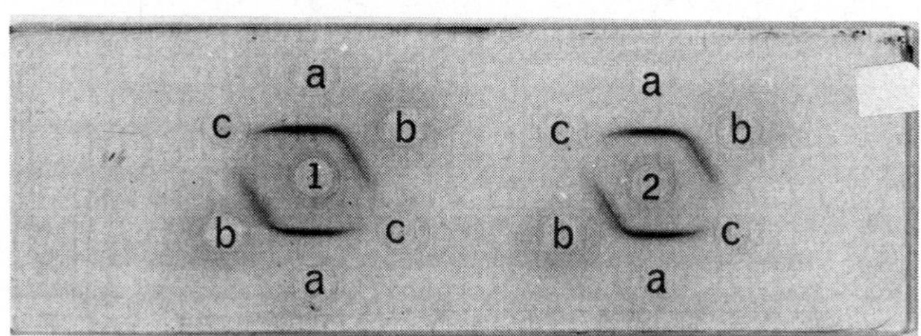
a
c
b
1
b
c
a
a
c
b
2
b
c
a

5. Discussion

Purification procedure of a reaction center, active in NADP photoreduction has been described. The source for the preparation were photosystem I particles which were depleted of cytochromes by digitonin treatment and differential centrifugation (20). The DEAE cellulose column that followed the Triton treatment was a crucial step which removed residual cytochromes and membrane fragments (33). Complete separation between photosystem I and residual photosystem II activities was obtained. Klein and Vernon (14) concluded that the major protein of photosystem II is expressed on SDS gels in the position of 23 Kilodalton and that of photosystem I in a region of 50-70 kilodaton. We observed correlation between a band in the pos-tion of 70 kilodalton and NADP photoreduction. Somewhat lower correlation was found with bands in the position of 20,18 and 16 kilodalton. Residual photosystem II activity appeared on the column in parallel with bands of 25,29 and 44 kilodalton. The band at 25 is probably analogous to a band of 23 kilodalton suggested by Klein and Vernon (14) as the major band of photosystem II. One of the bands in the position of 29 or 44 kilodalton might be cytochrome 559 which was detected in the fractions with photosystems II activity.

The purified active reaction center consists of five distinct polypeptide subunits with molecular weights of 70,000, 25,000, 20,000, 18,000 and 16,000. A diffused band in the 8,000 daltons position has a similar molecular weight as non-heme iron protein, recently discovered by Malkin et al (34). The 3.5 molecules of non-heme iron per one P700 we analyzed is in close agreement with the four irons per molecule they found. We cannot conclude yet if all of the subunits are an integral part of the active reaction center. The 25 kilodalton polypeptide, might be a contamination of a general light harvesting protein-chlorophyll complex and various amounts of it in different preparations did not affect the NADP photoreduction. The 70,000 molecular weight polypeptide is necessary for NADP photoreduction judged by the correlation between its amount and the specific activity of the preparation. As integrated from the scane of the SDS gel one of each of the low molecular weight subunits and two of the 70 kilodalton ones are present per reaction center. This might suggest that the 20,000, 18,000 and 16,000 dalton polypeptides play a role in NADP photoreduction. These polypeptides were present in the fractions from DEAE-cellulose column with NADP photoreduction activity.

Treatment with 0.5% SDS abolished the NADP photoreduction and enabled purification of the 70,000 molecular weight polypeptide. The low molecular weight subunits were removed while the P700 signal retained and its content doubled. This suggests that at least the oxidized site of photosystem I is associated with a polypeptide of 70,000 molecular weight. Disappearence of this band from a preparation with P700 signal might be due to its tendancy to aggregate.

The antibody we obtained against the purified reaction center is similar in its activity to that obtained by Regitz et al (35) by injecting broken chloroplasts. However the latter serum contained also antibodies against coupling factor I and ferredoxin-NADP-reductase which had to be neutrilized by excess of reductase in the NADP photoreduction experiments (35). The antibody described in this paper is specific for subchloroplasts preparations having a P700 signal, and does not interact with CF_1 and the reductase. The fact that this antibody inhibited most of the NADP reduction in isolated chloroplasts shows that the reaction center we purified, is on the main path of electron transport to NADP. After removal of the low molecular weight subunits, the antibody interacted with the 70 kilodalton polypeptide and not with the low molecular weight polypeptides. This might suggest that not only P700 is associated with this polypeptide but also the primary electron acceptor or at least the binding site for FRS (36) or ferredoxin.

References

1. Reed, D.W. and Clayton, R.K., 1968, Biochem. Biophys. Res. Commun. 30, 471-475.
2. Reed, D.W., 1969, J. Biol. Chem. 244, 4936-4941.
3. Feher, G., 1971, Photochem. Photobiol. 14, 373-387.
4. Clayton, R.K. and Haselkorn, R., 1972, J. Molec. Biol. 68, 97-105.
5. Wang, R.T. and Clayton, R.K., 1973, Photochem. Photobiol. 17, 57-61.
6. Okamura, M.Y., Steiner, L.A. and Feher, G., 1974, Biochemistry. 13, 1394-10403.
7. Clayton, R.K., Fleming, H. and Szuts, E.Z., 1972, Biophys. J. 12, 46-63.
8. Steiner, L.A., Okamura, M.Y., Lopes, A.D., Moskowitz, E. and Feher, G., 1974 Biochemistry. 13, 1403-1410.
9. Takamiya, A., 1971, Methods in Enzymology. 23, 603-613.
10. Thornber, J.P., 1969, Biochim. Biophys. Acta. 172, 230-241.
11. Thornber, J.P., 1971, Methods in Enzymology. 23, 682-687.
12. Shiozawa, J.A., Alberte, R.S. and Thornber, J.P., 1974, Plant Physiol. Sup. 63.
13. Vernon, L.P., Klein, S., White, F.G., Shaw, E.R. and Mayne, B.C., 1971, Proceedings II International Congress on Photosynthesis Research, eds. G. Forti, M. Avron and A. Melandri.PP. 801-812.
14. Klein, S.M. and Vernon, L.P., 1974, Photochem. Photobiol. 19, 43-49.
15. Crane, F.L., Arntzen, C.J., Hall, J.D., Ruzicka, F.J. and Dilley, R.A., 1971, in: Autonomy and Biogenesis of Mitochondria and Chloroplasts, eds. N.K. Boardman, A.W. Linnane and R.M. Smillie. P.P. 53-69. Elsevier, New York.
16. Remy, R., 1971, FEBS Lett. 13, 313-317.
17. Kung, S.D. and Thornber, J.P., 1971, Biochim. Biophys. Acta. 253, 285-289.
18. Lagoutte, B. and Duronton, J., 1971, Biochim. Biophys. Acta. 253, 232-239.

19. Levine, R.P., Burton, W.G. and Duram, H.A., 1972, Nature New Biol. 237, 176-177.
20. Nelson, N. and Racker, E., 1972, J. Biol. Chem. 247, 3848-3853.
21. Nelson, N. and Neumann, J., 1969, J. Biol. Chem. 244, 1926-1931.
22. Anderson, M.M. and McCarty, R.E., 1969, Biochim. Biophys. Acta. 189, 193-206.
23. Ouchterlony, O., 1968, Handbook of Immunodiffusion and Immunelectropharesis. p.p. 21-22 Ann Arbor-Humphrey Science Publishers, Ann Arbor, Michigan.
24. Lawry, O.H., Rosebrough, N.J., Farr, A.L. and Randall, R.J., 1951, J. Biol. Chem. 139, 265-275.
25. Arnon, D.I., 1949, Plant Physiol. 24, 1-15.
26. Brumby, P.E. and Massey, V., 1967, Methods in Enzymology. 10, 463-474.
27. Strain, H.H., Cope, B.T. and Svec, W.A., 1971, Methods in Enzymology, 23, 452-476.
28. Weber, K. and Osborn, M., 1969, J. Biol. Chem. 244, 4406-4412.
29. Nelson, N., Deters, D.W., Nelson, H. and Racker, E., 1973, J. Biol. Chem. 248, 2049-2055.
30. Ke, B., 1973, Biochim. Biophys. Acta. 301, 1-33.
31. Racker, E., Hauska, G.A., Lien, S., Berzborn, R.J. and Nelson, N., 1971, Proceedings II International Congress of Photosynthesis Research., eds. G. Forti, M. Avron and A. Melandri. pp. 1907-1113.
32. Hauska, G.A., McCarty, R.E., Berzborn, R.J. and Racker, E., 1971, J. Biol. Chem. 246, 3524-3531.
33. Nelson, N. and Neumann, J., 1972, J. Biol. Chem., 247, 1817-1824.
34. Malkin, R., Aparicio, P.J. and Arnon, D.I., 1974, Proc. Nat. Acad. Sci. U.S. 71, 2362-2366.
35. Regitz, G., Berzborn, R.J. and Trebst, A., 1970, Planta. 91, 8-17.
36. Siedow, J., Yocum, C.F. and San Pietro, A., 1973, in: Current Topics in Bioenergetics. 5, 107-123. eds. D.R. Sanadi and L. Packer. Acad. Press, New York.

M. AVRON, *Proceedings of the Third International Congress on Photosynthesis*, September 2-6, 1974, Weizmann Institute of Science, Rehovot, Israel
Elsevier Scientific Publishing Company, Amsterdam, The Netherlands, 1974

PREPARATION AND PROPERTIES OF PHOTOSYNTHETIC FRAGMENTS OF THE UNICELLULAR BLUE-GREEN ALGA *ANACYSTIS NIDULANS*

Claude SIGALAT and Yaroslav de KOUCHKOVSKY

Laboratoire de Photosynthèse, C.N.R.S., 91190 Gif-sur-Yvette (France)

ABSTRACT

A "mild" fractionation of *Anacystis* cells was obtained by lysozyme treatment followed by an osmotic shock. A double centrifugation on a sucrose gradient gives many fractions of which stand out : protoplasts, and light and heavy fragments. The *in vivo* extinction coefficients of chlorophyll a, C-phycocyanine and carotenoids (mainly β-carotene) are determined by comparison of absorption spectra of water and pyridine extracts with those of the whole algae and of the various fractions obtained, corrected for light-scattering thanks to an empirical equation and an abacus. An "excitation-transfer indicator" from phycocyanine to chlorophyll is computed on the basis of fluorescence data; only in the algae this transfer exists. The P700 content is measured and it is shown that the accompanying band at 675-685 nm is complex and is neither a simple satellite of it nor a System II-dependent signal. In all fractions, except protoplasts, the O_2 evolution is abolished, but the heavy fragments have still some System II activity (with artificial electron donors) and the light fragments, rich in carotenoids, present a fully functional System I (and no detectable System II). The high content of this light fraction in P700 (1 for *ca*. 36 chlorophylls) makes it comparable to the highly purified System I particles HP700 of Ogawa and Vernon, although a very simple procedure was used here.

INTRODUCTION

The fractionation of the photosynthetic apparatus of the blue-green algae (essentially : *Anabaena*)was already made several times, with a special attention paid to the properties of the different fragments and to the role of phycocyanine[1-7]. However , the fractions were often poorly active and some points still remained to be studied or required to be clarified. Thus, for Susor and Krogmann[2], the unique function of phycocyanine is to transfer the absorbed light energy to chlorophyll, whereas Thomas and De Rover[1] think that the loss of this accessory pigment during the cell disruption disturbs the membrane structure and hence lowers the photochemical activity. On the other hand, Ogawa and Vernon[6,7] succeeded in isolating fragments with either of the two photosystems properties but, very surprisingly, the light and the heavy fractions were respectively of SII and SI type. These different points prompted us t undertake the present work, applying a mildest as possible technique of fractionation to the alga *Synechococcus*(Indiana strain 625 of *Anacystis nidulans*). This species was chosen because it is an unicellular organism and is widely used in many laboratories. Only a short summary is given here, the detailed points (particularly on methodology) being presented elsewhere (ref. 8 and to be published general report).

Abbreviations : Chl = chlorophyll, Pcy = phycocyanine, Car = carotenoids, MV = methylviologen , DCPIP = dichlorophenolindophenol, DCMU = 3,4-(dichlorophenyl)-1,1-dimethylurea, SI = System I, SII = System II.

RESULTS

Obtention of fractions

The cell wall of the algae is firstly digested by incubating 45 min at 37° 3 ml of a concentrated suspension (10^{14} cells ml^{-1}) in mannitol 0.5 M + phosphate 0.03 M buffer pH 6.8 containing 25 mg of lysozyme (optimal conditions). The resulting spheroplasts (~2 µm diameter) are then diluted with the buffer and centrifuged, and the pellet resuspended in an equal volume of water to disrupt the protoplasts. After homogeneization, the mixture was put on top of a 12.5–50% linear gradient of sucrose in Tris 0.05 M + maleate 0.05 M buffer pH 7.5 (which preserves better the activity than mannitol-phosphate). Two 20 min centrifugation at 2 000 g (series A) and 40 000g (series B) are made, giving the various fractions illustrated in fig. 1 (the correspondance between series A and B was determined by separate centrifugation of A bands). A centrifugation at 60 000 g for 3 hours gives only a pellet and B1, which can be sedimented at 130 000 g in 1 hr. The additional bands (B4', B3' and below B1) are not always clearly seen - they are more abundant in 0.5 M NaCl - but have properties essentially similar to those of the standard bands. In general, for studying their properties, the fractions were concentrated by centrifugation, resulting in the loss of almost all of the phycocyanine - which is therefore only loosely bound- in the supernatant (that of B1, quite rich in phycocyanine, was called S and was sometimes used to supplement the fractions, particularly B3, almost devoid of this pigment). We will focus here our attention only on the bands A3 (= protoplasts which did not burst during the osmotic shock), B3 (= "heavy" fraction) and B1 (= "light" fraction), the other bands and the pellets being too heterogenous. On the basis of the chlorophyll content, A3, B3 and B1 represent respectively about 30%, 15% and <1% of the starting material.

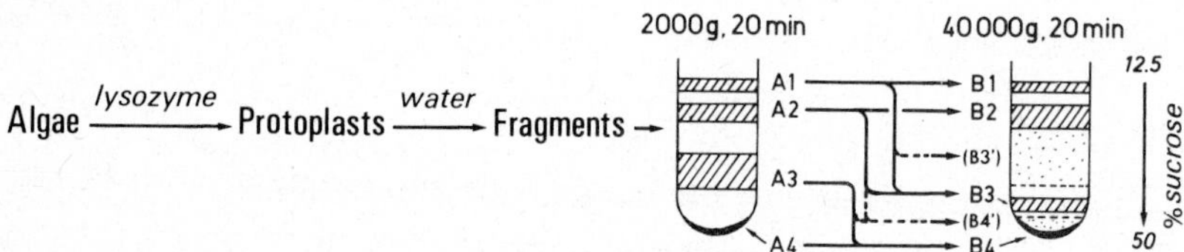

A3 (dark blue-green) = protoplasts; B3 (green-yellow) = heavy fraction
B1 (blue-violet) = light fraction [centrifuged → supernatant S]

Fig. 1. Schematic diagram of the preparation of *A. nidulans* photosynthetic fragments. See text for details.

Pigment characterization

The pigment concentration were obtained from the absorption spectra of water (for phycocyanine) or pyridine (for chlorophyll and carotenoids) extracts, using extinction coefficients described in the litterature and in this work[8]. The values obtained were used for the computation of the *in vivo* coefficients, given below, which we wanted to determine for a rapid but quantitative characterization of the pigment composition of the different fractions (therefore without the need of a solvent extraction).

To trace true *in vivo* absorption spectra, a correction for light-scattering must be applied. Using completely depigmented heterogenous cell suspensions as standards, with and without opal glasses of different thicknesses, we arrived at a practical equation of the turbidity D dependence *vs.* wavelength λ, which only requires the knowledge of the apparent absorbance - *i.e.* measured optical density - A'_o at a reference wavelength λ_o where there is no pigment absorption (A'_o is therefore a measure of the suspension concentration). This equation is : $D = k\ \lambda^{(\alpha A'_o - \beta)}$, where k is a "constant" (experimentally determined), $\alpha \approx 0.46$ and $\beta \approx 1.65$ (without opal glasses). It is thus possible to determine at any wavelength the true absorbance A from the recorded spectrum : A = A'-D. Some deviations (<5%) from the above indicated law may be observed and are eliminated with the use of an abacus and of an additional correction at 438 nm[8]. An illustration of the results is given in fig. 2.

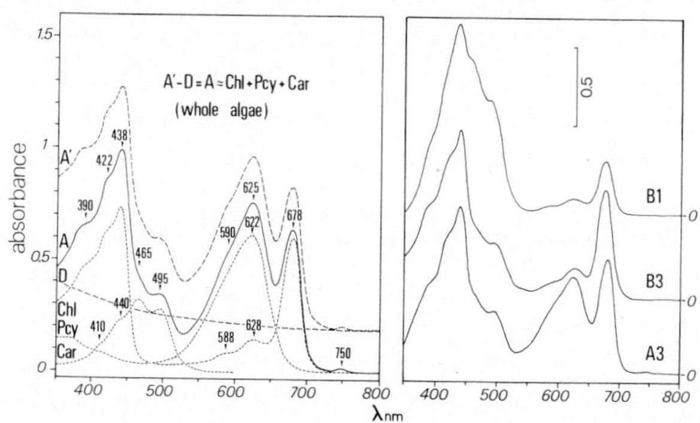

Fig. 2. *Left hand*: correction of a recorded absorption spectrum (A') for light-scattering (D) and calculated separation of the resulting spectrum (A) in the three indicated components (the sum of which reconstitute almost perfectly the curve A); algae : 5 μg Chl ml^{-1}. *Right hand* : absorption spectra, with light-scattering correction of protoplasts (A3), and heavy (B3) and light (B1) fractions (respectively : 5,6 and 2.5 μg Chl ml^{-1}).

To decompose the corrected spectrum into individual components, it was necessary to determine different equations giving the pigment concentrations from the total absorbance values at appropriate wavelengths and to dispose of reference spectra. We started with the relations established by Myers *et al.*[9,10] and used for absorption spectra of *in vivo* pigments the supernatant S for phycocyanine, a solution of β-carotene in benzene for carotenoids, and the washed B3 fraction *minus* carotenoids for chlorophyll a (satisfactory controls were obtained with two algae containing only chlorophyll a : *Botrydiopsis alpina* and a phycocyanine-less mutant of *Cyanidium caldarium*[11]). The *in vivo* extinction coefficients ε (in M^{-1} cm^{-1}) were calculated to be 103 500 for Chlorophyll a at 678 nm, 210 000 for the phycocyanine monomere at 625 nm and 94 500 for carotenoids at 465 nm. The equations giving the absorbance A at these peak wavelength, for a particular pigment *in vivo*,were finally determined (A/ε gives therefore the molar concentration) :

$$A_{465}Car = 1.1520\ A_{495} - 0.0244\ A_{678} - 0.0415\ A_{625}$$

$$A_{625}Pcy = 1.0114\ A_{625} - 0.2488\ A_{678}$$

$$A_{678}Chl = 1.0114\ A_{678} - 0.0465\ A\ 625$$

where the three A in the right part of equations are the apparent absorbance values minus scattering. These equations give an error $< 1\%$ and allow the tracing of the individual spectra as shown in fig.2 (unlike Jones and Myers[10],we did not need special adjustment). From fig. 2 it is clear that, for phycocyanine, the peak shift from 622 to 625 nm and the shoulder at 590 nm are actually due to the secondary bands of chlorophyll. In the right hand part of this figure are shown spectra of the main fractions (after washing out the unbound phycocyanine). It may be noticed that the special chlorophyll absorbing form at 750 nm[12] exists only in protoplasts (and in algae)-which means that it is essentially due to the physico-chemical environment - and that the light fraction is particularly rich in carotenoids. The numerical values of the pigment content are given in the Table.

Another pigment-state characteristics is obtained from fluorescence spectra analysis. The excitation transfer efficiency from phycocyanine to chlorophyll was estimated quantitatively with an indicator $t = K\ (f_{680}Chl\ /\ f_{650}Pcy)(a_{\lambda}Pcy\ /\ a_{\lambda}Chl)$, where K is an instrumental factor.$f_{680}Chl$ and $f_{650}Pcy$ are the fraction of fluorescence emitted at 680 and 650 nm by chlorophyll and phycocyanine, a_{λ} Pcy and a_{λ} Chl are the absorption ($= 1-e^{-2.3\ A}$), by phycocyanine and by chlorophyll, of the exciting light at the wavelength λ. The coefficient a are determined with equations of the type given above and for f, similar procedure than for a was used. The resulting equations are (f_{650} and f_{680} being the total fluorescence measured at these 2 wavelengths):

$$f_{650}Pcy = 1.025\ f_{650} - 0.109\ f_{680}$$

$$f_{680}Chl = 1.025\ f_{680} - 0.233\ f_{650}$$

The Table shows that t drops to almost zero already in protoplasts.

Electron-transfer chain

The O_2 evolution, already greatly diminished in protoplasts, is pratically absent in all fractions. The addition of an artificial electron donor such as p-phenylelenediamine or, better, hydroxylamine allows however the detection of a small System II activity in some fractions, especially in the heavy one B3 (where the whole chain is also functioning) : see the Table. The purely System I reaction

($DCPIPH_2$ → MV) is present in all fractions, the most active being the light fragments B1 (where no System II can be detected).

Another test of System II photochemistry was the effect of DCMU on fluorescence in weak light, which indicates the existence of a variable fluorescence. The results (see Table) confirms those above mentioned. No increase of activity was obtained by adding the supernatant S to the fractions.

P700

In fig. 3 are represented light-minus-dark (L-D) and ferricyanide-minus-hydrosulfite, *i.e.* dithionite, (F-H) difference spectra of an average fraction. In addition to the negative peak at 700-704 nm, the well-known[13,14] accompanying band around 682 nm ("P682") is seen, together with another one close to 620 nm, due to phycocyanine bleaching (indeed,ferricyanide oxidizes this pigment). A correlation between P700 and P682 and between the photoinduced and chemioinduced signals in the different fractions was searched and is illustrated in the right-hand part of fig.3. Roughly, photooxidizable and chemiooxidizable P700 vary in a parallel way, and photooxidizable P682, which is actually at 680-685 nm, follows a similar evolution. But chemiooxidizable P682, which is in fact at 678-683 nm, has a noticeably different behaviour. Also, addition of the supernatant S - an effect not easily understandable - increases more chemiooxidizable than photooxidizable P682 (6.3 times *vs.*2.5). All those facts underline the complexity of P682. The concentration of P700, as shown in the Table, is the highest for the light fraction B1 (1 for 36 chlorophylls), a result consistent with the redox activity measurements.

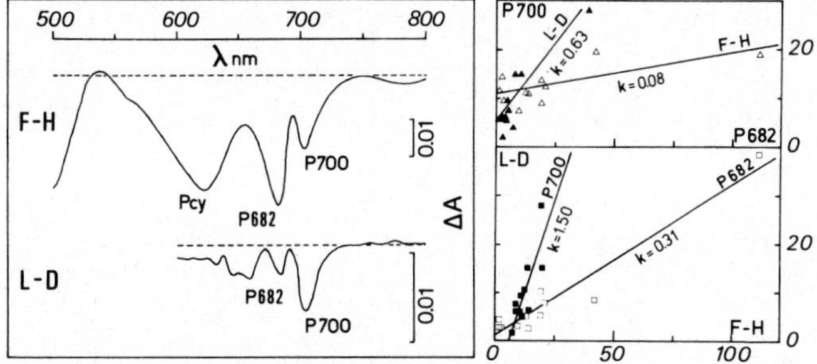

Fig. 3. *Left-hand* : ferricyanide-minus-hydrosulfite (F-H) and light-minus-dark (L-D) difference spectra of fraction A2 (8 μg Chl ml^{-1}). Ferricyanide and hydrosulfite : few crystals ; light (half of the maximum transmission from 410 to 520 nm) : 24 mW cm^{-2}. *Right-hand* : linear regression of P700 on P682, and of L-D on F-H signals (abscissa and ordinates in mmoles P $mole^{-1}$ bulk Chl) ; each point represents the data from a particular fraction of series A and B ; k represents the slopes, and the correlation coefficients (all significants within the confidence limits of 95 %) are : 0.94 (▲), 0.73 (△), 0.79 (■) and 0.97 (□).

CONCLUSION

Although the method of fractionation used was the "mildest" possible, two processes appear to be irreversibly destroyed and therefore to require a rather intact structure : the transfer of excitation from phycocyanine to chlorophyll - as could be expected if phycobilisomes are set wider apart from the chlorophyllous membranes during cell disruption - and the oxygen evolution. Since the latter drops less abruptly than the former and is completely insensitive to back addition of the supernatant S, it is highly improbable that phycocyanine has even an indirect protective function of the photochemical activity. Moreover, the System II as a whole seems very fragile, since artificial electron donors have but little restoring effect. On the contrary, the System I is largely preserved.

Table. Properties of the main fractions of *A. nidulans*.
C750 = chlorophyll form at 750 nm (identical extinction coefficients at their respective red peak are assumed for bulk Chl, C750, P700 and "P682"). L-D and F-H = light-dark and ferricyanide-hydrosulfite difference spectra ; t measured with 595 nm excitation. DCMU (50 µM) stimulation : on 680 nm fluorescence excited at 435 nm (no electron donor or acceptor added). Electron-transfer (rate in me$^-$g sec^{-1} mole^{-1} Chl), in saturating light : the numbers shown are good means, but the individual data are rather variable (H_2O donor : O_2 evolution measured, MV acceptor : O_2 uptake measured, DCPIP acceptor : dye reduction measured). The dashes in the "algae" column mean that the measurements are not made, because of the cell impermeability to substances. DCPIP = 50 µM, NH_2OH : 10 mM, MV : 500 µM, $DCPIPH_2$: DCPIP 50 µM + ascorbate 2 mM (PQ = endogenous plastoquinone) ; Chl : ∿ 10 µg ml^{-1}.

PIGMENTS :		Whole algae	Protoplasts (A3)	Heavy fragments (B3)	Light fragments (B1)
mmoles Chl/mg protein (other than Pcy)		40	40	70	10
mmoles pigment mole^{-1}Chl	Car	570	620	590	2660
	Pcy	490	300	10	20
	C750	20	20	0	0
Excitation transfer Pcy → Chl	(t)	1.41	0.03	0.03	0.08
ACTIVITY :					
moles Chl mole^{-1} P700	L-D	266	174	166	36
	F-H	-	90	105	52
Fluorescence stimulation by DCMU	SII	47%	10%	0%	0%
Electron transfer					
H_2O → PQ (burst)	SII	35	23	0	0
H_2O → CO_2 (photosynthesis)	SII+I	80	10	0	0
H_2O → DCPIP	SII	-	20	0	0
NH_2OH → DCPIP	SII	-	7	4	0
NH_2OH → MV	SII+I	-	15	7	0
$DCPIPH_2$ → MV	SI	-	200	70	700

The number of bands obtained on the gradients is rather important, especially in comparison to what is seen with higher plant chloroplasts. Many of them have however similar properties, and this shows how "crumbly" is the photosynthetic apparatus of these algae. The most noteworthy fractions in addition to the protoplasts A3

- which are an interesting material *per se* since they are very permeable and fully functional organisms - are the heavy B3 and the light B1 fragments. B3 exhibits the highest System II/System I activity and the lowest P700/Chl ratios, whereas B1 only presents a quite high System I functioning and is almost as rich in P700 as the highly purified HP700 fraction of Ogawa and Vernon[7], although we did not use detergents. However,it follows from our results that, contrarly to these authors but in agreement with Susor and Krogmann[3] and Shimony *et al.*[4], the lightest fraction is of System I type and the heaviest of System II, a "normal" situation if the reference taken is the chloroplast fractionation.

During this investigation, a particular point attracted our attention, the so-called P682. Our conclusion concerning its photoinducible form is in agreement with that of Murata[13,14], since we may rule out that it is related to System II. Indeed, the enrichment of the fractions in P700 is associated to a similar enrichment - and not to an impoverishment - in P682 and, because of the modulated detection method used, no fluorescence artifact could contribute to the measured signal. However it is unlikely that this photoinducible P682 is simply a secondary band of P700, the amplitude observed being too great, but it is impossible, at the present time,to give a precise signification to this component. The chemioinducible form is certainly complex : it may comprise the same pigment as that observed in the light-minus-dark spectra but seems mainly due to a non-biologically significant bleaching of the bulk chlorophyll (hence the peak wavelength position, high amplitude and asymmetry,on the one hand, and its very poor correlation with P700, on the other hand).

REFERENCES

1. Thomas, J.B. and De Rover, W., 1955, Biochim. Biophys.Acta 16, 391-395.
2. Susor, W.A. and Krogmann, D.W., 1964, Biochim. Biophys. Acta 88, 11-19.
3. Susor, W.A. and Krogmann, D.W., 1966, Biochim. Biophys. Acta 120, 65-72.
4. Shimony, C., Spencer, J. and Govindjee, 1967, Photosynthetica 1, 113-125.
5. Biggins, J., 1967, Plant Physiol. 42, 1442-1446 and 1447-1456.
6. Ogawa, T., Vernon, L.P. and Mollenhauer, H.H., 1969, Biochim. Biophys. Acta 172, 216-229.
7. Ogawa, T. and Vernon, L.P., 1969, Biochim. Biophys. Acta 180, 334-346.
8. Sigalat, C., 1974, Thesis in Biophysics, University of Paris-Sud.
9. Myers, J. and Kratz, W.A., 1955, J. Gen. Physiol. 39, 11-22.
10. Jones, L.W. and Myers, J., 1965, J. Phycol. 1, 6-13.
11. Volk, S.L. and Bishop, N.I., 1968, Photochem. Photobiol. 8, 213-221.
12. Govindjee, 1963, Naturwissenschaften 50, 720-721.
13. Murata, N. and Takamiya, A, 1969, Plant Cell Physiol. 10, 193-202.
14. Fork, D.C. and Murata, N., 1971, Photochem. Photobiol. 13, 33-44.

M. AVRON, *Proceedings of the Third International Congress on Photosynthesis*,
September 2-6, 1974, Weizmann Institute of Science, Rehovot, Israel
Elsevier Scientific Publishing Company, Amsterdam, The Netherlands, 1974

ENZYMES AND BOUND INTERMEDIATE INVOLVED IN PHOTOSYNTHETIC SULFATE REDUCTION OF SPINACH CHLOROPLASTS AND CHLORELLA

J.D. Schwenn and H. H. Hennies

Bochum, University, Department of Biology

In assimilatory sulfatereduction of intact chloroplasts (class I chloroplasts) sulfate is reduced to the level of sulfide in the light[1]. The reaction sequence proposed in this laboratory[2] includes a transfer of the activated sulfate from PAPS (in a cellfree system from APS[3]) on to an endogenous peptide forming a "bound sulfite". This compound is thought to serve as substrate for a thiosulfonate reductase[4]. The enzyme purified so far from Chlorella[4] requires reduced ferredoxin as electrondonor and yields a bound sulfide as already suggested in[2]. Further evidence of the proposed reaction sequence is put forward by Schiff[5].

An improved chloroplast system revealed new details on the formation of bound sulfite and bound sulfide. The reduction of analogous substrate by a thiosulfonate reductase from spinach will be given in this paper.

1. Results on the formation of "bound intermediates" by osmotically shocked chloroplasts

Walker et al. already proved the utility of ruptured class I chloroplasts[6] for CO_2-fixation hence the formation of bound sulfite and --sulfide has been followed with these chloroplasts using ^{35}S-sulfate as sole substrate. No free sulfite and sulfide are formed by this type of chloroplasts as observed earlier with intact class I chloroplasts. Both bound forms could therefore be measured after isotope exchange with unlabelled carrier as free $^{35}SO_3^{2-}$ and $^{35}S^{2-}$. The separation of all labelled product from ^{35}S - sulfate was performed by chromatography on ionexchange resin according to Iguchi[7].

Figure 1 demonstrates that the formation of bound sulfite and bound sulfide is proportional to the amount of chloroplasts. Because the attachment of these bound forms is assumed to occur on an endogenous thiol compound, therefore the effect of an additional thiol on the formation of bound intermediates was investigated. Figure 2 shows that there is no dependency on additional thiols in the ruptured class I chloroplast system. Only at higher concentrations there is a slight stimulation by the added mercaptoethanol. This finding supports the view of Schiff et al.[8] that various thiols like BAL, DTT and ß-MSH lead to the liberation of free sulfite or even thiosulfate from the bound sulfite and in this way possibly obscure the actual intermediate in assimilatory sulfate reduction of chloroplasts. Thiols were therefore omitted in all other experiments.

Fig. 1 Fig. 2

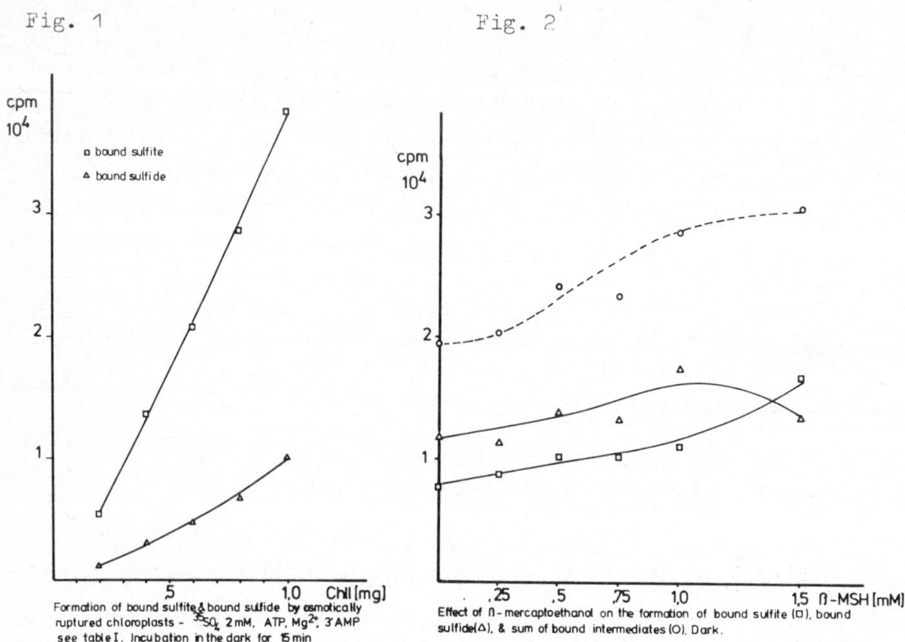

Formation of bound sulfite & bound sulfide by osmotically ruptured chloroplasts - $^{35}SO_4$ 2 mM, ATP, Mg^{2+}, 3'AMP see table I. Incubation in the dark for 15 min

Effect of ß-mercaptoethanol on the formation of bound sulfite (□), bound sulfide (△), & sum of bound intermediates (O). Dark.

The formation of bound sulfite and bound sulfide is dependent on the concentration of ATP as it could be expected from the participation of APS and PAPS as precursors of bound sulfite, figure 3. The optimal concentration of Mg^{2+} ion in the system varies with the amount of chloroplasts and the concentration of ATP used in the assay mixture. Generally 1,5 mM Mg^{2+} was saturating in the presence of 1 mM ATP per 100 ug Chll.

The role of 3'AMP is ambiguous and stimulation scarcely exceeds 10 to 20 % of the control. Anderson et al.[9] discuss its function as preservation of PAPS from degradation by nucleotidase activity associated with chloroplast preparations. The reduction of bound sulfite to --sulfide depends on light (Table 1) and on the concentration of NADP (fig. 4).

Fig. 3 Fig. 4

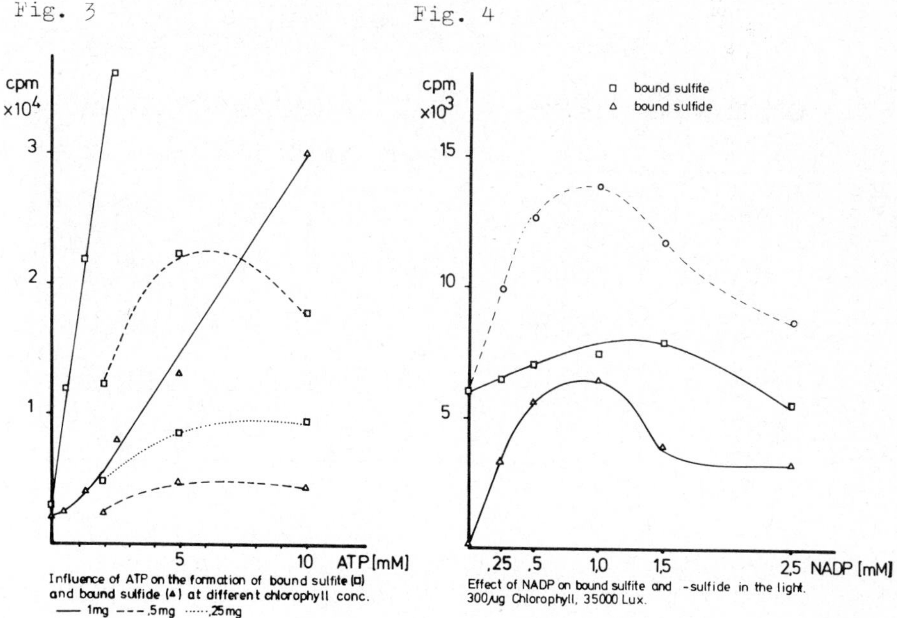

Influence of ATP on the formation of bound sulfite (□) and bound sulfide (▲) at different chlorophyll conc. —— 1mg - - - ,5mg ······ ,25mg

Effect of NADP on bound sulfite and -sulfide in the light. 300 µg Chlorophyll, 35000 Lux.

The occasionally observed reduction in the dark soon ceases and is probably due to remaining reducing power in the freshly prepared chloroplasts. It was found that 1 mM NADP is optimal under saturating light (for chloroplasts with a chlorophyll content of 300 ug). Under these conditions bound sulfide accumulates as product of the phothosynthetic sulfate reduction unless O-acetyl-serine is added. O-acetal-serine which provides the carbon sceleton for cysteine enhances the formation of cysteine whereas the amount of bound sulfide decreases.

Table 1 represents a comparison between light and dark reaction of the assimilatory sulfate reduction of chloroplasts in the presence of O-acetyl-serine. The addition of a protein fraction from chloroplast extract which consists only of peptides of low molecular weight leads to a twofold stimulation of cystein formation.

Table 1

Distributionpattern of S-labelled compounds from chloroplasts

compound	dark	illuminated chloroplasts minus frac. C	illuminated chloroplasts frac. C added
APS	138,5	142,9	166,8
PAPS	34,5	19,7	36,9
$\sim SO_3^-$	75,5	71,5	145,1
$\sim S^-$	19,9	33,5	56,9
cys	÷	15,5	28,2

x 10^3 tpm

50 mM Tris-HCl pH 8.4, 5 mM Mg^{2+}, 7.5 mM ATP, 2 mM $Na_2{}^{35}SO_4$ (specif. activ. 1.12×10^6 tpm/μM), 5 mM O-acetyl-serine, 1 mM NADP fraction C 0.22 mg, 300 μg chlorophyll in a volume of 3.0 ml. Time 20 min, 35000 lux under nitrogen. The low molecular weight fraction "C" from chloroplast extract was designated "C" according to the proposal of Bandurski et al.[11].

The protein fraction possibly contains the endogenous peptide which accepts the activated sulfate from PAPS. There is no stimulation of cysteine formation by this fraction in the dark. In the presence of acetyl-serine the bound sulfide occasionally found in the dark disappears and is completely incorporated into cysteine. No bound sulfide can be detected in this case whereas in the light even in the presence of acetyl-serine bound sulfide still remains in appreciable amounts. The acetyl-serine obviously pulls reduced sulfur out of the pool of bound sulfide, but under illumination this pool is constantly refilled.

2. Results on the enzyme thiosulfonate reductase from spinach leaf extract

Schmidt[4] has already published data on the purification of the enzyme, thiosulfonate reductase from chlorella. In his purification procedure he replaced the natural occuring bound sulfite by dithionite. Supplied with reduced ferredoxin the enzyme catalyses the reduction of S-sulfoglutathion and the reduction of dithionite. The reaction product in the first case is glutathion-persulfide[4] in the latter hydrogensulfide. Both substrates are considered to be analogous substrates to the natural occuring bound sulfite.

The enzyme from spinach has been purified to homogeneity as could be judged from polyacrylamide gelelectrophoresis. Table 2 gives details on the purification procedure employing dithionite as substrate.

Table 2
Purification of spinach thiosulfonate reductase

step	protein mg	rate	units	purification	yield
crude extract	49 400	60	2 964 000	--	100
ammonium-sulfate prec.	8 400	133	1 117 200	2.2	37.7
1. DE 52 column	2 880	203	584 640	3.3	19.7
2. DE 52 column	115	1 075	123 625	17.9	4.2
Sephadex G 200	22	3 910	86 020	65.0	2.9

Rates are expressed as mμmoles hydrogensulfide/mg^{-1} h^{-1}. Liberated one unit is defined as the amount of protein that forms 1 mμmol hydrogensulfide per hour at 37°C.

Assays were run with a ferredoxin reducing system using glucose 6 phosphate to regenerate NADPH and a Fd-NADPH-oxidoreductase from spinach. It is important to note that the highly purified enzyme does not reduce nitrite or sulfite, although both activities are present in the crude extract. The ability to reduce dithionite and S-sulfoglutathion are enriched by the purification procedure equally , until the final step of purification. The treatment with Sephadex G 200 removed the activity to reduce S-sulfoglutathion so that the remaining protein exclusively liberates hydrogensulfide from dithionite.

The two bands found in polyacrylamide gelelectrophoresis could be shown to be multiple forms of the same thiosulfonate-reducatse. Their partial identity was confirmed with antibodies.
The antibody precipitates both bands of the enzyme, but does not inhibit the dithionite reduction. This antibody was originally prepared against particles of the chloroplast thylakoid membrane. These particles (a byproduct of CF_1-preparations) have similar characteristics to the crude soluble sulfitreductase. Preparations of this membrane fractions[12] reduce sulfite, dithionite, and the natural occuring bound sulfite likewise.

Furthermore this enzyme deriving from the membrane fraction catalyses the reduction of bound sulfite to bound sulfide in vitro and the physiological importance of this protein is underlined by the fact that it reduces the same substrate in a reconstituted chloroplast system too. It appears therefore that parts of the assimilatory sulfate reduction in chloroplasts are membrane bound.

ACKNOWLEDGEMENTS

The support of Deutsche Forschungsgemeinschaft is gratefully acknowledged.

SUMMARY

Bound sulfite and bound sulfide are formed by osmotically ruptured class I chloroplasts in the light. A fraction of low molecular weight proteins from the chloroplast extract stimulates the formation of bound sulfite and bound sulfide.

A soluble thiosulfonate reductase has been purified from spinach. It is distinguished from a membrane fraction, which is capable of reducing the natural occuring bound sulfite to bound sulfide.

REFERENCES

1. Schmidt, A. and Trebst, A., 1969, Biochem. Biophys. Acta 180, 529.
2. Schmidt, A. and Schwenn, J.D., 1971, in: Proc. IInd Intern. Congr. Photosynth. Stresa, vol. 1, eds: G. Forti, M. Avron and A. Melandri (Dr. W. Junk, N.V. The Hague) p. 507.
3. Schmidt, A., 1972, Arch. Mikrobiol. 84, 77.
4. Schmidt, A., 1973, Arch. Mikrobiol. 93, 29.
5. Schiff, J.A. and Hodson, R.C., 1973, Ann. Rev. Plant Physiol. 24, 381.
6. Walker, D.A., 1974, in press.
7. Iguchi, A., 1958, Bull. Chem. Soc. Japan, 31, 600.
8. Abrams, W.R. and Schiff, J.A., 1973, Arch. Mikrobiol. 94, 1.
9. Burnell, J.N. and Anderson, J.W., 1973. Biochem. J. 134, 565.
10. Giovanelli, J. and Mudd, S.H., 1967, Biochem. Biophys. Res. Commun. 27, 150.
11. Torii, K. and Bandurski, R.S., 1967, Biochem. Biophys. Acta 136, 286.
12. Vambutas, K. and Racker, E., 1965, J. Biol. Chem. 240, 2660.

M. AVRON, *Proceedings of the Third International Congress on Photosynthesis*, September 2-6, 1974, Weizmann Institute of Science, Rehovot, Israel
Elsevier Scientific Publishing Company, Amsterdam, The Netherlands, 1974

AN ANALYSIS OF THE EFFECT OF ADDED THIOLS ON THE RATE OF OXYGEN UPTAKE DURING MEHLER REACTIONS WHICH INVOLVE SUPEROXIDE PRODUCTION

R. H. Marchant
Physiology Department, The Royal Veterinary College,
London NW1 0TU, England

1. Summary

The effect of thiols on oxygen uptake in a Mehler reaction conducted in the presence of azide and methyl viologen indicates that like ascorbate they react with the superoxide produced. The reduction in P/O measured in the presence of excess thiol indicates that the ratio of oxygen uptake to reducing equivalents transported is increased threefold by this reaction with superoxide. When the ratio of the nett rate of oxygen uptake in the presence of thiol to the corresponding rate in the absence of thiol is measured directly, however, the value obtained with high concns. of thiol tends to differ from three. This difference results either from an inhibition of electron transport or from uncoupling, or both, to an extent which varies according to which particular thiol is in use. These latter effects are minimal with glutathione, which has little effect on the photophosphorylation rate when assessed by uptake of inorganic phosphate.

2. Introduction

Following suggestions by Elstner et al[1] and Epel and Neumann[2] ascorbate has been shown to react with superoxide during a Mehler type reaction using methyl viologen, with the result that in the presence of azide a threefold increase in nett oxygen uptake occurs[3]. A reasonable explanation for this threefold change has been given by Allen and Hall[3]. Dithiothreitol has been shown to similarly increase the rate of oxygen uptake, and this increase, as in the case of ascorbate, is reversed by added superoxide dismutase[4]. Further experiments reported here have shown that this is also true of cysteine, glutathione and mercaptoethanol.

3. Materials and methods

The chloroplasts used were of Type C according to the classification of Hall[5] by the time that they were resuspended in the reaction medium. Examination of the stock chloroplasts under the phase-contrast microscope showed evidence of a moderate percentage of chloroplasts (approx. 40%) with a sickle shape and a refractive "halo" in spite of osmotic shock treatment; this halo disappeared completely, however, on suspension in the reaction medium at 15°C. The preparation of these spinach chloroplasts (from an original "type B" preparation by osmotic shock) and the reaction conditions used to measure oxygen uptake have been referred to elsewhere[4]. Measurements of oxygen uptake were made in the presence of azide and methyl viologen using a Clarke electrode (Yellow Springs Instrument Co., Yellow Springs, Ohio) connected to a pen-recorder. For measurements in the presence of a thiol, the thiol was added at the instant of turning on the actinic light (just after the addition of chloroplasts). ADP (0.01 M ; 30 µl) was added as soon as the state 2 rate had been clearly recorded. All rates and measurements of P/O were corrected for the oxygen uptake due to the autoxidation of the thiol. To allow for the decrease in control rates of oxygen uptake and P/O ratios with time, these were determined at intervals during the course of a day's experimenting values for a specific time could then be assessed graphically. Thiols were added from strong stock solns. which had been adjusted with KOH to pH 8. A separate check was made to confirm that there was no detectable effect on the pH of the reaction mixture as a result of the addition of the highest concn. of thiol used in each experiment. Aliquots of superoxide dismutase were added from a soln. of the crystalline Cu/Zn enzyme obtained from bovine erythrocytes and having an activity of 28,900 Fridovich units/ml. KCN was added from a 2.2 M stock soln. adjusted to pH 8 with HCl.

Photophosphorylation rates were assessed by the rate of uptake of inorganic phosphate and were measured in a reaction mixture as close in composition to that used for the study of rates of oxygen uptake as possible. The concns. of phosphate (2.9 mM instead of 9 mM) and of ADP (2.9 mM) were necessarily changed to make the estimation practicable, and the assay was conducted at a lower (but still saturating) light intensity of approx. 45,000 lux. The uptake of phosphate was assessed after 8 min to increase the precision of measurement (the measurements of oxygen uptake and P/O ratio on the other hand, were normally complete in 2 - 3 min). The inorganic phosphate was estimated in the presence of thiols as described earlier[6].

The relative rates of reaction of the thiols and of ascorbate with superoxide were measured by the effect of these substances on the rate of oxidation of sulphite in a chain reaction dependent on superoxide (generated in situ by the xanthine oxidase reaction). The reaction was assessed by measuring the rate of oxygen uptake using the Clarke electrode as developed by Tyler[7]. This is a method devised by Dr. Tyler[7] for assay of superoxide dismutase and involves determining the concn. of material under study required for 50% inhibition of the oxidation of sulphite (15 mM) in the presence of xanthine oxidase and xanthine (46 μM) and in a 0.05 M phosphate buffer medium containing 0.1 mM EDTA at pH 7.4 and 25°C. It was shown in separate experiments that the effect of ascorbate and of the thiols individually on the xanthine oxidase reaction was negligible at the concns. used.

4. Results

Table 1 shows the relative concns. of ascorbate and of the four thiols which were required for 50% inhibition of the sulphite oxidation rate (measured as described under materials and methods).

Table 1

Relative rates of reaction of ascorbate and thiols with superoxide measured by a sulphite oxidation assay system

Substance	Concn. required for 50% inhibition of the sulphite oxidation rate μM ± s.e.	Relative rate of reaction with superoxide (ascorbate = 100)
Ascorbate	2.1 ± 0.1 (4)	100
Cysteine	8.9 ± 0.8 (2)	24
Dithiothreitol	64 ± 6 (2)	3.3
Glutathione	63 ± 6 (2)	3.3
Mercaptoethanol	270 ± 24 (2)	0.8

This demonstrates that all of these compounds probably react with superoxide and by changing its steady-state concn. reduce the catalytically maintained sulphite oxidn. rate. Assuming that the effects are not due to inhibition of the sulphite oxidn. sequence for any other reason than a reaction with the catalytic superoxide radical, all the thiols would seem to react with superoxide and at an appreciably slower rate than ascorbate. This observation is in agreement with results already published on the Mehler reaction[4] where much higher concns. of dithiothreitol than of ascorbate were required to achieve a maximal reduction in P/O ratio. Of the thiols, cysteine was the most reactive and mercaptoethanol by far the least.

The effects of the thiols on uptake of phosphate during photophosphorylation are shown in Table 2. Glutathione had little effect on the rate of phosphate uptake even at a concn. of 22 mM On the other hand mercaptoethanol had an appreciable effect at this order of concn. (17 mM).

Table 2

The effect of thiols on the rate of uptake of phosphate during photophosphorylation

Thiol added	Concn. (mM)	Phosphate uptake (μ moles/mg chlorophyll/hr		Inhibition %
		No addition	Plus thiol	
Cysteine	16	192 ± 6 (5)	112 ± 14 (5)	42
Glutathione	10	147 ± 5 (4)	145 ± 2 (4)	1*
Mercaptoethanol	17	147 ± 5 (4)	98 ± 6 (2)	33
Dithiothreitol	16	246 ± 3 (5)	163 ± 7 (5)	34
Mercaptoethanol	35	176 ± 6 (5)	72 ± 8 (5)	60
Glutathione	22	140 ± 8 (5)	130 ± 6 (5)	8*

* Not significant ($P > 0.2$)

A number of experiments were conducted to determine the rates of oxygen uptake and P/O ratio in the Mehler type reaction with methyl viologen, carried out as described under materials and methods, with various concns. of each of the thiols added to the reaction mixture. Various parameters measurable from these recordings have been compared to their corresponding values obtainable without the thiol added (control). This comparison with the control has been made by expressing the values in the presence of thiol as a fraction or multiple of the control value. The following parameters were dealt with in this way: rates of oxygen uptake in states 2 and 3, P/O ratio and photosyn thetic control ratio. Selected curves showing the relative change in the rate of state 3 oxygen uptake and in the P/O ratio are shown in Figs. 1 and 2 respectively.

Only the effects of those thiols significant to the discussion have been shown in Figs. 1 and 2 to simplify the figures.

Finally Table 3 shows the reversal of the increase in oxygen uptake which occurred as a result of thiol addition during state 2, by the addition of superoxide dismutase. In the last column is shown the restoration of the stimulated oxygen uptake by KCN. High concns. of KCN are required at pH 8 to inhibit superoxide dismutase, but in each case this concn. had a negligible effect on the rates in the absence of S. O. D. (not shown).

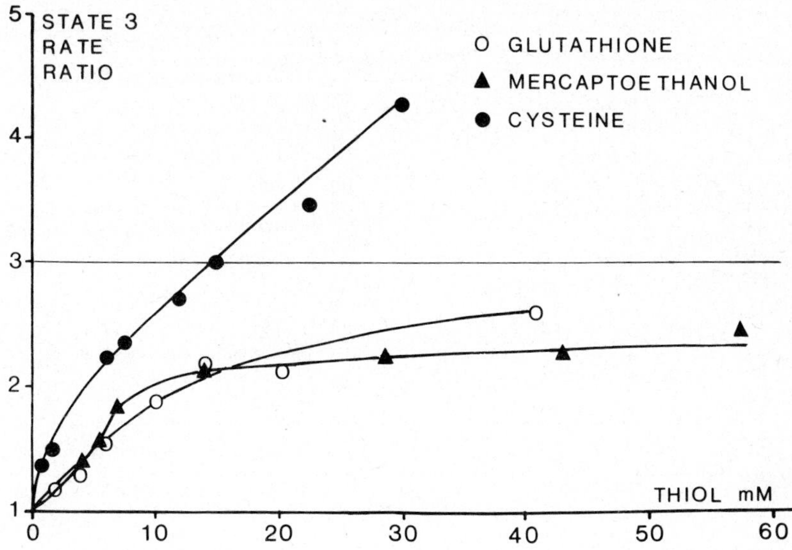

Fig. 1. Changes in the ratio of the state 3 rate of oxygen uptake to that of controls as the concn. of thiol is increased.

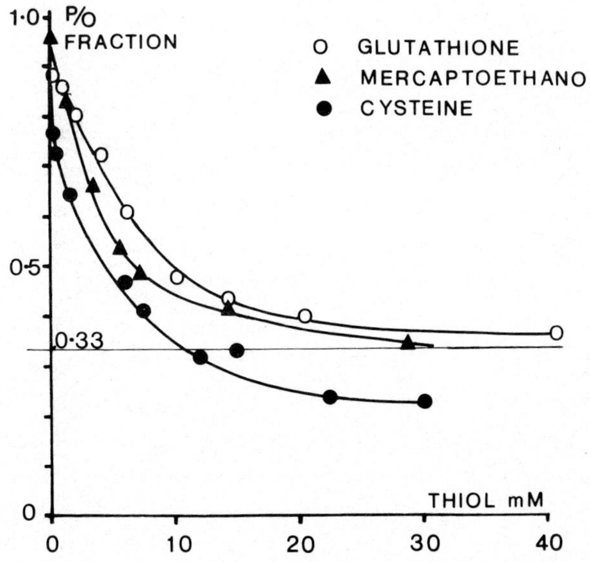

Fig. 2. Changes in P/O ratio (fraction of the control) as the concn. of thiol is increased.

Table 3

The reversal of thiol stimulation of nett oxygen uptake in the Mehler reaction by superoxide dismutase and its restoration by KCN

Thiol (Concn. in parenthesis - mM)	Rates of oxygen uptake in μ moles/mg chlorophyll/hr			
	State 2 (thiol abs.)	Plus thiol	Plus thiol plus S.O.D. (Fridovich U./ml in parenthesis)	Plus thiol, plus S.O.D. plus KCN (Concn.KCN in parenthesis-mM)
ME (7.5)	28	48	33 (1.7)	52 (17)
Cyst. (7.5)	31	75	29 (1.7)	68 (17)
Glut. (5.1)	25	43	29 (2.5)	47 (33)
DTT (20)	19	54	23 (1.7)	47 (33)

ME = mercaptoethanol; DTT = dithiothreitol; S.O.D. = superoxide dismutase

5. Discussion

All the thiols studied showed evidence of interaction with superoxide both separately from the chloroplast (Table 1) and during the Mehler reaction (Table 3).

Glutathione had a negligible effect on photophosphorylation rate (Table 2). This makes it probable that the changes in oxygen uptake and P/O ratio illustrated in Figs. 1 and 2 respectively, are due almost entirely to interaction with superoxide, in competition with the dismutation reaction. As in the case of ascorbate the two parameters concerned changed by a factor of three. In the case of cysteine it can be seen that the rate of oxygen uptake was stimulated as much as 4.5 fold at 30 mM cysteine and the P/O ratio reduced to a correspondingly greater degree to less than one third of the control. Thus in this case, the effects of interaction with superoxide are superimposed on an uncoupling effect. In the case of mercaptoethanol, although a considerable reduction in the rate of photophosphorylation occurred at 35 mM (Table 2), the P/O ratio was only reduced to about a third (Fig. 2). On the other hand the rate of oxygen uptake in state 3 at this concn. of mercaptoethanol was increased to a value appreciably less than three times the control (Fig. 1). This suggests that in the case of mercaptoethanol an inhibition of electron transport occurs as the primary complicating event rather than an uncoupling effect. The same type of effect was shown with dithiothreitol

In conclusion experiments with 0.8 M Tris-extracted chloroplasts and heat-treated chloroplasts have been attempted using methyl viologen as the electron acceptor in order to test whether the thiols and ascorbate are able to donate to system II in a partial reaction blocked by DCMU, as classically believed. To-date a clear demonstration of this partial reaction has not been achieved in this laboratory because of small residual activities in these systems in the absence of added "electron" donor. On adding ascorbate or one of the thiols, the oxygen uptake due to this residual activity was immediately stimulated threefold as a result of the superoxide interactions described. Whenever stimulation was greater than by a factor of three, it was then incompletely blocked by DCMU. Thus evidence for donation of reducing equivalents to system II by ascorbate and by the thiols when using electron acceptors like methyl viologen may need re-examining.

Thanks are due to the invaluable technical assistance given by Andrew Critchell. The University of London Central Research Fund supplied apparatus for thermostatic control.

References

1. Elstner, E.F., Heupel, A. and Vaklinova, S., 1970, Z. Pflanzenphysiol., 62, 184.
2. Epel, B.L. and Neumann, J., 1972, Abstracts VI International Congress on Photobiology, Bochum, 21 - 25th August, 1972, p. 237.
3. Allen, J.F. and Hall, D.O., 1973, Biochem. Biophys. Res. Communs., 52, 856.
4. Marchant, R.H., 1974, Biochem. Soc. Trans., 2(3), 532.
5. Hall, D.O., 1972, Nature New Biol., 235, 125.
6. Marchant, R.H., 1971, Arch. Biochem. Biophys., 147, 502.
7. Tyler, D.D., 1973, personal communication; being submitted for publication.

M. AVRON, *Proceedings of the Third International Congress on Photosynthesis*, September 2-6, 1974, Weizmann Institute of Science, Rehovot, Israel
Elsevier Scientific Publishing Company, Amsterdam, The Netherlands, 1974

THE CONTROL OF ELECTRON TRANSPORT BY TWO pH-SENSITIVE SITES

U. Siggel

Max-Volmer-Institut,
Technische Universität Berlin

1. Summary

The uncoupled electron transport rate is quantitatively described as being controlled by two reactions sensitive to the internal proton concentration which are associated with the oxidation of plastohydroquinone and the oxygen evolution respectively.

2. Introduction

Several years ago we brought out the theory that the rate of electron transport is controlled by the internal protons[1,2]. We concluded this from measurements of the electron transport rate and the reduction of oxidized chlorophyll-a_I. The effect on the electron transfer rate of some seconds of illumination with a probe without uncoupler turned out to be the same than that of lowering the pH of the suspending medium in an experiment with uncoupler. Concerning the correspondence between steady state electron transport and reciprocal half time of chl-a_I^+-reduction our notion was, that the oxidation of plastohydroquinone represents the main site of the electron transport chain sensitive to the internal pH. We used our curves for calibrating the internal pH.
The influence of the internal protons was later confirmed by Avron's group[3-6]. Additionally new ideas were introduced: the controlling effect of the pH-difference across the membrane or as an alternative the pH-value within the membrane as essential factor. The experiments also seemed to favour system II as responsible for the pH-effect. Recent results of other groups on this subject[7,8] are controversial. We now would like to give some more detailed information about our concept of rate control.

3. Materials and methods

Class II spinach chloroplasts were used prepared according to the method of Winget et al.[9]. Electron transport was monitored optically by observing the ferricyanide reduction at 420 nm. The absorption change of chlorophyll-a_I was measured at 703 nm with the single flash technique, that of plastoquinone at 265 nm with the repetitive flash technique.

4. Results and discussion

Qualitative results

1. Influence of the external pH.

In our view the totally uncoupled electron transport rate is determined by the internal protons only. This is true at least for $pH_i \leq 7$. The corresponding function is shown in fig. 1. For completeness of uncoupling see ref. 10. The decay of the rate for $pH_i > 7$ which normally occurs, but has been neglected for simplicity in our former publications, cannot undoubtedly be ascribed to the internal pH from our measurements.

In the absence of an uncoupler the internal proton concentration is raised by the action of electron transport, pumping protons into the thylakoids. The extent of the rise depends on the proton permeability of the membrane, which most probably is a function of the external pH. This may be concluded from the proportionality of the basal electron transport rate to the internal proton concentration[10], which seems to indicate the diffusion of protons out of the thylakoid. A simple evaluation of the proton permeability has already been given in ref. 11. More recent measurements in which care was taken to eliminate a diffusion potential by addition of valinomycin, yielded the dependence of the proton permeability on the external pH[12]. It is essentially responsible for the shape of the pH-function of the basal electron transport (fig. 1). Its rate is controlled by the internal protons, the stationary proton concentration depending on the external pH. So the influence of the internal protons is a direct, that of the external protons an indirect one.

With this concept we can directly read out the internal pH and the transmembrane pH-difference from a plot of basal and uncoupled electron transport rates as a function of the external pH (fig. 1). The result is similar to that of Avron's group[3-6], obtained with completely different methods.

2. Number of sites sensitive to the internal pH.

In view of the measured H/e-ratio of 2 [2,12] and the reasonable assumption that the controlling sites are identical with the proton pumps two pH-sensitive sites are to be expected. Protons are released into the inner phase of the thylakoids by the oxidation of plastohydroquinone and the water splitting reaction[13].

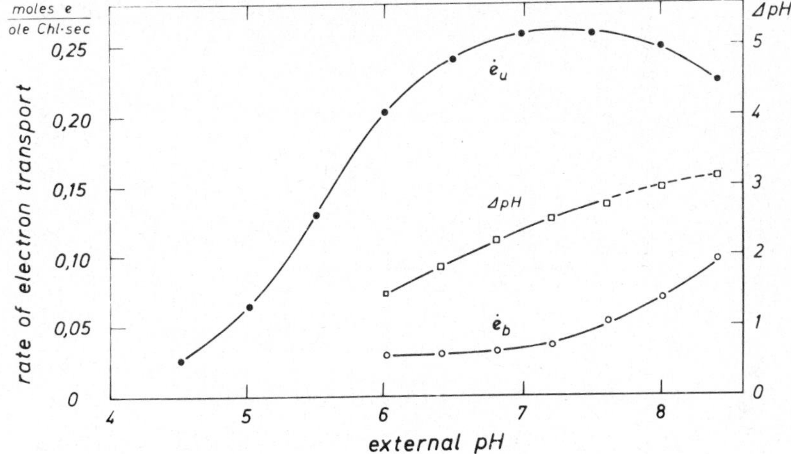

Fig. 1. Basal $\dot{e}_b$ and uncoupled $\dot{e}_u$ (with gramicidin) electron transport rates as a function of the external pH. In the presence of uncoupler internal and external pH are assumed to be equal. The values of internal pH corresponding to the basal curve are obtained by drawing horizontal lines to the uncoupled curve. The calculated pH-difference is plotted. The dashed part of the curve will only be correct if the decrease of $\dot{e}_u$ in this region will be due to the internal pH. Reaction mixture: 10 mM Tricine, 50 mM KCl, 2 mM $MgCl_2$, 0,1 mM ferricyanide, 0,01 mM chlorophyll, 0,005 mM gramicidin for $\dot{e}_u$.

Site I

The experimental result is that the electron transport rate and the reciprocal half time of chl-a_I^+-reduction are proportional for a number of values of the external pH in the presence of gramicidin[1]. This seems to indicate that the oxidation of plastohydroquinone PQH_2 is responsible for the pH-dependence. The published experiments with an illumination time of 0.2 sec for the chl-a_I-measurements are not yet conclusive. In the steady state which is attained after 0.2 sec approximately under uncoupled conditions all the transfer times of the electron transport chain should be equal. The mentioned proportionality reflects the existence of a pool for electrons. After half reduction of chl-a_I^+ the oxidation rate of PQH_2 is still nearly the same than in the steady state.
For the detection of the pH-sensitive site short flash experiments are appropiate. The pH-dependence of the chl-a_I^+-reduction is in fact very similar after a short and a long flash (fig. 2).

This proves that really the oxidation of PQH_2 is dependent on the internal pH.

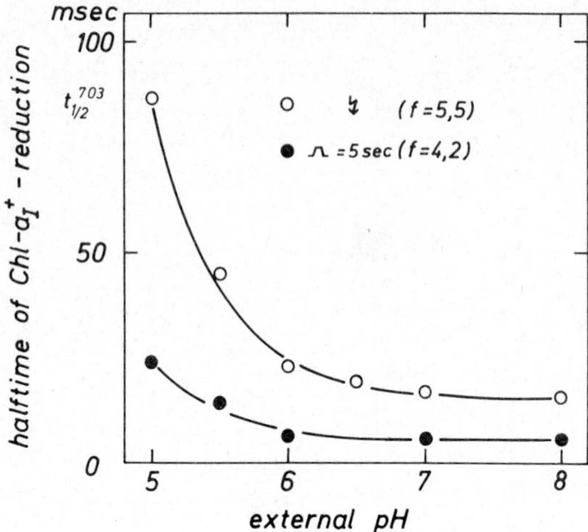

Fig. 2. Increase of the half time of the reduction of oxidized chlorophyll-a_I with decreasing external pH in the presence of gramicidin, interpreted as effect of the internal pH. The change is less pronounced after a long flash, indicated by a smaller enhancement factor f.

There is a twofold difference between short and long flash results. Firstly the minimal half time (at pH 8) is larger after a short flash. This is due to the different degree of PQ-reduction which is less after a short flash than in the steady state. Secondly the variation of half time is more pronounced in the short flash experiment. The reaction being approximately first order, the half time of chl-a_I^+-reduction is nearly independent of the degree of PQ-reduction. This does not hold for the long flash which will be seen from the quantitative treatment. The degree of stationary PQ-reduction is dependent on the efflux and influx of electrons. It will only be constant if both reactions are equally sensitive to the internal pH. So the long flash curve is a result of simultaneous variation in the PQH_2-concentration and its pH-dependent rate constant of oxidation. With higher internal proton concentration the oxidation of PQH_2 is slower. The amount of stationary electrons in the pool will be larger (relative to lower $[H_i^+]$) if the influx of electrons has no or a less effective pH-dependence than the efflux.

It might be added that the difference between the short and long flash curve cannot be a consequence of incomplete uncoupling. In that case the half time after a short flash should be less sensitive.

Site II

The discussion of chl-a_I^+-reduction has shown that it will not be possible to deduce a second pH-sensitive site from the evaluation of kinetic data. We have to look for the electron supplying reaction itself. This is the water splitting reaction the pH-dependence of which is well known. The oxygen yield of a short flash has to be measured (see e.g. ref. 14). Alternatively the amplitude of the positive absorption change of chlorophyll-a_I after a short flash may be observed. Following a short flash the reduction of oxidized chl-a_I very nearly reflects the number of electrons supplied by light reaction II if centres I and II are equal in number or centres I abundant. The decrease of the amplitude, that is of active centres II at low pH (fig. 3) is interpreted as the effect of the internal pH, although this cannot be proven with short flash experiments. The constant amplitude after a long flash shows that the reduction of chl-a_I^+ itself is independent of pH in the region observed.
The diminished amplitude at pH 8 indicates the existence of still another pH-sensitive site associated with the evolution of oxygen. It is probably related to the decrease of the uncoupled electron transport rate at the same pH. This effect shall however not be analysed further here.

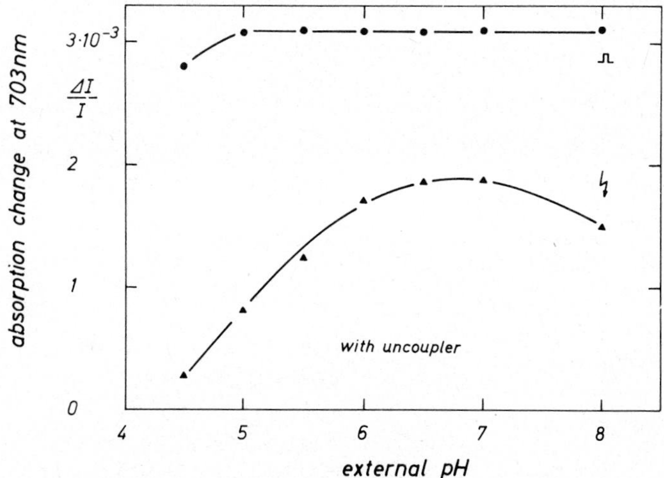

Fig. 3. Dependence of the amplitude of the positive absorption change of chlorophyll-a_I on the external pH in the presence of gramicidin for long and short flash excitation.

Quantitative results

It would be a nice support for our theory if the uncoupled rate of electron transport could be calculated taking into account the combined action of the two reactions shown to depend on the internal pH. We have to look for a mathematical formulation of the measured pH-dependencies as well as of the PQ-kinetics.

1. Oxidation of plastohydroquinone.

We have chosen this reaction because it is easier to be measured than the reduction of PQ. We measured the stationary amount of reduced PQ and the initial velocity of dark oxidation after a long flash, the degree of reduction being varied by DCMU. We were able to confirm the kinetic equation originally derived by Stiehl & Witt[15]. The rate of oxidation of plastohydroquinone is proportional to the product of concentrations of quinone and hydroquinone, the interpretation of which will be given in a forthcoming paper.

$$\left(\frac{dPQH_2}{dt}\right)^{ox} = -k_2 PQH_2 \cdot (PQ_o - PQH_2)$$

This equation in which PQ_o represents the total amount of reduced and oxidized plastoquinone, is visualized by a suitable plot of measured data (fig. 4). Additionally the stationary amount of plastohydroquinone has to be known. The formula of ref. 15 may be used:

$$PQH_2^{ss} = PQ_o \cdot \frac{k_1}{k_1 + k_2}$$

The ratio of the rate constants of reduction and oxidation k_1 and k_2 respectively is best deduced from electron transport measurements.

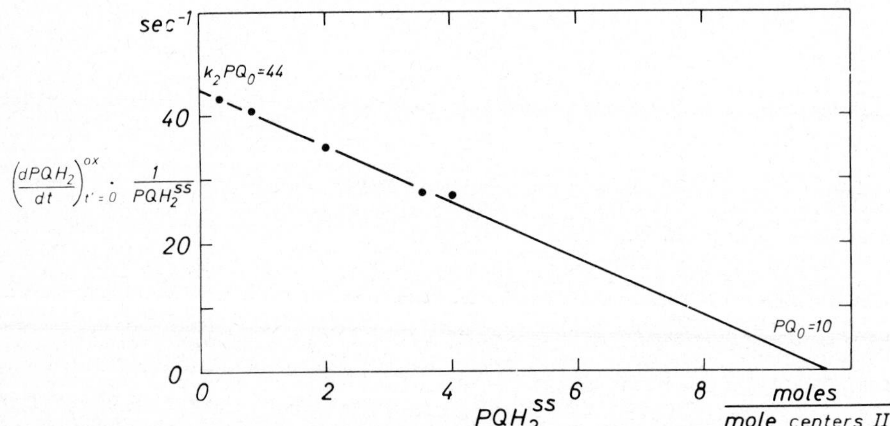

Fig. 4. Initial rate of oxidation of plastohydroquinone as dependent on its amount which was varied by different DCMU-concentrations.

2. Constant of PQH_2-oxidation (site I).

If the half time of chl-a_I^+-reduction is plotted as a function of the proton concentration a straight line is obtained (fig. 5). This means:

$$t_{1/2} \sim 1 + \frac{[H_i^+]}{K_A} \qquad \frac{1}{t_{1/2}} \sim \frac{A^-}{A_o} = \frac{K_A}{K_A + [H_i^+]}$$

The reciprocal half time seems to be proportional to the deprotonated form of a substance A with overall concentration A_o, characterized by the dissociation constant K_A. The identity of A is not clear, but not essential for our purpose. It might be the proton pump molecule delivering the protons into the inner phase. This formulation can at most be regarded as exact for the short flash experiment. For the long flash result it is then an approximation with an effective dissociation constant K_A'.

The pH-dependence of the constant of PQH_2-oxidation is proposed to be given by:

$$k_2 = k_2^o \cdot \frac{A^-}{A_o}$$

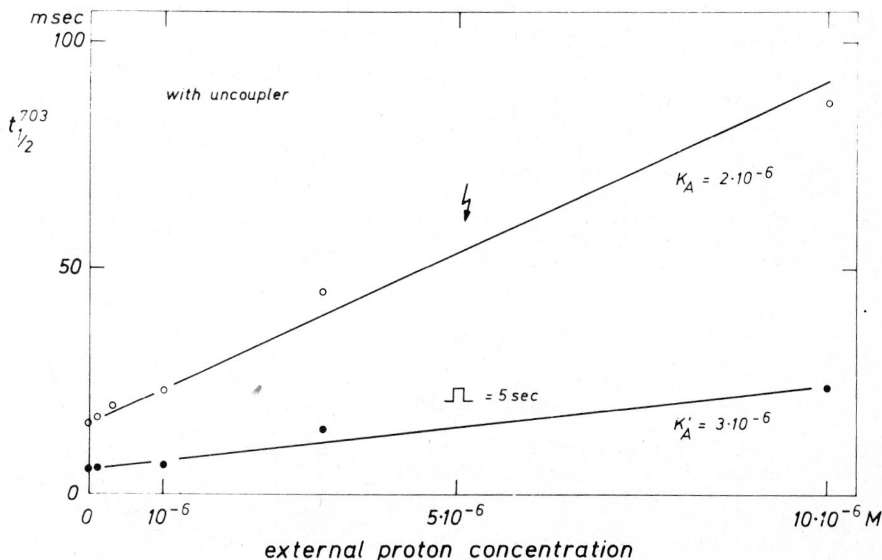

Fig. 5. Linear relationship between the halftime of chl-a_I^+-reduction and the proton concentration (interpreted as internal).

3. Constant of PQ-reduction k_1 (site II).

The evaluation of the amplitude of the chl-a_I-absorption change after a short flash is more complicated and therefore less sure.

The square root of the reciprocal amplitude ΔA seems to be a linear function of the proton concentration (fig. 6). This means:

$$\Delta A \sim \left(\frac{K_B}{K_B + [H_i^+]}\right)^2 = \left(\frac{B^-}{B_o}\right)^2$$

Accordingly the formula for the rate constant k_1 is proposed to be:

$$k_1 = k_1^o \cdot \left(\frac{B^-}{B_o}\right)^2$$

The amount of delivered electrons and the constant of PQ-reduction are proportional to the square of the deprotonated form of a substance B, characterized by the dissociation constant K_B. Perhaps a different formula might fit to the measured values equally well. The formulation presented was chosen because of the analogy to that of site I, with the square indicating a cooperation of two electron transport chains, as was already found by DCMU-experiments[16].

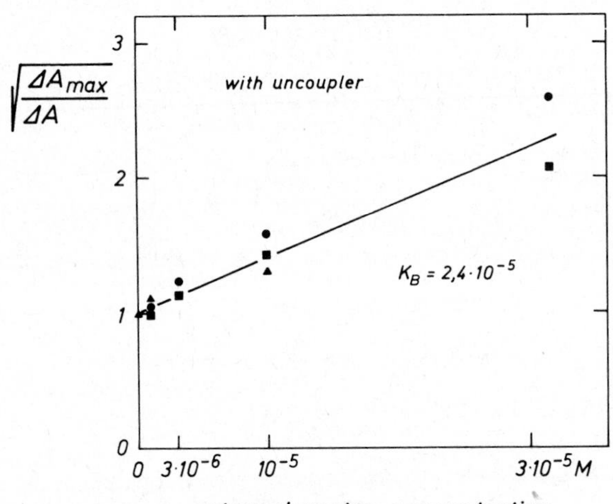

Fig. 6. Suitable evaluation of the chl-a_I-amplitude after a short flash as a function of proton concentration (interpreted as internal).

4. Overall electron transport.

Combination of the equations given above results in the final formula for uncoupled electron transport rate.

$$\dot{e}_u = 2k_1^o \cdot PQ_o^2 \cdot \left(\frac{B^-}{B_o}\right)^2 \cdot \left(\frac{k_2^o \cdot \frac{A^-}{A_o}}{k_1^o \cdot \left(\frac{B^-}{B_o}\right)^2 + k_2^o \cdot \frac{A^-}{A_o}}\right)^2$$

This function is calculated using the empirical rate and dissociation constants. The result is plotted in fig. 7 (curve 2) and fits to the measured rates for pH≤7. The decay at pH 8 is not included in the calculation.

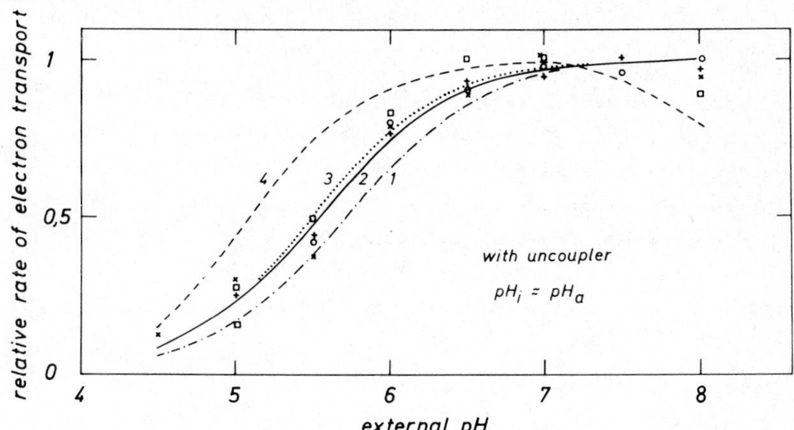

Fig. 7. Comparison between calculated and measured electron transport rates using the complete formula (curve 2) or the approximate formula (curve 3) with k_1^o/k_2^o=0.3. The pH-dependence of k_1 (water splitting) is given by curve 4 ($K_B=2\cdot 10^{-5}$) and that of k_2 (PQH_2-oxidation) by curve 1 ($K_A=2\cdot 10^{-6}$).

So our concept of rate control is consistent with our measurements. It may be summarized as follows. The uncoupled electron transport rate may be described as a function of the internal proton concentration only. The main pH-sensitive reaction is the oxidation of plastohydroquinone characterized by a dissociation constant of $K_A=2\cdot 10^{-6}$ M. The reduction of plastoquinone shows a less pronounced pH-dependence. The overall electron transport rate displays an intermediate sensitivity against lowering the internal pH. It can also approximately be described by the simple function

$$\dot{e}_u \sim \frac{K_A'}{K_A' + [H_i^+]} ,$$

where $K_A'=3\cdot 10^{-6}$ M is an empirical constant.

5. References

1. Rumberg,B. and Siggel,U.,1969, Naturwiss. 56,130
2. Rumberg,B., Reinwald,E., Schröder,H. and Siggel,U.,1969, in Progr. Photosynthesis Res. Vol. III,1374
3. Rottenberg,H.,Grunwald,T. and Avron,M.,1972,Eur.J.Biochem.25,54

4. Schuldiner,S.,Rottenberg,H. and Avron,M.,1972,Eur.J.Biochem.25,64
5. Rottenberg,H. and Grunwald,T.,1972, Eur. J. Biochem. 25,71
6. Bamberger,E.,Rottenberg,H. and Avron,M.,1973,Eur. J.Biochem.34,557
7. Gould,J.M. and Izawa,S.,1974, Biochim. Biophys. Acta 333,509
8. Harth,E.,Reimer,S. and Trebst,A.,1974, FEBS Lett. 42,165
9. Winget,G.D.,Izawa,S. and Good,N.E.,1965,Biochem.Biophys. Res. Commun. 21,433
10. Schröder,H.,Siggel,U. and Rumberg,B., this symposium
11. Reinwald,E.,Siggel,U. and Rumberg,B.,1968,Naturwiss. 55,219
12. Muhle,H., 1973, Thesis, TU Berlin
13. Junge,W. and Ausländer,W.,1973,Biochim.Biophys.Acta 333,59
14. Renger,G., 1969, THesis, TU Berlin
15. Stiehl,H.H. and Witt,H.T., Z. Naturf. 24b,1588 (1969)
16. Siggel,U.,Renger,G.,Stiehl,H.H. and Rumberg,B.,1972,Biochim. Biophys. Acta 256,328

Acknowledgements

We thank Mrs. Bathe for drawing the graphs.

M. AVRON, *Proceedings of the Third International Congress on Photosynthesis*, September 2-6, 1974, Weizmann Institute of Science, Rehovot, Israel
Elsevier Scientific Publishing Company, Amsterdam, The Netherlands, 1974

CONTROL OF THE PHOTOSYNTHETIC ELECTRON TRANSPORT BY FREE FATTY ACIDS AND Mn^{2+} SALTS

P.A. Siegenthaler and J. Horakova
Laboratoire de Physiologie végétale et biochimie
Université de Neuchâtel, 18 rue de Chantemerle
2000 Neuchâtel, Switzerland

1. Summary

C_{18}-unsaturated fatty acids seem to be mainly inhibitors of photosystem II (PS-II) electron transport in whole chloroplasts. The addition of increasing concentrations of linolenic acid caused an inhibition of DCPIP photoreduction and O_2 evolution. The activities were restored by Mn^{2+} ions (optimum, 5 mM) but not by 1,5-diphenylcarbazide (DPC) and Mg^{2+}. Tris-washed chloroplasts lost the photoreduction activity which was restored by DPC but not by Mn^{2+} and Mg^{2+} ions. The DPC-supported activity was inhibited by fatty acids. All DPC and Mn^{2+}-supported activities were inhibited by DCMU.

It is suggested that in the PS-II electron flow: $H_2O \rightarrow Y_1 \rightarrow Y_2 \rightarrow Y_3 \rightarrow$ PS-II $\rightarrow$ DCPIP, C_{18}-unsaturated fatty acids and Tris-washing inhibit the electron flow on the oxidizing side of PS-II, i.e. between Y_2 and Y_3 and between H_2O and Y_1, respectively. DPC feeds electrons into the chain through the Y_2-carrier and Mn^{2+} ions function as a catalyst that might establish a shunt between Y_1 and Y_3, bypassing Y_2 and the fatty acid block.

2. Introduction

Fatty acids, especially C_{18}-unsaturated acids, represent a substantial portion of the thylakoid lipids[1] and play an important role in the maintenance of the structural and functional integrity of the chloroplast membrane[2,3]. Interesting enough, free fatty acids, which accumulate in the membrane due to lipid hydrolysis under special conditions such as aging *in vitro*[4-7], seem to influence the structure and function of chloroplasts in a specific way.

A number of reactions and factors which may control the structure and function of chloroplasts in the course of aging have been summarized (see ref. 7, Fig. 5). The action of hydrolytic enzymes, namely galactolipases, alters the identity of the lipoprotein complexes of the thylakoid which, in turn, modify the physico-chemical properties of the membrane, namely the osmotic properties. These changes cause a swelling of the thylakoid membrane[8]. Also, free fatty acids themselves interact with the tertiary structure of protein, inducing conformational changes leading to chloroplast swelling[7].

Fatty acids released during plastid isolation or in the course of aging *in vitro*, or exogenous acids added to fresh chloroplasts, were also shown to inhibit electron flow[4-6,9-16] and energy linked reactions[4,7,14,17-18]. However, in spite of much evidence that fatty acids at appropriate concentrations are specific inhibitors of the photosystem II electron transport [9,14,15], their precise inhibition site is still unknown. In order to localize this site, several specific inhibitors of photosystem II (DCMU* and Tris-washing) and artificial electron donors (DPC and Mn^{2+} ions) were tested in various combinations, with and without fatty acids, for their action on the photoreduction of DCPIP and O_2 evolution.

The results indicate that the inhibition site of unsaturated fatty acids lies on the oxidizing side of photosystem II, probably between the hypothetical carriers Y_2 and Y_3 (see Fig. 4). In addition, we have found that Mn^{2+} ions interact with some unknown electron carriers of the electron transport chain located between H_2O and the pigment complex II, allowing a shunt between Y_1 and Y_3, bypassing thereby the fatty acid block (see Fig. 4). This is further proof for Mn^{2+} to be required in the photosystem II-mediated reactions[19] and we believe such a function to be postulated for the first time[15].

3. Materials and Methods

Chloroplasts were isolated from spinach (*Spinacia oleracea* L.) in 100 mM Tris-HCl (pH 8) and 175 mM NaCl, as described elsewhere[8]. The resulting stock suspension was adjusted to give 2 mg chlorophyll/ml in the same medium and kept at 0-4°C prior to use. This preparation, referred to as Tris-NaCl chloroplasts, contained mainly whole organelles (class II) and a few intact chloroplasts (class I). Chloroplasts were also prepared according to Kalberer *et al.*[20]: intact chloroplasts were first obtained in a solution containing 25 mM HEPES (pH 7.6) and 0.35 M sucrose and then shocked osmotically in the same medium diluted to 1:10. The resulting pellet containing whole chloroplasts was resuspended in the initial medium and adjusted to give 2 mg chlorophyll/ml. This preparation is referred to as HEPES-sucrose chloroplasts. Chlorophyll was measured by the method of Bruinsma[21] and

---

* Abbreviations: DAD, diaminodurene; DCMU, 3-(3,4-dichlorophenyl)-1,1-dimethylurea; DCPIP, 2,6-dichlorophenylindophenol; DPC, 1,5-diphenylcarbazide; HEPES, N-2-hydroxyethylpiperazine-N'-2-ethanesulfonic acid; PS, Photosystem; TMPD, N,N,N',N'-tetramethyl-p-phenylenediamine; Tricine, N-tris(hydroxymethyl) methylglycine; Tris, Tris(hydroxymethyl)aminomethane.

Tris-washed chloroplasts were obtained according to Yamashita and Butler[22].

The electron transport for photosystem II (from H_2O to DCPIP) was measured spectrophotometrically at 590 nm by the photoreduction of DCPIP as the oxidant, in the following reaction mixture: 50 mM Tris-maleate (pH 7), 35 mM NaCl, 0.15 mM DCPIP, 0.5% ethanol and chloroplasts (20 μg chlorophyll/ml). DCMU (10 μM), DPC, Mn^{2+}, Mg^{2+} ions and fatty acids at various concentrations were added to the basic reaction mixture, as indicated. The reaction was followed over a 1 min period at 20°C in the presence of white light of approximately 5.10^5 $ergs.cm^{-2}.sec^{-1}$ intensity.

The electron transport for photosystems II and I (from H_2O to $NADP^+$) was observed spectrophotometrically in a 0.2 mm cuvette by recording the absorbance changes of $NADP^+$/NADPH at 340 nm, in a Zeiss spectrophotometer modified as described by McSwain and Arnon[23]. The actinic light was supplied by a 1000 W iodine lamp passed through interference filters (Balzers Calflex and Balzers wide band DT-red). The measured intensity at the cuvette was 1.15 x 10^4 $ergs.cm^{-2}.sec^{-1}$. The reaction mixture contained 50 mM Tricine (pH 8.4), 35 mM NaCl, 2 mM $NADP^+$, ferredoxin in excess, chloroplasts (60 μg chlorophyll/ml) and, where indicated, linolenic acid. Ferredoxin was prepared from spinach leaves according to Buchanan and Arnon[24].

The electron transport for photosystem I (from reduced DAD, TMPD or DCPIP to $NADP^+$) was determined under the same conditions and in the same reaction mixture, except that it was supplemented with 4 μM DCMU, 4 mM ascorbate and any one of the three electron donors DAD, TMPD or DCPIP (30 μM).

Oxygen evolution was measured with a Clark-type oxygen electrode at 20°C in the reaction mixtures described in the legend of Fig. 3. The reactions were carried out over a 1 min period at a light intensity of approximately 5×10^5 $ergs.cm^{-2}.sec^{-1}$.

DPC was dissolved in methanol, while fatty acids, DCMU and DAD were dissolved in ethanol.

4. Results

The action of linolenic acid has been compared in different types of electron transport (Table I). It can be seen that the $H_2O/NADP^+$ system (i.e. photosystems II and I) was much more sensitive to the inhibitory action of linolenic acid than photosystem I measured with DAD, TMPD or DCPIP as electron donors and $NADP^+$ as the acceptor. This result prompted a more detailed study of the effect of C_{18}-unsaturated acids on photosystem II activities.

Fig. 1 shows the inhibitory effect of linolenic acid on the photoreduction of DCPIP and the ability of DPC to restore this activity. DPC has been

Table I

Comparison of the effect of linolenic acid on 4 different types of photosynthetic electron transport

Type of electron transport	Photoreduction of $NADP^+$		
	Control (μmoles NADPH /mg chl/h)	% of control	
		+ 100 μM $C_{18:3}$	+ 200 μM $C_{18:3}$
$H_2O \longrightarrow NADP^+$ (PS-II and PS-I)	74	40	0
DAD $\longrightarrow NADP^+$ (PS-I)	69	79	63
TMPD $\longrightarrow NADP^+$ (PS-I)	98	83	85
DCPIP $\longrightarrow NADP^+$ (PS-I)	32	148	215

shown to be an artificial electron donor to the oxidizing side of photosystem II[25]. The addition of increasing concentration of linolenic acid caused an inhibition of DCPIP photoreduction (Fig. 1 A). The activity was not restored by DPC; various concentrations of DPC in the presence of 100, 200 and 300 μM of linolenic acid did not change significantly the results (Fig. 1 B). These first observations suggested that the inhibition block caused by linolenic acid in the electron transport could be located after the entry point of DPC.

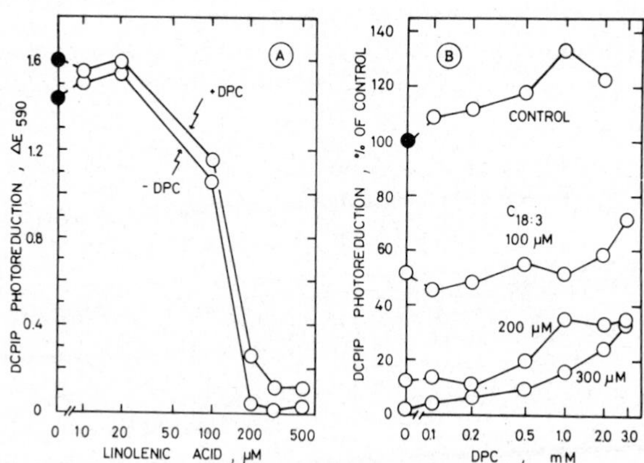

Fig 1. Inhibitory effect of linolenic acid on the photoreduction of DCPIP and ability of DPC to restore this activity. In controls, ΔE_{590}/min/20 μg chlorophyll per ml corresponded to 1.4. Chloroplasts were prepared by the Tris-NaCl procedure.

Fig. 2 shows the ability of $MnCl_2$ and $MgCl_2$ to restore the photosystem II electron flow activity which was inhibited by linolenic acid. Like DPC, $MnCl_2$ has been shown to be an artificial electron donor to the oxidizing side of photosystem II[26,27], whereas $MgCl_2$ is an important cofactor in numerous photosynthetic reactions[28]. The photosystem II activity was restored by $MnCl_2$ (Fig. 2 A) with an optimum at 5 mM (Fig. 2 B), whereas $MgCl_2$ had a much smaller effect.

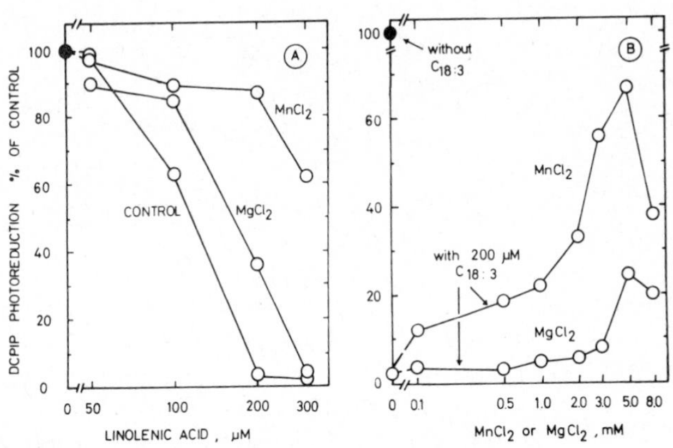

Fig. 2. Inhibitory effect of linolenic acid on the photoreduction of DCPIP and ability of $MnCl_2$ and $MgCl_2$ to restore this activity. The values 100% corresponded to ΔE_{590} = 1.72/min/20 μg chlorophyll per ml. Tris-NaCl chloroplasts were used.

The restoring abilities of various Mn^{2+}- and Mg^{2+}-salts on the DCPIP photoreduction are compared in the presence of three different C_{18}-unsaturated fatty acids (Table II).

First, it can be seen that the inhibition of PS-II electron flow caused by 300 μM of C_{18}-fatty acids depended on the degree of saturation of the molecule; for instance, linolenic acid ($C_{18:3}$) had a more inhibitory effect than linoleic ($C_{18:2}$) and oleic ($C_{18:1}$) acids. The restoration of the electron flow rate by Mn^{2+}-salts were less effective in restoring the activity in the presence of $C_{18:3}$ than in the presence of the other acids. Second, amongst all the Mn^{2+}-salts assayed, $MnCl_2$ and $MnSO_4$ were the most effective in restoring the activity (Table II). However, the anions did not appear to play a specific role. Third, the Mg^{2+}-salts had no restoring effect when $C_{18:3}$ was the inhibitor or had a much smaller effect than Mn^{2+}-salts when $C_{18:2}$ and $C_{18:1}$ were the inhibitors.

Table II

Restoration by Mn^{2+}-salts of the photoreduction of DCPIP inhibited by C_{18}-unsaturated fatty acids (FA)

Conditions	Photoreduction of DCPIP (% of controls)*			
	$+C_{18:3}$	$+C_{18:2}$	$+C_{18:1}$	without FA
Controls, without FA and ions	(100)	(100)	(100)	(100)
FA alone (300 μM)	4	8	18	--
" + $MnCl_2$ (5 mM)	53	64	70	93
" + $MnSO_4$ (5 mM)	50	79	74	98
" + $Mn(NO_3)_2$ (5 mM)	39	60	70	91
" + $Mn(CH_3COO)_2$ (5 mM)	39	64	66	89
" + $MgCl_2$ (5 mM)	3	28	47	99
" + $MgSO_4$ (5 mM)	4	12	27	99

* 100% corresponded to ΔE_{590} = 1.74/min/20 μg chlorophyll per ml.

These observations suggested that the inhibition block caused by the unsaturated fatty acids in the electron transport could be located after the entry point of DPC. Moreover, Mn^{2+} ions might either feed electrons into the system after the fatty acid block or bypass it.

In order to discriminate between these two possibilities, the photoreduction of DCPIP by Tris-washed chloroplasts was found to be crucial[15]. It is well known that washing chloroplasts with 0.8 M Tris at pH 8 caused an inhibition of the Hill reaction by blocking electron transport between water and photosystem II[22]. The results reported in column 1 (Table III) confirmed that such a treatment inhibited photosystem II activity and that DPC restored the electron flow to an extent of 65%. $MnCl_2$ on the other hand, did not have such a restoring effect indicating that it did not behave as an exogenous electron donor like DPC. Linolenic acid (300 μM) strongly inhibited electron flow in all cases (column 2).

Since DCPIP photoreduction is stoichiometrically related to O_2 evolution, it was expected that O_2 evolution should be inhibited by free unsaturated fatty acids and that Mn^{2+} ions should restore the lost activity when H_2O is the electron donor. Fig. 3 shows that this was indeed the case for all three C_{18}-unsaturated acids. As for the photoreduction of DCPIP, $MgCl_2$ and DPC (not shown) were ineffective in restoring the activity; again the Mn^{2+}-supported activity was inhibited by DCMU.

Table III

Ability of DPC and $MnCl_2$ to restore the photoreduction of DCPIP in Tris-washed chloroplasts in absence and presence of linolenic acid

Conditions	Photoreduction of DCPIP (% of control)	
	Controls	+300 μM $C_{18:3}$
Untreated chloroplasts	100*	2
" + DPC (0.5 mM)	107	10
Tris-washed chloroplasts	10	5
" + DPC	65	9
" + DPC + DCMU (10 μM)	3	2
" + $MnCl_2$ (5 mM)	5	2
" + DPC + $MnCl_2$	30	6

* The value of 100% corresponded to ΔE_{590} = 1.365/min/20 μg chlorophyll/ml.

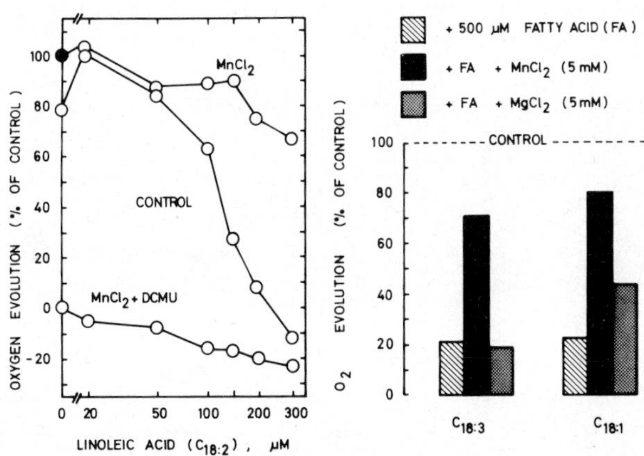

Fig. 3. Inhibitory effect of linoleic, linolenic and oleic acids on the oxygen evolution and ability of $MnCl_2$ and $MgCl_2$ to restore this activity.

Left hand: The reaction mixture contained 50 mM Tricine (pH 8), 35 mM NaCl, 0.3 mM DCPIP, chloroplasts (40 μg chlorophyll/ml) and where indicated linoleic acid (various concentrations), 5 mM $MnCl_2$ and 10 μM DCMU. Chloroplasts were prepared by the Tris-NaCl method. 100% corresponded to 96 μmoles O_2 evolved/mg chlorophyll/hour.

Right hand: The reaction mixture contained 50 mM Tris-maleate (pH 7), 35 mM NaCl, 0.15 mM DCPIP, chloroplasts (20 μg chlorophyll per ml) and where indicated 500 μM linolenic or oleic acid, and 5 mM $MnCl_2$ or $MgCl_2$. Chloroplasts were prepared by the HEPES-saccharose method. 100% corresponded to 123 μmoles O_2 evolved/mg chlorophyll/hour.

5. Discussion

The present findings show that C_{18}-unsaturated fatty acids are inhibitors of the photosystem II electron transport as reported earlier[9,14-16] and that the inhibition site of fatty acids is located on the oxidizing side of photosystem II.

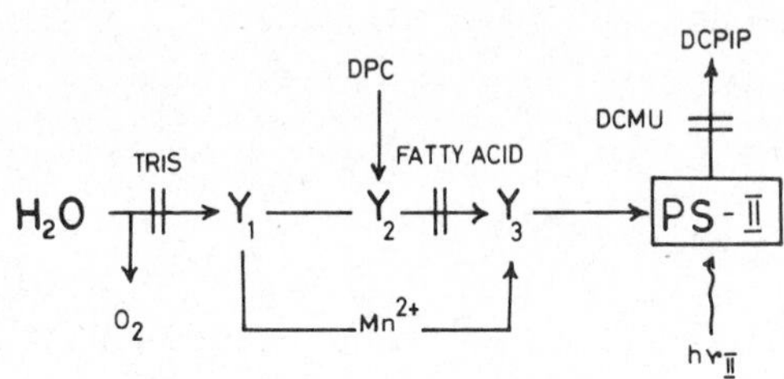

Fig. 4. Schematic representation of the action sites of C_{18}-unsaturated fatty acids, Tris-washing, DCMU, DPC, Mn^{2+} in the photosystem II electron transport in chloroplasts.

Taken together, our results suggest a schematic representation (see Fig. 4) of the action sites of C_{18}-unsaturated fatty acids, Tris-washing and DCMU and of the restoring effect of DPC and Mn^{2+} ions. According to Fig. 4, electrons from water are transported through three hypothetical carriers (Y_1, Y_2, Y_3) and boosted through the photosystem II pigments to DCPIP. C_{18}-unsaturated fatty acids and Tris-washing inhibit the electron flow between Y_2 and Y_3 and between H_2O and Y_1, respectively. DPC feeds electrons in the chain through the Y_2-carrier. Under our experimental conditions, Mn^{2+} did not serve as an electron donor to photosystem II but rather functioned as an electron carrier shunting Y_1 and Y_3, bypassing Y_2 and the fatty acid block. DCMU, a potent and specific inhibitor of the electron transport pathways as far studied. This suggested that the inhibition and restoration events take place at a site prior to the DCMU block.

The function of Mn^{2+} in relation to the fatty acid inhibition was found to be interesting. Although the absolute requirement of Mn^{2+} for a functional photosystem II is well documented[19], it is by no means certain whether exogenous Mn^{2+} participates directly as a carrier in the electron transfer reactions[19] and/or is an electron donor to some other carrier prior to photosystem II pigments[26,27,29,30]. Moreover, neither the exact position of Mn^{2+} as an electron carrier nor the entry point of the electrons provided by Mn^{2+} ions are well defined. Izawa[26] showed that in EDTA- and heat-treated chloroplasts, Mn^{2+} can be an efficient electron donor for photosystem II. Since Mn^{2+} failed to reactivate $NADP^+$ photoreduction in Tris-treated chloroplasts, Ben-Hayyim and Avron[27] concluded that the Mn^{2+} site of action not only precedes photosystem II but is rather close to the O_2 evolution step itself. Our results are in agreement with the above authors' observation since Tris-treated chloroplasts, either with or without added fatty acids, failed to show DCPIP photoreduction in the presence of Mn^{2+} (see Table III). However, when H_2O was the electron donor, Mn^{2+} was able to restore the electron flow activity in the presence of fatty acids, as measured by DCPIP reduction (Fig. 2) and O_2 evolution (Fig. 3). This suggested that exogenous Mn^{2+} ions might function as a catalyst (or an electron carrier) shunting Y_1 and Y_3 (see Fig. 4).

Finally, we would like to point out that compared to classical inhibitors of photosystem II, such as DCMU, Tris-washing, etc., we may consider fatty acids as physiological inhibitors. As a part of the lipids of the thylakoid membrane, any change in the ratio chlorophyll/(galacto-)lipids/free fatty acids, namely during aging, might greatly effect the structure and function in chloroplasts.

6. Acknowledgements

This investigation was supported by a grant (contract no. 3.566.71) to P.A. Siegenthaler from the Fonds National Suisse de la Recherche Scientifique. We thank Dr. P. Schürmann for kindly preparing the purified ferredoxin and building the Zeiss-spectrophotometer device, and Miss Françoise Prieur for arrangement and typing of the manuscript.

References

1. Allen, C.F., Good, P., Davis, H.F. and Fowler, S.D., 1964, Biochem. Biophys. Res. Commun. 15, 424.
2. Costes, C., Bazier, R. and Lechevallier, D., 1972, Physiol. Vég. 10, 291.
3. Benson, A.A., 1971, in: Structure and Function of Chloroplasts, ed. M. Gibbs (Springer-Verlag, Berlin, Heidelberg, New York), p. 129.
4. McCarthy, R.E. and Jagendorf, A.T., 1965, Plant Physiol. 40, 725.
5. Constantopoulos, G. and Kenyon, C.N., 1968, Plant Physiol. 43, 531.
6. Wintermans, J.F.G.M., Helmsing, P.J., Polman, B.J.J., Van Gisbergen, J. and Collard, J., 1966, Biochim. Biophys. Acta 189, 95.
7. Siegenthaler, P.A., 1972, Biochim. Biophys. Acta 275, 182.
8. Siegenthaler, P.A., 1969, Plant & Cell Physiol. 10, 801.
9. Cohen, W.S., Nathanson, B., White, J.E. and Brody, M., 1969, Arch. Biochem. Biophys. 135, 21.

10. Krogmann, D.W. and Jagendorf, A.T., 1959, Arch. Biochem. Biophys. 80, 421.
11. Katoh, S. and San Pietro, A., 1968, Arch. Biochem. Biophys. 128, 378.
12. Brody, S.S., Brody, M. and Döring, G., 1970, Z. Naturforsch. 25b, 367.
13. Brody, S.S., 1970, Z. Naturforsch. 25b, 855.
14. Siegenthaler, P.A., 1973, Biochim. Biophys. Acta 305, 153.
15. Siegenthaler, P.A., 1974, FEBS Letters 39, 337.
16. Terpstra, W., 1974, Plant Sci. Letters 3, 1.
17. Siegenthaler, P.A., 1970, Experientia 26, 1308.
18. Friedländer, M. and Neumann, J., 1968, Plant Physiol. 43, 1249.
19. Cheniae, G.M., 1970, Ann. Rev. Plant Physiol., 21.
20. Kalberer, P.P., Buchanan, B.B. and Arnon, D.I., 1967, Proc. Nat. Acad. Sci U.S.A. 57, 1542.
21. Bruinsma, J., 1961, Biochim. Biophys. Acta 53, 576.
22. Yamashita, T. and Butler, W.L., 1968, Plant Physiol. 43, 1978.
23. McSwain, B.D. and Arnon, D.I., 1968, Proc. Nat. Acad. Sci. U.S.A. 61, 989.
24. Buchanan, B.B. and Arnon, D.I., 1971, in: Methods in Enzymology, ed. A. San Pietro, vol. XXIII (Academic Press, New York and London), p. 413.
25. Vernon, L.P. and Shaw, E.R., 1969, Biochem. Biophys. Res. Commun. 36, 878.
26. Izawa, S., 1970, Biochim. Biophys. Acta 197, 328.
27. Ben-Hayyim, G. and Avron, M., 1970, Biochim. Biophys. Acta 205, 86.
28. Sun, A.S.K. and Sauer, K., 1972, Biochim. Biophys. Acta 256, 409.
29. Bachofen, R., 1966, Brookhaven Symp. Biol. 19, 478.
30. Kimimura, M. and Katoh, S., 1972, Plant & Cell Physiol. 13, 287.

M. AVRON, *Proceedings of the Third International Congress on Photosynthesis*,
September 2-6, 1974, Weizmann Institute of Science, Rehovot, Israel
Elsevier Scientific Publishing Company, Amsterdam, The Netherlands, 1974

THE SITES OF INDUCED LIMITING STEPS OF THE ELECTRON TRANSPORT REACTIONS IN CHLOROPLASTS ISOLATED WITH LOW-SALT MEDIUM

Yung-Sing Li

Institute of Botany, Academia Sinica,
Republic of China

Summary

Two rate-limiting sites are located on the photosynthetic electron transport chains for acidic media-inhibited or for $MgCl_2$-inhibited electron transport in chloroplasts isolated with low salt medium.

1. Introduction.

The rate of photosynthetic electron transport is low in acidic media[1], it is also limited by adding divalent cations to chloroplasts suspended in low salt media[2-4]. In this report, using electron donor reactions, two limiting sites are located; one situates between the two photosynthetic systems, the other exists on the water site of the photosystem II. It is the photosystem II limiting site which sets the lower limit for the overall electron transport.

2. Material and Methods

Three to four inches long green house grown oat seedlings were used for the experimental material. Ten grams of green part of seedling were chilled in the refrigerator in distilled water for 10 minutes. They were then cut into inch long pieces and hand ground in mortar (prechilled) with 20 ml cold buffer which consisted of sucrose (400 mM), tricine-NaOH (20 mM), pH8. The grinding was carried out at subtropical room temperature, ranging from 10°C to 30°C. After removal of cell debris by straining through eight layers of cheese cloth and one layer of nylon mesh, chloroplasts were sedimented at 4340 x g for five minutes, and resuspended in distilled water (~ one mg chl/ml). The resuspended chloroplasts were kept in ice and used half an hour to one hour later. Methyl viologen reduction was measured as O_2 uptake with a membrane covered Clark-type electrode.

3. Results and Discussion

Fig. 1 depicts the time courses of various photochemistry reactions at pH7 using methyl viologen as electron acceptor. The rate of water-supported electron transport is low, replacing water with hydroquinone and ascorbate as electron source stimulates the methyl viologen reaction. The donor reaction is DCMU (3-[3, 4-dichlorophenyl]-1, 1-dimethylurea) sensitive. Gramicidin increases the rate of the water-supported electron transport. These observations clearly place an uncoupler sensitive rate-limiting site between the H_2O splitting system and the photoact II(site II). The DCIP (2, 6-dichlorophenol-indophenol) and ascorbate donor couple reaction

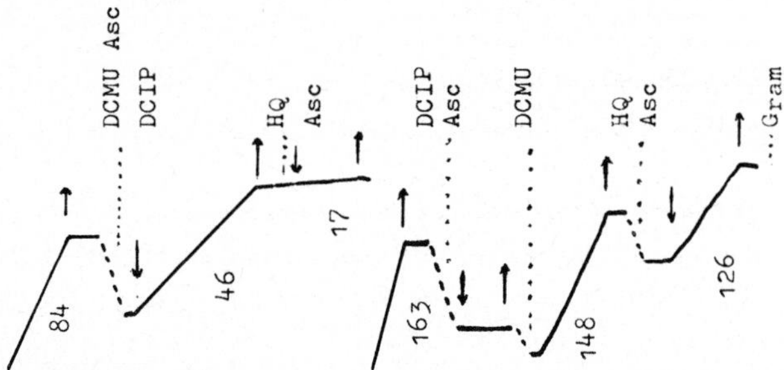

Fig. 1 Schematized time courses of the H_2O supported, the photosystems I and II donors supported electron transport. Stock chloroplasts suspended in distilled water. The reaction medium contained chloroplast, 22 μg chl/ml; TES (N-tris Hydroxymethyl methyl-2-aminoethane Sulfonic Acid) 20 mM, pH 7; sucrose 0.4 M; NaCl 10 mM; $MgCl_2$, 2.5 mM; KCN, 1 mM pH 7 and methyl viologen 50 μM. Other chemicals when added had the following cocentrations: hydroquinone (HQ) 100 uM; Na-ascorbate (Asc) (PS II donor), 300 μM; DCMU 50 μM, Na-ascrobate, 5 mM (PS I donor); DCIP 50 μM; Gramicidin (Gram) 5 x 10^{-7}M Reactions were run at 25°C. The broken lines showing the period while the reaction media were exposed to air. ↑ light on, ↓ light off. Numbers indicate rates in μmoles O_2 uptake/mg chl. hr.

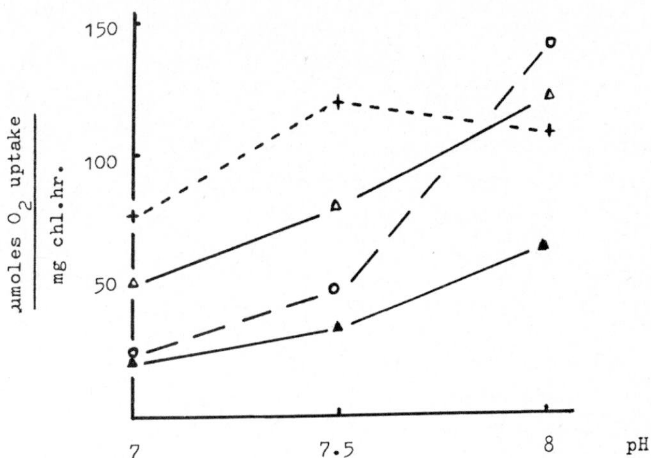

Fig. 2 $MgCl_2$ effects on the H_2O supported and the donor supported electron transport. Stock chloroplasts suspended in distilled water. The reaction media contained: chloroplasts, 23 µg chl/ml; sucrose, 0.4 M; NaCl, 10 mM; KCN, 1 mM, pH 7; methyl viologen, 50 µM; 20 mM appropriate buffer (tricine, pH 8 and 7.5; TES, pH 7) plus chemicals indicated in the figure; reactions were run at 25°C. (o) control; (▲) $MgCl_2$, 2.5 mM; (△), $MgCl_2$, ascorbate, 5 mM (including 10-20% DCMU insensitive rate); (+) $MgCl_2$, Gramicidin, $5x10^{-7}$M. The photosystem II donor rate was calculated assuming that the H_2O supported rate was not changed by adding ascorbate.

reveals that there is another uncoupler sensitive limiting site located between the two photosystems (site I). Site I may be less limitative to the electron flux than site II, because appreciable amount of electrons from ascorbate which bypass site II can pass through site I.

At pH 8 the rate of electron transport is high, still, two limiting sites can be induced by introducing $MgCl_2$. The $MgCl_2$ effect saturates near 1 mM in the presence of 10 mM NaCl (data not shown). Fig. 2 shows that one of the $MgCl_2$ induced limiting sites can be either bypassed by ascorbate which is DCMU sensitive, or relieved by Gramicidin D. Table 1 illustrates that, as acidic media, $MgCl_2$ induces another limiting site between the two photosystems. Incidentally, it is uncertain whether the reduced indophenol donor reaction supports phosphorylation or not[5-7].

Table 1

$MgCl_2$ effect on the electron transport of photosystem I. Stock chloroplasts suspended in 0.4 M sucrose and 20 mM tricine (pH8). The reaction medium contained: chloroplasts, 21 ug chl/ml; tricine, 20 mM (pH8); sucrose 0.4 M; KCN, 1 mM (pH7); DCMU, 50 μM; DCIP, 50 μM; ascorbate, 5 mM and methyl viologen 5 μM plus chemicals as indicated in the table. Rate was taken at about one minute after the onset of illumination. Reactions were run at 14°C

Additions	μmoles O_2 uptake / mgchl. hr
Control	80
NH_4Cl 5 mM	129
$MgCl_2$ 2.5 mM	40
$MgCl_2$ 2.5 mM; NH_4Cl 5 mM	123

However, table 1 seems to indicate that, in the presence of $MgCl_2$, at least 70% electrons from indophenols enter the electron transport chain prior to an uncoupler sensitive rate-limiting site. Below pH 7.5 $MgCl_2$ does not decrease the rate appreciably, and the ascorbate donor reaction rate (with $MgCl_2$) is far below the uncoupled rate (Fig. 2).

Knowing that site II is more limitative than site I, it may help explain the Gramicidin induced fluorescence change[8]. Chloroplasts which are not water-treated are only uncoupled partially, and $MaCl_2$ (Table 1) or $CaCl_2$ decreases the Hill reaction rate. Gramicidin D in the presence of $CaCl_2$ increases the rate of the Hill reaction as well as the minimal fluorescence while the ferricyanide Hill reaction is in effect; However, the Gramicidin effect on fluorescence disappears when the Hill oxidants are exhausted, therefore, the maximal fluorescence is not changed by adding Gramicidin. Fig. 3 reproduces from reference 8 two traces recorded from chloroplasts fluorescence excited by strong light in the presence of small amount of ferricyanide. Trace A represents the time course of fluorescence change in the absence of Gramicidin. After an intial spike, fluorescence drops to its minimal intensity and later rises to a maximal intensity as the ferricyanide Hill reaction is finished. Trace B is the recorded fluorescence change in the presence of Gramicidin. The intial spike is lowered and obscured by a higher minimal fluorescence. The maximal

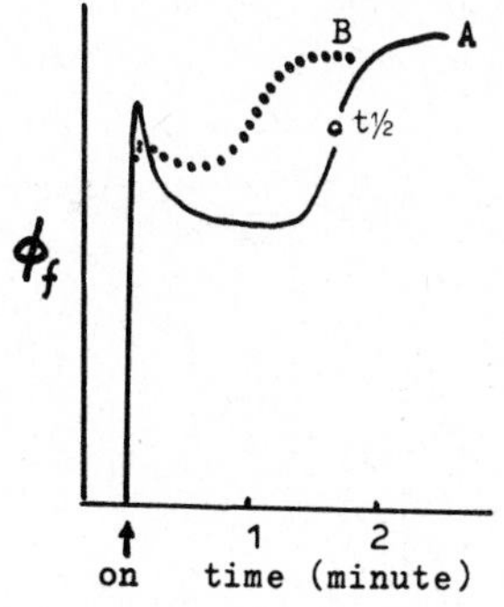

Fig. 3. Effect of Gramicidin on the fluorescence induction in the presence of ferricyanide. The reaction mixture consists of chloroplasts, 13 µg chl/ml; sorbital, 150 mM; tricine-NaOH, 50 mM, pH 8 and ferrciyanide, 2.8 x 10^{-5}M in a volume of 1 ml. A. calcium chloride, 5 mM; B. calcium chloride, 5 mM+gramicidin, 4.4 µM. t½: half time of fluorescence raise, indicated by a circle on trace A. ϕ_f : relative fluorescence intensity. Arrow indicates light on.

fluorescence is reached sooner because of faster Hill reaction rate[8]. The fluorescence change is more pronounced in terms of Δ fl-- the difference between maximal and minimal fluorescence. The fact that there is little change in the maximal fluorescence indicates that the increase in the minimal fluorescence is not due to Gramicidin per se. Rather, it is due to the interactions between Gramicidin and the electron transport process. The observed phenomena may be explained in terms of changed equilibrium ratio of Q^-/Q (the reduced and the oxidized form, respectively, of system II primary electron acceptor). In the absence of phosphorylation cofactors or uncoupler, site 2 limits the electron flow. The steady state ratio of Q^-/Q is kept low, hence a low intensity of minimal fluorescence. Upon the finishing of the Hill reaction, almost all Q is reduced for lack of oxidant and fluorescence rises to its maximum. In the process of Hill reaction, Gramicidin, which accelerates electron flow to Q, maintains a higher steady state ratio of Q^-/Q and hence a higher intensity of minimal fluorescence. An explanation proposed to deal with the presence of the initial fluorescence spike is offered in reference 8.

Both the donor reaction and the fluorescence measurement seem to indicate that there is a limiting site on the water side of the photosystem II; in fact, this site has been implied to exist in Keister's[7] and Böhme and Trebst's[10] reports, and the latter authors[10] and Trebst and Reimer[11] suggest that there is a coupling site on the water site of photosystem II. A second coupling site is also proposed by the group in Izawa's laboratory

12-17 and by Reeves and Hall[18]. What is the relationship between the limiting site II and this newly found coupling site is a matter of interesting.

Acknowledgements:

Research supported by Academia Sinica and The National Science Council, Republic of China.

4. References

1. Good, N.,Izawa, S. and Hind, G., 1966, in: Current topics in bioenergetics, Vol. 1 ed. Sanadi,D. R. (Academic Press, New York and London) p. 75.
2. Jagendorf, A. T. and Smith, M., 1962, Plant Physiol. 37, 135.
3. Spencer, D. and Unt, H., 1965, Australian J. Biol. Sci. 18, 197.
4. Gross, E., Dilley, R. A., and San Pietro, A., 1969, Arch. Biochem. Biophys. 134, 450
5. Trebst, A. and Eck, H., 1961, Z. Naturforsch. 166, 455.
6. Wessels, J. S. C., 1964. Biochim. Biophys. Acta 79, 640.
7. Keister, D. L., 1965, J. Biol. Chem. 240, 2673.
8. Li, Y. S.,1973, Bot. Bull. Acad. Sinica 14, 65.
9. Li, Y. S., 1973, ibid 14, 70.
10. Böhme, H. and Trebst, A., 1969, Biochim. Biophys. Acta 180, 137.
11. Trebst, A. and Reimer, S., 1973, Biochim. Biophys. Acta 305, 129.
12. Saha, S., Ouitrakul, R., Izawa, S. and Good, N. E., 1971, J. Biol. Chem. 246, 3204
13. Ouitrakul, R. and Izawa, S., 1973, Biochim. Biophys. Acta 305, 105.
14. Izawa, S., Gould, J. M., Ort, D. R., Felker, P. and Good, N. E., 1973, Biochim. Biophys. Acta 305, 119.
15. Gould, J. M. and Izawa, S., 1973, Biochim. Biophys. Acta 314, 211.
16. Gould, J. M. and Izawa, S., 1973, Eur. J. Biochem. 37, 185.
17. Ort, D. R. and Izawa, S., 1973, Plant Physiol. 52, 595.
18. Reeves, S. G. and Hall, D. O.,1973, Biochim. Biophys. Acta 314, 66.

Note: Paper No. 149 of the Scientific Journal Series, Institute of Botany, Academia Sinica

Addendum: A five-fold stimulation of electron flux by hydroquinone and ascorbate couple was observed in lettuce chloroplasts in the absence of uncoupler in acidic medium.

M. AVRON, *Proceedings of the Third International Congress on Photosynthesis*,
September, 2-6, 1974, Weizmann Institute of Science, Rehovot, Israel
Elsevier Scientific Publishing Company, Amsterdam, The Netherlands, 1974

POLYLYSINE AND CYANIDE INTERACTIONS WITH ELECTRON TRANSPORT CATALYSTS ON THE CHLOROPLAST MEMBRANE

R. Schneeman, S. P. Berg and D. W. Krogmann

Department of Biochemistry
Purdue University

1. Introduction

Large polycations inhibit Photosystem 1 on or near the site of plastocyanin function without diminishing Photosystem 2 activity.[1] Polycations are potentially instructive probes of other membrane bound enzymes and the ferredoxin:NADP oxidoreductase interaction with polylysine required detailed examination. This enzyme, when solubilized and purified, is stimulated in its catalytic activity by polylysine and the stimulation is greater than that experienced when the oxidoreductase is bound to the chloroplast membrane. Ouitrakul and Izawa have shown that cyanide is an inhibitor of plastocyanin function in the chloroplast.[2] We have developed supporting evidence on the details of this inhibitor-enzyme interaction.

2. Results

Ferredoxin:NADP oxidoreductase catalyses a number of reactions in addition to its central biological function in transferring electrons from ferredoxin to NADP in the chloroplast. We chose to examine the following three reactions catalysed by this flavoprotein.

Diaphorase

NADPH + oxidized indophenol ⟶ NADP + reduced indophenol

Transhydrogenase

NADPH + thionicotinamide NADP ⟶ NADP + thionicotinamide NADPH

Cytochrome c reductase

NADPH + oxidized cytochrome c + ferredoxin
⟶ NADP + reduced cytochrome c + ferredoxin

Each of these reactions is stimulated from three to seven fold by addition of polylysine. Nakamura et al. have described another reaction of ferredoxin:NADP oxidoreductase in which a complex of the oxidoreductase and ferredoxin catalyses the transfer of electrons from NADPH to oxygen.[3] This latter reaction is best stimulated by cadaverine and does not respond to polylysine while the three activities which we have studied are best stimulated by polylysine.

The stoichiometry of polylysine to enzyme in achieving maximum stimulation of activity is of interest and depends on the size of the polylysine molecule used. This is illustrated in Table 1.

Table 1

Polylysine size in relation to the stimulation of ferredoxin:NADP oxidoreductase

Polylysine molecular weight	n moles polylysine per n mole flavin of oxidoreductase	lysine residues per n mole flavin of oxidoreductase	% of maximum rate
control	0	0	13
1,000	65	520	28
4,000	15	480	100
30,000	2	480	89

Polylysine binding to the flavoprotein can be observed by alteration of the absorption spectrum of this enzyme. Polylysine causes a substantial increase in absorption at 275 nm. Titration of this spectral change shows that the maximum increase in absorption is achieved at a ratio of two moles of polylysine of molecular weight 30,000 to one mole of flavin equivalent of enzyme. The fluorescence of this enzyme is quenched by polylysine with the same stoichiometry.

Kinetic analyses of the effects of polylysine on both the diaphorase and transhydrogenase activities indicated that polylysine prevents the substrate inhibition of enzyme activity observed at high concentrations of NADPH. In addition to its effects on NADPH binding, polylysine causes a hyperbolic noncompetitive activation of diaphorase and transhydrogenase activities indicating that these reactions are allowed to proceed over a alternate, more active pathway by polylysine. In examining the relations between polylysine concentration, enzyme concentration and activity, we observed that the apparent affinity of enzyme for polylysine increased at higher enzyme concentrations. This suggests that the enzyme might exist in multiple forms of aggregation. A plot of activity as a function of enzyme concentration reveals that as the enzyme is diluted, the specific activity increases approximately two fold, indicating that the disaggregated species is more active than the aggregated species. Polylysine stimulates at all enzyme concentrations but the stimulation is greater at high enzyme concentration. Polylysine may stimulate activity by promoting the disaggregation of the enzyme into a smaller, more reactive species.

Fredricks and Gehl have reported on the existence of ferredoxin:NADP oxidoreductase in two size classes of molecular weights 117,000 and 50,000.[4] Since these enzymes were immunologically identical, two molecular weight forms of the same protein seem likely. We subjected a highly purified sample of ferredoxin:NADP oxidoreductase to gel filtration chromatography and obtained a molecular weight of 85,000. When the enzyme is chromatographed in the presence of a high concentration of urea, it is converted into a smaller molecular weight form. Disc gel electrophoresis with sodium dodecyl sulfate indicated two polypeptides of molecular weights 35,000 and 45,000. We have very preliminary evidence that polylysine may dissociate the enzyme but much more work is needed to put this on a firm basis.

Can these observations on polycation stimulation of ferredoxin:NADP oxidoreductase have any biological significance? One could imagine that this flavoprotein is bound *in vivo* to a cationic region on the chloroplast membrane which holds the enzyme in an optimally active state. An indication that this might be the case came from experiments in which we attempted to relate the appearance of soluble enzyme with the disappearance of particulate activity when chloroplasts are washed repeatedly in low ionic strength buffers. When 90% of the diaphorase activity has been washed off of the chloroplast membranes, only 30% of this activity can be found in the soluble proteins. It is, of course, very easy to loose enzymatic activity, so this kind of experiment does not prove that the enzyme is more active when bound to the chloroplast. After many unsuccessful attempts at enzyme-chloroplast reconstitution, the experiments of Dr. Berzborn suggested a fruitful approach.[5] Berzborn had found that an antibody to ferredoxin: NADP oxidoreductase would not agglutinate chloroplasts to which this enzyme antigen was bound. Agglutination was achieved when the coupling factor of photophosphorylation, CF_1, was removed from the chloroplasts. This experiment indicates that the ferredoxin:NADP oxidoreductase is held in a cleft below the coupling factor on the membrane surface. When the coupling factor is removed by the procedure of Lien and Racker[6] from the chloroplast membranes, much of the ferredoxin: NADP oxidoreductase is solubilized as well. When the depleted membranes are mixed with purified ferredoxin:NADP oxidoreductase, the activity of the combination is considerably greater than the sum of the activities of the membrane and enzyme when assayed individually. This is illustrated by data given in Table 2.

Table 2

Stimulation of diaphorase activity by CF_1 depleted chloroplasts

	A_{620nm}/90 sec.
soluble enzyme	0.080
0.02 ml depleted membranes	0.045
soluble enzyme + 0.02 ml depleted membranes	0.250
0.05 ml depleted membranes	0.100
soluble enzyme + 0.05 ml depleted membranes	0.320

This experiment suggests that the enzyme is activated by binding to the chloroplast membrane. It is noteworthy that the membrane induced increase in enzyme activity occurs only at pHs above 6.8. Thus changes in pH near the membrane surface may be influential in regulating ferredoxin:NADP oxidoreductase activity.

In another line of investigation, the interaction of plastocyanin with cyanide has been examined. Ouitrakul and Izawa had found cyanide to be a selective inhibitor of Photosystem 1 activity and they found that cyanide destroys purified plastocyanin. We have found that inhibitory concentrations of cyanide release copper from chloroplasts and from purified plastocyanin. Radioactive cyanide binds to plastocyanin in both the purified form and when plastocyanin is bound to the chloroplast membrane. Finally, purified plastocyanin can reverse cyanide inhibition of chloroplast Photosystem 1 most effectively only after the cyanide inactivated plastocyanin has been removed from the membranes. This experiment suggests that exogenous plastocyanin restores activity by returning to the original plastocyanin site on the membrane.

References

1. Brand, J., Baszynski, T., Crane, F. L. and Krogmann, D. W., 1972, J. Biol. Chem. 247, 2814.
2. Ouitrakul, R. and Izawa, S., 1973, Biochim. Biophys. Acta 305, 105.
3. Nakamura, S., Kimura, T. and Chu, J. W., 1972, FEBS Letters 25, 249.
4. Fredricks, W. W. and Gehl, J. M., 1973, Fed. Proc. 32, 477.
5. Berzborn, R. J., 1969, Z. Naturforsch. 23b, 1096.
6. Lien, S. and Racker, E., 1971, Methods Enzymol. 23, 547.

Journal Paper No. 5253 of the Purdue Agricultural Experiment Station.

This work was supported by research grant GB 36956X from the Metabolic Biology Program of the National Science Foundation.

M. AVRON, *Proceedings of the Third International Congress on Photosynthesis*, September, 2-6, 1974, Weizmann Institute of Science, Rehovot, Israel
Elsevier Scientific Publishing Company, Amsterdam, The Netherlands, 1974

INHIBITION OF THE PHOTOSYNTHETIC ELECTRON TRANSPORT BY BENTAZON*

K. Pfister, C. Buschmann and H.K. Lichtenthaler
Botanical Institute (Plant Physiology) University of Karlsruhe
D-75 Karlsruhe, FRG

Summary

1. The herbicide bentazon acts as a good inhibitor of Hill-reactions with electron acceptors like DCPIP, Fecy, p-BQ and MV.
2. Bentazon inhibits photo-reactions of PS II but does not affect reactions of PS I (DCPIP/Asc.→MV).
3. Bentazon also prevents the formation of the light induced pH-gradient and suppresses the variable fluorescence.
4. Noncyclic photophosphorylation is inhibited by bentazon, while cyclic phosphorylation is not affected.
5. The inhibition site of bentazon in the photosynthetic electron transport is assumed to be in the range of the reaction center of PS II and the PQ-shuttle.

1. Introduction

The herbicide bentazon (fig. 1) is known as an inhibitor of the DCPIP catalysed Hill-reaction. Furthermore, an inhibition of the photosynthetic CO_2-fixation was found and the effectiveness of herbicide on different kinds of plants was studied[1]. The question whether the effectiveness of bentazon is based on an unspecific damage of the photosynthetic membranes or on the interference in the electron transport has not yet been solved. Therefore thorough investigations on the influence of bentazon on the photo-reactions of pigment systems I and II and on photophosphorylation were carried out with the aim of localizing more precisely a possible inhibition site in the cyclic and noncyclic electron transport.

Abbreviations: Asc: ascorbate; bentazon: 3-isopropyl-2,1,3-benzothiadiazinone (4)2,2 dioxide; p-BQ: p-benzoquinone; DBMIB: dibromothymoquinone; DCMU: 3-(3,4-dichlorophenyl)-1,1-dimethylurea; Fecy: ferricyanide; MV: methylviologen; PD: phenylendiamine; PS I and PS II: pigment systems I and II.

---

* Basagran ® selective herbicide for soybeans, cereal, mayze and rice produced by BASF AG.

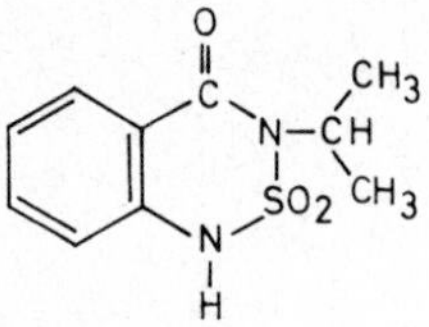

Fig. 1. Structure of bentazon 3-isopropyl-2,1,3-benzothiadiazinone(4)-2,2-dioxide.

2. Material and Methods

Chloroplasts were isolated from cotyledons of 6 day old Raphanus sativus L. var. Saxa Treib by differential centrifugation in a medium containing 0.4 M sucrose, 0.1 M potassium phosphate buffer pH 7.4, 35 mM NaCl, 5 mM $MgCl_2$. The O_2-evolution was measured polarographically with a Clark-type-electrode[2]. Simultaneous measurements of the variable fluorescence and O_2-evolution were carried out with a modified apparatus after Strasser[3]. Electron transport and photophosphorylation were measured in a medium after Strotmann and Gösseln[4]. The photosynthetic ATP formation was measured with the luciferase assay. Electron donors and -acceptors were added at the indicated concentrations. All experiments were carried out at a temperature of 17.5° C. Pure bentazon was kindly provided by the BASF AG, Ludwigshafen.

3. Results and Discussion

a) Inhibition of O_2-evolution. The influence of bentazon on the O_2-evolution was studied in four different Hill-reactions.

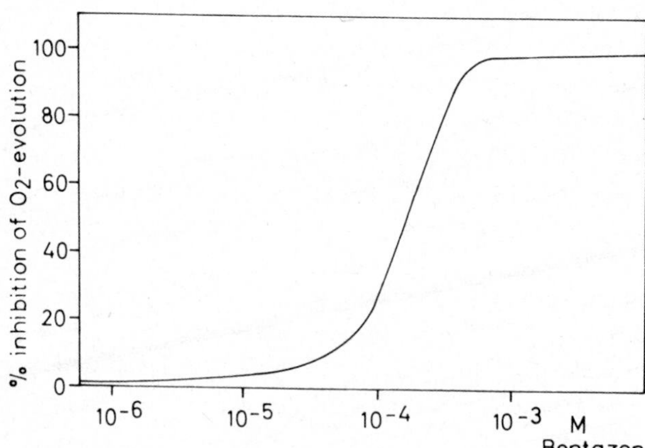

Fig. 2. Inhibition of O_2-evolution of the MV-reduction by different bentazon concentrations. Reaction rates measured after 5 min incubation time.

The results in table 1 show that Hill-reactions with electron acceptors like DCPIP, p-BQ and ferricyanide can be almost completely inhibited with concentrations of 4×10^{-4}M bentazon.

Table 1

Inhibition of O_2-evolution by bentazon in Hill-reactions with different electron acceptors. Reaction rate measured after 5 min incubation time

donor/acceptor system	bentazon concentration	µM O_2/µM Chl. + h	O_2-evolution in % of control
	0	35	100
$H_2O \rightarrow$ MV	4×10^{-5}M	30	86
	4×10^{-4}M	7	20
	0	26	100
$H_2O \rightarrow$ DCPIP	4×10^{-5}M	21	81
	4×10^{-4}M	1.5	5
	0	45	100
$H_2O \rightarrow$ p-BQ	4×10^{-5}M	24	53
	4×10^{-4}M	4	9
	0	28	100
$H_2O \rightarrow$ Fecy	4×10^{-5}M	19	67
	4×10^{-4}M	0.5	2

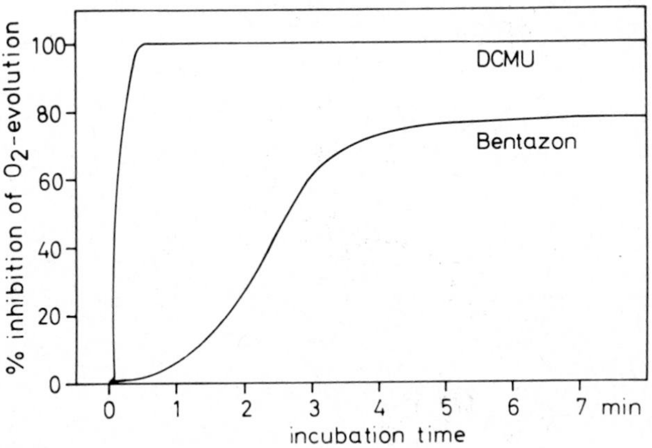

Fig. 3. Inhibition of O_2-evolution of MV-reduction by 5×10^{-4}M bentazon and by 5×10^{-5}M DCMU after different incubation times.

A 50% inhibition of the MV-reduction can be achieved with concentrations of 2×10^{-4}M. The required concentrations for a complete inhibition of the MV-reduction are a bit higher. A complete inhibition of the MV-reduction is reached with concentrations higher than 8×10^{-4}M (fig. 2). Since not only the concentration but also the incubation time of the inhibitor are important for the extent of the electron transport inhibition (fig. 3), we measured the reaction rate after 5 min incubation time for all data shown. After that time, irrespective of whether light or dark incubation, the inhibitory effect of bentazon is almost fully stabilized. After prolonged incubation time there is only a slight increase of the inhibition rate.

In contrast to DCMU, the classical inhibitor of the photosynthetic electron transport, it is apparent that in order to obtain maximal inhibition rates for bentazon, a prolonged incubation time is needed as well as considerably higher concentrations.

b) Effect on the reactions of PS I and PS II. The system $H_2O \rightarrow$ MV was used for the measurement of PS I and PS II activity, while for the measurement of PS II activity we used the systems $H_2O \rightarrow$ DCPIP and $H_2O \rightarrow$ p-BQ.

From these data it is not yet possible to determine whether bentazon inhibits only PS II or PS I as well. To characterize the inhibition of the electron transport more precisely, artificial donor systems were used. The donors ascorbate and phenylendiamine/ascorbate pass electrons into the transport chain before PS II.

Table 2

Inhibition of the electron transport by bentazon. Reaction mixture after Strotmann and Gösseln[4]. Electron donors: a) H_2O b) 0.5 mM Asc c) 33 µM PD, 330 µM Asc d) 5 mM Asc, 0.4 mM DCPIP. Bentazon concentration 7×10^{-4}M, DCMU 5×10^{-5}M. Results in $2e^-$/µM Chl. + h

donor/acceptor system	control	+ bentazon	+ bentazon + DCMU
a) $H_2O \rightarrow$ MV	36	0	0
b) Asc $\rightarrow$ MV	40	2	0
c) PD/Asc $\rightarrow$ MV	41	8	5
d) DCPIP/Asc $\rightarrow$ MV	47+)	46+)	

+) Reaction mixture contained DCMU.

The results (table 2) show that with ascorbate as electron donor there occurs no circumvention of the bentazon inhibition of the MV-reduction. With the donor couple PD/Asc only a slight reversion of the blocking of the

MV-reduction is possible. This small rate of the electron transport obtained under these conditions cannot be substantially diminished by subsequent addition of DCMU. This remaining MV-reduction rate must be considered as an activity of PS I, since firstly the system PD/Asc → NADP can only be inhibited by DCMU to 75% and, secondly, PD/Asc in higher concentrations can also act as donor for PS I. From this it follows that the donor couple PD/Asc does not eliminate the bentazon inhibition close to PS II.

To test whether bentazon shows an inhibition effect also in the region of PS I, the influence of the herbicide on the system DCPIP/Asc → MV was studied. By addition of DCMU to the reaction mixture the activity of PS II was eliminated. Since the electron transport in the system DCPIP/Asc → MV was not affected, even by high concentrations of bentazon, an inhibition site of bentazon in the region of PS I can be excluded. Therefore the effect of bentazon must be associated with the function of PS II.

c) Destruction of the light induced pH-gradient. Isolated chloroplasts show during illumination the formation of a light induced pH-gradient, which, as the energy conserving step, connects the electron transport to photophosphorylation[6].

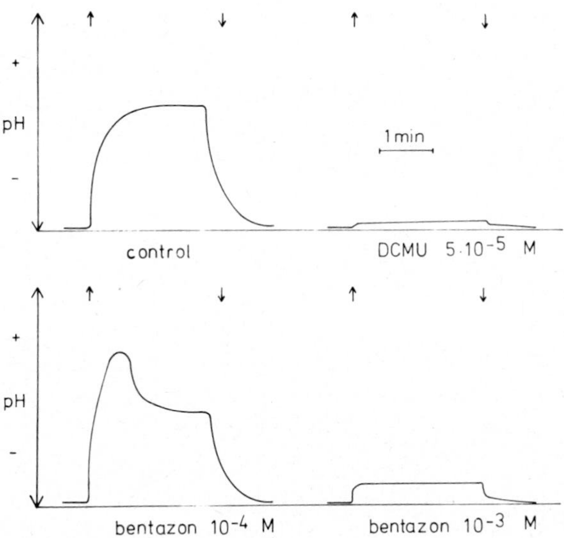

Fig. 4. Inhibition of the formation of the light induced pH-gradient by bentazon. Chloroplasts (0.7 μM chlorophyll) were suspended in 30 mM NaCl and 5 mM $MgCl_2$ at pH 8. (Arrows up or down = light on or off).

As expected from the data of the inhibition of electron transport obtained here, the formation of the light induced pH-gradient is strongly affected by bentazon. At low bentazon concentrations, which partially inhibit the electron transport, a characteristic signal is shown (fig. 4): at first a normally high pH-gradient is built up, which partially breaks down within a few seconds and settles down to a lower value. This effect cannot be explained by gradually increasing inhibition as shown in figure 3, because it needs a considerably shorter time and can be repeated after a dark phase. Higher bentazon concentrations, which are adequate for a complete blocking of the electron transport (10^{-3}M), suppress the initial high pH difference, but allow the formation of a lower gradient. While DCMU also totally eliminates the proton pump (fig. 4) by the blocking of the electron transport, this is not possible with bentazon even by complete inhibition of the PS II activity (10^{-3}M bentazon). The remaining low proton gradient may possibly indicate the proton pump capacity in connection with cyclic photophosphorylation.

d) Photophosphorylation. The effect of bentazon on noncyclic photophosphorylation was measured in the system $H_2O \rightarrow$ MV. As expected from the inhibition of Hill-reaction, there was no photophosphorylation detectable in the presence of 7×10^{-4}M bentazon. The cyclic photophosphorylation with phenazinemethosulphate as cofactor was not affected by bentazon. These data are presented elsewhere.

e) Suppression of the variable fluorescence. The variable amount of fluorescence measured on intact plant material with a maximum of 685 nm is usually attributed to the PS II[7]. If during illumination the reoxidation of the reduced quencher QH is prevented, a higher and constant fluorescence level occurs. Thus the variable fluorescence is a good and fast indicator of an undisturbed electron transport and the photolysis of water.

In our experiments whole Raphanus cotyledons were incubated in bentazon, DCMU and for control in water. After an incubation time of 5 or 15 min in 10^{-3}M bentazon the variable fluorescence is diminished and after prolonged incubation time ($>$ 25 min) it is almost completely suppressed (fig. 5). With an incubation in DCMU 8 minutes were sufficient to suppress the variable part of fluorescence completely. These results show that also on intact leaves with bentazon a fast inhibition of the photosynthetic electron transport is obtained, though DCMU even under these conditions works faster and in lower concentrations. From this data we assume that bentazon is transported fairly quickly to its inhibition site in the electron transport chain in isolated chloroplasts and in intact tissues, but slower than DCMU.

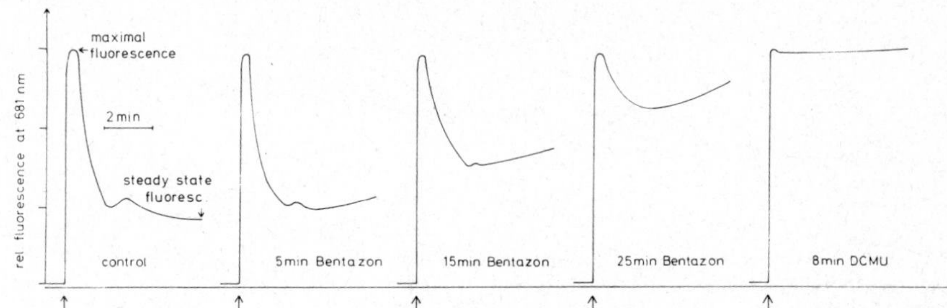

Fig. 5. Suppression of the variable fluorescence by bentazon. Raphanus cotyledons were incubated in 10^{-3} M bentazon for the indicated times. The fluorescence changes after onset of illumination were measured at 681 nm. (Upward arrows: light on).

Since bentazon is a very lipophilic substance, a transport to the more lipophilic inner part of the thylakoids, where photolysis of water occurs, seems to be fairly possible. From our results we assume that at this inner thylakoid side the inhibition site for bentazon is located in the range of the reaction centre of PS II and the PQ-shuttle. The place of action ought to be located before the inhibition site of DBMIB because the DCPIP/Asc catalysed electron transport through PS I is not affected by DBMIB[8] and bentazon. While after addition of DBMIB[9] p-BQ can be reduced via PS II, this is not possible after addition of bentazon. This again shows that the inhibition site is located before that of DBMIB.

This work was sponsored by the Deutsche Forschungsgemeinschaft.

References

1. Retzlaff, G. and Fischer, A., 1973, Mitteil. a. d. Biolog. Bundesanstalt f. Land- u. Forstwirtschaft 151, 179.
2. Lichtenthaler, H.K. and Tevini, M., 1969, Die Naturwissenschaften 56, 284.
3. Strasser, R.J., 1974, Experientia 30, 320.
4. Strotmann, H. and Gösseln, Chr. v., 1972, Z. Naturforsch. 27b, 445.
5. Yamashita, T. and Butler, W.L., 1968, Plant Physiol. 43, 1978.
6. Schwartz, M., 1971, In: Ann. Rev. Plant Physiol., 469.
7. Duysens, L.N.M. and Sweers, H.E., 1963, In: Studies on microalgae and photosynthetic bacteria (The University of Tokyo Press), 353.
8. Trebst, A. and Reimer, S., 1973, Z. Naturforsch. 28c, 710.
9. Gimmler, H. and Avron, M., 1971, In: Proc. II. Int. Congr. Photos. Res. (W. Junk N.V. Publishers, The Hague), 789.

M. AVRON, *Proceedings of the Third International Congress on Photosynthesis*, September 2-6, 1974, Weizmann Institute of Science, Rehovot, Israel
Elsevier Scientific Publishing Company, Amsterdam, The Netherlands, 1974

EFFECTS OF N-(PHOSPHONOMETHYL)GLYCINE ON PHOTOSYNTHETIC REACTIONS IN *SCENEDESMUS* AND IN ISOLATED SPINACH CHLOROPLASTS

J.J.S. van Rensen

Laboratory of Plant Physiological Research
Agricultural University, Wageningen, The Netherlands

Summary

The new herbicide N-(phosphonomethyl)glycine inhibited oxygen evolution in the green alga *Scenedesmus*. This inhibition increased with time. Incubation of the cells with a 7×10^{-4}M solution resulted after one hour in the light in 50% inhibition. After washing the cells the inhibiting effect could be partly removed. The inhibition increased both with higher light intensity and with temperature.

N-(phosphonomethyl)glycine had no immediate effect on electron transport in isolated spinach chloroplasts at concentrations below 10 mM. After pretreatment during 15 min. with the herbicide, inhibition was obtained below this concentration. The effect on electron transport was equal for $H_2O \rightarrow$ diquat and for $H_2O \rightarrow$ DCPIP. Since electron transport of ascorbate-DCPIP $\rightarrow$ diquat was not inhibited, it is concluded that the herbicide inhibits electron transport in or near Photosystem II.

Introduction

N-(phosphonomethyl)glycine is a new herbicide, abbreviated as glyphosate. It provides effective destruction of a broad variety of plants. Its mechanism of action is not known yet.

Upchurch and Baird[1] observed that light intensity has a significant influence on the response of plants to glyphosate. Therefore, we have studied the effects of this herbicide on photosynthetic oxygen evolution in the green algae *Scenedesmus* and on electron transport reactions in isolated spinach chloroplasts.

Materials and Methods

The herbicide glyphosate was formulated as the isopropylamine salt of N-(phosphonomethyl)glycine without a surfactant. The cultivation of the unicellular green alga *Scenedesmus* and the measurement of photosynthetic oxygen evolution were carried out as described previously[2,3].

Chloroplasts were isolated from fresh spinach leaves grown in a controlled climate room. Leaves (20 g) were ground in a Sorvall Omnimixer for 5 sec in 50 ml of an ice-cold medium containing 0.4 M sorbitol, 20 mM tricine-NaOH (pH 7.8), 10 mM NaCl, 2 mM sodium ascorbate and 2 mg/ml bovine serum albumin. The brei was filtered through four layers of cheese-cloth and centrifuged for 90 sec at 500 g. The supernatant was centrifuged during 8 min at 1000 g and the pellet was washed once with 50 ml of the grinding medium. The chloroplasts were taken up in 3 ml isolation medium and after chlorophyll determination according to Bruinsma[4] the suspension was diluted with the same medium to a suitable chlorophyll concentration. The whole procedure was carried out at 2°C.

For the study of electron transport these chloroplasts were osmotically shocked in the oxygen electrode chamber to give type D chloroplasts[5].

Three types of electron transport were measured: $H_2O \rightarrow$ diquat, ascorbate-DCPIP → diquat and $H_2O \rightarrow$ DCPIP. Oxygen uptake or release in these reactions were measured with a Gilson oxygraph provided with a YSI Clark oxygen electrode in a temperature-controlled reaction vessel at 25°C. The samples were illuminated with a projector lamp. The light beam passed a heat filter and 2.5 cm of a 1 M solution of $CuSO_4$ and reached the reaction vessel at an intensity of 15×10^5 ergs $cm^{-2}sec^{-1}$.

Results and discussion

The effect of glyphosate was studied on photosynthetic oxygen evolution in *Scenedesmus* cells. The inhibition increases with time and in table 1 the average effects for various time intervals after addition of the herbicide are given. Between 60 and 90 min after addition of the chemical, 7×10^{-4}M glyphosate causes 50% inhibition.

Table 1

Effect of glyphosate on oxygen evolution in *Scenedesmus* at various time intervals after addition

glyphosate concentration	averaged effect between 0-30 min after addition	averaged effect between 30-60 min after addition	averaged effect between 60-90 min after addition
4×10^{-4}M	102	95	89
10^{-3}M	83	38	21
2×10^{-3}M	44	9	7

20 µl cells per vessel, temperature 25°C, light intensity was saturating for photosynthesis, data in percent of control values.

In the following experiment oxygen evolution was first measured in two pairs of vessels. One hour after addition of water to the control pair of vessels and 2×10^{-3}M glyphosate to the other pair, the oxygen production was again determined.

Thereafter, aliquots of suspension of each pair of vessels were combined, centrifuged, washed once and resuspended in a fresh amount of buffer. The suspension density was made up to the same value as at the beginning of the experiment. Oxygen evolution was then measured in two vessels, one for each treatment. The data of table 2 show that the inhibiting effect can only be partly removed by this procedure.

Table 2

Effect of washing the cells on the inhibition of oxygen evolution by glyphosate

	control	2×10^{-3}M glyphosate
before addition	495	507
60 min after addition	514	10
after washing	530	168

20 µl cells per vessel, temperature 25°C, light intensity was saturating for photosynthesis, data in µl O_2 per hour.

Table 3 demonstrates that inhibition by glyphosate increases with higher light intensities. The herbicide inhibits both at light-limited and at light-saturated conditions. This shows that glyphosate belongs to the group of inhibitors affecting light dependent reactions as well as dark reactions in photosynthesis.

Table 3

Effect of light intensity on the inhibition of oxygen evolution in *Scenedesmus* by glyphosate

light intensity in 10^3 ergs $cm^{-2}sec^{-1}$	control	10^{-3}M glyphosate
2.5	221	186 (84)
8.8	270	200 (74)
41.5	461	168 (36)

20 µl cells per vessel, temperature 25°C, effect of glyphosate is measured 60 minutes after addition of the herbicide, data in µl O_2 per hour, numbers between parentheses: percentages of control.

Temperature too has a great influence on the inhibition of oxygen evolution by glyphosate (table 4). This is an indication that also dark reactions within the photosynthetic process are affected, since inhibition of photochemical reactions only should be largely temperature independent.

While we were investigating the effect of glyphosate on oxygen evolution in *Scenedesmus*, a communication of Croft et al.[6] appeared stating that N,N-bis(phosphonomethyl)glycine caused no significant inhibition of electron transport in isolated spinach chloroplasts. Preliminary studies on the effect of our compound,

N-(phosphonomethyl)glycine, on the Mehler reaction (electron transport $H_2O \rightarrow$ diquat) also showed no immediate effect up to a concentration of at least 10 mM. The inhibition increased with time.

Constant inhibition rates at lower glyphosate concentrations were obtained after incubation of the chloroplasts with the herbicide during 15 min at 0^oC. The slopes of the recorder trace when the inhibition was effective were compared with those obtained with chloroplasts incubated with the same volume of distilled water during the same time and at the same temperature.

Table 4

Effect of temperature on the inhibition of oxygen evolution in *Scenedesmus* by glyphosate

temperature	control	10^{-3}M glyphosate
20^oC	397	366 (92)
30^oC	784	193 (25)

20 µl cells per vessel, light intensity saturating for photosynthesis, effect of glyphosate is measured 60 min after addition of the herbicide, data in µl O_2 per hour, numbers between parentheses: percentages of control.

Table 5

Effects of glyphosate on three types of electron transport in isolated chloroplasts

glyphosate concentration	$H_2O \rightarrow$ diquat	$H_2O \rightarrow$ DCPIP	ascorbate-DCPIP $\rightarrow$ diquat
2×10^{-4}M	90	94	100
10^{-3}M	60	79	105
5×10^{-3}M	36	28	108

Chloroplasts were incubated either with glyphosate or with distilled water during 15 min at 0^oC. The reaction medium for electron transport $H_2O \rightarrow$ diquat contained 50 mM tricine-NaOH (pH 7.6), 0.1 M sorbitol, 5 mM $MgCl_2$, 5 mM NH_4Cl, 10 µM diquat, 2 mM dithioerythritol and chloroplasts equivalent to 150 µg chlorophyll in a total volume of 5 ml. The reaction medium for electron transport ascorbate-DCPIP $\rightarrow$ diquat was the same as that for $H_2O \rightarrow$ diquat, except the addition of 20 µM DCMU, 6 mM sodium ascorbate and 0.2 mM DCPIP. The reaction medium for electron transport $H_2O \rightarrow$ DCPIP contained 50 mM tricine-NaOH (pH 7.6), 0.1 M sorbitol, 5 mM $MgCl_2$, 5 mM NH_4Cl, 250 µM DCPIP and chloroplasts equivalent to 300 µg chlorophyll in a total volume of 5 ml. Data are in percentages of control.

Table 5 shows the effects of glyphosate on three types of electron transport. In the Mehler reactions $H_2O \rightarrow$ diquat and ascorbate-DCPIP $\rightarrow$ diquat the hydrogen peroxide was trapped with dithioerythritol and the net uptake of oxygen was measured. In the Hill reaction $H_2O \rightarrow$ DCPIP oxygen production was determined. In all cases electron transport was uncoupled by addition of NH_4Cl.

It is illustrated in table 5 that electron transport $H_2O \rightarrow$ diquat is inhibited to about the same extent as electron transport $H_2O \rightarrow$ DCPIP. Since electron transport ascorbate-DCPIP $\rightarrow$ diquat is not affected, it can be concluded that glyphosate inhibits photosynthetic electron transport in or near Photosystem II.

Acknowledgement

The herbicide glyphosate was kindly supplied by Mr. G.R. Schepens, Monsanto Europe S.A., Brussels.

References

1. Upchurch, R.P. and Baird, D.D., 1972. Herbicidal action of MON-0573 as influenced by light and soil. Proc. Western Society of Weed Science 25, 41-44.

2. Van Rensen, J.J.S., 1971. Action of some herbicides in photosynthesis of *Scenedesmus*, as studied by their effects on oxygen evolution and cyclic photophosphorylation. Meded. Landbouwhogesch. Wageningen 65-13, 1-80.

3. Van Rensen, J.J.S., 1974. Lipid peroxidation and chlorophyll destruction caused by diquat during photosynthesis in *Scenedesmus*. Physiol. Plant., in press.

4. Bruinsma, J., 1963. The quantitative analysis of chlorophylls a and b in plant extracts. Photochem. Photobiol. 2, 241-249.

5. Hall, D.O., 1972. Nomenclature for isolated chloroplasts. Nature 235, 125-126.

6. Croft, S.M., Arntzen, C.J., Vanderhoef, L.N., and Zettinger, C.S., 1974. Inhibition of chloroplast ribosome formation by N,N-bis(phosphonomethyl)glycine. Biochim. Biophys. Acta 335, 211-217.

M. AVRON, *Proceedings of the Third International Congress on Photosynthesis*,
September 2-6, 1974, Weizmann Institute of Science, Rehovot, Israel
Elsevier Scientific Publishing Company, Amsterdam, The Netherlands, 1974

THE EFFECT OF UNCOUPLERS ON ARTIFICIAL DONOR SHUTTLES FOR PHOTOSYSTEM I

G.A. Hauska

Lehrstuhl für Biochemie der Pflanzen, Ruhr-Universität Bochum, Germany.

1. SUMMARY

Photoreductions by photosystem I with weakly acidic donor compounds, like 2,6-dichlorophenolindophenol, are stimulated by uncouplers in phosphorylating chloroplast membrane preparations, while photoreductions with weakly basic compounds, like diaminodurene, are not. This uncoupler effect cannot be simply explained as a release of electron transport control by the high energy state, since the reaction with diaminodurene is also coupled to ATP formation. An explanation can be provided based on the topography of photosystem I with the oxidizing on the inner, the reducing side on the outer surface of the thylakoid membrane, and on the chemical properties of the donor compounds which determine their uncoupler sensitive distribution within illuminated, vesicular chloroplast membrane preparations.

The dependence of photoreductions in a particulate chloroplast preparation on plastocyanin is also shown.

2. INTRODUCTION

Electron transport through photosystem I in the presence of DCMU can be sustained by a number of artificial redox compounds[1,2], with varying efficiency of energy conservation[2]. A puzzling uncoupler effect on photoreductions has been known for quite some time[3], which cannot simply be explained by release of electron transport control: Although both are coupled to photophosphorylation only photoreduction with DPIP is stimulated by uncouplers like NH_4Cl, but photoreduction with DAD is not.

The experiments reported here allow the generalization, that donor compounds being able to form phenolate ions are stimulated by uncouplers while others, like the ones of the phenylenediamine type, are not. An explanation based on the topography of photosystem I is attempted.

3. METHODS and MATERIALS

Spinach chloroplasts[4] and plastocyanin from spinach[5] were prepared following the literature.

Digitonin subchloroplast fragments enriched in photosystem I were prepared according to Anderson and Boardman[6], using 0.5% for a

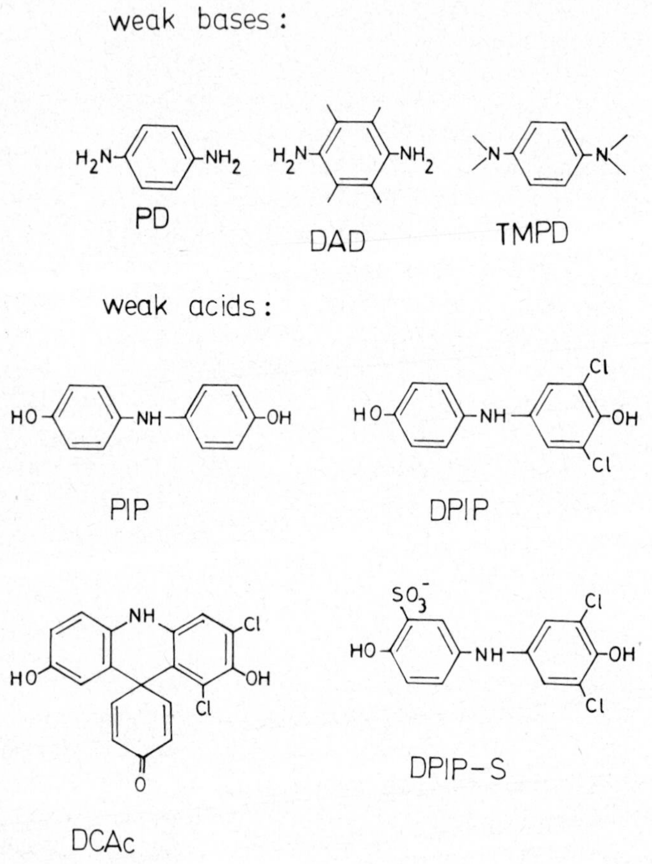

Fig. 1. Chemical formulae of the redox compounds used.
PD stands for p-phenylenediamone, DAD for diaminodurene, TMPD for N,N'-tetramethyl-p-phenylenediamine, PIP for phenolindophenol, DPIP for 2,6-dichlorophenolindophenol, DCAc for Dichloroacridan derivative, DPIP-S for 2,6-dichlorophenolindophenol-2'-sulfonate. All compounds have a pK around neutral pH.

"vesicular" preparation ($D200_v$) or 1.5% digitonin final concentration for a "particulate" preparation ($D200_p$); "vesicular" and "particulate" was discerned by ultrafiltration[7].

The assay for plastocyanin content in chloroplast preparations was carried out as previously described[8].

Photoreduction by photosystem I was followed by oxygen consumption with methyl viologen as the autooxidable electron acceptor, monitored by a Gilson oxygraph. The illuminating light intensity was 10^6 ergs/cm^2 per sec at the surface of the reaction vessel. The reaction mixture is described in the legend to Table 1.

Dichloroacridan was a kind gift of Dr.R. Hill.

4. RESULTS

In Fig. 1. the chemical formulae of the redox compounds used in photoreductions with photosystem I are depicted. They can be grouped into compounds with weakly basic (aromatic amines) and weakly acidic substituents (phenols). Because of its sulfonate group DPIP-S has the special property of being lipid impermeable.

Table 1. shows that photoreductions with PD, TMPD and DAD are not affected by the uncoupler NH_4Cl, but that the reactions with DPIP and DCAc are greatly stimulated in the vesicular systems, i.e. in chloroplasts and $D200_v$. Gramicidin gave similar results. A great number of additional redox compounds have been tested - they all fall into the line of the correlation of stimulation by uncouplers with the ability of the redox compound to form phenolate anions (s. DISCUSSION). Maximal rates are somewhat higher in $D200_v$ than in chloroplasts, reflecting some enrichment in photosystem I.

Plastocyanin and DPIP-S, both impermeable, showed only feeble reactions with chloroplasts, which is drastically increased in the digitonin fractions, expecially for plastocyanin, as has been known before (s.Ref 8). The reaction with both compounds is not stimulated by NH_4Cl. This suggests that DPIP-S and plastocyanin can react only if oxidation sites are exposed after fragmentation of chloroplasts and that the stimulation of the reaction with DPIP by uncouplers in vesicular systems might reflect the oxidation site inside the membrane (s. DISCUSSION).

Table 1.- Effect of NH_4Cl on Photoreductions by Photosystem I in chloroplasts, in Vesicular - and in Particulate Digitonin Subchloroplast Fragments.

The conditions and the assay of the reaction are described under METHODS. The reaction mixture contained: 50 mM Tricine-NaOH, pH 8.0, 50 mM NaCl, 5 mM $MgCl_2$, $2x10^{-5}$M DCMU and air saturated water to give a final volume of 1.5 ml. Ascorbate (5mM) and the electron donor compound (10^{-4}M, PC: $2x10^{-6}$M) were added before the reaction was started. NH_4Cl was added to 5 mM where indicated. The rate of photoreduction is expressed as µatoms O_2 reduced mg chlorophyll per hr. Eventual dark rates were subtracted from the rates in the light. PC stands for plastocyanin, $D200_v$ for vesicular and $D200_p$ for particulate digitonin subchloroplast fragments. The other abbreviations are the same as for Fig. 1.

Donor	Rate of Photoreductions					
	chloroplasts		$D200_v$		$D200_p$	
	-	$+NH_4Cl$	-	$+NH_4Cl$	-	$+NH_4Cl$
PD	660	625	-	-	-	-
DAD	1950	2020	1380	1380	150	165
TMPD	2450	2300	2860	2680	175	170
PIP	870	1530	-	-	-	-
DPIP	1290	2500	1980	3980	600	630
DCAc	1810	4500	4050	6100	6200	6350
DPIP-S	150	155	440	425	520	540
PC	50	50	1230	1100	925	900

DCAc is a very good donor also in D200p, followed by plastocyanin and the indophenols. Phenylenediamines showed a very poor reaction in D200p. D200p contained no detectable endogenous plastocyanin in contrast to $D200_v$ which retained about 30% on a chlorophyll basis compared to chloroplasts. Consequently phenylenediamines seem to react predominantly via plastocyanin, while indophenols, especially DCAc, can react with P_{700} directly.

Table 2. - The Dependence of Photoreductions by Photosystem I in particulate Digitonin Subchloroplast Fragments on Plastocyanin.

The conditions and the assay of the reaction are described under METHODS. The reaction mixture is given in the legend to Table 1. Plastocyanin was added to $2x10^{-6}$M were indicated. The abbreviations are given in the legends for Fig. 1 and Table 1. The rate of photoreduction is expressed as µatoms O_2 reduced/mg chlorophyll per hr.

Donor	Rate of Photoreduction	
	-	+ PC
-	50	1230
DAD	150	1630
TMPD	175	1550
DPIP	900	2150
DCAc	6200	6550
DPIP-S	700	1760

Table 2. shows this dependence of photoreductions on plastocyanin in $D200_p$ once more, with the expected pattern. Only the reactive DCAc is independent of plastocyanin, DPIP and DPIP-S give an appreciable rate without plastocyanin, while DAD and TMPD are almost inactive. From the data it cannot be estimated how much of the electron transport in the presence of plastocyanin is proceeding via the reduced donor compound and plastocyanin and how much via plastocyanin reduced directly by ascorbate.

5. DISCUSSION

The schemes in Fig. 2 provide a possible explanation for the different behaviour of redox compounds in photoreductions described here, and previously[3]. It is based first of all on the assumption that charged forms of the reduced donor - DH^- and DH_3^+ - are impermeable relative to the neutral form DH_2 (if charged forms are permeable - dotted arrows through the membrane in Fig. 2. - uncoupling by the donor itself is expected). Secondly it is assumed that phenolate anionic forms, being more electronegative, donate electrons more rapidly to photosystem I than neutral forms; ammonium forms are thought to be inactive. Thirdly we assume that the neutral forms of phenylenediamines react more readily than neutral forms of the phenol type. Of course the scheme rests entirely on the notion that the oxidizing end of photosystem I is located on the inner surface of the thylakoid membrane (s. 9 for a review). This is established here once more by the relative inactivity of the impermeable compound DPIP-S, as has been shown before[10].

According to Fig. 2. equilibration of the donor through the membrane is brought about by the neutral forms; its concentration is the same on both sides of the membrane, independent on pH. It forms charged species in the aqueous phases which is governed by the respective pH and pK's. A pH-gradient established by the oxidation of the donor will suppress formation of phenolate anions but will increase formation of ammonium ions inside the vesicles. In other words it will force active phenolate forms (scheme A) out, especially if the pK of the donor falls between the pH values outside and inside. It will, on the other hand, cause accumulation of inactive ammonium forms of the donors (scheme B). This is analog to the mechanism of uncoupling by NH_4Cl[11], but the concentration of the donor and the pK of aromatic amines are too low to cause uncoupling, but internal buffering does occur[12]. Uncouplers abolish the pH gradient and therefore increase the concentration of phenolate forms of a donor, but do not affect the distribution of neutral forms. Thus they increase the rate of photoreduction with reactive phenolate anions, but have no effect with donors of the phenylenediamine type. The results with external plastocyanine suggest that phenolate forms can react with the reaction center directly, while neutral forms preferentially react with plastocyanin.

This mechanism for the stimulation of electron transport by uncouplers is more specified than the classical view of uncoupler action by release of electron transport control. It is another example of how a pH-gradient might control electron flow, this time not via the pH-optimum of rate limiting electron transport enzymes inside the thylakoid[13], but by controlling the distribution of active forms inside-outside the membrane core of a mobile electron carrier.

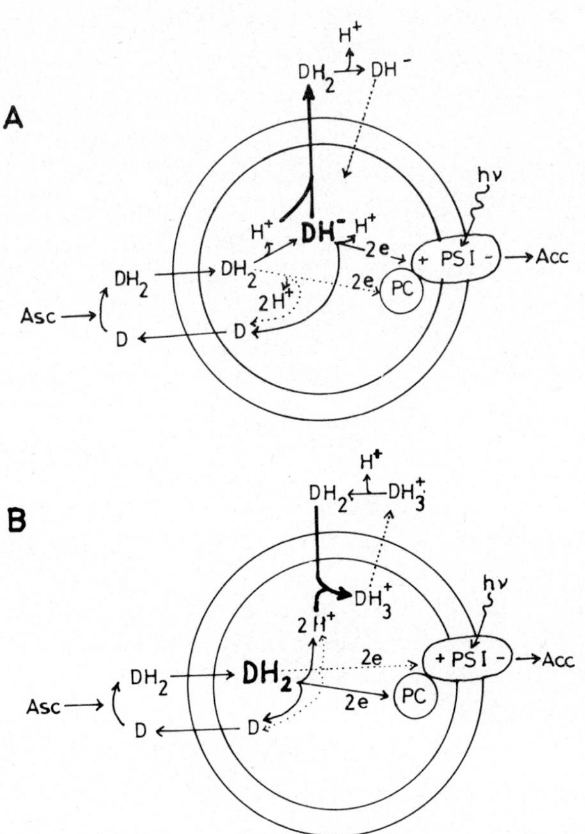

Fig. 2. Influence of a pH-gradient on the steady state distribution of acidic and basic electron donors for photosystem I in chloroplast membrane vesicles.
Asc stands for ascorbate, DH_2, DH^- and DH_3^+ for reduced forms of the electron donor, D for the oxidized form of the donor, PC for plastocyanin, PSI for photosystem I and Acc for electron acceptor.

Acknowledgements

I am grateful to Dr. A. Trebst and coworkers for their interest and numerous discussions, and to Deutsche Forschungsgemeinschaft for financial support.

REFERENCES

1. Vernon, L.P. and Zaugg, W.S., 1960, J. Biol. Chem. 235, 2728
2. Trebst, A. and Pistorius, E., 1965, Z. Naturf. 20b, 143
3. Izawa, S. et al., 1967, Brookhaven Symp. in Biology 19, 169
4. McCarty, R.E. and Racker, E., 1967, J. Biol. Chem. 242, 3435
5. Anderson, M.M. and McCarty, R.E., 1969, Biochim. Biophys. Acta 189, 193
6. Anderson, J.M. and Boardman, N.K., 1966, Biochim. Biophys.Acta 112, 403
7. Arntzen, C.J. et al., 1972, Biochim. Biophys. Acta 256, 85
8. Sane, P.V. and Hauska, G.A., 1972, Z. Naturf. 27b, 932
9. Trebst, A., these proceedings
10. Hauska, G.A., Draber, W. and Trebst, A., 1973, Biochim. Biophys. Acta 305, 632
11. Crofts, A.R., 1967, J. Biol. Chem. 242, 3352
12. Avron, M., 1972, Proc. 2nd Int. Congress Photosynthesis, vol.2, eds. G. Forti, M. Avron and B.A. Melandri (Dr. W. Junk, The Hague) p. 861
13. Bamberger, E.J., Rottenberg, H. and Avron, M., 1973, Eur. J. Biochem. 34, 557

M. AVRON, *Proceedings of the Third International Congress on Photosynthesis*, September 2-6, 1974, Weizmann Institute of Science, Rehovot, Israel
Elsevier Scientific Publishing Company, Amsterdam, The Netherlands, 1974

RESTORATION BY SILICOTUNGSTIC ACID OF DCMU-INHIBITED PHOTOREACTIONS IN SPINACH CHLOROPLASTS

J.M. Galmiche, G. Girault

Département de Biologie
Centre d'Etudes Nucléaires, Saclay, B.P. 1, Gif-sur-Yvette
France.

Introduction.

Uncoupled chloroplasts by removal of their coupling factor exhibit a very fast DCMU sensitive Hill reaction but do not show any reversible light-induced pH rise. A polyanion, the silicotungstic acid (STA), fixed by those chloroplasts, restores the light-induced reversible pH rise[1] although it changes the nature of the other photoreactions especially with respect to their sensitivity to DCMU[2]. We will discuss the effects of silicotungstic acid addition on three Photosystem II dependent reactions which were shown to be inhibited by DCMU : Hill reactions[3], photobleaching of the carotenoids[4] and pyocyanine photoreduction prior to the cyclic photoreactions mediated by the oxidized dye[5-6].

It is important to state that, during chloroplasts preparation[1], no protein like serum albumine was added and that silicotungstic acid was always used at pH below 7.5 as it decomposes at alkaline pH[7].

Hill reactions.

Hill reaction, measured as oxygen evolution in the presence of ferricyanide, p-benzoquinone and p-phenylenediamine or as 2,6-dichlorophenolindophenol (DCPIP) reduction, is inhibited by 1 μM DCMU with chloroplasts suspensions (10 to 80 μg of chlorophyll per ml). The Hill reactions are partly restored by silicotungstic acid addition (Fig. 1).

If 200 to 250 nmoles of silicotungstic acid per mg of chlorophyll are mixed with the chloroplasts suspension all the acid is fixed by the chloroplasts which display DCMU insensitive O_2 evolution even if they are spun down, washed and resuspended in a fresh medium without silicotungstic acid. The optimum rate of O_2 evolution is observed around pH 7[2] as Bamberger et al.[8] have reported for the electron transport rate as a function of the membrane pH (arithmetic mean of the internal and external pH). But the light dependent O_2 uptake with methylviologen is completely inhibited by DCMU even after addition of silicotungstic acid (Fig. 2).

Photobleaching of the carotenoids.

After treatment of the chloroplasts by silicotungstic acid ferricyanide reduction is accompanied by a bleaching of the carotenoids, insensitive to DCMU. Thus it is not possible to follow ferricyanide reduction by recording the change of the optical density difference between two wavelengths (420 nm minus 520 nm). There is not only a drop of the optical density near 500 nm but also a rise around 360 nm which might correspond to a shift of the absorption bands of the pigments (Fig. 3). Carotenoids bleaching is observed at the same extent under air and nitrogen, catalase being or not added. That photobleaching is comparable to this reported by Yamashita et al.[4] during Hill reaction in the presence of azide or hydroxylamine; that later photobleaching is inhibited by DCMU whereas this is not the case with silicotungstic acid.

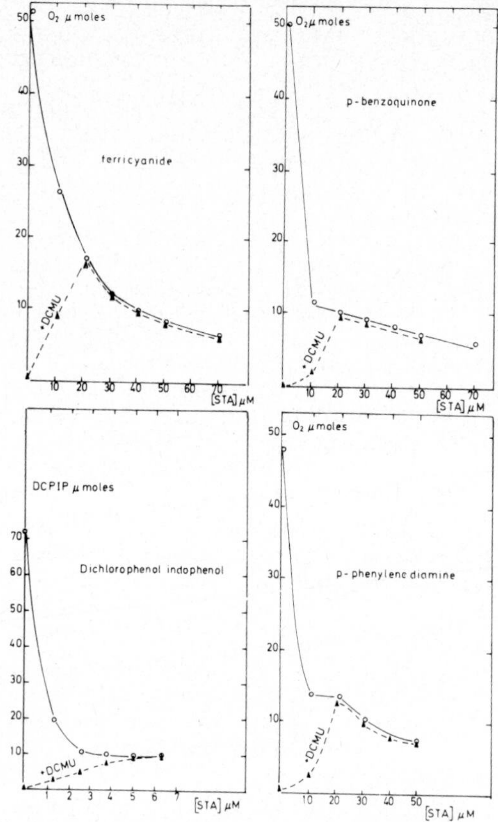

Fig. 1 : Rate of the Hill reactions with and without DCMU as a function of the silicotungstic acid concentration in µmoles per h and mg of chlorophyll.

EDTA-treated chloroplasts were suspended in the basic medium, 10mM tricine-maleate, 20 mM KCl, 10 mM $MgCl_2$. Chlorophyll concentration was 80 µg per ml except for the experiments with DCPIP where it was 10 µg per ml. Additives were 0.5 mM ferricyanide 0.5mM p-benzoquinone, 0.25 mM p-phenylenediamine and 0.5 mM ferricyanide, 40 µM DCPIP. pH was 6.5 and temperature 20°C. O_2 evolution was measured with a Clark electrode in white light (40 000 lux). DCPIP reduction was followed with the scattering attachment of the Cary 14 spectrophotometer in 480 nm light (22 000 ergs. $cm^{-2}.s^{-1}$).

Pyocyanine mediated proton uptake.

DCMU inhibits cyclic photoreactions mediated by oxidized pyocyanine[5-6]. Prior to those cyclic photoreactions oxidized pyocyanine must be partly reduced by chemicals or by a light Photosystem II dependent reaction which is inhibited by DCMU.

Broken chloroplasts suspensions display a light dependent reversible pH change mediated by oxidized pyocyanine. DCMU inhibits that reaction but addition of silicotungstic acid, roughly in the same ranges of concentration as those needed for restoration of the Hill reaction, restores completely the light dependent pH changes (Fig. 4). Silicotungstic acid addition depresses the diphenylcarbazone (DPCN) dismutation which is a Photosystem I dependent reaction[9]. Therefore that restoration is rather linked to the restoration of the Hill reaction (Photosystem II dependent) than to a stimulation of Photosystem I reactions.

Discussion and conclusion.

We have already shown that is not the non fixation of DCMU on the membranes which explains the lack of inhibitory effect of DCMU on Hill reaction with chloroplasts suspension in the presence of silicotungstic acid[2].

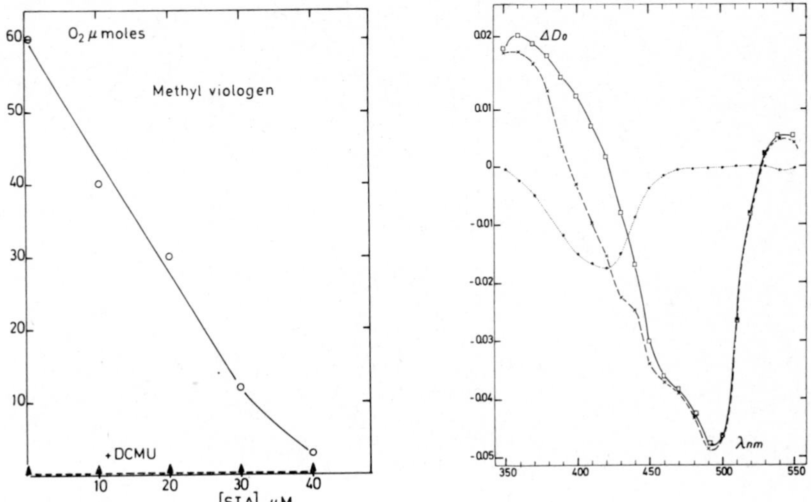

Figure 2 : Rate of oxygen consumption mediated by methylviologen with and without DCMU as a function of the silicotungstic acid concentration (left hand) : in µmoles per h and mg of chlorophyll.

The conditions were the same as these in Figure 1 except 50 µM methylviologen.

Figure 3 : Absorption changes of a broken chloroplasts suspension in white light 5000 lux , (right hand).

Broken chloroplasts (40µg of chlorophyll per ml) were suspended in a medium 2 mM tricine-maleate, 20 mM KCl, 10 mM $MgCl_2$, 0.1 mMferricyanide, 6.5 µM DCMU and 27 µM silicotungstic acid (STA). pH was 6.5. Sample was illuminated in the double beam spectrophotometer during 5 minutes at 20°C. Sample spectrum (x - - - x) was recorded against a standard kept in the darkness. Chloroplasts suspension was spun down and the supernatant spectrum was recorded (o - - -o), giving the part of the suspension spectrum which was alloted to ferricyanide reduction. By difference we drew the difference spectrum (□——□) giving the spectral changes of the pigments.

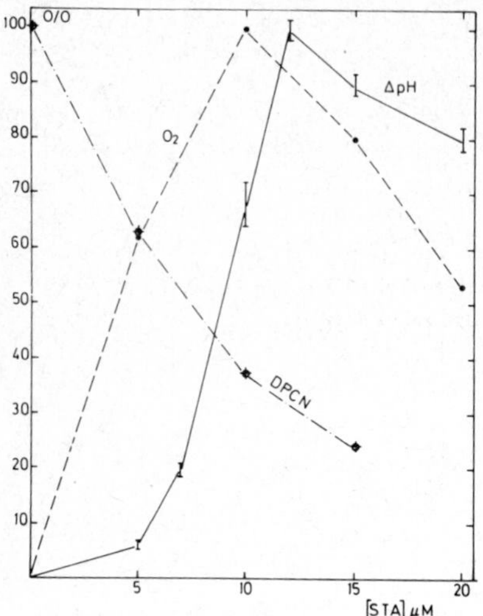

Figure 4 : Oxygen evolution (O_2), light induced proton uptake (Δ pH) and diphenylcarbazone dismutation (DPCN) in percent of the maximum rate as a function of the silicotungstic acid concentration.

Experiments were performed with broken chloroplasts. For O_2 an ΔpH measurements the basic medium was 1 mM tricine maleate, 20 mM KCl, 10 mM $MgCl_2$, 5 μM DCMU. Additives were 50 μM pyocyanine or 0.5 mM ferricyanide. For DPCN dismutation medium was 100 mM sodium phosphate, 0.3 mM DPCN, 5 μM DCMU. Chlorophyll concentration was 40 μg per ml, pH 7 and temperature 20° C. Oxygen evolution was measured with a Clark electrode in white light (20000 lux) the maximum rate was 25 μmoles per h and mg of chlorophyll. Proton uptake was determined by following the pH of the suspension in white light (40000 lux). The maximum rate was 600 nmoles of HCl per mg of chlorophyll. DPCN dismutation was measured by recording the absorption change at 480 nm with the scattering attachment of the Cary 14 spectrophotometer in a 680 nm light (40000 ergs.cm^{-2}. s^{-1}), the maximum rate was 300 μmoles per h and mg of chlorophyll.

Chloroplasts treatment with silicotungstic acid leads to a sharp decrease of the electron transport rate but it is possible to protect at least partly those chloroplasts against that drop. After silicotungstic acid has been fixed by the chloroplasts we add 0.2 % of bovine serumalbumine. The oxygen evolution rate of those chloroplasts is closer to that of the control. With that sequence of treatments (silicotungstic acid then serumalbumine) the chloroplasts suspensions display an oxygen evolution with ferricyanide almost insensitive to dibromothymoquinone (DBMIB[10]) or DCMU addition; in contrast the oxygen uptake with methylviologen is completely inhibited by DCMU (Table 1).

All the reactions we have shown to be insensitive to DCMU in chloroplasts after silicotungstic acid addition depend on the Photosystem II working. Pyocyanine like ferricyanide or quinones are reduced at the reducing site of the Photosystem II whereas carotenoids are photooxidized at the oxidizing site. Methylviologen has a too low redox potential to be reduced at the reducing site of Photosystem II but is reduced at the reducing site of the Photosystem I; so oxygen uptake with methylviologen needs the both Photosystems to be working together.

Therefore we conclude that after treatment with silicotungstic acid the Hill reagents with a high enough redox potential (at least higher than that of the methylviologen) can accept electrons from a part of the electron transport chain between the Photosystem II reaction center and the site of the block by DCMU and a fortiori by DBMIB. Those sites which are able to cede electrons are usually buried in the membrane. Treatment with silicotungstic acid allows those sites to be reached by lipid and water soluble Hill reagents[11]. Two prerequisites have to be fullfilled : redox potential of the Hill reagent must be higher than a lower limit which is fixed by the reducing capacity of the Photosystem II and the external pH must be the optimal pH for the Photosystem II photoreactions.

O_2 EVOLUTION + μmoles.h^{-1}
O_2 UPTAKE −

ADDITIVES	SILICOTUNGSTIC ACID	
	0 μM	25 μM
O	+ 48 − 43	+ 11 − 16
BSA	+ 60 − 46	+ 32 − 33
DCMU	+ 0 − 0	+ 10 − 0
BSA + DCMU	+ 0 − 0	+ 28 − 0
DBMIB	+ 24	+ 9
BSA + DBMIB	+ 28	+ 28

Table 1 : Effect of silicotungstic acid addition on the initial rate of oxygen evolution (+) with ferricyanide and of oxygen uptake (-) with methylviologen in μmoles per h and mg of chlorophyll.

The basic medium was 10 mM tricine maleate, 20 mM KCl, 10 mM $MgCl_2$, 0.5 mM ferricyanide or 50 μM methylviologen. Additives were 10 μM DCMU, 10 μM dibromothymoquinone (DBMIB), 0.2 % bovine serumalbumine (BSA). Chlorophyll concentration was 80 μg per ml, pH 6.5, temperature 20°C. Oxygen uptake (-) and oxygen evolution (+) were measured with a Clark electrode in white light (40000 lux).

References.

1 - Girault, G., Galmiche, J.M., FEBS Letters (1972), 19, 315.
2 - Girault, G., Galmiche, J.M., Biochim. Biophys. Acta, (1974), 333, 314.
3 - Duysens, L.N.M., Sweers, H.E., in Studies on Microalgae and Photosynthetic Bacteria, Japan Society Plant Physiology, University of Tokyo, p. 335 (1963).
4 - Yamashita, K., Konishi, K., Itoh, M., Shibata, K., Biochim. Biophys. Acta, (1969), 172, 511.
5 - Asahi, T., Jagendorf, A.T., Arch. Biochem. Biophys., (1963), 100, 531.
6 - Hauska, G.A., Mc Carty, R.E., Racker, E., Biochim. Biophys. Acta, (1970), 197, 206.
7 - Souchay, P., in Polyanions et Polycations, Gauthier Villars ed., (1963), p. 49.
8 - Bamberger, E.S., Rottenberg, H., Avron, M., Europ. J. Biochem. (1973), 34, 557.
9 - Haveman, J., Duysens, L.N.M., Van Der Geest, T.C.M., Van Gorkom, H.J., Biochim. Biophys. Acta, (1972), 283, 316.
10- Trebst, A., Harth, E., Z. Naturforsch. (1970), 25 b, 1157.
11- Saha, S., Ouitrakul, R., Izawa, S., Good, N.E., J. Biol. Chem., (1971), 246, 3204.

M. AVRON, *Proceedings of the Third International Congress on Photosynthesis*, September 2-6, 1974, The Weizmann Institute of Science, Rehovot, Israel
Elsevier Scientific Publishing Company, Amsterdam, The Netherlands, 1974

COMPARISON OF THE PHOTOSYNTHETIC MEMBRANE OF VEGETATIVE CELLS AND HETEROCYSTS OF THE BLUE-GREEN ALGA ANABAENA FLOS-AQUAE*

Sigrid M. Klein, Jesse M. Jaynes and Leo P. Vernon
Department of Chemistry and Research Division
Brigham Young University, Provo, Utah

Summary

Photosynthetic membrane fragments from vegetative cells and heterocysts of Anabaena flos-aquae were treated with Triton X-100 and the resultant fragments separated by discontinuous sucrose density gradient centrifugation. In both cases the major band was a dense blue-green band which concentrated in the 8% sucrose layer. This band contained Chl a, but little or no carotenoids, and exhibited photosystem 1 photoactivity as measured by the diphenylcarbazone reaction. With vegetative cells, another brown-orange band was observed in the 4% sucrose layer. This band contained most of the carotenoids and Chl a. With heterocysts, the 4% layer contained neither Chl a nor carotenoids, but did contain pheophytin a. In the case of heterocysts, the 20% layer contained carotenoids and some Chl a.

Analysis of the derived fractions with SDS gel electrophoresis showed that a 75 kilodalton polypeptide was the major one in the 8% sucrose layer which contained the majority of the Chl a and exhibited PS 1 activity. This was observed for both vegetative cells and heterocysts, and leads to the conclusion that this polypeptide is the one associated with Chl a in PS 1 of this alga. The photosynthetic membrane fragment exhibited other polypeptides of lower molecular weight, but has none of the polypeptides in the 20 to 30 kilodalton range which are associated with the light harvesting Chl-protein complex found in higher plants and which are associated with Chl b in higher plants.

Introduction

The blue-green algae perform photosynthesis in a manner generally similar to other plant species, but these algae have several distinct features which set them apart from higher plants in terms of their organization and mechanism of photosynthesis. Like the photosynthetic bacteria, the blue-green algae are prokaryotes and contain their photosynthetic pigments and functional protein components within a membrane system which is not organized into a specific

*This work was supported by Research Grant GB-28008X (L. P. V.) from the National Science Foundation.

organelle such as the chloroplast of higher plants, but is distributed throughout the cell[1,2]. The photosynthetic apparatus of blue-green algae resembles that of higher plant chloroplasts in that it contains chlorophyll a (Chl a) as a light harvesting and photoreactive pigment, produces oxygen and reduces NADP by coupling two photosystems functionally in electron transport[3], contains as the usual electron transport components cytochromes, plastocyanin and quinones[1,3] and performs both non-cyclic and cyclic phosphorylation[4,5]. The blue-green algae have no Chl b, however, utilize phycocyanin as an accessory light harvesting pigment, do not form stacked thylakoids, and lack the specific polypeptides which are part of the "light harvesting" Chl-protein complex of higher plants[6].

A unique feature of the blue-green algae is the formation of heterocysts under conditions of limiting nitrogen nutrition[1,2]. These heterocysts carry out nitrogen fixation at an elevated rate, and recent experiments show that the polypeptide components of the nitrogenase enzyme are the major polypeptides synthesized in forming heterocysts[7]. Changes in the pigment organization and perhaps the electron transfer components occur during the formation of the heterocysts, and analysis of the fluorescence yield and delayed light emission of heterocysts show that they contain an active photosystem 1 (PS 1) with a Chl/P700 ratio of 90, but they do not contain an active PS 2[8].

Treatment of *Anabaena variabilis* with Triton X-100 produces membrane fragments which have been separated by sucrose density gradient centrifugation[9]. PS 1, as measured by the presence of P700, was located as a distinct blue-green band. A lighter, orange band appeared to contain PS 2, since it showed a 685 nm fluorescence band, but photochemical activity relating to PS 2 was not shown. In the present experiments we report on the treatment of vegetative cells and heterocysts of *Anabaena flos-aquae* with Triton X-100 and the characterization of the fragments so produced.

Materials and methods

A culture of *Anabaena flos-aquae* was obtained from the Charles F. Kettering Laboratories, Yellow Springs, Ohio. Cells were grown on a medium containing per liter; 0.0176 M $NaNO_3$; 0.31×10^{-3}M $MgSO_4 \cdot 7H_2O$; 0.19×10^{-3}M Na_2CO_3; 0.24×10^{-3} M $CaCl_2 \cdot 2H_2O$; 0.03×10^{-3}M citric acid; 6 mg $FeNH_4$-citrate; 1 mg Na_2-EDTA; 0.22×10^{3}M K_2HPO_4; 2.86 mg H_3BO_3; 1.81 mg $MnCl_2 \cdot 4H_2O$; 0.222 mg $ZnSO_4 \cdot 7H_2O$; 0.39 mg $Na_2MoO_4 \cdot 2H_2O$; 0.79 mg $CuSO_4 \cdot 5H_2O$; 0.05 mg Co $(NO_3)_2$; $6H_2O$. The cultures were gassed with a mixture of 1% carbon dioxide in air and exposed to fluorescent light for 6-10 days. Vegetative cells of one liter culture of *A. flos-aquae* were suspended in 40 ml of 0.01 M Tris pH 7.5 and sonicated twice for one minute with a 5 minute cooling period in a Sonifier Cell Disrupter at 120 Watts. Larger cell debris was removed by centrifugation at

17,300 g for 10 minutes. The photosynthetically active membrane fragments were collected by further centrifugation of the supernate at 160,000 g for 45 minutes. The pellets were washed twice with the above buffer in order to remove soluble proteins and pigments. The membrane fragments were subjected to 5% Triton-X100 in 0.005 M Tris pH 7.5 and a grey debris removed by centrifugation at 17,300g for 10 minutes. The dark green supernate was further resolved on a non-continuous sucrose gradient ranging from 4-50% sucrose. The gradient was subjected to 131,000g for 16 hours.

Heterocysts were obtained by growing A. flos-aquae in the above medium, omitting sodium nitrate and substituting all other nitrogen containing compounds with appropriate derivatives, under 1% carbon dioxide in a nitrogen atmosphere. After 8-10 days the cells were harvested and suspended in a buffer containing 10^{-3} M Tris-HCl, pH 7.6; 2×10^{-3}M EDTA; 0.5M D-mannitol; 1 mg/ml. lysozyme; and incubated at 37°C overnight. The heterocysts were collected by centrifugation at 10,000g for 10 minutes. They were washed repeatedly with the above mannitol tris buffer in the absence of lysozyme until the vegetative cell debris had been completely removed. The heterocysts were fragmented by means of a RHO Scientific Inc. Glass Bead Cell Homogenizer at 3/4 speed for 12 minutes using glass beads of 0.10-0.11 mm diameter. The preparation was kept at 4°C. Large particles were separated by centrifugation at 10,000g for 10 minutes and discarded. The supernate was subjected to 160,000g for 45 minutes and the membrane fragments collected as a well-packed pellet. Further treatment of the membrane fragments from heterocysts was according to the procedure described for the vegetative system.

SDS-acrylamide electrophoresis was carried out essentially as described by Weber and Osborn[10]. The membrane preparations were suspended in 1% SDS:0.01 M sodium phosphate buffer pH 7.0; 1% mercaptoethanol, and heated at 94° C for 2 minutes. The ratio of SDS:protein was 5. The dimensions of the gel were 0.6x7 cm. Electrophoresis was carried out at 8 mA with the anode connected to the bottom of the gel. The gels were stained with Coomassie blue for 3 hours and destained electrophoretically at 6 mA per gel. Desitometer tracings were obtained with a Schoeffel Spectrodensitometer Model SD 3000.

Results

The membrane fragments produced with Triton X-100 were separated by discontinuous sucrose density gradient centrifugation. The vegetative cells of A. flos-aquae produced a centrifugation profile which was similar to that obtained earlier with A. variabilis[9], with the major blue-green band appearing at the 8% sucrose layer. The majority of the carotenoids appeared at the 4% sucrose layer, which was a brownish-orange in color. Some orange color was located in the 6% layer, and some faint blue-green color was located below the 8% layer, but the majority of the pigments sedimented to these two layers.

In the case of the heterocysts of A. flos-aquae, a somewhat different pattern was obtained, as shown in Fig. 1. In this case a major pigmented band was observed at the 20% sucrose layer, and the 4% band was more brownish in color.

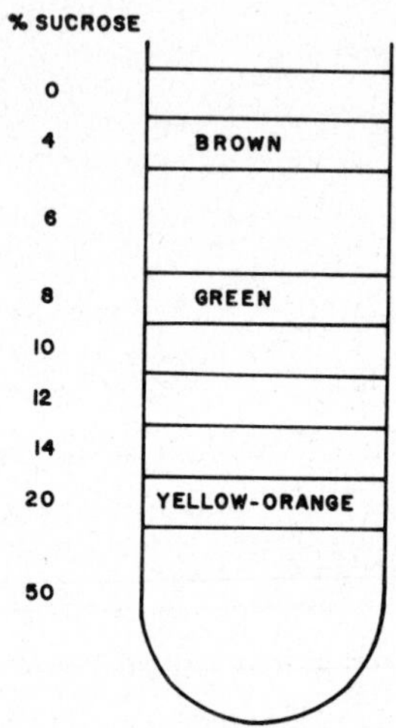

Fig. 1. Sucrose density gradient of membrane fragments obtained by the action of Triton X-100 upon photosynthetic membrane fragments of Anabaena flos-aquae heterocysts.

The absorption spectra of the fractions obtained from the vegetative cells are shown in Fig. 2. The pattern observed is similar to that shown earlier for A. variabilis, with the carotenoids and some Chl a located in the 4% layer while the major amount of Chl a and only small amounts of carotenoid are located in the blue-green band at 8% sucrose. The specific carotenoids have not been identified in the present investigation, but with A. variabilis, β-carotene was the major carotenoid in the fraction corresponding to the present 8% layer, while myxoxanthophyll was the predominent one in the 4% layer[9].

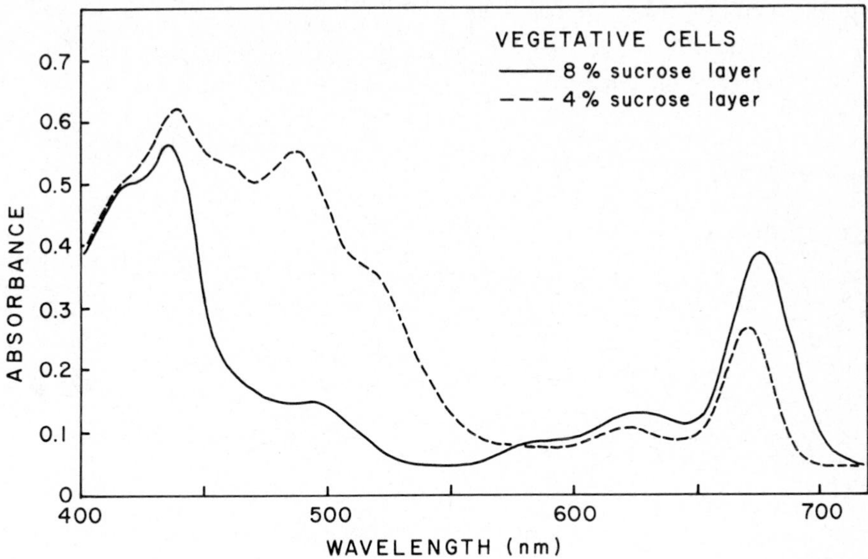

Fig. 2. Absorption spectra of the two major pigmented fractions obtained by the action of Triton X-100 upon photosynthetic membrane fragments of vegetative Anabaena flos-aquae. The fractions were isolated by discontinuous sucrose density gradient centrifugation.

Fig. 3 shows similar data obtained with heterocysts of A. flos-aquae. In this case the absorption spectrum of the 8% layer is not included, since it was essentially similar to that of the 8% layer of the vegetative cells. The exact locations of the absorption maxima are given below in Table 2. The data of Fig. 3 show a marked difference in the 4% layer obtained from heterocysts, when compared to the similar fraction from vegetative cells. The major Chl a peak at 438 nm is missing, and the main absorption in the blue region is shifted toward the blue, occurring at 418 nm. New bands appear at 508 and 540 nm. These data are consistent with the presence of pheophytin a in lieu of Chl a in the 4% fraction of heterocysts. Also, a strong yellow-orange band is observed at the 20% sucrose layer. The absorption spectrum shows the presence of carotenoid absorption in the 450 to 520 region, and also shows the presence of Chl a in this fraction.

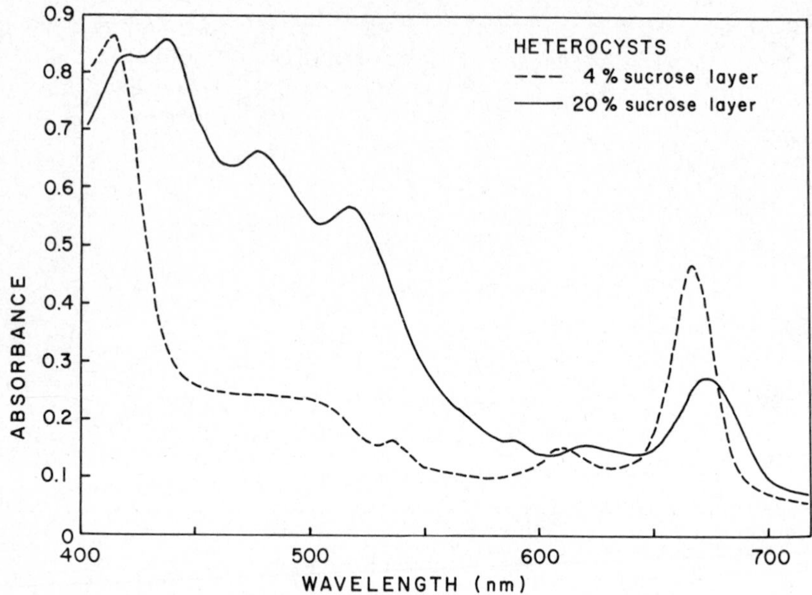

Fig. 3. Absorption spectra of the 4% and 20% bands obtained by sucrose density gradient centrifugation of photosynthetic membranes from *Anabaena flos-aquae* heterocysts treated with Triton X-100.

The fractions obtained from both vegetative cells and heterocysts of *A. flos-aquae* were assayed for photochemical activities, with the results shown in Table 1. Photosystem 1 activity was determined by the diphenylcarbazone (DPCO) method[11], and photosystem 2 activity was determined by use of the artificial electron donor to photosystem 2, diphenylcarbazide (DPC)[12]. As shown in Table 1, the 8% fraction from both the vegetative cells and heterocysts showed photosystem 1 activity, but a higher rate on a Chl basis was obtained in the 4% layer in each case. In terms of total activity, the majority of the activity is in the 8% layer. The assays for DPC-supported reduction of dichlorophenolindophenol (DPIP) were positive, but the reaction was not sensitive to either 3-(3,4-dichlorophenyl)-1, 1-dimethylurea (DCMU) or heating. Thus, it appears that the DPC-DPIP activity is not necessarily an indication of a functional PS 2, but could be catalyzed by some solubilized pigments. It should be noted also, that PS 1 activity, as measured by the DPCO reaction, is not always observed. In general, better DPCO activity is obtained with cells which have been ruptured by shaking with glass beads as opposed to sonication.

Table 1

Photochemical activities of sucrose density gradient fractions of Triton X-100 - treated Anabaena flos-aquae

Fraction	Photosystem 1 DPCO reaction*	Photosystem 2 DPC-DPIP reaction* Control	+ 10^{-5} M DCMU	Heated
Vegetatave				
4%	244	45	35	-
		70	-	87
8%	171	0	0	-
Heterocysts				
4%	123	42	-	-
8%	99	0	-	-

*Rates in µmoles/hr•mg Chl

Once the photosynthetic membrane fragments have been isolated from either vegetative cells or heterocysts, the resultant 8% sucrose band from density gradient centrifugation contains the majority of the Chl a and protein. We have examined the nature of the polypeptide composition of this 8% sucrose layer for both the vegetative cells and heterocysts, as shown in Figs. 4 and 5. The 4% sucrose band also contains protein, but it is present in much smaller amounts, and further work will be necessary to characterize this fraction in terms of its polypeptide composition.

As shown in Fig. 4, the membrane fragment of vegetative cells shows three major polypeptide components at 57, 63 and 75 kilodaltons. A higher molecular weight polypeptide is also present. It is important to note that these fragments do not contain any major polypeptides in the 20 to 30 kilodalton range, which are always found in the photosynthetic membranes of higher plants which contain Chl b. Three polypeptides which occur in this region are related to the presence of Chl b, and two of them are complexed to Chl b to serve as a light harvesting Chl-protein complex[6, 13-15]. The absence of these polypeptides in photosynthetic membranes of A. variabilis was noted in a previous study[6].

The 8% sucrose layer, which contains the photosynthetic apparatus associated with PS 1, as indicated by the presence of P700[9] and DPCO activity, shows the presence of the three polypeptides in the 57-75 kilodalton region, with the 75 component being the major polypeptide. The higher molecular weight component is not observed, and the concentrations of the lower molecular weight components are greatly reduced.

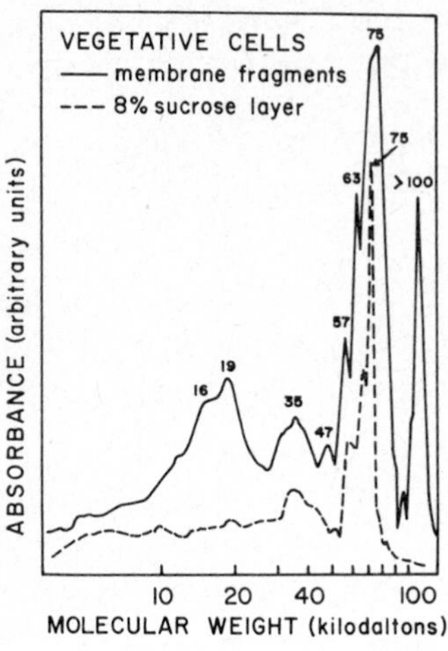

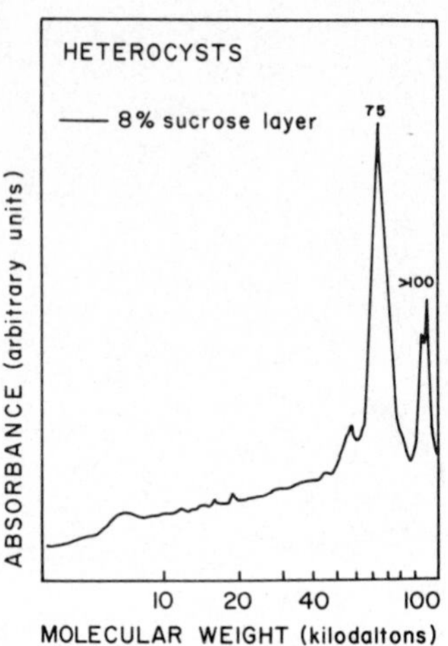

Fig. 4. SDS gel electrophoresis patterns of photosynthetic membrane fragments and the material located in the 8% sucrose layer following treatment of vegetative cells of A. flos-aquae with Triton X-100 and sucrose density gradient centrifugation.

Fig. 5. SDS gel electrophoresis patterns of the Chl a-containing fraction isolated in the 8% sucrose layer obtained by treating Anabaena flos-aquae heterocysts with Triton X-100 followed by discontinuous sucrose gradient centrifugation.

Discussion

The formation of heterocysts result in significant changes in the physiology of the cell, with PS 2 becoming nonfunctional[8] and the emergence of a prominent nitrogen fixation enzyme complex[7]. These events allow the heterocyst to shift functionally from a CO_2 fixation to a nitrogen fixation system utilizing the reducing power and ATP produced via PS 1 in the reduction of nitrogen. Such changes would be expected to be reflected in some corresponding changes in the structure of the photosynthetic membrane. Our data show that with both vegetative cells and heterocysts, a fraction containing most of the Chl a and exhibiting PS 1 activity is obtained. The polypeptide compositions of these two fractions are quite similar, indicating no significant change in this portion of the photosynthetic apparatus. Although in both cases a fraction is isolated which contains the majority of the carotenoids and some Chl a, and which by comparison with fluorescence data obtained previously with A. variabilis[9] would be expected to contain PS 2, it was not possible to identify a photochemically active PS 2, since the DPC-DPIP reaction used to assay PS 2 activity[12] was not sensitive to DCMU or heat.

Table 2 summarizes the spectroscopic data obtained, giving the locations of the absorption maxima of the fractions studied. The major bands are underlined.

Table 2

Absorption maxima of sucrose density gradient fractions of Triton X-100 - treated *Anabaena flos-aquae*

Fraction	Absorption maxima
4% sucrose	
vegetative	418, 438, 486, 514, 620, 670
heterocysts	418, 508, 540, 613, 670
8% sucrose	
vegetative	418, 438, 490, 625, 678
heterocysts	418, 438, 625, 675
20% sucrose	
heterocysts	418, 438, 475, 518, 620, 675

A major difference between vegetative cells and heterocysts was observed in the 4% sucrose layer. This fraction from vegetative cells contained carotenoids and some Chl a, but this fraction from heterocysts contained little or no carotenoids or Chl, but instead contained pheophytin a. Since this is a 4% sucrose layer obtained from density gradient centrifugation, this would not contain solubilized pheophytin, since the latter should accumulate in the 0% sucrose layer. In the case of the heterocysts, however, the 20% sucrose layer was found to resemble the 4% sucrose layer in vegetative cells, containing most of the carotenoids and some Chl a. These differences reflect a rather profound change in the structure of the PS 2 in the heterocysts, which is in agreement with the apparent inactivation of PS 2 during the formation of heterocysts.

Figs. 4 and 5 show several interesting features. Fig. 4 shows the absence of the polypeptides in the 20 to 30 kilodalton range which in higher plants form the Chl-protein complex which functions in light harvesting. In all experiments performed to date, the presence of these polypeptides in higher plants is associated with the presence of Chl b in the photosynthetic membrane[6, 13-15]. In the blue-green algae this light harvesting function is assumed by phycocyanin which is associated with the membrane in the form of phycobilisomes[1]. These polypeptides have also been implicated in the phenomenon of thylakoid stacking in the chloroplasts of higher plants[15], and this is again in agreement with the lack of stacking in the photosynthetic membranes of the blue-green algae.

The polypeptide profile of membrane fragments from vegetative cells shows that the major polypeptides are located in the 57 to 75 kilodalton range, with the major one being the 75 kilodalton polypeptide. Since these are also the major polypeptides found in the 8% sucrose layer, and this is the band which contains PS 1, we conclude that the 75 kilodalton is the one which binds the Chl a in PS 1 of blue-green algae. With only these preliminary data available, it is not possible to say anything definite about the role of each polypeptide, and only a circumstantial association of these polypeptides with PS 1 of blue-green algae can be made. It is important to note, however, that in higher plants and *Chlamydomonas*, the Chl a-protein complex of PS 1 migrates as a 110 kilodalton polypeptide with SDS gel electrophoresis under conditions where the lipids are not removed and some Chl remains complexed to the polypeptide[15]. Removal of the lipids results in the disappearance of this band in subsequent SDS gel electrophoresis and the appearance of new bands in the region of 60 kilodaltons. Recent work in our laboratory (manuscript in preparation)[16] shows that spinach subchloroplast fractions enriched in PS 1 show a strong band at 52 kilodaltons upon SDS gel electrophoresis, and fractions enriched in PS 2 show a strong band in 63 kilodaltons. From these data we have concluded that the 52 and 63 kilodalton bands are from polypeptides which are associated with Chl a in PS 1 and PS 2 respectively.

Recent data obtained by Nelson (Nathan Nelson, personal communication) show that with swiss chard, the purified PS 1 fraction contains a 70 kilodalton polypeptide which retains the reaction center P700 as well as the light harvesting Chl a. Thus it appears that in higher plants and in the blue-green algae a polypeptide in the range of 52 to 75 kilodaltons is associated with PS 1 and becomes prominent in PS 1 preparations prepared from the different plant species. When the Chl a remains on the polypeptide, higher molecular weights are obtained, with Anderson and Levine reporting a value of 110 kilodaltons[15]. In the experiments reported here, the Chl a was solubilized and removed during the electrophoresis procedure, while in the experiments of Nelson the Chl a was retained on the polypeptide, which allowed the determination of the P700 content. Thus, it appears that the presence of lipids, including Chl a, on the polypeptide will influence its migration in gel electrophoresis procedures and thus its apparent molecular weight. The general trend, however, of polypeptides in this region being associated with the Chl a of PS 1 is becoming well established.

References

1. Wolk, C. P., 1973, Bacteriological Rev. 37, 32.
2. Carr, N. G. and Whitton, B. A., 1973, The biology of the blue-green algae, Blackwell Scientific Publications, Oxford.
3. Knaff, D. B., 1973, Biochim. Biophys. Acta 325, 284.
4. Bedell, G. W. and Govindjee, 1973, Plant and Cell Physiol. 14, 1081.
5. Neumann, J., Ogawa, T. and Vernon, L. P., 1970, FEBS Letters 10,253.

6. Klein, S. M. and Vernon, L. P., 1974, Plant Physiol. 53, 777.

7. Fleming, H. and Haselkorn, R., 1973, Proc. Nat. Acad. Sci. USA 70, 2720.

8. Donze, M., Haveman, H. and Schiereck, P., 1972, Biochim. Biophys. Acta 256, 157.

9. Ogawa, T., Vernon, L. P. and Mollenhauer, H. H., 1969, Biochim. Biophys. Acta 172, 216.

10. Weber, K. and Osborn, M. 1969, J. Biol. Chem., 244, 4406.

11. Vernon, L. P. and Shaw, E. R., 1972, Plant Physiol. 49, 862.

12. Vernon, L. P. and Shaw, E. R., 1969, Plant Physiol. 44, 1645.

13. Thornber, J. P. and Highkin, H. R., 1974, Europ. J. Biochem. 41, 109.

14. Anderson, J. M. and Levine, R. P., 1974, Biochim. Biophys. Acta 333, 378.

15. Anderson, J. M. and Levine, R. P., 1974, Biochim. Biophys. Acta 357, 118.

16. Klein, S. M. and Vernon, L. P. Manuscript submitted for publication.

M. AVRON, *Proceedings of the Third International Congress on Photosynthesis*,
September 2-6, 1974, Weizmann Institute of Science, Rehovot, Israel
Elsevier Scientific Publishing Company, Amsterdam, The Netherlands, 1974

THE REGULATION OF ELECTRON FLOW IN SYNCHRONIZED CULTURES OF GREEN ALGAE

H. Senger and B. Frickel-Faulstich

Botanisches Institut
Universität Marburg

1. Summary

The unicellular green algae Scenedesmus obliquus and Chlamydomonas reinhardii were synchronized under a light-dark regime of 14 : $\overline{10}$ hrs. Photosynthetic capacity, photosystems II and I activities and steady state fluorescence were determined throughout the life cycles. It was concluded that the changes in photosynthetic capacity and photosystem II activity are caused by a rate limiting step in electron flow.

Changes in pool sizes of plastoquinone and cytochromes could not account for the variations in electron transport. Experiments with DCMU and DBMIB on cell free preparations of the synchronous cultures proved the reoxidation capacity of plastoquinone to be the life cycle dependent rate limiting step.

2. Introduction

Synchronous cultures of unicellular photosynthetic organisms represent an excellent tool to study changes in the photosynthetic apparatus of these organisms. Changes in the photosynthetic capacity are a frequently studied and generally accepted fact. The cause of these changes has not satisfactorily been explained. Changes in the pigment composition were ruled out by pigment analysis and in vivo spectra (SENGER & BISHOP, 1971). The changes in quantum yield of photosynthesis focussed the interest on the photochemical reactions (SENGER & BISHOP, 1967).

Abbreviations used: PCV = packed cell volume,
PS I = photosystem I, PS II = photosystem II,
PQ = plastoquinone, Cyt = Cytochrome,
DCMU = Dichloro-phenyl-dimethyl-urea
DBMIB = Dibromo-methyl-isopropyl-p-benzoquinone
DCPIP = Dichloro-phenol-indophenol
EDTA = Ethylene-diamine-tetraacetic-acid

In this study the pool sizes of several redox-carriers and the electron transport capacities of synchronized cultures of *Scenedesmus* and *Chlamydomonas* have been investigated in order to encircle the changing rate limiting step. An attempt to find a regulatory effect in the electron transport chain was made by SCHOR et al (1970). Since also changes between cyclic and non cyclic electron transport have been reported (SENGER, 1970b; GIMMLER et al., 1971) it was necessary to consider a possible balance between cyclic and non cyclic electron transport.

3. Material and methods

The eukaryotic unicellular green algae *Scenedesmus obliquus*, strain D_3 (GAFFRON) and *Chlamydomonas reinhardii*, strain 11-32 (Algal Collection, Göttingen) were grown in liquid culture medium (*Scenedesmus*, BISHOP & SENGER, 1971; *Chlamydomonas*, KUHL, 1962) in a light thermostate at 28°C, areated with 3% CO_2 in air. They were synchronized under a light-dark regime of 14 hrs light of 15.000 Lux and 10 hrs darkness and automatically diluted to 3.12×10^6 cells/ml (*Scenedesmus*) resp. 1.56×10^6 cells/ml (*Chlamydomonas*) at the beginning of each light period with a photoelectrically controlled dilution device (SENGER et al., 1972).

Chlorophyll was extracted with hot methanol and calculated as reported earlier (SENGER, 1970). - Oxygen evolution was measured polarographically with a micro-Clark-electrode (Gilson Medical Electronics) under saturating (10^6 erg/cm^2 sec) red light above 620 nm. The sample contained 10 µl PCV per ml phosphate buffer (0.05 M, pH 7.0). For Hill-reaction in whole cells p-benzoquinone (final concentration 0.4 mg/ml) was added.

Cell free preparations were obtained by shaking 500 µl PCV in 20 ml buffer (KCl 10^{-2} M, NaCl 10^{-2} M, $MgCl_2$ $2,5 \times 10^{-3}$ M, EDTA 10^{-5} M, Tricine/KOH 10^{-2} M pH 7.5) with 51 ml glass beats (0.7 mm Ø) for 3.5 min in a Vibrogen-Cell-Mill (Bühler, Tübingen) under cooling to 1.5°C (BERZBORN & BISHOP, 1973; SENGER & MELL, 1975). The breakage was 100% and microscopically controlled. Under addition of sucrose (5.6 M) the broken cells were separated from the glass beats and centrifuged for 10 min at 250 g. For Hill-reaction samples were diluted to 7 µg chlorophyll/ml and DCPIP (4×10^{-5} M) was added. Under actinic light above 630 nm (10^5 erg/cm^2 sec) the Hill reaction was measured as absorbance change at 578 nm in a Shimadzu MPS-50L spectrophotometer using an interference filter with a transmission peak at 578 nm (Schott IL 578) to shield the photomultiplier from the actinic light.

Photosystem I activity was determined as oxygen uptake during photoreduction of methylviologen in cell free preparations. The cells were broken as described in phosphate buffer (0.05 M, pH 6.5). The supernatant was brought to 20 μg chlorophyll / ml and DCPIP (100 μM), ascorbate (5 nM), methylviologen (1 mM) and DCMU 2 μM were added (STROTMANN & v. GÖSSELN, 1972). Oxygen consumption under red light above 620 nm (3 x 10^5 erg/cm^2 sec) was measured polarographically.

For cytochrome determination the cells (500 μl PCV) were broken in 20 ml buffer (0.1 M Tris/HCl, pH 7.5 for <u>Scenedesmus</u>; 0.1 M Tris/HCl, KCl 10^{-2} M, Mg Cl_2 10^2 M, EDTA 2.5 x 10^{-3} M, pH 7.5 for <u>Chlamydomonas</u>) for 10 min as described above. The supernatant was diluted to 90 μg chlorophyll / ml. According to BENDALL et al. (1971) cytochrome concentrations were determined in various difference spectra of reduced (sodium dithionite, ascorbate, hydroquinone) and oxidized (ferricyanide) samples using the following extinction coefficients:

cytochrome f = 17.7 cm^2 / mM
cytochrome b_{563} = 20.0 cm^2 / mM
cytochrome b_{559} = 20.0 cm^2 / mM

For details see FRICKEL-FAULSTICH (1974).

Plastoquinone determination was modified after BARR & CRANE (1971) and POWLS et al. (1969). After extraction with hot methanol the extract was shaken 1 min with ferricyanide (10^{-3} M) and subsequently 3 times washed with petrolether. The extract of the petrolether phase was chromatographed on thin layer (Silica gel G ; heptane/benzine) and developed with rhodamine B. The pastoquinone band (R_f 0.7) was eluated, transferred into ethanol and the absorption difference between the oxidized and reduced form (10 min borohydride) determined from the maxima around 255 nm. The extinction coefficient used is E = 14.9 cm^2 / mM.

Steady state fluorescence was measured with a modified Aminco DW2 spectrophotometer using monochromatic light of 444 nm (Schott DAL 444) and 5 x 10^3 erg/cm^2 sec for exciting cross illumination. The photomultiplier was shielded with a cut off filter transmitting above 665 nm (Schott RG 665). The samples were set to 5 x 10^{-3} mg chlorophyll/ml cell suspension.

4. Results and discussion

Cultures of the two green algae used for this study were completely synchronized. Thus each whole culture represents the life cycle of the respective organism. Samples taken from these cultures and brought to the same amount of chlorophyll demonstrate a life cycle dependent photosynthetic capacity (Fig. 1).

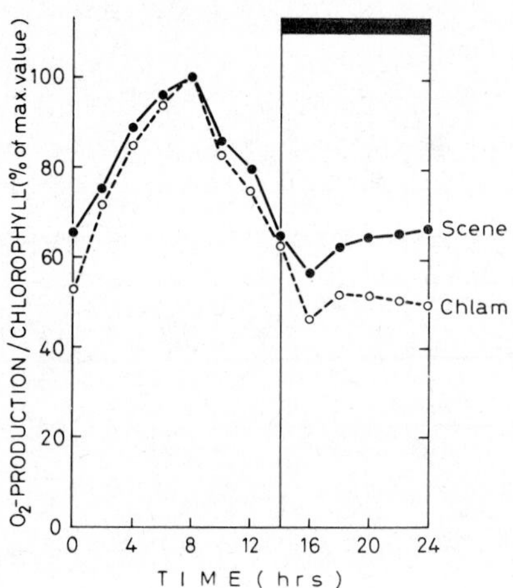

Fig.1. Photosynthetic oxygen evolution in synchronous cultures of Scenedesmus and Chlamydomonas. Oxygen evolution was measured polarographically under saturating red light (>620 nm), compensated for respiration and computed for equal amounts of chlorophyll for each organism. All values are expressed as percent of the maximum value.

This fact has been reported in detail for Scenedesmus (c.f. BISHOP & SENGER, 1971) and for a variety of other eukaryotic unicellular algae. The maximum of capacity is mostly reached in the middle of the growth phase i.e. in the middle of the light period and the minimum just prior to the separation in daughter cells. It is also established that changes in photosynthetic capacity are accompanied by changes in the quantum yield (SENGER, 1970b, 1971). - From the two photosystems, PS II follows closely the course of photosynthetic capacity (Fig. 2).

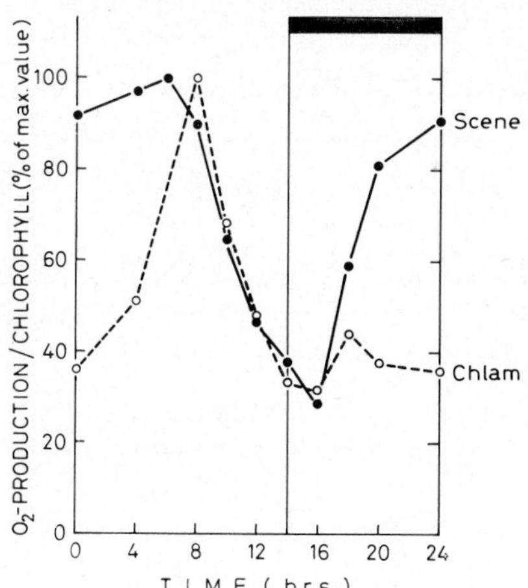

Fig.2. Hill-reaction in whole cells of synchronous cultures of Scenedesmus and Chlamydomonas. Oxygen evolution with p-benzoquinone as Hill-acceptor was measured polarographically under saturated red light (> 620 nm) and computed for equal amounts of chlorophyll for each organism. All values are expressed as percent of the maximum value.

All investigations on Hill-reaction with p-benzoquinone in whole cells (GERHARD, 1964; SENGER, 1971b) or with DCPIP in cell free samples (SCHOR et al., 1970) confirm these results. - Measurements of steady state fluorescence yield show an inverse course to the photosynthetic capacity (Fig. 3)

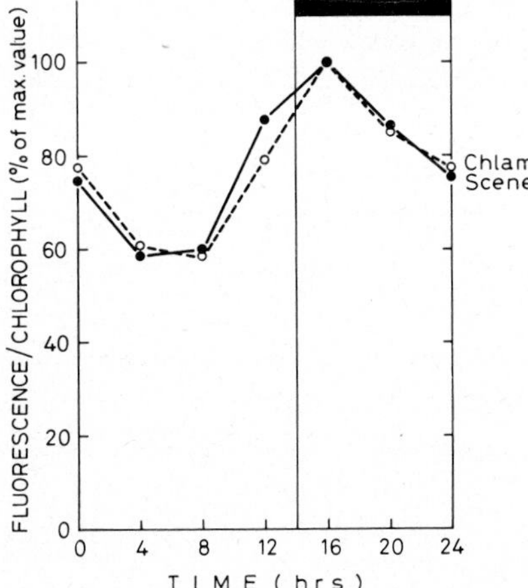

Fig. 3. Steady state fluorescence in synchronous cultures of Scenedesmus and Chlamydomonas. All samples contained cells with equal total amount of chlorophyll. Fluorescence was excited with light of 444 nm and measured above 665 nm.

This is in agreement with the data reported for synchronized cultures (DÖHLER, 1964; SENGER & BISHOP, 1971). The increase in fluorescence during the stage of decreasing photosynthetic and Hill-reaction activity cannot be explained with a decrease in PS II itself, like a degradation in pigments, but rather with a change in the electron flow linked to PS II. - The capacity of PS I seems to be sufficient throughout the life cycle to move on the electrons provided by PS II. In synchronous cultures of Scenedesmus PS I capacity measured as photoreduction of methylviologen (Fig. 4)

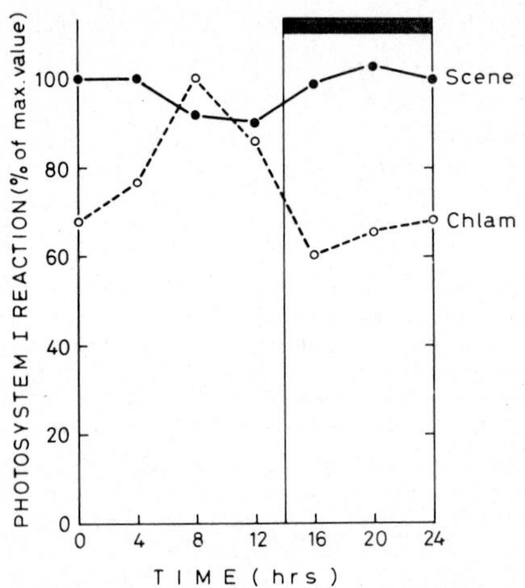

Fig. 4. Photosystem I reaction in cell free preparations from synchronous cultures of Scenedesmus and Chlamydomonas. Photosystem I reaction was measured polarographically as oxygen uptake during photoreduction of methylviologen under red light (> 620 nm). The values are expressed as percent of the maximum value.

or as photoreduction of CO_2 (SENGER & BISHOP, 1967) remains almost constant, whereas in synchronous cultures of Chlamydomonas PS I activity determined either by photoreduction of methylviologen (Fig. 4) or of methyl red (SCHOR et al., 1970) changes parallel to photosynthetic capacity but less pronounced.

From the data reported here, the changes in the Emerson - enhancement effect (SENGER & BISHOP, 1969) and the fact that there is no change in the pigment composition during the life cycle (SENGER & BISHOP, 1971) one has to conclude that no changes in PS I or PS II itself but rather in the linking electron transport chain are responsible for the changes in photosynthetic capacity. From the studies on Hill-reactions and on the 520 nm absorbance change (SENGER, 1970b) they do not reach the difference in photosynthetic capacity of more than 50%. Similar results were obtained by SCHOR et al. (1970). From these results it is quite improbable that the changes in pool sizes of redox-carriers could account for the changes in the photosynthetic capacity. The change in pool sizes of Chlamydomonas might just be coincidal or have a mutual cause with the other regulatory effect on the electron transport.

Special attention should be drawn to the fact that Cyt b_{563}, attributed to cyclic electron transport, remains constant during the life cycle of Scenedesmus and in Chlamydomonas it is the redox-carrier with the smallest change.

The application of the strictly localized inhibitors of photosynthetic electron transport DCMU (BISHOP, 1958) and DBMIB (TREBST et al., 1970) provided more information about the changes in electron transport. It is generally accepted that the site of inhibition of DCMU is before and that of DBMIB behind plastoquinone.

As demonstrated earlier the inhibitory effect of DCMU on whole cells undergoes considerable changes during the life cycle of Scenedesmus (FRICKEL-FAULSTICH & SENGER, 1972). This effect was very small in Chlamydomonas. From unpublished experiments we know that these effects are caused by a permeation barrier, strongest at the time of lowest photosynthetic capacity. Therefore, the experiments on the inhibitory effect of DCMU had to be carried out with cell free preparations from the synchronous cultures. The inhibitory effect of different concentrations of DCMU on Hill-reaction with DCPIP in cell free preparations from cells of the 8th and 16th hour of synchronous cultures of Scenedesmus and Chlamydomonas are shown in Fig. 5.

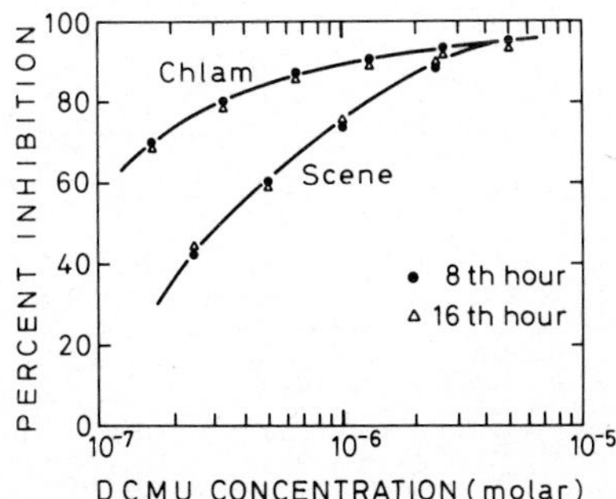

Fig. 5. Inhibitory effect of different concentrations of DCMU on DCPIP Hill-reaction in cell free preparations from cells of the 8th and 16th hour of synchronous cultures of Scenedesmus and Chlamydomonas. All samples contained equal amounts of chlorophyll and were irradiated with actinic light above 630 nm.

it is reasonable to expect a variation in electron transport close to PS II.

The pool size of the redox-carriers plastoquinone, cytochromes b_{559}, b_{563} and cytochrome f were determined for the stages of maximal (8th hour) and minimal (16th hour) photosynthetic capacity in the life cycles of Scenedesmus and Chlamydomonas (Tables I and II).

Redox-carrier	8th hour	16th hour	% decrease from 8th to 16th hour
Plastochinon $_{total}$	25.4 ± 2.5	22.5 ± 0.7	11
Cytochrom f	1.85 ± 0.08	1.66 ± 0.08	10
Cytochrom b_{563}	1.65 ± 0.02	1.74 ± 0.05	- 5
Cytochrom b_{559}	4.88 ± 0.15	4.40 ± 0.11	10
Cytochrom b_{559} HP	3.58 ± 0.20	2.28 ± 0.08	37
Cytochrom b_{559} LP	1.2 ± 0.09	1.78 ± 0.12	-48

Table I : Amounts of different redox-carriers in cells from the 8th and 16th hour of synchronous cultures of Scenedesmus. The data are expressed in 10^{-3} µM/mg chlorophyll. For analytical determination of plastoquinone and cytochromes see methods.

Redox-carrier	8th hour	16th hour	% decrease from 8th to 16th hour
Plastochinon $_{total}$	27.0 ± 0.3	17.3 ± 1.9	36
Cytochrom f	5.77 ± 0.46	3.67 ± 0.22	36
Cytochrom b_{563}	4.57 ± 0.41	3.18 ± 0.37	30
Cytochrom b_{559}	11.27	6.94	38
Cytochrom b_{559} HP	6.16 ± 0.50	4.00 ± 0.31	35
Cytochrom b_{559} LP	5.11 ± 0,33	2.94 ± 0.09	42

Table II : Amounts of different redox-carriers in cells from the 8th and 16th hour of synchronous cultures of Chlamydomonas. The data are expressed in 10^{-3} µM/mg chlorophyll. For analytical determination of plastoquinone and cytochromes see methods.

In the case of Scenedesmus (Table I) the small differences of PQ, Cyt f and total Cyt b_{559} could never be responsible for a change in photosynthetic capacity of more than 40%. Chlamydomonas demonstrates larger variations in its redox-carrier pool sizes (Table II). Changes between 36% and 38% occur parallel to photosynthesis, but

With the appropriate concentration electron flow is completely blocked and there is no difference between samples from cells with different photosynthetic capacity. This proves that the inhibitory site of DCMU is not influenced by the life cycle.

Scenedesmus demonstrates a less pronounced but similar life cycle dependent penetration barrier to DBMIB as to DCMU. Since Chlamydomonas did not show any penetration problem, the effect of different DBMIB concentrations on photosynthetic oxygen evolution in whole cells from the 8th and 16th hour on the life cycle was tested (Fig. 6).

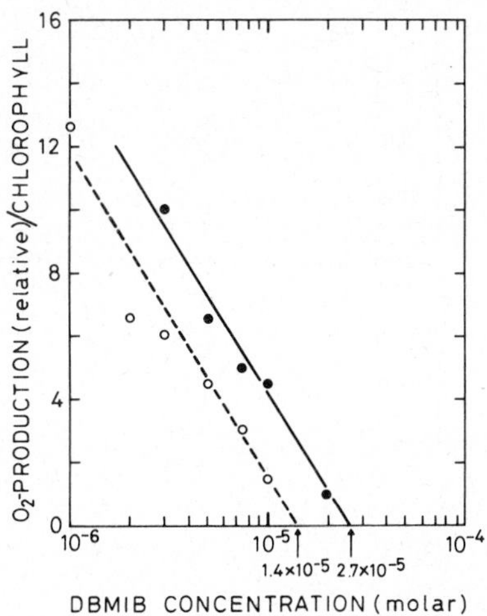

Fig. 6. The effect of different DBMIB concentrations on photosynthetic oxygen evolution in whole cells from the 8th and 16th hour of synchronous cultures of Chlamydomonas. Oxygen evolution was measured polarographically under saturating red light (> 620 nm) in samples containing equal amounts of chlorophyll.

DBMIB inhibits the electron flow completely in both types of cells, but needs the double concentration for those cells with the highest photosynthetic capacity. When amounts of DBMIB proportional to the photosynthetic capacity of the cells are necessary for complete inhibition the number of inhibitory sites should be equivalent too (c.f. BÖHME et al., 1971). This fact makes the inhibitory site of DBMIB a potential bottleneck of electron flow.

This conclusion is supported by experiments on the effect of DBMIB on Hill-reaction with DCPIP in cell free preparations from cells of the 8th and 16th hour from Scenedesmus and Chlamydomonas (Fig. 7 and 8).

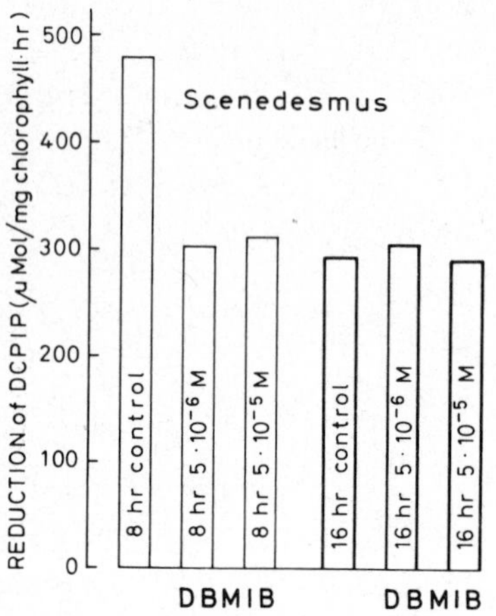

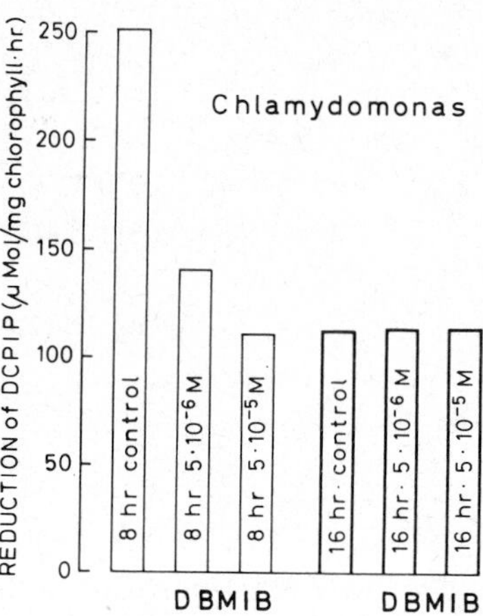

Fig. 7. The effect of DBMIB on DCPIP-Hill-reaction in cell free preparations from cells of the 8th and 16th hour from synchronous cultures of Scenedesmus. DCPIP-reduction was measured spectrophotometrically at 578 nm under saturating red actinic light (> 630 nm) in samples containing equal amounts of chlorophyll. The values are given in uM DCPIP reduced / mg chlorophyll x hour.

Fig. 8. The effect of DBMIB on DCPIP-Hill-reaction in cell free preparations from cells of the 8th and 16th hour from synchronous cultures of Chlamydomonas. DCPIP-reduction was measured spectrophotometrically at 578 nm under saturating red actinic light (>630 nm) in samples containing equal amounts of chlorophyll. The values are given in uM DCPIP reduced / mg chlorophyll x hour.

For both organisms the untreated controls demonstrate the same difference between the 8th and the 16th hour as the photosynthetic capacity. Addition of DBMIB to the preparations from the 8th hour cells (maximal capacity) suppress the Hill-reaction down to the level of the 16th hour (minimal capacity). On preparations from the 16th hour cells DBMIB had no effect.

In order to explain the results one has to postulate two sites of electron drainage by DCPIP, one before and one after the inhibitory site of DBMIB. This includes that the PS II side of the inhibitory site is constant throughout the life cycles of both organisms and thus the PS II capacity itself and the quencher Q too. It also clearly demonstrates that the inhibitory site of DBMIB is the location of the life cycle dependent rate limiting step.

Since DBMIB inhibits behind plastoquinone and the plastoquinone pool does not chenge to the necessary extent, we have to conclude that the reoxidation capacity of plastoquinone is the rate limiting step regulating the photosynthetic capacity during the life cycle. This result is in agreement with the finding that the reoxidation of PQ is generally the rate limiting step of photosynthetic electron transport (WITT, 1971).

The localization of the rate limiting step brings up a new question: what is regulating the reoxidation capacity of plastoquinone? Since photosynthetic capacity is proportionally inhibited in the light limiting and saturating region (SENGER, 1970a) the regulating mechanism has to be adapted to the respective light intensity. Life cycle dependent conformational changes (SCHÄFER & SENGER, unpublished) make structural changes the most probable cause. But potential gradients or photosynthetic intermediate products should as well be considered as possible candidates.

Another complication is raised by the cyclic electron flow. For *Scenedesmus* it was reported that under anaerobic conditions the decrease in non cyclic photophosphorylation is accompanied by an increase in cyclic photophosphorylation (SENGER, 1970b), whereas *Ankistrodesmus* demonstrates during its life cycle a change in the non cyclic and a constant cyclic photophosphorylation (GIMMLER et al., 1971). The constant pool size of Cyt b_5 in *Scenedesmus* and the small change in *Chlamydomonas* have also to be considered under the aspect of separately regulated cyclic and non cyclic electron flows.- These facts suggest to include a possible cyclic electron flow into the scheme of regulated electron flow during the extreme stages of photosynthetic capacity (Fig. 9).

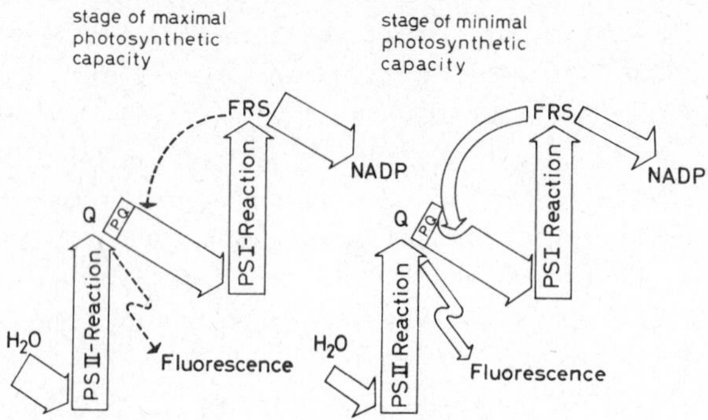

Fig. 9 . Schematic diagram of the potential electron flow during the stages of optimal and minimal photosynthetic capacity during the life cycle of synchronized unicellular green algae. The width of the arrows indicates the electron flow capacity.

The role of cyclic electron flow and its changes during the life cycles of synchronous algae is still under investigation.

5. Acknowledgements

We thank Prof. Trebst for kindly providing us the DBMIB and the Deutsche Forschungsgemeinschaft for supporting this study.

References

1. Barr, R. & Crane, F.L., 1971, In: Methods in Enzymology, Vol.XXIII, Photosynthesis (A. San Pietro ed.), 372-408, Academic Press, N.Y.
2. Bendall, D.S., Davenport, H.E. & Hill, R., 1971, In: Methods in Enzymology, Vol. XXIII, Photosynthesis (A. San Pietro ed.), 327 - 343, Academic Press, N.Y.
3. Berzborn,R.J. & Bishop, N.I., 1973, Biochim.Biophys. Acta 292, 700 - 714
4. Bishop, N.I., 1958, Biochim.Biophys. Acta 27, 205 - 206
5. Bishop, N.I. & Senger, H., 1971, In: Methods in Enzymology, Vol. XXIII, Photosynthesis (A. San Pietro ed.), 53 - 66
6. Böhme, H., Reimer, S. & Trebst, A., 1971, Z. Naturforsch. 26b, 341 - 352
7. Döhler, G., 1963, Planta 60, 158 - 165
8. Frickel-Faulstich, B., 1974, Ph.D.Thesis, University of Marburg, Germany, 1 - 119
9. Frickel-Faulstich, B. & Senger, H., 1972, Ber. Deutsch.Bot.Ges. 85, 401 - 408

10. Gerhard, B., 1964, Planta 61, 101 - 129

11. Gimmler, H., Neimanis, S., Eilmann, J. & Urbach, W., 1971, Pflanzen-Physiol. 64, 358 - 366

12. Kuhl, A., 1962, In: Beiträge zur Physiologie und Morphologie der Algen, 157 - 166, Fischer Verlag, Stuttgart

13. Powls, R., Wong, I & Bishop, N.I., 1969, Biochim.Biophys. Acta 180, 490 - 499

14. Schor, S., Siekevitz, P. & Palade, G.E., 1970, Proc. Natl.Acad. Sci. USA 66, 174 - 180

15. Senger, H., 1970a, Planta 90, 243 - 266

16. Senger, H., 1970b, Planta 92, 327 - 346

17. Senger, H., 1971, In: G. Forti, M. Avron and A. Melandri (Eds.), Proc. 2nd Intern.Congr. Photosynthesis Res., Stresa 723 - 730. Dr.W.Junk N.V., Publ., The Hague

18. Senger, H. & Bishop, N.I., 1967, Nature 214, 140 - 142

19. Senger, H. & Bishop, N.I., 1969, Nature 221, 975

20. Senger, H. & Bishop, N.I., 1971, In: G. Forti, M. Avron and A. Melandri (Eds.), Proc. 2nd Intern.Congr. Photosynthesis Res., Stresa 678 - 687. Dr. W. Junk N.V., Publ., The Hague

21. Senger, H. & Mell, V., In: Methods in Cell Physiology (D.M. Prescott ed.) Vol. XII, Academic Press, N.Y. (in press)

22. Senger, H., Pfau, J. & Werthmüller, K., In: Methods in Cell Physiology (D.M. Prescott ed.) Vol. V, 301 - 323, Academic Press, N.Y.

23. Strotmann & v. Gösseln, Ch., 1972, Z. Naturforsch. 27b, 445 - 455

24. Trebst, A., Harth, E. & Draber, W., 1970, Z. Naturforsch. 25b, 1157 - 1159

25. Witt, H.T., 1971, Quart.Rev. of Biophysics 4, 363 - 477

M. AVRON, *Proceedings of the Third International Congress on Photosynthesis*,
September 2-6, 1974, Weizmann Institute of Science, Rehovot, Israel
Elsevier Scientific Publishing Company, Amsterdam, The Netherlands, 1974

ISOLATED PHOTOREACTIONS IN CHLOROPLASTS PREPARED FROM PINUS SILVESTRIS

G. Öquist and B. Martin

Department of Plant Physiology,
University of Umeå, S-901 87 Umeå, Sweden

1. Summary

Photoactive chloroplasts have been isolated from the needles of Pinus silvestris using 20% PEG*. Sucrose and PEG stimulated photoreduction of DPIP by chloroplasts with optima at 1.2 M sucrose and 20 - 30% PEG. NADP reduction with ascorbate as electron donor was inhibited at all tested concentrations of sucrose and PEG. Also DPC-supported DPIP photoreduction by a 10,000 x g fraction of broken chloroplasts was inhibited by these compounds. Some possible modes of action of sucrose and PEG are discussed.

2. Introduction

There are only a few reports concerned with photosynthetic studies on the chloroplast or subchloroplast level in conifers[1,2,3,4,5,6]. The reason is probably that conifers contain considerable amounts of resins as well as other substances, tannin and lipids, which can inhibit the chloroplasts during the preparation[7,8,9,10,11]. By adding 25% PEG to the preparation medium Oku and Tomita[2] have managed to prepare photoactive chloroplasts from Pinus tunbergii. Using basically the same method active chloroplasts have also been prepared from Pinus silvestris and Picea abies and the reaction conditions have been optimized for DPIP photoreduction by the chloroplasts[6]. Both sucrose and PEG stimulated the rate of DPIP photoreduction. These compounds were also effective in protecting the isolated chloroplasts against ageing. The present work extends the studies of the effects and the modes of action of sucrose and PEG on the electron transport of isolated Pinus chloroplasts.

*Abbreviations used: CCCP, carbonyl cyanide m-chlorophenylhydrazone; Chl, chlorophyll; DCMU, 3-(3,4-dichlorophenyl)-1,1-dimethylurea; DPC, diphenylcarbazide; DPIP, 2,6-dichlorophenol indophenol; HEPES, N-2-hydroxyethyl-piperazine-N-2-ethanesulfonic acid; NADP, nicotinamide adenine dinucleotide phosphate; PEG, polyethylene glycol 4000; Tris, tris(hydroxymethyl)aminomethane.

3. Materials and Methods

Dormant seedlings of Pinus silvestris L., both a one-year-old crossing ($N64^{o}10'E20^{o}43'$, altitude 90 m X $N64^{o}37'E21^{o}06'$, altitude 50 m) and a two-year-old open pollinated progeny ($N64^{o}30'E17^{o}33'$, altitude 300 - 400 m), were placed in a controlled environment chamber: light/dark cycle 24 hours (19 h, 39 $W \cdot m^{-2}$/ 1 h, 28 $W \cdot m^{-2}$/ 1 h, 13 $W \cdot m^{-2}$/ 1 h, dark / 1 h, 13 $W \cdot m^{-2}$/ 1 h 28 $W \cdot m^{-2}$); max temperature in light $20^{o}C$ and min temperature in dark $11^{o}C$; min relative humidity in light 40% and max relative humidity in dark 90%. Fluorescent tubes (General Electric Power Grove, "Daylight" and "White de Lux") in the proportion of 1:1 were used for illumination. The plants were potted in peat and they were watered (100 ml) automatically every day . Wallco nutrient solution L-63/13 (Hasselfors Bruk AB, Sweden) was added to the water (10 ml per 10 l water).

When the dormancy of the plants had been broken chloroplasts were prepared from one-year-old needles as described elesewhere[6]. The preparation medium was composed of 0.4 M sucrose, 0.005 M Na_2HPO_4 - KH_2PO_4 buffer pH 7.6, 0.01 M NaCl, 0.005 M $MgCl_2$ and 20% (w/w) PEG - 4000. To disintegrate the needles a Colworth "Stomacher 80" (A.J. Seward. London) was used. This blendor works with two reciprocating paddles which repeatedly pound a plastic bag containing 2 g whole needles in 20 ml preparation medium. A disintegration time of 1 minute was sufficient to split up the needles and give a satisfactory yield of chloroplasts. After isolating the chloroplasts by fractionated centrifugation[6] the chloroplasts for DPIP reduction were suspended in 1.3 M sucrose and 0.05 M HEPES pH 6.8. The chloroplasts for NADP reduction were suspended in 0.05 M HEPES pH 6.8, 0.01 M NaCl and 0.005 M $MgCl_2$.

To obtain a grana preparation chloroplasts were incubated with 0.25% digitonin (British Drug House. Ltd.) for 30 minutes at $+4^{o}C$ in darkness together with 0.05 M HEPES pH 6.8, 0.15 M KCl, 0.01 M NaCl and 0.005 M $MgCl_2$ (Chl concentration 40 $\mu g/ml$). The detergent treated chloroplast suspension was passed once through a Yeda press at a pressure of 100 $kg \cdot cm^{-2}$ (Yeda Research and Development Co., Rehovot). The grana fraction was then isolated by differential centrifugation, first at 1,000 x g for 10 minutes and then at 10,000 x g for 30 minutes. The 10,000 x g fraction was finally suspended in 0.05 M HEPES pH 6.8, 0.15 M KCl, 0.01 M NaCl and 0.005 M $MgCl_2$.

The chloroplast catalyzed photoreduction of DPIP and NADP was measured with a Shimadzu spectrophotometer MPS-50L for 5 minutes at 587 and 337 nm, respectively. By use of Schott filters RG610 and KG3 the actinic light (600 - 800 nm, 64 $W \cdot m^{-2}$) was isolated from a 100 W/12 V Zeiss microscope lamp. The effects of different concentrations of sucrose (0 - 2.1 M) and PEG (0 - 40%) on DPIP and NADP reduction were investigated using water, DPC or sodium ascorbate as electron donors. For the DPIP reduction the reaction mixture also contained 0.05 M HEPES pH 6.8, 25 μM

DPIP and 4 μg/ml Chl. For the NADP reduction the basal reaction medium contained 0.05 M HEPES pH 6.8 (or 7.7), 6.7 mM sodium ascorbate, 67 μM DPIP, 50 μM DCMU, 6.7 mM NaCl, 3.3 mM $MgCl_2$, 0.33 mM NADP, 4 μg/ml Chl, ferredoxin and NADP-reductase from spinach[12,13] in excess. All reactions were performed at room temperature. The Chl concentrations were calculated according to Arnon[14]. All light intensities were measured with a calibrated thermopile (Higler-Schwarz FT 17.1/442).

4. Results

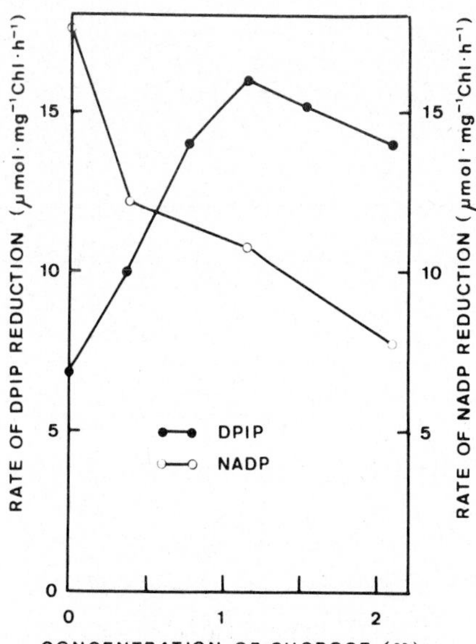

Fig. 1. Photoreduction of DPIP and NADP by Pinus chloroplasts at different sucrose concentrations.

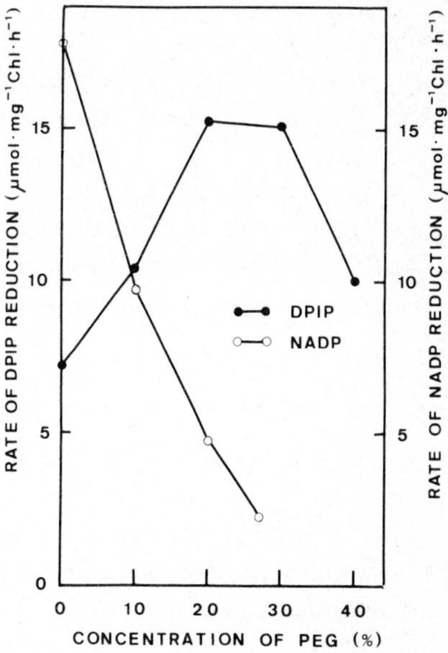

Fig. 2. Photoreduction of DPIP and NADP by Pinus chloroplasts at different PEG concentrations.

Photoreduction of DPIP by the Pinus chloroplasts was stimulated by sucrose up to 1.2 M and by PEG up to 20 - 30% (Figures 1 and 2), when no artificial electron donor to photosystem II was added. At higher concentrations inhibition occurred. Using ascorbate as electron donor to photosystem I photoreduction of NADP was inhibited by sucrose and PEG at all tested concentrations (Figures 1 and 2). NADP reduction was generally studied at pH 6.8, but increasing pH to 7.7 did not alter the activity. In another set of experiments (Table 1) it was found that DPIP photoreduction could also be stimulated by the artificial electron donor DPC. DPC was not as effective as 1.2 M sucrose in increasing the rate of DPIP reduction.

Table 1
The effect of sucrose and DPC on the rate of DPIP photoreduction by Pinus chloroplasts

Rate of DPIP reduction ($\mu mol \cdot mg^{-1} Chl \cdot h^{-1}$)		
0 M sucrose	1.2 M sucrose	0.5 mM DPC
6	18	13

Table 2
The effect of DPC on DPIP photoreduction by the 10,000 x g fraction of broken Pinus chloroplasts and the influence of sucrose and PEG on the DPC-supported DPIP photoreduction

Rate of DPIP reduction ($\mu mol \cdot mg^{-1} Chl \cdot h^{-1}$)			
0 M DPC	0.33 mM DPC	0.33 mM DPC	
		1.2 M sucrose	20% (w/w)PEG
3	9	6	5

The DPIP photoreduction rate in the 10,000 x g fraction of fragmented chloroplasts was low without artificial electron donor but it increased about three times when DPC was added (Table 2). Those concentrations of sucrose and PEG, which stimulated the DPIP reduction in chloroplasts without artificial electron donor (Figures 1 and 2), were instead inhibitory to DPIP reduction in the 10,000 x g fraction (Table 2). In all tested conditions DPIP photoreduction was totally inhibited by DCMU.

5. Discussion

A promotive effect of sucrose and PEG on the photoreduction of DPIP by isolated photosynthetic systems has been shown before[6,15,16,17]. Fujita and Suzuki[17] have proposed that these compounds adjust the hydratization of the photosynthetic membranes, thus protecting the active system from being disorganized by too high water concentrations. They have also reported that the lowest concentration of sucrose or PEG for stabilizing the DPIP reduction at a high level vary between different species of blue-green algae. Correspondingly, chloroplasts prepared from different progenies of Pinus silvestris show optima at different sucrose concentrations. This work reports an optimum for DPIP reduction at 1.2 M sucrose, while another progeny of Pinus silvestris[6] had the optimum around 2.4 M sucrose, when the reaction was followed for 20 minutes. Picea abies[6] showed an increasing rate of DPIP reduction up to 2.8 M sucrose.

Working with artificial electron donors to photosystem II in some blue-green algae Suzuki and Fujita[18] have come to the conclusion that the sensitive step to suboptimal sucrose or PEG concentrations is somewhere on the oxidizing side of photosystem II. This interpretation is supported by this work, where it is shown that DPC, which is an artificial electron donor to photosystem II[19], partly can substitute for sucrose in the photoreduction of DPIP (Table 1). Also the fact that sucrose and PEG inhibited the DPC-supported DPIP photoreduction by the 10,000 x g fraction of broken chloroplasts supports the idea that the positive action of these compounds is on the oxidizing side of photosystem II, probably before the input of electrones from DPC (Table 2). This interpretation agrees with the general experience that the oxidizing side of photosystem II is very sensitive to inactivation. Tris- and hydroxylamine-washing, heat-treatment and CCCP-addition are some examples of such inhibitions, which are more or less compensated for by artificial electron donors to photosystem II[19].

Both DPIP and NADP photoreduction were inhibited by sucrose and PEG when the reactions were supported by electrons from DPC and ascorbate, respectively (Table 2, Figures 1 and 2). If specific sites are blocked by sucrose and PEG or if the inhibition is the result of some more general effect by these compounds on the photosynthetic membranes is impossible to say from the results of this work. It must be mentioned, however, that high concentrations of PEG is known to inhibit Hill reactions[6,20,21]. Since Stocking[22] has shown that high concentrations of PEG can percipitate proteins, one possible explanation to the inhibitory effect of PEG is that some protein factor necessary for the photosynthetic electron transport is inactivated.

Sucrose and PEG thus seem to have both a stimulating and an inhibitory effect on photosynthesis in isolated Pinus chloroplasts. The electron donation from water to the reaction centre of photosystem II is promoted by sucrose and PEG, but the remaining part of the electron transport chain ending with NADP seems to be inhibited. The optimum curves for DPIP photoreduction (Figures 1 and 2) may be the result of these opposite effects of sucrose or PEG.

Acknowledgements: We are very grateful to Mrs. Karin Areblad for her skilful technical assistance. This investigation was supported by the Swedish Council for Forestry and Agricultural Research and by the Swedish Natural Science Research Council.

6. References

1. Firenzuoli, A.M., Ramponi, G., Vanni, P. and Zanobini, A., 1968, Life Science 7, 905.
2. Oku, T. and Tomita, G., 1971, Photosynthetica 5, 28.
3. Oku, T., Kawahara, H. and Tomita, G., 1971, Plant Cell Physiol. 12, 559.
4. Michel-Wolwertz, M.R. and Bronchart, R., 1974, Plant Sci. Lett. 2, 45.
5. Oku, T., Sugahara, K. and Tomita, G., 1974, Plant Cell Physiol. 15, 175.
6. Öquist, G., Martin, B. and Mårtensson, O., 1974, Photosynthetica 8, In press.
7. Clendenning, K.A., Brown, T.E., Walldov, E.E., 1956, Physiol. Plant. 9, 519.
8. Krogmann, D.W. and Jagendorf, A.T., 1959, Arch. Biochem. Biophys. 80, 421.
9. McCarty, R.E. and Jagendorf, A.T., 1965, Plant Physiol. 40, 725.
10. Constantopoulos, G. and Kenyon, C.N., 1968, Plant Physiol. 43, 531.
11. Harris, W.M., 1971, Can. J. Bot. 49, 1107.
12. Buchanan, B.B. and Arnon, D.I., 1971, in: Methods in Enzymology, vol. 23, eds. S.P. Colowich and N.O. Kaplan (Academic Press, New York - London) p. 413.
13. Shin, M., 1971, in: Methods in Enzymology, vol. 23, eds. S.P. Colowick and N.O. Kaplan (Academic Press, New York - London) p. 440.
14. Arnon, D.I., 1949, Plant Physiol. 24, 1.
15. McClendon, J.H., 1954, Plant Physiol. 29, 448.
16. Pavlova, I.A., 1972, Fiziol. Rast. 19, 877.
17. Fujita, Y. and Suzuki, R., 1971, Plant Cell Physiol. 12, 641.
18. Suzuki, R. and Fujita, Y., 1972, Plant Cell Physiol. 13, 427.
19. Trebst, A., 1974, Annu. Rev. Plant Physiol. 25, 423.
20. Fredricks, W.W. and Jagendorf, A.T., 1964, Arch. Biochem. Biophys. 104, 39.
21. Fujita, Y. and Myers, J., 1965, Arch. Biochem. Biophys. 111, 619.
22. Stocking, C.R., 1956, Science 123, 1032.

M. AVRON, *Proceedings of the Third International Congress on Photosynthesis*, September 2-6, 1974, Weizmann Institute of Science, Rehovot, Israel
Elsevier Scientific Publishing Company, Amsterdam, The Netherlands, 1974

INVOLVEMENT OF OXYGEN DURING PHOTOSYNTHETIC INDUCTION

R.C. Jennings and G. Forti

Istituto Botanico, Università di Napoli, Via Foria 223, Napoli, Italy

1. Introduction

In a recent review on the lag period of photosynthetic induction, Walker[1] has supported and emphasised earlier hypotheses that the process is fundamentally a reflection of the autocatalytic build up of Calvin cycle intermediates. While he stressed the fact that electron transport does not display a similar lag when provided with a terminal acceptor such as ferricyanide with broken chloroplasts and phosphoglycerate with intact chloroplasts, the problem of electron transport during the lag phase of CO_2-dependent photosynthesis remains. During this period net oxygen evolution, like carbon fixation, is extremely low, which must be the result either of a low rate of electron transport or else of oxygen being consumed in a Mehler-type reaction which would permit a high electron flux with low rates of net oxygen evolution. In this communication we provide evidence for the involvement of an oxygen-consuming Mehler-type reaction during the lag period in isolated intact chloroplasts. Furthermore we demonstrate that this reaction is involved in ATP synthesis, and in the development of a stromal pH value which is optimal for the operation of the Calvin cycle.

2. Methods

Chloroplasts were prepared from freshly harvested leaves according to the technique of Heldt and Sauer[2]. Oxygen evolution was measured with a Clark-type oxygen electrode (Radiometer, Copenhagen) in a 25°C thermostated cell. Illumination was provided by a 500W tungsten lamp, sometimes filtered through a 600 nm cut-off filter (Sovirel CS 2-60), and the intensity was saturating unless otherwise stated. The incubation medium was fundamentally that indicated by Heldt and Sauer[2], with $NaHCO_3$ 10 mM, unless otherwise indicated. Oxygen was removed from solutions by bubbling with oxygen-free nitrogen which had been passed through a saturated solution of $NaHCO_3$. Bubbling of chloroplasts was minimal in order to reduce damage, though on occasions some chloroplast breakage was inevitable. In such experiments, when the final rate of photosynthesis was lower than that of unbubbled chloroplasts, the necessary correction was applied to the nitrogen-bubbled sample on the basis of final photosynthetic rates. In no case was the correction greater than 1.4, though frequently no correction was required.

Cytochrome f photo-oxidation was induced by light, passed through a 620 nm cut-off filter (Sovirel CS 2-60), of 12 x 10^4 ergs cm^{-2}·sec. It was measured in a dual wavelength spectrophotometer at 555-540 nm. The light-dark difference spectrum showed the 554 nm maximum typical of cytochrome f, with an additional peak at 550 nm. Treatment with DCMU or DBMIB greatly increased only the 554 nm peak, providing additional evidence that this peak was in fact that of cytochrome f.

ATP was measured by the luciferin-luciferase assay, according to Stanley and Williams[3]. Chlorophyll was measured according to Arnon[4].

3. Results and Discussion

Oxygraph traces of a typical experiment are shown in figure 2, where the commonly observed lag in photosynthetic induction is clearly visible for 2 to 3 min after the initial illumination. However, as can be seen in fig. 1 cytochrome f oxidation was rapid, indicating the immediate onset of electron transport, usually attaining the steady state within a few seconds. It was also immediately responsive to DCMU thus indicating that electron transport between photosystem 2 and an acceptor on the reducing side of photosystem 1 was operative upon the initial illumination. In table 1 are given the data for cytochrome f photo-oxidation upon initial illumination and also after 3 minutes illumination. Though rates of photosynthetic oxygen evolution are very different, the steady state photo-oxidation levels seem to be the same. The experiments suggest that the rate of photosynthetic oxygen evolution during and after the lag may not reflect the electron flux; a conclusion supported by experiments on the rate of Q re-oxidation in the dark[5] immediately following illumination for either 15 sec or 4 min. Though the rates of oxygen evolution differed by more than a factor of thirteen, the rate of Q re-oxidation during the first 5 sec dark (when the re-oxidation rate was linear) was similar in both cases.

The foregoing observations are strongly suggestive that during the lag period of photosynthetic oxygen evolution, electron transport functions at rates comparable with these encountered during periods of rapid oxygen evolution. Presumeably then oxygen serves as a terminal electron acceptor during the lag, a proposition supported by the lower steady state level of cytochrome f photo-oxidation encountered in anaerobic chloroplasts (fig. 1). A similar conclusion was reached recently by Patterson and Myers[6], who demonstrated a burst of H_2O_2

production during the photosynthetic lag in Anacystis nidulans. The lower steady state levels of cytochrome f photo-oxidation under oxygen-deficient conditions indicate, furthermore, that oxygen interacts with the electron transport chain on the reducing side of photosystem 1, as has been indicated by others[1,8,9,10].

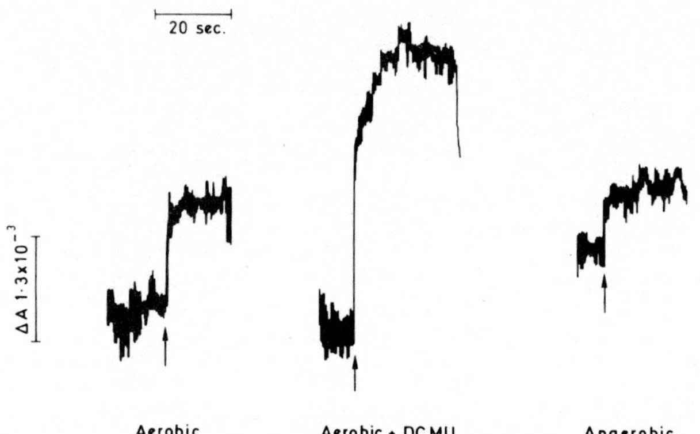

Fig. 1. Cytochrome f photo-oxidation in spinach chloroplasts, under aerobic and anaerobic conditions and in the presence of DCMU (6 μM). Measurements were made at 555-540 nm (see Methods). Chlorophyll content was 75 μg/ml and about 70% of the chloroplasts were intact as judged by the ferricyanide method[18]. Arrows indicate the onset of illumination (12 x 10^4 ergs/cm^{-2}·sec; Sovirel CS2-60 filter).

Table 1

Cytochrome f photo-oxidation and rates of photosynthetic oxygen evolution measured with the same light intensity during the first 15 seconds of illumination and after a subsequent 3 minutes illumination. The experimental scheme was similar to that illustrated in fig. 2 where after 3 min the light was turned off for 1 min. Sodium ascorbate (1 mM) was added to those chloroplasts being utilised for the cytochrome measurements. Treatment with ascorbate, added to rapidly reduce all the cytochrome, had no effect on its steady state oxidation level in the light. Measurements of both oxygen evolution and cytochrome f oxidation are those attained 10-20 seconds after illumination. Degree of chloroplast intactness, as indicated by the ferricyanide method[18], was around 70%. Chlorophyll content was 45 μg/ml for both measurements

	15 seconds light	3 minutes light - 1 minute dark - 20 seconds light
Oxygen evolution (nmoles/μg Chl/hr)	Not detectable	13
Cytochrome f ($\Delta A \times 10^{-3}$)	0.63 ± 0.02	0.60 ± 0.05

In order to understand the role of this oxygen-consuming, Mehler-type reaction, we examined the effect of anaerobiosis on the photosynthetic lag itself. A typical experiment is shown in fig. 2, where it is clear that anaerobic chloroplasts photosynthesised at much slower rates during the lag. A possible reason for this may have been due to decreased ATP levels as several reports indicate the possibility that spinach chloroplasts can phosphorylate with oxygen as the terminal electron acceptor[11,12]. From fig. 3 it is clear that anaerobic chloroplasts contained lower levels of ATP than aerobic chloroplasts, presumeably due to a lower rate of synthesis. The time taken to attain the maximal level was also somewhat longer in anaerobic chloroplasts. The lower ATP levels in 3-phosphoglycerate treated chloroplasts are expected as a consequence of an enhanced rate of the 3-PGA-kinase reaction during 3-PGA reduction[1] (see also table 2), here also lower ATP levels were encountered in oxygen-deprived chloroplasts. It can also be seen in table 2 that ATP addition stimulated photosynthetic oxygen evolution during the lag in both aerobic and anaerobic chloroplasts, though the final rate was little effected. The stimulation was proportionally greater in anaerobic chloroplasts, though the actual amount of stimulation was similar in both cases, a situation commonly observed and which does not unequivocably support the interpretation that ATP addition overcame an ATP deficiency in oxygen-deprived chloroplasts. However, in the presence of phosphoglycerate, where the effect of anaerobiosis was much reduced, ATP addition stimulated photosynthesis during the lag both proportionally and in terms of amount to a greater extent in oxygen deficient chloroplasts. It should also be pointed out that the inhibitory effect of anaerobiosis was largely overcome by the addition of phosphoglycerate and ATP, though in the first minute some difference was still apparent. Thus it is clear that while oxygen is necessary to attain high ATP levels during the lag, presumeably by a Mehler-reaction mediated increase in electron transport and hence phosphorylation, oxygen deprivation effects on the lag cannot be completely reversed by ATP even in the presence of phosphoglycerate. This failure is probably not due to limitations in the rate of ATP uptake, as interspersing the light with 30 sec dark periods at 1 min intervals, did not increase the amount of stimulation by ATP in the presence or absence of oxygen (unpublished observations), though ATP does penetrate spinach chloroplasts in the dark[13]. Another function for oxygen during the lag is therefore indicated.

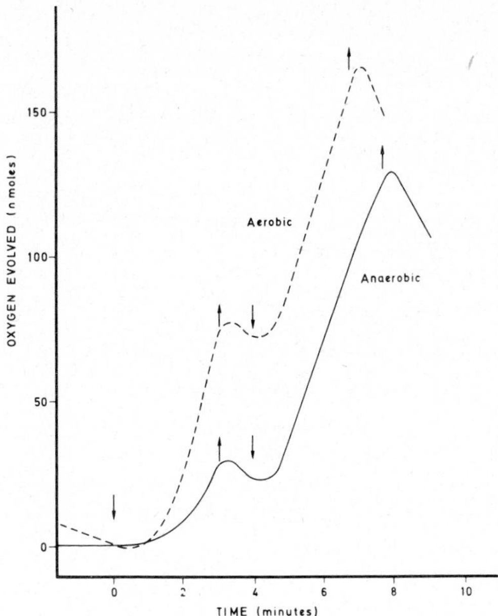

Fig. 2. Oxygen electrode traces of photosynthetic induction in spinach chloroplasts, under both aerobic and anaerobic conditions. Light intensity was 600,000 ergs/cm^2 ·sec (Sovirel CS2-60 filter) and the chlorophyll content was 50 μg/ml. Downward pointing arrows (↓) represent light on, and upward pointing arrows (↑) represent light off. For all other details, see Methods.

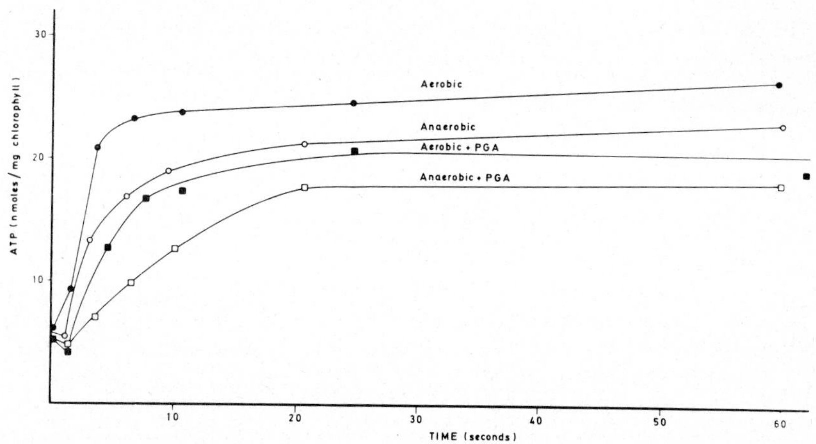

Fig. 3. Changes in ATP levels in spinach chloroplasts upon illumination under anaerobic and aerobic conditions and in the presence and absence of phosphoglycerate (PGA; 1 mM). The chlorophyll concentration was 100 μg/ml, and the final photosynthetic rate was 110 n moles of oxygen evolved/μg chlorophyll/hr. Illumination was with a 500 W tungsten lamp (see Methods).

Table 2

Effect of ATP (1 mM) and phosphoglycerate (PGA;1 mM) addition on the lag period of photosynthetic oxygen evolution. The experimental scheme was similar to that of fig. 2. The net rate of oxygen evolution was estimated for each minute during the early lag period of 2 to 3 min, when the light was turned off for 1 min in order to estimate the rate of oxygen exchange in the dark. The dark oxygen exchange values are given as negatives, as the pen deflection is downward, owing to oxygen consumption by photorespiration, and also equilibration reactions with the N_2 atmosphere in the anaerobic experiments. All additions were made 2 min before the commencement of illumination. The chlorophyll concentration was 50 μg/ml. Units are nmoles/μg Chl·hr

	Oxygen Evolution			Dark Oxygen Exchange	Maximal Rate of Photosynthesis
	0-1 min	1-2 min	2-3 min		
Anaerobic	0	2	6	-3	38
Anaerobic + ATP	0	4	13	-7	43
Aerobic	5	9	12	-3	48
Aerobic + ATP	6	13	17	-4	49
Anaerobic + PGA	7	28	-	-16	57
Anaerobic + PGA + ATP	11	36	-	-19	65
Aerobic + PGA	17	28	-	-5	65
Aerobic + PGA + ATP	20	30	-	-6	67

To investigate whether oxygen deprivation might inhibit electron transport, as suggested by Diner and Mauzerall[14],and hence inhibit NADP reduction during the lag, we measured the rate of oxaloacetate reduction in intact chloroplasts. This reaction is catalysed by malic dehydrogenase with photosynthetically produced NADPH[7], and so can be followed as light-dependent oxygen evolution. In order to reduce interference from Calvin cycle-dependent oxygen evolution the measurements were taken only in the first minute. From table 3 it can be seen that oxaloacetate dependent oxygen evolution was substantially greater in anaerobic than in aerobic chloroplasts, thus indicating that the anaerobic lag did not result from an NADPH deficiency.

Table 3

The effect of oxygen on oxaloacetate reduction, as measured by photosynthetic oxygen evolution. Oxaloacetate (1 mM) was added 2 min before illumination and the oxygen evolved after 1 min was estimated, corrections being made for the dark oxygen exchange. The chlorophyll concentration was 50 μg/ml, and the light was 600,000 ergs/cm^2·sec intensity (Sovirel CS2-60 filter). Units are nmoles of oxygen evolved/ μg chlorophyll/hr.

	Aerobic	Anaerobic
No addition	7.5	1.4
Plus oxaloacetate	9.2	8.2

Heldt et al.[15] have recently demonstrated that during photosynthesis, the stroma of intact spinach chloroplasts is alkalinised concurrently with thylakoid acidification. These authors speculated on the possible regulatory role of this alkalinisation on enzymes of the Calvin cycle, though no firm evidence was available at that time. We therefore felt that in view of the decreased electron flux in anaerobic chloroplasts, a decreased rate of stromal alkalinization may be partly responsible for the slower photosynthetic rates encountered in oxygen deficient chloroplasts during the lag. To this end we studied the effect of reaction medium pH on photosynthetic rates during the lag, as Heldt et al.[15] have demonstrated that the stroma becomes more strongly alkalinised as the pH of the buffer increases. From table 4 it is clear that while increasing the pH from 7.6 to 8.4 had only a very slight effect on aerobic chloroplasts, there was a substantial stimulation with anaerobic chloroplasts. This stimulation was only seen during the lag period and the maximal photosynthetic rate was lower at the elevated pH. From fig. 4 it can be seen that the optimal pH for steady-state photosynthesis was around pH 7.6, while in both aerobic and anaerobic chloroplasts the optimum was displaced to higher values for the lag period. Oxygen deficient chloroplasts were much more pH sensitive than aerobic chloroplasts, and the pH optimum was also higher.

Table 4

Effect of increased pH on the lag period of photosynthetic oxygen evolution in aerobic and anaerobic chloroplasts, and in the presence of phosphoglycerate (PGA; 1mM) and ribose-5-phosphate (R5P; 1 mM). All experiments contained ATP (1 mM). The experimental scheme is similar to that described in table 3. The chlorophyll concentration was 75 μg/ml. Units are nmoles of oxygen/μg chlorophyll/hr

	Oxygen Evolution		Dark Oxygen Exchange	Maximal Rate of Photosynthesis
	0-1 min	1-2 min		
Aerobic pH 7.6	15	32	-4	65
Aerobic pH 8.4	17	29	-4	46
Anaerobic pH 7.6	4	20	-12	73
Anaerobic pH 8.4	7	29	-16	52
Aerobic + PGA pH 7.6	31	47	-7	71
Aerobic + PGA pH 8.4	29	42	-7	52
Anaerobic + PGA pH 7.6	18	39	-18	80
Anaerobic + PGA pH 8.4	20	39	-18	61
Aerobic + R5P pH 7.6	16	45	-6	68
Aerobic + R5P pH 8.4	20	42	-6	60
Anaerobic + R5P pH 7.6	6	33	-16	70
Anaerobic + R5P pH 8.4	10	47	-18	63

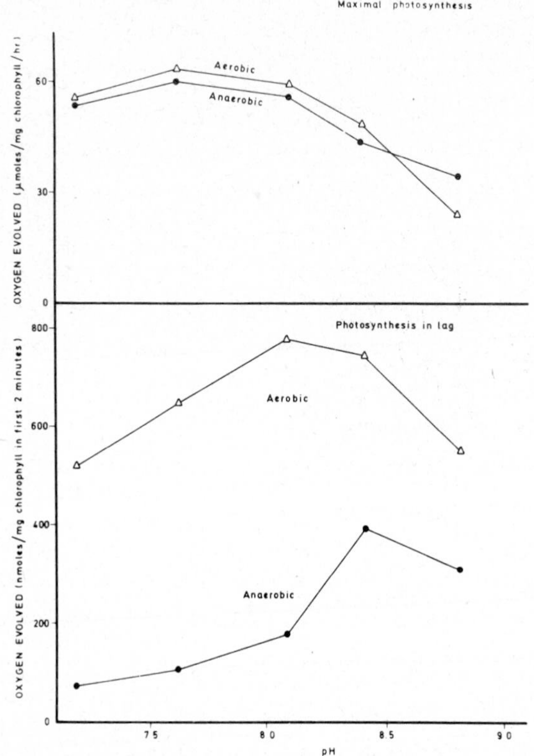

Fig. 4. The effect of pH on maximal rate of photosynthetic oxygen evolution and on the rate of oxygen evolution during the lag period. The experimental scheme was similar to that of fig. 2. The chlorophyll content was 75 µg/ml, and illumination was with a 500 W tungsten lamp.

In an attempt to define the pH sensitive site we compared the effect of pH in the presence of phosphoglycerate and ribose-5-phosphate (table 4). These experiments were performed in the presence of ATP, in order to minimise the complicatory effect of lower ATP levels in oxygen deprived chloroplasts. Clearly phosphoglycerate reduction was almost pH insensitive during the first 2 min, while in the presence of ribose-5-phosphate pH 8.4 was stimulatory and to an extent comparable with that encountered when no addition was made. Similar data have also been obtained when ribulose-5-phosphate was added and clearly indicate that the pH sensitive site is located between ribulose-5-phosphate and phosphoglycerate in the Calvin cycle, an entirely reasonable hypothesis in view of the published pH optima of the enzymes involved (phosphoribulokinase, pH 7.9[16]; ribulodiphosphate carboxylase, pH 7.6[17]) and the data for pH changes in the stroma upon illumination, determined by Heldt et al.[15].

These experiments indicate that the function of the Mehler-type reaction during the lag period of photosynthesis is to increase both ATP synthesis and the pH of the stroma, via the increased electron flux, which in turn leads to greater photosynthetic rates during the lag. However, we have not been able to completely reverse the anaerobic lag, even at optimal pH values in the presence of ATP, or in the presence of phosphoglycerate and ATP. It is therefore probable that oxygen deprivation inhibits yet another reaction during the early part of the lag, possibly associated with the expected lower pH values across the thylakoid membrane accompanying the decreased electron flux.

Acknowledgment

This work was supported by the Consiglio Nazionale delle Ricerche of Italy.

References

1. Walker, D.A., 1973, New Phytol. 72, 209.
2. Heldt, H.W. and Sauer, F., 1971, Biochim. Biophys. Acta 234, 83.
3. Stanley, H.W. and Williams, S.G., 1969, Anal. Biochem. 24, 381.
4. Arnon, D.I., 1949, Pl. Physiol. 24, 1.
5. Jennings, R.C. and Forti, G. Unpublished observations.
6. Patterson, C.O.P. and Myers, J., 1973, Pl. Physiol. 51, 104.
7. Vidaver, W. and French, C.S., 1965, Pl. Physiol. 40, 7.
8. Heber, U. and French, C.S., 1968, Planta 79, 99.
9. Heber, U., 1969, Biochim. Biophys. Acta 186, 302.
10. Osada, K., Kiso, K. and Yoshikawa, K., 1974, J. Biol. Chem. 249, 2175.
11. Forti, G. and Jagendorf, A., 1961, Biochim. Biophys. Acta 54, 322.
12. Heber, U., 1973, Biochim. Biophys. Acta 305, 140.
13. Heldt, H.W., 1969, FEBS Letters 5, 11.
14. Diner, B. and Mauzerall, D., 1973, Biochim. Biophys. Acta 305, 329.
15. Heldt, H.W., Werdan, K., Milovanev, M. and Geller, G., 1973, Biochim. Biophys. Acta 314, 224.
16. Hurwitz, J., Weisback, A., Horecker, B. and Smyrniotis, P., 1966, J. Biol. Chem. 218, 769.
17. Bassham, J.A., Sharp, P. and Morris, I., 1968, Biochim. Biophys. Acta 21, 438.
18. Heber, U. and Santarius, K.A., 1970, Z. Naturforsch. 25b, 718.

M. AVRON, *Proceedings of the Third International Congress on Photosynthesis*, September 2-6, 1974, Weizmann Institute of Science, Rehovot, Israel
Elsevier Scientific Publishing Company, Amsterdam, The Netherlands, 1974

CIRCADIAN RHYTHM IN THE HILL REACTION OF ACETABULARIA

T. Vanden Driessche

Département de Biologie moléculaire
Université Libre de Bruxelles

1. Summary

Evidence for a circadian variation in the Hill reaction has been obtained by determining twice a day the rate of DCIP reduction. The variation has been obtained with chloroplasts of the stalk and of the apex of the alga Acetabularia mediterranea. However, the extent of the variation is much larger in the stalk than it is in the apex. The Hill reaction itself is higher in the latter. The greater the amplitude of the circadian rhythm in photosynthesis, the larger the difference between the values of the Hill reaction measured at the beginning and at the middle of the light period. The results are discussed in view of the fact that a variation in the Hill reaction is of crucial significance for an understanding at the molecular level of the circadian rhythm of photosynthesis.

2. Introduction

Circadian variations in photosynthesis, irrespective of light intensity, are well known in the unicellular alga Acetabularia (1-6). Which part of the complex phenomenon of photosynthesis is directly modulated by circadian rhythmicity remains, however, unknown.

From their experiments conducted at various light intensities, Terborgh and McLeod (5) concluded that rhythmicity probably results either from changes in the light and dark reactions, or from changes in their coupling. No variation in activity was found in the 9 enzymes of the Cavin cycle assayed in the middle of the light and of the dark period (7). In Euglena, one of these enzymes, glyceraldehyde-3-phosphate dehydrogenase,in its two forms, NADH and NADPH dependent did sustain the variation in photosynthetic capacity (8).

In order to get more information concerning the support of rhythmicity in Acetabularia, experiments have been undertaken, with the aim of determining whether or not there is a circadian variation in the Hill activity of the chloroplasts. This implicates the determination of the concentrations of suspensions of chloroplasts and of electron acceptor appropriate to various Hill activity levels.

The alternative possibility of circadian variation of photosynthesis relying on changes in chlorophyll concentration has been ruled out by previous experiments (7 and 9).

It is also known that there is a physiological gradient as well as a morphological one along the stalk of the alga. Specifically, the photosynthetic activity (measured as ^{14}C incorporated from bicarbonate) has been found to be different, on a common chlorophyll weight basis, in the apex, the stalk and the rhizoid (10). Consequently, the different parts of the alga have been examined for their Hill activity. However, it has not been possible to obtain either a good yield in the Hill reaction nor reproducible values with the chloroplast suspensions obtained from the rhizoid, for some unknown reason.

This paper presents (1) the values of the Hill reaction measured twice a day on chloroplasts obtained from the apex and from the stalk of Acetabularia (at the beginning and at the middle of the light period) or from the stalk only. In several experiments, measurement of oxygen evolution has been determined in parallel with the Hill reaction.

3. Materials and methods

Culture of the algae and samples. The algae, Acetabularia mediterranea,are cultivated in standard medium(11)in culture rooms maintained at 20°C, under light-dark cycles of 12-12 h. The light was given by fluorescent daylight tubes "Phytor" (ACEC). The intensity at the level of the algae was of 1000-1200 lux.

75-200 algae of 15-30 mm in length have been used for each determination of the Hill reaction rate. When photosynthesis has also been measured, additional samples of 20-25 algae have been taken from the batch.

Fragmentation. The Hill reaction has been carried out on either stalk or apical chloroplasts. The fragments have been obtained in the following manner. Algae are aligned in groups of the same length on filter paper. The rhizoid is severed with a scalpel. The apical portion of approximately 5 mm is also cut off and used for the determination of the reducing activity of the chloroplasts (in smaller algae, a somewhat shorter apical fragment is taken whereas in longer algae, a somewhat longer fragment is taken off). In experiments 1-6, the next 2-5 mm (depending on the length of the alga) is also removed in order to ascertain a good discrimination between the two chloroplast populations.

The Hill reaction. The chloroplast suspensions have been prepared according to Goffeau and Brachet (12). An aliquot is used for the determination of the reduction of 2,6-dichlorophenol indophenol (sodium salt) (13). The reduction is allowed to take place for 6 mi at 20°C at 5000 lux in experiments 1-7 and 4200 lux in 8-12 (or in the dark for the controls). Each value is the mean of 2 or 3 determinations carried out on 2 or 3 different samples of algae. The chloroplast suspensions are either adjusted to the same chlorophyll content (3.62 µg) (exper. 1-8) or on variable chlorophyll content never exceeding 30 µg.

The pH of the buffer which has been used is 6.9 . Dood and Bidwell (14) have pointed out that the cytoplasmic pH of Acetabularia ranges from 8.0 to 8.4 whereas the optimum pH for CO_2 fixation or O_2 evolution of Acetabularia chloroplast suspensions lies between 7.6 and 7.7. In preliminary experiments a much better yield in the Hill reaction has also been obtained at pH 7.8 than at pH 6.9 (the yield was approximately twice as good). Since the preparation of the chloroplasts suspensions, until the Hill reaction itself, has to be carried out at 4°C in order to avoid chlorophyll degradation, and since precipitations occur in the buffer at pH 7.8 (and not 6.9) which interfer with the spectrophotometric readings, all experiments have been carried out in the 6.9 buffer.

Chlorophyll content. The determinations have been performed according to Arnon (15).

Photosynthesis. The oxygen evolution is measured manometrically by means of a recording Gilson respirometer (RR8). A constant CO_2 pressure is ensured by the Pardee medium (16). The light intensity is indicated for each experiment.
The photosynthesis determinations have been carried out on the same day as the Hill determination. Consequently, the first values came out late in the morning but the difference between the maximum which occurs in the middle of the light period and 4 h before can be estimated by its value 4 h later, since the circadian rhythm of oxygen evolution is symetrical.

4. Experimental results

In table 1 are reported the results obtained in 12 experiments. The rate of the Hill reaction on a chlorophyll weight basis is always increased in the middle of the day as compared with the beginning of the light period. Expressed as percent of the midday value, the reduction rate in the morning varies from 65 to 85 % for chloroplasts obtained from the stalk. It is a usual feature of circadian rhythms in Acetabularia that amplitude varies greatly from batch to batch.

The value of the Hill reaction of the chloroplasts obtained from apeces is always higher than that of chloroplasts obtained from the stalk (only one value is the same : in the middle of the light period in exper. 4). The circadian variation in Hill activity is observed in the stalk as well as in the apex (table 1). However, the importance of the circadian variation is always smaller than in the stalks. Since the rate difference between morning and middle of the day is smaller in the apex than in the stalk , the difference in the Hill reaction between these two fragments is more apparent in the morning.

Table I - Hill reduction of chloroplast suspensions obtained from stalks and apex in the beginning (a) and in the middle of the light period. The Hill value is expressed as the number of μM of DCIP reduced per μg of chlorophyll (means of duplicate determinations).

Exper.	STALK a	STALK b	a as % of b	APEX a	APEX b	a as % of b
1	6.70	9.65	70			
2	21.1			24.55		
3	11.75	13.85	85	15.25	16.2	94
4	12.85	18.8	68	17.2	18.55	91
5	10.55	16.2	65	14.8	17.85	83
6	14.8	19.7	75	16.45	21.1	78
7	12.8	18.3	70			
8	10.95	13.85	79			
9	11.1	14.55	76			
10	4.16	5.55	75			
11	4.55	6.50	70			
12	8.85	24.25	37			

The photosynthetic rate has been reported in table 2. The data show that the circadian rhythm is expressed in all cases. In experiment 1-7, the light intensity has been 5500 lux for photosynthesis and 5000 lux for the Hill reaction. If one compares the experimental results for both the oxygen evolution and the Hill reaction (stalk), a very similar picture is obtained : the greatest circadian differences are observed in experiments 5,4 and 1, the smaller one in experiment 3. However, in experiment 7, the difference is relatively smaller for photosynthesis than for the Hill reaction.

Table II - Oxygen evolution of samples of whole algae from the same batches as those used for the Hill determinations ($\mu l/h$).

Exper.	Lux	3h before maximum	maximum (b)	4h after maximum(c)	c as % of b
1	5500		24.25	14.75	61
3	5500		23.3	19.3	83
4	5500		57.4	35.4	62
5	5500		55.4	30.6	55
6	5500		48.0	32.0	67
7	5500		42.0	37.0	88
8	5500	18.0	25.0		72
10	5500		37.0	28.0	76
11	3500		43.0	31.0	72
12	3500		42.5	32.0	75

5. Discussion and conclusions

A circadian variation in the Hill reactions, measured twice a day is expressed by both stalks and apeces of Acetabularia. However, the amplitude is higher in the stalk. Since the chloroplasts of the apex are "young" and smaller than in the stalk (17,18) it could be suggested that the full expression of rhythmicity is a developmental process of chloroplasts.

It should also be remembered that intense cytoplasmic streaming takes place in the illuminated stalks of Acetabularia, chloroplasts moving longitudinally between cell wall and vacuole, but is not likely to involve apical chloroplasts embedded in much less vacuolized cytoplasm nor rhizoidal chloroplasts maintained in the network formed by the perinuclear maze (made up from ramifications of the vacuole and digitations of the cytoplasm).

An alternative explanation of the smaller difference between the two values of the Hill reaction in the apex as compared with the stalk would be a phase delay which would either completely mask the difference if either the maximum or the minimum occured at equal distance from either measurement or would reduce it if the maximum or the minimum occured at any time but the one between the two measurements. This however is not likely since the chloroplasts are synchronized by the same entraining agents and are located in one and the same cytoplasm.More direct evidence that it is not the case arises from the figures presented by Puiseux-Dao and her collaborators (18,19). The authors counted from hour to hour the number of chloroplasts containing one, two or several polysaccharide granules(19).In light-dark conditions, the number of choroplasts containing one granule decreases from 9 am (2 h after the beginning of the light period) until 2 pm, whereas the number of chloroplasts containing three granules (or more) increases. The number of chloroplasts containing 2 granules does not show dramatic change. In a second paper, the authors discreminate between apex, stalk and rhizoid and counted the different classes of chloroplasts from 9 am to 3 pm(18)The chloroplasts of both apex and stalk have a maximum of 3 granules at 2 pm and a minimum of 1 granule at 3 pm. However, the authors paid more attention to the inversion in the ratio : the number of chloroplasts containing one granule versus the number of chloroplasts containing 2 granules, inversion which occurs at different times in the apex and the stalk. From the first paper, this seems to result from fluctuations in the number of chloroplasts containg 2 granules, originating from compensation between two opposed curves. On the contrary, the determination of the minimum and maximum of chloroplasts containg one or several granules is be more relevant to the interpretation of the results presented here.

The determination of the photosynthetic rate has been carried out in most experiments in order to ascertain the expression of the rhythm. In all cases, variations in both oxygen evolution and Hill reaction have been manifested. Moreover, a general parallel in the importance of these variations is established. It should be kept in mind, however that, in this experimental series, the measurements have been carried out twice a day and that other experiments have to be designed in order to examine the phase relationship of the rhythms of photosynthesis and Hill reaction.

The experimental results have shown that the Hill reaction differs in the apex and in the stalk of the algae (the latter being lower). This is in agreement with the results of Issinger et al. (10) on the photosynthetic rate (measured by CO^2 incorporation) in these parts. Owing to the circadian rhythm of different amplitude in the two types of fragments, the difference between apex and stalk is larger in the morning than in the middle of the day.

The finding that, in contrast to Gonyaulax (20), the rate of the Hill reaction does vary with the time of the day opens two alternatives. Either the membranar state of the electron carriers is such that the transfer takes place at different rates depending on the time of the day or that the increase in the Hill reaction at midday reflects a change in the ATP pool(which is known to be lower at that time(21) or in other molecules. Futher research is needed in order to clarify this point.

Acknowledgment

The author is indebeted to Dr. P. Malpoix for kindly improving the English of the text. She thanks Miss M. Hayet and L. Lateur for technical assistance. Partial support of the International Biology Program is acknowledged.

References

(1) Sweeney B.M. and Haxo F.T. 1961, Science 134, 1361-1362.

(2) Richter G., 1963 - Z. Naturforsch. 18b, 1085-1089.

(3) Schweiger E., Walraff H.G., and Schweiger H.G., 1964, Z. Naturforsch. 19b, 499-505.

(4) Vanden Driessche T., 1966, Biochim. Biophys. Acta 126, 426-470

(5) Terborgh J. and McLeod G.C., 1967, Biol. Bull. 133, 659-669.

(6) Mergenhagen D., and Schweiger H.G., 1973, Exper. Cell. Res. 81, 360-364.

(7) Hellebust J.A., Terborgh J. and McLeod G.C., 1967, Biol. Bull. 133, 670-678

(8) Walther W.G. and Edmunds L.N. jr., 1973, Plant Physiol. 51, 250-258.

(9) Hellesbust, in Terborgh and McLeod, 1967 and Vanden Driessche, unpublished experiments.

(10) Issinger O, Issinger M. and Clauss H. 1971, Planta (Berl.) 101, 360-364.

(11) Lateur L., 1963, Rev. algol. 1, 25.

(12) Goffeau A. and Brachet J., 1965, Biochim. Biophys. Acta 95, 302

(13) Sober H.A., Harte R.A. and Sober E.K. (1970) Handbook of biochemistry chemical Rubber Co.

(14) Dood W.A. and Bidwell R.G.S., 1971, Plant Physical 47, 779.

(15) Arnon D.I., 1949, Plant Physical 24, 1.

(16) Umbreit W.W., Burris R.H. and Stanfer J.F., 1957, Manometric techniques, Burgess Publ. Co.

(17) Boloukhère M., 1972, J. Microsc. 13, 401

(18) Dazy A.C.,Mathys E. and Puiseux -Dao S., 1970, Soc. bot. Fr., Mémoires 117, 311.

(19) Puiseux-Dao S. and Gilbert A.M., 1967, C.R. Acad. Sc. Paris (série D) 265, 870.

(20) Sweeney B.M., 1965, Circadian Clocks, Aschoff J. Ed., North Holland Publ. Co, 190

(21) Vanden Driessche T., 1970, Biochim. Biophys. Acta 205, 526.

M. AVRON, *Proceedings of the Third International Congress on Photosynthesis*,
September 2-6, 1974, Weizmann Institute of Science, Rehovot, Israel
Elsevier Scientific Publishing Company, Amsterdam, The Netherlands, 1974

HILL-REACTION OF CHLOROPLASTS FROM RAPHANUS SEEDLINGS GROWN WITH ß-INDOLE-ACETIC ACID AND KINETIN

C. Buschmann and H.K. Lichtenthaler
Botanical Institute (Plant Physiology) University of Karlsruhe
D-75 Karlsruhe, FRG

Summary

IAA and kinetin, when given to the growing Raphanus seedlings, increase the formation of plastoquinone-9 and of P 700 per plant and on a chlorophyll basis. Both phytohormones also enhance the photosynthetic activity with DCPIP and methylviologen as Hill-reagents. As expected, addition of ammonium chloride as uncoupler increases Hill-activity in all cases; and even to a higher degree in plants grown with IAA and kinetin. Both phytohormones apparently induce in growing plants the formation of more photosynthetic units, which is valid for both photosystems PS I and PS II. Kinetin gives a higher promotion effect than IAA.

1. Introduction

The known natural and synthetic plant hormones have a considerable influence on growth processes in plants. They induce or regulate cell growth, cell multiplication and enzyme synthesis. In the following we describe the effect of the endogenous phytohormone ß-indole-acetic acid (IAA) and of the synthetic cytokinin kinetin on the formation and function of the photosynthetic apparatus. It has been shown before that application of IAA and kinetin to light grown Raphanus seedlings increases to a high extent the formation of plastoquinone-9[1,2], the well-known terminal electron acceptor of the photosynthetic light reaction II. The synthesis of α-tocopherol in turn, which is the main benzoquinone component in thylakoid-free etioplasts[3], is slightly (IAA) or strongly (kinetin) diminished by the phytohormones. There is also a low but insignificant increase in the formation of chlorophylls; however, little influence on the carotenoid level. Because of the specific promotion of the plastoquinone-9 synthesis, it was assumed that the phytohormones also influence the rate of photosynthetic activity by increasing the number of photosynthetic units on a chlorophyll basis. This has been proved in the present paper.

2. Methods

The seedlings of Raphanus sativum L. (Saxa Treib) were grown for 8 days on a 10% van der Crone nutrition solution (controls) in the light (fluora lamps, 2000 lux)[1,2]. Kinetin was added to the nutrition solution during germination in a concentration of 2ppm and IAA on the 3rd day in a level of 0.5ppm. The chloroplasts were isolated from the 8 day old, light grown plants in a buffer containing 0.5 M saccharose, 0.1 M potassium phosphate (pH 7.4), 75mM NaCl, 2mM $MgCl_2$. The light induced reduction of dichlorphenol-indophenol (DCPIP) was determined spectrophotometrically and that of methylviologen polarographically. NH_4Cl as uncoupler was applied in a concentration of 10^{-3}M. The concentration of P 700 was measured spectrophotometrically by chemical oxidation and reduction[4]. The chlorophyll content was used as reference point.

3. Results and discussion

Raphanus seedlings grown 8 days without addition of phytohormones exhibit hypocotyls of ca. 5cm length and possess fully green cotyledons. Treatment with IAA has little influence on the length or appearance of the seedlings. By application of kinetin the plants exhibit an initial lag phase in growth of about 2 days which is compensated by an increased growth thereafter. The hypocotyl and cotyledons of 8 day old kinetin treated plants are yet somewhat smaller than in the controls. The higher chlorophyll level of phytohormone treated plants, with a particular enrichment of the plastoquinone-9 level as compared to the controls, is shown in table 1.

Table 1

Pigment- and lipoquinone composition of Raphanus seedlings grown without (controls) and with addition of IAA and kinetin (in µg per 100 plants, according to Straub and Lichtenthaler 1973)

	control	IAA	kinetin
chlorophyll a	2700	3000	3100
chlorophyll b	900	850	900
chlorophyll a/b	3.0	3.5	3.4
carotenoids	572	592	480
plastoquinone-9	299	595	500
α-tocopherol (+ α-tocoquinone)	353	296	143
total benzoquinones	652	891	562

The chloroplasts isolated from 8 day old Raphanus seedlings are photosynthetically active and exhibit DCPIP reduction rates of 40-52 μ mol O/mg chlorophyll and hour. Plants grown with the addition of IAA and kinetin show a reduction rate which is higher by 10 and 35%, respectively than that of the controls. By the addition of the uncoupler ammonium chloride the DCPIP reduction rate is increased, to a much higher extent in the plants grown with IAA and kinetin (table 2). In order to exclude the possibility that IAA and kinetin increase photosynthetic activity by uncoupling the electron transport, we added IAA and kinetin directly to chloroplasts isolated from untreated plants. We did not find any uncoupling effects of the phytohormones. The Hill-activities measured lay in the range of the controls. From the stimulation of Hill-activity (DCPIP reduction) it is clear that IAA and kinetin not only change the chlorophyll/plastoquinone ratio but also increase the number of photosynthetic units (pigment system II) on a total chlorophyll basis.

Table 2

Rate of DCPIP reduction (μMol O/μMol chlorophyll and hour) by Raphanus chloroplasts isolated from plants grown with the addition of IAA and kinetin

		+ NH_4Cl	increase
control	51.2	64.8	1.26 times
IAA	56.4	85.7	1.52 times
kinetin	69.0	139.2	2.02 times

Similar results have been obtained from the polarographic measurement of methylviologen reduction which includes pigment system I and pigment system II activity (table 3). The reduction of methylviologen is increased by IAA (12%) and even to a higher extent by kinetin (36%). As in DCPIP reduction kinetin has also in this case a bigger promotion effect than IAA. Addition of the uncoupler ammonium chloride increases the reduction rate of methylviologen which again is higher by addition of phytohormones.

Table 3

Reduction of methylviologen (μMol O/μMol chlorophyll and hour) in Raphanus chloroplasts isolated from plants grown without (controls) and with addition of phytohormones)

		+ NH_4Cl	increase
control	42.8	58.8	1.38 times
IAA	47.9	68.2	1.42 times
kinetin	58.1	90.4	1.55 times

These results show that the phytohormones IAA and kinetin increase not only the number of pigment system II units per chlorophyll but also the number of pigment system I units.

In order to have an additional parameter for pigment system I activity, we determined the concentration of P 700 in the Raphanus chloroplasts. Under the influence of both phytohormones during growth the amount of P 700 is increased as expected. Per one P 700 we find after kinetin and IAA treatment only 161 and 192 moles of chlorophyll respectively, in the controls, however, 225 moles (table 4).

Table 4

Concentration of P 700 in chloroplasts isolated from Raphanus seedlings grown without (controls) and with the addition of phytohormones

	10^{-4} P 700/chlorophyll	P 700/chlorophyll
control	44.4	1/225
IAA	52.0	1/192
kinetin	62.3	1/161

Thus the concentration of P 700 is similarly increased by the phytohormones as the Hill-activity with DCPIP and methylviologen. Therefore it is clear that IAA and kinetin induce during growth of Raphanus seedlings not only an increased plastoquinone-9 and P 700 formation, but also an increased number of photosynthetic units on a chlorophyll basis, which gives rise to the increased DCPIP and methylviologen reduction. This means, consequently, that the amount of antenna chlorophyll per photosynthetic unit is reduced in the phytohormone treated plants as compared to the controls.

This work was sponsored by the Deutsche Forschungsgemeinschaft.

References

1. Straub, V. and Lichtenthaler, H.K., 1973, Z. Pflanzenphysiol., 70, 30-45.

2 Straub, V. and Lichtenthaler, H.K., 1973, Z. Pflanzenphysiol., 70, 308-321.

3. Lichtenthaler, H.K., 1973, Ber. Dtsch. Bot. Ges., 86, 313-329.

4. Marsho, T.V. and Kok, B., Detection and Isolation of P 700, in: Methods in Enzymology, A. San Pietro (ed.) Academic Press, New York, London 1971, Vol XXIII, S. 515-522.

M. AVRON, *Proceedings of the Third International Congress on Photosynthesis*,
September 2-6, 1974, Weizmann Institute of Science, Rehovot, Israel
Elsevier Scientific Publishing Company, Amsterdam, The Netherlands, 1974

ELECTRON TRANSPORT PATHWAYS IN CHROMATOPHORES

J.B. JACKSON

Max Volmer Institut fur Physikalische Chemie,

Technische Universitat, Berlin

Permanent address: Department of Biochemistry, University of Birmingham, England, U.K.

INTRODUCTION

The decay of the carotenoid shift of chromatophores from photosynthetic bacteria following a period of illumination depends on the ionic permeability of the membranes. In untreated chromatophores the carotenoid shift decays with similar kinetics to the release of protons which have been taken up in the light[1]. Low concentrations of uncoupling agents accelerate the decay of the shift and the H^+-release in parallel[2]. When chromatophores are doped with the ionophore, valinomycin the carotenoid shift decay following a short flash is enhanced due to the efflux of K^+[3,4]. These observations are consistent with the suggestion that the shift in the carotenoid absorption spectrum is an electrochromic response to an electric field generated across the chromatophore membranes during photosynthesis[5]. A kinetic analysis of the risetime of the carotenoid shift at different ambient redox potentials shows that it is generated at three sites of light-driven electron transport[6]. The fastest of these is the photochemical electron transfer between reaction centre bacteriochlorophyll (P) and the primary electron acceptor or photoredoxin (X). The two slower components may involve the oxidation/reduction reactions of the b and c type cytochromes.

Since electron transport in photosynthetic bacteria is most probably a cyclic process[7] the generation of the electric field in continous light will be limited by the rate of electron return from X to P. An indication of this rate may be taken from the kinetics of P re-reduction following a period of illumination. This does not necessarily give the rate of the limiting steps of electron transport, for example when the carriers in the cycle are mostly reduced, the off-rate may only indicate the rate of the reaction between the primary donors and P. The bleaching of several of the chromatophore absorption bands by light has been attributed to the photo-oxidation of the reaction centre bacteriochlorophyll. The recovery of the bands upon darkening an illuminated sample or after flash excitation has been studied in a number of laboratories[8-11]. There is general agreement that the re-reduction process is biphasic although there are some disagreements over the effects of specific inhibitors and uncoupling agents. The reason for the biphasic decay has not been adequately explained.

Non-standard abbreviations: FCCP carbonyl cyanide-p-trifluoromethoxy phenylhydrazone; R. Rhodospirillum; Rps. Rhodopseudomonas.

In this report the kinetics of the generation of the chromatophore electrochromic shifts in flashing light are compared with the recovery of the bleached bacteriochlorophyll signal centred around 600 nm after flash or continuous illumination. Although smaller than some of the other bacteriochlorophyll bands, the absorption change at 600 nm is relatively free from interfering signals ie. the bleaching at 430 nm contains components due to cytochrome oxidation and carotenoid shift; the signal at 800 nm is partly electrochromic and P870 is masked by bacteriochlorophyll fluorescence.

METHODS

Cells of *Rps. spheroids*, *Rps. capsulata* and *R. rubrum* were grown anaerobically in the light in the medium of Sistrom[12]. The harvested bacteria were broken in a Ribi Cell fractionater at 10000 psi and the chromatophore fraction, sedimenting between 25000 x 15 g min and 100000 x 90 g min was isolated in a medium containing 50 mM KCl, 50 mM tricine, 8 mM $MgCl_2$, 10% sucrose, pH 7.4. Bacteriochlorophyll was assayed using the *in vivo* extinction coefficients given by Clayton[13].

The absorption changes of the chromatophore carotenoids and P600 bacteriochlorophyll were measured with a single-beam spectrophotometer designed and built in the Max Volmer Institut (see reference 14 for details). The measuring beam was attenuated with interference and neutral density filters to an intensity that had no effect on the flash-induced signals. Each recording was an average signal of 32 or 64, 20 µs half-width flashes of saturating intensity at wavelenths $>$ 650 nm (see figure legends for details). Flash frequency was continuously variable down to 25 Hz.

RESULTS

Dark electron flow rate from the regeneration of the flash-induced carotenoid shift

A single short light pulse of sufficient intensity will oxidise all the reaction centre bacteriochlorophyll molecules of a chromatophore suspension with reduction of the corresponding acceptors. A second flash will produce P oxidation only if P^+ has been re-reduced and X^- re-oxidised either by the reverse reaction ($P^+ \leftarrow X^-$) or by electron flow from cytochrome *c* $\rightarrow P^+$ and $X^- \rightarrow$ ubiquinone. Subsequent flashes at short intervals exhaust these pools unless the oxidation/reduction equivalents can re-combine through cyclic electron flow. We have shown previously that the secondary donor and acceptor pools are exhausted after about five flashes and a steady-state is reached[6]. This is illustrated for *Rps. spheroides* chromatophores in figure 1 where the carotenoid shift response is shown for chromatophores exposed to light pulses at 20 Hz and 1.1 Hz. At the lower frequency, once the steady-state is established, the dark time between the flashes was sufficient for 70% of the electrons to return from X^- to P^+. At 20 Hz only 20% recombination of reducing equivalents took place in the dark period between the flashes.

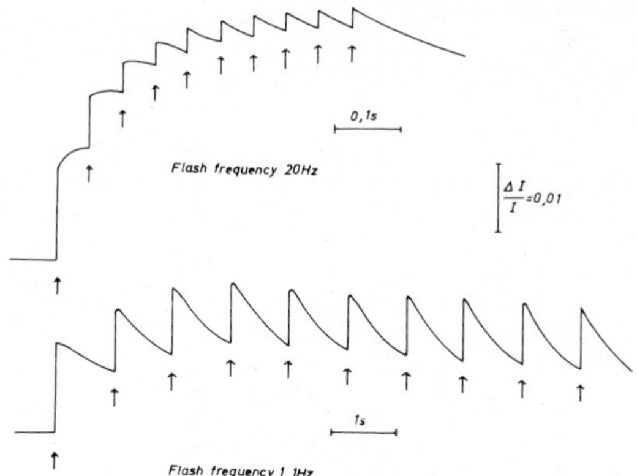

Figure 1 The attainment of a steady state electron flow pattern under repetitive flash conditions. Chromatophores of Rps. spheroides, containing 0.1 µmole bacteriochlorophyll were suspended in 10.0ml of a solution containing 50 mM tricine, 8 mM $MgCl_2$, 10% sucrose, pH 7.9 at 20°C in a 2 cm light path cuvette. An upward deflection of the trace represents a decrease in transmission of the 523 nm measuring light. At the time indicated by the arrows the chromatophore suspension was irradiated with a short light flash as described in the METHODS. Each trace is an averaged signal of 16 groups of 10 flashes with a dark time between the flash groups of 12.0 s.

Once the flash steady-state had been reached, the amplitude of the carotenoid shift remained undiminished for > 100 flashes (not shown). It is therefore unlikely that non-cyclic electron flow involving pools of endogenous oxidant and reductant makes any significant contribution.

The regeneration of the carriers required for the development of the carotenoid shift can be measured by this technique. Only the fast phase of the shift($t_{\frac{1}{2}} \leqslant 50$ ns) reported to respond to the P→X reaction,[6] has been considered. The chromatophore suspension was exposed to light pulses at a fixed frequency and measurement was started after five flashes. The carotenoid shift following each flash was stored on the signal averager operating at a sweep rate of 1 ms/cm. The average amplitude (between 0-100 µs after the flash) for 32 flashes was recorded for each frequency.

Figure 2 shows that complete recovery of the carotenoid shift fast phase of untreated Rps. spheroides chromatophores requires a dark period of approximately 10 s between the flashes. Log plots reveal that the recovery is biphasic, the half-times for the resolved components being approximately 55 ms and 920 ms. The electron flow pattern was unaffected by the medium pH in the range 6-8. Antimycin A is a potent inhibitor of cyclic electron flow in bacterial chromatophores. It interferes with oxidation/reduction reactions between b and c-type cytochromes[15,16], abolishes electric field formation and H^+ translocation associated with this site[2,6] and inhibits steady-state light induced ATP synthesis[17]. Figure 2 shows that only the fast cycling of electrons was inhibited by this antibiotic, the slow recovery phase remaining unaffected.

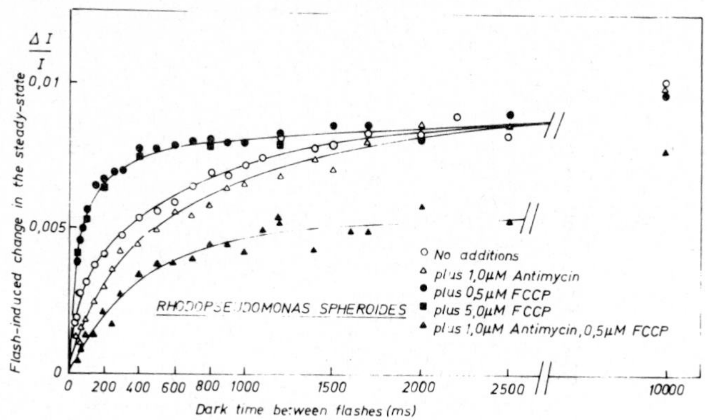

Figure 2 The effects of antimycin and FCCP on electron flow in Rps. spheroides chromatophores Conditions similar to figure 1 except that 0.084 μmole bacteriochlorophyll and 50 mM KCl were present. Each point is an average of 32 flashes and measurement was begun after the first five flashes. The vertical axis represents the change in transmission at 523 nm per flash at the flash repetition frequency indicated by the horizontal axis. Antimycin A and FCCP were present where shown.

Low concentrations of the uncoupling agent FCCP stimulated the rate of electron flow in Rps. spheroides chromatophores (figure 2). The half time of the uncoupler stimulated rate was $<$ 45 ms. Further resolution was limited by the 25Hz maximum frequency of the flash system. Concentrations of FCCP above 5×10^{-7} M did not produce further stimulation of the electron flow rate and a slow pathway remained. This contrasts with the effect of FCCP on the decay of the electrochromic shifts of Rps. spheroides. In this case increasing concentrations of FCCP, even up to 10^{-5} M produced further stimulation of the carotenoid shift decay (reference 2 and unpublished observations).

Similar effects on electron transport were observed with the uncoupling combination of nigericin, valinomycin and K^+ (figure 3).

In chromatophores treated with antimycin A, uncoupling agents failed to stimulate the recovery of the carotenoid shift in flashing light (figures 2 and 3) There is a small inhibition of the maximum amplitude (ie. at low flash frequency) but log plots show that the half-time for the recovery is not affected.

Rps. spheroides and Rps. capsulata show several significant differences in the thermodynamic properties of their photosynthetic electron transport components[16,18] and R. rubrum is quite different[16]. The midpoint potential of its primary acceptor is some 130 mV lower than in the two Rhodopseudomonas species. Nevertheless the pattern of electron transport, indicated by the recovery of the fast phase of the carotenoid shift in flashing light, is similar in chromatophores

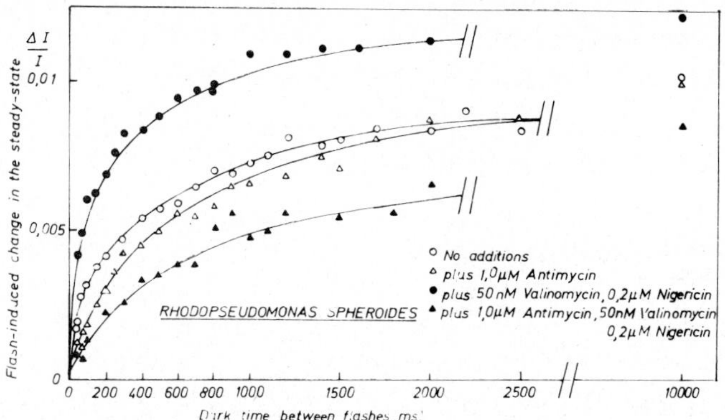

Figure 3 The effects of antimycin, nigericin and valinomycin on electron flow in Rps. spheroides chromatophores. Conditions as in figure 2 except that nigericin and valinomycin were present where shown.

from each of these species (compare figures 2,4 and 5). All three showed a bi-phasic recovery with approximate $t\frac{1}{2}$ values of the fast phase of 55 ms, 70 and 60 ms and of the slow phase 920 ms, 1300 ms and 1050 ms for Rps. spheroides, Rps. capsulata and R. rubrum respectively. In each case, antimycin A inhibited the fast phase without affecting the slow and FCCP increased the amount of electron return through the fast pathway at the expense of the slow. Uncoupling agent had a slight inhibitory effect on the antimycin insensitive electron transport.

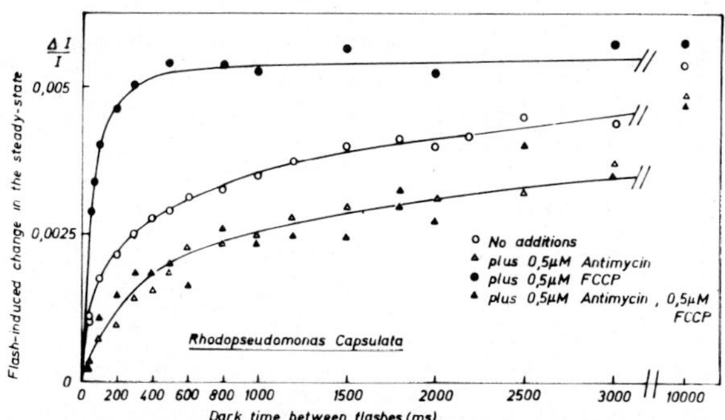

Figure 4. The effects of antimycin and uncoupling agent on electron flow in Rps. capsulata chromatophores Conditions as in figure 2, except using Rps. capsulata chromatophores. The extinction of the final chromatophore suspension at 856 nm was 0.96.

Dark electron flow rate from the reduction of reaction centre bacteriochlorophyll

The recovery of the absorption band of Rps. spheroides chromatophores at 605 nm. bleached by flash-light or a 5 s period of illumination is shown in figures 6 and 7. The decay of the photo-oxidation was approximately biphasic with half-times of the two components, 60 ms and 1500 ms. This is consistent with the

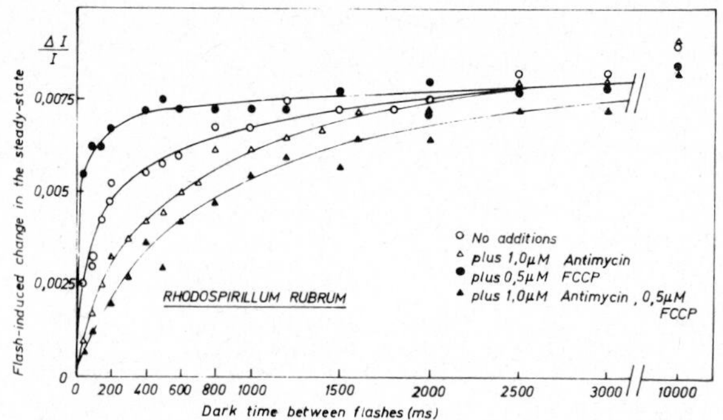

Figure 5. The effect of antimycin and uncoupling agent on electron flow in R. rubrum chromatophores. Conditions as in figure 2 except using R. rubrum chromatophores containing 0.14 μmole bacteriochlorophyll

description of the decay of the bleaching at 435 nm and 861 nm in R. rubrum chromatophores reported by Geller[9] and Parsons[10] respectively except that the 5-20% decay in 20 μs[10] was not resolved in the present experiments.

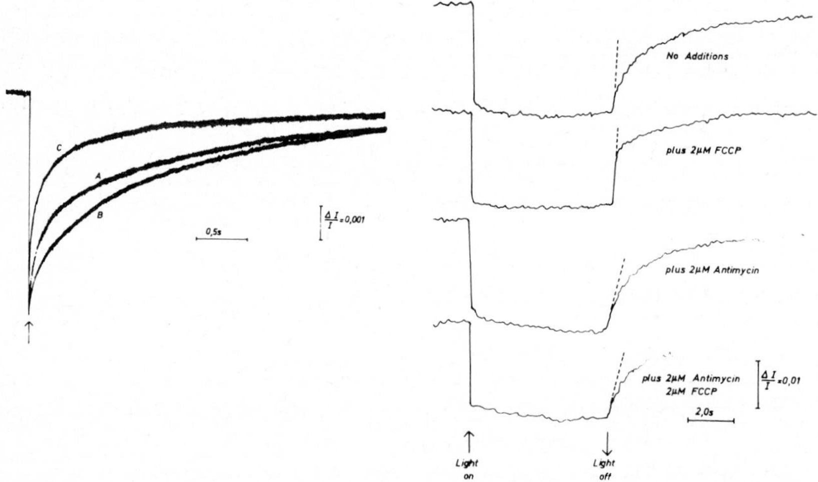

Figure 6 (left) The decay kinetics of P605 following single flash activation of Rps. spheroides chromatophores. Chromatophores containing 0.13 μmole bacteriochlorophyll were suspended in the medium described in the legend to figure 1. Each trace shown is an average from 64 flashes at a frequency of 0.083 Hz with the measuring beam set to 603 nm. Decreasing the measuring light intensity or the flash repetition rate did not affect the result. A: no additions B: plus 1 μM antimycin A. C: plus 0.5 μM FCCP.

Figure 7 (right) The kinetics of P605 during and after a period of continuous illumination in Rps. spheroides chromatophores. Conditions as figure 6 except using 0.26 μmole bacteriochlorophyll. Each trace shows the effect of a single period of illumination on the kinetics of the bleaching and recovery at 603 nm (not an averaged signal). The actinic light was attenuated with a filter transmitting at wavelengths $\rangle$ 650 nm.

In agreement with Geller,[9] antimycin inhibited and uncoupler stimulated the fast decay phase of the bacteriochlorophyll signal a part of the slow phase remaining unaffected. Parsons observed the antimycin inhibition but found no effect with uncoupling agent[10]. It seems likely that Parson's chromatophores, prepared by sonication, were already partly uncoupled.

The recovery of P605 bleaching after a period of continuous illumination is also biphasic and is similarly affected by antimycin A and by FCCP (figure 7). The relative amplitudes of the fast and slow recovery phases were significantly different after flash and continuous illumination. The fast phase accounts for approximately 45% of the decay following a short flash (76% in the presence of uncoupler) and 35% following a 5 s period of illumination (60% in the presence of uncoupler). The decay kinetics in the presence of a combination of antimycin A and FCCP are similar to those with antimycin alone but the amplitude of the bleaching is slightly inhibited (compare figures 2 and 3).

Similar biphasic decay of flash-induced bleaching of P605 and sensitivity to uncouplers and antimycin A were observed in chromatophores from R. rubrum (not shown. In Rps. capsulata the absorbance changes around 600 nm were complex and could not be attributed to the oxidation/reduction reactions of a single pigment.

Stimulated electron flow under anaerobic conditions

When chromatophores from Rps. spheroides were supplemented with substrate under anaerobic conditions, the net rate of cyclic electron flow estimated from the carotenoid shift recovery was considerably increased (figure 8) as a result of the diversion of electrons from the slow pathway to the fast. The fast phase was still completely inhibited by antimycin A and the antimycin insensitive pathway was similar in aerobic and anaerobic suspensions. The redox potential dependence of this phenomenon was not investigated; the time-course of the experiment

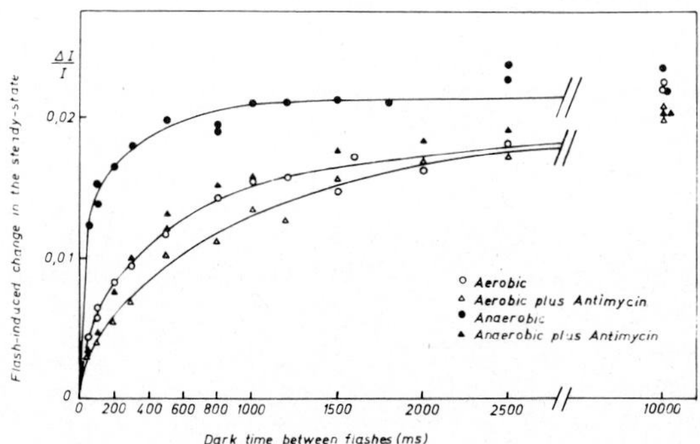

Figure 8. The effect of anaerobic conditions on electron flow in Rps. spheroides chromatophores. Experimental conditions as in figure 2 except using 0.14 µmole bacteriochlorophyll. The anaerobic samples were incubated in the dark under nitrogen for 15 min in the presence of 5 mM succinate and 5 mM ascorbate prior to each experiment.

(several seconds or minutes) prohibited the use of redox mediators[16]. It is well known that dyes such as phenazine methosulphate and diaminodurene, etc. can act as artificial electron donors and acceptors to create bipasses around inhibited sites.

DISCUSSION

The rate of regeneration of reduced P and oxidised X measured from the carotenoid response in flashing light clearly corresponds to the decay of the light induced reaction centre bacteriochlorophyll bleaching and illustrates the pattern of dark electron flow in chromatophores from purple non-sulphur bacteria. The differential response of the two phases of electron transport to inhibitors and uncouplers suggests that two different pathways are involved. Current opinion favours the co-operation of a cyclic and a non-cyclic or open electron transport chain operating through the same or different reaction centre bacteriochlorophyll (see review by Frenkel)[23]. Antimycin A sensitive electron transport is clearly cyclic[7]. The slow antimycin insensitive pathway could conceivably be a non-cyclic process between large endogenous pools (more than 100-fold the size of P) of reductant and oxidant. This possibility is rendered unlikely however by the finding (figure 8) that provision of exogenous substrate increases electron flux through the antimycin-sensitive cyclic pathway.

The bi-phasic decay of light-induced bacteriochlorophyll bleaching has received much attention. Parsons[10] and Loach and Sekura[11] have suggested that inhomogeneity in the chromatophore preparation, perhaps through damaged electron transport chains, would lead to different rates of electron flow into the reaction centre. This however seems unlikely because the amplitude of the fast phase increases at the expense of the slow phase under uncoupled conditions (reference 9 and figure 6) and at low redox potentials[11,16]. The increased amplitude of the fast phase in aged chromatophores[11] is probably an uncoupling effect (cf. the effect of FCCP or nigericin/valinomycin/K^+) rather than progressive damage to the electron transport reactions _per se_.

A more attractive possibility suggested by Loach and Sekura[11] and also by Fleischman and Cooke[19] is that one phase of the decay of the bacteriochlorophyll bleaching represents the reversal of the P$\rightarrow$X reaction from electrons in the secondary acceptor pools. One phase in the profile of delayed light emission was found to have a similar half-time to the slow phase of the decay of the chlorophyll reaction[11]. Two observations are difficult to reconcile with this model (i) the slow decay phase of the decay of the bacteriochlorophyll bleaching is insensitive to FCCP (figure 6) whereas delayed light emission in _Rps. spheroides_ chromatophores is strongly inhibited by uncoupling agents[20]; (ii) the reversal of P$\rightarrow$X would be expected to be more extensive when cyclic electron flow is inhibited with antimycin because the secondary acceptors would be more highly reduced. The slow phase of the bacteriochlorophyll re-reduction, however remains unchanged under these conditions (figure 6). The proportion of electrons returned to P with delayed light emission may therefore be negligable compared with cyclic electron flow.

The recovery of the light-induced g=2,0025 ESR signal in bacterial chromatophores has similar kinetics to the decay of the spectroscopically measured bacteriochlorophyll bleaching and has been identified with the P^+ free radical[11,12]. Cost _et al_[22] showed that a fast decay phase of the ESR signal in _R. rubrum_ chromatophores persists at $-150^{\circ}C$. They suggested that the fast recovery of P oxidation is a temperature independent, non-enzymic reversal of the photoact. This, however, is not consistent with the data of Parsons[10] who showed that the rate of the fast decay phase of P890 is decreased 8-fold between $23^{\circ}C$ and $0^{\circ}C$. It is probable that the fast decays observed at liquid nitrogen and room temperature are caused by two unrelated processes; the former being P⟵X and the latter antimycin sensitive flow through the cyclic electron transport chain.

Another explanation for the biphasic decay is suggested by the results reported in this communication: two parallel cyclic electron transport pathways operate between a common reaction centre with common secondary electron donors and acceptors. One pathway is fast ($t\frac{1}{2} \simeq 60$ ms) is inhibited by antimycin and conserves energy; the other is slow ($t\frac{1}{2} \simeq 1500$ ms) and is not significantly affected by antimycin or uncoupling agents. In untreated chromatophores in aerobic suspension, in continuous or repetitive flash light about one third of the electrons take the fast, uncoupled pathway.

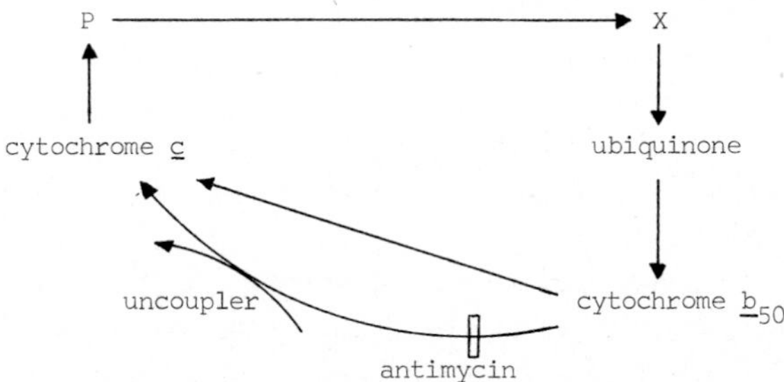

Figure 9 Branched electron transport chain in chromatophores from Rps. spheroides

It is proposed that the two pathways compete for electrons arriving at the branch point. Two corollaries to this model are supported by the data presented above. First, it is clear that following several turnovers of the reaction centre, more electrons will be "trapped" in the slow pathway. This is consistent with the finding that a higher proportion of the P605 recovery was fast following a short flash than after a period of continuous illumination, in the presence or absence of FCCP - see figures 6 and 7. Second, when the electron carriers in the slow pathway are chemically reduced before illumination, reducing equivalents will be transported preferentially through the fast pathway. These conditions are evidently met when the redox potential of the suspension is lowered by providing substrate in the absence of oxygen - see figure 8. According to this model the switch between the alternative pathways is operated by the redox potential of the chromatophore environment.

In figure 9 the parallel pathways are placed tentatively between cytochromes $\underline{b}_{50}$ and $\underline{c}$, a region of the chain which had presented some difficulties following a recent kinetic analysis of the electron transport reactions of Rps. spheroides[16]. In one formulation[2] a pathway short-circuiting the site II electrogenic and H^+-translocating reactions had been suggested between the $\underline{b}$ and $\underline{c}$-type cytochrome The data presented here gives some experimental support to this scheme.

ACKNOWLEDGEMENTS

I should like to express my gratitude to the members of the Max Volmer Institut for their hospitality and for many stimulating discussions. I am particularly indebted to Professor H.T. Witt for introducing me to repetitive flash spectrophotometry and I also thank the electronic and mechanical workshop staff and Miss C. Fleckenstein for their assistance. Financial support was from the European Molecular Biology Organisaion.

REFERENCES

1 Jackson, J.B., Saphon, S. and Witt, H.T., manuscript in preparation.
2 Cogdell, R.J., Jackson, J.B. Crofts, A.R., 1972, Bioenergetics, 4, 413.
3 Jackson, J.B. and Crofts, A.R., 1971, Eur. J. Biochem. 18, 120.
4 Nishimura, M., 1970, Biochim. Biophys. Acta, 197, 69.
5 Jackson, J.B. and Crofts, A.R., 1969, FEBS Lett. 4, 185.
6 Jackson, J.B. and Dutton, P.L., 1973, Biochim. Biophys. Acta, 325, 102.
7 Bose, S.K. and Gest, H., 1963, Proc. Nat. Acad. Sci. U.S., 49, 337.
8 Kuntz, I.D., Loach, P.A. and Calvin, M., 1964, Biophys. J., 4, 227.
9 Geller, D.M., 1967, J. Biol. Chem., 242, 40.
10 Parsons, W.W., 1967, Biochim. Biophys. Acta., 131, 154.
11 Loach, P.A. and Sekura, D.L., 1967, Photochem. Photobiol., 6, 381.
12 Sistrom, W.R., 1970, J. Gen. Microbiol. 22, 778.
13 Clayton, R.K., 1963, in Bacterial Photosynthesis, editors Gest, H., San Pietro, A. and Vernon, L.P., Antioch Press, Yellow Springs, Ohio, p. 498.

14 Ruppel, H. and Witt, H.T. 1970, Methods in Enzymol. 16, 316.
15 Nishimura, M., 1963, Biochim. Biophys. Acta, 66, 17.
16 Dutton, P.L. and Jackson, J.B., 1972, Eur. J. Biochem. 30, 495.
17 Geller, D.M. and Lipmann, F., 1960, J. Biol. Chem. 235, 2478.
18 Evans, E.H. and Crofts, A.R., 1974, Biochim. Biophys. Acta, 357, 78.
19 Fleischman, D.E. and Cooke, J.A. 1971, Photochem. Photobiol. 10, 251.
20 Fleischman, D.E. and Clayton, R.K. 1968, Photochem. Photobiol. 8, 287.
21 Bolton, J.R., Clayton, R.K. and Reed, D.W., 1969, Photochem. Photobiol., 9,209.
22 Cost, K., Bolton, J.R. and Frenkel, A.W., 1969, Photochem. Photobiol. 10, 251.
23 Frenkel, A.W., Biol. Rev. 45, 469-616

M. AVRON, *Proceedings of the Third International Congress on Photosynthesis*, September 2-6, 1974, Weizmann Institute of Science, Rehovot, Israel
Elsevier Scientific Publishing Company, Amsterdam, The Netherlands, 1974

ASYMMETRY OF AN ENERGY-COUPLING MEMBRANE. THE IMMUNOLOGICAL LOCALISATION OF CYTOCHROME $\underline{c}_2$ IN RHODOPSEUDOMONADS.

R.C. Prince, G.A. Hauska,[a] A.R. Crofts, A. Melandri[b] and B.A. Melandri.[b]

Department of Biochemistry, University of Bristol, Bristol. BS8 1TD. England, and [a]Lehrstuhl fur Biochemie der Pflanzen, Ruhr-Universitat Bochum, Germany, and [b]Instituto Botanico, Universite di Bologna, Italy.

INTRODUCTION

During the primary light reaction of photosynthetic electron flow in Rhodopseudomonas spheroides and Rps. capsulata, the excited reaction centre bacteriochlorophyll is oxidised, and the primary electron acceptor is reduced. The reaction centre is subsequently re-reduced by a c-type cytochrome[1,2]. In the work reported in this paper, we have used monospecific antibodies prepared against cytochrome $\underline{c}_2$ prepared from both Rps. spheroides and Rps. capsulata to determine the location of these proteins in the bacterial cell.

The very specific interactions between antibodies and antigens have been widely used in the elucidation of the functional roles of the proteins used as antigens, and their intracellular location. This latter approach has been of great interest in the study of energy conserving organelles. Racker et al.[3] have used antibodies prepared against mammalian cytochrome c to show that in mitochondria this protein is outside the inner membrane, while in sub-mitochondrial particles the antigen is inaccessible to the antibody. Similarly, antibodies have been used to demonstrate that spinach plastocyanin is situated inside the chloroplast thylakoids,[4] as is Euglena gracilis cytochrome c_{552}.[5]

The results presented here show that in both Rps. spheroides and Rps. capsulata, cytochrome $\underline{c}_2$ is located between the cell wall and the cell membrane, in the periplasmic space. When chromatophores are prepared from whole cells, this cytochrome becomes trapped inside the vesicles.

METHODS

Bacteria were grown anaerobically in the light, and chromatophores prepared as previously described[6]. Spheroplasts were prepared as described by Karunairatnam et al.[7] Bacterial cytochromes were purified from the supernatants remaining after the preparation of chromatophores using data from reference 8, and reaction centres and bulk bacteriochlorophyll-protein complexes were isolated as described earlier.[9,10,11.] The coupling factor from Rps. capsulata was also prepared as described previously.

Antibodies were prepared against cytochrome $\underline{c}_2$ from Rps. spheroides Ga and Rps. capsulata St. Louis grown anaerobically in the light, using procedures similar to those used earlier.[4] The antisera were shown to be pure by immuno-electrophoresis, and both antisera reacted only with cytochrome $\underline{c}_2$ from the same species. Moreover no cross-reaction occurred with reaction centres, bulk bacteriochlorophyll-protein complexes or coupling factor, nor with a wide range of c-type cytochromes. Immunoglobulins were routinely used instead of whole sera, and these were prepared as described earlier.[4]

Spectroscopy was performed on the dual spectrophotometers described earlier,[12,13] and 9-amino acridine fluorescence was monitored as before.[14] Bacteriochlorophyll was assayed after extraction into acetone: methanol,[15] and cytochrome $\underline{c}_2$ from ascorbate reduced minus ferricyanide oxidised difference spectra, assuming an extinction coefficient of $20mM^{-1}$ cm^{-1} at 551-540nm.[16]

RESULTS

When chromatophores are prepared from whole cells, considerable quantities of soluble proteins are liberated into solution, including cytochromes $\underline{c}_2$ and cc'[8]. However, the mole ratio of cytochrome $\underline{c}_2$: bacteriochlorophyll in the chromatophores is very similar to that of whole cells (approximately 1:100 in Rps. spheroides Ga and 1:120 in Rps. capsulata Ala pho+), showing that chromatophores are not depleted in cytochrome $\underline{c}_2$ with respect to whole cells, but rather that the soluble cytochromes come from partially broken cells which do not liberate chromatophores.

In contrast, when spheroplasts are prepared, cytochrome $\underline{c}_2$ is liberated into free solution - in some preparations from Rps. spheroides Ga as much as 85% of the c-type cytochrome was lost, with the concommitant loss of more than 90% of cytochrome photo-oxidation. Similar results were obtained with Rps. capsulata. Both Rps. spheroides and Rps. capsulata possess several c-type cytochromes. Rps. spheroides has one with an E_{m7} of +120mV[17] in addition to $\underline{c}_2$ (E_{m7}=+346mV[8]), while Rps. capsulata has three with E_{m7} values of +340mV (cytochromes $\underline{c}_2$), +120mV and 0mV[2], where the latter two may account for up to 50% of the total.[2] Both organisms also liberate cytochromes $\underline{c}_3$ and $\underline{c}_{554}$ when broken in a French pressure cell,[8] although these have not yet been detected in chromatophores. Several of these cytochromes would be reduced by ascorbate, and are hence included in the "cytochrome $\underline{c}_2$" estimated above. However, there is no evidence that any cytochrome other than $\underline{c}_2$ can be directly photo-oxidised by the reaction centres, so the amount of photo-oxidation in saturating light is a good estimate of the amounts of cytochrome $\underline{c}_2$ present. Very little of this cytochrome remains associated with spheroplasts.

The effects of the antisera on sub-cellular fractions.

(i) Photochemical reaction centres. Those used in this work contain no endogenous cytochromes[9,10], but nevertheless can rapidly photo-oxidise added c-type cytochromes.[13] The photo-oxidation of added cytochrome $\underline{c}_2$ could be completely

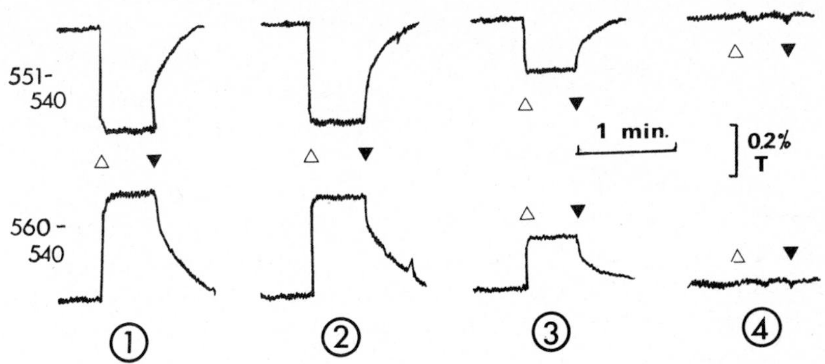

FIGURE 1 THE EFFECT OF ANTI-c_2 ON CYTOCHROME c_2 PHOTO-OXIDATION AND CYTOCHROME b REDUCTION IN Rps. spheroides Ga

The reaction mixture contained 3ml 10mM Tris Cl,100mM KCl,pH 7.5,0.1mg BChl/ml,2μM Antimycin and Valinomycin. ① No additions ② +200μl anti-c_2 ③ as 2 +1% cholate ④ as 3 but +500μl anti-c_2.

inhibited by the addition of the appropriate antiserum. The control serum had no effect, and neither serum had any effect on the photo-oxidation of mammalian cytochrome c.

(ii) Chromatophores. Chromatophores possess a cyclic photosynthetic electron transport chain, but in the presence of Antimycin A this becomes an essentially linear pathway; illumination causes an oxidation of c-type cytochrome, and a reduction of a b-type cytochrome.

When anti-c_2 was added to chromatophores of Rps. spheroides Ga it had little effect, but in the presence of 1% sodium cholate, anti-c_2 had a marked inhibitory effect on both cytochrome c_2 photo-oxidation, and cytochrome b photo-reduction (Fig. 1). Sodium cholate is a mild detergent, and at concentrations of up to 1% it solubilises only a negligible amount of the bacteriochlorophyll, in marked contrast to Triton X-100 or lauryl dimethylamine oxide, which at such concentrations completely solubilise the membrane. Nevertheless, 1% sodium cholate completely inhibits the generation of a transmembrane pH gradient, as monitored with 9-amino acridine fluorescence,[14] and when chromatophores were washed five times in 50mM HEPES, 150mM KCl, 1% sodium cholate, pH 7, all the light induced c-type cytochrome photo-oxidation was lost, as was most of the cytochrome. The solubilised cytochrome cross-reacted with anti-c_2.

Qualitatively similar results were obtained using chromatophores from Rps. capsulata Ala pho+ or St. Louis and the appropriate antiserum, except that in the presence of 1% cholate, c-type cytochrome photo-oxidation was never completely inhibited by anti-$\underline{c}_2$; the residuum varied from 10-35% in different preparations, and cytochrome $\underline{b}$ reduction was either slightly inhibited or unaffected. In addition, not all cytochrome photo-oxidation was lost after five washes with 1% cholate. However, a redox titration of the cytochrome photo-oxidation occurring in the presence of 1% sodium cholate plus excess anti-$\underline{c}_2$ indicated that the remaining cytochrome photo-oxidation was indistinguishable by its redox characteristics from the total cytochrome photo-oxidation occurring in the presence of 1% sodium cholate (D. Crowther and R.C. Prince, unpublished observation), suggesting that 1% cholate does not render chromatophores of Rps. capsulata as permeable to large molecules as those of Rps. spheroides.

An interesting anomaly of Rps. spheroides is the fact that the c-type cytochrome which donates electrons to the oxidised reaction centre in chromatophores has an E_{m7} = +295mV[1], while that of the soluble cytochrome $\underline{c}_2$ liberated into solution during the preparation of chromatophores has an E_{m7} of +346mV[8]. We have been able to confirm these results, and in addition to show that in the presence of 1% cholate, the midpoint potential of the c-type cytochrome of the chromatophores rises to +340mV. Since,(when this cytochrome is solubilised by cholate treatment,)it cross reacts with anti-$\underline{c}_2$ prepared against the cytochrome liberated during the preparation of chromatophores, and it has the same redox potential, the two forms are the same protein species, and the difference in midpoint potential when bound to the chromatophore must be due to this binding. The midpoint shift observed indicates that the oxidised form is bound approximately six times more tightly than the reduced form. The midpoint potential of both the soluble and the chromatophore associated cytochrome $\underline{c}_2$ in Rps. capsulata are the same, + 340mV[2].

Chromatophores will also photo-oxidise added cytochrome $\underline{c}_2$or cytochrome c . The former reaction can be inhibited by anti-$\underline{c}_2$ in the absence of cholate. However we find (G.A. Hauska & A.B. Melandri, unpublished observation) that this photo-oxidation is not coupled to phosphorylation, in contrast to the earlier work of Horio and Yamashita[18] on Rhodospirillum rubrum. Our results are similar to those of Racker et al.[3] with sub-mitochondrial particles.

(iii) Spheroplasts. As discussed above, during the preparation of spheroplasts, cytochrome $\underline{c}_2$ is liberated into solution, and this cytochrome is precipitated by the appropriate anti-$\underline{c}_2$. Spheroplasts will readily photo-oxidise added cytochrome $\underline{c}_2$ or cytochrome c, although the rate of rereduction is slow in an aerobic cuvette in the absence of added reducing agents. As with reaction centres, ant-$\underline{c}_2$ inhibited the photo-oxidation of cytochrome $\underline{c}_2$, but not of cytochrome c.

DISCUSSION

Earlier workers[19] have produced antibodies against cytochrome $\underline{c}_2$ from R.

rubrum, but this was not used to localise the antigen in vivo. Interestingly, their antiserum cross reacted with other bacterial cytochromes, while neither of our antisera did, indicating that the various cytochromes may share antigenic determinants.

The results presented here indicate that the cytochrome $\underline{c}_2$ liberated by French pressure cell treatment of whole cells is identical to that retained in chromatophores, in both Rps. spheroides and Rps. capsulata, and that in vivo the cytochrome is trapped in the periplasmic space. Since cytochrome $\underline{c}_2$ can be lost from spheroplasts, chromatophores cannot be sealed vesicles in vivo, and their contents must be a continuum of the periplasmic fluid. This implies that the chromatophores have no autonomy in vivo, as is shown in figure 2.

Thus cytochrome $\underline{c}_2$ is located inside chromatophores. Rapid proton uptake[20] occurs on the outside of chromatophores and we have recently shown that part of this is due to the reduction of ubiquinone[21]. Thus the topography of the chromatophore may be represented as in figure 3. The primary electron transfer step is the electrogenic movement of an electron from P, the primary donor, to X, the primary acceptor. This primary electrogenic event is then followed by the neutral transport of a hydrogen across the membrane in the opposite direction, leaving a charge separation across the membrane.

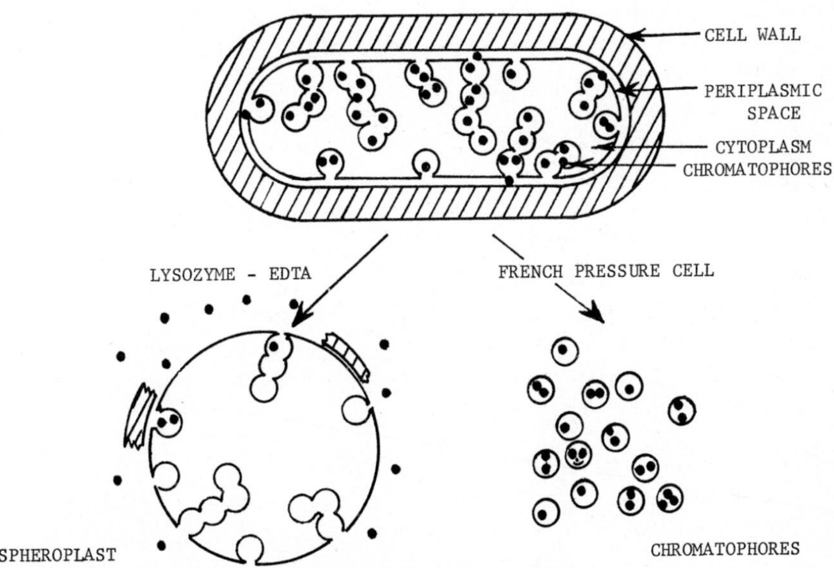

FIGURE 2 DIAGRAMMATIC REPRESENTATION OF Rps. spheroides and Rps. capsulata

The solid circles represent cytochrome $\underline{c}_2$.

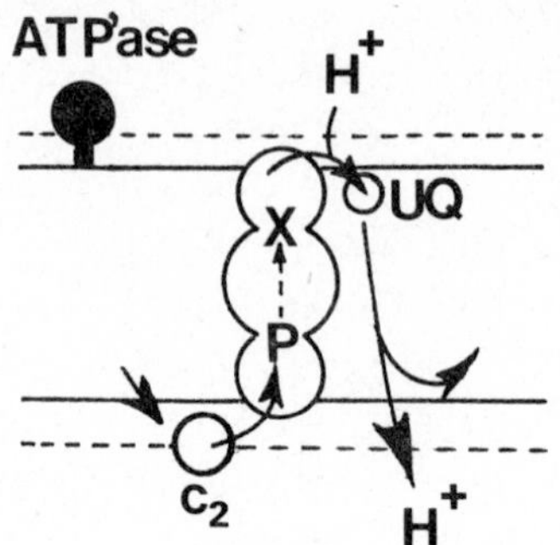

FIGURE 3 THE TOPOGRAPHY OF THE LIGHT-REACTION OF *Rps. spheroides* and *Rps. capsulata*

Analogous situations occur in other energy transducing systems. Photosystem II in chloroplasts appears to be very similar,[22] except that the secondary donor is water and the secondary acceptor is plastoquinone (Fig. 4). Likewise, photosystem I in chloroplasts of spinach and *E. gracilis*, and site III in mitochondria, have a similar topography (Fig. 4). The donors for these latter "coupling sites" are small soluble proteins,[5,4,3] and antibodies have been prepared against all of them. They differ from the bacterial photochemical reaction in that the secondary acceptor is in the aqueous phase, either $NADP^+$ or O_2. Nevertheless the analogy between the sites is striking.

Recent work by Case and Parson[22] on *Chromatium* has suggested that the primary electron transport event in this organism may not occur across the membrane, but that instead the rereduction of P^+ by cytochrome C_{555} is an electrogenic step. This cytochrome is tightly bound to the membrane, and a possible model for the topography of this system is included in figure 4.

The photo-oxidation of external cytochromes *c* and c_2 by chromatophores of *Rps. spheroides* and *Rps. capsulata* could be due to the presence of non-vesicular membrane fragments[24] or to "membrane scrambling"[25]. The fact that this photo-oxidation is uncoupled, and relatively slow, leads us to believe that it has no physiological significance.

REFERENCES

1. Dutton, P.L. and Jackson, J.B. (1972). Eur. J. Biochem. 30, 495-510.
2. Evans, E.H. and Crofts, A.R. (1974). Biochim. Biophys. Acta. in press.
3. Racker, E., Burstein, C., Loyter, A. and Christiansen, P.O. (1970). in "Electron Transport and Energy Conservation" (Tager, J.M., Papa, S., Quagliariello, E. and Slater, E.C., eds). Adriatica Editrice, Bari. p. 235-252.
4. Hauska, G., McCarty, R.E., Berzborn, R. and Racker, E. (1971). J. Biol.Chem. 246, 3524-2531.

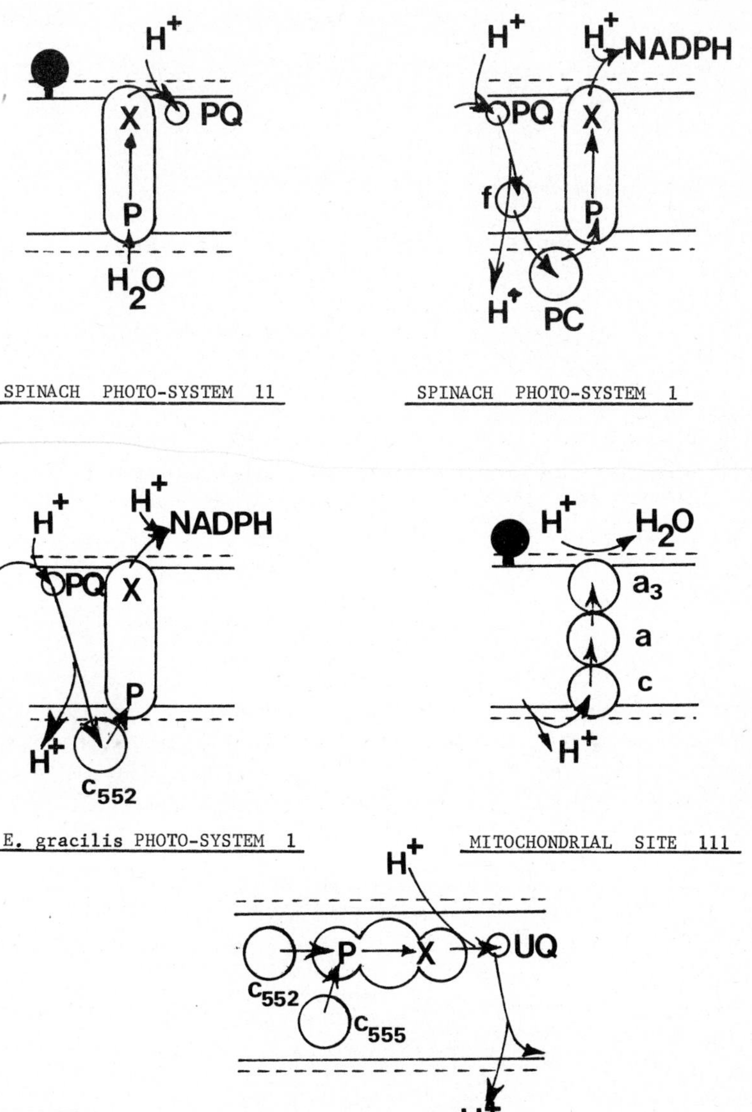

CHROMATIUM LIGHT-REACTION

FIGURE 4 THE TOPOGRAPHY OF SOME OTHER ELECTRON TRANSPORT SYSTEMS

5. Wildner, G.D. and Hauska, G. (1974). Arch. Biochem. Biophys. in press.

6. Jackson, J.B., Crofts, A.R. and von Stedingk, L.V. (1968). Eur. J. Biochem. 6, 41-54.

7. Karunairatnam, M.C., Spizizen, J. and Gest, H. (1958). Biochim. Biophys. Acta. 29, 649-650.

8. Bartsch, R.G. (1971). in Methods in Enzymology 23A (San Pietro, A., ed), 344-363.

9. Clayton, R.K. and Wang, R.T. (1971). in Methods in Enzymology 23A (San Pietro, A. ed), 696-704.

10. Prince, R.C. and Crofts, A.R. (1973). FEBS Lett. 35, 213-216.

11. Clayton, R.K. and Clayton, B.J. (1972). Biochim. Biophys. Acta. 283, 492-504.

12. Jackson, J.B. and Crofts, A.R. (1969). Eur. J. Biochem. 10, 226-237.

13. Prince, R.C., Cogdell, R.J. and Crofts, A.R. (1974). Biochim. Biophys. Acta. 347, 1-13.

14. Deamer, D.W., Prince, R.C. and Crofts, A.R. (1972). Biochim. Biophys. Acta. 274, 323-335.

15. Clayton, R.K. (1963). in Bacterial Photosynthesis (Gest, H., San Pietro, A. and Vernon, L.P. eds) Antioch Press, Yellow Springs, Ohio, p. 498.

16. Reed, D.W. (1969). J. Biol. Chem. 244, 4936-4941.

17. Saunders, V.A. and Jones, O.T.G. (1974). Biochim. Biophys. Acta. 333, 439-445.

18. Horio, T. and Yamashita, J. (1964). Biochim. Biophys. Acta. 88, 237-250.

19. Smith, W.R., Sybesma, C., Litchfield, W.J. and Dus, K. (1973). Biochemistry. 12, 2665-2671.

20. Cogdell, R.J. and Crofts, A.R. (1974). Biochim. Biophys. Acta. 347, 264-272.

21. Prince, R.C., Cogdell, R.J. and Crofts, A.R. (1974). Biochem. Soc. Trans. 2, 162-164.

22. Junge, W. and Auslander, W. (1973). Biochim. Biophys. Acta. 333, 59-70.

23. Case, G. and Parson, W.W. (1973). Biochim. Biophys. Acta. 325, 441-453.

24. Crofts, A.R. (1970). in "Electron Transport and Energy Conservation" (Tager, J.M., Papa, S., Quagliariello, E. and Slater, E.C. eds) Adriatica Editrice, Bari, p.221-228.

25. Chance, B., Erecinska, M. and Lee, C.P. (1970). Proc. Natl. Acad. Sci. U.S. 66, 928-935

ACKNOWLEDGEMENTS.

We are grateful to NATO for travel grants to allow this co-operation between our laboratories, and to the S.R.C. and D.F. for general support.

M. AVRON, *Proceedings of the Third International Congress on Photosynthesis*, September 2-6, 1974, Weizmann Institute of Science, Rehovot, Israel
Elsevier Scientific Publishing Company, Amsterdam, The Netherlands, 1974

PHOTOSYNTHETIC ELECTRON TRANSPORT IN RHODOPSEUDOMONAS CAPSULATA[1]

A. Hochman[2] and C. Carmeli[3]

Department of Biochemistry
The George S. Wise Life Sciences Center
Tel-Aviv University, Tel-Aviv, Israel

1. Introduction

It is generally accepted that photosynthetic electron transport in most photosynthetic bacteria occurs in two distinct pathways [1], a cyclic one [2] which is coupled to ATP formation and a non cyclic one which is coupled to substrate oxidation. The electron transport pathways include reaction center bacteriochlorophyll, quinone, and cytochromes. Vernon [3] observed a cytochrome c in R.rubrum [3] and Duysens showed [4] that this cytochrome was oxidized by light. Cytochromes of the c type are found in all photosynthetic bacteria. They were isolated from many species including Rhodopseudomonas capsulata [5] and their properties were studied. Several investigators [4,6] suggested that cytochrome c was the primary donor to the photo-oxidized bacteriochlorophyll. Most photosynthetic bacteria possess more than one kind of cytochrome c. The relation of the various c type cytochromes to the electron pathways is not clear yet. In R.rubum, cytochromes of the c type were assumed to mediate cyclic and non cyclic electron transfer from succinate or DCIP to either NAD or methyl viologen, or to O_2 [7]. A different experimental approach was used by Horio and Kamen [8]. They observed that long incubation followed by centrifugation of R.rubum chromatophores resulted in a loss of photophosphorylation activity which could be restored with cytochrome C_2 in the presence of ascorbate.

There are very few data published about the properties and the role of cytochrome C_2 in Rhodopseudomonas capsulata. Cytochrome C_2 was isolated [5] and the oxidized minus reduced spectra of c type cytochromes were observed in chromatophores [9]. It was the purpose of this work to further study the role of cytochrome C_2 in Rhodopseudomonas capsulata.

A different approach to study the photosynthetic electron pathway was the use of mutants. Sistrom and Clayton [10] selected a mutant strain of Rps. spheroides that lacked reaction center bacteriochlorophyll. Mutants of the photosynthetic bacteria which can not grow aerobically in the dark [11] are useful

---

[1]We are grateful to Dr. H. Gest for supplying the inoculum of Rps capsulata.

[2]This is part of a work presented by A. Hochman towards a Ph.D. Degree at the Tel-Aviv University.

[3]Supported by the United-States - Israel Binational Science Foundation.

for study of the oxidative electron transport pathway. In this work photosynthetic mutants were used for the study of the components of electron transport in Rhodopseudomonas capsulata.

2. Materials and Methods

Growth Conditions - Rhodopseudomonas capsulata was grown anaerobically in the light or aerobically in the dark in the medium described by Ormerod, et al [12], supplemented with 0.0002% thiamine hydrochloride. Both cultures were harvested after 5 days, washed once with 100mM tricine-NaOH pH 7.2 containing 8mM $MgCl_2$ (tricine-Mg buffer) and stored under nitrogen at -20°C.

Mutations- Photosynthetic mutants were induced with Nitrosoguanidine [13]. The culture was enriched with the mutants by the penicillin method. Colonies which failed to grow anaerobically in the light were selected.

Preparation of chromatophores by sonication - The frozen cells were thawed and suspended in the tricine - Mg buffer (1:1 v:w) and sonicated for 1 min. The whole cells and cell debris were removed from the suspension by centrifugation at 12,000 x g for 15 min. The chromatophores were sedimented by centrifugation at 144,000 x g for 90 min and resuspended in a minimal volume of tricine - Mg buffer.

Preparation of heavy chromatophores - The cells were suspended in tricine-Mg buffer (0.75:1 v:w) and broken by two passages through the Yeda Press cell under nitrogen pressure of 1800 psi. The suspension was centrifuged for 16 min at 12000 x g. The supernatant was composed of two distinct layers. A lower viscous layer and a clear upper layer. These two layers were separated by decantation and the pellet which contained cell debris was discarded. The chromatophores from the lower layer were sedimented by centrifugation at 48,000 x g for 50 min. The heavy chromatophores were resuspended in tricine-Mg buffer. The chromatophores in the upper layer of the supernatant were sedimented by centrifugation at 144,000 x g for 90 min and resuspended in a tricine-Mg buffer. The supernatant which remained after sedimentation of the chromatophores was called Sup I.

Preparation of washed chromatophores and fractionation of Sup II - Chromatophores suspension was diluted with tricine-Mg buffer to a concentration of 250 μg BChl/ml and centrifuged at 144.000 x g for 90'. The sediment consisted of washed chromatophores while the remaining supernatant was called Sup II. The washed chromatophores were suspended in a minimal volume of tricine-Mg buffer. Sup II was fractionated on columns of DEAE cellulose according to the method of Bartch and Meyer [14]. Large quantities of cytochrome C_2 were prepared from Sup I [14]. Bacteriochlorophyll content was measured by extracting the chromatophores with acetone methanol according to the method of Clayton [15].

Photophosphorylation was assayed in a reaction mixture containing: tricine - NaOH (pH 8.0) 33mM; $^{32}P-K_2HPO_4$ (pH 8.0) 3.3 mM (containing 10^6 cpm); $MgCl_2$ 8mM; ADP, 1.6 mM and chromatophores containing 70-80 μg bacteriochlorophyll in a total volume of 3 ml. The reaction was carried out in Warburg vessels, [^{32}P] ATP formation was measured according to the method of Avron [16]. Protein concentration was measured according to the method of Hartree [17]. Absroption spectra were recorded in a Cary moled 14 spectrophotometer. Light induced absorbance changes were measured with Aminco Chance dual wavelength spectrophotometer.

3. Results

Light induced pH changes in heavy chromatophores - In electron micrographs of a negatively stained preparation (unpublished) heavy chromatophores looked like clusters of vesicles, sometimes appearing in closed structures. Chromatophores prepared by sonication and the light fraction of the Yeda Press preparation appeared as separate small vesicles. Illumination of chromatophores suspended in a medium of low buffer capacity caused an increase in the pH (Fig. 1). The pH change reached a steady state and was reversed when light was turned off.

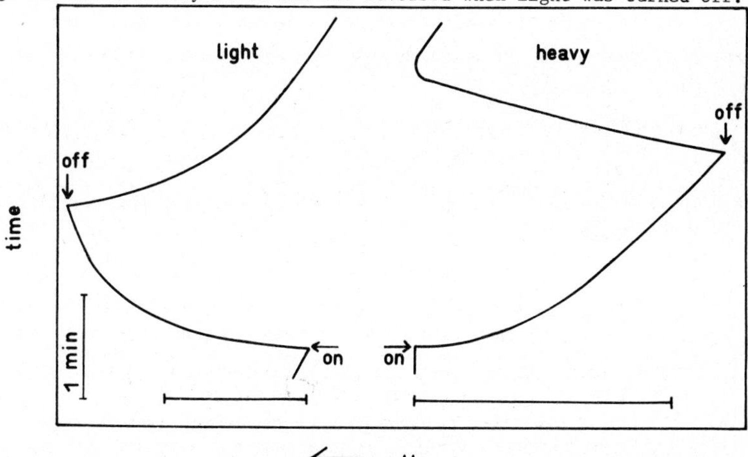

Fig. 1. pH changes induced by light in chromatophores and heavy chromatophores. The reaction mixtures contained: 300 μmoles KCl, 1 μmoles PMS and 198 μg bacteriochlorophyll in a final volume of 2 ml at pH 6.06. The pH changes were measured by a recording pH meter. The bar indicates 50 nmoles of H^+ ions.

Illumination of the heavy chromatophores resulted in a pH decrease which was also reversed when light was turned off. The pH changes were probably due to proton movement across the membrane. In the heavy chromatophores this movement was in the same direction as in whole cells and was opposite to the direction of the movement of protons in the chromatophores. This might indicate that membranes of the two types of chromatophores were opposite in their polarity.

Effect of washing on photophosphorylation activity in chromatophores. - The main properties of photophosphorylation were similar in the two kinds of chromatophores.

Table 1

Photophosphorylation activity in chromatophores and heavy chromatophores. The reaction was carried out under nitrogen unless indicated otherwise.

Additions	Heavy Chromatophores	Chromatophores
	μmoles Pi esterifes x mg bacteriochlorophyll^{-1} x h^{-1}	
None	30.0	105.0
PMS 10^{-4}M	126.0	240.4
Antimycin A 3.3x 10^{-6}M	0	0
Antimycin A 3.3 x 10^{-6}M + 5 x 10^{-4}M PMS	98.0	192.0
Ascorbate 8.3 x 10^{-3}M	138.0	208.0
Air	17.5	63.0
Air + 5 x 10^{-4}M PMS	118.0	221.0

The activity in the heavy chromatophores was 30-70% of the activity in the chromatophores. In both cases photophosphorylation was enhanced by PMS and by ascorbate (Table 1). Antimycin A and air inhibited this activity but PMS restored most of it. Washing of the heavy chromatophores caused a loss of their photophosphorylation (Table II). The activity was completely restored either by the addition of PMS or by the addition of the supernatant which was obtained from their washing, Sup II (Table II).

Table II

Effect of washing on anaerobic phosphorylation in heavy chromatophores. The reaction mixture as described under Methods. One equivalent of Sup II was obtained from the washing of the same amount of chromatophores which were used in the assay.

Additions	Heavy chromatophores	Washed heavy chromatophores	Washed heavy chromatophores +5 equ. Sup II
	μmoles Pi esterified x mg bacteriochlorophyll^{-1} x h^{-1}		
None	65.8	7.5	55.3
5 x 10^{-4}M PMS	139.7	132.9	161
6 x 10^{-6}M antimycin A	2	2	6
5 x 10^{-4} PMS +6 x 10^{-6}M antimycin A	104	100	-

Although both treatments restored the activity there was a difference in the mode of restoration. The activity restored by Sup II was inhibited by antimycin A whereas the one restored by PMS was not inhibited. PMS enhanced photophosphorylation in the untreated chromatophores giving maximal activity at 5×10^{-4}M. In the washed chromatophores restoration of the activity by PMS was dependent on its concentration but the maximal activity was similar to that of the untreated chromatophores. The photophosphorylation activity which was lost by washing of the heavy chromatophores was also restored by ascorbate (Fig. 2). The optimal

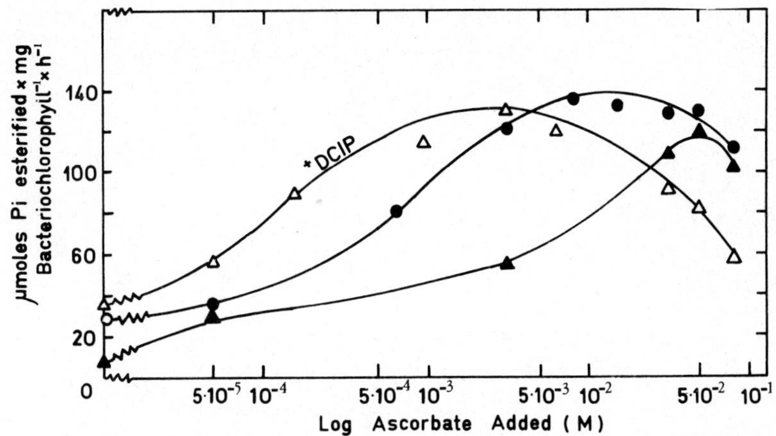

Fig. 2. Effect of ascorbate on photophosphorylation in heavy chromatophores (●). Washed heavy chromatophores were assayed in the absence (▲) or in the presence of 10^{-5}M DCIP (Δ). For experimental details see under Methods.

concentration for restoration of the activity was at 5×10^{-2}M. However, in the presence of DCIP the optimal concentration of ascorbate was reduced to 3.3×10^{-3}M. Ascorbate enhanced phosphorylation in untreated chromatophores but the optimal concentration here was 8.3×10^{-3}M. Washing of the chromatophores caused only 19.5% loss of the photophosphorylation activity and this loss could not be restored by PMS (Table III). Ascorbate enhanced phosphorylation in the chromatophores and in the washed chromatophores at the same optimal concentration of 6×10^{-3}M (Fig 3). Differences in the properties of the membranes could explain the greater sensitivity of the heavy chromatophores to washing.

Electron transport from Ascorbate-DCIP to O_2 in heavy chromatophores.

The heavy chromatophores as well as the chromatophores catalyzed an electron transport in the dark from Ascorbate-DCIP to O_2. This activity was inhibited by 5×10^{-4}M N_3. However, in the presence of N_3, light induced electron transport, The light induced electron transport was stimulated by 6.6×10^{-6}M antimycin A.

Table III

Effect of washing on photophosphorylation catalyzed by chromatophores and by heavy chromatophores.

Additions	Heavy chromatophores	Washed heavy chromatophores	Chromatophores	Washed Chromatophores
	μmole Pi esterified x mg bacteriochlorophyll^{-1} x h^{-1}			
None	52.3	7	84.4	67.8
5 x 10^{-4}M PMS	84.5	88.5	171.1	134.7

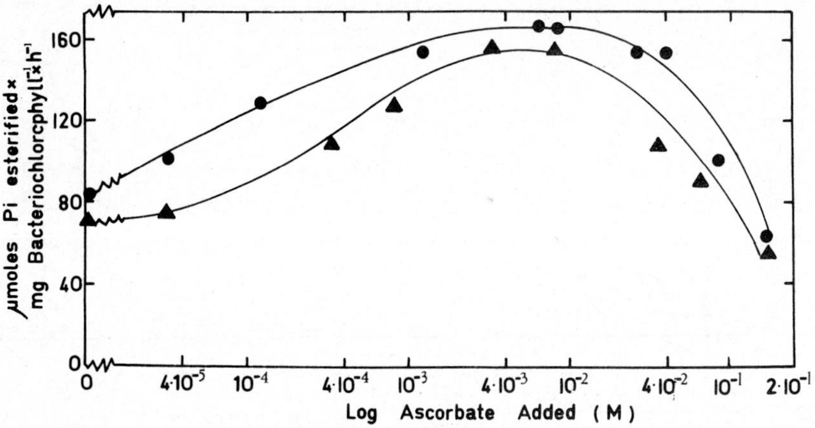

Fig. 3. Effect of ascorbate on chromatophores (●) and on washed chromatophores (▲).

Table IV

Effect of washing on electron transport from DCIP to O_2 in heavy chromatophores.

Additions	Condition	Heavy chromatophores	Washed heavy chromatophores	Washed +5 equivalent Sup II
		μmoles O_2 x mg bacteriochlorophyll^{-1} x h^{-1}		
None	dark	135	81.2	130.8
5 x 10^{-4}M N_3	dark	0	0	0
5 x 10^{-4}M N_3	light	81.5	24.8	98.9
5 x 10^{-4}M N_3 + 6.4 x 10^{-6}M antimycin A	light	289	97.1	183.8

Washing of the chromatophores resulted in some loss of the dark electron transport and in 70% loss of the light activity. Like the activity in the untreated chromatophores, the residual activity in the washed chromatophores was also enhanced by antimycin A. Sup II restored all the activities to their original values (Table IV). Since the electron transport experiments were performed under aerobic conditions we examined the photophosphorylation activity under the same conditions (Table V).

Table V

Effect of washing on aerobic photophosphorylation in heavy chromatophores.

Additions	Heavy chromatophores	Washed heavy chromatophores	Washed heavy chromatophores +5 equ. Sup II
	μmoles Pi esterified x mg bacteriochlorophyll^{-1} x h^{-1}		
None	38.6	7.8	42.8
5 x 10^{-4}M PMS	128.5	90.6	-
1.3 x 10^{-3}M Ascorbate +2.5 x 10^{-5}M DCIP +5 x 10^{-4}M N_3	183.9	85.8	103.4
1.3 x 10^{-3}M Ascorbate +2.5 x 10^{-5}M DCIP 5 x 10^{-4}M N_3 +6 x 10^{-6}M antimycin A	4	4	6

The photophosphorylation assayed under aerobic conditions was washed from the chromatophores and could be restored by Sup II, PMS and partly by DCIP Ascorbate. Antimycin A inhibited phosphorylation under all conditions except for the activity catalyzed in the presence of PMS.

Fractionation of Sup II

Bartch and Meyer's procedure for separation of Rhodopseudomonas capsulata cytochromes on DEAE columns was used for factionation of Sup II. Except for ferredoxin which composed about half of the protein in Sup II, another major component was found (Fig. 4). This component was identified according to its

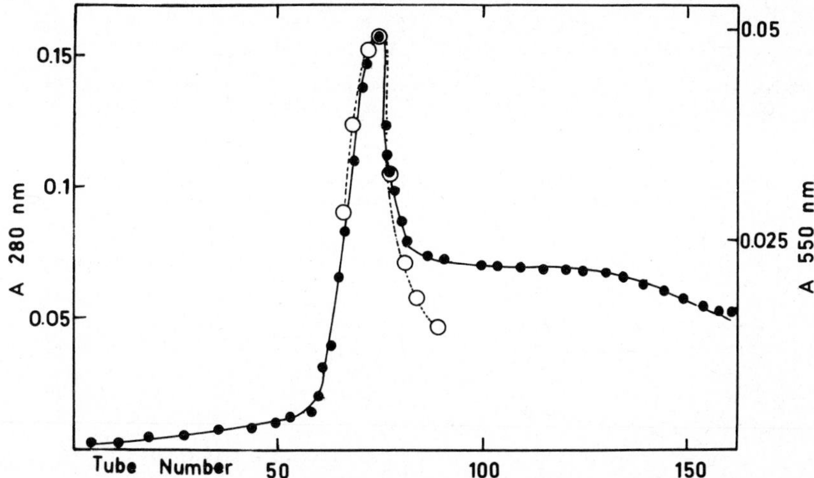

Fig. 4. Components in Sup II. The eluent obtained after the separation of ferredoxin from Sup II was applied to a second DEAE cellulose column. Protein content (-) given as A_{280}nm and cytochrome C_2 content (---) measured at A_{550}nm were determined in each fraction. The elution was done by a gradient of NaCl in 10mM tris-NaOH (pH 8) from 0 to 0.16 M. The peak of cytochrome C_2 was eluted by 0.04 M NaCl.

spectrum as cytochrome C_2 (Fig. 5). Cytochrome C_2 in Sup II which in its reduced

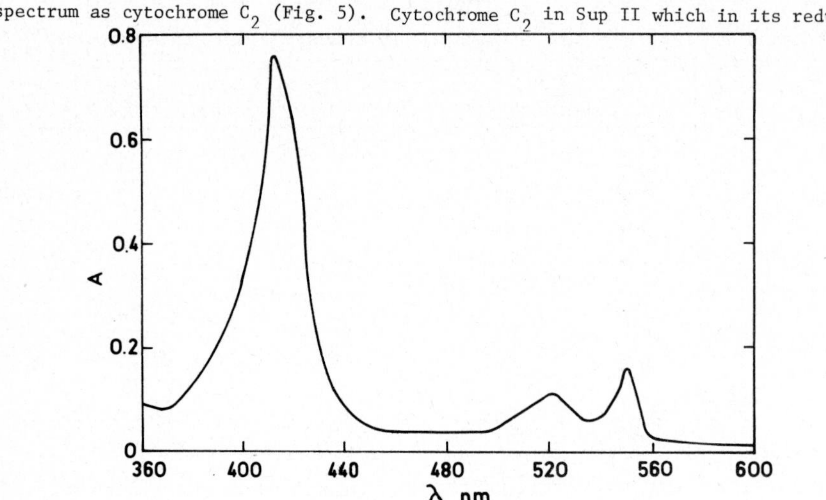

Fig. 5. Spectrum of the active fraction separated from Sup II.

form restored 100% activity to the washed chromatophores at a concentration of 2.8×10^{-6}M. The concentration was determined by measuring ΔA at 550nm of the oxidized minus reduced form using an extinction coefficient of 29. Oxidized C_2 in the same concentration restored only 25 percent of the washed activity. Oxidized cytochrome C_2 added together with ascorbate lowered the optimal ascorbate concentration which was required for restoration of the photophosphorylation activity to the washed chromatophores (Fig. 6). The concentration of ascorbate required in the presence of cytochrome C_2 was similar to the concentration of ascorbate which gave optimal activity in untreated chromatophores.

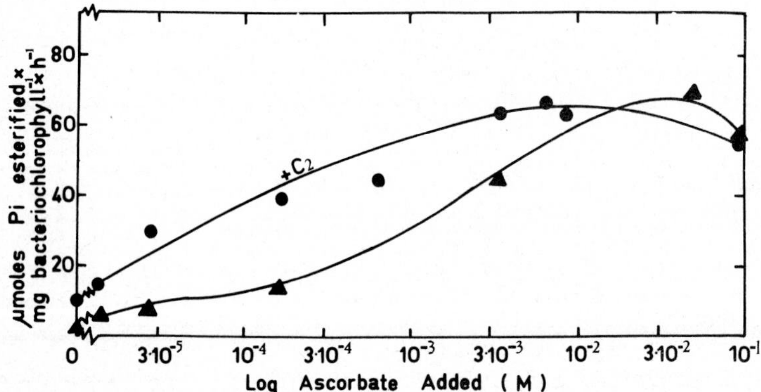

Fig. 6. Effect of ascorbate on photophosphorylation in washed heavy chromatophores. Activity was assayed in the presence of ascorbate (▲) and in the presence of ascorbate and 2.8×10^{-6}M cytochrome C_2 (●).

A photosynthetic mutant of Rhodopseudomonas capsulata.

We have selected a mutant strain of Rhodopseudomonas capsulata which could not grow under anaerobic conditions in the light but grew aerobically in the dark. The same optima were observed in the spectra of the bacteriochlorophylls both in the mutant grown aerobically in the dark and in the wild type grown under the same conditions. The ratio of bacteriochlorophyll to protein in the chromatophores was lower in the mutant than in the wild type (Table VI). A reaction center could be observed in a difference spectrum of dithionite-reduced against ferricyanide-oxidized chromatophores obtained from the mutant.

Table VI

Ratio of bacteriochlorophyll to protein in the chromatophores

Wild type anaerobic	Wild type aerobic	Y 10 aerobic
	μg bacteriochlorophyll x mg protein^{-1}	
47.2	65.0	36.2

Photophosphorylation activity in chromatophores obtained from the mutant was very low under anaerobic conditions but was enhanced by air. Light induced absorbance changes of cytochromes b and c could be observed in this mutant. A clear difference spectrum of the two cytochromes was obtained in the presence of antimycin A, ascorbate and FCCP (Fig. 7). Cytochrome b was reduced by light while cytochrome c was oxidized. In the wild type cytochrome b reduction was also inhibited by DBMIB (Fig. 7). The small absorbance changes in the region of

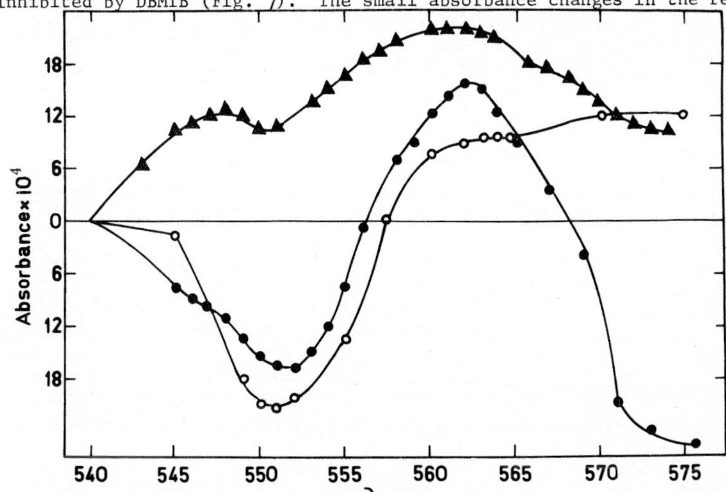

Fig. 7. Light induced absorbance changes in chromatophores from aerobically grown Rps. capsulata. The absorbance changes were measured in a reaction mixture containing 30 mM tricine-NaOH (pH 8), 6 μM FCCP and 30 mM Na-ascorbate (▲) either in the presence of 1 μM antimycin A (●) or in the presence of 5 μM DBMIB (o).

b was mainly due to the carotenoid shift since it appeared with no distinct spectral maxima (in collaboration with A. Gofer). Cyclic photophosphorylation was inhibited by DBMIB but the inhibition was bypassed either by PMS or by ascorbate (in collaboration with Dr. G. Ben Hayyim).

Chromatophores of both Y 10 and of the aerobically grown wild type catalyzed electron transport in the dark from DCIP to O_2 (Table VII). This activity was inhibited by N_3. In the presence of N_3 light induced electron transport. The rate of the light induced electron transport was lower in the mutant chromatophores than in the wild type.

Table VII

Electron transport from DCIP to O_2 in chromatophores. Both the mutant and the wild type were grown aerobically.

Additions	conditions	wild type	Y 10
		μmoles O_2 x mg bacteriochlorophyll^{-1} x h^{-1}	
None	dark	496	357
10^{-3}M N_3	dark	26.4	1
10^{-3}M N_3	light	202.5	64.2
4.4 x 10^{-6}M Antimycin A	dark	342	357
10^{-3}M N_3 + 4.4 x 10^{-6}M Antimycin A	dark	19.8	7
10^{-3}M N_3 +4.4 x 10^{-6}M Antimycin A	light	337.5	54

The activity was stimulated by antimycin A in the wild type but not in the mutant chromatophores.

4. Discussion

In the study of the photosynthetic electron transport in Rhodopseudomonas capsulata two approaches were used. In the first set of experiments a component was removed from the chromatophores and put back and in the second set we used mutants selected in our laboratory.

It is accepted that the bacterial photosynthetic apparatus is an essential part of the intracytoplasmic membrane system [18,19]. Electron micrographs of whole R.rubum cells show a membrane bound vesicular system in the cytoplasm [20]. It was suggested [21,22] that these vesicles were the cellular elements from which the chromatophores arise following mechanical breakage of the cell. It was assumed that during the process of the preparation of chromatophores the vesicles detached from the membrane were resealed with their inner membrane facing the outside medium [23]. The change in the direction of the light induced proton movement is one expression of the change in the polarity of the membrane in cells of R.rubrum and R.ps. spheroides.

In whole bacteria cells proton translocation associated with light dependent electron flow was directed outward through the cytoplasmic membrane. In isolated chromatophores [24] the proton translocation was directed into the vesicles [25]. Heavy chromatophores were somewhat similar in their structure to the vesicular structures inside the cell as seen in electron micrographes. They also catalyzed proton movement in the same direction as the intact cells. Therefore, it was suggested that heavy chromatophores retained the native structure of the intracytoplasmic membranes. This structure was very unstable and was damaged even by moderate stirring or dilution. The single chromatophores with their inside out membrane were more stable.

Chromatophores and heavy chromatophores catalyzed photophosphorylation activity, the rate of which was faster in the former than in the latter chromatophores. This fact was attributed to the unstable structure of the heavy chromatophores and to the fact that cytochrome C_2 was readily washed out of them. However, in both types of chromatophores phosphorylation was enchanced by PMS and by ascorbate and inhibited by antimycin A and air. The inhibited activities were restored almost to their original values by PMS.

Light induced electron transport from DCIP to O_2 in the heavy chromatophores. This electron transport was not inhibited by N_3. Antimycin A stimulated the oxygen uptake presumably by blocking the cyclic electron flow. The cyclic and non cyclic electron transport could pass through a common carrier of the electron transport chain. If this were the case there might be a competition on the electrons between O_2 and the cyclic pathway. Antimycin A blocked the cyclic electron flow, as indicated by the complete inhibition of photophosphorylation. Therefore, in its presence all the electrons from DCIP flow only to O_2. The stimulation of non cyclic electron flow by antimycin A could not be caused by uncoupling since PMS could restore up to 85% of the activity. It is suggested that there is no phosphorylation on the way from DCIP to O_2. A careful washing of chromatophores did not affect their activity but inhibited photophosphorylation activity and most of the light induced electron transport from DCIP to O_2 in the heavy chromatophore. This difference was attributed to the opposite polarity of the two membranes. If the washable component was facing the inside of the vesicle in the chromatophores it could not be readily washed. In the heavy chromatophores this compound faced the surrounding medium and since it was not tightly bound it was detached from the membrane.

Photophosphorylation was completely resotred to the washed particles by PMS which is known to bypass some of the carriers of electron transport [26]. This finding also suggests that the component washed from the chromatophores was an electron carrier rather than a coupling factor. The lost activities in the washed heavy chromatophores was also restored by Sup II and by ascorbate. In Sup II ferredoxin and cytochrome C_2 were identified as the two major proteins but only C_2 could restore the photophosphorylation to the depleted chromatophores. It is suggested that cytochrome C_2 was required for both cyclic and non cyclic transport. The fact that oxidized cytochrome C_2 was not as efficient as the reduced cytochrome in the restoration of the activity was attributed to its oxidizing effect. This was in accordance with the effect of ferricyanide which at the same concentration inhibited 40% of the photophosphorylation in the untreated chromatophores.

Horio and Kamen [8] observed that photophosphorylation in R.rubrum chromatophores prepared by sonication was lost following overnight incubation and washing. The activity could be restored either by a supernatant fluid which was obtained during the preparation of the chromatophores before aging, or by ascorbate or by PMS. Our findings are in agreement with these data. We also showed that cytochrome C_2 reduced the optimal ascorbate concentration for restoration of photophosphorylation in the washed chromatophores. We suggested that the heme protein obtained by Horio and Kamen was a C_2 cytochrome since this cytochrome was found in our preparation although in the present experiments the active fraction was obtained from the depleted chromatophores.

In a second set of experiments we selected photosynthetic mutants one of which is Y 10 strain. This strain seems to possess an unimpaired complement of the photosynthetic pigments including the reaction center bacteriochlorophyll. The fact that it did not grow anaerobically in the light could be due to a defect in its photosynthetic electron transfer or in the phosphorylation coupled with it. From the fact that the mutant had normal photophosphorylation rates under aerobic conditions we rule out the possibility of a defect in the coupling mechanism. The Y 10 strain had light induced spectral changes in the region of cytochrome c and cytochrome b under aerobic condition. The reduction of b type cytochrome as well as the photophosphorylation activity were inhibited under anaerobic conditions. From this we suggest that the mutant possibly had a defect in the electron transport chain between the reaction center bacteriochlorophyl and cytochrome b, and that this defect was partially overcome by O_2.

References

1. Hind, G., Olson, J.M., 1968, Ann. Rev. Plant. Physiol. 19, 249.
2. Ormerod, J.G., Gest, H., 1962, Bacteriol. Rev. 26, 51-66.
3. Vernon, L.P., 1953, Arch. Biochem. Biophys. 43, 492-493.
4. Duysens, L.N.M., 1954, Nature. 173, 692-693.

5. Bartch, R.G., 1971, Methods in Enzymol. Vol 23. pp. 344-363, Colowich and Kaplan ed.

6. Chance, B., Nishimura, M., 1960, Proc. Natl. Acad. Sci. U.S. 46, 19-24.

7. Feldman, N., Gromet-Elhanan, Z., 1971, IInd. Intern. Cong. of Photosynthesis, Stressa. pp. 1217-1220.

8. Horio, T., Kamen, M.D., 1962, Biochemistry. 1, 144-156.

9. Klemme, J.H., Schlegel, H.G., 1969, Arch. Mikrobiol. 68, 326-354.

10. Clayton, R.K., Sistron, W.R., 1964, Biochim. Biophys. Acta. 88, 61-73.

11. Marrs, B., Gest, H., 1973, Jour. of Bact. 114, 1045-1051.

12. Ormerod, J.G., Ormerod, K.S., Guest, H., 1961, Arch. Biochem. Biophys. 94, 449.

13. Hochman, A., Yagil, E., Carmeli, C. In preparation.

14. Personal communication.

15. Clayton, R.K., 1963, Biochim. Biophys. Acta. 75, 312.

16. Avron, M., 1960, Biochim, Biophys, Acta. 40, 257-272.

17. Hartree, E.F., 1972, Anal. Biochem. 48, 422-427.

18. Lascelles, J., 1968, Adv. Microbiol. Physiol. 2, 1.

19. Pfenning, N., 1967, Ann. Rev. Microbiol. 21, 285.

20. Cohen-Bazire, G., Riyo Kunisawa., 1963, J. Cell. Biol. 16, 401-419.

21. Vatter, A.E., Wolfe, R.S., 1958, J. Bact. 75, 480.

22. Bergerson, J.A., Fuller, R.C., 1961, Macromolecular Complexes (M.V. Edds Jr. ed.) pp. 179. New-York,Ronald Press.

23. Jackson, J.B., Crofts, A.R., 1969, Eur. J. Biochem. 10, 226.

24. Von Stedingk, L V., Baltscheffsky, M., 1966, Arch. Biochem. Biophys. 117, 400.

25. Melandri, B.A., Baccarini-Melandri, A., San Pietro, S., Gest, H., 1970, Proc. Natl. Acad. Sci. 67, 477-484.

26. Geller, D.M., Lipmann, F., 1960, J. Biol. Chem. 235, 2478.

M. AVRON, *Proceedings of the Third International Congress on Photosynthesis*, September 2-6, 1974, Weizmann Institute of Science, Rehovot, Israel
Elsevier Scientific Publishing Company, Amsterdam, The Netherlands, 1974

EFFECT OF AUROVERTIN ON ENERGY CONVERSION REACTIONS IN RHODOSPIRILLUM RUBRUM CHROMATOPHORES

Z. Gromet-Elhanan

Department of Biochemistry, Weizmann Institute of Science, Rehovot, Israel

Summary

Aurovertin has been found to act as an energy transfer inhibitor in chromatophores. Like oligomycin it inhibited photophosphorylation but not NAD^+ photoreduction or light-induced proton uptake, but unlike oligomycin it could not inhibit the phosphorylation completely and the residual rate remained resistant even to very high aurovertin concentrations.

The particle bound ATPase, although inhibited by oligomycin, was in chromatophores as well as in mitochondria quite resistant to aurovertin. However, unlike in mitochondria, no increase in the fluorescence of aurovertin was observed upon addition of either coupled chromatophores or their solubilized coupling factor.

1. Introduction

Photophosphorylation in chromatophores of photosynthetic bacteria was found to respond to various kinds of inhibitors of respiration in a manner similar to that of oxidative phosphorylation[1]. This similarity included also the effect of the energy transfer inhibitor, oligomycin, which was shown to inhibit photophosphorylation in R. rubrum chromatophores at the same concentrations that were reported to be effective in mitochondria. Photophosphorylation in chloroplasts on the other hand, was not inhibited at all by similar concentrations of oligomycin[2], while at much higher concentrations it was found to be a weak uncoupler[3].

The antibiotic aurovertin was found by Lardy et al[4] to share most of the effects of oligomycin on mitochondrial reactions. Furthermore, the titration curves for inhibition of oxidative phosphorylation, Pi-ATP and Pi-H_2O exchange reactions were identical with oligomycin and aurovertin. Subsequent workers[5,6] have, however, reported important differences between the effects of the two agents. Thus, while oligomycin blocked completely the mitochondrial ATPase, aurovertin inhibited it only partially or not at all. In fact the forward reaction of ATP synthesis in oxidative phosphorylation was in general much more sensitive to aurovertin inhibition than the reverse energy-transfer reactions, such as ATP dependent NAD^+ reduction by succinate, whereas their sensitivity to oligomycin inhibition was almost identical[6]. These results indicate that oligomycin and aurovertin cannot react at the same site, but as yet no convincing explanation was given to the strange effect of aurovertin, namely to the discrepancy of aurovertin sensitivity between forward and reverse energy transfer reactions.

In this communication the effect of aurovertin was tested on a variety of chromatophore photoreactions and, like oligomycin, it was found to act as an energy transfer inhibitor. The above stated discrepancy in aurovertin effectiveness was observed also in these particles, but in other tests its effect in chromatophores was found to differ from that observed in mitochondria.

2. Experimental Procedure

The growth of R. rubrum cells was as previously described[7]. Chromatophores were isolated according to Gromet-Elhanan[8,9] and stored in liquid air.

Phosphorylation experiments were carried out as previously described[8] and ATP formation was measured according to Avron[10]. The reaction mixture contained, in 3 ml, 30 mM Tricine-NaOH, pH 8.0; 3.3 mM $MgCl_2$; 1.6 mM ADP; 3.3 mM sodium phosphate containing ^{32}P (2×10^6 cpm); chromatophores containing 20 μg bacteriochlorophyll and when indicated, 66 μM PMS or 0.3 mM diaminodurene (DAD).

NAD^+ reduction was assayed as described by Gromet-Elhanan[11]. The reaction mixture contained, in 3 ml, 30 mM Tricine-NaOH, pH 8.0; 3.3 mM $MgCl_2$; 3.3 mM ascorbate; 0.13 mM DAD; 0.33 mM NAD^+; 1 mg bovine serum albumin, and chromatophores containing 40 μg bacteriochlorophyl. Where indicated, 1.6 mM ADP and 3.3 mM sodium phosphate containing ^{32}P(2×10^6 cpm) were added.

Proton uptake was measured as previously described[12]. The reaction mixture contained, in 2.25 ml, 18 mM NaCl; 3.3 mM $MgCl_2$; 300 μM succinate; 66 μM PMS; chromatophores containing 60 μg bacteriochlorophyll and 0.1 mM Tricine-NaOH to a final pH of 6.9.

ATPase was assayed for 10 min at 30° in a reaction mixture containing, in 3 ml, 30 mM Tricine-NaOH, pH 7.8; 2 mM $MgCl_2$; 2 mM ATP and chromatophores containing 20 μg bacteriochlorophyll. The reaction was started by addition of ATP and stopped by the addition of perchloric acid to a final concentration of 1.1%. Liberated Pi was determined using the colorimetric method of Taussky and Shorr[13].

Aurovertin was a gift of Dr. H. Lardy and Dr. T. A. Out. Both samples gave identical results. Oligomycin was purchased from Sigma.

3. Results and Discussion

The titration curves of inhibition of PMS-catalyzed photophosphorylation by oligomycin and aurovertin were found to be quite similar, so that at 1 μM both antibiotics leftover about 10 % of the control activity (Fig. 1). At this level of residual phosphorylation a doubling of oligomycin concentration resulted in complete inhibition of ATP formation, whereas even a 20 to 50 fold increase of aurovertin concentration did not lead to any additional inhibition.

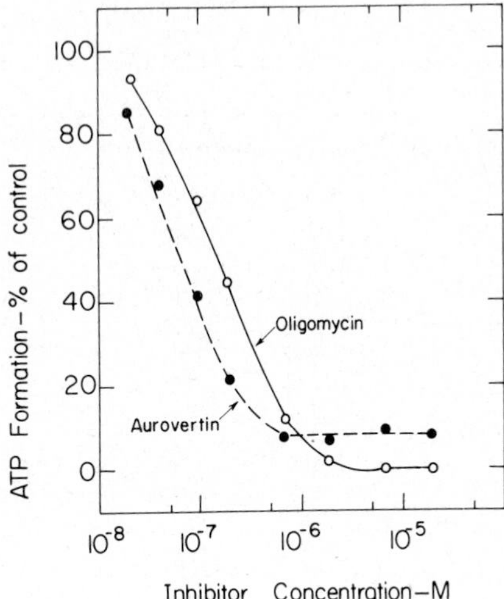

Fig. 1: Comparison of the effect of oligomycin and aurovertin on PMS- catalyzed photophosphorylation. Control activity was 715 μmoles ATP formed per mg bacteriochlorophyl per h.

The difference between oligomycin and aurovertin was even more pronounced when the effect of both inhibitors was compared in other phosphorylating systems (Table I). Oligomycin at 3 μM completely inhibited all the systems, but the effect of aurovertin leveled off at 0.7 μM leaving a residual rate of phosphorylation of between 30 to 60 μmoles per mg bacteriochlorophyll per h.

Inhibition of photophosphorylation in chromatophores cannot be used as a test to differentiate between uncoupling and inhibition of energy transfer, since this phosphorylation is coupled to a cyclic electron flow. The conclusion that oligomycin acted in chromatophores, like in mitochondria, as an energy transfer inhibitor was reached from its effect on a number of other photoreactions, such as light-induced NAD^+ reduction[11,14] and light-induced proton uptake[15]. These photoreactions were inhibited by uncouplers, but rather stimulated by oligomycin.

Table 1: Effect of Oligomycin and Aurovertin on Various Phosphorylation Systems

Numbers in parenthesis indicate % of control

Inhibitor concentration	ATP synthesis with system tested		
	PMS	Endogenous	DAD
	µmoles/mg bacteriochlorophyll/h		
None	570(100)	109(100)	223(100)
Oligomycin			
3×10^{-7}M	205(36)	72(66)	133(60)
7×10^{-7}M	68(12)	36(33)	-
3×10^{-6}M	6(1)	6(6)	6(3)
Aurovertin			
7×10^{-8}M	358(52)	47(43)	133(60)
7×10^{-7}M	57(10)	25(23)	67(30)
3×10^{-6}M	46(8)	27(25)	58(26)

The initial effect of aurovertin on NAD^+ photoreduction resembled that of oligomycin (Table II). With ascorbate-DAD, as with other electron donors[11,14], NAD^+ reduction was inhibited by addition of ADP and Pi and this inhibition was relieved by aurovertin as well as by oligomycin. However, the difference reported above between oligomycin and aurovertin was observed also in this system (Table II). Thus, with 0.7 µM oligomycin, which enabled photophosphorylation to proceed at a third of its control rate, the inhibition of NAD^+ reduction by the concomitant phosphorylation was not completely relieved. But at 7 µM when practically no ATP was formed, oligomycin completely relieved the inhibition of NAD^+ reduction. With aurovertin, on the other hand, a residual rate of phosphorylation at 25% of control and a parallel 20 to 25 % of inhibition of NAD^+ reduction remained even when its concentration was raised from 0.4 to 20 µM (Table II).

Oligomycin was reported to have a slight stimulatory effect on light-induced proton uptake in R. rubrum chromatophores[15]. As can be seen in Table III, this effect was not shared by aurovertin. The stimulation by oligomycin was quite pronounced, about 150 % of control, but it disappeared when proton uptake was tested in the presence of either permeant anions, such as SCN^-(12) or permeant cations (KCl in the presence of valinomycin) which by themselves already stimulated the proton uptake by 2 to 5 fold. Aurovertin was ineffective in all three cases (Table III). This inactivity indicated, however, that in this test system, as in NAD^+ reduction, aurovertin behaved as an energy transfer inhibitor rather than an uncoupler, since uncouplers were shown to inhibit proton uptake[15].

Table II: Effect of oligomycin and aurovertin on ascorbate-DAD linked NAD^+ photoreduction and photophosphorylation

Numbers in parenthesis indicate % of control

Inhibitor concentration	Reaction tested		
	NAD^+ photoreduction		Photophos-phorylation
	(-ADP -Pi)	(+ ADP + Pi)	
	µmoles/mg bacteriochlorophyll/hour		
None	66 (100)	29 (44)	244 (100)
Oligomycin			
7×10^{-7}M	70 (106)	52 (80)	80 (33)
7×10^{-6}M	71 (108)	67 (102)	5 (2)
Aurovertin			
4×10^{-7}M	61 (93)	52 (80)	54 (22)
4×10^{-6}M	64 (97)	52 (80)	61 (25)
2×10^{-5}M	61 (92)	50 (76)	68 (28)

Table III: Proton uptake characteristics of *R. rubrum* chromatophores. Effects of oligomycin and aurovertin.

Where indicated 20 mM NaSCN or 0.1 µM valinomycin and 100 mM KCl were added.

Inhibitor concentration	Extent of proton uptake with		
	control	+ NaSCN	+ valinomycin + KCl
	µeq. H^+/mg bacteriochlorophyll		
None	0.167	0.360	0.780
Oligomycin			
2×10^{-6}M	0.240	0.330	0.795
Aurovertin			
4×10^{-6}M	0.175	0.340	0.765

The most pronounced difference between oligomycin and aurovertin reported in mitochondria was their effect on ATPase activity[5,6]. This difference was observed also in chromatophores (Table IV). While oligomycin inhibited the control as well as the uncoupler stimulated rate, the effect of aurovertin levelled off when 60 to 70 % of the activity was still present in both systems. The inhibition by aurovertin although much less pronounced than with oligomycin was, in contrast to the stimulation by uncoupler (Table IV), enough to demonstrate that here as in other test systems (Tables II and III) aurovertin was acting as an energy transfer inhibitor.

Table IV: ATPase activity in the absence and presence of FCCP as affected by oligomycin and aurovertin.

Where indicated 1 μM FCCP was added. Numbers in parenthesis indicate % of control.

Inhibitor concentration	ATPase activity	
	-FCCP	+FCCP
	μmoles Pi released/mg Bchl/h	
None	120 (100)	341 (100)
Oligomycin		
2×10^{-6}M	54 (45)	123 (36)
2×10^{-5}M	30 (25)	37 (11)
Aurovertin		
4×10^{-6}M	72 (60)	208 (61)
4×10^{-5}M	84 (70)	232 (68)

Although aurovertin was found to be relatively inactive on mitochondrial ATPase, it has been found to inhibit the soluble mitochondrial ATPase[16] and to form stoichiometric fluorescent complexes with it[17] as well as with coupled mitochondria[18] and submitochondrial particles[19,20]. The increase in fluorescence of aurovertin observed upon addition of soluble ATPase, or of mitochondria, was interpreted as binding of the aurovertin to the ATPase[17,20]. During the course of this study, no increase in the fluorescence of aurovertin was, however, observed upon addition of either coupled or depleted chromatophores or a solubilyzed preparation of coupling factor prepared according to reference 21.

The data summarized above indicate that, although oligomycin and aurovertin act as energy transfer inhibitors in chromatophores as well as in mitochondria, their site of inhibition can not be identical. Oligomycin was found to inhibit completely all types of phosphorylation as well as the ATPase reaction whereas with aurovertin a residual rate of phosphorylation and even more so of ATPase was always retained. It is therefore concluded that the various sites of phosphorylations, suggested to be coupled to the cyclic electron transport in chromatophores, differ in their sensitivity to aurovertin but not to oligomycin.

Acknowledgment:

Thanks are due to Miss R.Wolfowitch for running the ATPase assay.

References

1. Baltscheffsky, H. and Baltscheffsky, M., 1960, Acta Chem. Scan. 14, 257-263.
2. Baltscheffsky, H., 1960, Acta Chem. Scan. 14, 264-272.
3. Avron, M. and Shavit, N., 1965, Biochim. Biophys. Acta 109, 317-331.
4. Lardy, H. A., Connelly, J. L. and Johnson, D., 1964, Biochemistry 3, 1961-1968.
5. Lenaz, G., 1965, Biochem. Biophys. Res. Commun. 21, 170-175.
6. Lee, C.P. and Ernster, L., 1968, Eur. J. Biochem. 3, 391-400.
7. Briller, S. and Gromet-Elhanan, Z., 1970, Biochim. Biophys. Acta, 205, 263-272.
8. Gromet-Elhanan, Z., 1970, Biochim. Biophys. Acta 223, 174-182.
9. Gromet-Elhanan, Z., 1974, J.Biol. Chem. 249, 2522-2527.
10. Avron, M., 1960, Biochim. Biophys. Acta 40, 257-272.
11. Gromet-Elhanan, Z., 1969, Arch. Biochem. Biophys. 131, 299-305.
12. Gromet-Elhanan, Z. and Leiser, M., 1973, Arch. Biochem. Biophys. 159, 583-589.
13. Taussky, H. and Shorr, E., 1953, J. Biol. Chem. 202, 675-685.
14. Keister, D. L. and Yike, N. J., 1967, Arch. Biochem. Biophys. 121, 415-422.
15. Von Stedingk, L. V. and Baltscheffsky, H., 1966, Arch. Biochem. Biophys. 117, 400-404.
16. Robertson, A. M., Beechey, R. B., Holloway, C. T. and Knight, I. G., 1967, Biochem. J., 104, 54C-55C.
17. Lardy, H. A. and Lin, C. H. C., 1969, in Inhibitors, Tulls for Cell Research (Buecher, T., ed) pp. 279-281, Springer Verlag, New York.
18. Bertina, R. M., Schrier, P. I. and Slater, E. C., 1973, Biochim. Biophys. Acta 305, 503-518.
19. Chang, T. and Penefsky, H. S., 1973, J. Biol. Chem. 248, 2746-2754.
20. Chang, T. and Penefsky, H. S., 1974, J. Biol. Chem. 249, 1090-1098.
21. Binder, A. and Gromet-Elhanan, Z., 1974, Third International Congress on Photosynthesis, Rehovot, pp. 1161-1166.

M. AVRON, *Proceedings of the Third International Congress on Photosynthesis*,
September 2-6, 1974, Weizmann Institute of Science, Rehovot, Israel
Elsevier Scientific Publishing Company, Amsterdam, The Netherlands, 1974

THE EFFECT OF DIBROMOTHYMOQUINONE ON LIGHT INDUCED REACTIONS IN CHROMATOPHORES FROM RHODOSPIRILLUM RUBRUM

M. Baltscheffsky

Department of Biochemistry, Arrhenius Laboratory,
University of Stockholm, S-104 05 Stockholm, Sweden

1. Summary

The effect of dibromothymoquinone (DBMIB) has been studied on various electron transport linked reactions in Rhodospirillum rubrum chromatophores. When present in concentrations nearly stoichiometric with the concentration of endogenous quinone, DBMIB strongly inhibits the rereduction of photooxidized cytochrome $\underline{c}_2$, the reoxidation of photoreduced cytochrome $\underline{b}$, the reduction of cytochrome $\underline{c}'$ by succinate, photophosphorylation, the light induced H^+ uptake and the reversed electron flow induced by inorganic pyrophosphate (PP_i). The results indicate that the endogenous quinone constitutes an additional, but obligatory, link between cytochromes $\underline{b}$ and $\underline{c}_2$.

2. Introduction

DBMIB was first found by Trebst et al.[1] to act as an inhibitor of photosynthetic electron transport in chloroplasts, and was proposed to be an antagonist against plastoquinone.

Since ubiquinone in chromatophores from the photosynthetic bacterium Rhodospirillum rubrum has been assumed rather generally during the last decade to participate in the cyclic electron transport, it could be expected that DBMIB would act as an inhibitor for the cyclic electron flow in the bacterial chromatophores. If this was the case we would also possibly gain some insight in the actual site of ubiquinone in this cyclic electron transport which has been uncertain. For some time quinone was thought to be the primary acceptor, x in the photoact[2], it has later been suggested to act as the secondary electron acceptor, the unknown y[3,4]. Our data suggest that the role of quinone in the electron transport may be quite different from that of both x and y.

3. Methods and materials

Rhodospirillum rubrum, strains S-1 and G-9 were grown and chromatophores prepared as described previously[5].

The spectrophotometric measurements were carried out in a dual beam spectrophotometer designed and built at Johnson Foundation, Philadelphia, USA. Actinic illumination was applied at a 90° angle from the measuring beam and was filtered through double layers of Wratten 88A gelatin filters. The light intensity was adjusted to be saturating.

Photophosphorylation was measured with the ΔpH method[6].

The dibromothymoquinone was a gift from Dr. R. Berzborn, which is gratefully acknowledged. Antimycin A was purchased from Sigma Chemical Co. All other reagents were of analytical grade.

The rate of reduction of cytochrome *c*´ was measured spectrophotometrically at 428-450 nm with succinate (5 mM) as substrate. The amount of reduced cytochrome *c*´ was determined using a mM extinction coefficient of 112[7].

4. Results

Photophosphorylation is strongly inhibited by DBMIB (Fig. 1) although at higher concentration than is required in chloroplasts. The concentration is also higher than that needed of antimycin, the well known inhibitor of electron transport at the level of cytochrome *b*. Antimycin almost completely blocks electron transport and photophosphorylation in less than μM concentration. It was found that the concentration of DBMIB needed to inhibit the chromatophore system was directly related to the concentration of chlorophyll in the experiment, and that the stoichiometry was usually in the neighbourhood of 0.4-0.5 DBMIB per chlorophyll. This is very close to the endogenous content of quinone in these chromatophores, which ranges from 0.3-0.45 quinone per chlorophyll in a number of preparations, as analysed by the quinone specialist in the Biochemistry department, Dr. Birgitta Norling. The concentration of quinone is according to the data of Kakuno *et al.*[7] more than hundredfold that of cytochrome c_2 or cytochrome *b*. Fig. 1 also shows that apparently all of the quinone pool participates in the electron transport coupled phosphorylation, since partial inhibition is obtained already at low concentrations of DBMIB, with no plateau observed before the inhibition occurred. It also shows that the inhibition is not

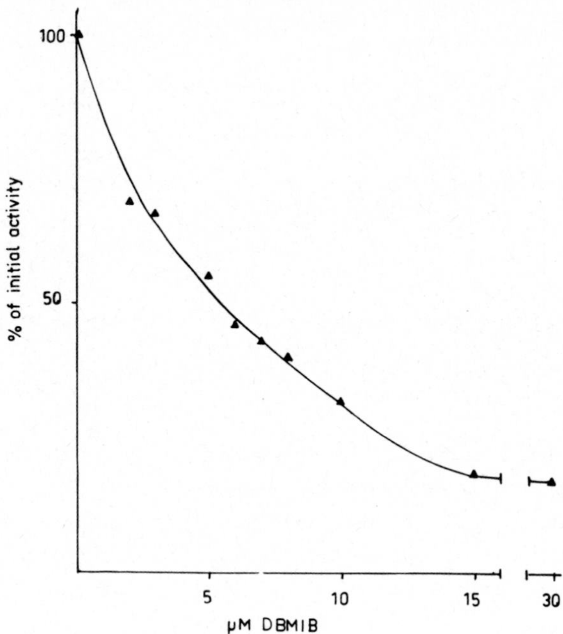

Fig. 1. Inhibition of photophosphorylation by DBMIB. Chromatophores equal to 27.2 μM bacteriochlorophyll were suspended to a final volume of 2.5 ml in a reaction mixture consisting of 0.1 M KCl; 0.4 mM succinate; 12 mM $MgCl_2$; 8 mM KH_2PO_4 and 2 mM ADP. The initial pH was 7.4.

complete, even at high concentrations of DBMIB. There may be several explanations for this, one is that DBMIB itself takes over the role of quinone, but at a much slower rate, another that a small part of the quinone pool is not reached by the inhibitor. The inhibition of phosphorylation is completely reversed by addition of phenazine methosulfate (PMS), well known to reverse also the antimycin inhibition of electron transport. So, DBMIB inhibits, in all likelyhood at the quinone level also in bacterial chromatophores. The next question was, where did it inhibit, what other reactions were affected? Fig. 2 shows that the light induced proton uptake in the chromatophores is strongly inhibited by DBMIB. This is quite in agreement with the proposal by Trebst[8], that in chloroplasts the quinone pool acts as the translocating proton shuttle. Also here,

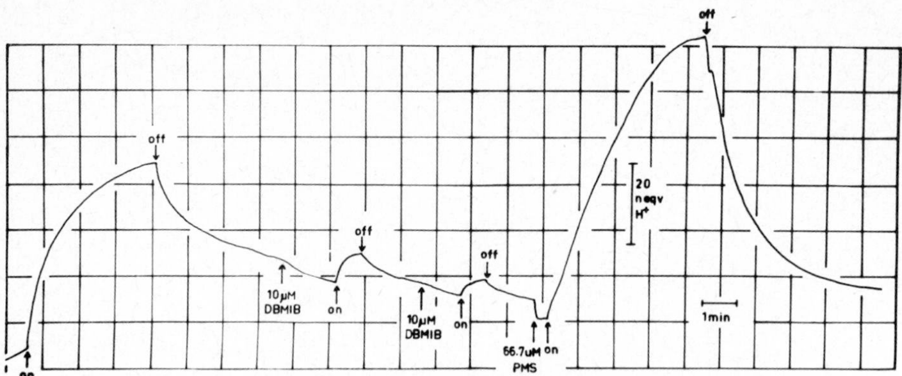

Fig. 2. Inhibition by DBMIB of the light-induced H^+ uptake. Chromatophores equivalent to 21 µM bacteriochlorophyll were suspended to a total volume of 3 ml in 0.35 M NaCl; 13 µM Na-succinate. Starting pH 6.2.

PMS will provide a bypass around the site of inhibition and again allow the reaction to proceed efficiently.

If the quinone pool is identical with y, the secondary acceptor in the cyclic electron transport, one would expect that the first cytochrome accepting electrons from the reducing end, cytochrome b with E_m = -100 mV, would be severely inhibited in its light induced reduction both to rate and extent. In Fig. 3 is seen that this is really not the case. The extent of reduction is slightly decreased, but the rate of reduction is not slowed down. This has been further examined in single flash experiments, not shown here, where measurements on a faster time scale still give no evidence of a slower rate

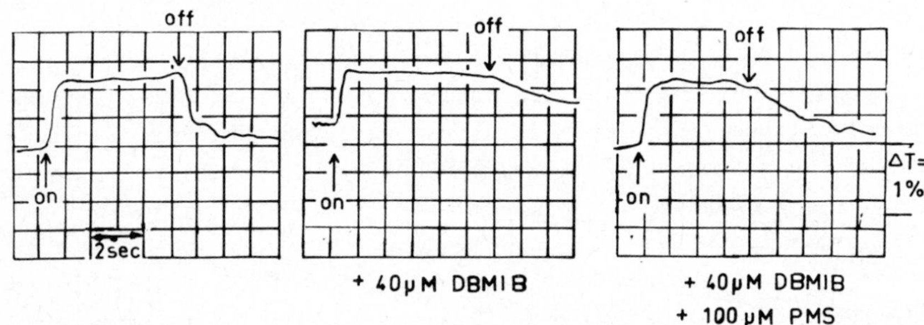

Fig. 3. Effect of DBMIB and PMS on light induced reduction of cytochrome b measured at 562-572 nm. Chromatophores corresponding to 78 µM bacteriochlorophyll were suspended to a final volume of 1.5 ml in 0.2 M glycyl-glycine buffer pH 7.4. Additions as indicated in the figure.

of reduction. On the other hand, the reoxidation of cytochrome b after the illumination has been turned off is inhibited in the presence of DBMIB, and this inhibition is reversed by PMS. The unusual feature of this is that the extent is slightly decreased although the reoxidation is inhibited, suggesting that there is another reoxidation pathway which is the rate limiting reaction that determines the steady state extent of reduction. This would be the pathway inhibited by antimycin. In the presence of antimycin, the extent of the light induced reduction is greatly enhanced as is seen in Fig. 4 and the reoxidation is also inhibited. Addition of DBMIB to the antimycin inhibited system causes a further inhibition of the reoxidation. The reoxidation rates, relative to each other, are approximately 1:0.5:0.25, for respectively the control, in the presence of antimycin, and in the presence of both antimycin and DBMIB.

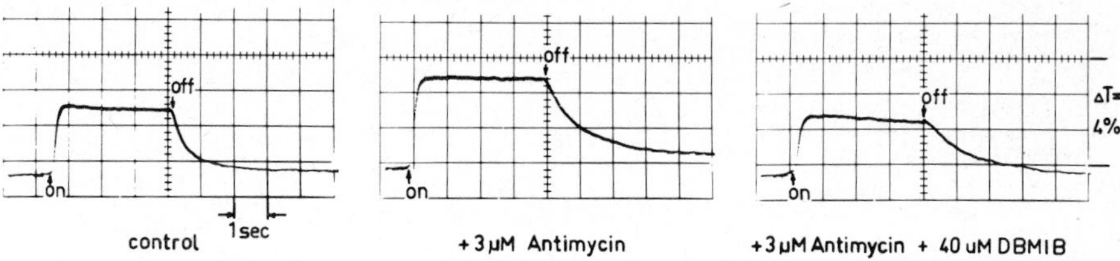

Fig. 4. Effect of antimycin on light-induced reduction of cytochrome b in the absence and presence of DBMIB. Measuring wavelength 428 nm with 450 nm as reference. Conditions as in Fig. 3 except that bacteriochlorophyll concentration was 30 µM, and 3.3 µM Na-succinate was present.

Similar results are obtained when the light induced oxidation of cytochrome c_2 is studied. In this case the effects are even more pronounced. In Fig. 5 is shown that DBMIB alone here also increases the extent of the steady state oxidation as well as decreases the

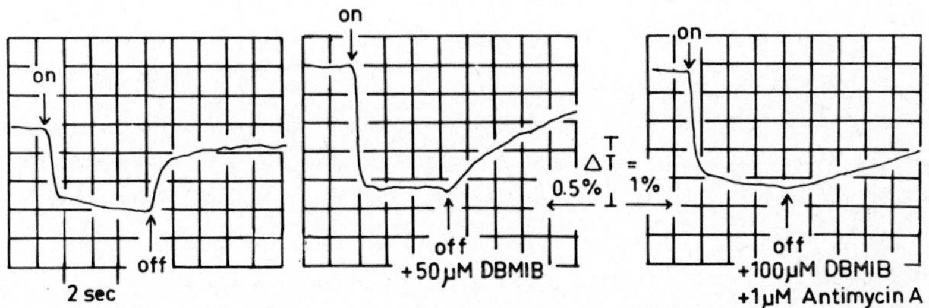

Fig. 5. Effect of DBMIB and antimycin on the light induced oxidation of cytochrome c_2 measured at 552-540 nm. Conditions as in Fig.3.

rate of rereduction. The effect is similar to that obtained with antimycin alone, but when the two inhibitors are added together the inhibitions are synergistic, the rereduction becomes extremely slow, and the extent of oxidation is approximately doubled. This synergistic effect clearly points to different sites of action for the two inhibitors.

The succinic dehydrogenase system in animal mitochondria is well known to reduce ubiquinone, and since it is known that succinate in these chromatophores reduces cytochrome c´ but not cytochrome b it was expected that the rate of reduction of cytochrome c´ would be inhibited by DBMIB. Fig. 6 shows that this is the case. With increasing concentrations of the inhibitor, the rate of reduction is strongly decreased until almost complete inhibition.

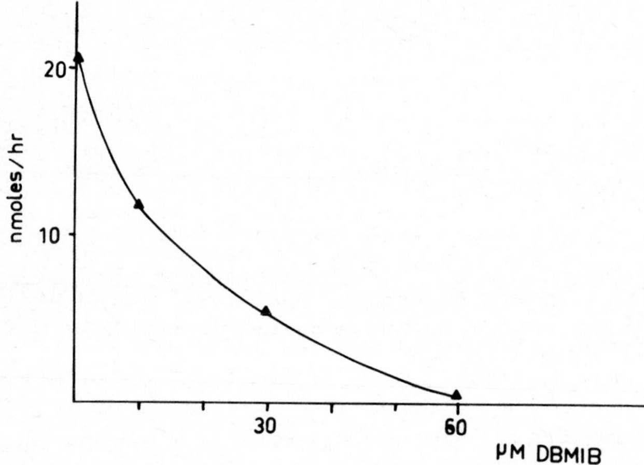

Fig. 6. Inhibition by DBMIB on reduction of cytochrome c´ by succinate measured at 428-540 nm. Conditions as in Fig. 3. Succinate was added in each experiment to a final concentration of 5 mM.

The reversed reactions of phosphorylation, the reduction in the dark of cytochrome b, induced by PP_i is also inhibited by DBMIB, but only in the absence of the exogenous electron donor succinate. If, as seen in Fig. 7, succinate is allowed to reduce the system before the addition of PP_i, the reversed flow, utilizing the energy from the coupling site, will proceed although the usual enhancement of the reduction[9] in the presence of succinate is not seen.

In this case, where the electron transport at the quinone level is partially inhibited, electrons fed into the system from succinate will still be donated under energy pressure to cytochrome b. The

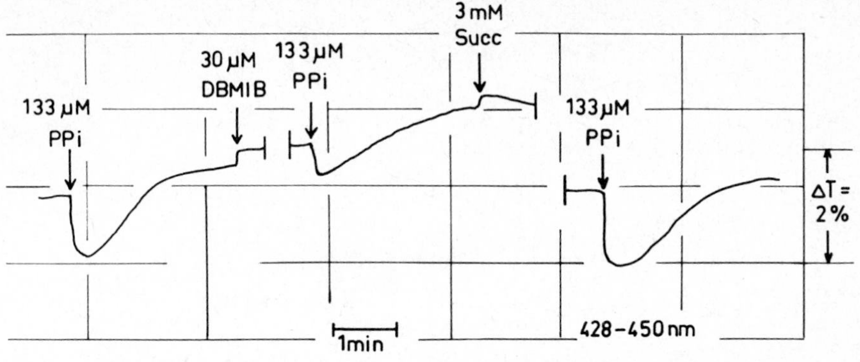

Fig. 7. Effect of DBMIB on PP_i induced reduction of cytochrome b measured at 428-450 nm. Chromatophores equivalent to 79 μM bacteriochlorophyll; 0.2 M glycyl-glycine buffer pH 7.4 and 5 mM $MgSO_4$. Total volume 1.5 ml.

immediate electron donor here is probably reduced cytochrome $\underline{c}_2$.

5. Discussion

In a recent paper Loschen and Azzi[10] found that also in animal mitochondria DBMIB is a potent inhibitor of electron transport. This makes the inhibitor an extremely useful tool for the study of electron transport and energy conversion in different systems, and makes it possible to try to extrapolate the results from one system to another. The proposed role of Q as the secondary acceptor[4] seems difficult to reconcile with these results since the photoreduction of b is very little inhibited by DBMIB, not at all in rate and only slightly in extent. On the other hand it seems clear that electrons are accepted from cytochrome b and donated to cytochrome $\underline{c}_2$ by the quinone pool since their respective reoxidation and rereduction are inhibited by DBMIB. That this is not the only pathway is also evident, especially in the case of cytochrome $\underline{c}_2$ since antimycin causes an extra inhibition. Fig. 8 gives a highly simplified scheme for a possible "branched" electron transport at the b-$\underline{c}_2$ coupling site.

I would at this point like to speculate about the possibility that when a pair of electrons pass from cytochrome b to cytochrome $\underline{c}_2$ the electron transfer directly betwen the two cytochromes is a one electron transfer, and that the other electron passes in a concerted fashion *via* the quinone pool, resulting in the formation of the semiquinone radical. This would agree with the old observation by Redfearn[11], that quinone extracted under illumination is always oxidized, since it is well known that the semiquinone form is rapidly

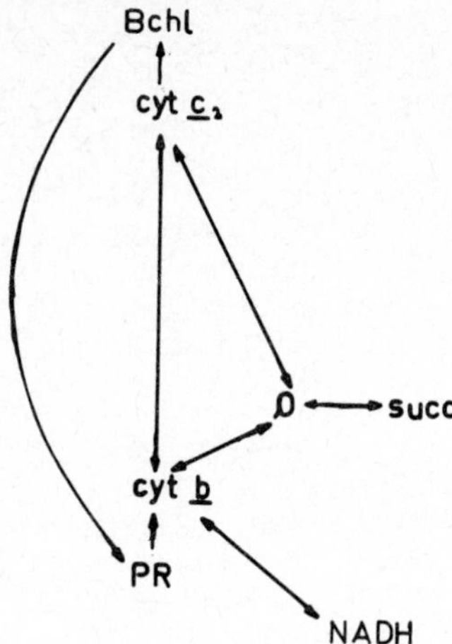

Fig. 8. Proposed scheme for electron transport pathways at the cytochromes *b*-c_2 coupling site.

autooxidized. The new concept that may have emerged from these studies is the splitting up of a pair of reducing equivalents in two discrete but obligatorily coupled electron transfer steps *via* the two proposed pathways. The fact that inhibition of only one of these two pathways, either by antimycin or by DBMIB, nearly eliminates the photophosphorylation indicates that this splitting up phenomenon may be a necessary prerequisite for energy conservation.

References

1. Trebst, A. et al., 1970, Z. Naturforsch. 25 b, 1157.
2. Ke, B. et al., 1968, Biochemistry 7, 311.
3. Parson, W.W., 1969, Biochim. Biophys. Acta 189, 384.
4. Reed, D.W. et al., 1969, Proc. Natl. Acad. Sci. 63, 42.
5. Baltscheffsky, H., 1960, Biochim. Biophys. Acta 40, 1.
6. Nishimura, M. et al., 1962, Biochim. Biophys. Acta 59, 177.
7. Kakuno, T. et al., 1971, J. Biochem. Tokyo 70, 79.
8. Trebst, A., 1974, Ann. Rev. Plant Physiol. 25, 423.
9. Baltscheffsky, M., 1969, in: Progress in Photosynthesis Research, ed. H. Metzner (Tübingen) p. 1306.
10. Loschen, G. and Azzi, A., 1974, FEBS Lett. 41, 115.
11. Redfearn, E., 1967, Biochim. Biophys. Acta 131, 218.